Handbook of Power Quality

Handbook of Power Quality

Edited by

Angelo Baggini
University of Bergamo, Italy

John Wiley & Sons, Ltd

Telephone (+44) 1243 779777

Email (for orders and customer service enquiries): cs-books@wiley.co.uk
Visit our Home Page on www.wiley.com

Other Wiley Editorial Offices

John Wiley & Sons Inc., 111 River Street, Hoboken, NJ 07030, USA

Jossey-Bass, 989 Market Street, San Francisco, CA 94103-1741, USA

Wiley-VCH Verlag GmbH, Boschstr. 12, D-69469 Weinheim, Germany

John Wiley & Sons Australia Ltd, 42 McDougall Street, Milton, Queensland 4064, Australia

John Wiley & Sons (Asia) Pte Ltd, 2 Clementi Loop #02-01, Jin Xing Distripark, Singapore 129809

John Wiley & Sons Canada Ltd, 6045 Freemont Blvd, Mississauga, ONT, L5R 4J3

Wiley also publishes its books in a variety of electronic formats. Some content that appears in print may not be available in electronic books.

Library of Congress Cataloging in Publication Data

Handbook of power quality / Edited by Angelo B. Baggini.
p. cm.
Includes bibliographical references and index.
ISBN 978-0-470-06561-7 (cloth)
1. Electric power transmission—Reliability. 2. Distributed resources (Electric utilities)—Reliability.
I. Baggini, Angelo B.
TK3001.H34 2008
621.31—dc22

2007050178

British Library Cataloguing in Publication Data

A catalogue record for this book is available from the British Library

ISBN: 978-0-470-06561-7

Typeset in 10/12pt Times by Integra Software Services Pvt. Ltd, Pondicherry, India

FSC
Mixed Sources
Product group from well-managed forests and other controlled sources
Cert no. SGS-COC-2953
www.fsc.org
© 1996 Forest Stewardship Council

Contents

List of Contributors

Hermina ALBERT
Institute for Studies and Power Engineering
Lacul Tei Blvd1–3
Bucharest 020371
Romania

Alan ASCOLARI
ECD Engineering Consulting and Design
via Maffi 21
27100 Pavia
Italy

Angelo BAGGINI
Dipartimento di Ingegneria Industriale
Università degli Studi di Bergamo
Via Marconi 5
24044 Dalmine (BG)
Italy

Araceli HERNÁNDEZ BAYO
Departamento de Ingenieria Eléctrica
Universidad Politécnica de Madrid
Escuela Técnica Superior de Ingenieros Industriales
C/José Gutiérrez Abascal, 2
28006 Madrid
Spain

Joan BERGAS-JANÉ
CITCEA
Universitat Politécnica de Catalunya
Av. Diagonal, 647. Planta 2
08028 Barcelona
Spain

Janusz BROŻEK
AGH-University of Science and Technology
30-059 Kraków
Al. Mickiewicza 30
Poland

Franco BUA
ECD Engineering Consulting and Design
via Maffi 21
27100 Pavia
Italy

Francesco BURATTI
ECD Engineering Consulting and Design
via Maffi 21
27100 Pavia
Italy

Maurizio CACIOTTA
Università degli Studi
Facoltà INGEGNERIA
Dipartimento Ingegneria Elettronica
via della Vasca Navale, 84
00146 Roma
Italy

David CHAPMAN
Copper Development Association
5 Grovelands Business centre
Boundary Way
Hemel Hempstead
HP2 7TE
UK

Mircea CHINDRIS
Technical University of Cluj-Napoca
Power Systems Dept.
15, C. Daicoviciu St.
400020 Cluj-Napoca
Romania

Johan DRIESEN
Katholieke Universiteit Leuven
Dept. Electrical Engineering (ESAT),
Div. ELECTA
Kasteelpark Arenberg 10, bus 2445
3001 Heverlee
Belgium

Stefan FASSBINDER
Beratung elektrotechnische
Anwendungen
Deutsches Kupferinstitut
Am Bonneshof 5
D-40474 Düsseldorf
Germany

Andrzej FIRLIT
AGH-University of Science
and Technology
30-059 Kraków
Al. Mickiewicza 30
Poland

Samuel GALCERAN-ARELLANO
CITCEA
Universitat Politècnica de Catalunya
Av. Diagonal, 647. Planta 2
08028 Barcelona
Spain

Nicolae GOLOVANOV
Institute for Studies and Power
Engineering
Lacul Tei Blvd1–3
Bucharest 020371
Romania

Krish GOMATOM
EPRI
942 Corridor Park Blvd
Knoxville
TN 37932
USA

Zbigniew HANZELKA
AGH-University of Science
and Technology
30-059 Kraków
Al. Mickiewicza 30
Poland

Andrzej KANICKI
Technical University of Łodź
Institute of Electrical Power
Engineering
18/22 Stefanowskigo Str.
90-924 Łodź
Poland

Antoni KLAJN
Wroclaw University of Technology
Institute of Electrical Power Engineering
ul. Wyb. Wyspianskiego
27 50–370 Wroclaw
Poland

Aleksander KOT
AGH-University of Science
and Technology
30-059 Kraków
Al. Mickiewicza 30
Poland

Jonathan MANSON
JEL Consulting Limited
6 Staveley Road
Chiswick
London
W4 3ES
UK

Johan RENS
School for Electrical Engineering
North-West University
Potchefstroom Campus
Potchefstroom
South Africa

Tom SHORT
EPRI
1462 Erie Blvd
Schenectady
NY 12305
USA

Antoni SUDRIÀ-ANDREU
CITCEA
Universitat Politècnica de Catalunya
Av. Diagonal, 647. Planta 2
08028 Barcelona
Spain

Andreas SUMPER
CITCEA
Universitat Politècnica de Catalunya
Av. Diagonal, 647. Planta 2
08028 Barcelona
Spain

Piet SWART
School for Electrical Engineering
North-West University
Potchefstroom Campus
Potchefstroom
South Africa

Roman TARGOSZ
Polish Copper Promotional Centre
Plac Jana Pawla II 1-2
50-136 Wroclaw
Poland

Vu Van THONG
Katholieke Universiteit Leuven
Dept. Electrical Engineering (ESAT),
Div. ELECTA
Kasteelpark Arenberg 10, bus 2445
3001 Heverlee
Belgium

Pieter VERMEYEN
Katholieke Universiteit Leuven
Dept. Electrical Engineering (ESAT),
Div. ELECTA
Kasteelpark Arenberg 10, bus 2445
3001 Heverlee
Belgium

Roberto VILLAFÁFILA-ROBLES
CITCEA
Universitat Politècnica de Catalunya
Av. Diagonal, 647. Planta 2
08028 Barcelona
Spain

Irena WASIAK
Technical University of Łodź
Institute of Electrical Power Engineering
18/22 Stefanowskiego Str.
90-924 Łodź
Poland

Preface

Power quality (PQ) issues are relatively new: years ago this was a problem concerning only power stations and arc furnace engineers. It is only recently that the electrical engineering community has had to deal with the analysis, diagnosis and solution of PQ problems, even if it has not become a major topic in the industry. Professionals are being confronted with PQ issues on a daily basis, yet only the latest generation of engineers has been trained to face and solve these issues.

The main reason is probably due to the fact that PQ is a complex area covering many different topics. This is also something that makes a comprehensive book difficult – each PQ topic can warrant an entire book, but time is more and more a constraint.

If PQ issues applied to utility networks are relatively new, the same concepts applied to customer installations and equipment have attracted the attention of the electrical world only in really recent times.

The problems related to PQ are often difficult to solve, and may allow different solutions, so the choice is not always simple for those engineers and professionals who are not trained in PQ. The optimal solution to a PQ problem is usually a mix of solutions for a specific situation. In such a situation, it is necessary to identify that problem and propose different solutions to allow the technicians to make the optimal choice.

Evaluation of solutions is probably the key element in PQ problem solving, chiefly for economic reasons. Actually, some solutions require higher investments, and thus the necessary management approval, but managers usually lack the knowledge to evaluate the problems properly.

For these reasons, in 2000 a group of academics and industrialists launched a cultural programme (www.lpqi.org) co-founded by the European Commission and fully dedicated to PQ from the perspective of not just power suppliers but electricity users too. Seven years later this program has more than 100 partners around the world and numerous sub-projects focused on specific issues related to PQ. It was at one of these, LPQIves (LPQI Vocational Educational System), during the Berlin meeting in April 2005, that the idea for this book was born.

Basically, the aim of the LPQIves project was to develop a system of vocational training consisting of methodology and content and, in some countries, expert certification. The members of this project came to the conclusion that the book should be on system components. The authors felt that the book should be a manual for participants on educational courses. Furthermore, the book can serve as a reference book for teachers and a handbook for students on regular university courses, and as a guidebook for people who seek background information on practical solutions to PQ problems.

The unique character of the book is a well-balanced one between a scientific approach and practical knowledge which can be used in everyday situations by people who have only a fundamental electrical engineering background. To reflect this, one of the first decisions taken by the authors was to illustrate each chapter with a case study of a practical application, its measurements and solution.

This multi-use approach makes the book very comprehensive, practice oriented and attractive for a relatively broad audience: namely, scientists looking for links between their specific domain and other PQ domains; engineers seeking a methodology and information on the identification, analysis and solution of a PQ problem; electricity users who need explanations of different PQ terms and definitions; managers looking for background information on the economic consequences of PQ; and students who require a comprehensive manual covering the whole spectrum of PQ.

In order to consider PQ from different perspectives and topics, this book has been organized to cover five 'themes'. The first is dedicated to power system issues. The second is fully dedicated to PQ phenomena in terms of physics, parameters, measurements, sources and mitigations. The last three are dedicated to PQ in practice, PQ problems and economical aspects of PQ. The case studies and other specific content from each chapter are also available on the companion website, *www.wiley.com/go/powerquality*

Before you begin what I hope is interesting reading, let me mention that, although I was still young at the time of the 2005 Berlin meeting, prior to coordinating this group of creative authors around the world, I have been indicated as the main author of this book.

Angelo Baggini
Pavia, Italy

1

Frequency Variations

Hermina Albert, Nicolae Golovanov, Aleksander Kot and Janusz Brożek

Frequency is one of the most important parameters in the assessment of a power system's operational characteristics. Being shared by all the points in the power network, it requires centralized control or at the zone power system levels.

The frequency control and maintenance within allowed limits requires the existence at the system operator level of important power reserves that can be called automatically to assure at any moment a balance of the set-point frequency value of power consumption and generation. An ample and reliable information network that provides the system operator with necessary data is a prerequisite for control of the system frequency in real time.

Failures in the interconnected power system are events that are felt throughout the system, while returning to the stable operation point, and arise from both the automated response of generators in the affected area and through the contributions of neighboring areas via interconnection lines. To this effect, the system operators of the various zone systems use special help procedures to achieve recovery of the normal operational status.

Effective control of the generation and consumption in each zone of the system by the system operator, and also good collaboration with system operators in neighboring zone systems, ensures that the frequency can be maintained within the entire system at the set-point value.

Handbook of Power Quality Edited by Angelo Baggini

One of the most important indices in the operation of alternating voltage systems is the supply voltage frequency, defined as the repetition frequency of the fundamental voltage curve, measured over a specific time interval.

In each power system, the operational moment of the frequency value depends on the extent to which the demand is met by the power sources.

Setting up a nominal power system operational frequency – the frequency value supposed to assure this balance – is a matter of optimizing the possibilities of equipment manufacture and the requirements of specific producers and customers. The selection of a 50 Hz nominal frequency in Europe and 60 Hz in the USA relies on a complex process in which the technical aspects, historical matters and companies' interests all play an important role.

The use of frequency below 30 Hz (the frequency above which the human eye can no longer distinguish a separate succession of images) was accompanied by disturbing variations of the luminous flux of incandescent lamps and these frequencies had to be abandoned. The frequency of 25 Hz was in use on the Cote d'Azure up to 1955. During the interwar period, 42 Hz was widely used in Europe, then it switched to 50 Hz, beginning in 1930. The frequency of 42 Hz was used locally up to 1964.

The use of transformers is advantageous at high frequency. Currently, industry uses on a large scale transformers operating at 30–50 kHz frequencies (welding transformers, lighting transformers, etc.). High frequencies which, due to the skin effect, increase linearly with the frequency of inductive reactance and also increase the dielectric displacement currents through parallel-connected capacitors, are not recommended for power transmission that provides economic parameters, at frequencies as low as possible (the use of transmission lines at direct voltage is an example).

Often, the customer will use direct voltage and in order to obtain it, complex circuits rectifying the alternative voltage are used.

Electric traction, during its first development stage, used alternative voltage single-phase motors with a collector that required a low frequency. For this purpose, 16 2/3 Hz (in Europe) and 20 Hz (in the USA) frequency systems were developed and are still in operation.

The increasing development of power electronics nowadays allows the use of frequency converters in industrial processes, which provides optimum frequency for various processes.

The requirement of power system interconnection determines the standardization of frequency.

Frequency monitoring of the public network and its conservation within required limits is the duty of the system operator, who is supposed to have at his or her disposal sufficient reserves of active power and adequate power frequency control in order to keep frequency deviations within allowed limits.

All equipment (installations) in the European power network are projected to operate at a rated frequency of 50 Hz. Actually, due to the fact that under normal operating conditions the frequency in the power system varies in terms of power variation and according to the response speed of its control systems, while under fault and post-fault conditions its variation depends on the efficiency of the measures adopted to clear the fault, the electric power quality normally requires limits that allow for the frequency variations.

1.1 FREQUENCY QUALITY INDICES

In order to characterize the power system frequency, under normal operating conditions, the following indices are used:

- Δf, the frequency deviation allowing evaluation of slow frequency variations:

$$\Delta f = f - f_r \tag{1.1}$$

where f_r is the rated frequency (50 Hz or 60 Hz) and f is the real frequency (Hz);

- relative frequency deviation, ε_f (%):

$$\varepsilon_f(\%) = \frac{f - f_r}{f_r} \cdot 100 \tag{1.2}$$

- the integral deviations during the day, required to ensure appropriate operation of clocks synchronized to the electrical network frequency:

$$I_f = \int_0^{24} \Delta f \cdot dt \tag{1.3}$$

According to standard EN 50160/2006 [1], the rated frequency of the supply voltage is 50 Hz. Under normal operating conditions, the mean value of the fundamental frequency measured over 10 s stays within the following range:

- for systems with synchronous connection to an interconnected system:

$50\,\text{Hz} \pm 1\,\%$	(i.e. 49.5–50.5 Hz) for 99.5 % of the year;
$50\,\text{Hz} + 4\,\%/-6\,\%$	(i.e. 47–52 Hz) for 100 % of the time;

- for systems with no synchronous connection to an interconnected system (e.g. supply systems on certain islands):

$50\,\text{Hz} \pm 2\,\%$	(i.e. 49–51 Hz) for 95 % of the week;
$50\,\text{Hz} \pm 15\,\%$	(i.e. 42.5–57.5 Hz) for 100 % of the time.

Figure 1.1 shows the curve of frequency variation under normal operating conditions, the values indicated being contained within allowable standard limits.

Regarding transient regimes, it is required that the extensive variations of frequency must be rapidly decreased in order to fall within the frame of the trumpet curve (Figure 1.2) set in compliance with system safety conditions [2]. That is,

$$H(t) = f_0 \pm A \cdot e^{-t/T} \tag{1.4}$$

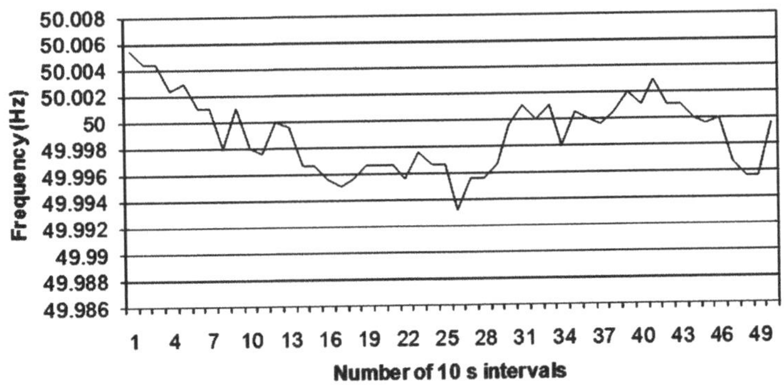

Figure 1.1 Frequency variations within 500 s

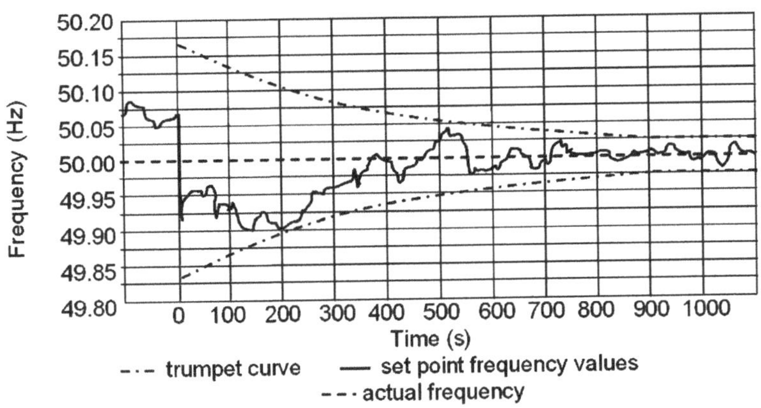

Figure 1.2 Frequency variation during a fault in a power network

where A is the experimental value; f_0 is set frequency value; and T is a time constant resulting from the relation

$$T = \frac{900}{\ln \frac{A}{d}} \quad \text{for} \quad T \leq 900\,\text{s} \quad \text{and} \quad |d| = 20\,\text{mHz} \tag{1.5}$$

Relation (1.5) considers that the post-transient regime begins within 900 s. Consequently, the frequency must be in the range $f_0 \pm 20\,\text{mHz}$.

1.2 FREQUENCY MEASURING

The frequency reading is obtained every 10 s. As power frequency may not be exactly 50 Hz within the 10 s time interval, the number of cycles may not be an integer number. The fundamental frequency output is the ratio of the number of integral cycles counted

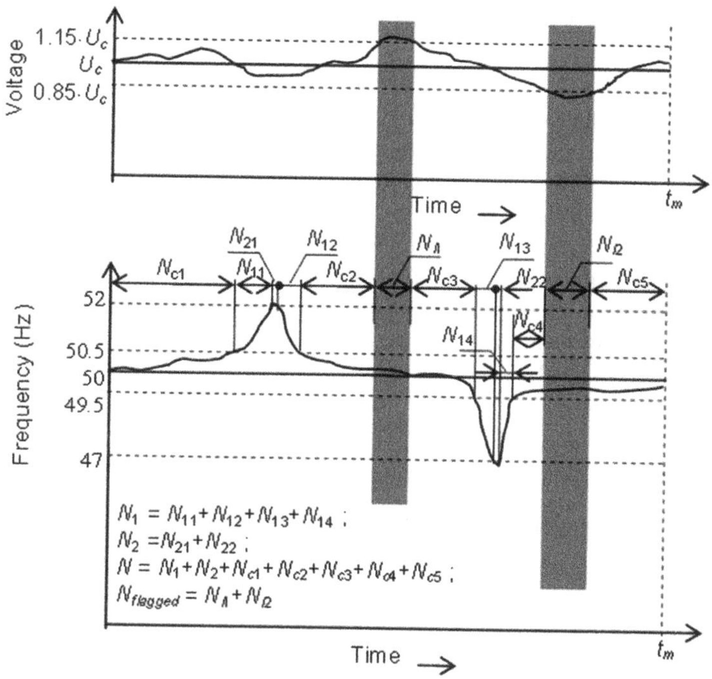

Figure 1.3 Assessment of frequency quality in a power system

during the 10 s time interval, divided by the cumulative duration of the integer cycles. In order to avoid determination errors, it is required to ensure mitigation of harmonics and interharmonics, thus limiting the possibility of unwanted voltage passage through zero. The 10 s measuring intervals should not overlap.

The frequency measurement is performed by class A equipment, with ε_r error, that does not exceed 50 mHz and is not affected by a variation of the total harmonic distortion (THD) of the voltage up to 20 %.

The evaluation of system frequency quality relies on the following procedure:

- monitoring the duration t_m (Figure 1.3) over one week, based on the data obtained on 10 s measuring windows;
- determination of N (the number of 10 s intervals) in which the supply voltage had no deviation larger than $\pm 15\,\%$ from the contracted voltage;
- determination of N_1 (the number of 10 s intervals) in which the frequency differs by more than 0.5 Hz from the rated value while the voltage is within $\pm 15\,\%$ of the contracted voltage;
- determination of N_2 (the number of 10 s intervals) in which the frequency is below 47 Hz or over 52 Hz while the voltage is within $\pm 15\,\%$ of the contracted voltage;
- checking conditions $N_1/N \leq 0.05$ and $N_2 = 0$.

1.3 LOAD–FREQUENCY CHARACTERISTICS

The power system interconnection and the steps taken to maintain the frequency within required limits render deviations from the normalized values very rare phenomena. In this way, an analysis of the influence of frequency variations on the final customers is only performed for a reduced interval about ± 3 Hz of the rated value and for rather short periods.

Within this reduced variation field, a considerable number of static customers (about 40 % of total consumption) are not affected by the frequency variations (rectifier installations, resistance ovens, electric arc ovens, etc.).

On setting the frequency control steps, and lacking further information, it is generally considered that load self-control is 1 % per Hz; that is, the load decreases by 1 % if the frequency goes down by 1 Hz.

The static safety limit is the 20 mHz difference, identical to the one for primary control action.

1.3.1 Influence of the Frequency Variation on the Actuation Motors

The asynchronous and synchronous driving motors, *connected directly to the supply network*, and used extensively in industrial actuation, have a power–frequency characteristic $P = T \cdot \Omega$ dependent on the mechanical characteristic of the load involved, $T = f(\Omega)$, where T is the coupling torque to the motor shaft and Ω is the motor's speed of rotation. Figure 1.4 shows the variation curves of the power according to frequency for various types of customers. Curve 1 corresponds to receivers that in the analyzed frequency field have a consumption independent of frequency. Curve 2 corresponds to types of hoisting installations, mine lifts, beet conveyers, etc., that with a uniform load have practically a speed-independent coupling and therefore the power consumption is proportional to the frequency. Curve 3 is characteristic of viscous loads (calenders for paper fabrication, plastic mass hot processing machines, textile industry machines, etc.). Curve 4 corresponds to a large number of receivers with a parabolic mechanical characteristic (ventilation pumps) and thus a cubic characteristic power–frequency [5].

The speed of asynchronous or synchronous motors connected directly to the electric power supply network varies in proportion to the applied voltage frequency. The frequency variation leads to the corresponding modification of the process productivity throughout the supply with a reduced frequency.

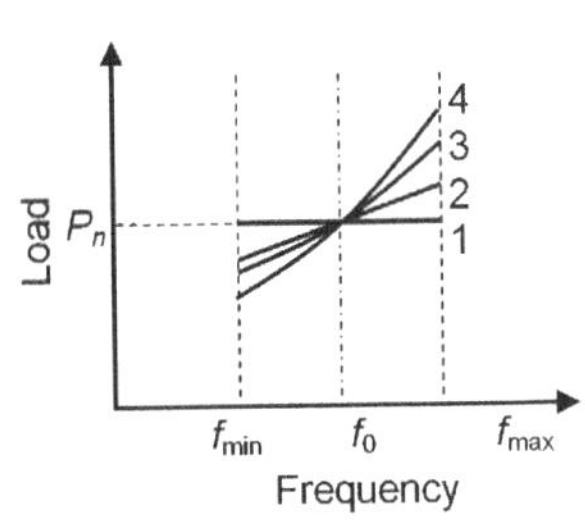

Figure 1.4 Power consumption of various types of receivers versus frequency variation

1.3.2 Capacitor Bank and Harmonic Filters

The reactive power Q generated by the capacitor bank is directly proportional to the supply voltage frequency

$$Q = 2 \cdot \pi \cdot f \cdot C \cdot U^2 \tag{1.6}$$

where C is the capacitor bank capacity and U is the voltage at its terminals.

From the relation (1.6) one can notice the fact that variation of the supply voltage frequency modifies the reactive power determined by the capacitor battery and thus it can influence the value of the power factor at the supply busbars. In most cases, for the allowed range of frequency variation, the influence on the power factor is not important. The frequency variation effects are particularly felt when the capacitor batteries are parts of the harmonic filters. Under normal operating conditions (for a frequency equal to the rated one) the parameters L_h and C_h of the resonant circuit are tuned on harmonic h so that the circuit impedance for this frequency is actually zero, i.e.

$$2 \cdot \pi \cdot f_h \cdot L_h - \frac{1}{2 \cdot \pi \cdot f_h \cdot C_h} \cong 0 \tag{1.7}$$

On operation at any other frequency than the rated frequency, the filter will have an impedance other than zero, which shows a certain degree of discord (out of time tuning) as referred to the h rank harmonic corresponding to the new fundamental frequency.

With decreasing supply voltage frequency, the input impedance of resonant circuits becomes capacitive and can determine overloading of a rank h circuit due to inferior rank harmonics which are not completely filtered.

1.3.3 Transformers and Coils in the Power Network

The maximum value of the magnetic flux Φ_μ determined by the application of voltage U to the terminals of a magnetic circuit winding is obtained through the relation

$$\Phi_\mu \cong \frac{\sqrt{2} \cdot U}{2 \cdot \pi \cdot f \cdot w} \tag{1.8}$$

where U is the actual value of the voltage applied to the winding terminals; f is frequency; and w is the number of coils of the winding.

Analysis of the relation (1.8) reveals that for a certain configuration of the magnetic circuit, the frequency cut leads to an increase in magnetic flux and therefore of magnetic induction accompanied by possible operation in the non-linear zone of the magnetic characteristic. In this way, the transformer's no-load losses increase. It can also result in a distortion of the electric current in the circuit. So, the magnetic circuit operating at reduced frequency becomes a non-linear element of the network. The cut of the voltage frequency to the transformer's terminals also results in cutting of its leakage reactance, $X_\sigma = 2 \cdot \pi \cdot f \cdot L_\sigma$, as well as magnetizing reactance, $X_\mu = 2 \cdot \pi \cdot f \cdot L_\mu$ (Figure 1.5), so that it results in an

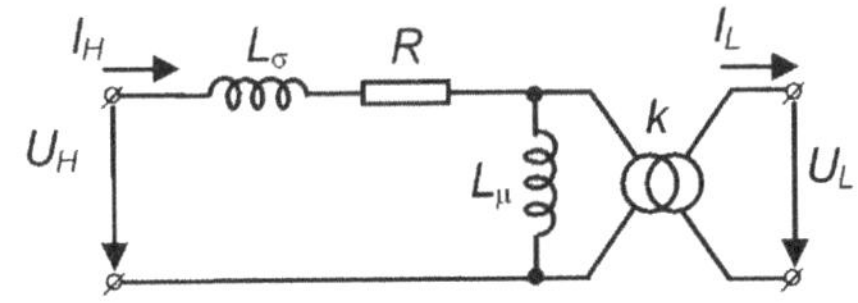

Figure 1.5 Transformer simplified electric circuit

increase in the voltage to the transformers' output and an increase in the no-load operating current. At the same time, the coils' reactance decreases, affecting the characteristic of the network in which the coil is connected.

1.4 INFLUENCE OF FREQUENCY ON USERS' EQUIPMENT

The supply voltage frequency in the public electrical network, under current conditions in Europe, has very low variations in regular operation. Under emergency conditions, however, transient conditions emerge in which the frequency variations range within the limits of the so-called trumpet curve.

Practical calculations can consider the electrical network supply voltage frequency to be constant and equal to 50 Hz.

An industrial electrical network mostly uses driving systems with asynchronous motors supplied via frequency converters which allow control of the asynchronous motor's rotational speed within large limits in practical cases from zero to double the rotational speed obtained with 50 Hz supply (Figure 1.6).

1.4.1 Influence of Frequency Variations on Asynchronous Motors

An analysis of the behavior of an asynchronous motor on variation in frequency can be performed based on a simplified equivalent diagram (Figure 1.7). This simplified diagram uses the following notation:

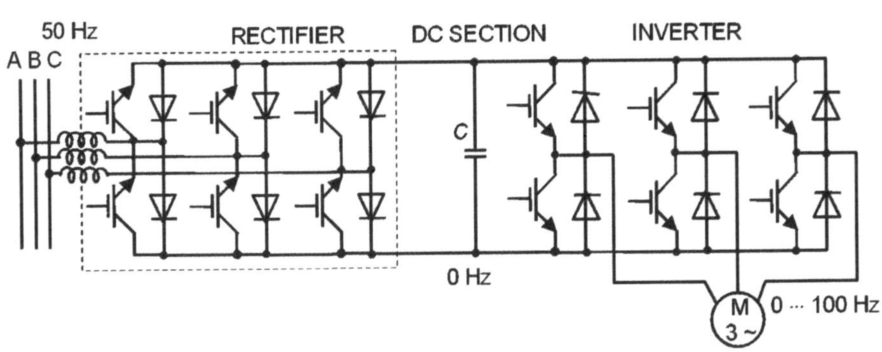

Figure 1.6 Asynchronous motor variable frequency supply

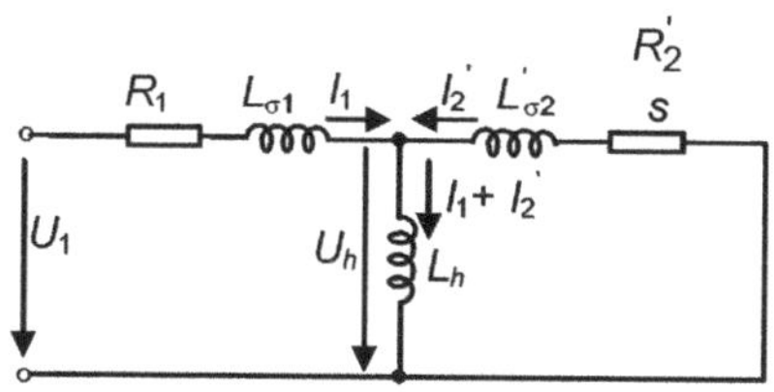

Figure 1.7 Simplified equivalent electric circuit for an asynchronous motor

- $\mathbf{U}_1$ is the supply voltage per phase;
- R_1 is the stator winding electric resistance;
- $L_{\sigma 1}$ is the stator winding leakage inductivity as compared to the rotor winding;
- $\mathbf{I}_1$ is the stator winding electric current value;
- $\mathbf{U}_h$ is the air gap equivalent voltage;
- L_h is the inductivity corresponding to the machine magnetization flux;
- $\mathbf{I}_2'$ is the rotor electric current value, related to stator voltage;
- L_{s2}' is the leakage inductivity of the rotor winding as compared to the stator winding, related to stator voltage;
- R_2' is the rotor winding electric resistance related to stator voltage;
- s is machine slip.

Assuming that a magnetization current $\mathbf{I}_1 + \mathbf{I}_2'$ is actually negligible, torque M_{el} developed by the machine can be written as

$$M_{el} = \frac{3 \cdot R_2' \cdot U_1^2 \cdot p}{s \cdot \omega \cdot \left[\omega \cdot L_\sigma^2 + \left(R_1 + \frac{R_2'}{s} \right)^2 \right]} \tag{1.9}$$

where $L\sigma = L\sigma_1 + L\sigma_2$ is the machine leakage inductivity; p is the number of pole pairs; and $\omega = 2 \cdot \pi \cdot f$ is the applied voltage pulsation (f – supply voltage frequency).

The machine's mechanical feature is $T_{el} = f(\Omega)$, where Ω is the speed of rotation indicated in Figure 1.8 for various supply voltage frequencies.

Analysis of the curves in Figure 1.8 outlines the reduction of supply frequency compared to increased rated frequency $f_r (f_1 < f_r)$, breakdown torque T_k and starting torque

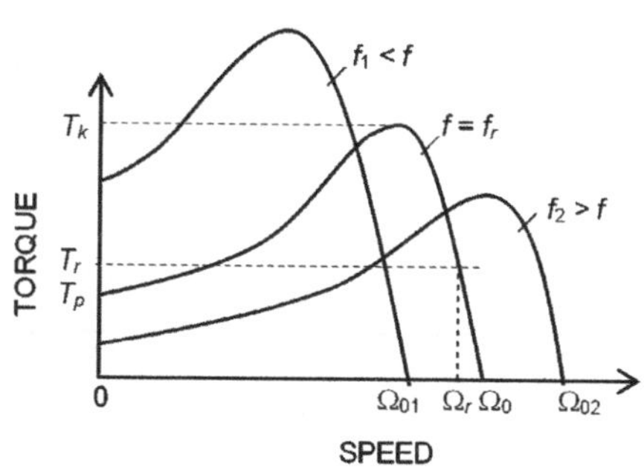

Figure 1.8 Asynchronous machine mechanical features for various supply voltage frequencies

T_p, which stresses the advantage of using asynchronous motors in starting up working machines. For a frequency above the rated frequency, the working machines' driving speed increases.

Actually, the curves in Figure 1.8 cannot be used because, for reduced frequencies, the machine magnetization current leads to magnetic circuit saturation.

To keep the induction constant within the machine magnetic circuit, the following condition has to be met:

$$\frac{U_1}{f} = \text{constant} \tag{1.10}$$

For frequencies above the rated frequency, the condition (1.10) would allow operation with a voltage above the rated one, but this is not allowed due to the inadmissible electrical requirement of the machine's electric insulation. Thus, the relation between machine supply voltage and frequency is as shown in Figure 1.9.

Relation (1.9) indicates that, to meet condition (1.10), the breakdown torque is actually constant, which determines the mechanical characteristics used to obtain the form shown in Figure 1.10.

Analysis of the curves in Figure 1.10 underlines the fact that the use of asynchronous motors fed with variable frequency voltage ensures control within a large range of the speed of rotation without causing further losses throughout the control process.

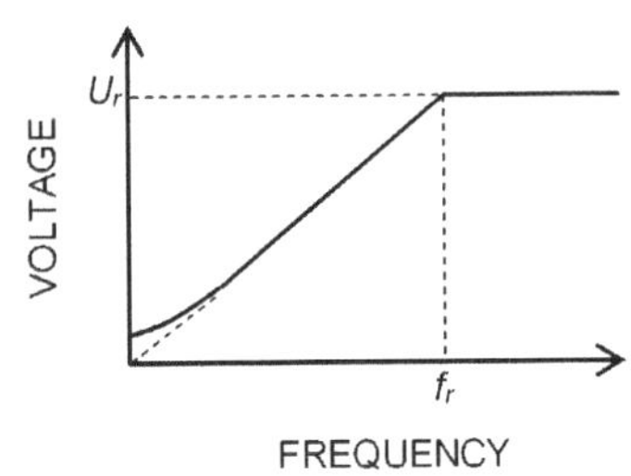

Figure 1.9 Voltage–frequency characteristic of asynchronous motor supply with variable frequency

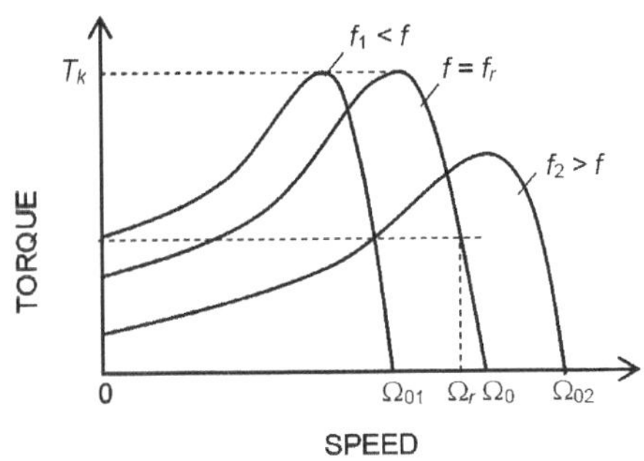

Figure 1.10 Mechanical characteristics of an asynchronous motor fed with variable frequency that meets the constant magnetic induction condition for frequencies under the rated frequency

Obviously, the use of an asynchronous motor fed with variable frequency voltage has important technological advantages which triggered the widespread use of this rotational speed control system in modern industry. Still, it is necessary to consider that, at low speeds, further machine cooling is required (the fan on the machine shaft determines a low air flow due to the low driving speed) and also that lacking appropriate measures to limit the electromagnetic disturbances, the system can lead to the occurrence of harmonics and interharmonics which could affect the quality of the electric power supplied to other consumers in the area.

1.4.2 Influence of Frequency Variations on Parallel-Connected Condensers and Coils

Until now, the use of the control diagram in Figure 1.6 was only possible for motors fed from a low-voltage network due to the lack of technical and economic solutions acceptable for the medium-voltage zone.

Currently, these solutions exist, and therefore the medium-voltage motors too can be fed directly with variable frequency voltage.

Existing installations for medium-voltage motors supplied with variable frequency use an intermediate reduced voltage circuit and an increased frequency converter outlet voltage (Figure 1.11).

Transformer T_2 operation at variable frequency requires the magnetic induction to be kept at a value that should not exceed the saturation limit. As in the case of the asynchronous machine, to keep the induction equal to the rated induction, for frequencies under the rated frequency it is necessary to meet the condition (1.10) between the terminal voltage and supply voltage frequency. In the specific case shown in Figure 1.11, the condition (1.10) is met, being required by the asynchronous motor operation.

Obviously, the condition (1.10) has to be met by the transformers of another type that can operate at frequencies other than the rated frequency.

One case often met is that where pieces of equipment are designed for 60 Hz operating systems. Their utilization in 50 Hz frequency systems requires the supply voltage to be reduced some 20 % against their rated voltage.

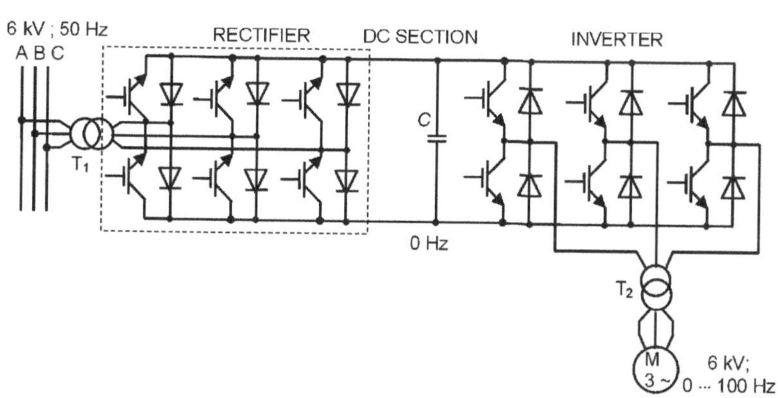

Figure 1.11 Medium-voltage asynchronous motor supply with variable frequency

The pieces of equipment with magnetic circuits designed to operate both to 50 Hz and 60 Hz have larger losses with the 50 Hz supply. For low-power equipment the losses are irrelevant. For equipment of larger power it is required to consider operation at reduced frequency.

Identically, the utilization of parallel-connected coils, at a frequency other than the rated frequency, requires analysis of the conditions when saturation limits are exceeded and losses increase.

Induction keeping to the rated value requires control of the voltage to terminals as per condition (1.10).

1.4.3 Influence of Frequency Variations on Series-Connected Condensers and Coils

It is customary in variable frequency circuits of asynchronous motors to use filters consisting of parallel-connected condensers and series-connected coils.

Reactive power Q generated by a parallel-connected condenser of capacity C and fed with variable frequency voltage U is given by

$$Q = C \cdot \omega \cdot U^2 = C \cdot \omega^3 \cdot \frac{U^2}{\omega^2} \tag{1.11}$$

The asynchronous motor supply by meeting condition (1.10) determines that, at low-frequency operation, the reactive contribution of the condenser connected to the motor terminals must be greatly reduced.

Coils connected in series with the motor fed with variable frequency have an inductive reactance proportional to the actual frequency going through the coil, which is to be considered on sizing the filtering circuit connected in series with an asynchronous motor.

1.5 GOVERNING OF TURBINE SPEED

The frequency variations are determined by variations of the active power in the system, due mainly to variations in the final users' consumption. Frequency stabilization at the set-point value requires a permanent adjustment of the electric power required at the generators' terminals with the mechanical moment at their shafts (performed by the actuation equipment).

The frequency control (generator rotational speed) is performed by means of the automated speed regulator (ASR). The main function of an ASR is to provide a constant speed to the electric generator rotation using the valves of the turbine for control, according to the deviation $\Delta\Omega$ of the rotational speed Ω versus the Ω_0 set-point value, determined by the active power modification

$$\Delta\Omega = \Omega - \Omega_0 \tag{1.12}$$

by

$$\Omega_0 = \frac{\omega_0}{p} \tag{1.13}$$

where ω_0 is the pulsation ($\omega_0 = 2 \cdot \pi \cdot f_0$) and p is the number of pole pairs of the electric generator.

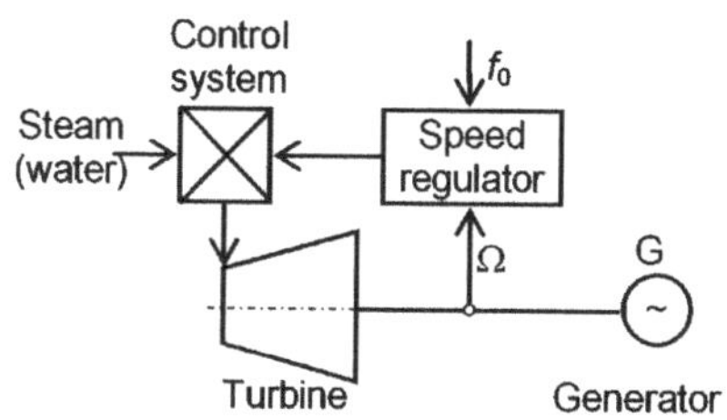

Figure 1.12 Speed regulator in the power plant

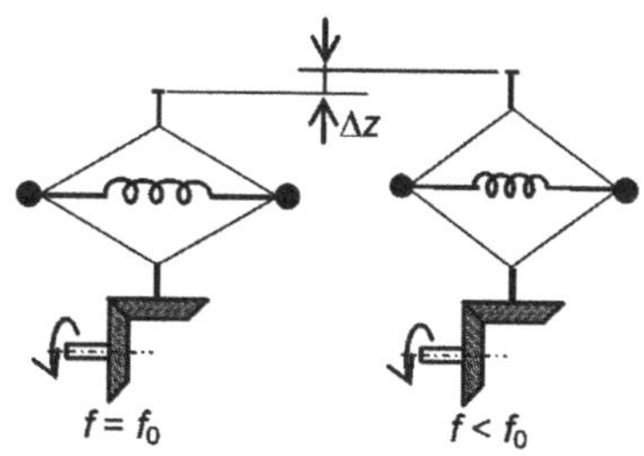

Figure 1.13 Centrifugal regulator operation system

The speed regulator should determine the steam (water) flow increase to the turbine inlet for $\Omega < \Omega_0$ and fluid flow decrease for $\Omega > \Omega_0$ (Figure 1.12).

The first speed regulators were of mechanical–hydraulic type and changed the signal for turbine shaft rotational speed into a linear displacement with a negative slope via a centrifugal system (Figure 1.13), which imposed a linear speed of rotation characteristic with a slightly negative slope (Figure 1.14, curve P_{d2}) on the variation of the torque for the actuation turbine shaft. The speed regulator actuates on admitting steam (water) into the turbine by a value Δz proportional to the variation Δf of the frequency (speed of rotation of the generator shaft) that controls the coupling with the turbine shaft and therefore its

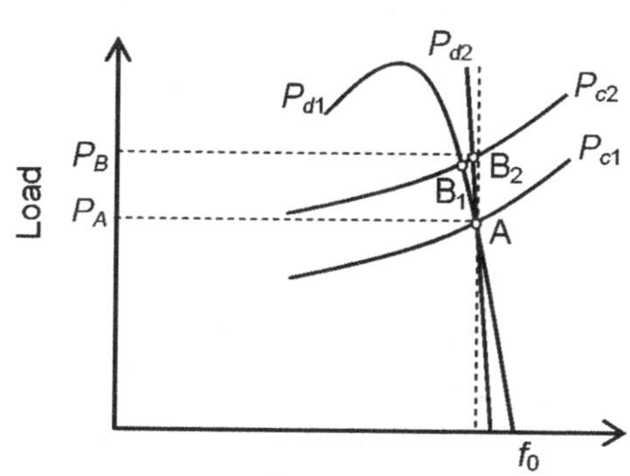

Figure 1.14 Frequency modification on load increase within power system

rotational speed. Without the speed regulator, the rotational speed variation on variation of the coupling with the turbine shaft would correspond to the curve P_{d1}.

Consider set-point frequency f_0, while the power required at the generator terminals has a characteristic P_{c1}. Characteristics P_{d1} and P_{d2} are controlled so that operation point A corresponds to the set-point frequency value and to the power required P_A.

If the power requested increases corresponding to characteristic P_{c2}, for the case of default of the automated speed regulator, the new operation point is established at point B_1 for a rotational speed (frequency) that is generally unacceptable (beyond allowed variation limits). The use of ASR ensures a new operation point B_2 corresponding to a frequency lower than the frequency set-point value but close enough to it. The power transmitted by the generator increases the value P_B, thus complying with the power demand in the system.

The main disadvantage of mechanical–hydraulic regulators is the low sensitivity (no reaction unless the amplitude of the $\Delta\Omega$ deviation exceeds a certain value) and variations in accuracy due to wear.

Currently, mechanical–hydraulic regulators have been replaced by electric–hydraulic regulators having the same operating principle yet with higher accuracy through replacing the speed measuring system and mechanical systems with high-performance electronic systems that are very reliable and accurate. These systems also allow control by means of additional signals for effective control of the electric power generation system.

Figure 1.15 presents the diagram of such a speed regulator.

The deviation signal $\Delta\Omega$ is amplified and then integrated to obtain the control signal ΔY that will command the fluid turbine admission system. The integrated-type regulator generates a new balance condition between the active power released by the generator and the rotational speed when deviation $\Delta\Omega = 0$.

The response in time of the generating unit, with an active power increase of ΔP, is shown in Figure 1.16.

An active power increase from an initial value P_1 to a value P_2 should trigger a reduction in the rotational speed and consequently the frequency. The regulator should command an increase in steam (water) flow at the turbine inlet which would lead to an increase in mechanical power and then rotational speed up to the set speed of Ω_0.

The astatic regulators are specific to the isolated systems considering that, within an interconnected system, the recommended specification of the same set speed for the different unit regulators leads to an unstable operation of the frequency control process within the system.

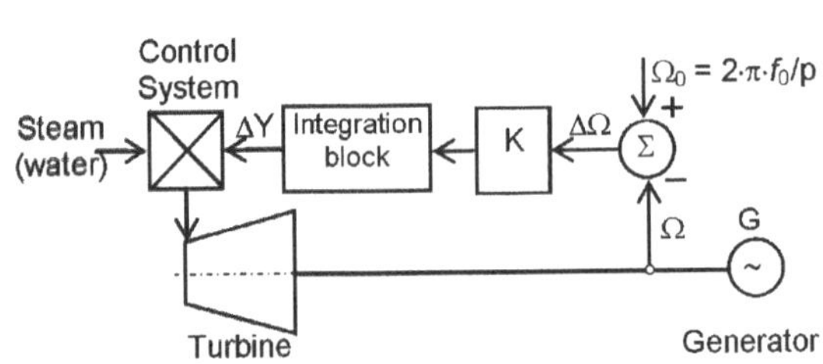

Figure 1.15 Schematic diagram of an astatic speed regulator

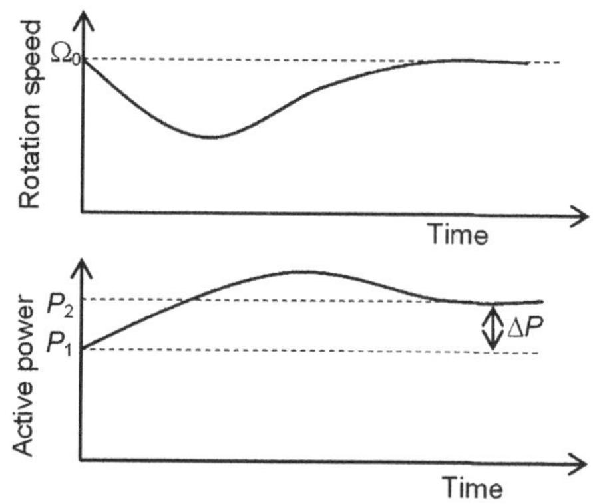

Figure 1.16 Response of an astatic speed regulator to a change in active power

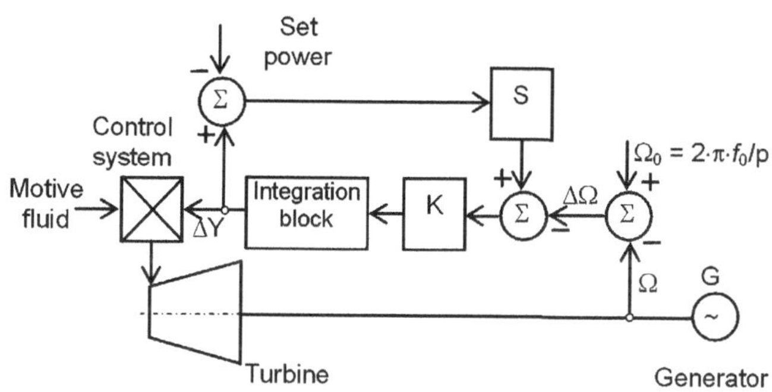

Figure 1.17 Schematic diagram of a static speed generator

To obtain a stable partition of the power variations between generating units trying to cover the power deviations, the speed regulators can be allowed a small frequency reduction on increase of active power (static speed regulators).

Figure 1.17 presents the schematic diagram of a static speed regulator, while Figure 1.18 shows the system response to the use of this control adjustment system.

The interconnection between the rotational speed (proportional to frequency) and turbine mechanical power represents the static feature of the unit. Ideally, this feature is a line that has a slope R named **statism**:

$$R = \frac{\Delta f^*}{\Delta P^*} = \frac{\dfrac{f_{s\min} - f_{s\max}}{f_0}}{\dfrac{P_{\max} - P_{\min}}{P_0}} \tag{1.14}$$

where Δf^* is the relative frequency variation; $f_{s\min}$, $f_{s\max}$ are the frequencies corresponding to minimum unit maximum load, respectively; $P_{\max}$, $P_{\min}$ are powers under unit maximum and minimum load condition; and P_0 is unit reference power (currently rated power).

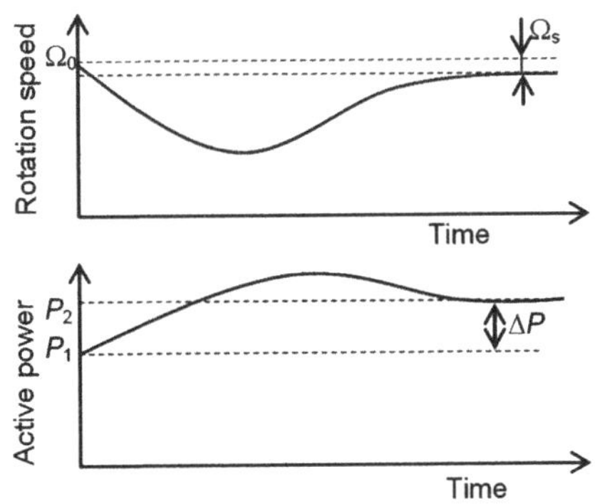

Figure 1.18 Response of a static speed regulator to a change in active power

Habitually, the generating unit statism ranges between 1 and 7 % so that for a unit with 3 % statism there is a 1.5 % frequency variation when a 50 % load variation occurs.

For the linear static feature, the relation (1.8) can also be written as follows:

$$P = P_0 - \frac{P_0}{R} \cdot \frac{f - f_0}{f_0} = P_0 - \frac{P_0}{R} \cdot \frac{\Delta f}{f_0} \tag{1.15}$$

where f is the frequency within the system and P is the unit active power.

In an interconnected electric power system, where the frequency deviation $\Delta f = f - f_0$ is the same at all units, each unit will take power depending on its statism.

1.6 FREQUENCY CONTROL IN POWER SYSTEMS

The electric frequency f in the network (the system frequency) is the only parameter that is common for synchronously interconnected systems. It is a power quality parameter whose steady-state value is identical at every point of the interconnected power system.

The system frequency f appears in relationships with other, fundamental quantities which characterize power system operation (active power P, reactive power Q, voltage V). Particular relationships take place (Figure 1.19) between the active power P and frequency f (P–f), the reactive power Q and voltage V moduli (Q–V), and Q–V parameters and P–f parameters.

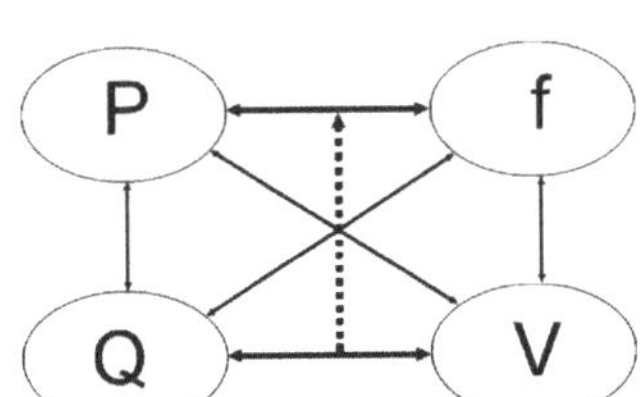

Figure 1.19 Relationships between basic electrical quantities in the power system

The system frequency is determined by the rotational speed of synchronized generators. In the steady state all connected generators are synchronized, i.e. the rotors of all generators have the same rotational speed and rotate in the same direction. According to Newton's second law for rotational motion, the rotors' motion can be expressed by the following equations:

$$M\frac{d^2\delta}{dt^2}+D\frac{d\delta}{dt}=P_m-P_e$$
$$\Delta\omega=\frac{d\delta}{dt} \tag{1.16}$$

where M is the inertia coefficient, D the damping factor, P_m the mechanical power delivered by the turbine, P_e the generator electric power, δ the rotor angle with respect to the synchronously rotating axis, and $\Delta\omega=(\omega-\omega_s)$ the change in the rotor angular speed with respect to synchronous speed ω_s.

If the power system is balanced (the frequency is constant), then $\Delta\omega=(\omega-\omega_s)=0$, i.e. $P_m=P_e$. That means the generated power is balanced by the power of loads and active power losses in the network. Unbalance of power in the system results in an increase or decrease of the system frequency (the rotational speed of rotors). The rate of frequency changes depends on the power unbalance and the inertia of rotating masses of generators' rotors.

As follows from the above, *the frequency is directly related to the balance of power generated and consumed in the system* and frequency variation is a sensitive **indicator** of the active power balance.

Since electric power cannot be stored easily or economically on a large scale, it has to be generated and delivered at the same time as it is consumed. Distribution of electric energy in power systems takes place with a speed near to that of electromagnetic wave propagation. In order to keep the power balance, the power system is operated according to the principle that active power generation follows the varying demand. The frequency variation is therefore a signal for the system control. Thus a control system, based on the direct relationship between the active power and frequency, is the fundamental control in the power system.

The voltage level at power system nodes is controlled by alteration of reactive power flow (V–Q control, Figure 1.19). This is a local control, in contrast to the frequency control performed for the entire system. Due to the significant difference between the speed of V–Q control and P–f control, the system analyses should take into account the influence of V–Q control on the P–f control (the voltage increase causes an increase in consumed power, which results in the frequency deviation).

1.6.1 Composite Load

A large number of various loads are being connected to electric power systems. In terms of the power system analysis it is impracticable to follow the behavior of each load. Therefore, for practical reasons, the notion of composite load has been introduced. It represents the resultant behavior of a large group of diverse loads connected to a node of a high-voltage network.

Variation of system parameters (including the frequency f) results in a change of the active power consumption of composite loads. The dependence of consumed active power

P_L on the frequency f is expressed by the static power–frequency characteristic for the consumed active power $P_L = P(f)$.

The static characteristics are usually unitized with respect to the nominal frequency f_n, nominal power P_n and nominal voltage V_n.

In the neighborhood of the nominal frequency value (e.g. 48–52 Hz) the static power–frequency characteristic for consumed active power can be expressed by a linear relationship. This relationship is numerically expressed by the **frequency susceptibility of a load k_{Pf}**, defined as the quotient of the relative power deviation and the relative frequency deviation:

$$k_{Pf} = \frac{\Delta P_{pu}}{\Delta f_{pu}} = \frac{P_{npu} - P_{pu}}{f_{npu} - f_{pu}} = \frac{(P_n - P)/P_n}{(f_n - f)/f_n} \tag{1.17}$$

Figure 1.4 shows power–frequency static characteristics for active power consumption of various types of loads.

The value of susceptibility factor k_{pf} depends on the structure of a composite load, i.e. on the contribution of a given load group. In theory (assuming all loads are connected in parallel at a common node) the resultant frequency susceptibility factor k_{pf} can be determined from the formula

$$k_{Pf} = \frac{\sum_{i}^{n} k_{Pfi} \cdot P_{Li}}{P_L} \tag{1.18}$$

where k_{Pfi} is the frequency susceptibility factor for the ith group of loads; P_{Li} is the power of the ith group of loads; and P_L is the total power of loads.

Analytical calculations are difficult because the contribution P_{Li}/P_L of a given load group to the composite load cannot be determined. The values of k_{Pf} are usually obtained by measuring $P_L = P(f)$ characteristics for typical groups of loads.

For the convenience of practical calculations the frequency susceptibility factor c_f expressed as a percent of power deviation per 1 Hz is used. Converting the relationship (1.17), it is possible to obtain

$$k_{Pf} = \frac{\Delta P_{pu}}{\Delta f_{pu}} = \frac{\dfrac{\Delta P}{P_n}}{\dfrac{\Delta f}{f_n}} \rightarrow \frac{\Delta P}{P_n} = k_{Pf} \frac{\Delta f}{f_n} \tag{1.19}$$

That is,

$$\Delta P\,\% = \frac{\Delta P}{P_n} \cdot 100 = k_{Pf} \cdot \frac{100}{f_n} \Delta f = c_f \cdot \Delta f \tag{1.20}$$

For a typical composite load, comprising industrial, commercial and residential loads, the c_f factor values are of the order of 1–6 % per 1 Hz.

1.6.2 The Generation Characteristic

The preceding subsection describes the loads behavior under frequency variations. Now, let us discuss the resultant generation characteristics.

The rotational speed of generators in the power system is proportionally related to the system frequency:

$$\frac{\Delta f}{f_n} = \frac{\Delta \omega}{\omega_n} \tag{1.21}$$

From relationship (1.21) it is possible to formulate an equation for a single generating unit:

$$\frac{\Delta f}{f_n} = -R_i \frac{\Delta P_{mi}}{P_{ni}} \quad \text{or} \quad \frac{\Delta P_{mi}}{P_{ni}} = -K_i \frac{\Delta f}{f_n} \quad \text{or} \quad \Delta P_{mi} = -K_i \cdot P_{ni} \frac{\Delta f}{f_n} \tag{1.22}$$

where R_i is the droop of the ith generating unit as determined by (1.14); ΔP_{mi} is the change in mechanical power of the ith generating unit; and P_{ni} is the nominal power of the ith generating unit.

The overall power change ΔP_G for N generators in service will be

$$\Delta P_G = \sum_{i-1}^{N} \Delta P_{mi} = -\frac{\Delta f}{f_n} \sum_{i=1}^{N} K_i P_i \tag{1.23}$$

At steady state the equilibrium between the power generation P_G and the system load P_L (i.e. total power consumed, including the system losses) occurs:

$$P_G = \sum_{i-1}^{N} P_{mi} = P_L \tag{1.24}$$

Dividing Equation (1.23) by P_L, the system static characteristics are obtained which express the dependance of the change of generated power and frequency variation for N operating generators:

$$\frac{\Delta P_G}{P_L} = -\frac{\sum_{i=1}^{N} K_i P_{mi}}{P_L} \frac{\Delta f}{f} = -K_G \frac{\Delta f}{f} \quad \text{or} \quad \frac{\Delta f}{f} = -R_G \frac{\Delta P_G}{P_L} \tag{1.25}$$

where $R_G = 1/K_G$ is referred to as the system droop.

Figure 1.20 shows a summation of characteristics of generating units which yields a resultant generation characteristic of the system.

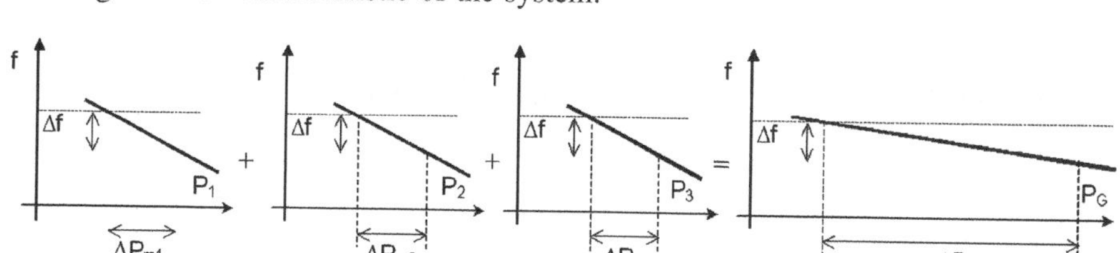

Figure 1.20 Summation of characteristics of generating units

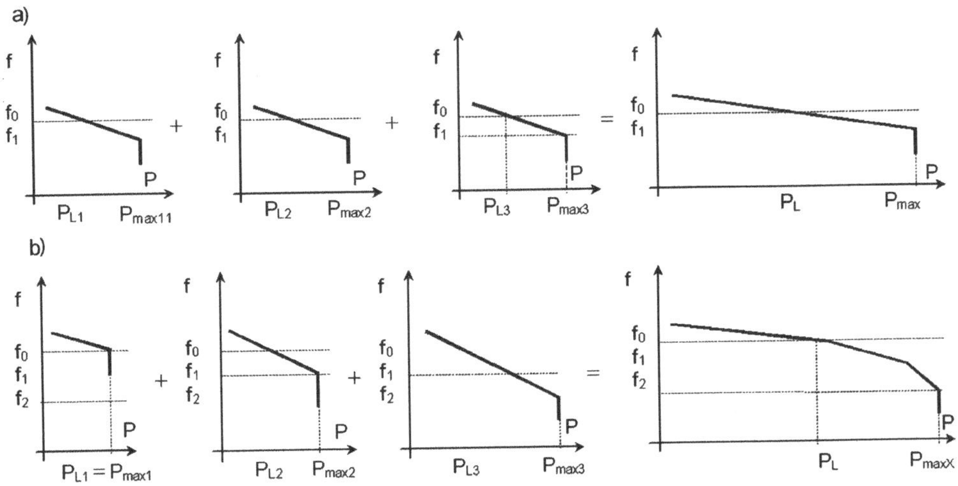

Figure 1.21 The influence of generator loads on the resultant generation characteristic

The summation is made at a constant frequency and under assumption that generation units' droop characteristics are linear and with no limitations. The resultant characteristic has a very small slope – the sum of power changes of all generating units ΔP_G corresponds to the frequency increment Δf.

The above discussion leading to a resultant linear generation characteristic (Figure 1.21) assumed no constraints in power generation.

Under real conditions the generation units must not be loaded with a power greater than their maximum power $P_{\max}$ resulting from technical constrains. Reaching the unit maximum power appears as a vertical cut-off on its characteristic (Figure 1.21) which corresponds to infinite droop.

The presence of generators loaded with the maximum power in the set of generating units in service changes the droop of the resultant generation characteristic, noticeable in Figure 1.21(b) as an increasing slope of the characteristic.

The spinning reserve of a system is the difference between the maximum power of generating units in use and their actual load. The amount of spinning reserve and its allocation within the system has a significant influence on the resultant generation characteristic, which is evident in Figure 1.21.

1.6.3 The System Properties and Control Basics

The frequency properties of a system are determined by the resultant characteristics of both power generation and demand versus the system frequency f, combined together. The essence of frequency control can be derived from the analysis of the system behavior as a whole.

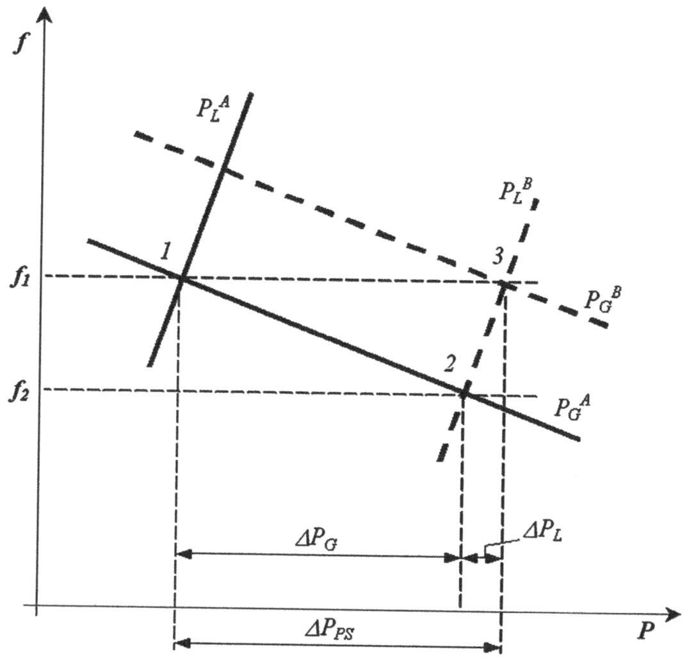

Figure 1.22 The equilibrium points of the system for two different power demand values

Figure 1.22 [6] shows the system generation characteristic P_G, and the characteristic of the change in active power P_L, versus the system frequency. The point of system equilibrium is the point of intersection of characteristics $P_G = P_L$.

The point of system equilibrium 1 at the intersection of characteristics P_A^L and P_G^A corresponds to a certain load P_A^L. The system operates at the frequency f_1. An increase in the system load shifts the demand power characteristic from P_A^L to P_L^B. The new point of the system equilibrium will be the point 2 at the intersection of characteristics P_L^B and P_G^A. The system now operates at frequency f_2, lower than f_1.

As follows from Figure 1.22, the system frequency has decreased by $\Delta f = f_1 - f_2$ due to the increase of load. According to the generation characteristic, the generated power has increased by ΔP_G. As a result of the frequency decrease the actual power demand has decreased by ΔP_L (the so-called self-regulating effect of load, being the consequence of the proportional relationship between the consumed power and frequency). The overall value of these changes corresponds to the segment 1–3 and can be computed from the relationship

$$\Delta P_{PS} = \Delta P_G - \Delta P_L = -(K_G + K_L) \cdot P_L \cdot \frac{\Delta f}{f_n} = -K_{PS} \cdot P_L \cdot \frac{\Delta f}{f_n} \tag{1.26}$$

or

$$\frac{\Delta P_{PS}}{P_L} = -K_{PS} \cdot \frac{\Delta f}{f_n}$$

Coefficient K_{PS} is referred to as the **system stiffness**. It determines the resultant frequency variation caused by the change in power demand in the system.

On increasing the load, the system frequency decreases. The system, whose turbines are equipped only with speed governors, is not able to return unaided to the initial frequency f_1. The new point of equilibrium 2 is below the initial point 1. The frequency returns to the f_1 value if the generation characteristic P_G^A is shifted to the position indicated by the dashed line (P_G^B) in Figure 1.22. Then the new point of equilibrium will be point 3.

The position of the generation characteristic can be changed by means of altering the settings of speed governors of generating units operating in the system.

Each turbine speed governor allows the reference rotational speed ω_{ref} to be set. Turbine governing systems are equipped with a supplementary control node where the set power value P_{ref} can be introduced in order to avoid regulation by means of the rotational speed set value (cf. Section 1.5). In consequence of a change in the set value of ω_{ref} or P_{ref}, the generation characteristic of a generating unit is shifted.

The system generation characteristic is composed of individual generating unit characteristics. Therefore, shifting the generating unit characteristics also shifts the entire system characteristic. The modification of reference values does not need to be applied to all the operating units in a given system.

1.6.4 Frequency Control in an Islanding System and in Interconnected Systems

A power system can be referred to as an islanding system when it is disconnected from other systems and does not exchange power through tie-lines.

Due to the fact that frequency is the same within the entire system, frequency control in an islanding system can be achieved in a relatively simple way. The turbine speed governors should be provided with supplementary elements that change the settings according to frequency variations. More information on turbine governing systems can be found in Section 1.5.

After a load change, the frequency reaches a new steady-state level, different from the initial value. The frequency (or rotational speed) error causes an additional integral term to generate a signal which modifies the value of the power setting of a generating unit. With a sufficiently large number of generating units supplied with control systems of this type, the power system yields such a change in the generated power that frequency returns to its initial value.

This concept of frequency control can be referred to as decentralized since it is performed by regulators in power stations situated at various locations within the system.

Such a solution cannot be applied to cooperating power systems. In interconnected systems the goal of frequency control is to provide a power control also in the tie-lines. The frequency and power exchange control can only be achieved by means of centralized control.

The central control system also allows for other goals, such as automation of optimal load flow.

1.6.5 Frequency Control: Primary, Secondary and Tertiary

1.6.5.1 Primary Control

Primary control consists of changing a generating unit's power versus the frequency, according to its static generation characteristic as determined by the speed governor settings.

The objective of primary control is to re-establish a balance between generation and demand within the synchronous area at a frequency different from the nominal value. This is done at the expense of the kinetic energy of rotating masses of generating sets and connected motors.

The primary control action time is 0 to 30 s after disturbance of the balance between generation and demand.

Under normal conditions the system operates at nominal frequency, maintaining the condition of equality of generated power and demand. Each disturbance of this balance, due to, for example, disconnection of a large generating unit or connection of a large load, causes a change in frequency. At first, the frequency varies rapidly, practically linearly, and attains the maximum deviation from the nominal value, referred to as the dynamic frequency deviation (Figure 1.23).

This deviation in the system frequency will cause the primary controllers of all generators subject to primary control to respond within a few seconds. The controllers alter the power delivered by the generators until a balance between the power output and consumption is re-established. At the moment when the balance is re-established, the system frequency stabilizes and remains at a quasi-steady-state value, but differs from the frequency set point because of the generators' droop.

The magnitude of the dynamic frequency deviation depends on: the amplitude and development over time of the disturbance affecting the balance between power output and consumption; the kinetic energy of rotating machines in the system; the number of generators subject to primary control; the dynamic characteristics of the machines (including controllers); and the dynamic characteristics of loads. The quasi-steady-state frequency deviation is governed by the amplitude of the disturbance and the system stiffness.

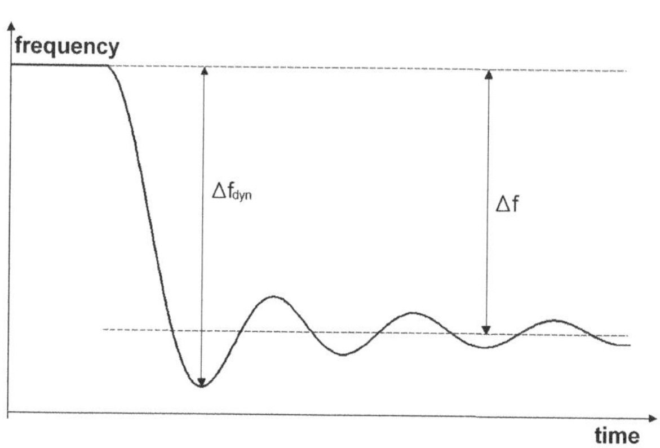

Figure 1.23 Definition of the dynamic (Δf_{dyn}) and quasi-steady-state frequency (Δf) deviation

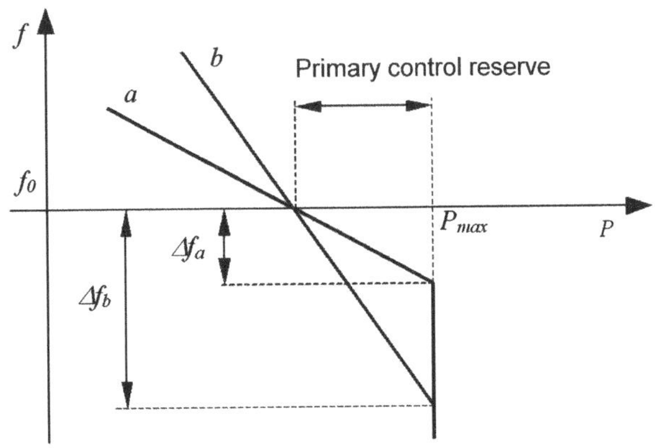

Figure 1.24 The contribution of two generators, with different droops, to primary control

The contribution of a generator to primary control depends upon the droop of the generator and the primary control reserve of the generator concerned. Figure 1.24 shows the characteristics of two generators *a* and *b* and of different droops under equilibrium conditions, but with identical primary control reserves. In the case of a minor disturbance, for which the frequency offset is smaller than Δf_a, the contribution of generator *a* (with the smaller droop) will be greater than that of generator *b* (the one with the greater droop).

The primary control reserve of generator *a* is exhausted (i.e. where the power generating output reaches its maximum value) earlier (at the frequency offset Δf_a) than that of generator *b* (which will be exhausted at the frequency offset Δf_b), even when both generators have identical primary control reserves.

For an adequate operation of frequency control it is crucial that the system has a proper level of primary control reserve at any instant of time allocated in a possibly large number of generating units and activated within a few seconds of detecting the frequency deviating from its nominal value.

1.6.5.2 Secondary Control

Secondary control makes use of a central regulator, modifying the active power set points of generating sets subject to secondary control, in order to restore power interchanges with adjacent control areas to their programmed values and to restore the system frequency to its set-point value at the same time. By altering the operating points of individual generating units, secondary control ensures that the full reserve of primary control power activated will be made available again.

Secondary control operates slower than primary control, in a timeframe of minutes. Its action becomes evident about 30 s after a disturbance/event, and ends within 15 min.

Since under normal operating conditions of a power system the power demand varies continuously, secondary control takes place continually in such way as not to impair the action of primary control.

Secondary control requires: a central regulator; a system measuring the interchanged power in tie-lines; measurements of the system frequency; and a system for transmission of regulator signals to the relevant generating units.

The central regulator minimizes in real time the system control error G, expressed as

$$G = P_{measure} - P_{program} + K \cdot (f_{measure} - f_0) \tag{1.27}$$

where $P_{measure}$ is the sum of the instantaneous measured active power transfers on the tie-lines; $P_{program}$ is the resulting programmed exchange with all the neighboring control areas; K, the system factor, is a constant in MW/Hz set on the secondary controller; $f_{measure}$ is the measured instantaneous value of system frequency; and f_0 is the set-point (nominal) frequency.

The desired behavior of secondary control over time will be obtained by assigning a proportional–integral (PI) characteristic of the central regulator, according to the following equation:

$$\Delta P_d = -B \cdot G - \frac{1}{T}\int G dt \tag{1.28}$$

where P_d is the correction signal of the central regulator governing the generating units subject to secondary control; B is the gain (proportional term) of the central regulator; T is the integral time constant of the central regulator; and G is the system control error, described by the Equation (1.27)

A disturbance of the balance between power generation and demand in synchronous systems gives rise to variations in the system frequency observed over the entire system despite the location of the disturbance. In such cases a joint reaction of primary control of all interconnected systems is foreseen in order to re-establish the balance between generation and demand. The result will be achieved at a frequency differing from its set-point value by Δf, and the power interchanges on tie-lines will be different from the scheduled values.

Whereas during primary control all systems provide mutual support, only the system in which the unbalance occurred is required to undertake secondary control action. The controller of this system activates appropriate secondary control power restoring the nominal frequency and scheduled power exchanges.

This concept of control in the interconnected systems will be discussed in the example of two systems A and B, connected with a tie-line. The disturbance will be disconnection of generator B with generated power P_X (Figure 1.25).

The systems are assumed to operate at nominal frequency f_n, and the actual power exchange is assumed to be equal to the scheduled exchange before the disturbance, thus the deviation of power exchanges $\Delta P_{AB} = 0$. The systems A and B are characterized by stiffness factors K_{PSA} and K_{PSB}, respectively.

Due to a sudden loss of generated power, the frequency begins to change (decrease) and the frequency deviation Δf can be determined from the relationship

$$\Delta f = \frac{P_X}{K_{PSA} + K_{PSB}} \tag{1.29}$$

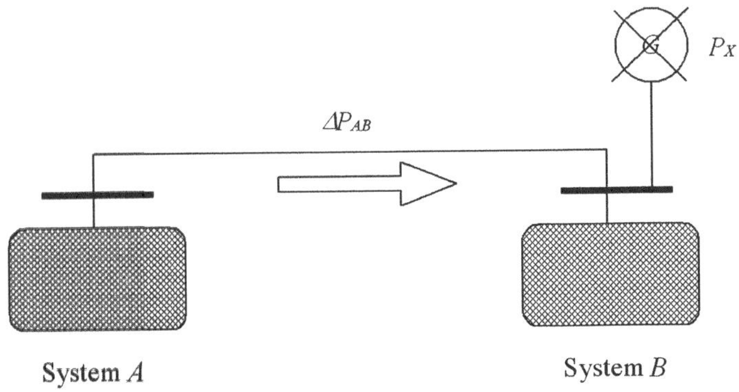

Figure 1.25 Two power systems interconnected with a tie-line

Since generating capacity is lost, P_X will have a negative value. Hence, Δf will also be negative.

In response to the frequency deviation, appropriate power values will be activated by primary control in both systems:

$$\begin{aligned} \Delta P_A &= -K_{PSA} \cdot \Delta f \\ \Delta P_B &= -K_{PSB} \cdot \Delta f \end{aligned} \tag{1.30}$$

The loss of power will be offset by the power values ΔP_A and ΔP_B in such a manner that the sum $\Delta P_A + \Delta P_B$ will be equal to the disconnected generating capacity P_X. The frequency will be stabilized at the lower level, equal to the nominal value reduced by Δf. At this point the action of primary control comes to the end.

Now the action of secondary control begins.

The power exchange P between systems A and B will no longer correspond to the programmed value, therefore ΔP_{AB} will not equal zero but will assume the value $\Delta P_{AB} = \Delta P_A$ of primary control power activated in system A and transferred to system B in order to bring the situation under control. In terms of system A the transferred power is exported power, hence its value is considered positive. In terms of system B it is imported power, thus $\Delta P_{BA} = -\Delta P_{AB}$.

If the value of $K = K_{PSA}$ is set on the secondary controller of system A, then the control error for this system is

$$G_A = \Delta P_{AB} + K_{PSA} \cdot \Delta f = \Delta P_A + (-\Delta P_A) = 0 \tag{1.31}$$

That means the central regulator of system A does not react and no secondary control will be activated in system A. Primary control in system A will be maintained as long as the frequency deviation from the nominal value persists in the interconnected systems.

If the value of $K = K_{PSB}$ is set on the secondary controller of system B, then the control error for this system is

$$G_B = \Delta P_{BA} + K_{PSB} \cdot \Delta f = -\Delta P_A + (-\Delta P_B) = P_X, \text{ while } P_X < 0 \tag{1.32}$$

It is self-evident that the control error determined by the central regulator of system B will be negative and its value equal to the power P_X, lost due to the tripping of the generating unit. The secondary control in system B is activated. The loss of generating capacity, due to the disturbance, will be offset by the action of the secondary control in system B, the frequency in the systems will be restored to the nominal value, and power exchange will be re-established at its scheduled value (i.e. before the disturbance). The action of primary control in system B will decline when the frequency deviation Δf approaches zero.

In order to provide effective secondary control, the generating units that contribute to this control process must have sufficient power reserve to be able to respond to the regulator signal with both the required change in generated power and the required rate of change.

The rate of change in the power output at the generator terminals significantly depends on the generation technique. Typically, for oil- or gas-fired power stations this rate is about 8 % per min, for lignite-fired and hard-coal-fired power stations it is up to 2 % and 5 % per min, respectively, and for nuclear power stations this rate is up to 5 % per min. Even in the case of reservoir power stations the rate is 2.5 % of the rated plant output per second.

The secondary control range is the range of adjustment of the secondary control power, within which the central regulator can operate automatically, in both directions (positive and negative) from the working point of the secondary control power. The secondary control power is the portion of the secondary control range already activated at the working point. The secondary control reserve is the positive part of the secondary control range between the working point and the maximum value.

1.6.5.3 Tertiary Control

Tertiary control is any automatic or manual change in the working points of the generating units participating, in order to restore an adequate secondary control reserve or to provide desired (in terms of economic considerations) allocation of this reserve within the set of generating units in service.

Tertiary control may be achieved by means of: changing the set operating points of thermal power plant generation sets, around which the primary and secondary control acts; connection/disconnection of pump storage hydro power stations operated the an intervention mode; altering the power interchange program; and load control (centralized tele-command system or controlled load shedding).

The timing of the primary, secondary and tertiary control ranges is shown in Figure 1.26.

The discussed frequency control system, comprising primary, secondary and tertiary control, ensures the frequency control under normal operating conditions of a power system. In such cases frequency remains within the range of permissible variation.

Where the frequency variation exceeds the permissible range, due to a significant loss of generation or consumed power, the system conditions are deemed impaired (emergency)

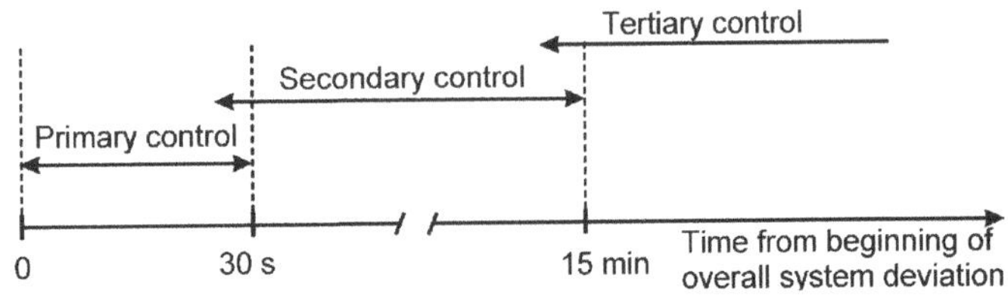

Figure 1.26 The timing of the primary, secondary and tertiary control ranges in a power system

conditions. In such circumstances supplementary actions are needed in order to re-establish the active power balance. These include:

- emergency load tripping (system load shedding) in case of a major frequency drop;
- emergency disconnection of generators in case of a large frequency increase.

1.6.6 Technical and Organizational Aspects of Load–Frequency Control

For the purposes of frequency and power control, each power system must be equipped with adequate technical means, as well as organizational structures which provide all the necessary procedures.

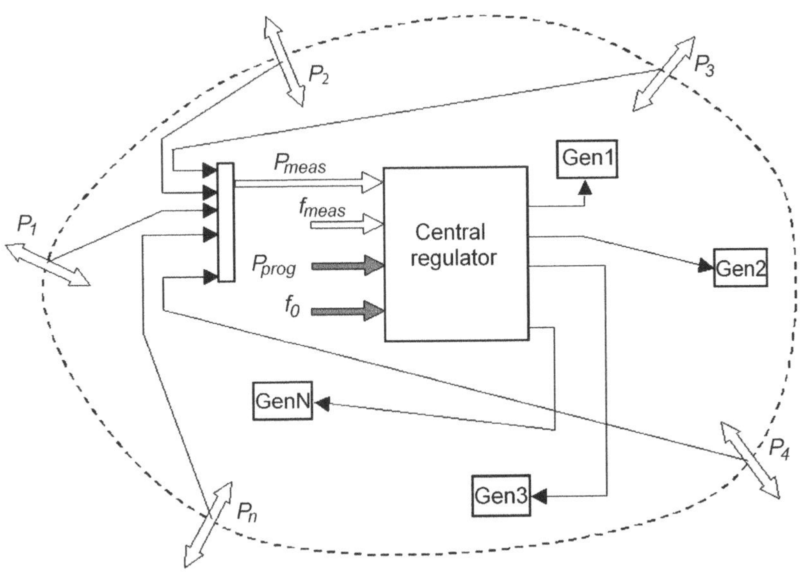

Figure 1.27 A system for frequency and power exchange central control

The necessary technical means include:

- generating units with their governing systems, whose parameters ensure primary control, and suitable for cooperating with the central control system;
- a central regulator suitable for providing secondary control in real time;
- a system for power measurement in tie-lines and a system transmitting the measured values in real time to the central regulator;
- a system for distribution of the central regulator signals to governing systems of generating units subject to secondary control
- frequency measurement systems for both the central regulator and generator controllers.

All equipment and systems used for frequency and power control should comply with appropriate requirements concerning the speed and accuracy of operation in order to ensure the required quality of the control process. Figure 1.27 shows the concept of a central system for frequency and power exchange control.

Organizational operations regarding power and frequency control in each power system are the responsibility of the transmission system operator (TSO). The scope of responsibilities of the TSO include, *inter alia*, ensuring reliable operation of the central regulator together with the necessary measuring and IT systems, preparing specific requirements for generating units connected to the transmission system, monitoring cross-border exchanges, and contracting and accounting ancillary services necessary for frequency control.

For a case study see web address

BIBLIOGRAPHY

[1] EN 50160/1999, *Voltage characteristics of electricity supplied by public distribution systems.*
[2] *UCTE Operation Handbook*, www.ucte.org.
[3] IEC 61000-4-30/2003, *Electromagnetic compatibility (EMC), Part 4-30: Testing and measurement techniques – Power quality measurement methods.*
[4] Dugan R. C., *Electrical Power Systems Quality*, Second Edition, McGraw-Hill, New York, 2002.
[5] Mircea E., *Electric Power Systems: Electric Networks*, Academia Romana, Bucharest, 2005.
[6] Machowski J., Białek J. W., Bumby J. R., *Power system dynamics and stability*, John Wiley & Sons, Ltd, Chichester, 1997.

2

Continuity of Supply

Krish Gomatom and Tom Short

Distribution reliability is the ability of the distribution system to perform its function under stated conditions for a stated period of time without failure. Distribution reliability is becoming significantly important in the current competitive climate because the distribution system feeds customers directly. Transmission and generation events can cause interruption to customers but events on these systems are much less likely to affect customers than those on the distribution system.

Many utilities across the world today use reliability indices to track the performance of the utility or a region or a circuit. Regulators require most investor-owned utilities to report their reliability indices. The regulatory trend is moving to performance-based rates where performance is penalized or rewarded based as quantified by reliability indices. Some utilities also pay bonuses to managers or others based in part on reliability achievements. Some commercial and industrial customers ask utilities for their reliability indices when planning to find a location for their facility [15].

2.1 DISTRIBUTION RELIABILITY [15]

To understand distribution reliability, consider a basic distribution feeder. There are many protective devices (fuses, reclosers, sectionalizers, breakers), overhead and/or underground line segments, three-phase and single-phase line elements, several different distribution voltage levels (4–35 kV), and in general many places for failure of components.

Distribution systems generally consist of radial feeders, which are not looped. The consequence of radial feeders is that many customers can be affected by the failure of

Handbook of Power Quality Edited by Angelo Baggini

any single component. Modified radial system designs with normally open tie points have become popular to minimize the reliability impact of radial design.

The distribution system is greatly affected by weather and vegetation. In most European countries, snow and ice are major problems. In other areas, lightning is a major cause of interruption in service. Utilities typically track weather such as wind, rain, lightning, snow, salt accumulation and tornadoes in order to predict the performance of the distribution network and for planning purposes. Tree branches are by large one of the most common causes for distribution system interruptions.

Better reliability performance can be achieved with better tree-trimming schedules, regular maintenance schedules, effective crew placement, better design schemes which can include the use of expensive equipment (i.e. reclosers), and distribution automation such as sensors, monitors and advanced technologies for condition monitoring. Utilities which feed only urban centers can achieve the highest reliability performance because they can feed their customers via a looped underground network rather than overhead radial feeds. Hence designing and maintaining a system which is as impervious as possible to failure can improve reliability. The utilities with the most reliable systems are those that have mastered the art of juggling the above-mentioned factors which affect reliability.

There are several ways to define the reliability or the quality of the electric supply. The approach described here extends the traditional utility measures of reliability to include short-duration power quality events. The two main ways used to quantify the reliability of the electric supply (they are also used to quantify the reliability of many other systems) are based on:

1. Frequency – The rate of interruptions of the electric supply is often quantified in terms of the mean time between failures (MTBF). The failure rate is often denoted by λ in units of average failures per unit of time ($\lambda = 1/\text{MTBF}$).
2. Duration – An interruption of the electric supply can be quantified as the availability or as the unavailability. Availability (and also unavailability) can be specified in per unit or percent. Even though the unavailability is a unit-less quantity, it is often referred to in units of time such as minutes per year. The unavailability is $r\lambda$, where r is the repair time and λ is the failure rate.

A common way of defining reliability is in terms of customer- and load-based indices.

2.1.1 Customer-Based Indices

Utilities commonly use the following two reliability indices for frequency and duration to quantify the performance of their systems [12]:

SAIFI The *System Average Interruption Frequency Index* (sustained interruptions) is designed to give information about the average frequency of sustained interruptions per customer over a predefined area:

$$\text{SAIFI} = \frac{\text{Total number of customer interruptions}}{\text{Total number of customers served}}$$

SAIDI The *System Average Interruption Duration Index* is commonly referred to as customer minutes of interruption or customer hours, and is designed to provide information about the average time that customers are interrupted:

$$\text{SAIDI} = \frac{\sum \text{Customer interruption durations}}{\text{Total number of customers served}}$$

These indices are used by utilities to quantify the performance of their systems and circuits (they are usually system-wide averages). The utility SAIFI and SAIDI numbers are indirectly useful by providing useful data points on the performance of existing electrical services. SAIFI is comparable to λ, and SAIDI is the unavailability in units of time.

CAIDI The *Customer Average Interruption Duration Index* is another customer-based index which is comparable to the repair time r:

$$\text{CAIDI} = \frac{\text{SAIDI}}{\text{SAIFI}} = \frac{\sum \text{Customer interruption durations}}{\text{Total number of customer interruptions}}$$

ASAI The *Average Service Availability Index* = 1− (SAIDI in hours)/8760 hours. (Use 8784 hours per year for a leap year.)

The utility indices have traditionally only included long-duration interruptions (usually defined as interruptions longer than five minutes). Many other disturbances referred to as power quality disturbances can cause customer equipment to malfunction. The two major causes are voltage sags and momentary interruptions. The following indices have been defined to track these:

MAIFI The *Momentary Average Interruption Frequency Index* is very similar to SAIFI, but it tracks the average frequency of momentary interruptions:

$$\text{MAIFI} = \frac{\text{Total number of customer momentary interruptions}}{\text{Total number of customers served}}$$

MAIFI_E The *Momentary Average Interruption Event Frequency Index* is very similar to SAIFI, but it tracks the average frequency of momentary interruption events:

$$\text{MAIFI}_\text{E} = \frac{\text{Total number of customer momentary interruption events}}{\text{Total number of customers served}}$$

A momentary interruption event is defined as [6]: *An interruption of duration limited to the period required to restore service by an interrupting device.*

SARFI$_x$ SARFI$_x$ is the average number of specified r.m.s. variation measurement events that occurred over the assessment period per customer served from the assessed system. The specified disturbances are those r.m.s. variations with a voltage magnitude less than X for voltage drops or a magnitude greater than X for voltage increases [8], where X = r.m.s. voltage threshold.

Possible values are 140, 120, 110, 90, 80, 70, 50 and 10.

SARFI$_X$ is calculated in a manner similar to SAIFI, but SARFI$_x$ is used to assess the frequency of occurrence of sags, swells and shortduration interruptions. Furthermore, the inclusion of the index threshold value, X, provides a means for assessing sags and swells of varying magnitudes. For example, SARFI 70 represents the average number of sags below 70 % experienced by the average customer served from the assessed system.

As can be seen, it is much more difficult to characterize power quality disturbances. No one index (or even two or three indices) is going to quantify accurately the reliability of electric power for particular users. Users have different needs, and interruptions affect them differently. As examples, some users (and the equipment they have) are only affected from 8 a.m. to 6 p.m. on weekdays. Others are affected only when there are short-duration interruptions.

2.1.2 Load-Based Indices

Residential customers dominate SAIFI and SAIDI since these indices treat each customer the same. Even though residential customers make up 80 % of a typical utility's customer count, they may only have 40 % of the utility's load. To weight larger customers more fairly, load-based indices are available; the equivalent of SAIFI and SAIDI, but scaled by load, are ASIFI and ASIDI:

ASIFI The *Average System Interruption Frequency Index*:

$$\text{ASIFI} = \frac{\text{Connected kVA interrupted}}{\text{Total connected kVA served}}$$

ASIDI The *Average System Interruption Frequency Index*:

$$\text{ASIDI} = \frac{\text{Connected kVA interruption duration}}{\text{Total connected kVA served}}$$

Most utilities in Europe and in North America do not track ASIFI and ASIDI, mainly since they are hard to track (knowing load interrupted is more difficult than knowing number of customers interrupted). Utilities also feel that commercial and industrial customers have enough clout that their problems are given due attention.

Most companies calculate a subset of the indices listed above and use them for planning and reporting to the regulatory commission. There are many factors which affect these calculations. Some utilities calculate one set of indices and include every type of interruption.

Others calculate the same set of indices but for different subsets of interruption types. For example, some calculate a set with storm data excluded, a set with planned interruptions excluded, and a set with nothing excluded. This helps planners make good decisions about system improvements.

2.1.3 Variation in the Utility Indices

The main reasons utilities calculate indices differently is based on the method used to count storms, planned outages, momentary interruptions, the number of customers in the service territory and the duration of events. Each of these factors can affect different indices and give misleading information.

The definition of major storm has been very difficult to formulate such that all utilities are willing to use the definition. Each company seems to have a different definition of a storm event. Many of the definitions have been mandated by the regulatory commission and most of them seem to have some relationship to the National Weather Service, the amount of mechanical damage to the system and an extended duration of customer interruption. Since the concept of storm was so difficult to define due to the many different climactic regions across a country, the IEEE Task Force settled on a definition for major event. This definition includes data about earthquakes and other natural disasters which are not necessarily weather. Some utilities include the interruption duration and frequency of planned interruptions in their index calculation.

Momentary interruptions are a topic of much debate in the industry today. The major area of concern has been their duration. Companies count them as for a variety of durations: 1 min, 2 min, 3 min and 5 min. Regardless of their length, be it one minute or five minutes, the number of momentaries will not affect the overall reliability numbers significantly. Each utility has a set time requirement for operation of automatic reclosing schemes. This time is based on the type of protection scheme used by the utility. The reason for using five minutes instead of a lesser number is to allow for the maximum duration that an automatic reclosing scheme needs to clear a fault. It is important to calculate the number of these interruptions due to the increasing sensitivity of customer loads such as computers.

Utility indices vary widely because of many differing factors, mainly [15]:

- Weather
- Physical environment (mainly the amount of tree coverage)
- Load density
- Distribution voltage
- Age
- Percent underground
- Methods of recording interruptions

Within a utility, performance of circuits varies widely for many of the same reasons causing the spread in utility indices: circuits have different lengths necessary to feed different areas of load density, some are older than others, and some areas may have less tree coverage.

Customer reliability is not normally distributed. A skewed distribution such as the log-normal distribution is more appropriate and has been used in several reliability applications

[1],[2]. A log-normal distribution is appropriate for data that is bounded on the lower side by zero. The skewed distribution has several ramifications:

- The average is higher than the median. The median is a better representation of the typical customer.
- Poor-performing customers and circuits dominate the indices (which are averages).
- Storms and other outliers easily skew the indices.

Note that SAIFI and SAIDI are weighted performance indices. They emphasize the performance of the worst-performing circuits and the performance during storms. SAIFI and SAIDI are not necessarily good indicators of the typical performance that customers have.

2.2 QUALITY OF SUPPLY [7]

According to the Council of European Regulators (CEER) Working Group on Quality of Supply, the following definitions apply:

- Customer service.
- Continuity of supply (typically referred to as reliability in the USA).
- Voltage quality (typically referred to as power quality in the USA).

2.2.1 Customer Service

Customer service typically relates to the nature and quality of the service provided to electricity consumers. Customer service (referred to as *commercial quality* in Europe) indicators are typically quantitative operational measures that are used extensively in existing performance-based ratemaking (PBR) schemes throughout the world. Some examples of customer service indicators utilized include:

- Customer satisfaction.
- Customer complaints.
- Customer meter and billing accuracy.
- Response to customer enquiries.
- Appointments met.
- Customer call/wait times.
- Time required for new service connections.
- Emergency/storm response.
- Safety/health.
- Estimating charges for work.

Although important in assessing the overall quality of supply, customer service is not the focus of this research and will largely not be addressed in this chapter. There is a variety of information and resources available to policy makers and utilities regarding this component of the quality of supply, some of which are referenced at the end of this chapter.

2.2.2 Continuity of Supply

Continuity of supply is characterized by the number and duration of interruptions. Several indicators are used to evaluate the continuity of supply in transmission and distribution networks. Regulation can aim to compensate customers for very long supply interruptions, keep restoration times under control and create incentives to reduce the total number and duration of interruptions (and disincentives to increase them). Different methods and accuracies of measuring interruptions and in assigning liability for each of them create problems in regulating continuity of supply.

2.2.3 Voltage Quality

The third and final component of the quality of supply is voltage quality. Voltage quality is the quantitative form of describing power quality and includes both steady-state power quality variations and momentary disturbances that may impact loads. Categories of voltage quality include:

- Power frequency
- Magnitude of the supply voltage
- Harmonics and interharmonics
- Voltage unbalance
- Flicker
- Voltage sags and momentary interruptions (r.m.s. variations)
- Transients

The International Electrotechnical Commission (IEC) also has defined a set of parameters to quantify power quality variations. These include voltage harmonics and interharmonics, voltage flicker, voltage unbalance, voltage sags (dips), forced interruptions, voltage regulation, frequency variation, voltage surges (swells) and switching disturbances. A series of IEC 61000 documents provides the general description, environment, measurement method, and emission and immunity targets for most of these power quality parameters. For instance, IEC 61000-2-2 provides the basic compatibility levels for power quality parameters at low-voltage locations. IEC 61000-4-30 provides overall guidelines for measuring and characterizing these power quality characteristics. Similar to the IEEE 1159-1995 definition, the IEC definition for voltage interruption is considered as part of the quality of supply parameters. See Table 2.1.

In Europe, the standard EN 50160:2000 *Voltage characteristics of electricity supplied by public distribution systems* that provides the limits and tolerances of various phenomena that can occur on the mains is widely used by utilities and regulators in order to define power quality in European distribution grids. The EN 50160 standard defines the main voltage

Table 2.1 Categories of power quality variation [10]

Categories			Typical spectral content	Typical duration	Typical voltage magnitude
Transient	Impulsive	Nanosecond	5 ns rise	< 50 ns	
		Microsecond	1 μs rise	50 ns to 1 ms	
		Millisecond	0.1 ms rise	> 1 ms	
	Oscillatory	LF	< 5 kHz	0.3–50 ms	0–4 pu
		MF	5–500 kHz	20 μs	0–8 pu
		HF	0.5–5 MHz	5 μs	0–4 pu
Short-duration variations	Instantaneous	Sag (dip)		0.5–30 cycles	0.1–0.9 pu
		Swells		0.5–30 cycles	1.1–1.8 pu
	Momentary	Interruption		0.5 cycles-3 s	< 0.1 pu
		Sag		30 cycles-3 s	0.1–0.9 pu
		Swells		30 cycles-3 s	1.1–1.4 pu
	Temporary	Interruption		3 s to 1 min	< 0.1 pu
		Sag		3 s to 1 min	0.1–0.9 pu
		Swells		3 s to 1 min	1.1–1.2 pu
Long-duration variations	Interruption sustained			> 1 min	0.0 pu
	Undervoltages			> 1 min	0.8–0.9 pu
	Overvoltages			> 1 min	1.1–1.2 pu
Voltage unbalance				Steady state	0.5–2 %
Waveform distortion	D.C. offset			Steady state	1–0.1 %
	Harmonics		0–100th H	Steady state	0–20 %
	Interharmonics		0–6 kHz	Steady state	0–2 %
	Notching			Steady state	
	Noise		Broadband	Steady state	0–1 %
Voltage fluctuations			< 25 Hz	Intermittent	0.1–7 %
Power frequency variations				< 10 s	

parameters and the permitted ranges of derivations in public low-voltage and medium-voltage networks at the customer's point of common coupling. Note that this standard defines only voltage parameters. Table 2.2 contains a summary of the criteria for the low-voltage side of the supply network.

Table 2.2 Limits defined by EN 50160 (low voltage) [5]

Supply voltage phenomenon	Acceptable limits	Measurement Interval	Monitoring period	Acceptance percentage
Grid frequency	49.5 Hz to 50.5 Hz 47 Hz to 52 Hz	10 s	1 week	95 %
Slow voltage changes	230 V ± 10 % −15 % − 230 V ± 10 %	10 min	1 week	95 %
Fast voltage changes	5 % Max. 10 %	10 ms	N/A	100 %
Flicker	Plt < 1	2 h	1 week	95 %
Voltage dips (sags) (<1 min)	10 to 1000 times per year ($90\,\% > U_n > 1\,\%$)	10 ms	1 year	100 %
Short interruptions (< 3 min)	10 to 100 times per year ($U_n < 1\,\%$)	10 ms	1 year	100 %
Accidental, long interruptions (> 3 min)	10 to 50 times per year ($U_n < 1\,\%$)	10 ms	1 year	100 %
Temporary overvoltages (line-to-ground)	Mostly < 1.5 kV	10 ms	N/A	100 %
Transient overvoltages (line-to-ground)	Mostly < 6 kV	N/A	N/A	100 %
Voltage unbalance	Mostly 2 % but occasionally 3 %	10 min	1 week	95 %
Harmonic voltages	8 % total harmonic distortion (THD) Even/non-multiples of 3: 5 6.0 %; 7 5.0 %; 11 3.5 %; 13 3.0 %; 17 2.0 %; 19 1.5 %; 23 1.5 %; 25 1.5 % Odd multiples of 3: 2 2.0 %; 3 5.0 %; 4 1.0 %; 9 1.5 %; Rest: 6–24 0.5 %	10 min	1 week	95 %
Interharmonic voltages	N/A	10 min	N/A	N/A
Signaling	N/A	3 s	1 day	99 %

For a case study see web address

2.3 FACTORS AFFECTING RELIABILITY PERFORMANCE [9]

As mentioned above, reliability performance varies dramatically from one system to another and this is not necessarily an indication that one system has poor performance. Many factors influence the expected reliability at a particular location or for an entire system. Some of these factors are indicated in Table 2.3.

Reliability indices that reflect reliability performance differ with data definitions and data classifications. Most utilities define separate indices for planned and unplanned events. Interruption data may or may not be considered in reliability performance because interruptions generally occur due to a major event like a storm, forest fire or a blizzard – which the utility has least control over. Distribution and transmission events are considered separately for reliability performance evaluation, due to the data classes. Utilities could use different reliability indices for transmission and distribution due to the nature of the events and the type of data.

The service territory of the utility determines the nature of events that could be expected which effect reliability performance. Geography of the service territory such as a wooded area, mountainous terrains, etc., are likely to cause reliability issues. Weather is an important factor that can severely affect reliability levels. Failures due to lightning, wind and rain occur during storms, and the failure rate during storms can be much higher than normal. Vegetation and animal activity could affect reliability performance. The effects of vegetation such as tree-falls, branch intrusions and animal activity from birds, squirrels and pests causing ground faults affect reliability levels. Maintenance practices such as tree-trimming programs and installing animal guards could help achieve higher levels of reliability.

Obviously, a rural, overhead distribution system with significant exposure to trees, animals and lightning cannot be expected to have the same reliability as an underground network. An important area of current research is attempting to quantify the effect of some of these parameters on expected reliability levels. However, for the purposes of this chapter it is enough to recognize that different reliability levels should be expected at different locations.

Table 2.3 Summary of important factors that can affect reliability levels

Cause	Definition and data classification	Service territory	Data collection process	System design
1	Major events	Geography	Outage notification	Urban/rural/downtown
2	Interruption	Weather	Outage reporting	Load characteristics
3	Planned/ unplanned	Vegetation	Step restoration process	Underground/overhead
4	Distribution/ transmission	Animal activity	Customer to network connectivity	Voltage level
5				Protection scheme

2.3.1 Reliability Indices Reporting [15]

Many utilities track reliability indices. Indices provide some data on the frequency of long-duration interruptions and the availability of a utility supply (availability is dominated by long-duration events). Figure 2.1 and Figure 2.2 show survey results of American utilities for SAIFI, SAIDI and MAIFI. The SAIFI and SAIDI numbers help quantify a utility's performance for Quality Level 4 (long-duration interruptions), and MAIFI helps quantify Quality Level 3 (momentary interruptions).

Utility indices vary widely because of many differing factors, mainly:

- Weather
- Physical environment (mainly the amount of tree coverage)
- Load density
- Distribution voltage
- Construction (age, overhead or underground)
- Methods of recording interruptions

Whether storms or major events are included in indices makes a big difference in the interruption duration; SAIDI and availability are much worse when storms are included. Figure 2.1 shows differences in the Edison Electric Institute (EEI) Survey of utilities reporting SAIDI

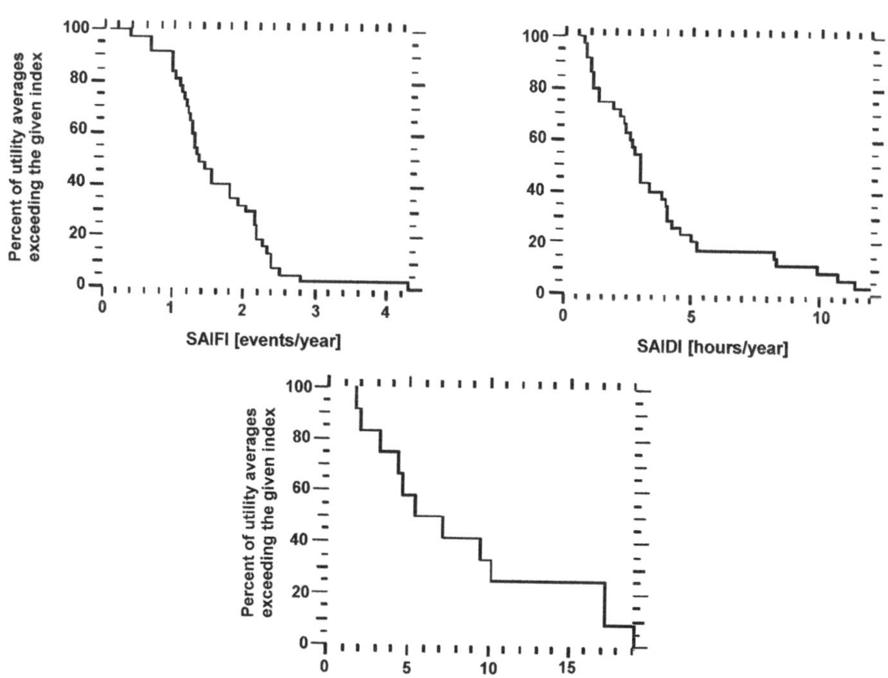

Figure 2.1 The 1998 Edison Electric Institute Survey results (including storms) [4] (Reproduced from *EEI Reliability Survey 1998, Edison Electric Institute*)

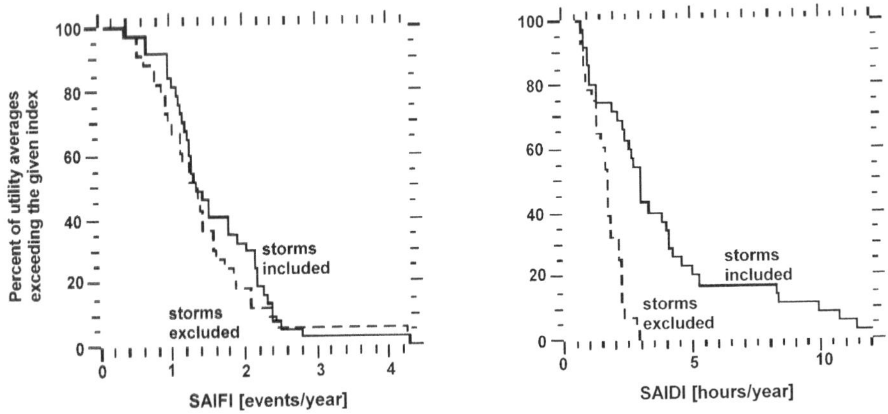

Figure 2.2 Effect of including storms on utility reliability indices based on the 1998 Edison Electric Institute Survey [4] (Reproduced from *EEI Reliability Survey 1998, Edison Electric Institute*)

and SAIFI with and without storms. During storms, equipment and personnel shortages increase the repair time significantly.

Within a utility, the performance of circuits varies widely for many of the same reasons causing the spread in utility indices: circuits have different lengths necessary to feed different areas of load density; some are older than others, and some areas may have less tree coverage.

When using a utility's SAIFI and SAIDI to estimate the customer's quality, keep in mind the following points:

- Poor-performing customers and circuits dominate the indices (which are averages).
- Storms and other outliers easily skew the indices.
- Residential customers dominate the indices. Many industrial and commercial customers have better reliability (although they may be more sensitive to short-duration events).

Treat SAIFI and SAIDI as weighted performance indices. They stress the performance of the worst-performing circuits and the performance during storms. SAIFI and SAIDI are poor indicators of the typical performance that customers have. From the point of view of individual customers, they ignore the most problem-causing events – voltage sags and momentary interruptions.

2.3.2 Differences Based on Type of Supply

The distribution supply greatly impacts reliability. Long radial circuits provide the poorest service; grid networks provide exceptionally reliable service. Table 2.4 gives estimates of the reliability of several common distribution supply types developed by New York City's Consolidated Edison. Note that the repair time (CAIDI) increases for the configurations used in more urban areas. Having underground lines and having to deal with traffic increase the time for repairs.

Table 2.4 Comparison of the reliability of different distribution configurations [13]

	SAIFI (interruptions/year)	CAIDI (min/interruption)	MAIFI (momentary interruptions/year)
Simple radial	0.3 to 1.3	90	5 to 10
Primary auto-loop	0.4 to 0.7	65	10 to 15
Underground residential	0.4 to 0.7	60	4 to 8
Primary selective	0.1 to 0.5	180	4 to 8
Secondary selective	0.1 to 0.5	180	2 to 4
Spot network	0.02 to 0.1	180	0 to 1
Grid network	0.005 to 0.02	135	0

It is easier to provide higher reliability in urban areas: circuit lengths are shorter; more reliable distribution systems (such as a grid network) are more economical. Figure 2.3 and Figure 2.4 show reliability for different distribution services in several Commonwealth countries. The delineations used for this comparison for Victoria (in Australia) are:

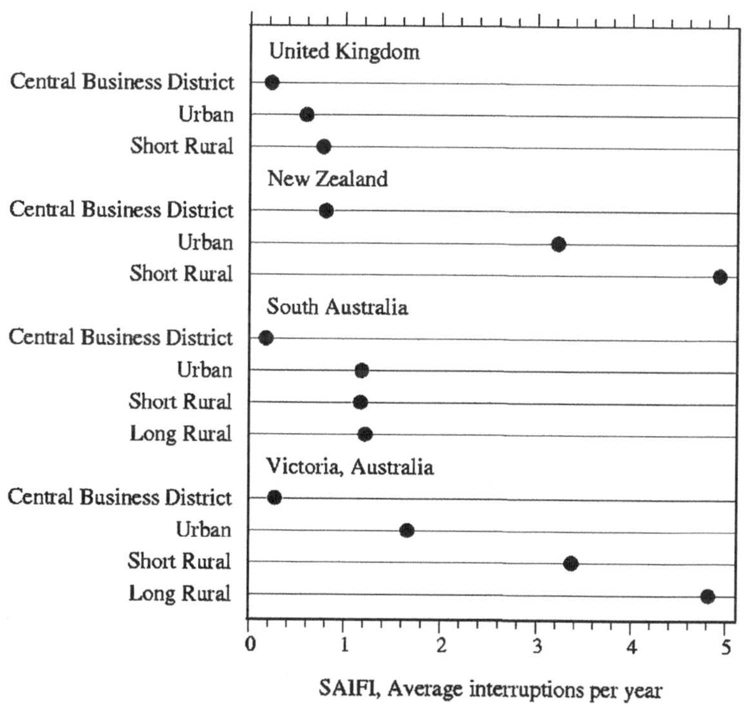

Figure 2.3 Comparison of SAIFI by load density for several former colonies of the British Empire [3] (Reproduced from 2001 Electricity Distribution Price Review Reliability Service Standards, Office of the Regulator-General and Service Standards Working Group)

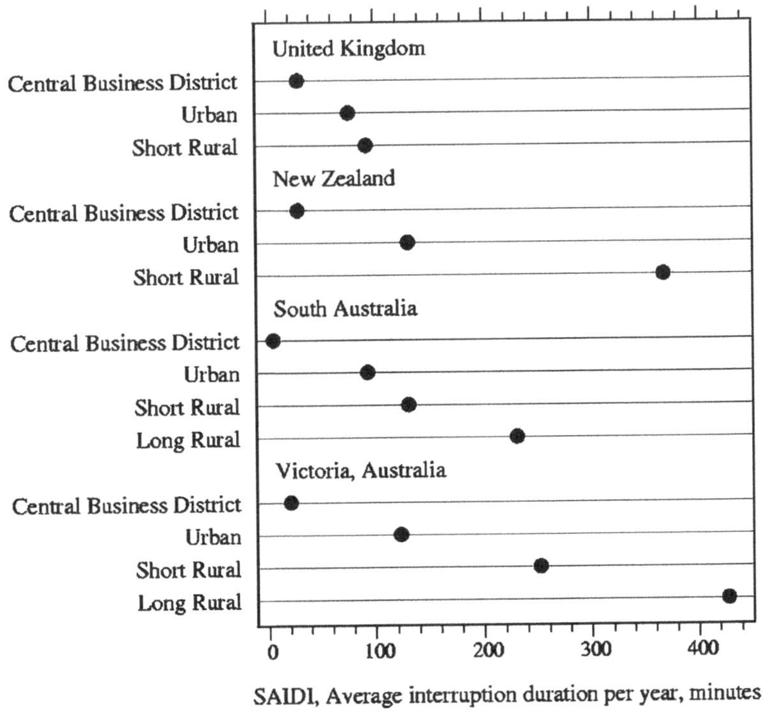

Figure 2.4 Comparison of SAIDI by load density for several former colonies of the British Empire [3] (Reproduced from 2001 Electricity Distribution Price Review Reliability Standards, Office of the Regulator-General and Service Standards Working Group)

- Central business district: used map boundaries.
- Urban: greater than 0.3 MVA/km.
- Short rural: less than 200 km.
- Long rural: greater than 200 km.

For shorter-duration disturbances, the analysis in Section 2.3 showed that urban sites have fewer events than suburban or rural sites. The difference is not as dramatic as the difference in long-duration interruptions. The distribution power quality (DPQ) data provides information on whether the sites are urban or rural, but it does not differentiate between configurations (such as radial and network supplies). Some data on voltage sags is available for network systems in New York City.

2.4 IMPROVING RELIABILITY

2.4.1 Utility-Side Improvement Options [15]

2.4.1.1 Tree Trimming

The main way to improve reliability is to have an effective tree-trimming program. Trees very often cause temporary faults on laterals which result in nuisance fuse blowings and sustained interruptions to customers. Spacer cable or covered wire helps the situation, but trimming is a very effective way to reduce indices.

2.4.1.2 System Improvements

Reliability can be improved by system improvements. Adding protective equipment such as reclosers, sectionalizers and fuses can affect reliability very positively. There is always a law of diminishing returns and more is not necessarily better. Studies should be performed to determine the number of devices per feeder and the best location for devices. Changes in system protection philosophy can also have a positive effect on reliability. Many utilities are using the practice of fuse saving on feeders known as feeder selective relaying (FSR). This practice allows the breaker one instantaneous operation to try to clear any temporary fault without having a fuse blow. It is intended to reduce nuisance fuse blowings which can result in high SAIFI and SAIDI. The only problem with this philosophy is that if the protective devices do not coordinate, then the breaker will operate, giving the whole feeder a momentary and the fuse will still blow and cause a sustained interruption for customers on the lateral. FSR should only be applied where devices coordinate. Changing the practice for certain feeders can have positive affects on reliability.

System improvements add to costs. Theoretically, utilities should use reliability indices and their trends to affect cost decisions. Reliability data should be coupled with customer trouble reports to help determine the best location to spend money on system improvements.

There are several costs a utility must be aware of – mainly customer cost, utility liability cost, utility labor cost and utility lost energy cost. If interruptions to customers are costing the customers thousands of dollars each time they are subjected to an interruption, then there needs to be a bridge between the utility and the customers to possibly share the cost or rectify the problem. Utilities need to keep their large customers and customers need to have a certain level of service which should be paid for in the rate structure. Utilities may also have to spend some capital on reliability to avoid lawsuits or liability. Certain reliability improvements will also save the utility money in labor costs.

One example of how reliability indices can be used to make cost-related decisions is to consider a feeder which has an unusually high SAIDI. Adding a recloser to this feeder may reduce the customer minutes significantly.

2.4.1.3 Crew Placement and Management

This measure can also affect reliability. Centralizing crews often results in lower reliability because it takes the crews longer to arrive at a trouble spot. Decentralizing crew placement so they can reach any part of the system within a set time period will improve reliability.

2.4.1.4 Maintenance Practices

Lack of maintenance will negatively affect reliability. Reliable systems typically have well-established maintenance procedures which are not the first thing to be cut from the budget each year.

2.4.1.5 Advanced Technology and Monitoring Systems

Advanced technology initiatives such as distribution automation could help improve reliability performance. Sensors, better protection systems and outage management systems and software could lead to reducing downtime and improve reliability. In today's electricity market, utilities are expected to provide highly reliable power supplies at competitive prices. Distribution automation is the way for intelligently using advanced technology to increase the reliability at reduced system operation cost. During recent years there has been an increasing focus on the application of automation in distribution systems. Distribution automation applied to distribution feeders can result in rapid service restoration and provide effective feeder performance at relatively low operation costs [16].

Most of these are directed at reducing long-duration interruptions, and while they will provide improvements in SAIFI and SAIDI, large improvements (meaning an order of magnitude) cannot be obtained using these methods.

Realistically, only two utility-side approaches increase the overall quality and reliability to a customer:

1. Provide a more reliable distribution configuration (like a spot network).
2. Use a custom power device.

More reliable configurations are a well-developed means to greatly reduce the number of long-duration interruptions and possibly short-duration interruptions. Each of these schemes provides significant improvement:

- Automated loop distribution system (see Figure 2.5) – reduces sustained interruptions, but does not help with momentaries or voltage sags.
- Primary or secondary selective scheme (switch between two utility sources) – reduces sustained interruptions, but does not help with momentaries or voltage sags.
- Spot or grid network – greatly reduces momentary and sustained interruptions, but does not help much with voltage sags.

2.4.2 Custom Power Devices [15]

Custom power devices are medium-voltage devices that improve voltage sags (and possibly momentary interruptions) to a load. They are generally configured to supply improved quality to a fixed location – either one customer or a power quality park. The three main custom power configurations (Table 2.5) are:

1. Static series compensator – provides ride-through for sags by injecting a signal to offset the voltage lost during the sag.
2. Backup stored energy – provides ride-through for sags and momentary interruptions by using stored energy (e.g. batteries, superconducting coil, ultracapacitors).
3. Static transfer switch – provides ride-through for momentary interruptions and most voltage sags by quickly switching between two different utility feeders.

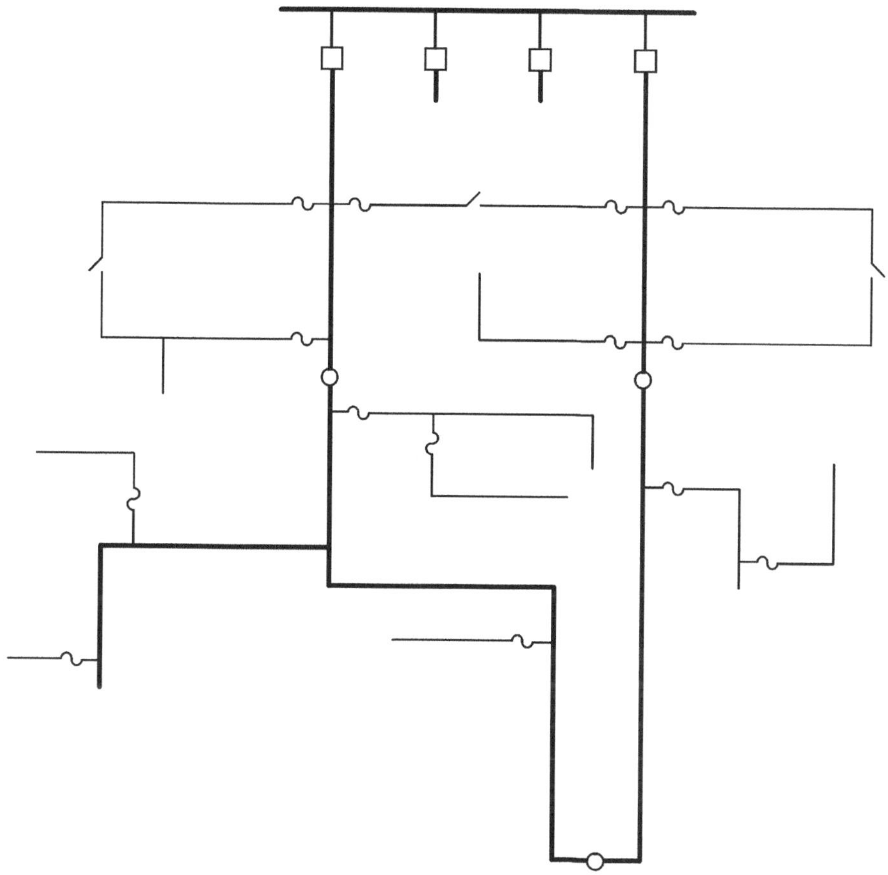

Figure 2.5 Automation scheme using three reclosers to provide a backup power path in the event of a feeder interruption close to the substation

The most common series device is a dynamic voltage restorer (DVR) that injects voltage onto the faulted phase from the unfaulted phases, and it also contains a small amount of stored energy in the form of capacitors (see Figure 2.6). The dynamic voltage restorer can protect against most sags to 40 % of nominal indefinitely.

The most prominent stored energy device is an off-line uninterruptible power supply (UPS) (Figure 2.7). An off-line UPS allows the utility to power the equipment until a disturbance is detected and a switch transfers the load to the battery-backed inverter. The transfer time from the normal source to the battery-backed inverter is important. Because a very short-duration interruption exists during the time it takes to detect a mains failure, start the inverter and transfer the load to battery power, a load with some inherent ride-through capability is required for the interruption to go unnoticed.

Table 2.5 Various medium-voltage ride-through devices

Generic name	Device model names	Typical mitigation level	Typical cost (US $)	Typical size
Static series compensator	Dynamic voltage restorer (DVR) PureWave VSS SIPCON-S PQ-IVR	Medium-voltage/utility-side/facility level	150–300 $/kVA	2–20 MVA
	http://www.abb.com/us http://www.powerquality.de/ http://www.sandc.com/ped/pwproducts.htm http://www.geindustrial.com/industrialsystems/powerquality/catalog/pqivr/index.htm			
Backup stored energy supply devices	Dynamic UPS (DUPS) PQ AC PureWave UPS	Medium-voltage/utility-side/facility level	250–400 $/kVA	2–20 MVA
	http://www.abb.com/us http://www.sandc.com/ped/pwproducts.htm http://www.amsuper.com/			
MV source transfer switch	PureWave Source Transfer System ParaDigm Subcycle Transfer Switch Cyberex Static Transfer Switch	Medium-voltage/utility-side	300–600 000, 15 kV, 600 A	< 34.5 kV, 400–6000 A
	http://www.cyberex.com/ http://www.meppi.com/html/power_electronics.html http://www.sandc.com/ped/pwproducts.htm			

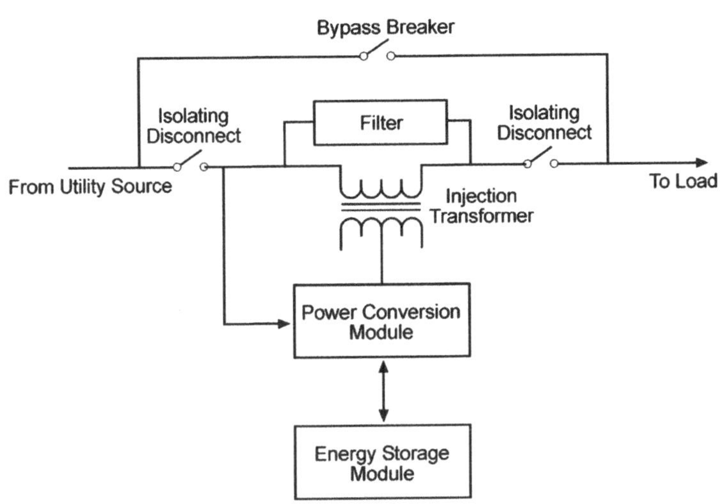

Figure 2.6 Basic configuration of a dynamic voltage restorer

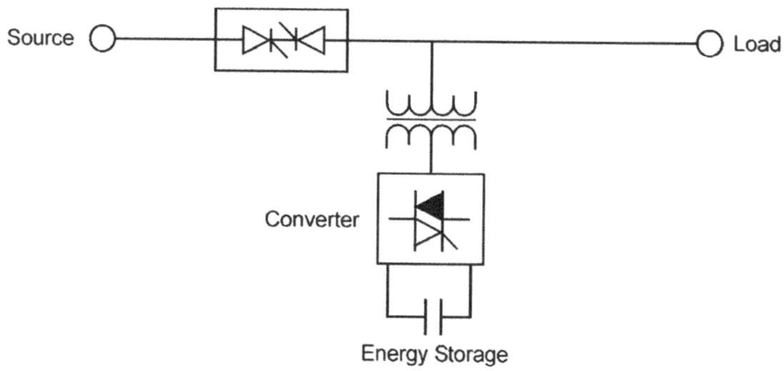

Figure 2.7 Medium-voltage standby uninterruptible power supply

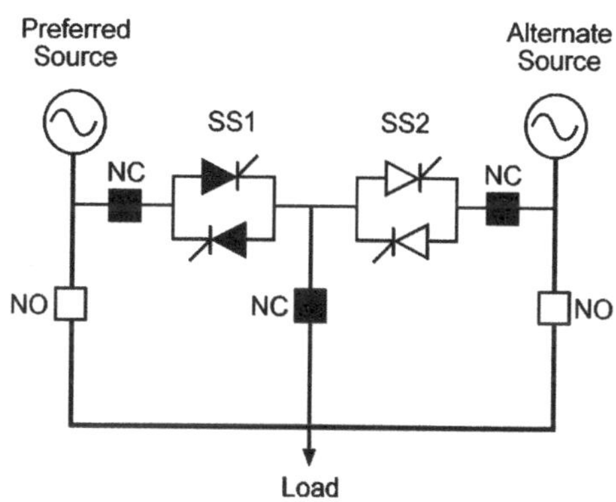

Figure 2.8 Preferred/alternate configuration of a static switch

A static source transfer switch (SSTS) uses solid-state switches to provide an almost seamless transfer from one source to another. The most commonly used solid-state device in SSTSs is the thyristor or silicon-controlled rectifier (SCR). The transfer time can be a quarter cycle. It may take a half cycle in some cases because the SCR cannot turn off until the current crosses zero (to change polarity). Figure 2.8 shows a common static switch configuration.

2.5 COSTS, MARKETS AND VALUE FOR RELIABILITY

2.5.1 Size of the End-User Load and Duration Affect Cost

The size of an electric power end user's electric load is an important metric for understanding the market for reliability. Not only will different-sized solutions be required to solve problems at large versus small facilities, entirely *different* technologies may be appropriate. Figure 2.9 shows cost of interruptions versus duration for end users of different load sizes in Canada.

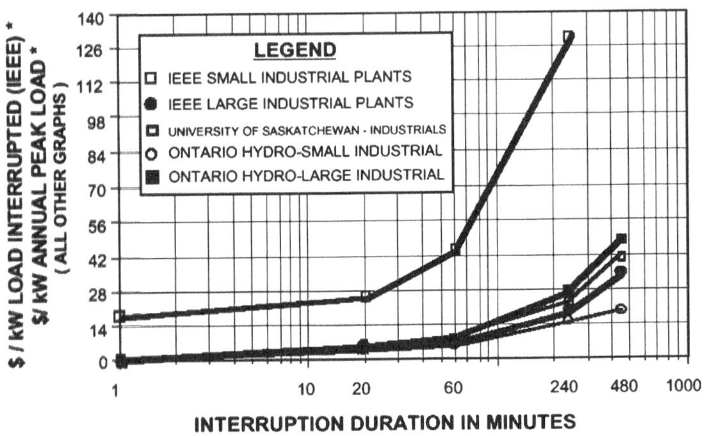

Figure 2.9 Cost of interruptions vs. duration [11] (Reproduced from IEEE standard 493-1997, IEEE)

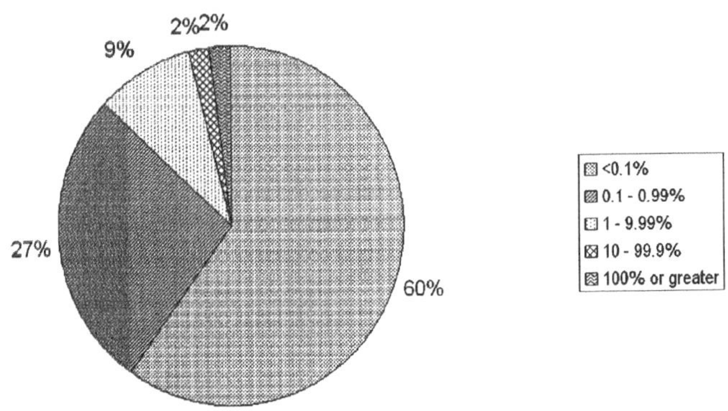

Figure 2.10 End-user reliability costs vs. revenues

Shown in Figure 2.10 are the self-reported costs due to reliability problems of all US electric power end users in the digital economy, continuous process manufacturing, and fabrication and essential services sectors. Costs are presented as a percentage of the facility's revenues. Making the assumption that facilities experiencing reliability costs in excess of 1 % of their revenues are being injured by reliability costs would lead us to conclude that approximately 12.7 % of industrial end users are ripe for the purchase of reliability-enhancing measures.

2.5.2 Market for Reliability [6]

Estimating the market for reliability requires a two-pronged approach, focusing both on the economic impact of reliability issues (the macro picture) and on identifying who among

Table 2.6 Sample base for CEIDS Value of Reliability Market Survey

Number of employees	Digital economy	Continuous process manufacturing	Fabrication and essential services	Total
1 to 19	179	166	159	504
20 to 249	101	87	101	289
250 or more	62	74	56	192
Total	342	327	316	985

end users of electric power are most likely to purchase reliability-enhancing solutions (the micro picture). The macro picture can provide one with a rough approximation of the overall value of improved reliability to the US economy (and, thereby, a rough estimate of what this overall market might be worth to an industry of reliability solutions providers), while the micro picture can help to quantify what percentage of electric power end users may actually consider purchasing reliability-enhancing products.

The latest, most definitive view of both the macro and micro reliability market pictures is enabled by recent market research completed by CEIDS, where over 980 commercial and industrial end users were surveyed and asked their opinions and circumstances vis-à-vis electric power reliability (see the survey sample in Table 2.6) [14]. The study enabled quantification of both the overall value of reliability to the US economy, while also allowing some intelligent approximations of what percent of end users are likely to purchase reliability-enhancing solutions.

2.5.3 Value of Reliability: a Macro View

The importance of reliable, high-quality electric power continues to grow as society becomes ever more reliant on digital circuitry for everything from e-commerce to industrial process controllers to the onboard circuitry in toasters and televisions. With this shift to a digital society, business activities have become increasingly sensitive to disturbances in the power supply. Such disturbances include not only *power outages* (the complete absence of voltage, whether for a fraction of a second or several hours), but also *power quality phenomena* (all other deviations from perfect power, including voltage sags, surges, transients and harmonics). Three sectors are particularly sensitive to power disturbances:

1. **The digital economy (DE).** This sector includes firms that rely heavily on data storage and retrieval, data processing or research and development operations. Specific industries include telecommunications, data storage and retrieval services (including collocation facilities or Internet hotels), biotechnology, electronics manufacturing and the financial industry.
2. **Continuous process manufacturing (CPM).** This sector includes manufacturing facilities that continuously feed raw materials, often at high temperatures, through an industrial process. Specific industries include paper; chemicals; petroleum; rubber and plastic; stone, clay and glass; and primary metals.

3. **Fabrication and essential services (F&ES).** This sector includes all other manufacturing industries, plus utilities and transportation facilities such as railroads and mass transit, water and wastewater treatment, and gas utilities and pipelines.

These three sectors account for roughly 2 million business establishments in the USA. Although this is only 17 % of all US business establishments, these same three sectors account for approximately 40 % of US gross domestic product (GDP). Moreover, disruptions in each of these sectors – but especially DE and F&ES – have an almost immediate effect on other sectors that depend on the services they provide.

Power outages cost each of the roughly 2 million establishments in these three sectors more than $23 000 a year. The bulk of this loss ($29.2 billion) is concentrated in the F&ES sector, which is particularly vulnerable to equipment damage. DE firms lose $13.5 billion to outages annually, primarily from lost productivity and idled labor. The greatest losses per establishment are among CPM firms, which suffer the loss of raw materials as well as the costs incurred by other sectors. Industrial and digital economy firms are collectively losing $45.7 billion a year to outages.

Costs vary with the length of the outage, but even short outages are costly. Even a one-second outage can damage equipment and disrupt highly sensitive operations to the point where labor becomes idled as systems are reset and brought back on-line. The average cost of a one-second outage among industrial and digital economy firms is $1477 versus an average cost of $2107 for a three-minute outage and $7795 for a one-hour outage. Brief outages are also more frequent than outages of an hour or more; industrial and digital economy establishments report that 49 % of the outages they experience last less than 3 min.

Industrial and digital economy companies lose another $6.7 billion each year to power quality (PQ) phenomena. Digital economy firms have lower PQ-related losses per establishment than either of the industrial sectors. The F&ES sector seems to be particularly sensitive to PQ phenomena, losing more than $9600 annually per establishment and accounting for 85 % of the aggregate losses across all three sectors. Once again, equipment damage seems to play a large role in the costs to industrial facilities.

Data suggests that across *all* business sectors, the US economy is losing between $104 and $164 billion a year to outages and another $15 to $24 billion to PQ phenomena. California has the highest costs for both outages and PQ phenomena (between $13.2 and $20.4 billion), followed by Texas ($8.3 to $13.2 billion) and New York ($8.0 billion to $12.6 billion). California's costs are based on a typical year of power disturbances; costs are likely to increase dramatically if the state experiences high levels of rolling blackouts. Projections to all business sectors are extrapolations from the survey data based on the assumption that per-establishment costs from outages and PQ phenomena for firms outside the DE and industrial sectors are anywhere from 25 % to 50 % as high as the costs reported by these sectors, and are statistically valid as long as this assumption is correct.

Figure 2.11 shows the percentage of end users in different load-size tiers reporting various levels of cost (presented as a percentage of revenue) due to reliability problems. The data makes clear that all sizes of companies suffer from losses due to reliability. The smallest companies (quartile 1) offer something of a paradox in that the greatest percentage of all the quartiles report no costs to reliability (65.3 %), but the greatest percentage report losses greater than 1 % of revenues (18.8 % vs. 13.3 %, 11.3 % and 11.9% respectively for

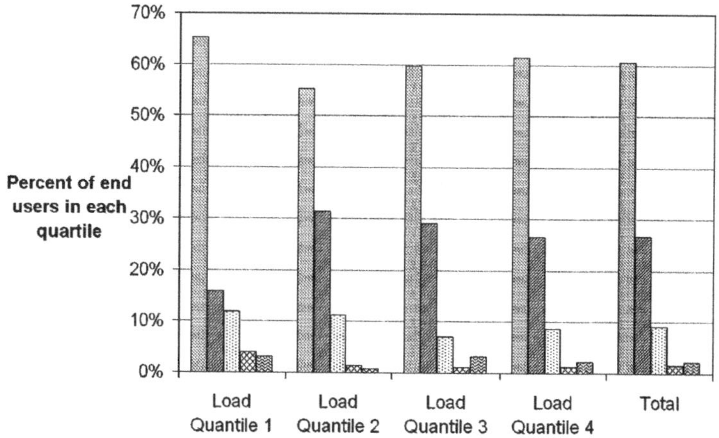

Figure 2.11 End-user reliability costs vs. revenues by electric load

quartiles 2, 3 and 4). For quartiles 2, 3 and 4, 40–50 % of respondents report no losses due to reliability.

2.5.4 Impact of Reliability Events on End-User Productivity [6]

The CEIDS Value of Reliability Study asked electric power end users about the impact on their operations of a number of different reliability events, including outages of 1 s, 3 min and 1 h. Paramount to understanding the market for reliability is first gaining an appreciation for how the impact of comparable reliability events varies from end user to end user.

Figure 2.12 presents results of the impact of a 1 s interruption in utility-supplied power on end users. One out of four end users are able to get their facilities up and running almost

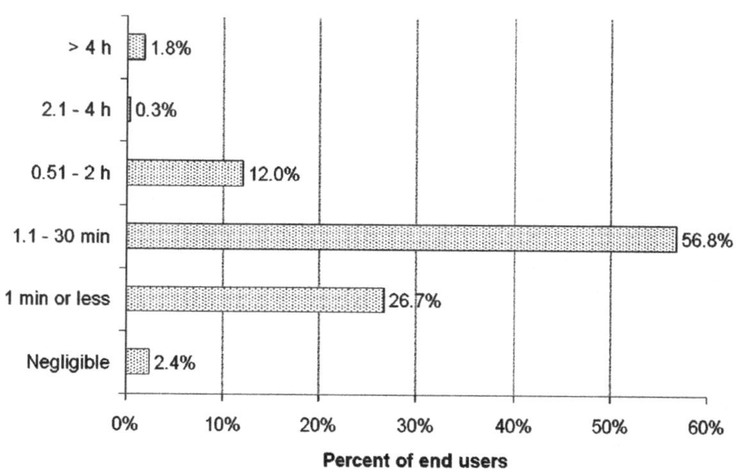

Figure 2.12 Duration of facility outage following a 1 s power interruption

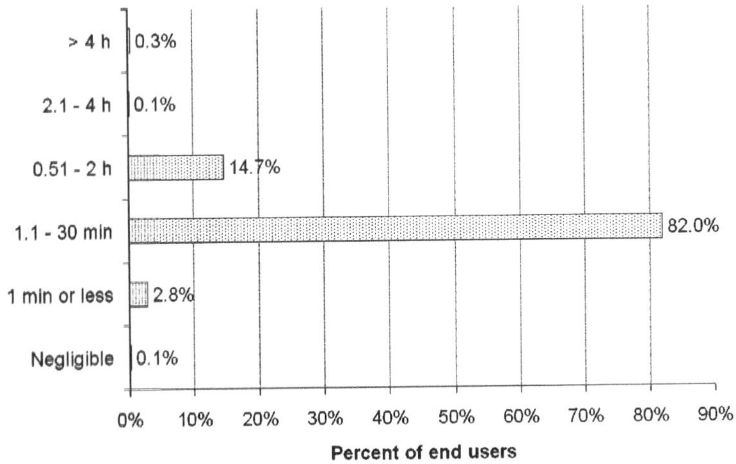

Figure 2.13 Duration of facility outage following a 3 min power interruption

instantly after a 1 s outage, and over 80 % are down less than 30 min. However, just over 14 % of end users' facilities are down more than 30 min, and just over 2 % are down for more than 2 h. The average outage duration across all companies is approximately 21 min.

The study used data to understand the differences in losses following a 1 s duration and following a 3 s duration. As illustrated in Figure 2.13, the only significant change is that fewer companies can weather a 3 min outage with only a minute or less of downtime. The average outage duration across all companies is 22 min, essentially unchanged from the 1 s outage. The portion of companies experiencing an outage of 30 min or more increases by a percentage point to just over 15 %.

It is also interesting to note that those reporting very long interruptions (in excess of 2 h) actually decline for 3 min power interruptions vs. 1 s interruptions. This can logically be attributed to equipment damage and psychological effects. Manufacturing equipment, motors and electronics are much more likely to be damaged by quick cycling of power than by outages that allow all spinning loads to stop and capacitors to discharge. Psychologically, manufacturing facilities are much more confident that resumption of power after a 3 min interruption is likely to hold, while brief outages are perceived as more likely to reoccur, making restarting complicated processes too quickly counterproductive.

When the outage duration is increased to 1 h, the impact on end-user productivity is more profound. With this type of outage, nearly 90 % of end users experience a process interruption exceeding 30 min, and close to 6 % are down for more than 2 h. The average duration of facility interruption nearly trebles to 62 min (see Figure 2.14).

2.5.5 Mapping Reliability to End-User Facility Operating Hours [6]

Figure 2.15 shows reliability losses vs. facility revenues for five different operating schedules. Interestingly, those segments with the greatest percentage of end users claiming no reliability losses are the 8×5 and the 24×7 operators – the two extremes in terms of hours

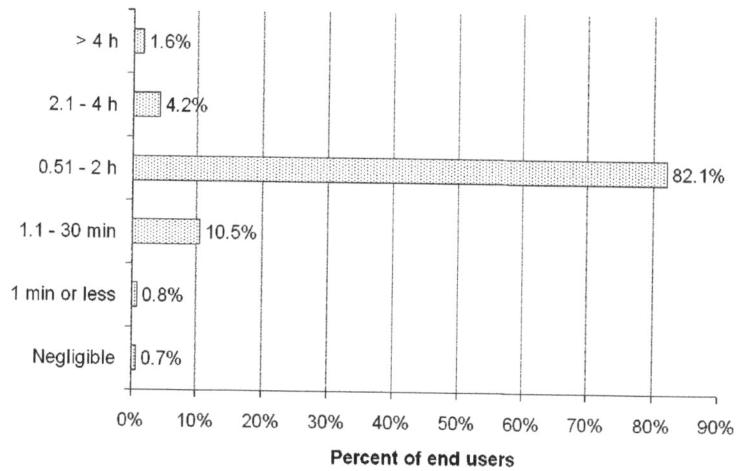

Figure 2.14 Duration of facility outage following a 1 h power interruption

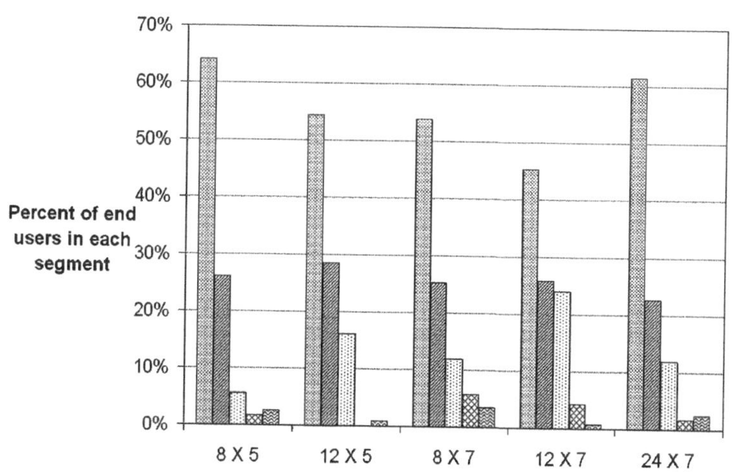

Figure 2.15 End-user reliability losses vs. revenues by operating hours

open for business. By one measure, the segment suffering the most is the 12×7 segment, where nearly 29 % of these operators report reliability losses that are greater than 1 % of revenues. The 8×7 segment is second, with 21 % reporting similar losses.

A more comprehensive weighted comparison of average end-user losses due to reliability problems is presented in Figure 2.16. Overall, one-shift operations open seven days per week report the highest average reliability costs of any segment, perhaps because of the difficulty of such an operation to make up lost production. On average, 24×7 and 8×5 businesses suffer equally from reliability problems, although at a level about 40 % lower than that reported by 8×7 operations.

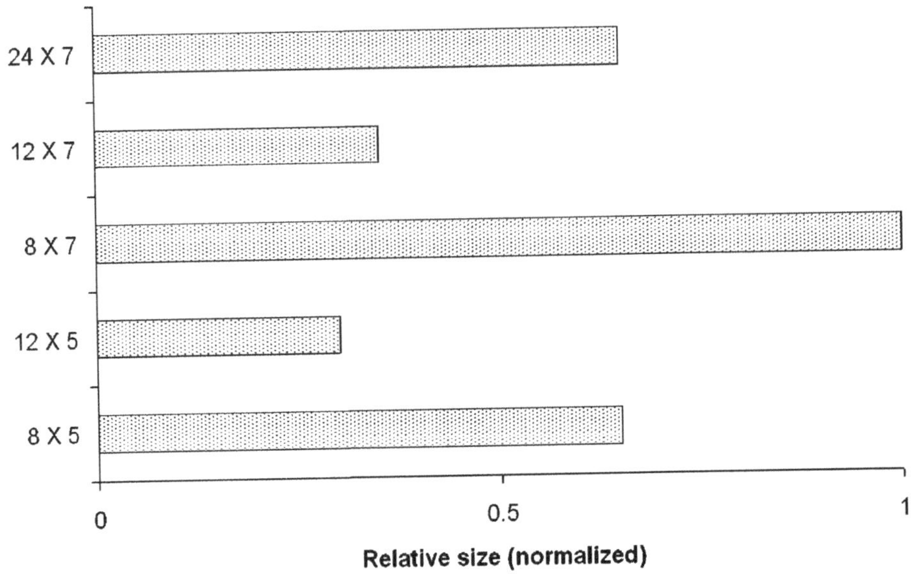

Figure 2.16 Comparison of average end-user losses by operating hours

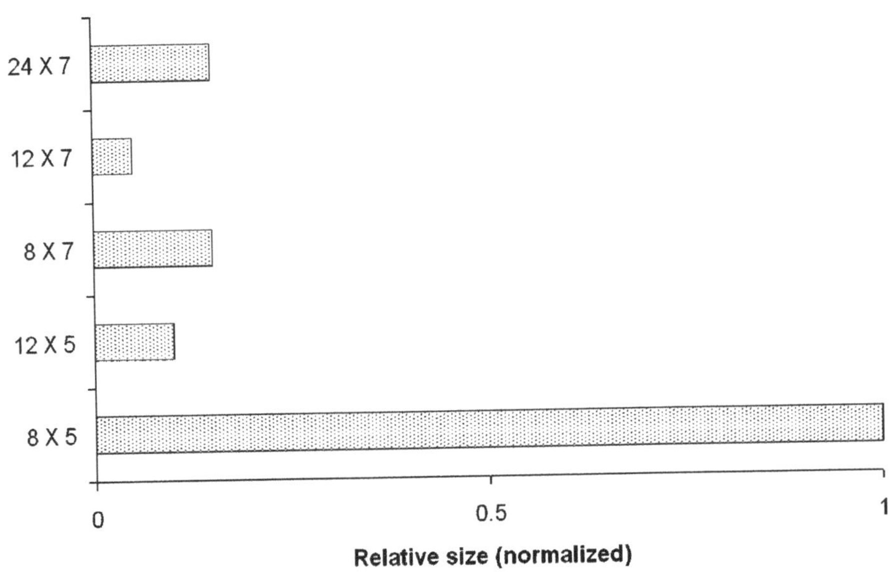

Figure 2.17 Comparison of overall losses for all end users by operating hours

The pain suffered by individual companies is, of course, important to assessing reliability markets, but of even greater importance, perhaps, is overall market value. The sheer number of 8×5 establishments makes this market segment the most potentially lucrative among the various operating hours scenarios (see Figure 2.17).

For a case study see web address

BIBLIOGRAPHY

[1] Brown R. E., Burke J. J., Managing the risk of performance-based rates. *IEEE Transactions on Power Systems*, vol. 15, pp. 893–898, 2000.

[2] Christie, R. D., Statistical classification of major event days in distribution system reliability. *IEEE Transactions on Power Delivery*, vol. 18, no. 4, pp. 1336–1341, 2003.

[3] Coulter R. T., 2001 Electricity Distribution Price Review Reliability Service Standards. Prepared for the Office of the Regulator-General and Service Standards Working Group, Victoria, Australia, 1999.

[4] *EEI Reliability Survey*, Minutes of the 8th Meeting of the Distribution Committee, 28–31 March 1999.

[5] EN 50160, *Voltage characteristics of electricity supplied by public distribution systems.*

[6] EPRI, Analysis of Extremely Reliable Power Delivery Systems: A Proposal for Development and Application of Security, Quality, Reliability, and Availability (SQRA) Modeling for Optimizing Power System Configurations for the Digital Economy, Palo Alto, CA, 2002. Product ID 051207.

[7] EPRI, Global Survey of Regulatory Approaches for Power Quality and Reliability, Palo Alto, CA, 2005. 1008589.

[8] EPRI, Reliability Benchmarking Methodology, TR-107937, May 1997.

[9] EPRI, Service Quality Index – Example Application, Palo Alto, CA, 2006. 1010199.

[10] IEEE 1159-1995, *Recommended practice for monitoring electric power quality*, 1995.

[11] IEEE Standard 493-1997, Figure 2.5.

[12] IEEE Standard 1366-1998, *Trial-Use Guide for Electric Power Distribution Reliability Indices.*

[13] Settembrini R. C., Fisher J. R., Hudak N. E., Reliability and Quality Comparisons of Electric Power Distribution Systems. IEEE T&D Conference, 1991.

[14] Shawn McNulty, Primen, The Cost of Power Disturbances to Industrial and Digital Economy Companies. Report for CEIDS, 2001.

[15] Short T. A., *Electric Power Distribution Handbook*, CRC Press, Boca Raton, FL, 2003.

[16] Third Benchmarking Report on Quality of Electricity Supply. Council of European Energy Regulators (CEER), 2005.

3

Voltage Control in Distribution Systems

Andrzej Kanicki

Voltage, unlike frequency, varies at different points of a power system. The voltage level and changeability are influenced by many factors occurring in the generating, transmission and distribution processes. The main factor, however, is load changeability. Significant and accidental load changeability in a distribution system imposes requirements concerning voltage stability maintenance. Another factor affecting the voltage is a balance between the reactive power consumed and generated. This balance has to consider reactive losses which may be a considerable part of reactive load demand. When the reactive power consumed is bigger than generated, voltages decrease and vice versa.

Some of the devices and strategies available for reactive power compensation and voltage control are as follows:

- Voltage control by means of reactive power flow change:
  - Synchronous generators
  - Synchronous compensators
  - Transformers with on-load or off-load tap-changers
  - Shunt capacitors
  - Shunt reactors
  - Static var compensators and STATCOM
- Voltage control by means of network impedance change:
  - Network configuration changes
  - Series capacitors
- Voltage control by means of undervoltage load shedding.

Handbook of Power Quality Edited by Angelo Baggini

Selecting the optimal voltage control strategy is a difficult process and should be meticulously analyzed in three stages, as described below.

Inadequate control of voltage and/or reactive power flows results in:

- unsuitable bus voltages;
- inappropriate operation of all electrical devices;
- augmenting the active power losses of a power system;
- limiting the capacity of system devices;
- limiting generating capacity.

3.1 DESCRIPTION OF THE PHENOMENA

In order to limit voltage deviation on the consumers' side and maintain it within acceptable limits, voltage regulation is used in power systems. In networks that lack voltage control, voltage drop values tend to be higher than the acceptable deviation values. Voltage changes that violate the acceptable limits are detrimental to all types of receivers and disturb their performance. Lowering the voltage value causes increased transmission losses in lines, transformers and other devices. It may even lead to stability violation of the whole system, called voltage avalanche. An excessive increase in the voltage value, on the other hand, leads to the increase of magnetizing currents of transformers and engines as well as damage or weakens the insulation of devices. Voltage control can regulate both the voltage value and voltage phase. In radial and tree networks only the former is of any importance, as changes in the voltage phase do not cause any load flow change. In grids, however, a change in the voltage module as well as in the phase module causes a load flow change and, as a result, a voltage change in a system. There are many voltage control methods that concern:

- excitation generator control;
- active power control;
- network scheme optimization;
- voltage level control by means of control transformers.

The voltage control process is divided into three stages:

Stage 1 An introductory stage which involves planning a distribution system scheme taking into account distribution system safety regulations and technical requirements. There may be different planning timeframes. In this stage network reliability and required electric energy quality parameters (including node voltage values) are established. As a result of the analysis, the following aspects are decided:

- a distribution system scheme for a given period of work;
- planned investments in a power system ensuring the maintenance of appropriate technical requirements;
- node voltage values, which are preset voltage control values in voltage regulators.

Stage 2 An execution real-time stage which involves automatic voltage and reactive power control together with overlapping performance of other control types and system operators. Three types of voltage control can be distinguished: primary, secondary and tertiary. Primary voltage control is executed by excitation generator voltage controllers which initiate quick changes in generator excitation in a situation when they detect a voltage change on the generator terminals. Primary voltage control includes other fast regulators such as static var compensators (SVCs). Secondary voltage control coordinates the operation of voltage and reactive power regulators in a given distribution network zone in order to maintain the required voltage levels. Tertiary voltage control involves voltage-level optimization using on-line calculations. This control type aims to modify settings of voltage and reactive power regulators.

Stage 3 A stage which involves technical and statistical analyses of voltage and reactive power control as well as the analysis of power system operators.

Voltage drop calculations and voltage control methods have already been discussed in various publications, e.g. [1],[2],[3],[5],[6],[7], but considering the fact that new devices, mainly electronic ones, are being installed in electrical networks, the issues are still present in the literature.

This chapter emphasizes voltage drop calculations and voltage control methods in distribution networks.

3.2 DISTURBANCE SOURCES

The main reason why voltage control should be used in a distribution system is to prevent voltage values exceeding acceptable limits. These limits are exceeded due to the following reasons:

- slow load changes connected mainly with daily and annual load curves;
- switching on/off essential power system elements;
- generator self-excitation occurring when the generator works with capacitance load;
- swinging of synchronous machines.

The first two reasons do not require any further explanation. Generator self-excitation while working with capacitance load is fortunately a rare phenomenon. In spite of the fact that generator voltage controllers eliminate this phenomenon, it should only be taken into account during switching operations and not during voltage control. The phenomenon of synchronous machine swinging is connected with frequency control, therefore it will not be analyzed in this chapter.

3.3 DISTURBANCE EFFECTS

3.3.1 Load Models

Load modeling has an impact on load flow studies and so has a robust impact on a voltage drop. A composite load, i.e. a consumer load consisting of heating, air-conditioning, lighting, induction motors and television, is approximated by combining load models. The load models can be normalized to rated voltage V_r, rated power P_r and rated frequency f_r. The exponential load models are:

$$P = P_n \cdot \left(\frac{V}{V_n}\right)^{\alpha_P} \cdot \left(\frac{f}{f_n}\right)^{\alpha_f} \tag{3.1}$$

$$Q = Q_n \cdot \left(\frac{V}{V_n}\right)^{\alpha_Q} \cdot \left(\frac{f}{f_n}\right)^{\alpha_f} \tag{3.2}$$

The exponential factors depend on the load type. The frequency dependence on loads is ignored for load flow and voltage drop studies and is applicable to transient and dynamic stability studies. In such a case three typical load models are used:

- a constant power load $\alpha_P = \alpha_Q = 0$;
- a constant current load $\alpha_P = \alpha_Q = 1$;
- a constant impedance load $\alpha_P = \alpha_Q = 2$.

A composite load is a mixture of these three types and can be simulated with values between 0 and 2. For typical loads the exponential load models have $\alpha_P = 0.4$–0.6 and $\alpha_Q = 1.4$–1.6.

Another form of a load model is the polynomial model. Reference [2] provides more details concerning the polynomial model, as follows.

For air-conditioning:

$$P = 2.18 + 0.268V - 1.45V^{-1} \tag{3.3}$$

$$Q = 6.31 - 15.6V + 10.3V^2 \tag{3.4}$$

For fluorescent lighting:

$$P = 2.97 - 4.00V + 2.00V \tag{3.5}$$

$$Q = 12.9 - 26.8V + 14.9V^2 \tag{3.6}$$

For an induction motor:

$$P = 0.720 + 0.109V + 0.172V^{-1} \tag{3.7}$$

$$Q = 2.08 - 1.63V - 7.60V^2 + 4.08V^3 \tag{3.8}$$

To recapitulate, the constant power load model is a simple model by means of which the most conservative results are obtained. Its main drawback, however, is that the model of the whole power system is non-linear. In load flow studies, the constant power load model is the most frequently used type. In calculations of low- and middle-voltage networks, the constant current load model is often used because its algorithm of calculating load flow and voltage drop is very easy.

3.3.2 Voltage Drop

Slow load changes and switching on/off essential power system elements cause voltage changes and may in some cases violate power system stability (voltage avalanche). Sudden load impacts (e.g. starting up a large motor) or load demands under contingency operating conditions, when one or more tie-line circuits may be out of service, result in voltage dips and, in some cases, also voltage avalanche.

The difference between the sending V_S and receiving end voltage V_R as a result of power flows P_L, Q_L can be expressed in the following way:

$$\Delta \mathbf{V}_{SR} = \mathbf{V}_S - \mathbf{V}_R = \frac{P_L R_L + Q_L X_L}{\mathbf{V}_R^*} + j\frac{P_L X_L - Q_L R_L}{\mathbf{V}_R^*} \tag{3.9}$$

Assuming that the receiving voltage is a real number, the following is obtained:

$$\Delta \mathbf{V}_{SR} = \mathbf{V}_S - \mathbf{V}_R = \frac{P_L R_L + Q_L X_L}{\mathbf{V}_R} + j\frac{P_L X_L - Q_L R_L}{\mathbf{V}_R} \tag{3.10}$$

The voltage drop can be approximately expressed as:

$$\delta V_{SR} = \frac{P_L R_L + Q_L X_L}{V_R} \tag{3.11}$$

When the constant current load model is used, it is easier to express Equation (3.11) as a function of the active and reactive load current component:

$$\delta V_{SR} = I'_L \cdot R_L - I''_L \cdot X_L \tag{3.12}$$

Switching on/off power system elements or load demand changes cause a voltage step that is directly proportional to the reactive power change and inversely proportional to the short-circuit power according to:

$$\frac{\Delta V}{V} = \frac{Q_L}{S_{SC}} \tag{3.13}$$

where S_{SC} is the short-circuit power in the bus where reactive power changes and Q_L is the reactive power change.

The larger the short-circuit power, the smaller the voltage drop caused by the reactive power change. If the short-circuit power were equal to infinity, the node voltage would be constant regardless of the reactive power change. Switching on/off essential power system elements affects the voltage drop similarly to the reactive power change.

3.3.3 Voltage Stability

The following major power system blackouts have caused the studies of voltage stability to be one of the most important problems concerning power system control [6]:

- France, 19 December 1978
- Belgium, 4 August 1982
- Sweden, 27 December 1983
- Florida, 2 September, 26 November, 28 and 30 December 1982
- France, 12 January 1987
- Japan, 23 July 1987

All the above blackouts were caused by voltage collapse, also referred to as voltage avalanche. There are many mathematical models dealing with this phenomenon. Some of them use dynamic generator and load models but the majority use static power system models. Static models based on the solvability analysis of load flow equations prove to be useful while planning and controlling large power systems. These models are characterized by relatively short computation time and ability to identify quickly bottlenecks in a system. Voltage collapse may result from both a sudden load increase and an outage of a transmission branch that is important to the transmission system. During voltage avalanche five characteristic time periods resulting from different system phenomena may be distinguished:

- The period from 0 to 1 s results from transient electromagnetic states and budding electromechanical phenomena.
- The period from 1 to 20 s results from automatic excitation generation control.
- The period from 20 to 60 s results from primary control systems preventing the generator from violating the upper and lower limits of reactive power generation.
- The period from 1 to 10 min results from secondary control and transformer ratio control.
- The period longer than 10 min results from voltage restoration and switching operations of a transmission system operator.

An approximated curve of voltage avalanche and voltage restoration is presented in Figure 3.1. The following characteristic time intervals are marked in the figure:

- 0–t_1, dynamic interaction of loads;
- t_1–t_2, interaction of transformer ratio control;
- t_2–t_3, voltage avalanche in load nodes;
- t_3–t_4, voltage restoration resulting from switching off loads.

In order to show the reasons for voltage instability in a power system, reactive power balancing in a node will be analyzed. Figure 3.2 focuses on power system performance at one point of work, point A or B, where load reactive power is equal to the generator reactive power.

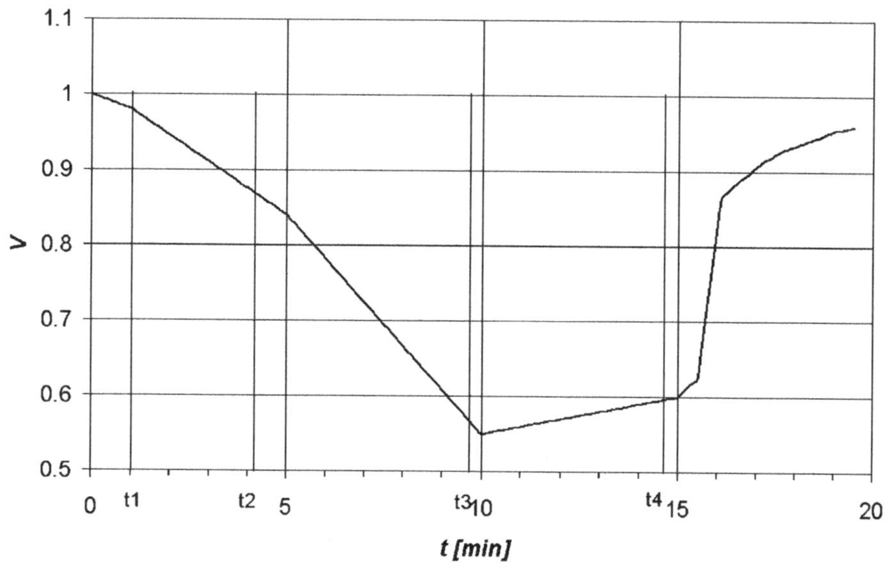

Figure 3.1 An approximated curve of voltage avalanche and voltage restoration based on the blackout incident in France in 1978

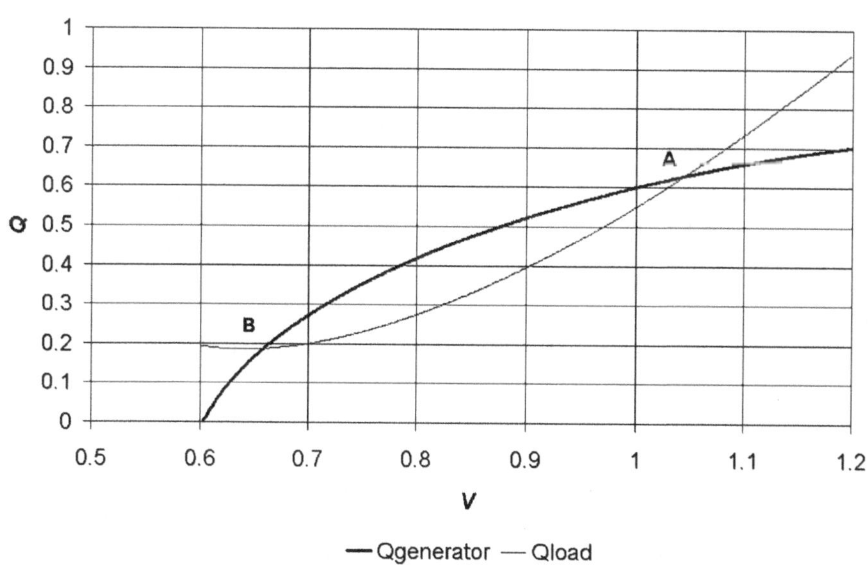

Figure 3.2 Reactive generator and load power according to voltage changes

Work at point A:

1. Decreasing voltage by ΔV causes Q_G (generator) to increase, and Q_L (load) to decrease (see Figure 3.2), so $Q_G > Q_L$, which in turn results in a voltage increase and return to the equilibrium point A.
2. Increasing voltage by ΔV causes Q_G to decrease and Q_L to increase, which in turn results in a voltage decrease and return to the equilibrium point A.

Work at point B:

1. Decreasing voltage by ΔV causes Q_G to decrease and Q_L to increase, which results in a further voltage decrease. The power system is going away from the equilibrium point while the voltage is constantly decreasing. This phenomenon is called voltage avalanche.
2. Increasing voltage by ΔV causes Q_G to increase and Q_L to decrease, which results in a further voltage increase and return to the equilibrium point A.

The voltage where $dQ/dV = 0$ is called the critical voltage. This phenomenon is particularly dangerous when asynchronous motors are linked to the analyzed node. Voltage avalanche causes the motors to stop, which results in the rapid growth in reactive power demand, which in turn causes a rapid voltage decrease. The resulting voltage decrease may cause a voltage avalanche in a neighboring node.

The above analysis proves that point A is a stable point and point B is an unstable point of work. Let us consider two values: $\Delta Q = Q_G - Q_L$ and ΔV. For point A:

$$\frac{d\,\Delta Q}{d\,\Delta V} < 0 \Rightarrow \text{stable point} \tag{3.14}$$

$$\frac{d\,\Delta Q}{d\,\Delta V} > 0 \Rightarrow \text{unstable point} \tag{3.15}$$

$$\frac{d\,\Delta Q}{d\,\Delta V} = 0 \Rightarrow \text{stability limit} \tag{3.16}$$

Switching on an additional load causes the load characteristic to move upwards. At a certain point of this movement there is only one intersection point, where $\Delta Q = 0$ (Figure 3.2). It is an unstable point and any further reactive power increase results in a definite stability loss.

In the next stage a relationship between the receiving end voltage and load of a line (with reactance X_L) is established. In order to do that, let us assume that the generator maintains a constant voltage V_R. After a series of transformations the following equation can be obtained:

$$(V_R)^4 + (2\,P_L\,(\tan\varphi)\,X_L - V_R^2)\,(V_R)^2 + \left(1 + (\tan\varphi)^2\right)\,P_L^2\,(X_L)^2 = 0 \tag{3.17}$$

The above equation is a biquadratic equation from the point of view of the receiving end voltage V_R and it may be solved analytically. The relationship between the receiving end voltage V_R and power P_L is presented in Figure 3.3. It can be seen that there is a maximum power, which can be sent also in a radial system when the load characteristic is not of a quadratic type. In Figure 3.3 horizontal lines denote the voltages when there is maximum power flow. These voltages are called critical voltages. If the power reaches

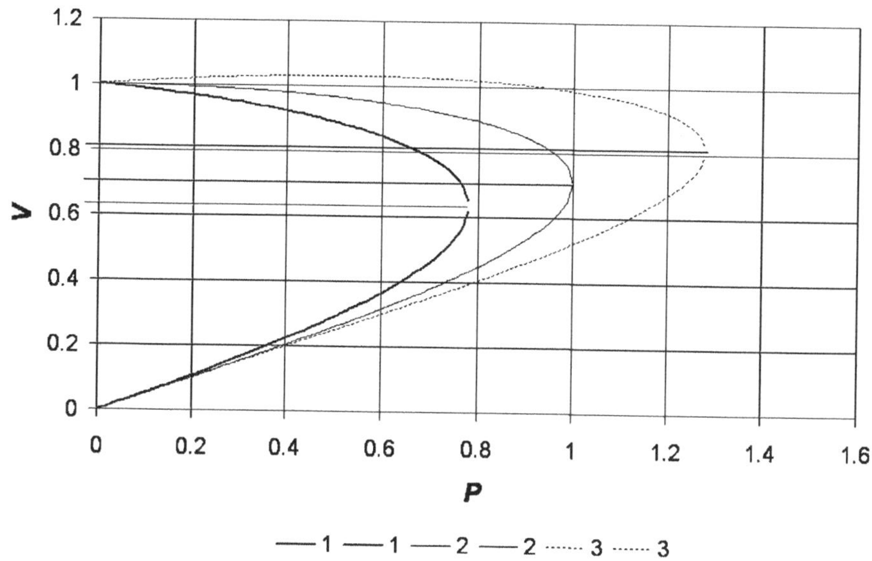

Figure 3.3 Relationship between voltage V_R and power P_L for: 1, $\cos\varphi = 0.97$ lag; 2, $\cos\varphi = 1.0$; 3, $\cos\varphi = 0.97$ leading

its maximum value, any further increase in its value results in voltage collapse and the voltage will eventually reach a value of zero. This phenomenon is called voltage avalanche. The curves in Figure 3.3 have been drawn assuming that active and reactive load power is constant. Any change in load characteristic will result in altering the curves. For loads of a quadratic characteristic type (impedance loads) this phenomenon will not occur.

For the more general case of a non-stiff load, the power demand will depend on voltage as described by the load voltage characteristics. The shape of the solution area of Equation (3.17) will vary depending on the actual voltage characteristics as shown in Figure 3.4. In general the less stiff the load, the more open the solution area. For the

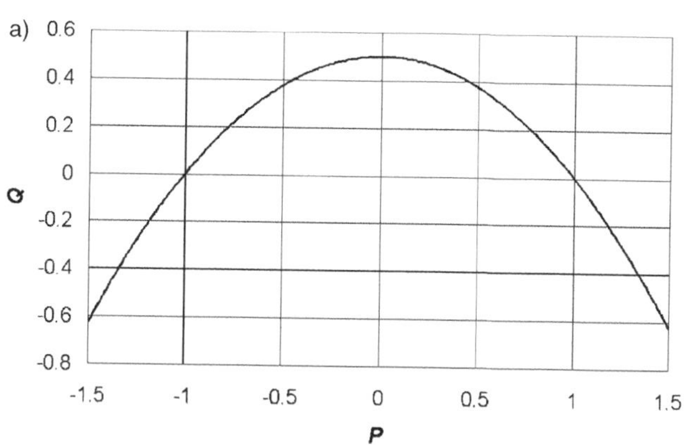

Figure 3.4 Dependence of the solution area on the shape of the load characteristics

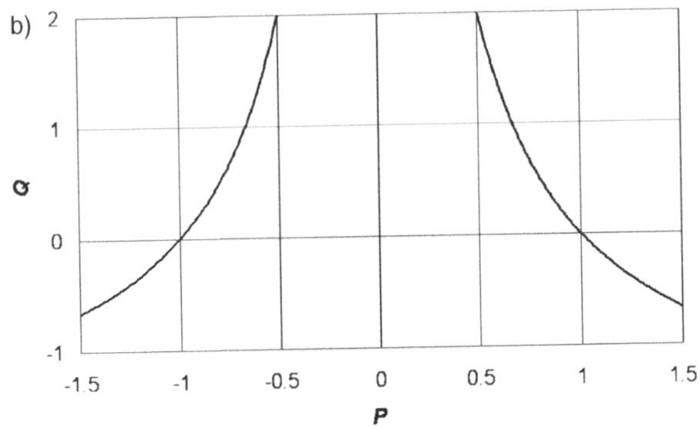

Figure 3.4 (Continued)

constant load, the solution area corresponds to a parabola (Figure 3.4a). If the reactive power characteristic is a square function of the voltage, then the solution area opens up from the top (Figure 3.4b).

3.4 METHODS OF EFFECT ELIMINATION

The effects described above can be eliminated by means of appropriate voltage and reactive power control. The control is maintained by different devices to provide at all take-off points supplies of reliable electric energy with good-quality parameters, and to provide at other points appropriate working conditions resulting from both parameters of the installed devices and load flow. Voltage control can be realized by changing:

- generator electromotive forces and the transformer ratio;
- network impedance;
- active and reactive load flow.

Voltage control, by changing generator electromotive forces and the transformer ratio, is direct control; the remaining two control types can be called indirect control. To provide voltage and reactive power control, the following ancillary services are ordered:

- voltage and reactive power control realized by means of automatic voltage control system of generating units;
- work done by hydroelectric sets in the compensator mode.

3.4.1 Generator Excitation Control

Generator excitation control is a basic way to influence the system voltage. A generator enabled to operate in a power system has to contain an appropriate excitation system and a regulator. Generators nowadays have expanded excitation systems that ensure proper work in both normal and emergency states. Generator excitation voltage control can ensure not only appropriate voltage but also desired reactive power generation. The value and type (inductive or capacitive) of the generated reactive power depend also on voltages in other parts of the system. It has to be noted here that work with underexcitation decreases the generator stability margin and work with overexcitation increases it.

Excitation control systems operate in such a way so as to:

- increase the stability margin;
- stabilize the work of consumer devices controlling voltage values;
- quickly and smoothly reduce voltage fluctuations;
- ensure proper reactive power generation and its suitable distribution during generator parallel work.

Generator voltage control is not conducted by the direct measurement of generator voltage but by current compensation of the voltage. Adopting this compensation as a preliminary stage of conversion of the measured generator voltage decreases/increases the voltage value. The resulting voltage is then compared to the reference regulator voltage and, as a result, a constant voltage value in a chosen network point is obtained. There are two types of current compensation process:

1. **Non-phase current compensation process** – where the generator voltage value depends also on the generator current magnitude.
2. **Phase current compensation process** – where the generator voltage value depends on the generator current magnitude and the angle between the generator voltage and current.

The drawback of non-phase current compensation control is the possible occurrence of voltage avalanche when the generator works with a capacitive load and when there is excessive reactive power in the system. Instead of consuming the reactive power, the generator produces it in increasing quantities, resulting in a growing excess of reactive power and a further voltage increase.

Generator excitation systems should contain the following items of equipment which set:

- the reference voltage locally or remotely;
- the dead zone of voltage control;
- the droop of voltage control;
- the voltage control interlocking time;
- maximum reactive power and voltage limiters;
- the maximum ceiling excitation time;
- the minimum reactive power limiter (power angle);
- the maximum stator current limiter;
- the maximum rotor current limiter;
- the induction limiter;
- the system stabilizer.

Generator excitation regulators should be characterized by:

- appropriate performance speed;
- appropriate regulation error;
- the ability not to cause voltage and reactive power fluctuations.

3.4.2 Transformer Ratio Control

Voltage control conducted by means of a transformer ratio change is another commonly used way of voltage control in power systems. This type of control involves a change in the number of turns in one of the transformer windings. Transformer and autotransformer ratio control does not change the reactive power balance, but altering voltage levels at different network points changes the reactive power flow. Taking tap changes into consideration, the following transformer types can be distinguished:

- Transformers whose ratio is changed after switching off the energy supply. Typical values of their ratio are: $\pm 5\,\%, 2 \times (\pm 2.5\,\%), \pm 2.5\,\%$. Such transformers are not used in transmission grids.
- Transformers whose ratio is changed on-load in wide ranges, e.g. $\pm 20\,\%$ and with the regulation step equal to or higher than $0.5\,\%$.

On-load ratio transformers must be equipped with special devices enabling on-load tap changes. Due to the high cost of such devices, they are used only in transformers whose rated power is higher than 5 MVA. Transformer ratio control can be conducted by changing the turn number in both transformer windings (primary and secondary) or only in one of them. The former produces better results because:

- the change in the turn number in the primary winding makes it possible to set appropriate work parameters (rated magnetic flux in the core);
- the change in the secondary winding makes it possible to obtain appropriate voltage levels, but is very expensive.

Therefore the transformer ratio change is conducted only on one side of the transformer. The choice of transformer side is determined by the values of voltage fluctuations, voltage and rated winding current. The chosen control windings are equipped with a number of taps, i.e. the so-called zero tap (corresponding to the transformer rated ratio) and additional taps. In some cases the control windings are divided into coarse control and accurate control windings. In most cases transformer rated ratios are not equal to the network rated voltage quotient, e.g. the transformer ratio is equal to 121 kV/6.3 kV, while the network rated voltages are 110 kV and 6 kV. As a result, during work in the zero tap the following voltage increase (expressed below as a percentage) can be obtained:

$$\delta V_{nT} = \left(\frac{\vartheta_s}{\vartheta_r} - 1 \right) \cdot 100\,\% \tag{3.18}$$

where ϑ_r is the transformer rated ratio and ϑ_s the network rated voltage quotient.

Apart from the above voltage increase, a voltage increase resulting from the ratio change δV_{zT} is obtained. The total voltage increase in the transformer is then equal to:

$$\delta V_T = \delta V_{nT} + \delta V_{zT} \tag{3.19}$$

Tap changing must be conducted without disturbing the system, i.e. it is unacceptable to disconnect the load current and resistance-less shortcuts of transformer turns. Therefore it is common to adopt on-load changing by means of shortcuts in control transformer turns using resistors or reactors. Two main types of tap-changers can be distinguished:

1. **Snap tap-changers** – changing time equals 0.01–0.1 s, characterized by shortcuts of turns by means of resistors, small overall dimensions but lower reliability.
2. **Slow tap-changers** – changing time a few seconds, characterized by shortcuts of turns by means of reactors and resistors.

The transformer ratio can be changed manually or automatically. For the latter a control-type tap-changer must be equipped with a voltage regulator similar to a generator voltage regulator. In order to synchronize tap-changer and transformer repairs and overhauls and to maintain appropriate contact life of the tap-changer, it is essential to set properly the dead zone and interlocking time of the regulator, which in turn causes a reduction in the excessive number of changes in a time unit. The interlocking time is usually assumed to be between 20 and 90 seconds. The regulator dead zone should be set within the following range:

$$\Delta V \leq \varepsilon \leq 2\,\Delta V \tag{3.20}$$

where ΔV is the voltage control degree and ε the regulator dead zone.

The optimum value of the regulator dead zone is $\varepsilon = (1.2 \div 1.4)\,\Delta V$. The regulator interlocking time can be a function of the voltage difference value: the bigger the difference, the smaller the interlocking time. Regulators are equipped with safety blockades preventing the following:

- Regulator performance during voltage differences that are too big and are impossible to eliminate by means of a ratio change. In this way unnecessary tap-changer performance is avoided and when a transformer is loaded it additionally prevents it from being overloaded.
- Unnecessary tap-changer performance during short circuits.
- Overshoot incidences.

3.4.3 Voltage Control by Means of Reactive Power Flow Change

Voltage control by means of reactive power flow change consists of a change in the reactive power value in network branches, which in turn results in a change of voltage drop. The above statement is also proved in Equation (3.11), but if the reactive power were negative (capacitive), the voltage would increase. Reactive power control is conducted by means of:

- Synchronous generators – described in the previous section.
- Shunt capacitors.
- Synchronous compensators.
- Shunt rectors.
- Static var compensators and STATCOM.

3.4.3.1 Shunt Capacitors

Depending on where they are installed, shunt capacitor banks affect different quantities such as:

- power coefficient improvement – in low-voltage networks and in individual consumers;
- power coefficient improvement and voltage drop decrease – in distribution networks;
- reactive power balance improvement and voltage control – in substations of the distribution network.

Figure 3.5 presents the voltage control method by means of switching on shunt capacitor banks. Shunt capacitor banks are not used in transmission grids, but they influence considerably the reactive power balance in distribution networks.

When analyzing a schematic diagram of a certain network it can be seen that using shunt capacitor banks may cause long-lasting supply voltage distortions resulting from resonance of the transformer and shunt capacitor banks. Such resonance occurs only when converters generate high-current harmonics and it increases phase voltage (Figure 3.6).

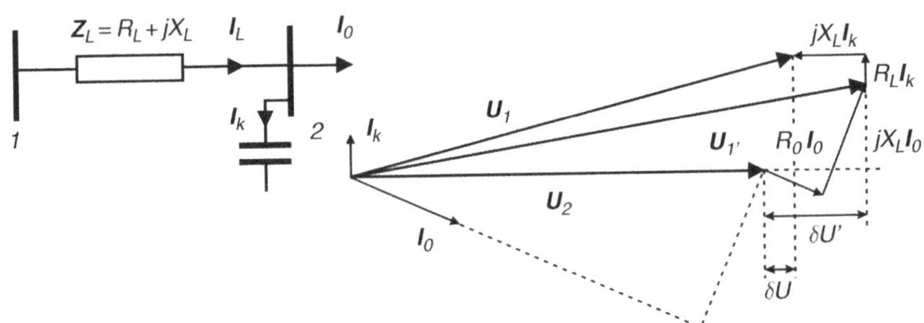

Figure 3.5 Voltage control by means of switching on shunt capacitor banks

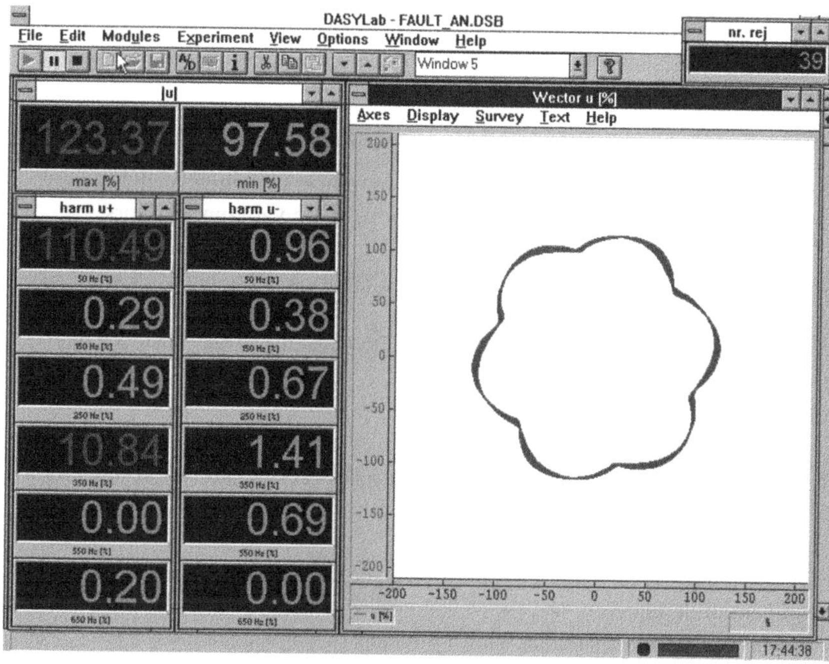

Figure 3.6 Three-phase voltage hodograph during a disturbance characterized by long-lasting, stable distortions caused by high harmonics

3.4.3.2 Synchronous Compensators

Nowadays synchronous compensators are virtually not used in a electric network. They used to be incorporated in compensating windings of high-voltage transformers. A hydroelectric set working in the compensating mode behaves similarly to a synchronous compensator. Reactive power generation and control rules in the compensation mode of the hydroelectric set are described in the previous section.

3.4.3.3 Shunt Reactors

Overhead lines in transmission grids consume a substantial amount of lagging reactive power to compensate for series power losses (proportional to power squared) and at the same time they generate leading reactive power (proportional to voltage squared). A 400 kV overhead line generates about 60 Mvar/100 km worth of leading reactive power. In the case of long overhead lines with a weak load, i.e. during off-peak hours, voltage rises substantially. To prevent this, it is essential to switch off some of those lines. Such action is detrimental, however, as it reduces supply reliability and increases power loss. Therefore, devices consuming lagging reactive power, e.g. shunt reactors, are installed in chosen substations of transmission grids. Switching on such a shunt reactor during weak load of a network reduces the leading reactive power excess. Shunt reactors are incorporated in compensating windings of high-voltage transformers or as primary reactors. Such reactors are controlled using a bang–bang method (switch-on/off).

3.4.3.4 Static Var Compensators and STATCOM Compensators

Quick load changes should be compensated for by quick switching of the reactive power source (receiver), which is possible by using various static var compensators:

- a compensator consisting of thyristor-controlled shunt capacitor banks (thyristor-switched capacitor, TSC);
- a compensator consisting of thyristor-controlled reactors (thyristor-switched reactor, TSR);
- STATCOM compensator.

From the point of view of the whole voltage control process, static var compensators have the following advantages:

- they increase the stability limit by increasing the critical power which depends on the line compensation degree;
- they increase the voltage stability limit by means of voltage stabilization.

3.4.4 Voltage Control by Means of Network Impedance Change

Two ways of control can be distinguished:

1. Network configuration changes.
2. Series capacitors.

3.4.4.1 Network Configuration Changes

This method is used mainly where there are network elements working in parallel and involves switching on additional lines or transformers during a significant load. During a small load the elements working in parallel are switched off. The main purpose of the network configuration change, however, is to adjust rated power of the elements to the load and reduce no-load losses; the voltage change is only a side effect.

3.4.4.2 Series Capacitors

This second method involves compensating line reactance by means of series capacitors (Figure 3.7). A voltage drop in the capacitors changes together with the load similarly to a voltage drop in the line reactance. Several series capacitors are installed in a line and they can be bypassed in order to change the compensation degree of the line reactance. Theoretically, only circuit-breakers can be bypassing devices, but nowadays thyristor-controlled series capacitors (TCSCs) are commonly in use.

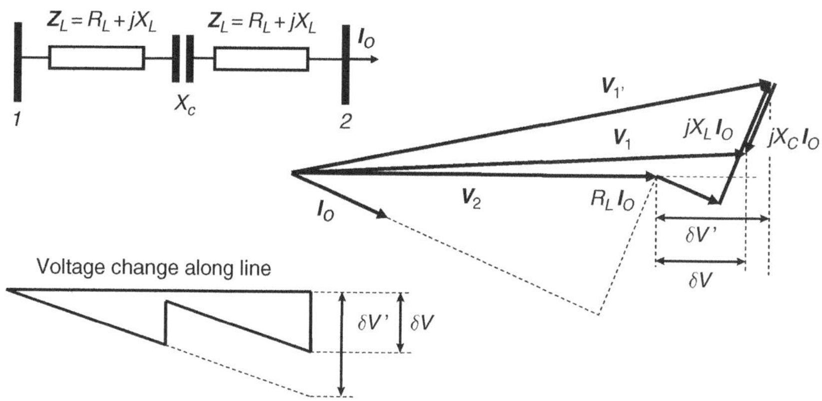

Figure 3.7 Voltage control by means of series capacitors

TCSCs have the following advantages:

- They increase the stability limit by means of the critical power increase resulting from the line reactance decrease. The stability limit increase depends mainly on the compensation degree of the line reactance. The critical power increment equals approximately:

$$\Delta P = \frac{V^2 \cdot \sin\delta}{X_L}\frac{k}{1-k} \tag{3.21}$$

$$k = \frac{X_c}{X_L} \tag{3.22}$$

where k is the compensation degree of line reactance.
- The above increment may eliminate the necessity of building an additional, parallel line.
- They can dump electromechanical oscillations occurring in a power system.
- They can increase the voltage stability limit by means of decreasing line reactance and maintaining constant voltage irrespective of quick load changes. Constant voltage can be maintained only in a certain range of power changes.
- They can decrease power losses and voltage drop.

TCSCs have some disadvantages:

- investment costs;
- the possibility of subsynchronous oscillations (series capacitors increase the R/X ratio);
- the possibility of ferroresonance in a circuit consisting of a series capacitor and an unloaded transformer.

Nowadays TCSC overvoltage protection is feasible. In distribution networks hundreds of TCSCs have already been installed, whereas in transmission grids only a few of them have been applied. It is predicted, however, that their number in the latter network type will grow.

3.4.5 Node Voltage Optimization

It is necessary to carry out constant on-line voltage optimization when a distribution network operates. The optimization process uses calculations based on on-line measurements and it aims to modify the reference values of a control system. The control system then affects reactive power flow. As a result of this optimization active power losses in a power system can be reduced by at least 10 %. The optimization process should be an integral part of a supervision control and data acquisition system (SCADA).

3.5 STANDARDS

3.5.1 Voltage Standards in Grid Normal Operating Conditions

The transmission system operator remains obliged to secure for each individual grid take-off point and grid supply point such power supply conditions that facilitate the maintenance of voltage standards. At meshed grid nodes where the end-use customers receive power which does not exceed the connection capacity, with tan φ smaller or equal to 0.4, the mean 15-minute rated voltage deviation should remain within the following ranges:

- Voltage standards for the grid supply points, applicable to normal grid operating conditions:
  - 400–420 kV for the 400 kV grid nodes
  - 220–245 kV for the 220 kV grid nodes
  - 110–123 kV for the 110 kV grid nodes.
- Voltage standards for the grid take-off points, applicable to normal grid operating conditions:
  - 380–420 kV for the 400 kV grid nodes
  - 210–245 kV for the 220 kV grid nodes
  - 105–123 kV for the 110 kV grid nodes.

3.5.2 Voltage Standards in Grid Disturbed Operating Conditions

These are as follows:

- Voltage standards for the grid supply points, applicable to disturbed grid operating conditions:
  - 380–420 kV for the 400 kV grid nodes
  - 210–245 kV for the 220 kV grid nodes
  - 105–123 kV for the 110 kV grid nodes.

- Voltage standards for the grid take-off points, applicable to disturbed grid operating conditions:
  - 360–420 kV for the 400 kV grid nodes
  - 200–245 kV for the 220 kV grid nodes
  - 100–123 kV for the 110 kV grid nodes.

3.5.3 Voltage Standards in Middle- and Low-Voltage Distribution Networks

In compliance with standard [4], slow voltage changes in middle- and low-voltage distribution networks are the reason why, in the normal state, 95 % of 10-minute mean values of measured r.m.s. voltage must be $U_n \pm 10\,\%$ for each week.

For a case study see web address

BIBLIOGRAPHY

[1] Anderson P. M., Fouad A. A., *Power System Control and Stability*, The Iowa State University Press, Ames, 1977.
[2] Das J. C., *Power System Analysis: Short-Circuit, Load Flow and Harmonic*, Marcel Dekker, New York, 2002.
[3] Dugan R. C., McGranaghan M. F., Beaty H. W., *Electrical Power Systems Quality*, McGraw-Hill, New York, 1996.
[4] EN 50160, *Voltage characteristics of electricity supplied by public distribution networks.*
[5] Hellmann W., Szczerba Z., *Frequency and Voltage Control in Power Systems*, WNT, Warsaw, 1978 (in Polish).
[6] Machowski J., Bialek J. W., Bumby J. R., *Power System Dynamics and Stability*, John Wiley & Sons, Ltd, Chichester, 1997.
[7] Taylor C. W., *Power System Voltage Stability*, McGraw-Hill, New York, 1994.
[8] Yong Hua Song, Johns A. T., *Flexible ac transmission systems (FACTS)*, IEE Press, Stevenage, 1999.

4

Voltage Dips and Short Supply Interruptions

Zbigniew Hanzelka

Two kinds of electromagnetic disturbances are dealt with in this chapter: voltage dips and short supply interruptions. Their sources and effects, as well as possible methods of their mitigation and measurement, are briefly described within this scope, which is sufficient for the formulation of contractual provisions. The issues that should be taken into account in concluding a contract between the supplier and consumer of electric power are considered on the basis of the existing standards and regulations.

4.1 DESCRIPTION OF THE PHENOMENA

A voltage dip[1] is a sudden reduction of the voltage at a particular point of an electricity supply system below a specified dip threshold[2] (within a time period no shorter than 10 ms), followed by its recovery after a brief interval. A voltage dip is in most of the regulations

[1] The term dip is normally preferred by the IEC, while sag is used mainly in US technical publications.
[2] The threshold value is the voltage r.m.s value defined in order to determine the start and end of a voltage dip. It can be expressed in volts or as per unit (percentage) value of the reference voltage. The voltage dip start threshold is specified for the purpose of defining the start of a voltage dip. Typically this value ranges from 0.8 to 0.9 of the reference voltage. The voltage dip end threshold is specified for the purpose of defining the end of a voltage dip. Typically, it has a value of 1–2 % of the reference voltage above the start threshold (Section 4.6.1.2).

Handbook of Power Quality Edited by Angelo Baggini

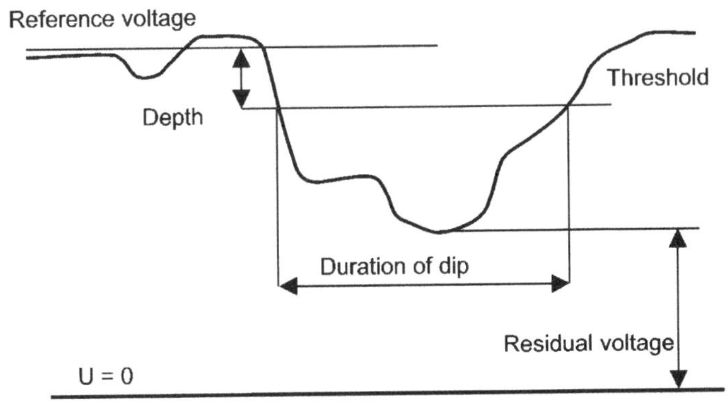

Figure 4.1 Characteristics of voltage dip

a two-dimensional electromagnetic disturbance, the level of which is determined by both voltage (the residual voltage[3] or the depth) and time (the duration – Figure 4.1).

The **depth of the voltage dip** is defined as the difference between the reference voltage and the residual voltage during voltage dip often expressed as a value in volts or as a percentage or per unit value of the reference voltage.

The **reference voltage** (U_{ref}) is the value specified as the base on which depth, thresholds and other values, which characterize the disturbance, are expressed in per unit (percentage) terms. Frequently the term depth (also 'magnitude' or 'value') is used in a descriptive, non-quantitative sense, in order to characterize a voltage dip without the intention of specifying whether it refers to the residual voltage or the magnitude in the meaning of definitions specified in this chapter.

The **duration of a voltage dip** is the time between the instant at which the voltage at a particular point on a supply system falls below the voltage dip start threshold and the instant at which it rises to the voltage dip end threshold.

In polyphase systems it has been commonly assumed that a three-phase dip starts when the voltage in the first disturbed phase falls below the dip start threshold and ends when voltages in all phases are equal to or above the dip end threshold (Figure 4.2). The duration of a voltage dip depends on the specified threshold value (Figure 4.3).

A **short supply interruption** is a particular case of a voltage dip. It is a reduction of a voltage on all phases at a particular point of an electric supply system below a specified interruption threshold[4] followed by its restoration after a brief interval (Figure 4.4).

[3] The residual voltage is the minimum value of r.m.s. voltage recorded during a voltage dip or short interruption (Figure 4.1). The residual voltage may be expressed as a value in volts or as a percentage or per unit value relative to the reference voltage.

[4] An **interruption threshold** is an r.m.s. value of the voltage specified as the minimum value of the residual voltage of a voltage dip.

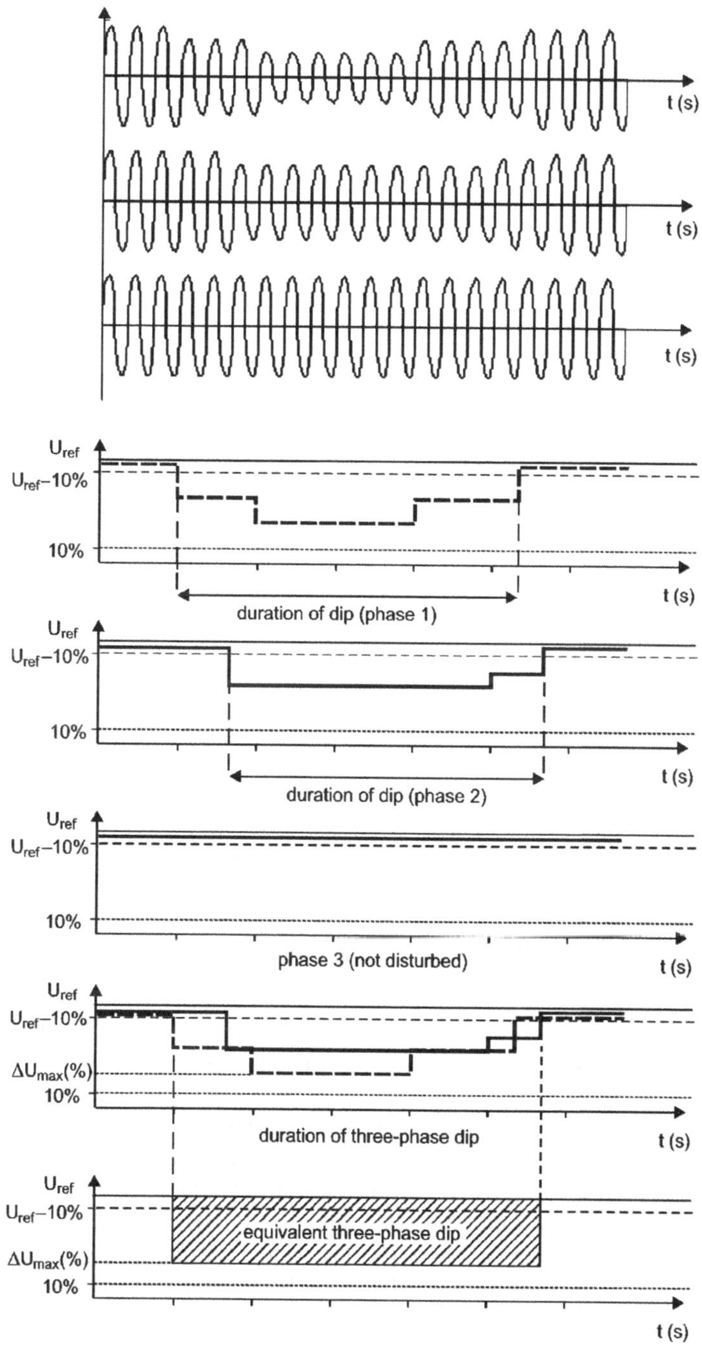

Figure 4.2 The duration of a voltage dip in a three-phase system; the threshold value has been assumed to be 90 % of U_{ref} [65]

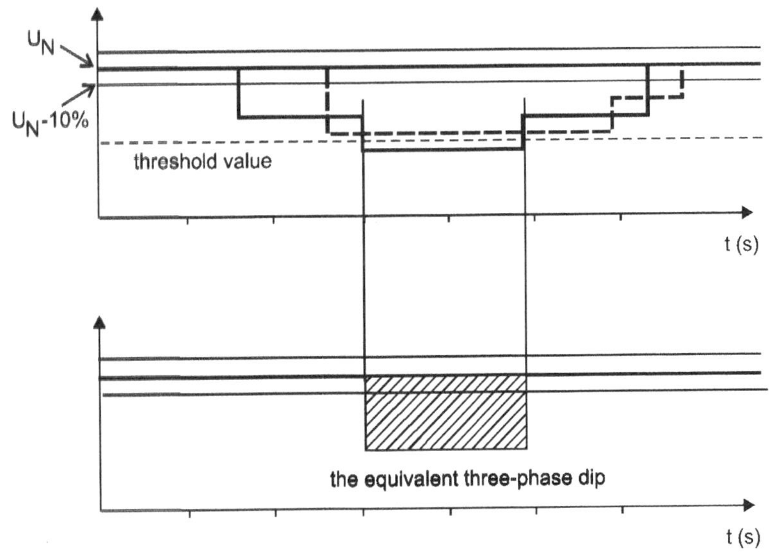

Figure 4.3 The influence of a threshold value on a voltage dip duration (cf. Figure 4.2) (Reproduced from Voltage Dips Measurement, UIEPQ-9559, October 1995)

4.2 PARAMETERS

The typical power system including generators, loads and coupling impedances is a single, integrated and dynamic system – any change of voltage, current, impedance, etc., at one point immediately brings about a change at every other point in the system.

4.2.1 Voltage Dip Duration

The duration of voltage dips is mainly determined by the operating time of the device which acts to disconnect the short circuit from the system, mainly fuses, circuit-breakers and protection relays. Protection relays often are designed to have an inverse time characteristic, so the lower the short-circuit current, the longer the fault clearance time. Fuses have similar characteristics. The time characteristics and settings of both the fuses and relays are graduated and coordinated, so that a fault detected by several devices is cleared at the most appropriate point of the system (normally, the closest to the fault location).

Many faults are cleared within the wide time range. Faster times may be achieved for short circuits on transmission lines (60–150 ms), while the fault clearing on distribution circuits may be considerably slower (MV, 0.5–2 s; LV, depending on the fuse characteristic).

When a disturbance other than a short circuit is the source of the voltage dip, the duration is governed by that of the causative event.

Some loads, e.g. electric motors, draw a large inrush current as the voltage recovers at the end of a disturbance. This results in extending the duration of the voltage dip.

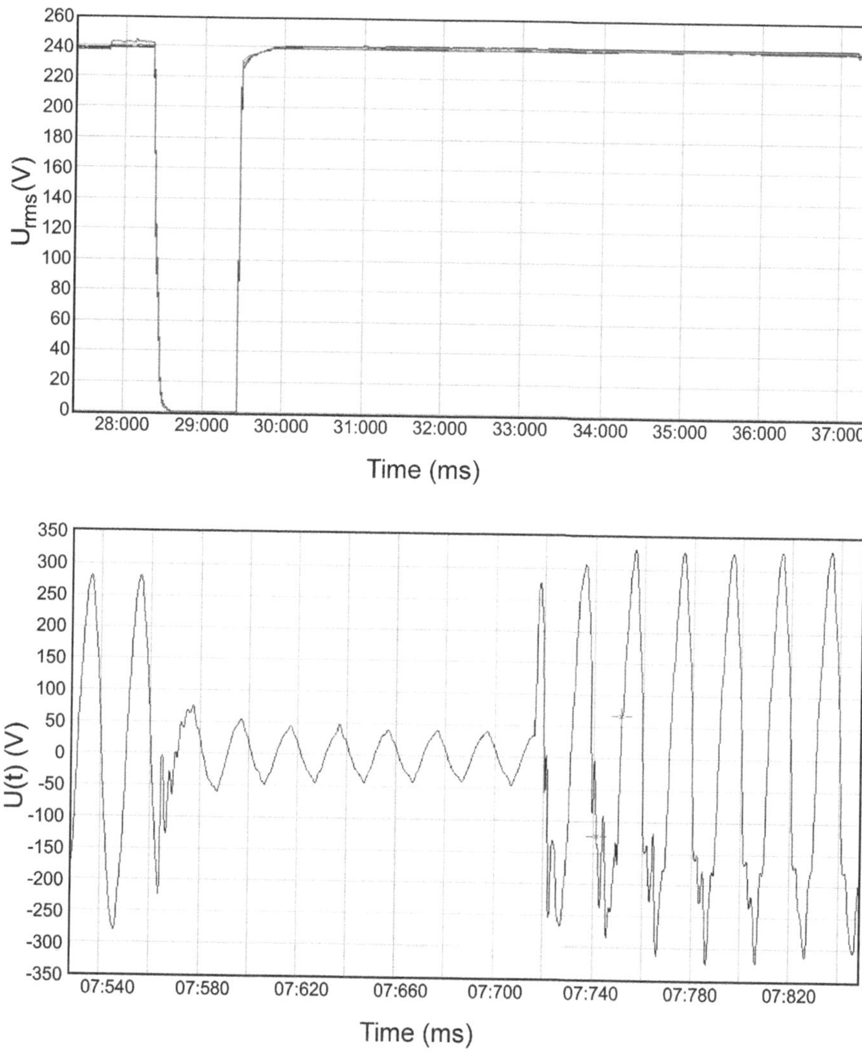

Figure 4.4 Short supply interruption: upper, r.m.s. voltage characteristics; lower, voltage waveform (the value of an interruption threshold has been assumed to be 90 % of U_{ref}) (Reproduced from Voltage Dips Measurement, UIEPQ-9559, October 1995)

4.2.2 Magnitude of a Voltage Dip

The magnitude of the voltage dip is governed by the electrical distance of the observation point from the site of the short circuit and the source(s) of supply. The level to which the voltage falls at a particular observation point during the dip is a random value, depending on its position in the network relative to a short circuit.

In the case of a zero-impedance short circuit, the system can be represented by a single-phase equivalent circuit as in Figure 4.5. The voltage in the considered point

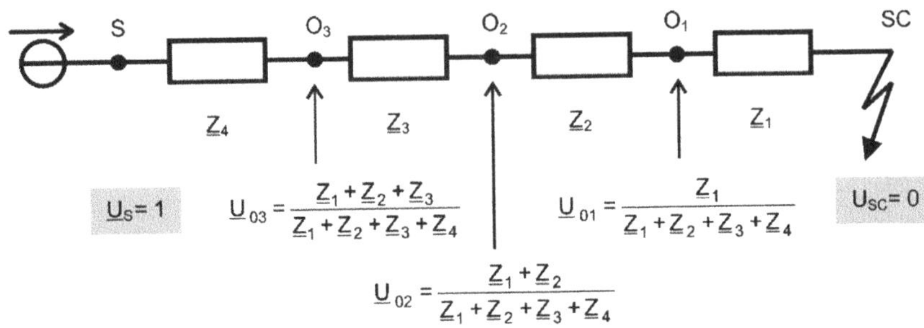

Figure 4.5 Voltage dips at points O_1, O_2 and O_3 for short circuit at the point SC and single equivalent source S (expressed in terms of residual voltage pu); Z, equivalent impedances

(e.g. O_1, O_2 or O_3) depends on the equivalent impedances connecting that point to the short-circuit location (SC) and the source. Depending on the relative magnitudes of these impedances, the depth of the voltage dip can vary over the range 0–100 %.

The nearer the short-circuit location is to the considered point, the lower is the residual voltage. On the other hand, the nearer the considered point is to the supply source (generally, a source of energy, which can also be a capacitor bank, battery, rotating machine, etc.), the lesser is the voltage reduction during the disturbance.

A short circuit on the transmission system is likely to result in a voltage dip that is observed over a very wide area, even at a distance of some hundreds of kilometers. A short circuit in a distribution circuit has a much smaller field of influence. The severity of the disturbance will be moderated considerably by neighboring circuits.

Given that the observation point is located near to a consumer, a short circuit within the consumer's installation can result in a voltage dip with a magnitude exceeding the dip caused by short circuits on the transmission or distribution system.

The depth of a voltage dip depends on the kind of short circuit and connection of windings of the transformer(s) between the short circuit's location and the considered point of the system. The phasing of the short circuit or other disturbance, the cause of a dip, the connection methods of the primary and secondary transformer windings are significant factors of the negative impact of the disturbance. For instance, considering a step-down transformer connected as Dyn or Dy, a single line-to-neutral fault on the primary side (initially a voltage dip of 0 V residual voltage on one phase) results in voltage dips on the secondary side in two phases, each 58 % of the pre-existing voltage (Table 4.1, based on [59]).

In practice, loads that are sensitive to voltage dips (power converters, adjustable-speed drives, motors, control equipment, etc.) are often connected line to line in industrial installations. They would therefore be subjected to line-to-line voltage dips rather than line-to-neutral dips.

By using the symmetrical components of a three-phase power system it is possible to categorize the dip characteristics in all three phases simultaneously. Such an approach has been developed by Bollen [8]. He developed the concept of voltage dip 'types' to describe the different voltage dip characteristics that can be experienced at the end-user terminals for different fault conditions and system configurations. Equipment is normally connected at a

Table 4.1 Transformer secondary voltages with a single line-to-neutral fault on the primary side (based on [59])

Connection of transformer windings[a]	Line-to-neutral voltage			Line-to-line voltage		
	U_1	U_2	U_3	U_{12}	U_{23}	U_{31}
Ynyn or Yny	0.0	1.0	1.0	0.58	1.0	0.58
Yy, Yyn or Dd	0.33	0.88	0.88	0.58	1.0	0.58
Ynd or Yd	–	–	–	0.33	0.88	0.88
Dyn or Dy	0.58	1.0	0.58	0.88	0.88	0.33

[a] Capital letters refer to primary winding connection (supply network side) and lower-case letters refer to secondary winding connection (load side). N or n designates a grounded primary or secondary transformer neutral, respectively.

lower voltage level than the level at which the fault occurs. The voltages at the equipment terminals, therefore, depend not only on the voltages at the point of common coupling (PCC) but also on the winding connection of the transformers between the PCC and the equipment terminals (phase shift). In this method, dips are divided into seven types, from A to G (Table 4.2). Type A refers to symmetrical three-phase dips, whereas single-phase and phase-to-phase faults cause type B, C or D dips. Three additional types correspond to two-phase-to-ground faults. These fault types can be used conveniently to summarize the expected performance at a customer's location for different types of faults on the supply system. In addition to the dip type, a complex phasor called the characteristic voltage is all that is needed to describe a voltage dip in a three-phase system without losing any

Table 4.2 Seven types of dip (based on [8]); h, residual voltage (pu)

Type of dip	Phasor diagram (dashed lines, before; solid lines, during fault)	Type of fault	Phase-to-neutral voltages
A		Three-phase faults	$\mathbf{U}_a = hU$ $\mathbf{U}_b = -\frac{1}{2}hU - j\frac{\sqrt{3}}{2}hU$ $\mathbf{U}_c = -\frac{1}{2}hU + j\frac{\sqrt{3}}{2}hU$
B	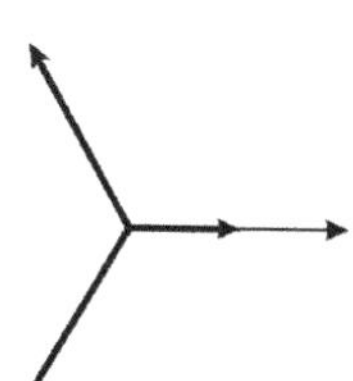	Single-phase faults	$\mathbf{U}_a = hU$ $\mathbf{U}_b = -\frac{1}{2}U - j\frac{\sqrt{3}}{2}U$ $\mathbf{U}_c = -\frac{1}{2}U + j\frac{\sqrt{3}}{2}U$

Table 4.2 (Continued)

Type of dip	Phasor diagram (dashed lines, before; solid lines, during fault)	Type of fault	Phase-to-neutral voltages
C	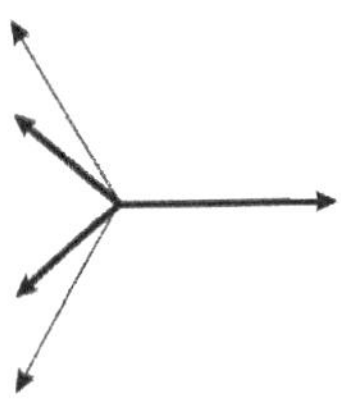	Phase-to-phase faults	$\mathbf{U}_a = U$ $\mathbf{U}_b = -\frac{1}{2}U - j\frac{\sqrt{3}}{2}hU$ $\mathbf{U}_c = -\frac{1}{2}U + j\frac{\sqrt{3}}{2}hU$
D	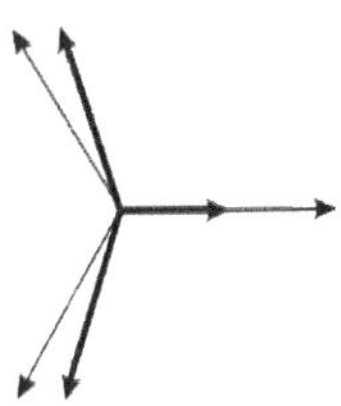	Phase-to-phase faults	$\mathbf{U}_a = hU$ $\mathbf{U}_b = -\frac{1}{2}hU - j\frac{\sqrt{3}}{2}U$ $\mathbf{U}_c = -\frac{1}{2}hU + j\frac{\sqrt{3}}{2}U$
E	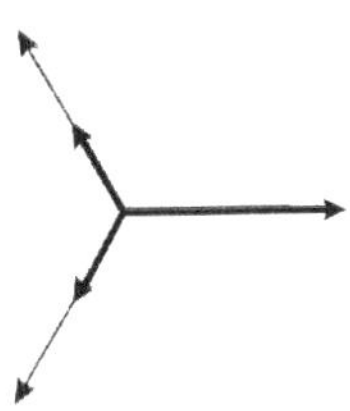	Two-phase-to-ground faults	$\mathbf{U}_a = U$ $\mathbf{U}_b = -\frac{1}{2}hU - j\frac{\sqrt{3}}{2}hU$ $\mathbf{U}_c = -\frac{1}{2}hU + j\frac{\sqrt{3}}{2}hU$
F	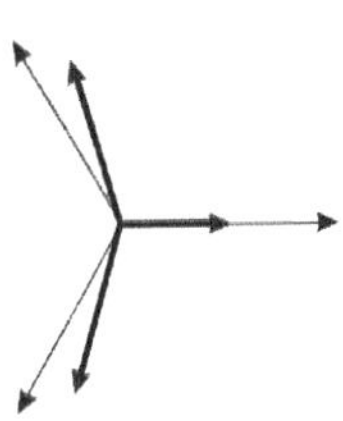	Two-phase-to-ground faults	$\mathbf{U}_a = hU$ $\mathbf{U}_b = -\frac{1}{2}hU - j\frac{1}{\sqrt{12}}(2+h)U$ $\mathbf{U}_c = -\frac{1}{2}hU + j\frac{1}{\sqrt{12}}(2+h)U$
G	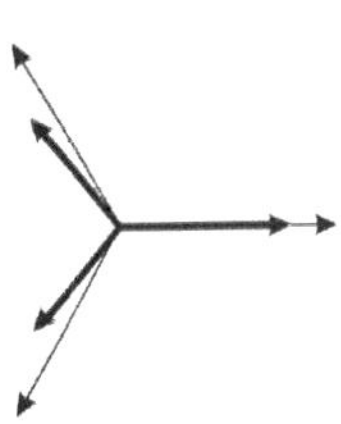	Two-phase-to-ground faults	$\mathbf{U}_a = \frac{1}{3}(2+h)U$ $\mathbf{U}_b = -\frac{1}{6}(2+h)U - j\frac{\sqrt{3}}{2}hU$ $\mathbf{U}_c = -\frac{1}{6}(2+h)U + j\frac{\sqrt{3}}{2}hU$

essential information. For systems where the positive and negative impedances are not equal, an additional PN factor (Positive–Negative factor, [8]) is needed in order to ensure accurate results. Any of the phase-to-ground or phase-to-phase voltages during a dip can be retrieved when the three parameters are known: namely, dip type, characteristic voltage and PN factor.

This approach is valid also from one voltage level to another because it takes into account transformer and load connections (star–delta), and is based on per unit calculations.

4.3 SOURCES

4.3.1 Sources of Voltage Dips

The primary source of voltage dips is the electrical short circuit occurring on the power supply system. The short circuit causes a very large current, and this, in turn, gives rise to large voltage drops across the impedances of the supply system. Short-circuit faults are an unavoidable state of system operation. They have many causes, but basically they are caused by exceeding the insulation level, mostly due to insulation breakdown or in effect to overvoltages caused by switching operations or atmospheric discharges.

The example of a voltage dip on an LV system caused by an unbalanced remote fault is shown in Figure 4.6.

Most supply systems are three-phase systems. The short circuit can occur between phases, phase and neutral, or phase and earth; one or more phases can be involved.

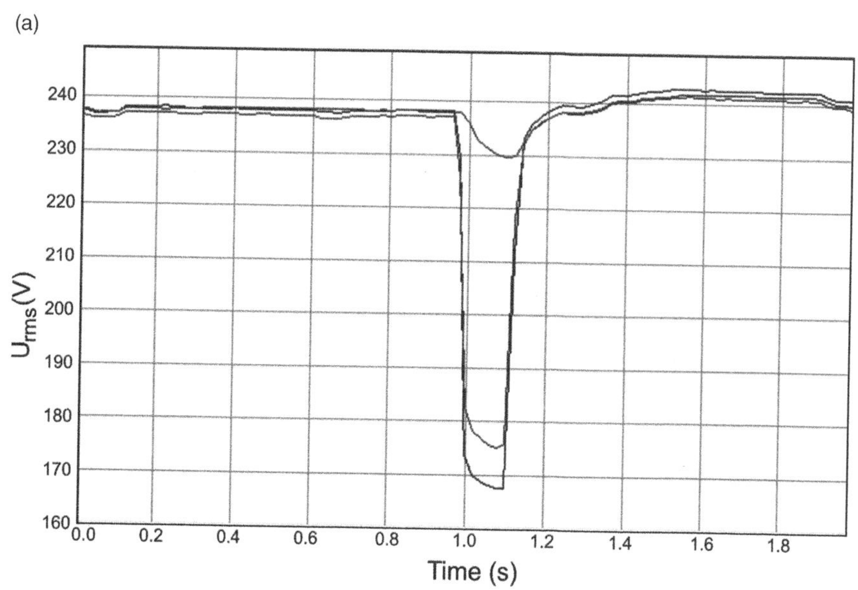

Figure 4.6 Voltage dips caused by an unsymmetrical remote power fault

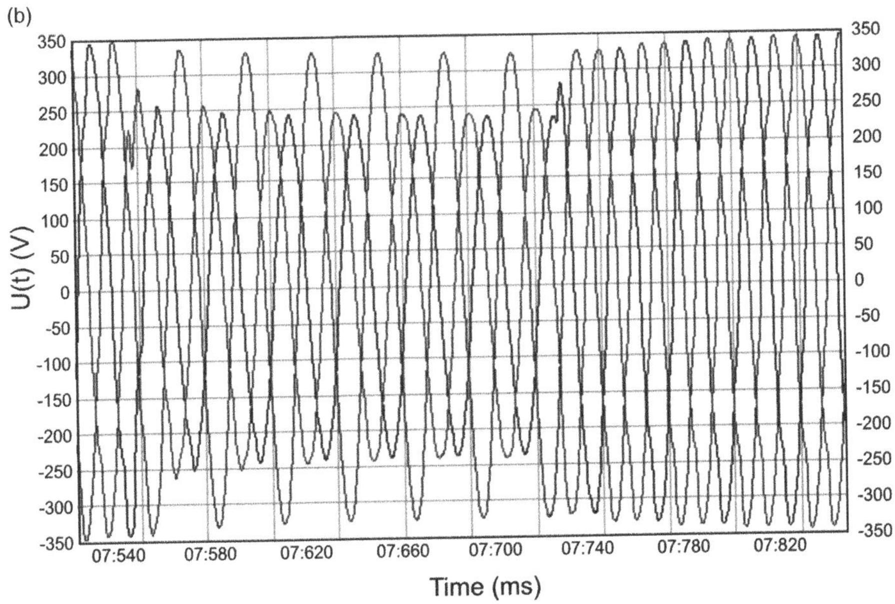

Figure 4.6 (Continued)

Supply systems are equipped with protective devices to disconnect the short circuit from the source of energy. As soon as the disconnection takes place, there is an immediate recovery of the voltage, approximately to its previous value, at every point except disconnected ones. Some types of faults are self-clearing: the short circuit disappears and the voltage recovers before line disconnection can take place.

The switching of large loads, starting of large motors connected to the end of a long supply line (Figure 4.7), power fluctuations of great magnitude (particularly of reactive power) are characteristic of some categories of loads and installations, such as variable load and/or speed drives, arc furnaces, welding equipment, etc., and all can produce large

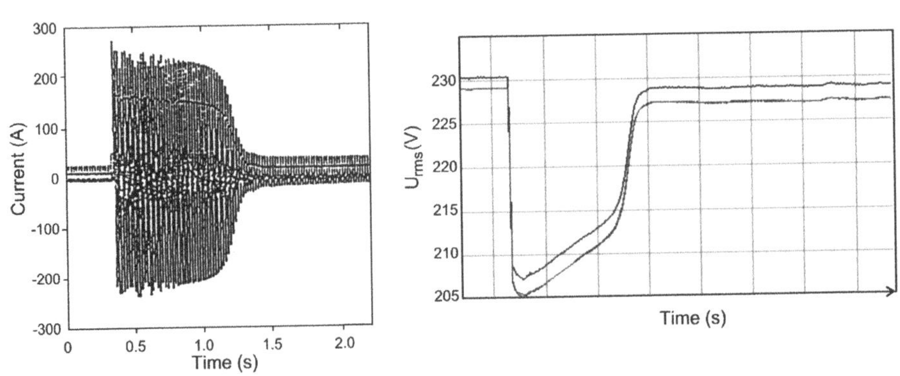

Figure 4.7 Voltage dip caused by starting of large motor

changes in current, similar in effect to those of the short-circuit states. The influence of this category of loads should be limited to an acceptable level in the conditions of connection issued by the system operator, depending on the actual state of the supply system.

4.3.2 Sources of Short Supply Interruptions

Short interruptions are typically associated with switchgear operation related to the occurrence and termination of short circuits in the system or installations connected to it The operation of a circuit-breaker or fuse disconnects part of the system from the source of energy. In the case of a radial circuit, this interrupts the supply to all downstream parts of the system. In the case of a meshed network, disconnections at more than one point are necessary in order to clear the fault. Electric power users within the disconnected segment of network suffer an interruption of supply.

Automatic reclosing sequences are often applied in the supply system. Their purpose is to restore the circuit to normal operation with the minimum of delay in the event that the fault proves to be a transient one. The reclosing operation may be attempted several times (depending on the adopted practice of fault clearing) until self-clearance of the fault, or the circuit-breaker remains in the open position if the fault is a permanent one. It should be noted that almost each reclosure results in an additional voltage dip.

In addition to the actual isolation of the fault, further switching operations are often carried out, either automatically or manually, in order to reduce the number of users interrupted as a result of the fault. Thus, a single fault can result in a complex series of switching operations, observable to users as a series of interruptions of various durations. Depending on the structure of the network, on the position of individual users relative to the location of the fault and the time of operation of relevant protection devices, some users will experience only minor dips, while others may even have to wait for repairs to be completed before supply can be restored.

Interruptions having a duration up to 1 min (or, in the case of some reclosing systems, up to 3 min) are classified typically as short interruptions.

4.4 EFFECTS

The energy sources which under normal conditions supply energy to the equipment do not perform their function during the disturbance or perform it only in a limited range. The reduction or loss of voltage causes the energy needed for normal operation not to be supplied to the equipment. This leads to degradation in the performance of the equipment and in extreme cases to a complete cessation of operation. Protective systems are often implemented for the purpose of disconnecting the supply in the event of the voltage falling below a set level. Such protection can have the effect of converting a voltage dip into a long supply interruption. This long interruption is not caused by the voltage dip, but is the intended result of a protective device's response to the reduced voltage.

A direct technical effect depends, among other factors, on the magnitude and duration of a voltage dip.[5]

If the voltage attains too low a value, or the duration of a dip is excessively long, the equipment may be disconnected by a protective system or may operate in an improper manner. The economic consequences of such an event can be of considerable significance. They include loss of production, costs of restarting the technological process (this is particularly significant for continuous processes, where the time needed for restarting is, as a rule, very long), damaged equipment and materials, delayed delivery, reduced customer satisfaction, a decrease in the power delivered to the user, etc. Also the dissatisfaction of employees (if their wages depend on the production output) should be taken into account. These costs may, and probably will, have an impact on the position of utility companies in the energy market as the users can seek alternative energy sources, e.g. a local one.

The immunity of equipment to voltage dips and short interruptions is described in several standard documents. IEC 61000-4-11 [38] defines the immunity test methods and range of preferred test levels for electrical and electronic equipment with a rated input current of 16 A or less per phase (connected to LV power supply networks). IEC 61000-4-34 [40] concerns equipment with an input current of more than 16 A per phase. Additional information can be found in standard documents IEC 61000-6-1 [41] and 61000-6-2 [42].

4.4.1 IT Equipment and Control Systems

Microprocessor systems, widely used for process control, are particularly sensitive to voltage dips. Irregularities in their operation can interrupt the process, even if drives and other 'power' equipment are immune to these disturbances. The most common effects of voltage dips are loss of transmission and errors in signal transmission. IT equipment is usually provided with built-in detectors of failures and external disturbances for the purpose of protecting internal memory data (including the software procedure for the response to voltage dips and short supply interruptions that guarantees data protection and correct operation after voltage recovery) or for safety reasons (the loss of transmission or erroneous commands in the process control).

This kind of equipment is more sensitive to gradual changes (decrease) in the voltage than to the sudden interruption of supply. Some failure detectors do not react sufficiently fast to the gradual change in the supply voltage. In this event the d.c. power supply voltage may drop below the permissible operating voltage before the failure detector is actuated. In consequence, the data will be lost or erroneous. After voltage recovery the equipment might not be capable of restarting properly and may need reprogramming. Therefore, detailed

[5] A sudden phase shift is associated with the voltage dip (Figure 4.9), as well as the voltage unbalance in polyphase systems that can have a significant effect on some equipment. This kind of influence is not discussed further in this chapter; however, it is a topic of discussion within the circle of standardization experts. The suggested approach involves: the determination of a complex value of the voltage fundamental harmonic during a dip; employing the symmetrical components method, etc.; and determination of the maximum value of the phase shift angle (also its rate of change).

All the suggested methods yield correct results in the voltage steady state, whereas in transient states significant discrepancies do occur.

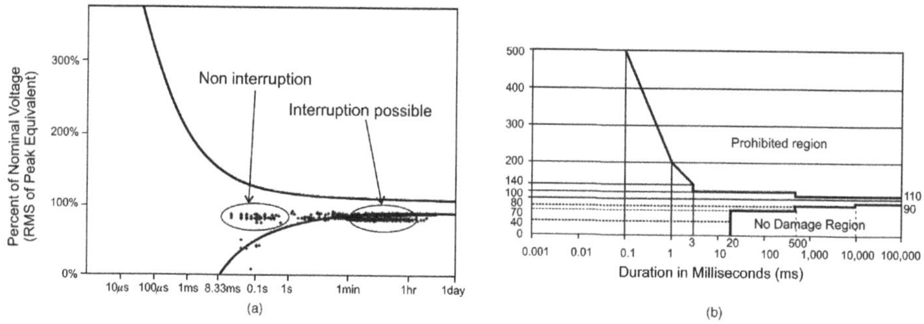

Figure 4.8 CBEMA (a) and ITIC (b) curves. These dip-withstand performance curves represent standards that computer and other equipment manufacturers are required to comply with by many customers

procedures for testing IT equipment for immunity to the discussed disturbances are given in relevant standards.

In the simplest description, a dip can be represented as a point on a voltage vs. duration graph. At the same level of simplification, the equipment voltage tolerance characteristic can be represented on the same axes (see Figure 4.11). In spite of neglecting the phase jumps and other aspects of dip behavior, this approach is a practicable means of comparing the performance of different equipment.

The information on the immunity of IT equipment to changes in r.m.s. voltage value provides so-called the ITIC[6] (formerly CBEMA)[7] curve, shown in Figure 4.8.[8]

The duration of disturbance is indicated on the abscissas in seconds (s) and cycles (c) of the power–frequency voltage fundamental component, the r.m.s. voltage value in percent of nominal voltage, is represented on the ordinates. As seen from the curve, the immunity of equipment (guaranteed for disturbances within the curve branches) is strongly dependent on the duration of a dip. According to these characteristics, the IT equipment (computers, components of computer networks, etc.) should obviously withstand steady-state voltage changes within 90–110 % of the nominal value.

The operation of a programmable logic controller (PLC) can be considered in four basic functional steps: reading the input data (input module), execution of the control program (CPU), self-diagnostics (CPU) and modification of output states according to the program (output module). Voltage dips can affect the CPU and I/O cards and influence the PLC logical levels during execution of each step. The continuity of technological process may be interrupted at any of these points of possible disturbance. The execution cycle, i.e. the time required for execution of all the four steps, should not exceed a dozen or so milliseconds and thus can be commensurate with the dip duration.

One of the most vulnerable modules of the PLC is its power supply. It is a typical unit (usually switch mode) which converts the a.c. supply to d.c. voltage supplying the other

[6] Information Technology Industry Council – http://www.itic.org.
[7] Computer Business Manufacturers Association [33].
[8] Other known characteristics are SEMI F-42 and F-47 developed for semiconductor equipment by the Semiconductor Equipment and Materials International Group [58].

PLC modules. The dip-withstand performance of a power supply module depends mainly on the required stabilization of d.c. output voltage and the energy stored in its capacitors. Sometimes, in order to minimize cabling, the I/O modules are installed in close proximity to actuators and used as data concentrators. In this event, their power supplies become a critical element of the system, especially as most CPU installations are provided with an uninterruptible power supply (UPS), which seldom applies to concentrators.

I/O modules provide an interface between the controller and peripherals. Input devices, like pushbuttons or sensors, are hard-wired to the controller. In general, the inputs are discrete. Threshold voltages used for defining logical signal 0 and 1 levels are not standardized. Thus, if the input signal value is decreased due to a voltage dip for a time of several cycles, it will cause a problem in the correct definition of logic states.

Each process control system is equipped with an emergency stop button. It may be the cause of inadvertent tripping, if configured in such a manner that a voltage dip can lead to an action analogous to its intended activation.

4.4.2 Contactors and Relays

Contactors and relays are used for switching of both power and control circuits. Irrespective of their application, a problem occurs when a contactor or a relay disconnects in an unpredictable manner during the electromagnetic disturbance. This usually leads to uncontrolled interruption of the process. Many manufacturers declare that their relays drop out at 50 % of nominal voltage U_N for a duration of more than one cycle [47]. These specifications vary depending on the manufacturer; in practice, irregular operation may already occur at 70 % or more of U_N.

Most European contactor manufacturers have designed their products according to IEC 60947-4-1 [43]. The standard gives the following limits for electromagnetic contactors, if used separately in motor starters. They:

- shall close satisfactory at any value between 85 % and 110 % of their rated control supply voltage;
- shall drop out and open fully between 75 % and 20 % of rated voltage for a.c., 75 % and 10 % for d.c.

The limits refer to steady-state conditions. No time limits are given and thus event-type phenomena, e.g. voltage dips, are not specifically considered [55].

4.4.3 Induction Motors

Induction motors, as a rule, are protected by their inertia (and that of the driven machine) against the effects of short-duration changes in the supply voltage, except for dips and interruptions of larger depth and duration, which can cause disturbances in motor operation. As a result of a voltage dip, the electromagnetic torque initially decreases, reducing the motor speed. A new point of equilibrium between the motor torque and the load torque is reached. Voltage dips with a depth of less than 30 % have normally no influence on asynchronous motor operation. During such a disturbance the motor torque remains greater than or equal to the load torque. Conversely, for voltage dips with a depth larger than 30 %, the motor torque may be smaller than the load torque. Then the motor speed decreases, and its reduction depends on the depth and duration of the dip, as well as on the inertia of rotating masses.

The motor restart after the disturbance involves a large inrush current and extends the duration of the dip beyond that of the disturbance. This results from the voltage reduction due to the large current and sometimes may prevent restarting. The greater the motor slip is at the end of a voltage dip, the closer the current value is to the starting current.

4.4.4 Synchronous Motors

In industrial applications synchronous motors are employed almost exclusively in constant speed drives. Because of their power they are mainly supplied from MV systems.

The disturbance, depending on its depth and duration, may result in overcurrent and, in an extreme case, a loss of synchronism. In that event the start-up procedure must be carried out.

The synchronous machine can tolerate short-duration voltage changes (with amplitude up to almost 40 %) due to the inertia associated with its large power, possibility of overexcitation and proportionality of the motor torque to the supply voltage. The operation of a synchronous motor is defined on the output side by torque and speed, and on the input side by voltage and active power. Flux, active power and power angle are variables that are linked to the voltage and torque. The voltage reduction can lead to a new stable operating point established in response to the voltage dip.

The ride-through capability of these machines can be improved by means of appropriately setting the current-protection tripping level to allow greater overcurrent and by providing properly controlled field excitation current to maintain the machine in synchronism.

4.4.5 Variable Speed Drives

Variable speed drives produce one of the most serious problems associated with voltage dips and short supply interruptions. They are particularly sensitive to this kind of disturbance and their large individual power causes the methods of mitigation of the disturbance effects to be technically difficult and costly. The problem concerns not only the adverse impact of these disturbances on electric drives, but also the influence on their whole electromagnetic and technological environment. Unlike other disturbances, such as harmonics, unbalance, etc., the effect of dips and short interruptions is instantaneous.

In the case of variable speed drives, characterizing the voltage dip solely in the coordinate system of residual voltage and duration becomes very often an oversimplification, though it is a common method of description and the basic purpose of measurements. This approach does not take into account the difference in the phase voltage values (their unbalance), as well as the change in their phase angles (Figure 4.9). Moreover, this simplified approach neglects the non-sinusoidal nature of the voltage waveform during the disturbance.

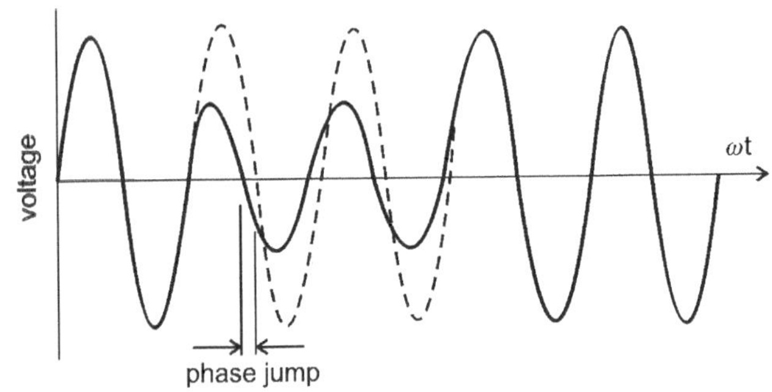

Figure 4.9 The influence of a voltage dip on the phase angle of supply voltage

D.C. and a.c. drives differ in the topology of their power part and control systems (both in hardware and software) and therefore they respond to voltage dips in different ways. There are three main reasons for their susceptibility to voltage dips:

1. The power supply of the control system. If the power supply cannot maintain the supply voltage at a sufficient level, the drive must be switched off for fear of loss of control.
2. Possible irregular operation or failure of the power part of the drive resulting in the disturbance (e.g. commutation failure in a d.c. drive).
3. In many processes the loss of precise speed or torque control is not tolerated for technological reasons, even for a very short time. The drive's response to a voltage dip is a function of the drive parameters and type of load, as well as the quantities which define the disturbance. Some loads (fans, blowers, etc.) can tolerate significant reduction of speed and torque, others do not allow such changes. A precise control of parameters, like pressure, temperature, flow, is required in many industrial processes. As most of these processes employ electric drives, the motor torque and speed directly influence the process variables.

4.4.6 High-Pressure Discharge Lamps

High-pressure sodium lamps – actually the most popular – are extinguished by a supply interruption of about two cycles' duration, or by a voltage dip that reduces the voltage to less than 45 % of the nominal value. The lamp requires time, from one to several minutes, for cooling and restart. In the case of wear-out lamps a voltage dip of a much lesser depth (residual voltage of about 85 % of U_N) is sufficient to extinguish the lamp [54], [55].

4.5 MITIGATION

There are many solutions to prevent damage due to voltage dips. Typically, these solutions can be categorized into three classes:

- solutions in the manufacturing process itself;
- solutions between the process and the public electric grid;
- solutions in the grid.

The standard approach to electromagnetic compatibility consists of coordinating emission and immunity limits. The attempt is made, on the one hand, to prevent electromagnetic disturbance from being emitted at an excessive level, and, on the other hand, to provide the equipment and installations exposed to disturbances with an adequate level of immunity; that is, the level that enables them to operate as intended. Some equipment and installations have an inherent increased immunity to voltage dips due to their inertia or energy storage capacity. This property can be provided at the design stage.

In the case of voltage dips defined in the two-dimensional coordinate system of residual voltage (depth) and duration, the emission and immunity limits are defined for both coordinates.

4.5.1 Reduction of the Number of Faults

Short circuits cannot be entirely eliminated. There are, however, methods for reducing their number and, in consequence, the frequency of voltage dips and occurrence of short interruptions. This is a very effective way of improving the quality of supply and many customers suggest that it is an obvious remedy to be taken in the case of the considered disturbances.[9] Such actions are: replacing overhead lines with cables; the use of insulated conductors on overhead lines; regular tree cutting in the area of the transmission line; fencing against animals; shielding overhead conductors with additional shield wires; increased insulation levels; increased frequency of overhaul and periodic maintenance, cleaning insulators, etc.

4.5.2 Reduction of the Fault Clearance Time

The duration of a voltage dip is largely determined by the speed at which short circuits are cleared. A necessary feature of short-circuit protection is the graduation of the operating times of switches, fuses, etc., in order to ensure that a short circuit is cleared at the most appropriate point in the supply system. This means that the clearance time and, consequently, the duration of voltage dips and short interruptions depend on the location where the short circuit has occurred.[10]

A reduction in fault-clearance time does not mean a decrease in the number of faults but only a mitigation of their effects. It also does not influence the number or the duration of supply interruptions, for the duration depends solely on the speed of voltage recovery. Fast fault clearing does not influence the number of voltage dips, but can significantly reduce their duration.

[9] Such options are, however, limited. Because of other effects of faults, power system operators reduce, on a continuous basis, the risk of fault occurrence to an economically reasonable level.

[10] If a disturbance results from another causative event than the short circuit, the duration of a dip depends on the individual characteristic of its cause.

The basic method for reducing fault duration consists of the use of current-limiting fuses. These are capable of clearing a fault in a very short time. Decreasing the short-circuit current and shortening its duration significantly limit the duration of a voltage dip to rarely exceeding one cycle.

Some devices (using power electronic devices), though they do not provide such a short reaction time, can significantly reduce a short-circuit current during 1–2 cycles. The essential limitation of their use (in LV and MV systems) is their maximum operating voltage – several tens of kilovolts. The actual state of semiconductor switch development creates real prospects for their application to HV circuits.

The fault-clearance time is not just the time needed for a circuit-breaker to open its contacts. It is also the time required for the process of decision making on disconnection. The fault-clearance time in a transmission system is often determined by the limitations of system's transient stability. They are much more restrictive than thermal limitations in distribution networks, and require a shorter disconnection time, rarely exceeding 100 ms. This makes further reduction of the fault-clearance time very difficult.

Not only is the fault-clearance time reduced, but also the graduation margin of distance protection decreases. The loss of selectivity cannot be accepted in most power systems, for it leads to disconnection of two or more lines/subsystems at the same time.

To clear the most common, single-phase faults in HV and EHV systems, single-phase reclosing systems are also used. Upon detection of a fault of this kind, only the faulty phase is disconnected and after some time has elapsed an autoreclosing cycle follows.

Transmission system protection is based on distance relays, whose principle of operation consists in the dependence of their actuation time on the distance to the fault in the system. This distance can be determined by calculating the impedance. Depending on the result of calculations, a command for instantaneous disconnection is generated if the fault is located within the first part of the protected line length, or a delayed disconnection command if the fault has occurred at a longer distance from the considered point.

4.5.3 Modification of the Supply System Configuration

These operations allow for a reduction in the severity of the phenomenon, but at a high cost, particularly in HV systems. The basic method of preventing voltage dips is to install elements of redundancy, as follows:

- Installing generators close to sensitive loads. They support the voltage during distant dips. The voltage reduction equals the percentage share of the generator current in the short-circuit current.
- Increasing the number of substations and busbars in order to limit the number of customers, who potentially may be affected by the disturbance.
- Installing current-limiting reactors at strategic points of the system in order to increase electrical distance to the fault. It should, however, be remembered that this action may make a voltage dip deeper for other customers.
- Supplying sensitive customers busbars from several substations. The effects of a voltage dip on one substation will be reduced by the influence of the others. The more independent

these substations are, the more effective the measure is. The best reduction effect can be achieved by providing a power supply from two different supplying systems. The second supply increases the number of dips but reduces their duration and depth.

The number of short interruptions can be reduced by connecting a smaller number of customers to a single circuit-breaker (to put it another way – by increasing the number of breakers).

4.5.4 Voltage Stabilizers

A more sophisticated way to eliminate the negative effects of dips is called custom power technology. This technology is mainly based on power electronics and also, on some occasions, electrical energy storage.

The most common method for mitigating the effects of the considered disturbances is the use of additional equipment, namely voltage stabilizers. They can be installed on both the supplier's or the customer's side but, as experience shows, the customer is the one who much more frequently does it, since the improvement in supply conditions and increasing the equipment's immunity are beyond the customer's control.

These systems can be generally termed as systems of improved power parameters. There is a wide variety of practical solutions in this group. The operation of this equipment - connected between the disturbed supply source and sensitive loads – consists, in essence, of quickly supplying the energy from an alternative source, or on the adaptation of their operating mode to a short supply interruption or to the limited amount of energy supplied, at the same time providing adequate supply conditions for the critical load.

Two kinds of technical solution can be distinguished:

1. **Energy storage systems.** The stored energy is utilized to supply a critical load during the disturbance. These systems can be used in the case of voltage dips with arbitrary residual voltage, as well as during short supply interruptions. The immunity level of equipment depends on the amount of energy stored and on the energy requirements of the protected process. In many cases the reaction time of the compensation equipment should be considered critical. Since the energy storage process is, as a rule, very costly, it is applied only to particularly sensitive equipment. Examples of energy storage systems are: uninterruptible power supplies (UPSs), superconducting magnetic energy storage (SMES), rotating machines with flywheels, motor–generator systems, etc.
2. **Systems having no energy-storing capability.** These can only be used to reduce the effects of voltage dips (typically up to a maximum of 50 %) but not of supply interruptions. They differ in depth of the voltage dip, which they are able to compensate. The duration of a dip is not a critical parameter in these systems. Their cost, as a rule, is smaller than that of the energy-storing systems.

   Example of such solutions are:

- the constant voltage transformer (CVT);
- static fast transfer switching (SFTS);
- static generators of the fundamental harmonic currents and voltages.

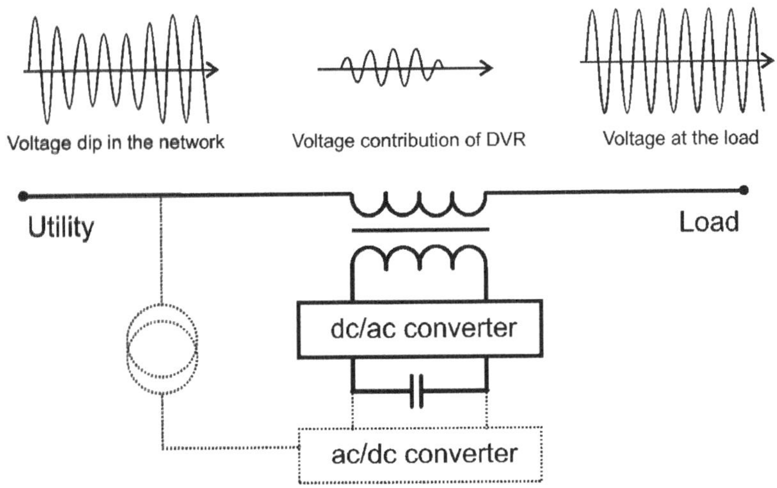

Figure 4.10 Dynamic voltage restorer (DVR)

Three groups of these are the following:

1. *Series* dynamic voltage restorer, DVR[11] (for LV networks, the series voltage restorer, SVR) – Figure 4.10. Apart from voltage stabilization by connecting a series voltage source between the critical load and disturbed power supply source, these systems can also alter the equivalent reactance of a power system, function as a phase shifter, provide balancing and active elimination of voltage distortion at the load terminals, etc.
2. *Shunt* static VAR compensator, SVC.[12] By their effect on the character (inductive or capacitive) of the reactive power consumed, they reduce or increase the magnitude of the voltage at the considered point of a supply system. STATCOM systems are also categorized in this group.
3. *Series–shunt* unified power quality conditioners.

4.5.5 Improvement in Equipment Immunity

One of the most advantageous solutions, in both technical and economical terms, is the use of equipment of a sufficient immunity level that is adequate for the intended operational environment. This is an effective method which eliminates unwanted disconnections due to voltage dips (short interruptions to a lesser extent). More and more frequently the immunity

[11] In some solutions (particularly in MV and HV systems), also with a separate energy storage system.

[12] The common feature of recent equipment is that it comprises inductances and/or capacitances, while thyristor switches are employed to change its equivalent impedance. Previously this equipment was designed solely as line-commutated converters; at present, fully controlled semiconductor devices are used to a large extent.

to a voltage dip of a specified depth and duration becomes the basis of a manufacturer's offer, determining its commercial success.

The level of compatibility of a sensitive load with the supply network is assessed prior to connection. The possible procedure includes three stages:

1. **Acquiring information on system operation.** That is, the prospective number of voltage dips. There are a number of ways to get such data: contacting the electric power supplier, monitoring the power supply over an extended period of time, analysis of faults, etc. Obtaining credible information requires the measurements to be performed for a long time. An alternative is the use of statistical methods of prediction.
2. **Acquiring information on equipment sensitivity.** This information can be obtained from the manufacturer, by conducting tests or assuming typical sensitivity characteristics. In practice, it frequently happens that the user learns about the limited immunity of the equipment only after installing it.
3. **Determination of the potential effect.** If the foregoing information is available, there is the possibility to assess the potential threat of equipment failure (failure rate) and evaluate the economic effect of its occurrence (Section 4.6.1). On that basis a method of proceeding can be chosen: improvement of supply conditions, better (i.e. less sensitive) equipment, application of a stabilizer or acceptance of the existing situation.

Generally, it is difficult to obtain information from the electricity supplier and from the equipment manufacturer. There are a number of reasons for this. The main causes are diverse, difficult to compare methods for describing the phenomenon (the lack of a standard format for the characteristic), and manufacturers seldom carry out relevant tests. Their knowledge of the equipment's immunity, e.g. to a phase jump or unbalanced voltage dip, is usually limited.

Due to the fact that various categories of equipment respond in different ways to voltage dips, it is not possible to develop a single standard which would define the sensitivity of the equipment used in industry. The closest to such unification is the CBEMA curve with its subsequent modifications (Section 4.4.1).

Since the design and manufacture of equipment that is more immune to the considered disturbances are possible, the effects of voltage dips and short supply interruptions can be taken into account at equipment design stage. The above-mentioned information allows, in technical and economic terms (i.e. no excessive cost incurred), adequate methods to be employed obtain the required immunity level.

There are several embedded solutions, which should be followed in order to increase the immunity of equipment. They are attractive, since, in theory, they do not require any additional power conditioning equipment, but instead involve the use of more robust or improved components in the tool design. They include (based on [60]):

- For single-phase equipment, the use of d.c./d.c. power supplies of a more sophisticated structure, which tolerate a wider range of supply voltage changes, maintaining proper equipment operation. The use of universal input switching power supplies at every location possible, wired phase to phase. The universal input-type power supply has a voltage range of 110–400 V a.c. typically.

- D.C. power supplies should not be overloaded. Since the amount of voltage dip ride-through time available from d.c. power supply is directly related to the loading, d.c. power supplies should not be running at their maximum capacity. Oversizing the expected load by at least two times will help the power supply to ride through voltage dips.
- Consolidation of control of the power sources. When designing the layout of a tool control circuit, the control power supplies should be consolidated to be fed from a common source or breaker where possible. In this way a small power conditioner is required to make the tool insensitive to voltage dip; this will make the implementation less painful.
- Equipment design employing components of a high, specified immunity to disturbances, e.g. utilizing certified relays, contactors, motor starters, etc., and avoiding the use of a.c.-powered general purpose relays. The use of robust inverter drives. The use of a.c. inverter drives in the tool design requires units that have good voltage dip ride-through. Flying restart, kinetic buffering and the ability to have a low d.c. bus-level trip point (50 % of rated value nominal is ideal) are essential.
- Mismatched equipment voltages should be avoided. If the equipment does not match the expected nominal input voltage, the tool will be more susceptible to voltage dips. This can occur when a transformer's secondary voltages do not match the rated voltage for the connected equipment.
- Circuit-breakers and fuses should be selected to allow for higher inrush currents due to power quality within the voltage dip range. Where possible, do not select breakers that have instantaneous trip characteristics.
- Phase monitoring relays should not be used in the interlock circuit. These devices will easily trip during a voltage dip and can lead to tool shutdown.
- A non-volatile memory should be utilized. This type of backup technique for tool controllers ensures that the control system will not lose its position in the event of a voltage dip.
- Software and control program issues should be considered. The tool or system designer should consider process variable fluctuations during voltage dips. It might be essential that the bandwidth for certain process variables be widened or a filter with time delays be added to avoid tripping of the process.

4.6 MEASUREMENT

4.6.1 Principles of Measurement

For the assessment of power quality with respect to voltage dips and short interruptions the following, five-stage procedure should be applied:[13]

Stage 1. Obtain sampled voltages with a certain sampling rate and resolution. The IEC 61000-4-30 [39] standard does not give any direct information on sampling rate and resolution for voltage dip measurements. However, in most cases the same

[13] Standard IEC 61000-4-30 [39] is helpful for stages 1–3, while IEC 61000-2-8 [36] is better for stages 4 and 5.

equipment will be used for obtaining information on the harmonic spectrum, which will require a sampling rate of at least 80 samples per cycle – 128 or 256 being typical values. Voltage sampling is synchronized with the system frequency. In this way the sampling frequency is not expressed as a constant number of samples per second but as the constant number of samples per cycle.

Stage 2. Calculate the event characteristic[14] as a function of time from the sampled voltages. IEC 61000-4-30 prescribes the use of a one-cycle r.m.s. voltage updated every half cycle ($U_{rms(1/2)}$).

Stage 3. Calculate single-event indices from the event characteristics. For example: (i) duration; (ii) the minimum value attained by the voltage during a disturbance (residual voltage), or the value by which the voltage is reduced with respect to the reference voltage.[15] Single-event indices are used mainly for troubleshooting and diagnostics. They should not be compared to any objectives, although a comparison to equipment tolerance curves is possible. The calculation of single-event indices is an intermediate step in the calculation of single-site indices.

Stage 4. Calculate single-site indices from the single-event indices of all events measured during a certain period of time. Different ways of representation are given in different standard documents and proposed in many papers.

Site indices are used for compatibility assessment between sensitive equipment and the power supply. Site indices for different locations can provide an aid in the choice of a suitable location for an installation containing sensitive equipment. They can also be used to provide information to local customers on the voltage quality, e.g. to choice of mitigation method or for following up premium power contracts.

One possible method for reporting site information from the magnitude and duration of an event is described in standards [29] and [30]. The method results in the so-called 'voltage dip coordination chart'. An example of such a chart is shown in Figure 4.11.

The chart, as defined in these standards, contains the performance of the supply at a given site and the voltage tolerance of one or more devices. The chart gives the number of events (dips and interruptions) as a function of the severity of the event acquired from utility data, measurements, monitoring or prediction. Using this dip data, the supply dip performance contours are drawn. Each contour represents the number of voltage dips per year. This contour map gives complete information on the supply performance, as long as the magnitude and duration of the events can be uniquely defined.

Next, equipment sensitivities (voltage tolerance curves) are overlaid on the supply dip performance contours. According to Figure 4.11, device A is susceptible

---

[14] A time-dependent parameter characterizing a change in voltage before, during and after a voltage event. Event characteristics are used to detected the event, to determine the type of event and to calculate event indices.

[15] Further single-event indices can be calculated to extract additional information about the event. Examples are phase angle jump, symmetrical component voltages, direction (upstream or downstream) and cause (e.g. motor starting, transformer energizing or fault) of the event.

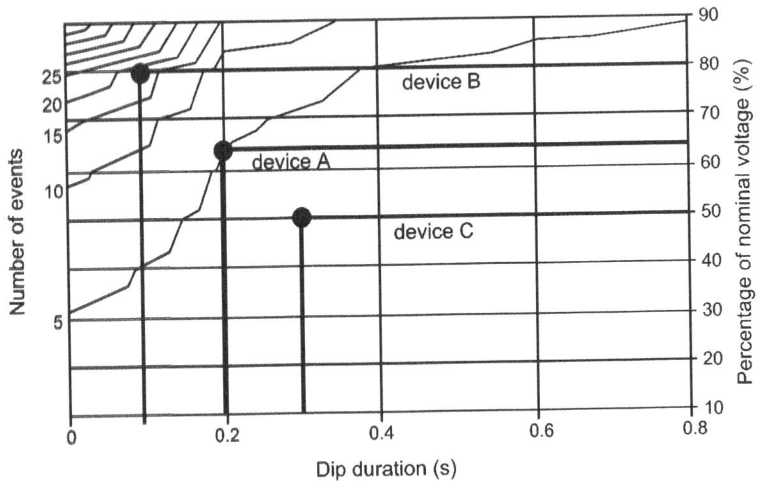

Figure 4.11 Voltage sag coordination chart (based on [66]) (Reproduced from Voltage Sag Indices – Draft 4, Working Document, IEEE)

to a dip in retained voltage of 65 % and duration 0.2 s. Hence it will be susceptible to any dip in retained voltage of 65 % or less and duration 0.2 s or more. The chart shows also that the supply would expect five such dips per year. Similarly device B is susceptible to a dip in retained voltage 80 % or less and duration 0.1 s or more and the supply voltage would expect about 17 of these per year. Device C is not influenced by dips at the considered point in the supply network. The sensitivity of the whole process is defined by the most sensitive component, with a knee point located at the uppermost left-hand portion of the chart [66].

The advantage of this method is that equipment behavior can be directly compared to system performance, for a wide range of equipment. The disadvantage of the method is that a two-dimensional function is needed to describe the site.

Comparison of site indices for different sites is not recommended other than as an aid in the choice of a suitable location for an installation containing sensitive equipment. Indices can be used to identify typical levels of disturbances for different types of sites and as such provide feedback for improvements to the network operator. An example of the use of site indices is in comparing the voltage dip frequency in underground, overhead and mixed distribution networks.

It is very difficult to give objectives for site indices due to the very large difference between the number of voltage dips to be expected at different sites. Year-to-year comparison of the site indices for one site may be used to see trends in the voltage quality of this site. For locations with a strong seasonal variation in the event frequency, a three- to five-year monitoring period is recommended to incorporate year-to-year variations in the seasonal effects.

The calculation of site indices is an intermediate step in the calculation of system indices.

Stage 5. Calculate system indices from the site indices for all (monitored) sites over a certain region or a part of the power system. These kinds of indices require a longer monitoring period – a minimum of one year. A good basis for prediction requires measurements over several years.

System indices are used by the network operator to assess the performance of the whole system. They can be used to compare year-to-year results, where the effect of weather variations should also be considered. The results of such a performance assessment or comparison can be used as a basis for improvements in the system. The indices can be used to identify typical levels of disturbances for various types of network and for ongoing monitoring of any one network. System indices can be:

- given as information to customers to enable them to compare the performance of their local site to the performance of the system as a whole;
- used by manufacturers of sensitive equipment to choose the immunity requirements of their equipment against voltage dips.

Many such indices have been proposed in technical bibliographies. They can be divided into two groups.

System indices are typically a weighted average of the single-site indices obtained for all or a number of sites within the system. Weighting factors can be introduced to take into account the sites not monitored and the difference of substations, the number of customers or the rated power represented by each site.

According to other conceptions, the weighting coefficients can be proportional to the number of customers who have experienced effects of a disturbance, or to the number of lodged complaints. The weighting coefficient is often assumed to be unity. A simple approach is to give a weighing factor of 0 if it lies above the CBEMA (Figure 4.8) curve and 1 if it lies below it. The index then becomes the number of dips lying below the CBEMA curve.

According to the second conception, system indices can be calculated either from the average of the site indices or from the value not exceeded by 95 % of the sites (the percentile of 95 % defined over the number of measurement points considered). It implies that a considerable number of points should be monitored.

Table 4.3 presents the conception of equivalent weighted voltage dips (calculated per year) according to [48] and [64], as the basis for the determination of mutual financial commitments between the electricity supplier and the customer.

The weighting coefficients provided in Table 4.4 are the product of the duration and amplitude of average dips for each cell.

For example, for the given time interval 0.5–0.75 s and amplitude 0.3–0.6U_{ref}, the value of the weighting coefficient is $0.281\,25 = 0.625 \times 0.45$. Given that an adverse impact of voltage dips with a duration longer than 1 s depends almost entirely on the amplitude of the disturbance, the same average duration of 0.875 s of a voltage dip has been assumed for the last four columns.

In order to obtain comparability of results it is necessary to take some arbitrary decisions on measurements. The decisions taken in selected published measurement projects are presented below.

Table 4.3 The conception of weighting the voltage dips with respect to the example number of disturbances according to [48] and [64]

Category of disturbance	Example weighting coefficient for the purpose of economic analysis	Example number of disturbances within a year	Equivalent disturbances within a year
Interruption	1	7	7.0
Voltage dip with depth < 50 %	1	2	2.0
50–70 %	0.7	10	7.0
70–80 %	0.4	8	3.2
80–90 %	0.1	27	2.7
Total		54	**21.9**

Table 4.4 The conception of weighting coefficients according to [57]

Voltage dip amplitude u (%)		Duration					
		$20\,\text{ms} \le t < 100\,\text{ms}$	$100\,\text{ms} \le t < 500\,\text{ms}$	$500\,\text{ms} \le t < 1\,\text{s}$	$1\,\text{s} \le t < 3\,\text{s}$	$3\,\text{s} \le t < 20\,\text{s}$	$20\,\text{s} \le t < 60\,\text{s}$
	Mean value	0.06	0.3	0.75	0.875	0.875	0.875
$15 > u \ge 10$	0.125	0.0075	0.0375	0.0938	0.1094	0.1094	0.1094
$30 > u \ge 15$	0.225	0.0135	0.0675	0.1688	0.1969	0.1969	0.1969
$60 > u \ge 30$	0.45	0.0270	0.1350	0.3375	0.3938	0.3938	0.3938
$99 > u \ge 60$	0.795	0.0477	0.2385	0.5963	0.6956	0.6956	0.6956
$0 > u \ge 99$	0.995	0.0597	0.2985	0.7463	0.8706	0.8706	0.8706

4.6.1.1 Reference Voltage for Measurement Purposes

Dip threshold can be a percentage of either nominal or declared voltage or a percentage of the sliding voltage reference which takes into account the actual voltage level prior to the occurrence of a dip. The choice of a dip threshold is essential for determining the duration of the event and is also important for counting events.

The voltage variation range in HV systems is much wider than that in LV and MV networks. In these cases it is advisable to measure the voltage dip with respect to the preceding voltage that is continuously determined over a preset time interval – measuring window – that is significantly longer that the duration of a voltage dip.

The sliding reference voltage poses some difficulties in predicting the reaction of equipment, whose immunity is often expressed in absolute values. For example, a voltage dip of $d\,\%$ of the sliding reference value can mean an actual voltage as low as $0.9dU_N$ or as high as $1.1dU_N$ (U_N is the nominal voltage), depending on the voltage level preceding the dip. This is not sufficiently precise enough to assess the equipment's response to the disturbance.

4.6.1.2 Voltage Thresholds Marking the Start and End of the Disturbance

The selection of threshold values depends on the type of reference voltage: whether the reference voltage is determined in a sliding window or is a fixed value [36].

4.6.1.2.1 Fixed Reference Voltage

At any given point in the power system the voltage is constantly subjected to changes in response to the load variations and switching of loads. The network is designed to maintain these voltage variations within a certain range that defines normal supply conditions. In the event of a short circuit or a large increase of current, a significant change in voltage with respect to the pre-existing voltage may occur. At certain measurement points, however, the voltage can remain high enough to be still contained within the permissible range of variation, especially if the pre-existing voltage was close to the maximum permissible value. In measurements at a given site, such a voltage reduction, which is often caused by a fault at a distant part of the network, should not be considered a voltage dip.

For that reason, many dip measurements are based on the lower limit of the permissible range of voltage variation. Only those disturbances during which the voltage decreases below this threshold are recorded as a disturbance.

Each voltage dip begins when the voltage falls below this threshold and ends when the voltage reaches, at least, the same level.[16]

If, however, the voltage value at the measuring point is near to this threshold it may happen that normal load changes or load switching operations induce voltage oscillations around this value. These load-induced voltage fluctuations around the set threshold can significantly increase the number of recorded dips. They can be excluded by setting an additional margin below the permissible band of voltage variation. In this case a voltage change is classified as a voltage dip only if the voltage falls below the threshold corresponding to a value lower than the bottom of the permissible range. This threshold is used to mark both the start and end of a voltage dip in measuring its duration.

An alternative method for excluding load-induced voltage oscillations near the bottom of the permissible band of variation consists of classifying as voltage dips only those events in which the voltage, having fallen below a threshold, recovers above a new threshold which is greater than the first threshold. The term *hysteresis* has been applied to the margin between the two thresholds (typically no more than 1–2 %). In the original development of this approach the second threshold was used only for the purpose of classifying a disturbance as a voltage dip – the first threshold was used to mark both its start and end. A modification of this method is to adopt the second threshold as marking the end of a recorded voltage dip.

Adopting the nominal voltage as the fixed reference voltage is recommended for measurements at the load terminals or close to them, for the correct operation of a load is determined by the residual voltage value expressed in volts.

---

[16] According to IEC 61000-4-30 [39] (also IEC 61000-2-8 [36]), dip thresholds are for example in the range 80–90 % of the fixed voltage reference, for troubleshooting or statistical applications, and 70 % for contractual applications.

The mean value of a voltage, which varies with the location of measuring points along a long line, has a significant influence on the description of disturbance in measurements on distribution systems. Close to a substation this mean value is, as a rule, greater than at the more distant points. For that reason two customers may experience significantly different residual voltages during a voltage dip caused by the same event, e.g. a short circuit. Voltage dip indices, determined close to a substation, may not give a correct value for a customer at the end of a long line. It is assumed that the correction of this effect is not made during the measurements since the determination of representative indices is not possible for all customers. Nominal voltage is therefore recommended as the reference voltage. The correction of measurement results can be made later, if required, at the stage of processing the results, depending on the individual needs of a customer.

The method of sliding time window allows most of the above difficulties to be avoided.

4.6.1.2.2 Sliding Reference Voltage

The sliding reference window may be used in HV and EHV systems with a relatively large variation in normal operational voltage, when HV/MV transformers are equipped with on-line tap-changers. The use of a reference voltage value determined in a sliding time window immediately before the occurrence of a voltage dip[17] has a smoothing effect, which automatically eliminates most of the voltage variations due to local load fluctuations and transformer tap changing. In this case, therefore, the start and end thresholds can be selected at a value quite close to the reference voltage. In the event that there is a downward trend in the voltage, independent of the dip, the value to which the voltage recovers at the end of the dip is somewhat less than the value immediately before the disturbance. Therefore, in order to ensure that the end of the dip is recognized in measurement, it may be necessary to set the end threshold at a value slightly below the reference voltage, e.g. 99 % of the reference voltage. For uniformity, the start threshold can be set at the same value.

4.6.1.3 Distinguishing Between Voltage Dips and Short Interruptions

Notionally, an interruption implies complete isolation from all sources of supply and, therefore, zero voltage. In practice, however, the isolated portion of network can include significant sources of stored energy, preventing the voltage from reaching zero value during a very short interruption. Furthermore, the theoretically most severe voltage dips can reach zero voltage. Such a voltage dip is, effectively, an interruption, although a connection to the voltage source remains. Thus, it can be difficult for the measuring instrument to distinguish a voltage dip from a short interruption. For this reason it is necessary to adopt a limit voltage of several percent (e.g. 1, 5, 10 %, cf. Section 4.8) in order to distinguish between these two

[17] As a pre-fault reference, the so-called 'sliding window reference' is used in IEC 61000-4-30 [39]. It is calculated from 200 ms r.m.s. voltages using a first-order filter with a 1 min time constant. In equation form $U_{SR(n)} = 0.9967U_{SR(n-1)} + 0.033U_{(10)rms}$, where $U_{SR(n)}$ is the present value of the sliding reference voltage; $U_{SR(n-1)}$ is the previous value of the sliding reference voltage; and $U_{(10)rms}$ is the most recent 10-cycle r.m.s. value. The declared voltage is assumed as the initial value of the reference voltage. The sliding reference voltage is updated every 10 cycles. If a 10-cycle value is flagged the sliding reference voltage is not updated and the previous value is used.

disturbances. For instance, a short circuit can result at different observation points in both voltage dips and short interruptions, depending on whether the recorded voltage is higher or lower than the selected limit level.

4.6.1.4 Reporting Measurements Results

Various methods for reporting dips have been proposed in the literature. Such methods may be classified here in two categories:

1. Methods to characterize site or system performance.
2. Methods most suitable to estimate the compatibility between equipment and supply.

The two-dimensional character of the description of disturbance implies a two-dimensional matrix or table, with residual voltage values in rows and duration of the disturbance in columns. Based on the experience of European power systems, UNIPEDE has proposed a method for the classification of disturbances in the form of two alternative tables, i.e. Table 4.5 and Table 4.6, determined over a set interval of time, e.g. 30 days, half a year, one year. If the characteristics of equipment sensitivity are known, Table 4.6 predicts how many irregularities may occur in its operation. It should be emphasized that the proposed tables are not the disturbance indices in the exact meaning of the words, but rather a way of presenting the indices. Each element of the table can be treated as a separate index.

The method of presenting considered disturbances, given in the tables, has a particular advantage, for it provides a large amount of data that can be used to create in coming years a database for the assessment of supply quality. This will allow more complex indices to be introduced and to be proposed in contracts concluded in the future.

Different publications use different values of voltage and duration values for disturbance classification. Their selection is the subject of discussion, and many customers apply other classifications than those given in Table 4.5 and Table 4.6 (e.g. [30]).

Table 4.5 Classification of voltage dips – version 1[18]

Duration / Residual voltage u (%)	$10\,\text{ms} \leq t < 20\,\text{ms}$	$20\,\text{ms} \leq t < 100\,\text{ms}$	$100\,\text{ms} \leq t < 500\,\text{ms}$	$500\,\text{ms} \leq t < 1\,\text{s}$	$1\,\text{s} \leq t < 3\,\text{s}$	$3\,\text{s} \leq t < 20\,\text{s}$	$20\,\text{s} \leq t < 60\,\text{s}$	$60\,\text{s} \leq t < 180\,\text{s}$
$90 > u \geq 85$								
$85 > u \geq 70$								
$70 > u \geq 40$								
$40 > u \geq 10$								
$10 > u \geq 0$								

[18] Measurement results in the first column and first row are increased by transients and load fluctuations, respectively.

Table 4.6 Classification of voltage dips – version 2

Duration / Depth u^* (%)	$10\,\text{ms} \leq t < 20\,\text{ms}$	$20\,\text{ms} \leq t < 100\,\text{ms}$	$100\,\text{ms} \leq t < 500\,\text{ms}$	$500\,\text{ms} \leq t < 1\,\text{s}$	$1\,\text{s} \leq t < 3\,\text{s}$	$3\,\text{s} \leq t < 20\,\text{s}$	$20\,\text{s} \leq t < 60\,\text{s}$	$60\,\text{s} \leq t < 180\,\text{s}$
$u^* \geq 10$								
$u^* \geq 15$								
$u^* \geq 30$								
$u^* \geq 60$								
$u^* \geq 90$								

The main advantages of this method for the presentation of results are [66]:

- highly detailed data;
- clear and easy interpretation;
- the same form of presentation can be used with respect to mean values, percentiles of 50 %, 95 %, and to maximum values;
- a broad range of disturbances are considered, from very short voltage dips to long supply interruptions.

Disadvantages are:

- The selected durations of disturbance are not very useful in terms of typical values encountered in practice. Since the duration of many dips is close to 100 ms they can be categorized into two adjacent time classes. The time interval 100–500 ms (the most common for the majority of cases) is too wide for the prediction of equipment performance (in Table 4.7 it has been split). Also a unique decision on a dip belonging to a specified class is required when the result of measurement occurs at the boundary of time intervals (categorizing a dip to a higher class – a more severe disturbance).
- Within the time interval of 1 s to 3 min, voltage dips seldom occur. Since this time interval is particularly important in terms of short supply interruptions, these columns in the tables can be combined into one or two columns.

Table 4.7 Classification of voltage dips according to IEC 61000-2-8 [36]

Duration / Residual voltage u (%)	< 0.1 s	0.1–0.25 s	0.25–0.5 s	0.5–1 s	1–3 s	3–20 s	20–60 s	1–3 min
$90 > u \geq 80$								
$80 > u \geq 70$								
$60 > u \geq 50$								
$50 > u \geq 40$								
$40 > u \geq 30$								
$30 > u \geq 20$								
$20 > u \geq 10$								
$10 > u \geq 0$								

Table 4.7 shows other proposals adopted in various standardization documents. Splitting the reference voltage values into nine ranges may in actual measurements result in many empty cells – their number increases from 35 according to the UNIPEDE proposal to 72. For the purpose of comparison between various measurement points and power systems this table is probably too detailed.

Standard document IEC 61000-4-11 [38] recommends duration values different from those recommended by UNIPEDE for testing equipment immunity to voltage dips and short interruptions: namely, ½ cycle, 1 cycle, 10 cycles, 25 cycles and 250 cycles; residual voltage values 0, 40, 70 and 80 %.

Tables, analogous to those presented above, are used for preparing reports from measurements carried out at many points. In this case each cell can for example contain:

- a percentile (typically 95 %, CP95) of all recorded disturbances;
- the maximum value of recorded disturbances;
- the average number of recorded disturbances;
- weighted averages from all measurement points (weighting can also be applied to CP95) etc.

When measurements are carried out on several types of network (cable, overhead, mixed, LV, MV, HV, etc.) a separate table can be prepared for each type.

The 2D or 3D histograms in Figure 4.12 are also well established as a means of reporting site dip behavior.

4.6.1.5 Measurement Remarks

The majority of voltage dips are simple in shape – the voltage drops to approximately constant value and, after some time, recovers to its previous level. It can be assumed that the voltage shape takes an approximately rectangular form. In this case the residual voltage of the dip is the lowest value to which the voltage falls during the disturbance. A single pair of numbers – residual voltage and duration – provide a full description of the disturbance.

In the case of complex, non-rectangular dips, during which the voltage takes on several levels, the description of such a disturbance by its minimal residual voltage and predefined duration may sometimes lead to its oversizing. The shape of the envelope may be also assessed using several dip thresholds set within the range of voltage dip and voltage interruption threshold detection. In this case a single dip is treated as a series of subsequent rectangular dips of different durations and different residual voltages. This method of description considerably increases the number of recorded dips.

A distinction can be also made between *multi-channel measurements* for which the IEC method is recommended and *three-phase measurements* for which an alternative method is recommended.

For multi-channel measurements (according to IEC 61000-4-30 [39]) the worst channel, i.e. the channel with the lowest r.m.s. voltage and the longest duration, is taken for further analysis when calculating single-event indices. This is not necessarily the same channel

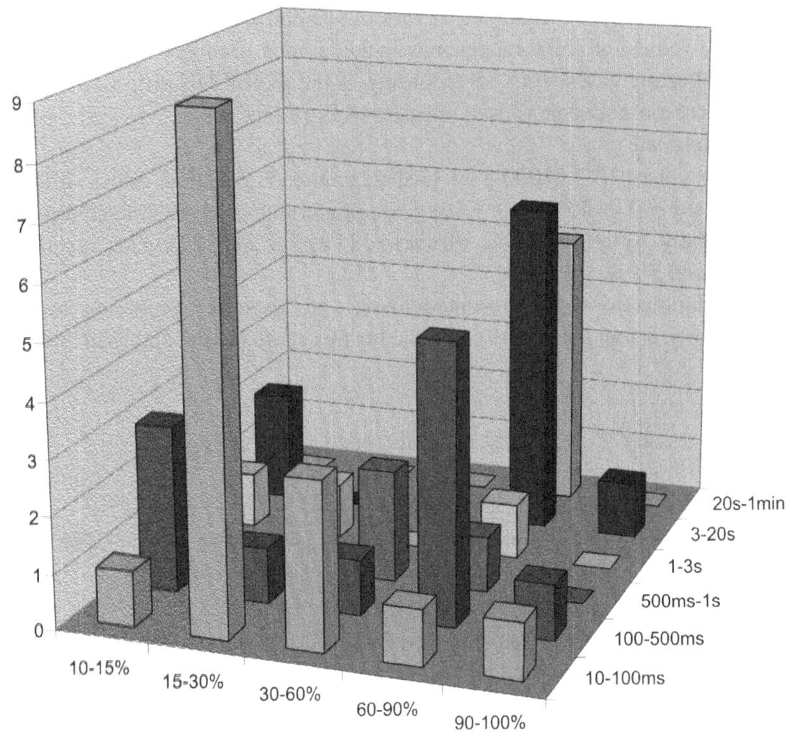

Figure 4.12 A common way of presenting annual dip distribution. Number of dips per year as a function of dip magnitude (remaining voltage) and duration (under preparation)

throughout the recording (three channels may correspond to line-to-line or line-to-neutral measurements).

From the three sampled waveforms one *characteristic voltage magnitude as a function of time* is obtained in accordance with the method for the characterization of three-phase unbalanced voltage dip. In this case voltage dips, whose durations overlap in time, are conventionally counted as a single disturbance. Frequently the duration of such a disturbance is measured from the start of a dip in the first disturbed phase to the end in the last phase (Figure 4.2). This decision is of some consequence. Firstly, a single-phase dip is treated in the same way as the three-phase one, despite the fact that its effects on equipment are typically more serious. Secondly, voltage dips caused by an earth fault on systems earthed through a large impedance are treated as being equally severe (or even more severe) as the dips caused by a line-to-line short circuit, despite the former having practically no effect on the equipment. The consequence of the adopted description of a disturbance is that recognizing the type of a short circuit and its location is difficult – hence the attempts made to describe three-phase short circuits by other methods.

Direct measurements require continuous recording of voltages in case their value exceeds preset limit levels. Measuring phase currents (where justified by analysis of other disturbances, e.g. harmonics) can sometime provide helpful information, e.g. on the location of a disturbance. In the case when the limit values are exceeded, the r.m.s. values of voltages on all measuring channels are recorded during the disturbance and also the instantaneous voltage values are recorded over the set number of cycles preceding and following both the start and end of the disturbance. When the duration of a disturbance exceeds a user-defined time interval, then r.m.s. values of monitored voltages, averaged over several cycles, are recorded in order to protect memory resources. It is advisable that the maximum and minimum values are recorded within the averaging interval.

It is advantageous when the data acquired this way is transferred to a server, processed by support software, and the results are accessible on-line to the user. Technical requirements regarding measuring instruments are typical for the equipment measuring r.m.s. voltage values. Particular attention should be paid to matching the instrument's input voltage level to output voltages of measuring transducers (transformers, voltage dividers) and to matching the instrument input impedance. In measuring voltage dips (in general, the voltage magnitude) both inductive instrument transformers and capacitive voltage dividers can be used.

The measurements can be performed directly on a transmission network (high cost, difficulties in connection of instruments) as well as on a distribution system. In the latter case it is necessary to develop methods for identifying the source of disturbance; that is, answering the question of whether the disturbance originated on the transmission or distribution network.

The essential benefits of measuring methods are the actual data obtained in this way. It should not, however, be forgotten that they reconstruct the past state of events, thus their usefulness in forecasting is limited, particularly when they result from monitoring over a limited period of time. The disadvantage is the high cost of equipment installation, which means that measurements on distribution networks are often used as a basis for assessment of supply quality on a transmission system.

For a case study see web address

4.6.1.6 Conclusions from the Measurements[19]

In fact the possibility of comparing the measurement results is limited. There are significant differences between procedures, associated with:

- the criteria of selecting measurement points and their individual environmental characteristics;
- the number and location of measurement points on the supply network;
- the threshold values for voltage dips and supply interruptions;
- the duration of measurements;
- the type of measuring equipment.

Further differences concern the analysis of data and method of their presentation, in particular:

---

[19] These conclusions have been formulated from the results presented here and other results that are not presented in this chapter [36],[56],[66].

- presentation of the data as absolute or per unit value, or in statistical terms;
- assigning voltage dips to the customer or to a measurement point;
- presentation of maximum, average values, percentiles, or other statistical indices;
- the method used for the aggregation of results (cf. Section 4.7.7.) etc.

In spite of these differences, there are some common traits:

- Results are presented in the same system of coordinates: namely, duration and voltage (residual or depth).
- There is a strong similarity in the distribution of relative probability density, according to which the dips are situated in the adopted system of coordinates.
- The obtained results confirm the thesis that the type of network has an influence on the number of disturbances, and their occurrence on an overhead network is more frequent. These results show that the number of voltage dips experienced by an individual user can range from some tens per year, where the networks are largely underground, to several hundreds per year, where overhead lines are involved [36].
- A large number of events, of very short duration and with voltage values close to nominal, indicate their source in transients and load fluctuations.
- For a larger number of measurement points, a considerable difference in the obtained results has been found. This indicates the effect of differences in supply networks – their type and configuration, climatic and environmental conditions, etc.

4.6.2 Statistical Methods of Analysis

A statistical approach to the considered disturbances leads to the conclusion:

> *A customer can expect on average* X *voltage dips per year, with a depth no greater than* Y *and of duration no longer than* Z.

In order to substitute actual numerical data for the symbols X, Y, Z, a statistical analysis of data collected from the past is necessary. This, as a rule, requires information on the supply system, in the form of its model, and information about the history of disturbances (preferably as long as possible). It is advisable to have information for each circuit-breaker in system lines (for if possible a long period of time) about the history of its operations, location of short circuits, number of disturbed phases, impedances, durations of short circuits, etc. More advanced cases of analysis take into account models of switching equipment and time characteristics of protective devices. When this information is available, a reliable analysis and forecasting can be undertaken. The lack of this information implies that simplifying assumptions have to be made, diminishing the likelihood of prognosis.[20]

[20] Very useful instruments for the acquisition of such data are digital short-circuit recorders. The majority of them facilitate the recording of r.m.s. voltage values, and some of them are provided with the option for power quality assessment.

A typical model of a system comprises a schematic diagram, the length and type of supply lines (cable or overhead), transformer data, and short-circuit capacity at given points of the network. The data is necessary for most short-circuit analyses.

The advantages of statistical methods are: rapid collection of information, the possibility of estimating the magnitude of dips from the short-circuit analysis, the possibility of determining disturbance duration (if models of protection equipment are used), usefulness in designing new systems, etc.

It should not be forgotten that the accuracy of the results of the applied method is determined by the model accuracy. If the model is erroneous, the estimations obtained with it will be inadequate. The second factor which limits the accuracy of prediction is the random nature of input data, e.g. the frequency of occurrence of short circuits, which depends on the season, weather conditions, network operating practices, etc. This approach is further hampered since some important parameters for estimating dip magnitude, such as the power output of power plants and the position of switching equipment, change frequently in time

Among various methods of analysis the fault position method should be distinguished. It allows the analysis of voltage dips and supply interruptions at selected points where a sensitive load is planned to be connected. The use of the method consists of analyzing systematically the system short circuits occurring at an increasing distance from the considered point of connection (PC), in order to determine the magnitude and duration of disturbances occurring at this point. The following stages can be distinguished in this algorithm:

- Determination of the maximum distance from the PC, within which short circuits are considered. In algorithms, where this procedure is executed automatically, this distance depends on the assumed voltage threshold at the PC, which is critical for sensitive equipment.
- Determination of the resolution of short-circuit analysis; this practically means dividing a supply line into sections of different lengths.
- For each selected, potential short-circuit location the rate of occurrence is determined. This is the prospective number of short circuits, which may occur within a period typically assumed as one year, on the model portion represented by the short-circuit location. It is determined for a line section, for example, by multiplying the line length by the average number of short circuits per unit of length.
- Short circuits are cyclically repeated for each location, and each time the depth and duration of a voltage dip at the PC is determined.

Increasing the number of short-circuit locations taken into account in this method will improve the accuracy of the results of analysis. Three decisions have to be taken when selecting the short-circuit location:

1. At which points of a system should short circuits be considered? For each voltage level the number of lines should be determined and busbars should also be taken into account as potential locations of the short circuit.

2. How long should the distance be between short-circuit locations? This should be determined for each voltage level.
3. What values should be determined for each short-circuit location? Typically this is the depth of a voltage dip, often the duration of a short circuit. It is also necessary to specify the type of short circuit: single-phase, three-phase or both. The analysis should be carried out for various configurations of a supply system, generating units and loads.

The advantages of both the statistical method and direct measurements are united in hybrid methods, which combine short-circuit analysis with limited data acquired by means of measurement [63].

4.6.3 Area of Vulnerability

This concept is used to visualize the residual voltage at the considered point in an electrical network (e.g. point of perspective connection of an adjustable-speed drive (ASD)) due to a three-phase short circuit somewhere in the network. Figure 4.13 shows this area of vulnerability for symmetrical three-phase short circuits. For example, a cable or busbar in this network situated in the gray area of 50–75 % indicates that a three-phase short circuit at this cable or busbar will lead to a voltage dip at the extrusion company with a residual voltage between 50 and 75 %. (This portion of the text is reproduced by permission of European Copper Institute) [21].

4.7 CONTRACT

The final user experiences the consequences of voltage dips, whose source is located somewhere on the system, on the supplier's side. The same user causes voltage dips, resulting for example from short circuits in the user's internal installations, which in turn, through the power system, affect other users. Who should bear responsibility, and in what way, for the losses caused by these disturbances, which can be huge in many continuous process industries like refineries or chemical plants?

The customer considers both voltage dips and short interruptions to be disturbances which affect the operation of equipment. The difference between these two disturbances is perceived in terms of their effects, i.e. will the operation be interrupted, lighting extinguished, etc.? Such an approach can lead to erroneous conclusions: in the case of light sources a voltage dip with a depth greater than 50 % (the typical threshold of extinguishing high-pressure discharge lamps) produces the same effect as a short supply interruption.

The supplier[21] classifies disturbances in terms of their cause. The supplier wants to be responsible only for the duration of a voltage dip, for it depends upon the supplier – and also on the protection systems on the supplier's side. The supplier does not wish to assume responsibility for the depth of a voltage dip since the supplier's influence on it is limited.

[21] Under the conditions of the competitive electricity market the producer has to be clearly distinguished from the operator/supplier, who deals with transmission or distribution and with whom the contract is concluded. The operator/supplier is the one who guarantees the quality of supply.

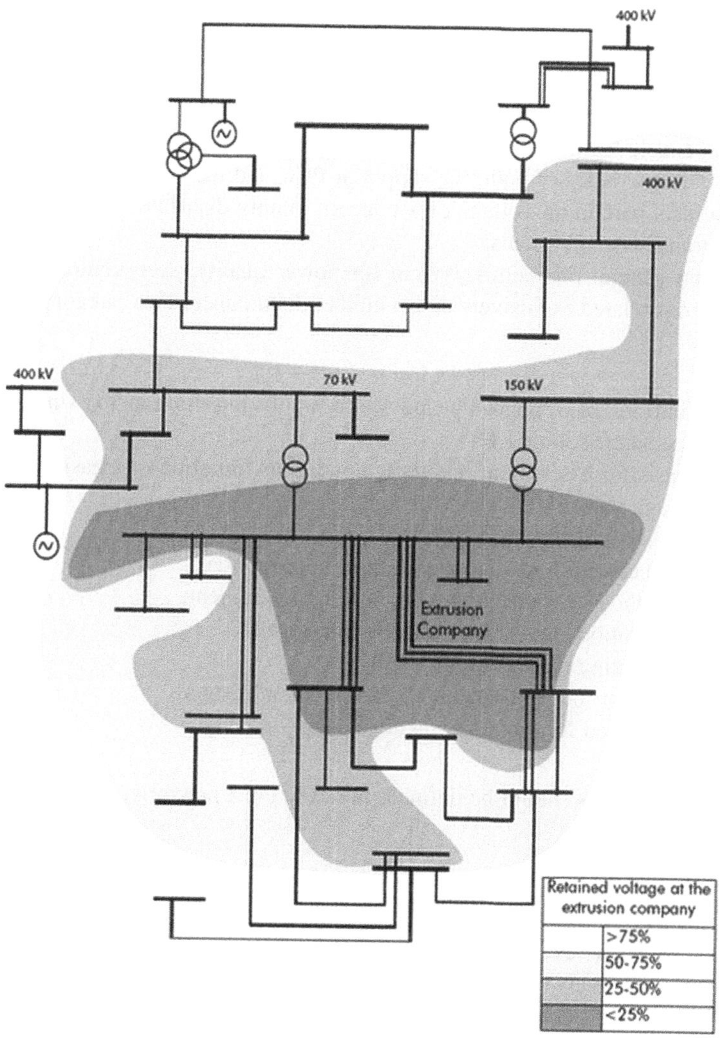

Figure 4.13 Area of vulnerability [21] (Reproduced from Power Quality and Utilisation Guide, Leonardo ENERGY)

Discussions are going on about how voltage dips should be taken into account in the contractual provisions. Many countries are introducing the electricity market. In consequence, customers and suppliers have begun to negotiate the conditions of electric power supply. Most often, the supplier defines a basic offer (quality standard), which is accepted by the majority of customers. There is, however, an increasing number of customers who are using equipment that is less immune to disturbances. This applies to both industrial and residential customers.

Requirements related to the better-than-standard quality of supply for certain categories of customers are expressed in contracts where the supplier guarantees for an extra price the limit level for disturbances which shall not be exceeded in the supply network. The criteria adopted in these contracts usually characterize the quality of power delivery to a certain area of the network, according to international recommendations and standards.

For voltage dips these criteria are sometimes formulated in a different way. In this case the contract is concluded between the electricity supplier and an individual customer taking into account characteristics of both: the supply at PCC and the customer.

The contracts pertaining voltage dips concern mainly distribution networks, and only in rare cases transmission systems.

Apart from general provisions, typical for power quality, such contracts should also provide provisions related exclusively to this kind of disturbance. This category of provisions includes:

- adopted definitions of voltage dip and short supply interruption for single-phase and polyphase systems (Section 4.1);
- time as the basis for assessment of supply conditions (duration of measurements);
- reference voltage value;
- threshold values for disturbance detection;
- time boundary between long and short supply interruptions;
- location and method of connection of measuring instrument;
- technical specifications for measuring instrumentation;
- technique of reporting measurement results;
- the method used for the aggregation of measurement results;
- other techniques used for the assessment of supply quality.

The foregoing provisions should be defined, possibly most precisely, in the contract.

4.7.1 Duration of Measurements

Whereas it is possible to evaluate system performance against a harmonic, flicker or unbalance index over a relatively short time period (e.g. a week), voltage dip performance must be evaluated over a longer period of time. In the large majority of already-concluded contracts the duration of measurement has been adopted as one year, due to the character of disturbance. The likelihood of prognosis based on the obtained data increases with time (number of years) for recording the analyzed disturbance.

4.7.2 Reference Voltage Value

The reference value for the determination of disturbance detection thresholds is the voltage declared in the contract for electric power supply, which in LV and MV networks is usually equal to the nominal voltage. In MV and HV systems the declared voltage may differ from nominal. As practical experience shows, adopting as a reference the voltage value determined over a sliding time window is in most cases advantageous for the supplier. The duration of a time window should be longer than the duration of a voltage dip. The main effect of the choice of threshold or reference value will be in the number of voltage dips detected. This will affect the single-site and system indices.

4.7.3 Location and Method of Connection of Measuring Instrument

There is no correct measurement connection – any connection scheme that records all of the instantaneous voltages between one reference terminal and each of the other terminals in the system provides sufficient information to calculate the voltage between any pair of terminals. So it does not matter if all voltages are recorded with respect to neutral, or if they are recorded with respect to earth, or if they are recorded with respect to the first phase. It just should be made clear that the measurements include all terminals, including the earth terminal.

If not all voltages are measured, then the choice of a connection method (e.g. connection to phase or phase-to-phase voltages) should result from a jointly made decision by the supplier and customer, taking into account:

- the method of supplying sensitive equipment;[22]
- the location of the connection with respect to sensitive equipment, taking into account the system elements, e.g. transformers, which may influence the transfer of voltage dips.

Recording the voltages and currents at the PCC does not always allow reconstruction of the supply conditions on the terminals of sensitive loads in LV networks.[23] In some public power systems phase-to-neutral voltages are measured on solidly earthed systems, and phase-to-phase voltages on impedance-earthed systems. Measurement of phase voltages provides more information on the type of short circuit and its location.

The results of measurements show that the number of dips in phase-to-phase voltages, as compared to phase voltages, is smaller and their residual voltage is higher. To some extent

[22] For example, measurements of phase voltages in MV networks are not representative from the point of view of the customer, because at this voltage level the customer's equipment is mainly phase-to-phase connected.

[23] According to the investigation presented in [53], phase-to-phase measurements are adequate for voltage dips caused by earth faults or phase-to-phase short circuits occurring in MV systems, whereas phase voltage measurements are adequate for phase-to-phase short circuits occurring in HV systems and Yd-connected HV/MV transformers. UNIPEDE recommends the recording of phase voltages. The majority of short circuits in MV and LV systems are phase-to-earth faults. Hence a dip in a phase voltage is always greater than that in phase-to-phase voltage, particularly in systems with isolated neutral points.

The method of earthing, or the fact of not earthing, a neutral point has particular influence on the result of measurements. It has been found that a number of dips depend on the system earthing impedance, i.e. most dips occur in a system with isolated neutral points and fewer in a system with large short-circuit current (small earthing resistance of a neutral conductor). Voltage dips with a greater depth occur in solidly earthed systems, though fault location is easier and, in consequence, fault-clearance times shorter.

Short-circuit durations are longer in networks with isolated neutral points. In the case of networks with isolated neutral or resonant-earthed systems, intermittent phase-to-earth faults are interpreted by measuring equipment as multiple dips.

it is the effect of the phase aggregation method used, where the least residual voltage value has been adopted for describing the disturbance (Section 4.7.7). The largest differences occur for voltage dips caused by phase-to-earth faults, during which a voltage drop in phase-to-phase voltages is significantly smaller than in phase voltages. On phase-to-earth faults in impedance-earthed systems connection to phase voltages results in very small residual voltage, whereas connection to phase-to-phase voltages in most cases yields very shallow dips. Another reason for the phase-to-phase connection is the fact that it can be made under any supply conditions. The method of connection has no significant influence on the duration of the disturbance.

4.7.4 Technical Specifications for Measuring Instrumentation

Detailed specifications for measuring instrumentation are given in standard IEC 61000-4-30 [39]. The measuring instruments used for the purpose of a contract are to comply with technical requirements set forth in this document for class A instruments.

4.7.5 Threshold Values for Disturbance Detection

Threshold value is usually assumed to be 90 % of the reference voltage[24] for a voltage dip and 10 % for short supply interruption.

It can be assumed, and it is advantageous for the supplier, that disturbances are the subject of a contract only when their duration exceeds a limit value, e.g. 1 s for a short supply interruption, 600 ms for a voltage dip [44].

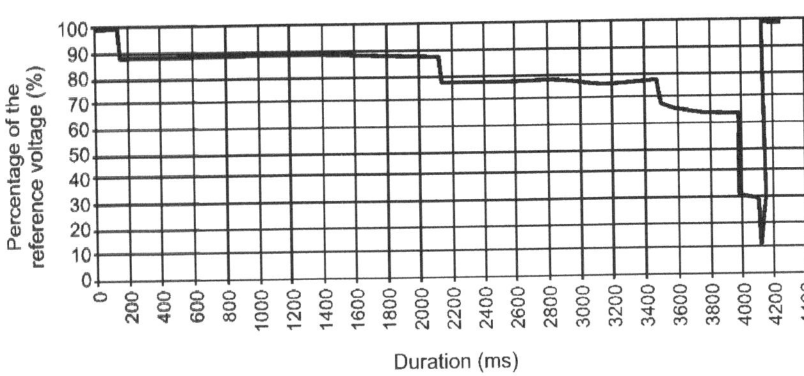

Figure 4.14 The influence of threshold voltage on dip duration [15] (Reproduced from Recommendations for Tabulating RMS Variation Disturbances with Specific Reference to Utility Power Contracts, CIGRE)

[24] If the immunity of the customer's sensitive equipment is known, a greater value of the voltage dip amplitude can be assumed as the threshold (i.e. detection of a dip can be started for $-d\,\%$, where $d \geq 10$). The reason for this attitude is to avoid recording disturbances of no practical significance.

It should be remembered that the voltage threshold has a fundamental influence on dip duration as shown in Figure 4.14 (see also Figure 4.3). Depending on the adopted threshold voltage value, the duration of individual dips takes the following values: $T_{90\%} = 4$ s, $T_{85\%} = 2$ s, $T_{70\%} = 600$ ms, $T_{40\%} = 150$ ms, $T_{10\%} = 50$ ms and $T_{1\%} = 0$.

For this reason, in some proposals several threshold voltages are found and the duration of a disturbance is determined separately for each threshold, e.g. 85, 75, 40 and 1 % of voltage value [66].

4.7.6 Techniques of Reporting the Measurement Results

The need for developing a method for counting voltage dips is imperative. Such a method should be clear and unambiguous and will balance the interests of both the supplier of electric power and the customer and will be taken into account in the contract concluded between them. Normally on three-phase systems phase voltages are monitored independently, r.m.s. half-cycle values calculated, and on this basis a summary of disturbances is obtained at the end of the measuring period. Each event record contains information on the date of occurrence of the disturbance, time of start and end, and on the depth or residual voltage of a dip. There is a considerable degree of acceptance of the tabulation developed by UNIPEDE (Section 4.6.1.4).

In many contracts the supplier guarantees that voltage dips with depth greater than X (e.g. 70 %) will not occur more than Y (e.g. 15) times over a period of one year. In the event that this number is exceeded, the customer is entitled to compensation to an amount agreed upon.

4.7.7 Methods for Aggregation of Measurement Results

In spite of the fact that voltage dips are measured and recorded independently for each measuring channel, they can be aggregated[25] according to contractual provisions. Various aggregation procedures are considered, among them the most extensively used are:

- voltage-level aggregation
- phase aggregation
- time aggregation
- space aggregation

4.7.7.1 Voltage-Level Aggregation

As the measurements show, the great majority of voltage changes during a dip have rectangular shape, i.e. they can easily be described in two coordinates: amplitude (residual

[25] In this case aggregation means a transition from describing a single disturbance to a higher level of data generalization; that is, the formulation of indices that describe a longer interval of time and/or selected part of the supply system.

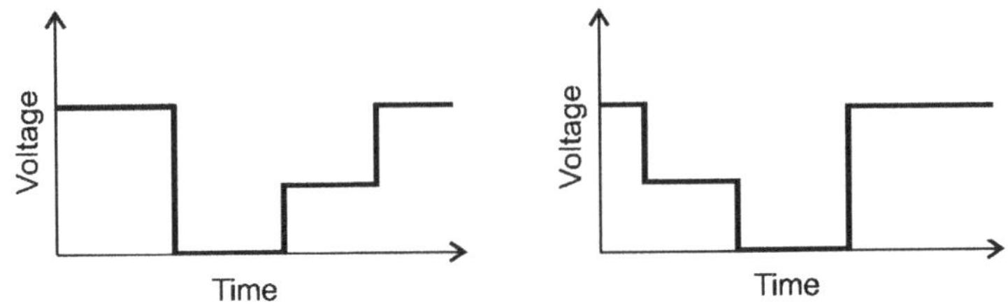

Figure 4.15 Example of two non-rectangular voltage dips of the same duration and amplitude (according to the voltage-level aggregation presented in this section)

voltage) and duration. However, events of multiple voltage changes may occur, as shown in Figure 4.15. This involves the description of a much greater complexity.

Usually, for contractual purposes, the amplitude of a non-rectangular, single-phase voltage dip is the maximum voltage change during the disturbance. This method is recommended by EPRI, UIE, CIGRE, UNIPEDE and is commonly accepted without major reservations.

This method of description of disturbance has, however, certain drawbacks. As an example, two non-rectangular voltage dips, shown in Figure 4.15, which in terms of contractual provisions are indiscernible, may substantially differ in their impact on sensitive equipment.

The duration of a voltage dip is defined as the time over which the voltage is lower than the set threshold value. The choice of the threshold value is the subject of negotiation between the parties to the contract.

4.7.7.2 Phase Aggregation

In the phase aggregation method, voltage dips occurring simultaneously in more than one phase are considered as a single disturbance described by the pair of numbers: amplitude–duration. Voltage dips in different phases are treated as simultaneous if they occur at least in a single common measuring window. This requires synchronization of measurements, e.g. with respect to the voltage zero crossing in the reference phase.

There is no international acceptance for the method of preparation of multi-phase measurement results. The most popular method – presented in Figure 4.2 – consists of determining the amplitude of disturbance as the maximum amplitude of individual phase dips and its duration as the time from the start of the dip in the first phase to the end of the dip in the last disturbed phase. According to this method, the amplitude of the voltage dip for the three-phase disturbance shown in Figure 4.16 is 100 %.

In some cases this method may be inadequate for the nature of the disturbance. In addition, if stipulated in the contract, it may excessively penalize the supplier. An example of such a case is shown in Figure 4.17. The duration of a three-phase dip is the time during which the voltage is less than 90 %. The duration of the dip in Figure 4.17 is indicated according to this definition. The three-phase dip from Figure 4.17 has amplitude 97 % and

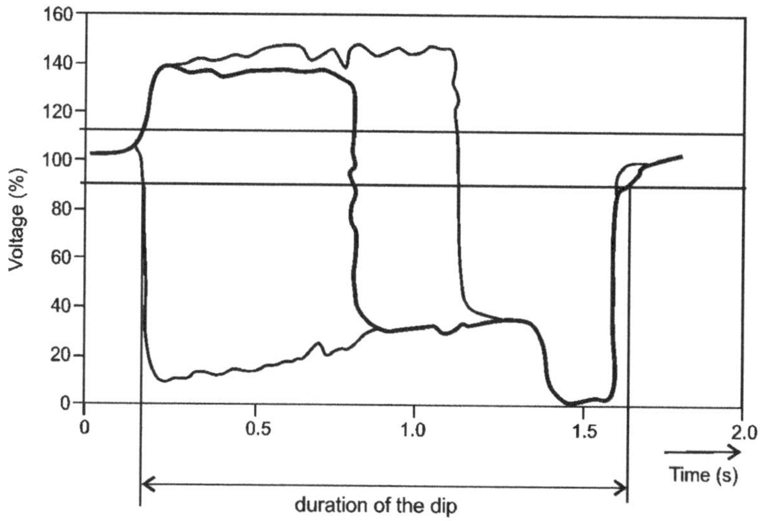

Figure 4.16 Example of determination of a three-phase voltage dip according to the UNIPEDE method [15] (Reproduced from Recommendations for Tabulating RMS Variation Disturbances with Specific Reference to Utility Power Contracts, CIGRE)

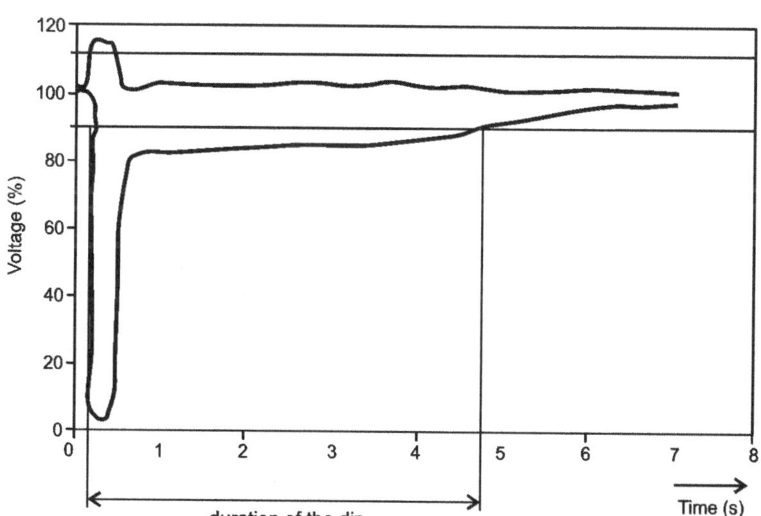

Figure 4.17 Example of parameters determination of a three-phase voltage dip [15] (Reproduced from Recommendations for Tabulating RMS Variation Disturbances with Specific Reference to Utility Power Contracts, CIGRE)

duration about 4.8 s. The actual voltage decrease below 80 % lasts only about 500 ms. Many sensitive loads are insensitive to such disturbances, whereas they are not immune to a dip of depth 97 % and duration 5 s. This interpretation is therefore unfavorable to the supplier.

Other methods adopt the depth and duration of the worst phase dip, the average of voltage dips' amplitudes in individual phases, weighted average, etc., as the value which characterizes the disturbance.

4.7.7.3 Time Aggregation

When collecting voltage dip data it often happens that the data has a tendency to occur in clusters. This observation is the basis for the concept of time aggregation. In this method a sequence of voltage dips separated by specified short time intervals (over a specified time interval) is treated as a single disturbance[26] (Figure 4.18).

The choice of aggregation time for measurements can be made easier when information is available on the autoreclosure time used at the voltage level where the monitor is located and higher voltage levels. A review of reclosing times and practices for some countries shows that autoreclosure time, and thus the window in which one may expect a second event, varies significantly even within one country. It goes from a fraction of a second to a few tens of seconds, but it is generally less than 1 min in the case of multiple reclosing attempts.[27] The discussion on the window width is still ongoing. The answer depends on the purpose that this type of aggregation is employed for. If the intention is to avoid (remove from statistics) multiple disturbances caused by an automatic reclosing sequence, then time aggregation over several seconds should be sufficient. If the purpose of time aggregation were more precise characterization of the influence on equipment, a window of several minutes would be acceptable. What is most often used is a fixed cycle of observations, starting with the beginning of a disturbance in the first phase.

From that instant the time is counted, whose value is conditional on the sequence of operation of the protective systems. Practically, that time is defined by the contract provisions, individually for each customer. For the chemical industry it can amount to several days. If a voltage dip were to occur within this time, it would be of no significance for the process has already been interrupted. The customer who needs, say, five days to restart the process will not incur additional costs only to guarantee that another voltage dip will not occur 5 min after the one that interrupted the process. Thus its is the customer who should mainly define the time interval relevant to the specific situation.

The time-aggregated disturbance can be represented by:

---

[26] The time between events is defined here as the time elapsed between the end of an event and the start of the next event.

[27] An example of a longer aggregation time is seen in the Detroit Edison special manufacturing contract (SMC) with its automotive manufacturers. The contracts require the worst voltage dip in 15-minute aggregation intervals to be archived (see Case Study 4).

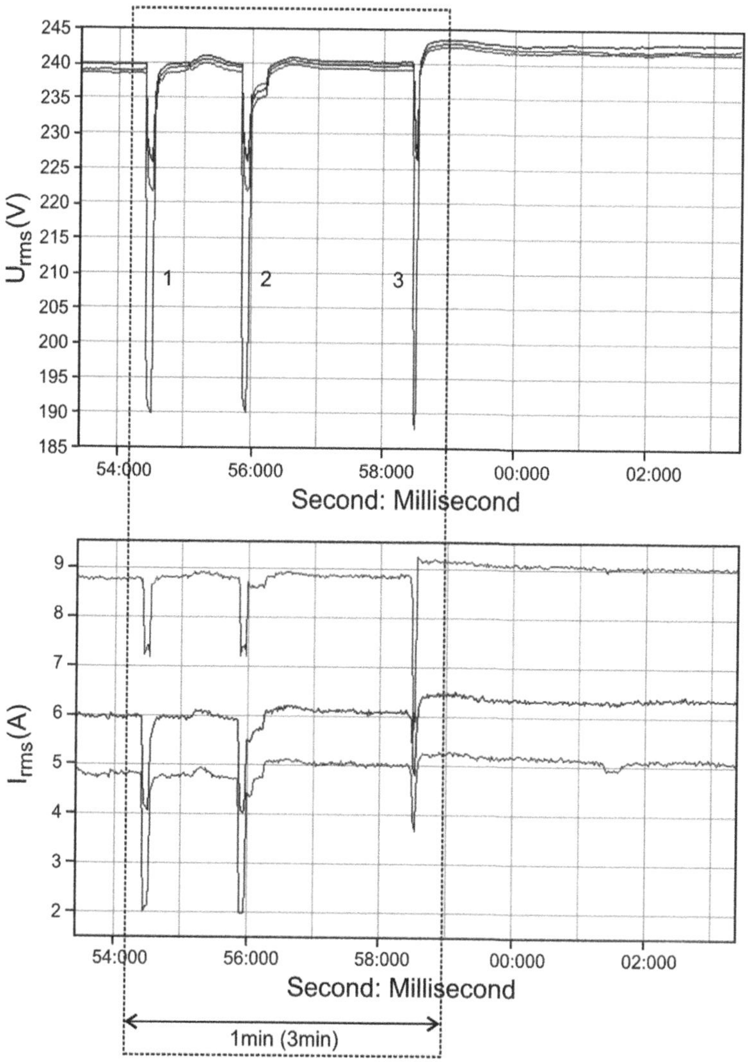

Figure 4.18 Example of time aggregation of three separate dips

- the depth and duration of the first dip;
- the depth and duration of the first dip which has disturbed, or can disturb, the operation of equipment;[28]
- the maximum depth and maximum duration of all dips which occurred within the aggregation time;
- the depth and duration of the voltage dip of the largest area;
- the total duration of aggregated dips, etc.

[28] In this case, knowledge of the immunity level of the sensitive equipment is required.

This type of aggregation is of particular importance, since resigning from use of it would lead to a deterioration in supply reliability due to desisting from automatic reclosing (Section 4.3.2).

It should be emphasized here that, due to adopting short aggregation times, this method does not falsify the assessment of repeatable dips caused by weather conditions, for example. Numerous measurements show that voltage dips which occur over a one-year period, for instance, are concentrated mostly within several days of what undoubtedly indicates their meteorological source. This particularly concerns MV and HV systems with a considerable share of overhead lines.

As the aggregation time is extended the following regularities can be observed:

- for an aggregation time longer than several hours, the number of voltage dips remains relatively constant;
- for an aggregation time of 1–1.5 h, the number of recorded dips changes significantly due to the duration of atmospheric phenomena, which cause short circuits. Similarly, the number of recorded disturbances changes when the aggregation time exceeds about 10 s. This is caused by reclosing and simultaneous faults caused by atmospheric phenomena [66].

4.7.7.4 Space Aggregation

This type of aggregation means the grouping of voltage dips (i.e. treating them as a single disturbance) simultaneously measured on different lines supplying the same user or at one substation when monitoring several busbars. A disturbance of the greatest depth or the longest duration of all disturbances, which occurred at the same time on all lines supplying the same user, can be adopted as the representative one. In this way measurements carried out at several points are reduced to a single-site measurement.

For a case study see web address

4.8 STANDARDS

On the basis of the comparative analysis of existing normative documents, technical recommendations, etc., differences in these documents have been found in the classification and definitions of the considered disturbances (see Table 4.8).

Many of them do not contain unambiguous information on the basic parameters of voltage dips, (Tables 4.8 and 4.9) i.e.:

1. The reference voltage value.
2. Threshold values.
3. The duration of limits of disturbance.
4. A method for assessing the quality of supply in terms of the analyzed disturbance (data processing for contractual purposes, type of aggregation, etc.).
5. A method of estimation of a three-phase voltage dip.
6. A method of connecting a measuring instrument.

Table 4.8 Voltage dip amplitude and duration values in various standardization documents, regulations and publications

Standardized quantity	Amplitude	Min. duration	Max. duration
IEC 1000-2-1	10–100 % of U_N	0.5 cycle	Several seconds
IEC 1000-2-2		10 ms	3 s
IEC 1000-2-5	10–99 % of U_N	10 ms	Several seconds
IEC 61000-2-12		10 ms	3 s
EN 61000-4-11	10–95 % of U_N	0.5 cycle	Several seconds
IEC 1000-6-1	10–95 % of U_N		
IEC 1000-6-2	10–95 % of U_N		
EN 50160	10–99 % of U_N	10 ms	1 min
GOST 13109-97[29]	more than 10 % of U_N	10 ms	Several tens of seconds
UNIPEDE	10–99 % of U_N		
UIE	10–99 % of U_N	10 ms	1 min
IEC 61000-4-30	All threshold values are the subject of a contract		
IEEE Std. 1159-1995	10–90 %	0.5 cycle	1 min
CENELEC	10–90 %	10 ms	1 min
EPRI	< 95 %	1 cycle	1 min
Brazilian classification [18]	10–90 %	1 cycle	1 min

Table 4.9 Short supply interruptions

Standardized quantity	Amplitude	Min. duration	Max. duration
IEC 1000-2-1	Voltage loss (100 %)		1 min
IEC 1000-2-2		10 ms	60 s (180 s)
IEC 1000-2-5	Less than 1 % of U_N		1 min.
IEC 61000-2-12		10 ms (inf.)	60 s (180 s)
IEC 61000-4-11	More than 95 % of U_N 80–100 % of U_N (measuring practices)		1 min (inf.)
IEC 1000-6-1 IEC 1000-6-2	More than 95 % of U_N		5 s (inf.)
EN 50160	More than 99 %		3 min
UIE	More than 90 %		1 min
UNIPEDE	More than 99 %		
IEC 61000-4-30	All threshold values are the subject of a contract		
IEEE Std. 1159-1995	More than 90 % of U_N	0.5 cycle	1 min
Emerald Contract	More than 90 % of U_N	1 s	3 min
Brazilian classification [18]	More than 90 % of U_N		3 s

[29] Russian standard.

4.9 ALTERNATIVE VOLTAGE DIP INDICES

The formulation of numerical indices of voltage dips is a compromise between the simplicity of calculations, their mathematical correctness and representation of the physical complexity of the phenomenon. Some examples are given below.

4.9.1 Indices Based on the Voltage Change [66]

4.9.1.1 Integral of Voltage Loss

That is,

$$L_U = \int (1 - U(t))\, dt$$

where U is a time-varying r.m.s. voltage value (pu) integrated over the duration of a disturbance. This time interval should be defined. Different integration times (important for slow-rising voltages at the end of a voltage dip) would yield significantly different results. If the dip is rectilinear with retained voltage R and duration T then

$$S = (1 - R)T$$

A disadvantage of this index is that a single voltage dip with a long duration can dominate many short-duration disturbances.

4.9.1.2 Voltage Sag Severity Index

The voltage sag severity S_e is defined from the retained voltage U in pu and the duration d by comparing these values with the SEMI[30] (CBEMA or ITIC) curve:

$$S_e = \frac{1 - U}{1 - U_{ref}(d)}$$

where U is the residual voltage value of a voltage dip with duration d and $U_{ref}(d)$ is the amplitude of a voltage dip on the reference curve for disturbance duration d (residual voltage).

The index takes the value 1 for disturbances on the reference curve – Figure 4.19. For those disturbances above the curve, the index value is less than 1, and for those below the

[30] The algorithm for calculating the voltage sag severity proceeds as follows [66]:

Duration range	Calculation of voltage sag severity
$d \leq 20$ ms	$S_e = 1 - U$
20 ms $< d \leq 200$ ms	$S_e = 2(1 - U)$
200 ms $< d \leq 500$ ms	$S_e = 3.3(1 - U)$
500 ms $< d \leq 10$ s	$S_e = 5(1 - U)$
$d > 10$ s	$S_e = 10(1 - U)$

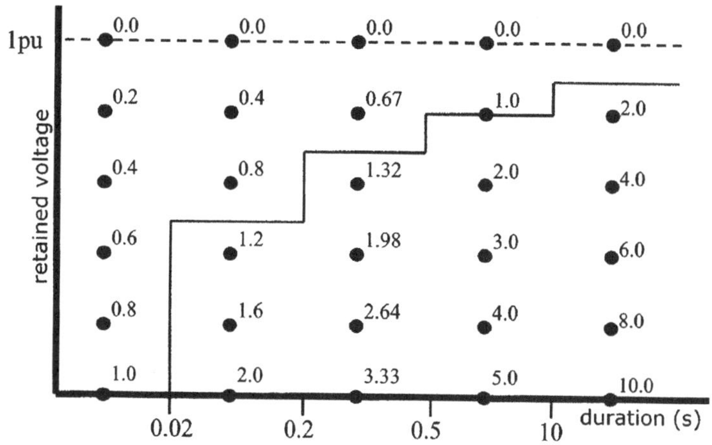

Figure 4.19 Voltage dip *severity index* related to the reference curve (solid line) for disturbances of different durations and with different residual voltages [11] (Reproduced from Ongoing Standard Work on Statistical Presentation of Voltage Dips, Discussion in the CIGRE Working Group)

curve it is greater than 1. For the disturbances with pu voltage value equal to 1 the index takes the value 0. The longer is the disturbance duration and the lower the residual voltage, the higher is the value of the voltage dip severity index.

4.9.1.3 RPM Power Quality Index Approach

RPM has developed a technique for determining a single dip number for each dip event as presented in [26]. Suppose a dip event has coordinates (T, V). Define the corresponding CBEMA voltage $V_{\mathrm{CBEMA}}(T)$ as a voltage on the CBEMA curve corresponding to duration T. The RPM 'PQ Index' for the dip then given by

$$\text{PQ Index} = \left(\frac{100\,\% - V}{100\,\% - V_{\mathrm{CBEMA}}(T)}\right) \times 100\,\%$$

A 'Dip Index' can be formed by the addition of the RPM indices over the dip monitoring period. It is pointed out that the index can be significant for small long dips that may have little effect on equipment. Many such dips may contribute to a substantial dip index although there are no real problems. Similarly, there are problems with very large dips giving very large values of the PQ Index.

4.9.2 Energy-Related Indices [66]

4.9.2.1 Voltage Dip Energy

The voltage sag energy index is given by

$$E_{VS} = \int_0^T \left\{1 - \left[\frac{U(t)}{U_{nom}}\right]^2\right\} dt$$

where $U(t)$ is the time-varying r.m.s. voltage value during a disturbance and U_{nom} is the nominal value. The integration is taken over the duration of the voltage dip. A unit of the index is a unit of time.

In the case when only the residual voltage and duration of a voltage dip are available, the r.m.s voltage value is assumed constant over the dip duration and the relation below applies:

$$E_{VS} = \left[1 - \left(\frac{U}{U_{nom}}\right)^2\right] T$$

The voltage sag energy E_{VS} can be interpreted as the energy (or lack of it) in the voltage dip event. For multi-channel events the voltage sag energy is defined as the sum of the voltage sag energy in the individual channels:

$$E_{VS} = (E_{VS-A} + E_{VS-B} + E_{VS-C})$$

It should be noted that in this method a single voltage dip of a longer duration would dominate numerous short disturbances.

This index can also be used to describe a measurement point within a given interval of time. For this purpose the so-called 'Sag Energy Index' (SEI) has been proposed, which is the sum of voltage sag energies for all qualified disturbances, n, at the given site over the given time interval. These indices are normally determined over a one-month or one-year interval:

$$\text{SEI} = \sum_{i=1}^{n} E_{VS-i}$$

SEI, when expressed in units of time, can be interpreted as the length of the equivalent interruption with the same loss of energy as all dips together that occurred during the observation period.

The 'Average Sag Energy Index' or ASEI is the average of voltage sag energies for all qualified events measured at a given site during a given period:

$$\text{ASEI} = \frac{1}{n}\sum_{i=1}^{n} E_{VS-1}$$

The ASEI is dependent on the triggering of the monitor. A sensitive setting will result in a large number of shallow events (with a low sag energy) and thus in a lower value of the ASEI. The SEI on the other hand will increase for a sensitive setting of the monitor. To compare results from site to site and from one period to another, a standardized trigger setting needs to be defined.

When using voltage sag energy indices it is recommended not to include momentary interruptions, as one momentary interruption may have a larger contribution to the index than all voltage dips together.

Also when using voltage sag energy indices, system indices are calculated by taking the average value of the site indices. The SEI values for the system are obtained by dividing the sum of the site values by the number of sites:

$$\mathrm{SEI}_{system} = \frac{1}{N}\sum_{s=1}^{N} SEI_s$$

where N is the number of sites.

4.9.3 Others [15]

4.9.3.1 System Average R.M.S. (Variation) Frequency Index$_{Voltage}$ (SARFI$_X$)

This index is used to describe a single measurement point within a specified time, e.g. one month or a year. It represents the average number of measured events of voltage change (dips, swells, interruptions), with a duration between half a cycle and 1 min, experienced by the user supplied from the assessed measurement point (a different duration, e.g. 5 min, may also be adopted):

$$\mathrm{SARFI}_X = \frac{\sum N_i}{N_T}$$

where:

- X is the r.m.s. voltage threshold (minimum voltage value during a dip). For a voltage dip it can be 90, 80, 70, 50 or 10 % of the reference voltage. The value of X is determined individually depending on the customer's equipment immunity characteristic, e.g. the dominant sensitive load. In the case of a three-phase dip it is recommended to take into account the phase on which the dip of the largest amplitude has occurred.
- N_i is the number of customers who experience voltage changes of residual voltage less than $X\%$ for $X < 100$, for the ith case.
- N_T is the number of customers supplied from the system section being assessed.

According to the foregoing definition, the calculation of SARFI is possible when all customers who have experienced the voltage dip are known. This index, when not calculated per single customer (or a single monitored line), can also be regarded as the number of disturbances in a given measurement point, over a specified period of time.

Determination of SARFI for a single customer is equivalent to the determination of the value of aggregated events defined by a voltage value lower than the specified threshold.

If, for instance, eight disturbances have been recorded during 92 days of measurement (at the assumed threshold value X), then after calculation to a standardized one-month period (30 days) SARFI for a given location will be $\mathrm{SARFI} - 8 \cdot (30/92) = 3.93$.

A modification of SARFI$_X$ is the so-called *SARFI-curve* index, e.g. *SARFI-CBEMA* (also *SARFI-ITIC* and *SARFI-SEMI*). Voltage dips, whose position on the plane defining the CBEMA curve is below the lower line of the characteristic, are taken into account for determining this index. It can limit the duration of a disturbance to 1 min, for example, taking into account all voltage dips of duration longer than 10 ms.

SARFI can also be applied to the system description, e.g. as an average value of indices obtained from different measurement points. In this way, the average supply quality index for the whole system is obtained. Because not all the system locations are monitored and not all measurement points are equivalent, a certain type of weighting can be used. By using the weighting coefficients, more weight can be assigned to a location with a more important load. A simple solution is the weighting over the number of customers. If the MVA supplied by each site is known, and are very different in value, than it can be weighted by the appropriate MVA to emphasize the power quality experienced by most of the load demand. In fact the weighting coefficients are in most cases taken as equal for all locations.

When a number of monitored lines are large, the percentile CP95 can be applied to the system description.

Advantages of the proposed index are [66]:

- the small number of SARFI indices makes it easy to compare different sites, different systems and year-to-year variations;
- due to its simplicity, the method is widely used;
- the index depends only on the total number of events. When the indices are used to quantify system performance there will be a strong incentive to reduce the number of faults. This also has a positive effect on the supply reliability (number of interruptions).

Disadvantages:

- the index is not relevant for assessing equipment immunity;
- all information on the duration of a disturbance is lost. When the indices are used to quantify system performance, there is no incentive to reduce fault-clearing times. Duration can be introduced as a parameter if it is assumed that all dips with a residual voltage greater than X and duration longer than Y are taken into account.

According to the EPRI proposal [63], three other indices have been defined. They are defined analogously to SARFI indices and are applied to the quantitative assessment of instantaneous, momentary and temporary voltage dips (Table 4.10).

Table 4.10 Summary of indices according to EPRI [63]

<table>
<tr><th></th><th>0.5 cycle to 60 s</th><th>0.5 cycle to 0.5 s</th><th>0.5–3 s</th><th>3–60 s</th></tr>
<tr><td>< 90 %</td><td>$SARFI_{90}$</td><td>$SIARFI_{90}$</td><td>$SMARFI_{90}$</td><td>$STARFI_{90}$</td></tr>
<tr><td>< 80 %</td><td>$SARFI_{80}$</td><td>$SIARFI_{80}$</td><td>$SMARFI_{80}$</td><td>$STARFI_{80}$</td></tr>
<tr><td>< 70 %</td><td>$SARFI_{70}$</td><td>$SIARFI_{70}$</td><td>$SMARFI_{70}$</td><td>$STARFI_{70}$</td></tr>
<tr><td>< 50 %</td><td>$SARFI_{50}$</td><td>$SIARFI_{50}$</td><td>$SMARFI_{50}$</td><td>$STARFI_{50}$</td></tr>
<tr><td>< 10 %</td><td>$SARFI_{10}$</td><td colspan="2">$SMARFI_{10}$</td><td>$STARFI_{10}$</td></tr>
</table>

4.9.3.2 *System Instantaneous Average R.M.S. (Variation) Frequency Index*$_{\text{Voltage}}$ *(SIARFI$_X$)*

THIS IS for voltage dips with duration of 0.5-30 cycles and $x = 90$, 80, 70 and 50 %.

4.9.3.3 *System Momentary R.M.S. (Variation) Frequency Index*$_{\text{Voltage}}$ *(SMARFI$_X$)*

THIS represents the average number of measured cases of voltage change with duration of 30 cycles to 3 s and $x = 90$, 80, 70, 50 and 10 %, which have occurred over a specified period of time at a given customer's site.

4.9.3.4 *System Temporary Average R.M.S. (Variation) Frequency Index*$_{\text{Voltage}}$ *(STARFI$_X$)*

THIS represents the average number of measured cases of voltage change with duration of 3-60 s and $x = 90$, 80, 70, 50 and 10 %.

ACKNOWLEDGEMENT

A portion of text in this chapter is reproduced by permission of European Copper Institute.

BIBLIOGRAPHY

[1] Ahang L. D., Bollen M., Characteristics of voltage dips (sags) in power systems. *IEEE Transactions on Power Delivery*, vol. 15, no. 2, pp. 827–832, 2000.
[2] Amentagui J., Power quality measurement & control procedures in Spain. CIRED 2003, Round table on Power Quality at the Interface T&D.
[3] Ball R. F., *Thoughts on voltage dip severity index*. The National Grid Company plc.
[4] Ball R. F., *Voltage dips: indices and objectives*. The National Grid Company plc.
[5] Beaulieu G., Borloo G., Bollen M., Koch R., Malgarotti S., Mamo X., Recommending power quality indices and objectives in the context of an open electricity market. CIGRE/IEEE-PES on Quality and Security of Electric Power Delivery Systems, Montreal, 7–10 October 2003.
[6] Beaulieu G., Bollen M., Koch R., Malgarotti S., Mamo X., Sinclair J., Power quality indices and objectives for MV, HV and EHV systems. 17th International Conference on Electricity Distribution, CIRED, Barcelona, 12–15 May 2003.
[7] Bollen M., Characterization of voltage sags experienced by three-phase adjustable-speed drives. *IEEE Transactions on Power Delivery*, vol. 12. no. 4, pp. 1666–1671, 1997.
[8] Bollen M. H. J., *Understanding power quality problems – voltage sags and interruptions*, IEEE Press Series on Power Engineering, 2000.
[9] Bollen M., Time aggregation of voltage dips. Discussion in the CIGRE Working Group.
[10] Bollen M., Comparision of the different indices. Discussion in the CIGRE Working Group.
[11] Bollen M. H. J., Ongoing standard work on statistical presentation of voltage dips. Discussion in the CIGRE Working Group.
[12] Bollen M. H. J., Goossens P., Robert A., Assessment of voltage dips in MV-networks: deduction of complex voltage from the measured rms-voltages. CIRED, Barcelona, 2003.

[13] Bollen M., Sabin D., Thallam R., Voltage-sag indices – recent developments in IEEE P1564 Task Force. CIGRE/IEEE-PES on Quality and Security of Electric Power Delivery Systems, Montreal, 7–10 October 2003.
[14] Bollen M., Sabin D., Thallam R., Sannino A., Voltage sag indices: recent developments in IEEE P1564 Task Force. EPRI PQIG, Paris, 2003.
[15] Brooks D. L., Gunther E. W., Sundaram A., Recommendations for tabulating rms variation disturbances with specific reference to utility power contracts. CIGRE 36.05/CIRED 2 CC02 9923.
[16] Conrad L., Little K., Griff G., Predicting and preventing problems associated with remote fault clearing voltage dips. *IEEE Transactions on Industry Applications*, vol. 27, no. 1, pp. 167–172, 1991.
[17] Davenport F.W.T., Voltage dips and short interruptions in medium voltage public electricity supply systems. UNIPEDE/DISDIP Report, 1990.
[18] Delmo de Macedo Corriera *et al.*, Pilot project for evaluation of the sag performance of some Brazilian network busbars. CIGRE/IEEE-PES on Quality and Security of Electric Power Delivery Systems, Montreal, 7–10 October 2003.
[19] Desquilbert G., Foucher C., Fauquembergue P., Statistical analysis of voltage dips. 3rd International Conference on Power Quality: End-Use Applications and Perspectives, 24–27 October 1994, Amsterdam.
[20] Dettlof A., Sabin D., Power quality performance component of the special manufacturing contracts between power provider and customer. 0-7803-6499-6/2000 IEEE.
[21] Didden M., Voltage sags in continuous process-case study (5.5.1). Power Quality and Utilisation Guide, Leonardo ENERGY (www.lpqi.org).
[22] Dugan R., McGranghan M.F., Santoso S., Beaty H.W., *Electrical power systems quality*, McGraw-Hill, New York, 2002.
[23] EN 50160, *Voltage characteristics of electricity supplied by public distribution systems.*
[24] Geiveldinger C. *et al.*, New power quality contracts in France. PQA, Amsterdam, 1994.
[25] Gosbell V. J., Robinson D., Barr R., Smith V. W., How should power quality be reported? Electricity 2002 – Winners and Losers from Reforms and Innovations.
[26] V. J. Gosbell, D. Robinson and S. Perera, The analysis of utility voltage sag data. Proceedings of the International Power Quality Conference, Singapore, October 2002, pp. 479–490.
[27] Häger M., Wahlström B., Friman E., Limits and responsibility sharing with regard to power quality for the Swedish 220- and 400- transmission system. STRI Report S98–171.
[28] Higgings M., Mansoor A., Sundaram A., Understanding premium power grades. *Electrical Power Quality and Utilisation*, vol. 2, no. 7, 2001.
[29] IEEE Standard 493, *Recommended practice for the design of reliable industrial and commercial power systems.*
[30] IEEE Standard 1159, *Recommended practice for monitoring electric power quality.*
[31] IEEE Standard 1346, *Recommended practice for evaluating electric power system compatibility with electronics process equipment.*
[32] IEEE Standard 1366, *Guide for electric power distribution reliability indices.*
[33] IEEE *Orange Book.*
[34] IEC 61000-2-2, *Electromagnetic compatibility (EMC) – Part 2-2: Environment compatibility levels for low-frequency conducted disturbances and signaling in power supply systems.*
[35] IEC 61000-2-5, *Classification of electromagnetic environments.*
[36] IEC 61000-2-8, *Voltage dips and short interruptions on public supply systems with statistical measurement results.*
[37] IEC 61000-2-12, *Electromagnetic compatibility (EMC) – Part 2-12: Environment compatibility levels for low-frequency conducted disturbances and signaling in public medium-voltage power supply systems.*

[38] IEC 61000-4-11, *Electromagnetic compatibility (EMC) – Testing and measurement techniques – Voltage dips, short interruptions and voltage variation immunity tests.*

[39] IEC 61000-4-30, *Electromagnetic compatibility (EMC) – Part 4-30: Testing and measurement techniques – Power quality measurement methods.*

[40] IEC 61000-4-34, *Electromagnetic compatibility (EMC) – Part 4-34: Testing and measuring techniques – Voltage dips, short interruptions and voltage variations immunity tests for equipment with input current more than 16 A.*

[41] IEC 61000-6-1, *Electromagnetic compatibility (EMC) – Part 6-1: Generic standards – Immunity for residential, commercial and light-industrial environments.*

[42] IEC 61000-6-2, *Electromagnetic compatibility (EMC) – Part 6-2: Generic standards – Immunity for industrial environment.*

[43] IEC 60947-4-1, *Low voltage switchgear and control gear – Part 4-1: Contactors and motor-starters – Electromechanical contactors and motor starters.*

[44] International Telecommunication Union, Geneva, 1989, *Emerald Contract for the supply of electric at the Green Rate*, EdF (Electricité de France).

[45] Javerzac J. L., *Contracting the quality of electricity: the French experience*, 0-7803-64499-6/2000, IEEE.

[46] Matusz G., Maj T., Hanzelka Z., Loziak W., Voltage dips and short supply interruptions in contracts for the electric power supply from high voltage transmission systems. Power Quality PCIM Europe, Nurnberg, 20–22 May 2003.

[47] McGranaghan M. F., Effects of voltage sags in process industry applications. Stockholm Power Technical Conference, Stockholm, Sweden, 18–22 June 1995.

[48] McGranaghan M., Gunther E., The economics of custom power. Power Quality and EMC in Power Quality, Colloquium of Czech National Committee and 36 Study Committee of CIGRE, Dept. of Power Engineering, CTU Prague, 28 September 2001, Prague.

[49] Mueller D., Understanding power quality measurements. Power Quality, August 1999.

[50] NRS 048-2, *Electricity supply – Quality of supply. Part 2: Voltage characteristics, compatibility levels, limits and assessment methods.* Published in the Republic of South Africa by the South African Bureau of Standards (http://www.sabs.co.za).

[51] NRS 048-4, *Electricity supply – quality of supply. Part 4: Application guidelines for utilities.* Published in the Republic of South Africa by the South African Bureau of Standards.

[52] Olguin G., Stochastic assessment of voltage dips caused by faults in large transmission system. Department of Electric Power Engineering, Chalmers University of Technology, Goteborg, Sweden, 2003.

[53] Pazos F. J., Measuring voltage dips oriented towards power quality contracts. CC02 9837, 1998.

[54] Pierz P., Rosołowski E., Quality of electric power in industrial plants (in Polish). Elektroinstalator 9/2000.

[55] Pohjanheimo P., A probabilistic method for comprehensive voltage sag management in power distribution systems. Doctoral Thesis, Helsinki University of Technology, Power System Laboratory, 2003.

[56] *Power quality indices and objectives*, Joint Working Group CIGRE C4.07/CIRED.

[57] Robert A., Hoeffelman J., De Jaeger E., CIRED 2001 – Session 2: Power Quality & EMC.

[58] SEMI F47-0999, *New standard – specification for semiconductor processing equipment voltage sag immunity.*

[59] Smith J., The impact of voltage sags on industrial plant loads. Paper C-24, PQA'91, First International Conference on Power Quality: End-Use Applications and Perspectives, Paris 1991, pp. 171–178.

[60] Stephens M., PQ in continuous manufacturing (1.5). Power Quality and Utilisation Guide, Leonardo ENERGY (www.lpqi.org).

[61] Thallam R. S., Power quality indices based on voltage sag energy values. Power Quality Proceedings, September 2001.
[62] The 3rd CEER Benchmarking Report on Quality of Electricity Supply – C05-QOS-0103, December 2005.
[63] Transmission Power Quality Benchmarking Methodology. EPRI Final Report, December 2000.
[64] Understanding premium power grades. EPRI Final Report, November 2000.
[65] Voltage dips measurement. UIEPQ-9559, October 1995.
[66] Voltage sag indices. IEEE P1564, August 2003.
[67] www.lpqi.org.

5

Voltage Fluctuations and Flicker

Araceli Hernández Bayo

The term flicker is used to refer to the subjective impression that is experienced by human beings when subjected to changes occurring in the illumination intensity of light sources. One of the reasons for the complexity in the evaluation of this phenomenon is the human factor involved in its definition, since it forces one to take into account the characteristics of the physiological process of perception.

From an electrical point of view, flicker is caused by voltage fluctuations with an amplitude which is generally much lower than the threshold of immunity for electrical equipment. So, it can be said that the major effect of rapid voltage fluctuations is flicker. Voltage variations on the order of only a few tenths of a percent can produce a very significant malaise, especially if the frequency of repetitive deviations is between 8 and 10 Hz.

Fluctuating loads such as arc furnaces, welders, etc., whose power demand experiences wide and rapid variations, can be potential sources of voltage fluctuations producing flicker. Different mitigation methods oriented to decrease voltage variations caused by this type of loads can be applied in order to reduce flicker levels.

5.1 DESCRIPTION OF THE PHENOMENON

5.1.1 Voltage Changes, Voltage Fluctuations and Flicker

Voltage changes, **voltages fluctuations** and **flicker** are related terms and very often are used indistinctly. However, it is important to understand that these terms designate different

Handbook of Power Quality Edited by Angelo Baggini

effects, although very often they can occur simultaneously. Before going deeper into this, it is convenient to establish clearly the distinction between these expressions:

- A voltage change is defined as a deviation in the r.m.s. voltage value with respect to a steady-state value averaged over some period of time. This voltage deviation may or may not be periodical. Since voltage changes can be easily measured, historically, curves relating periodic rectangular voltage changes to flicker have been used (see Figure 5.1). These curves represent the threshold of irritability for a periodic series of rectangular voltage changes. Some shape factors have been also proposed in order to adapt this curve to other types of voltage changes: sinusoidal voltage variations, triangular voltage variations and pulse-type disturbances. However, the applicability of these curves is restricted to the above-mentioned periodic voltage waveshape variations. In other cases, a more general measurement method is necessary as will be described in Section 5.3 of this chapter.
- Voltage fluctuations are defined as the cyclic variation of voltage with amplitude that does not exceed 10 %. This variation in magnitude is usually much lower than the sensitivity threshold of most equipment and, consequently, operating problems are experienced only in rare cases. Except for these very particular cases, the main disturbing effect of voltage fluctuations is producing changes on the illumination intensity of light sources.
- Finally, flicker is defined as the *impression of unsteadiness of visual sensation induced by a light stimulus whose luminance or spectral distribution fluctuates with time* [8]. In other words, flicker is defined as the unpleasant sensation experienced by the human visual system when subjected to changes occurring in the illumination intensity of

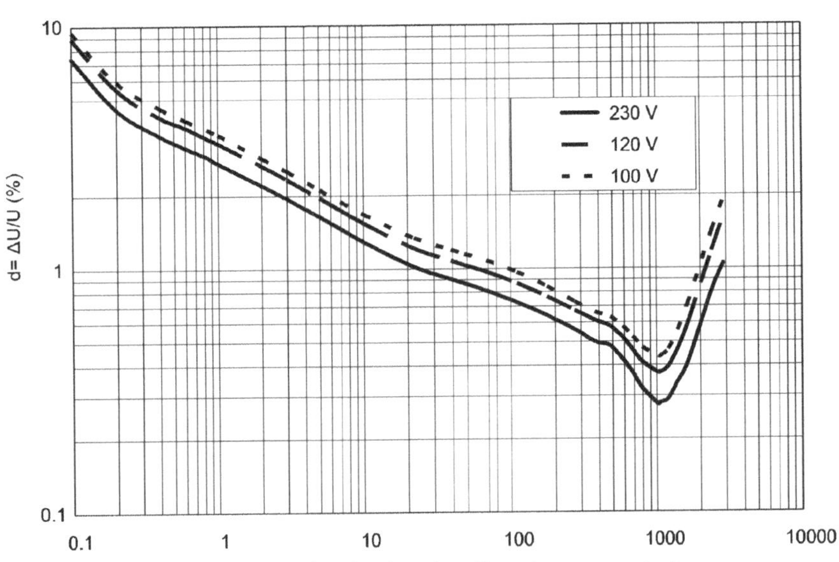

Figure 5.1 Rectangular voltage changes for the threshold of flicker irritability

light sources [26]. These illumination variations appear as a consequence of voltage fluctuations, so that a clear relationship can be established between these disturbances (flicker and voltage fluctuations). However, in spite of this, an important nuance must be pointed out between them: a physiological aspect is involved in the definition of flicker while voltage fluctuations are defined exclusively from an electrical point of view. This means that flicker is a phenomenon due to the combination of two factors: voltage fluctuations causing changes in light intensity and a person exposed to these changes. Thus, flicker is an effect strictly related to human perception and reaction and, for that reason, an appropriate characterization and measurement of flicker implies thorough knowledge of the physiological process involved in its perception.

5.1.2 Physiology of Flicker Perception

As previously stated, flicker is essentially a visual phenomenon that can cause significant malaise to the person exposed to it. The degree of annoyance experienced by the individuals depends on several factors: namely, the amplitude of the fluctuations, the frequency of the fluctuation, the duration of the disturbance or even the activity of the subject and the color of the lighting source, among others. Sensitivity to flicker is also dependent to a high degree on the individual, although it is recognized that, for a large population, it is normally distributed.

The malaise caused by flicker is more noticeable for two important domestic activities: reading and watching TV[1] [26]. Flicker may induce discomfort in the form of nausea, headaches, annoyance, distraction, feeling of tiredness and problems with concentration for the people exposed to it. It is also known that flicker, in high doses, can trigger epileptic fits in those susceptible to it. Sometimes it can be subtle enough so as not to be consciously detected by those affected.

Bearing in mind that a perfect compensation of load variations cannot be achieved in the supply system, a certain amount of voltage fluctuations must be tolerated, up to the limit where malfunctions or intolerable flicker can occur. Lamps, due to the flicker problem, are by far one of the most sensitive loads to voltage fluctuations. Since they are also one of the most common loads, concern about this disturbance developed very early and different investigations were carried out in order to characterize the response of the physiological perception of flicker. One of the first experiments on this phenomenon was performed by Simons [25]. In his experiments, Simons performed several tests on individuals producing light changes by a rotating sectored wheel which gave sinusoidal light fluctuations.

In the 1950s, de Lange studied the frequency sensitivity of the visual system. His research work [16], [17] showed that flicker can be observed for repetitive voltage fluctuations up to a frequency where the fusion of images in the eye makes it impossible to detect them. There is also a lower critical frequency below which flicker is again not perceived. This effect is called the **flicker fusion boundary** and was included in the flicker perception model developed by de Lange.

[1] The voltage fluctuation is generally too low to disturb operation of the TV; however, the flicker of the light source in the room where the TV is watched is perceived as very annoying by TV viewers.

More recently, in the 1970s, Rashbass carried out extensive experiments in which the perception threshold of light flashes of various duration was determined [22]. As a result of these experiments, he proposed incorporating into the perception model two further elements, namely a squaring multiplier block and a low-pass filter that simulated, respectively, the quadratic law of the visual amplitude response and the memory effect in the brain perception.

As will be discussed later, all these findings represent the basis for the development of flicker measuring equipment intended to simulate the flicker perception process and to provide an output expressed in terms of perceptibility level.

5.2 PARAMETERS

Flicker is basically characterized by two parameters obtained by means of the flicker measurement apparatus that will be described in detail below. These parameters are:

- The parameter P_{st} which is a measure of short-term flicker severity obtained for a 10-minute interval. This basic observation period of 10 minutes is a good compromise that is long enough to prevent too much weight to isolated voltages changes and to allow observation of the persistence of the disturbance. At the same time, it is short enough to characterize the voltage fluctuations produced by equipment with short duty cycles [27]. P_{st} is a dimensionless quantity. A unit of P_{st} corresponds to the threshold of irritability: that is, to the severity limit that should not be exceeded in order not to create malaise for an observer to any type of flicker, independently from the source of the disturbance.
- The parameter P_{lt} is a measure of long-term flicker severity obtained for a two-hour period. P_{lt} is deduced from 12 consecutive values of P_{st}, as given by the following equation:

$$P_{lt} = \sqrt[3]{\frac{\sum_{i=1}^{12} P_{st,i}^3}{12}} \tag{5.1}$$

The parameter P_{lt} provides a criterion for the long-term flicker assessment when flicker sources with long and variable duty cycles have to be considered or when several disturbing loads operate simultaneously in a random way. The threshold value for P_{lt} is 0.8 units.

The limits imposed on the flicker severity corresponding to the two time intervals (P_{st} and P_{lt}) are necessary in order to ensure that the flicker is not annoying. It is important to notice that if P_{st} is low, then P_{lt} will also be low. However, the opposite is not necessarily true.

In addition to P_{st} and P_{lt}, which are parameters defined in terms of flicker severity, another related parameter can also be mentioned, namely the relative voltage change. The relative voltage change is usually expressed in percent value and is designated as d or $\Delta V/V$. This parameter expresses the deviation in the r.m.s. voltage value with respect to a steady-state value averaged over some period of time and, as previously indicated, can be related to flicker severity by means of different approximate approaches. Figure 5.2 shows an example of a sinusoidal wave with rectangular voltage changes of amplitude 40 % and a frequency 8.8 Hz in order to illustrate how to interpret these values correctly.

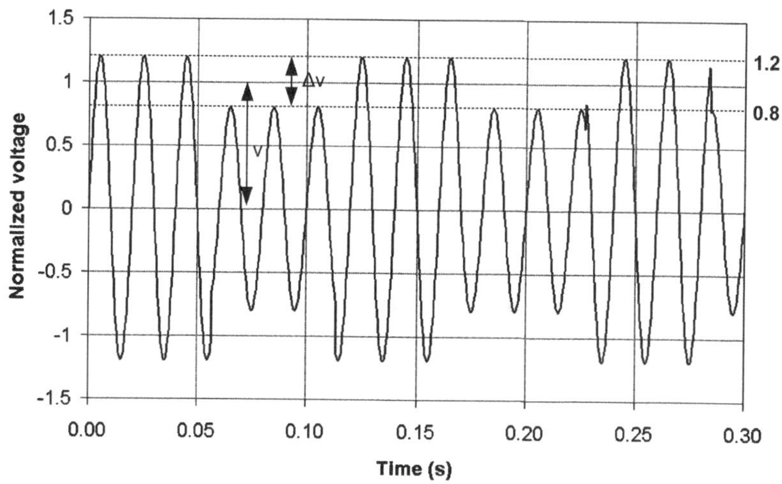

Figure 5.2 Rectangular voltage change $d = \Delta V/V = 40\,\%$, modulation frequency 8.8 Hz, 17.6 changes/second

5.3 MEASUREMENT

Standard IEC 61000-4-15 provides the functional and design specification for flicker measurement apparatus intended to indicate the correct flicker perception level for all practical voltage fluctuation waveforms [6].

Before the standardization of the flickermeter, several different instruments were used around the world to evaluate voltage flicker. These instruments, although based on the relation between voltage fluctuations and the flicker effect expressed in terms of visual sensation, did not yield comparable results. The necessity to achieve an internationally accepted flickermeter motivated the work of the International Union for Electricity Applications (UIE) on a standardized flickermeter over the last two decades. In 1986 the UIE/IEC flickermeter was internationally agreed by the IEC [5].

It is important to bear in mind that the primary objective of the IEC flickermeter is not to provide an evaluation of voltage fluctuations but of flicker perception caused by these fluctuations. To achieve this goal, the equipment must be designed so that it can transform the input voltage fluctuations into an output parameter proportionally related to flicker perception. This is possible by simulating the process of physiological visual perception, that is the so-called *lamp–eye–brain chain.*

5.3.1 The IEC Flickermeter

The UIE/IEC flickermeter was originally based on a 230 V, 60 W incandescent lamp [6]. In North America, lamps operate at 120 V and have to conduct higher currents. It is for this reason that they are built with thicker filaments which present a higher thermal inertia and, consequently, these lamps show a different frequency characteristic which is not as sensitive to voltage fluctuations as compared to a 230 V lamp.

Thus, the original IEC flickermeter was not directly applicable to measurement of flicker caused in 120 V systems. However, over the last few years, the IEC flicker measurement standard has been amended to include the characteristics of 120 V lamps for measurement in North America [7].

The description given in IEC 61000-4-15 is based on an analogue implementation although an instrument that is totally or partially digital is also valid under the condition of having the same functional characteristics [6].

The architecture of the flickermeter is depicted in Figure 5.3. As shown in this figure, the flickermeter can be divided into two parts, each performing one of the following tasks:

- Simulation of the response of the lamp-eye-brain chain
- Online statistical analysis of the flicker signal and presentation of results.

The first task is performed by blocks 2, 3 and 4 of Figure 5.3 while the second task is accomplished by block 5. The response of each block will be analyzed in the following subsections.

5.3.1.1 Block 1: Voltage Adapting Circuit

Block 1 of the flickermeter contains a voltage scaling circuit that accepts the supply voltage as an input and derives the relative voltage change. This is performed by adjusting the gain of the block to the average of the input voltage with a time constant. The time constant of the voltage adapter has been chosen equal to one minute as a compromise between a value longer enough to reproduce correctly the voltage changes which are relevant for flicker, but that still allows following quite closely slow variations produced by regulation of the supply system [27].

By performing this scaling, flicker measurements can be carried out independently of the actual input carrier voltage level and expressed as a percent ratio.

In addition, block 1 contains a signal generator to check the calibration of the flickermeter on site.

5.3.1.2 Block 2: Square Law Demodulator

Flicker is caused by the voltage fluctuation superimposed on the 50 or 60 Hz wave. The purpose of the demodulating block is to recover this modulating signal from the sinusoidal carrier.

Among the possible methods for demodulating the signal, a quadratic demodulator is used in the IEC flickermeter since it shows some advantages over other methods as a classical envelope detector. An envelope detector operates with peak values while the squaring demodulator considers the r.m.s. voltage, which is the relevant quantity for the variations in the illumination intensity of light sources. Moreover, an envelope detector would provide the fluctuation of the fundamental component of voltage ignoring an eventual

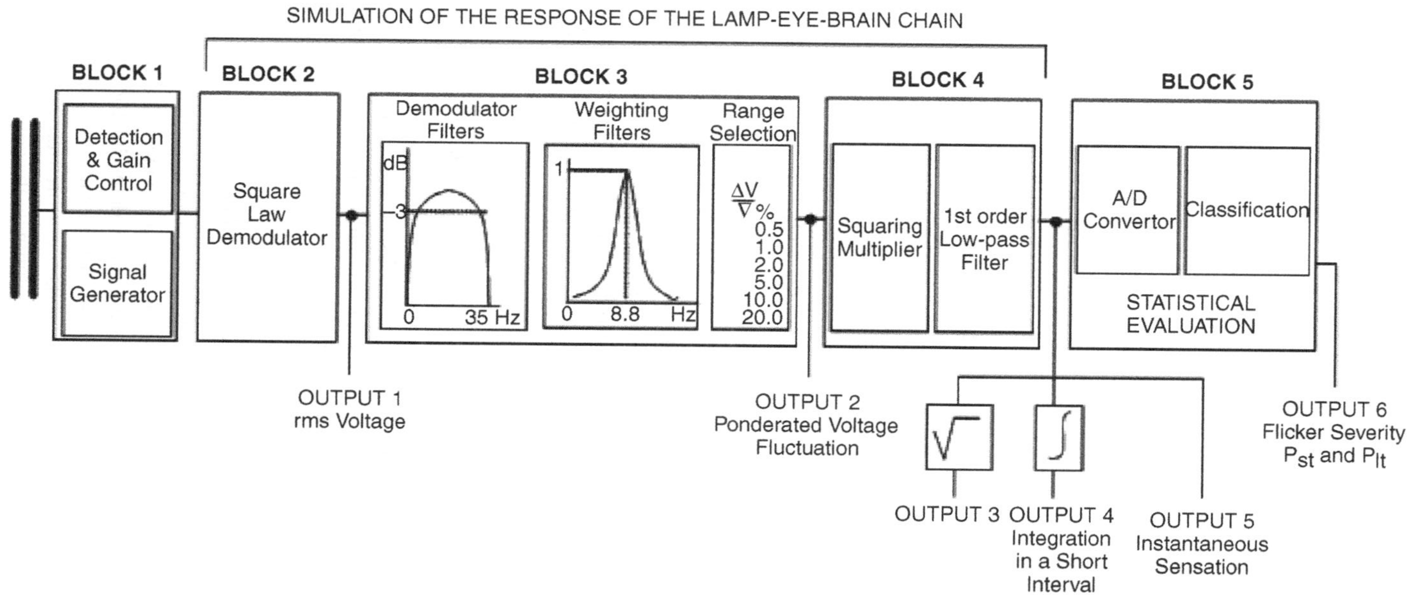

Figure 5.3 Schematic description of the IEC flickermeter

harmonic content, whereas the square demodulator reproduces the fluctuations of the square of the r.m.s. voltage, which is directly related to the electric power absorbed by the lamp and, therefore, is representative of flicker. One further advantage is that the quadratic demodulator can be quite easily realized.

To gain a better understanding of the operation of a quadratic demodulator, typically an example of a sinusoidal voltage modulated by a sinusoidal voltage fluctuation is proposed. This type of signal is expressed as follows:

$$v(t) = A\cos\omega_p t \cdot (1 + m\cos\omega_m t) \tag{5.2}$$

where $v(t)$ is the supply voltage with amplitude A at angular frequency ω_p, and m is the amplitude of the sinusoidal voltage fluctuation that modulates the carrier with angular frequency ω_m.

Applying the quadratic demodulator to this signal, its output would be of the form

$$\begin{aligned} v_S(t) = [v(t)]^2 = {} & \frac{A^2}{2}\left(1+\frac{m^2}{2}\right) + \frac{A^2}{2}\left(1+\frac{m^2}{2}\right)\cos 2\omega_p t + \frac{m^2A^2}{8}\cos 2(\omega_p+\omega_m)t \\ & + \frac{m^2A^2}{8}\cos 2(\omega_p-\omega_m)t + \frac{m^2A^2}{2}\cos(2\omega_p+\omega_m)t + \frac{m^2A^2}{2}\cos(2\omega_p-\omega_m)t \\ & + mA^2\cos\omega_m t + \frac{m^2A^2}{2}\cos 2\omega_m t \end{aligned} \tag{5.3}$$

By an appropriate filtering that will be described in the next block, it is possible to suppress some of the components obtained in expression (5.3). Specifically, the direct current component and the components with frequency higher than ω_p will be filtered. With this, the previous expression could be reduced to

$$v_F(t) = mA^2\cos\omega_m t + \frac{m^2A^2}{2}\cos 2\omega_m t \tag{5.4}$$

This equation is formed by two terms. It can be observed that the demodulation process introduces, even in this case of a purely sinusoidal modulation, a component with a sideband frequency different from that of the modulation signal.

5.3.1.3 Block 3: Demodulator Filters and Weighting Filter

Block 3 is composed of a cascade of two filters: a demodulator filter and a weighting filter.

The demodulator filter aims to attenuate the d.c. component and the components with frequency higher than ω_p of the output of the square-law demodulator of block 2 (that would be the components of Equation (5.3), in the case of sinusoidal modulation). To this end, it consists of a first-order high-pass filter, with a 0.05 cut-off frequency for suppressing the d.c. component, and a low-pass filter intended to eliminate the components with a frequency greater than or equal to the fundamental frequency of the carrier voltage. This low-pass filter has a cut-off frequency of 35 Hz.[2] The final necessary attenuation for the ripple components with a frequency greater than or equal to the fundamental is obtained by the combined effect of this low-pass filter and the weighting filter that is described next.

The weighting filter aims to simulate the frequency-selective behavior exhibited by the human eye. Clearly, this behavior is influenced by the response of the lamp that is subjected to voltage fluctuations. Therefore, in the IEC flickermeter, a reference lamp is considered.

The weighting filter exhibits a band-pass response with a maximum gain for frequencies between 8 and 10 Hz that correspond to the maximum perceptibility of light intensity variations. The upper frequency limit is 35 Hz taking into account the upper limit of the flicker fusion boundary and the thermal inertia of lamps. Furthermore, the weighting filter considers also the low-pass behavior of the reference lamp which has a cut-off frequency of about 6 Hz.

Finally, taking into account all these aspects, the transfer function of the weighting filter for a reference incandescent lamp of 230 V, 60 W is of the following type:

$$F(s) = \frac{k\omega_1 s}{s^2 + 2\lambda s + \omega_1^2} \times \frac{1 + {}^{s}/_{\omega_2}}{\left(1 + {}^{s}/_{\omega_3}\right)\left(1 + {}^{s}/_{\omega_4}\right)} \tag{5.5}$$

where s is the Laplace complex variable and, for a 230 V, 60 W bulb, the parameters take the values listed below [6]:

$$\begin{aligned} k &= 1.748\ 02 \\ \lambda &= 2\pi \times 4.059\ 81 \\ \omega_1 &= 2\pi \times 9.154\ 94 \\ \omega_2 &= 2\pi \times 2.279\ 79 \\ \omega_3 &= 2\pi \times 1.225\ 35 \\ \omega_4 &= 2\pi \times 21.9 \end{aligned}$$

The following values have been proposed for a 120 V, 60 W bulb [7]:

$$\begin{aligned} k &= 1.6357 \\ \lambda &= 2\pi \times 4.167\ 375 \\ \omega_1 &= 2\pi \times 9.077\ 169 \\ \omega_2 &= 2\pi \times 2.939\ 902 \\ \omega_3 &= 2\pi \times 1.394\ 468 \\ \omega_4 &= 2\pi \times 17.315\ 12 \end{aligned}$$

The amplitude response of the filters of block 3 is shown in Figure 5.4 where the weighting filter represented corresponds to the 230 V lamp. The combined effect of these filters gives a strong attenuation at frequencies out of the band of 0.05–35 Hz.

---

[2] A sixth-order Butterworth filter is suggested [27]. The choice of the Butterworth filter is justified because this filter is the flattest on the pass band and shows no damping in the considered frequency range.

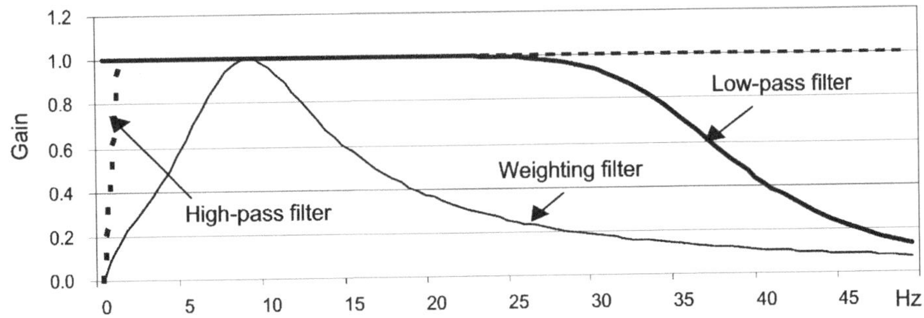

Figure 5.4 Response of the block 3 filters of the IEC flickermeter

5.3.1.4 Block 4: Non-Linear Variance Estimator

Considering Rashbass's findings about flicker perception [22], this block, called the non-linear variance estimator, consists of a squaring multiplier that simulates human non-linear visual perception, and a first-order low-pass filter with a time constant of 300 ms. This filter simulates the storage effect of the human brain.

It is very important to highlight that the output of block 4 of the flickermeter, after simulating the lamp–eye–brain chain, is an instantaneous signal proportionally related to the visual sensation of flicker. That was exactly the objective of the instrument: to provide a measurement method that could relate voltage fluctuations to units of visual sensation.

The absolute value of this signal must be converted to per units of perceptibility by scaling it down to a value of the perceptibility threshold. The perceptibility threshold level is given by a sinusoidal voltage fluctuation of 0.25 % amplitude and 8.8 Hz frequency, so that such a voltage fluctuation provides one unit of perceptibility as the output of block 4. Therefore, one unit of output corresponds to the visual perceptibility threshold of flicker occurrence. Higher output values mean that flicker is more than perceptible and can become annoying or intolerable.

5.3.1.5 Block 5: Statistical Evaluation

Considering the mechanism of vision and flicker annoyance, it seems clear that flicker evaluation must be done over a sufficiently representative period of time. In addition, it is necessary to take into account that flicker usually has a random nature and its instantaneous value can vary widely.

Therefore, to characterize the flicker level correctly it is necessary to determine for what percentage of a selected observation period a given flicker level has been exceeded.

The agreed evaluation method is based on calculating P_{st} and P_{lt}.

To calculate P_{st}, the cumulative probability function of the instantaneous flicker sensation (normalized output of block 4) over the 10-minute observation period must be evaluated. This analysis can be performed by block 5 on-line or off line on a recording of output 4. The sampled values of the flicker sensation must be distributed in a certain number of classes. Every time that a value occurs, the counter of the class corresponding to that value is incremented by one unit. In this way, the distribution function of the input values is obtained and the cumulative probability function used in block 5 can be obtained. The construction

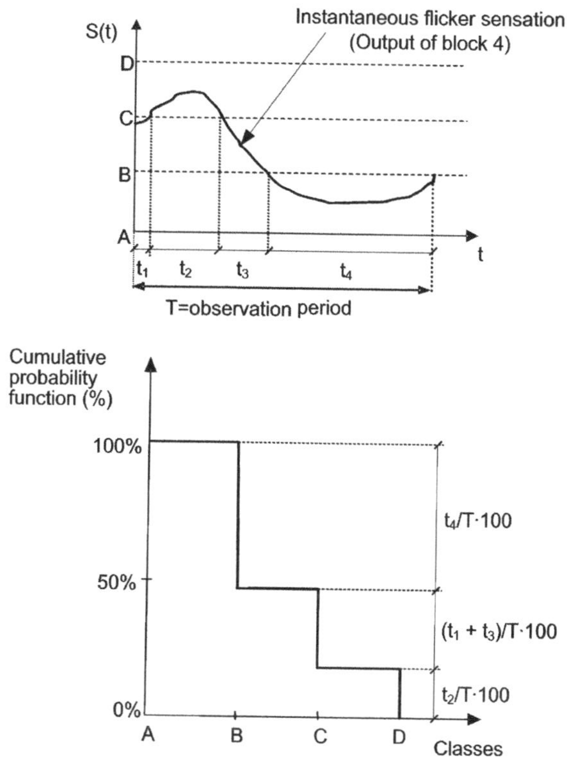

Figure 5.5 Cumulative probability function calculation

of this probability function is explained by means of Figure 5.5 where a simplified situation has been assumed and the values are distributed in only four classes. A more realistic and accurate calculation should consider a much larger number of classes.

After the cumulative probability function has been calculated, a number of points of this function are selected to calculate P_{st}. In accordance with IEC specifications [6], the following formula is used for such calculation:

$$P_{st} = \sqrt{0.0314P_{0.1} + 0.0525P_{1s} + 0.0657P_{3s} + 0.28P_{10s} + 0.08P_{50s}} \tag{5.6}$$

where the percentiles $P_{0.1}$, P_{1s}, P_{3s}, P_{10s}, P_{50s} are the flicker levels exceeded during 0.1, 1, 3, 10 and 50 % of the time of the observation period. The suffix s in the formula indicates that the *smoothed* values should be used; these are obtained using the following equations:

$$P_{1s} = (P_{0.7} + P_1 + P_{1.5})/3 \tag{5.7}$$

$$P_{3s} = (P_{2.2} + P_3 + P_4)/3 \tag{5.8}$$

$$P_{10s} = (P_6 + P_8 + P_{10} + P_{13} + P_{17})/5 \tag{5.9}$$

$$P_{50s} = (P_{30} + P_{50} + P_{80})/3 \tag{5.10}$$

It can be seen that the maximum flicker sensation level observed during the interval is not considered for the P_{st} calculation. This is due to the fact that a single peak level of very short duration cannot be representative of a flicker occurrence and could lead to overestimated values.

5.3.1.6 Performance Tests

IEC 61000-4-15 [6] defines two ways to check the output of flickermeters. Firstly, the overall response from the instrument input to the output of block 4 (called output 5 of the instrument) is given for a table of sinusoidal and rectangular voltage fluctuations at frequencies between 0.5 and 25 Hz. For all these input signals, the aforementioned response must be one unit.

Secondly, a performance test defines a set of rectangular voltage changes of different frequencies and depths for which the short-term flicker severity, P_{st}, must be $1.00 \pm 5\%$. For each frequency, the fluctuation amplitude must be increased and decreased and the resulting flicker severity must increase or decrease by the same factor.

The IEC specifications for blocks 1 to 4 are given for an analogue design. However, most flickermeters in the market nowadays are implemented in a digital or semi-digital way. Signal processing issues such as sampling rate, quantifying effects and windowing method play an important role in determining the performance of the meter, but are not covered in the IEC specification. Recently, certain investigations have shown that considerable differences in the readings might result from the different digital implementations [3]. So, further work must be done on this subject in order to characterize fully the response of the flickermeter [20].

5.3.2 Use of the Flickermeter

Flicker is a disturbance directly related to voltage variations and, therefore, flicker measurement must always take voltage as an input. When the measurements are aimed at assessing the emission level of an individual load connected to the network, this method can lead to errors, since the measurement of voltage includes the background disturbance that is not caused by the considered load. To avoid this influence, several approaches can be taken. The first one is based on measuring the flicker level at the point of common coupling (PCC) under two different conditions:

- with the fluctuating load of the consumer connected;
- with the fluctuating load disconnected.

The flicker level measured in this second situation is not caused by the load connection and, therefore, it can be assumed as the background flicker level. Therefore, this value should be subtracted from the first one which was measured when the load was connected. This subtraction must be done on the basis of the appropriate flicker aggregation law that will be described below.

This method can be applied with acceptable accuracy when the existing background P_{st} level at the PCC is low.[3] However, it is important to notice that measurements are not performed simultaneously and, therefore, changes in the network topology or in the operating conditions can influence the results. In these circumstances, a more elaborate approach must be applied.

An alternative method is to record the current absorbed by the load, and calculate the voltage by injecting this current into a model of the supply system (which must include the system impedance) by means of a simulation.

IEC 61000-3-3 [11], IEC 61000-3-5 [12] and IEC 61000-3-11 [13] define testing methods based on the utilization of a reference impedance for assessing the individual emission level of equipment connected to LV systems.

According to IEC 61000-4-30 [14], flicker measurements intended to verify the flicker limits' compliance should be carried out with a minimum duration of one week. From the P_{st} values measured during the observation week, the cumulative probability function of P_{st} and P_{lt} values must be obtained. The 99 % percentiles of P_{st} and P_{lt}, called $P_{st99\%}$ and $P_{lt99\%}$, respectively, must be derived from this curve.

In a supply system, the percentiles $P_{st99\%}$ and $P_{lt99\%}$ should not exceed the planning levels specified by the supply utility for all voltage levels of the system. These planning levels, which can be considered as internal quality objectives of the utility, may take different values from case to case. For this reason, only indicative values can be given. Table 5.1 shows the indicative planning levels provided by IEC 61000-3-7 [9]. Planning levels must be always equal to or less than the compatibility levels which are shown in Table 5.2 for flicker in LV and MV systems [9].

Table 5.1 Indicative planning levels for P_{st} and P_{lt} in MV, HV and EHV systems

	Planning levels	
	MV	HV and EHV
P_{st}	0.9	0.8
P_{lt}	0.7	0.6

Table 5.2 Compatibility levels for P_{st} and P_{lt} in LV and MV systems

	Compatibility levels
P_{st}	1.0
P_{lt}	0.8

[3] IEC 61000-3-7 considers a P_{st} equal to or less than 0.5 as an indicative value to apply this method with reasonable accuracy.

The limits of flicker emission for an individual fluctuating load must be determined in order to guarantee that the total flicker injection from all the consumers does not result in flicker levels exceeding the planning levels. IEC 61000-3-7 [9] provides the basis for determining the requirements for connecting large fluctuating loads to MV and HV levels. IEC 61000-3-3 [11], IEC 61000-3-5 [12] and IEC 61000-3-11 [13] provide emission limits for equipment connected to LV systems.

In polyphase supply systems, attention must be paid to the method of connection of the flickermeter. If the system is perfectly balanced, flicker measurements will be the same in the three phases, and line-to-line or line-to-neutral measurements will provide equal results. However, if the flicker source is unbalanced (for instance, a single-phase welding machine) it is necessary to measure the flicker level in the three phases and the worst case must be considered.

In general, if the flicker measurement is intended to assess whether the flicker severity is acceptable or not at an LV site, the flicker measurement should be performed to a signal representative of the voltage (line-to-line or line-to-neutral) that is finally feeding the lighting systems.

5.3.3 Simplified Methods for P_{st} Assessment

Ideally, the most direct way to assess the flicker level is the use of a flickermeter. However, this is a valid method only when the analyzed flicker severity is caused by loads that are already in operation and connected to the supply system. On the contrary, the direct measurement cannot be applied in the design or planning stage of an installation. In these cases, some predictive methods are needed in order to ensure that limits for flicker severity are not exceeded and to determine the possible mitigation strategies in case they are necessary.

The present section provides an overview of some practical approaches that can be taken to predict flicker. These approaches, which are based on indications provided in [9],[27],[11], generally consist of three steps:

1. Calculation of the maximum relative voltage change, d, produced by the fluctuating load.
2. Calculation of the flicker severity caused by that voltage change.
3. Addition of flicker originating from multiple sources.

Next, these three steps are explained in more detail.

5.3.3.1 Assessment of the Relative Voltage Change

A basic prerequisite for the calculation of flicker is the determination of the relative voltage change produced by the load expressed as a percent ratio of the nominal voltage. Where it cannot be measured, it must be calculated from the supply and load data.

For balanced three-phase loads, the relative voltage change d % of the phase-to-neutral voltages and of the phase-to-phase voltages is the same and can be calculated approximately as [9]

$$d = \frac{\Delta U}{U_N} 100\,\% \cong \frac{\Delta S_i}{S_{SC}} 100\,\% \tag{5.11}$$

where ΔS_i is the apparent power load change and S_{sc} is the short-circuit power of the feeding network at the point of common coupling.

If the active as well as the reactive part of the load change are known, the relative voltage change can be calculated more accurately by using the resistive and inductive part of the network impedance:

$$d = \frac{R_s \cdot \Delta P_i + X_s \cdot \Delta Q_i}{U_N^2} 100\,\% \tag{5.12}$$

where R_s is the resistive part of the network impedance; X_s is the reactance of the network impedance; U_N is the nominal voltage; ΔP_i and ΔQ_i are, respectively, the active and reactive power changes of the load.

Equation (5.12) can be particularized for the specific characteristics of some types of loads. For instance, mainly resistive loads (like electric boilers, resistance furnaces, etc.) cause a voltage drop given by

$$d = \frac{R_s \cdot \Delta P_i}{U_N^2} 100\,\% \tag{5.13}$$

Analogously, the voltage drop caused by reactive elements (for instance, connection and disconnection of capacitors) can be calculated by

$$d = \frac{X_s \cdot \Delta Q_i}{U_N^2} 100\,\% \tag{5.14}$$

When the voltages changes are caused by a two-phase load (e.g. a welding machine), the drop in the phase-to-neutral voltage can be calculated by conveniently correcting the previous expressions.

5.3.3.2 *Calculation of Flicker Severity*

Once the relative voltage change caused by the load is known, P_{st} can be estimated by means of two methods, as follows.

5.3.3.2.1 Unity flicker severity curve method

This method makes use of the fact that short-term flicker severity is a linear parameter with respect to the magnitude of the voltage change that causes it. To apply this method, the percent relative voltage change d (that can be estimated as described previously) and the repetition rate r of this voltage change must be known. If the voltage fluctuations are described by a frequency, this means that the repetition rate of the voltage fluctuations, r, is twice the value, i.e. 1 Hz corresponds to two changes per second. Substituting r, expressed as the number of voltage changes per minute, into the severity curve (Figure 5.6) for rectangular steps provides on the ordinates the voltage change $d_o = \Delta U/U$ (%) which produces $P_{sto} = 1$. If the voltage step change calculated for the load has a value d, the corresponding flicker severity is

$$P_{st} = P_{st\,o} \frac{d}{d_o} = \frac{d}{d_o} \tag{5.15}$$

This method can be applied for rectangular voltage changes with the same magnitude and regularly distributed in time [11].

5.3.3.2.2 Analytical method

This method is presented in IEC 61000-3-3 [11]. It is based on calculating the so-called *flicker time*, t_f, which is a magnitude with a time dimension that describes the flicker impression caused by a single voltage variation. This value is calculated by means of the following expression:

$$t_f = 2.3\,(d \cdot F)^{3.2} \tag{5.16}$$

where d is the relative voltage change expressed in percent ratio with respect to the nominal voltage and F is an equivalence factor, depending on the shape of the voltage fluctuation. This factor is one unit for rectangular voltage changes and its value for other shapes (sinusoidal periodical fluctuations, ramps, double steps, rectangular and triangular pulses) can be derived from the curves provided in IEC 61000-3-7 [9].

The factor 2.3 is included in order to achieve compliance with the flicker curve (Figure 5.6).

The addition of all the flicker times, $\sum t_f$, inside a total time interval, T_p, is the base of the P_{st} assessment. P_{st} is therefore derived from considering independently the disturbance created by each voltage variation and then adding their effects by an appropriate law of flicker addition. The proposed expression for the calculation of P_{st} is

$$P_{st} = \left(\frac{\sum t_f}{T_p}\right)^{1/3.2} \tag{5.17}$$

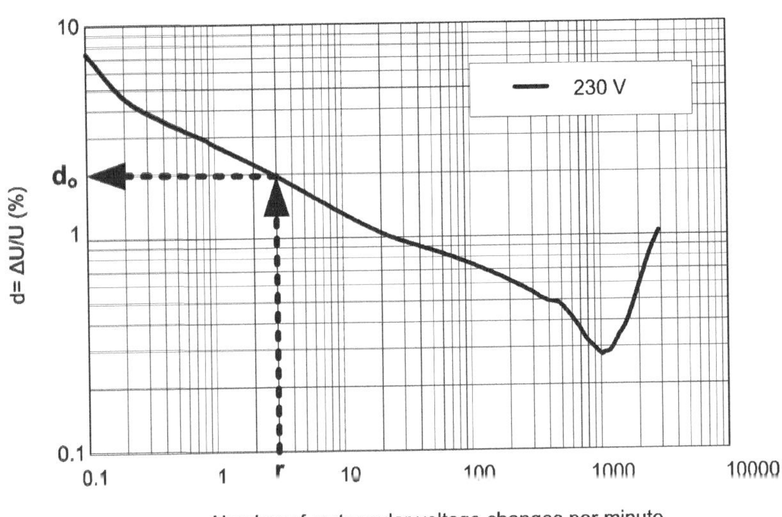

Figure 5.6 Utilization on the curve $P_{st} = 1$ for rectangular voltage changes for 60 W, 230 V incandescent lamps

Time T_p must be expressed in seconds and, for the calculation of P_{st}, this observation period should be 10 minutes.

The exponent 3.2 of (5.17) is derived from one of the proposed coefficients for flicker addition that will be explained in the following section.

This analytical method is recommended for the above-mentioned shapes of voltage variations and only when the time elapsed between voltages changes is over 1 s (i.e. the time between the end of a voltage variation and the beginning of the following one is over 1 s). Otherwise, there is a great possibility of coincident voltage changes and the flicker summation law involved in (5.17) will be incorrect.

5.3.3.3 Addition of Flicker Originating from Multiple Sources

The method described earlier estimates the flicker severity caused by a single item of equipment. IEC 61000-3-7 [9] provides rules to consider the summation effect of P_{st} values caused by multiple disturbing sources. The P_{st} originated by various loads operating simultaneously can be obtained from the following expression:

$$P_{st} = \sqrt[m]{\sum_i P_{st_i}^m} \tag{5.18}$$

where the P_{sti} are the individual levels of flicker severity emitted by each of the disturbing loads.

The value of the coefficient m depends upon the characteristics of the sources of fluctuation. Generally m is classed into the following five categories:

- $m = 4$: This value is used uniquely for the summation of voltage fluctuations produced by arc furnaces which are operated specifically to avoid coincident melts.
- $m = 3$: This is the value used in most of the types of voltage variations where the risk of coincident voltage changes is small. Most of the studies combining the effect of unrelated disturbing sources belong to this category. This is also the default value that must be used when there is any doubt about the risk of simultaneous occurrence of voltage variations.
- $m = 3.2$: This value corresponds to disturbances matching the linear part of the $P_{st} = 1$ curve (Figure 5.6). It is applied in expression (5.17).
- $m = 2$: This value is used when there is coincident stochastic noise (e.g. when several arc furnaces perform cycles in a synchronized manner or in the case of continuous operation of several wind turbines connected to the same PCC [10])
- $m = 1$: The resultant value of P_{st} will be the arithmetic sum of the individual flicker severity values produced by each disturbing source. This value must be used when there is a high probability of simultaneous occurrences of voltage changes.

5.4 SOURCES

In general, it can be said that the origin of flicker is in the devices connected to the electrical system (mainly loads) that produce rapid voltage fluctuations. Usually, these voltage fluctuations are caused by variations in the consumed power and, especially, by reactive power fluctuations. Variations of power can be due either to switching on or off large-capacity loads, or to loads with an intrinsically fluctuating behavior (i.e. arc furnace loads). Voltage reductions greater than 10 % are considered voltage dips and, therefore, they are not included in the common definition of voltage fluctuations of interest in this chapter.

5.4.1 Industrial Loads

The main source of voltage variations is large industrial loads. These loads are usually connected in HV or MV networks, so, in case they provoke noticeable flicker, they can affect a large number of consumers connected to the same network.

5.4.1.1 Arc Furnaces

Arc furnaces are a very large industrial load with ratings that can range from a few MVA to more than 100 MVA. Possibly, arc furnaces are also one of the major sources of perturbations connected to a HV network from the point of view of flicker. Arc furnaces are used in the steel industry for the production of steel from recycled scrap. The most common type of arc furnace is the a.c. arc furnace. In this kind of arc furnace, three electric arcs are created between the electrodes and the scrap. The heat dissipated in the process by the Joule effect melts the metal. The beginning of the heating cycle consists of boring the electrodes into the scrap. The arc established at this initial stage is very unstable, the length of the arc is very variable and the current can oscillate from short-circuit condition to zero (open circuit). These drastic power variations can cause very large voltage fluctuations at the PCC containing frequencies in the bandwidth of interest for flicker [4]. Figure 5.7 and Figure 5.8 show the apparent power demanded by an a.c. arc furnace during the beginning of the heating process and the r.m.s. current absorbed [18]. Very severe variations can be observed in these figures.

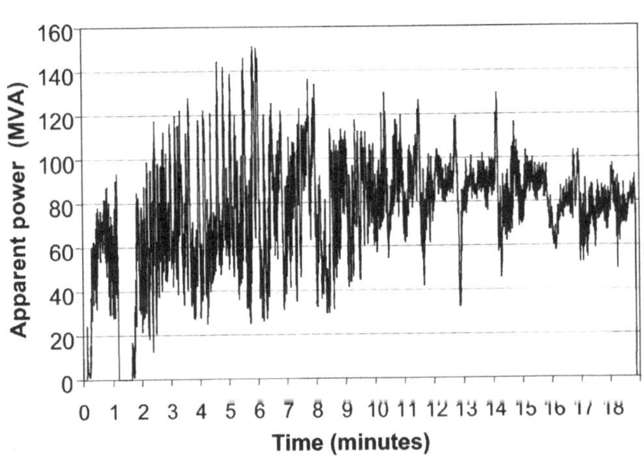

Figure 5.7 Apparent power demanded by an 80 MVA a.c. arc furnace in the first minutes of the melting process [18]

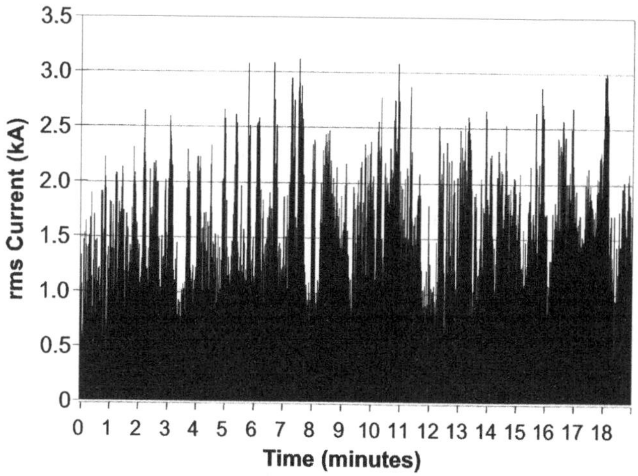

Figure 5.8 Current absorbed by an 80 MVA a.c. arc furnace in the first minutes of the melting process

Recently, the spectacular progress of power electronics has contributed to an increasing popularity of d.c. arc furnaces. This kind of furnace is formed by an a.c./d.c. rectifier. A single electrode is placed on the d.c. side acting as a cathode. The current returns through a conductive bottom lining or conductive pins in the base of the vessel that form the anode. One of the advantages of d.c. arc furnaces is that they have lower electrode consumption per tonne of steel produced, since only one electrode is used. In addition, it is generally accepted that voltage fluctuations caused by d.c. arc furnaces are significantly lower than that of an equivalent a.c. arc furnace.

Very often the connection of an arc furnace requires an exhaustive study that includes complex simulation methods together with knowledge of the background flicker level in order to predict the likelihood of potential problems. However, a simplified approach can be very helpful in providing an initial estimate of the flicker emission. This approach is based on considering the ratio of the arc furnace power to the available short-circuit power. As a general rule of thumb, some authors have proposed that a ratio of 80 or larger can be used as a guideline to avoid voltage flicker problems [19].

UIE proposed the following approximate formula to estimate P_{st} due to arc furnaces at the point of common coupling [27]:

$$P_{st} = K_{st} \frac{S_{sc_{four}}}{S_{sc_{net}}} \tag{5.19}$$

where:

- K_{st} is the coefficient of disturbance produced. It is constant for a certain furnace. If it is not known, then a furnace of similar size and technology has to be found to be able to get the value. The array proposed by UIE is $85 \geq K_{st} \geq 48$.
- $S_{sc_{net}}$ is the short-circuit power for the feeding network at the point of common coupling. The minimum value which is likely to occur should be chosen.

$S_{sc_{four}}$ is the short-circuit power of the furnace. In general it is higher than the furnace transformer rated power. It is given by the current that circulates when the electrodes are short-circuited, so it can be calculated as the short-circuit current multiplied by the rated open-circuit voltage, for the maximum tap employed on the furnace transformer. This value is generally known by the client for furnaces already in service. As a first approximation, the current can be evaluated as twice the furnace's transformer rating (in the absence of the series inductance for a.c. furnaces).

The inconvenience of expression (5.19) is the dependence on factor K_{st} whose range of values is too large, leading to flicker limits going from single to almost double the value.

5.4.1.2 Welding Machines

Welding machines are based on the Joule effect; that is, on using the heat dissipated by a current circulation in order to assemble two metallic pieces. This requires the passage of a very high current value and, for this reason, welders usually employ a step-down transformer in order to reduce the applied voltage to very low values and to drive the necessary high-intensity current. Welders can be directly connected in LV distribution networks or through a supply transformer that can feed several welders from an HV network.

In welding operation, short intervals of welding (with high current flow) are followed by others of no-load operation. In each welding operation, two voltage steps are produced, one at the time of establishing current and another at the instant of disconnection. These voltage variations can cause problems in the nearby loads producing voltage magnitudes outside the voltage tolerances and can even produce flicker.

To assess P_{st} produced by welders, the expected number of voltage changes produced per minute must be known. The amplitude of these voltage changes can be estimated by means of the expression (5.12) and then the method described earlier for flicker prediction can be applied.

For single-phase and two-phase welders, the transformer connection feeding the welding machine and the type of connection of the welding machine (line-to-line or phase-to-neutral) have to be taken into account.

5.4.1.3 Electric Boilers

Electric boilers are used in industrial applications for generating steam. They can work by heating a surface that transmits heat to evaporate water or, in other cases, the water can be used as a conductor being heated by the Joule effect. In both cases, the required control of the temperature together with the varying need of steam lead to periods of high power demand followed by periods without energy consumption. This can cause voltage steps whose magnitude, considering the boiler as a purely resistive load, can be estimated by means of expression (5.13). The predictive method explained previously in this chapter can help to estimate the expected level of flicker severity caused by these types of loads.

5.4.1.4 Capacitor Banks

Capacitors banks are connected in parallel with inductive loads in order to compensate for the power factor of the installation. Switching on and off a capacitor bank produces a voltage step proportional to the reactive power capacity of the capacitor bank and inversely proportional to the short-circuit capacity of the point of connection. Therefore, the magnitude of the voltage step can be estimated by means of the expression (5.14). By applying this expression and knowing the expected number of switching operations per minute, the flicker severity level can be estimated.

In general, it is accepted that the most common power quality problems related to capacitor banks are the appearance of resonance phenomena between the capacitor and inductive elements of the system and the possible overvoltage transients due to the capacitor's connection. It is only in very rare cases that voltage variations caused by capacitors switching can lead to a noticeable flicker level.

5.4.2 Electrical Appliances Supplied from LV Networks

Electrical appliances supplied from LV networks like drives for lifts, pumps, fans, electric boilers, refrigerator chambers, electric cookers or air-conditioning equipment, having a considerable high power and being switched on an off cyclically or irregularly are potential flicker sources.

Most of these disturbances are of the *motor-start* type. The voltage drop can be calculated by considering the ratio between the current absorbed by the load in the starting process and the rated current.

Other potential flicker generators in LV systems are X-ray equipment and large copying machines. These devices are within the scope of IEC 61000-3-3 [11] if they absorb a current lower than 16 A or IEC 61000-3-7 [9] if their rated current is between 16 and 75 A.

5.4.3 Wind Turbine Generation Systems

In recent years, power systems have experienced a significant increase in wind energy penetration. Presently, the trend is for planning large wind farms with a capacity of hundreds of MVA. This large-scale utilization of wind energy has caused an increasing concern about its influence on the voltage quality of the power system.

Wind turbines generate fluctuating power during continuous operation that can lead to voltage fluctuations at the PCC. In this regard, variable speed turbines have some advantages concerning flicker emission. In addition to continuous operation, switching operations of wind turbines are also a potential cause of voltage variations.

IEC 61400-21 [10] provides procedures for assessing the expected flicker emission levels caused by wind farms connected to the grid. This publication proposes a flicker coefficient for continuous operation of wind turbines, which is used to provide a flicker estimation under these circumstances, together with a flicker step coefficient which provides an estimate of flicker due to start-up and shutdown operations of the turbines. These procedures are designed to be as non-site specific as possible, so that the measurements taken in a turbine for the calculation of the coefficients at a site can be used to predict flicker at any other location.

5.5 EFFECTS

Most of the equipment is designed to be insensitive to voltage fluctuations within the statutory limits. So, the most important effect of voltage fluctuations is flicker produced by the variations in the fluctuation of illumination of light sources.

The physiological effects of flicker on individuals have already been discussed in Section 5.1.2, so the purpose of this section is to provide a brief overview of the response to voltage fluctuations of different types of lighting equipment.

Although a reference lamp is used for standardization purposes in the IEC flickermeter, different types of lamps present a different behavior when subjected to voltage changes. In some situations, replacement of the existing lighting system can be an efficient and economical alternative to attenuate the flicker level.

Tungsten filaments of incandescent lamps of different nominal wattage behave differently when subjected to the same voltage fluctuations because of differences in construction. The added thickness of a filament increases thermal inertia, so an incandescent lamp of higher wattage is less sensible to voltage fluctuations. For the same reason, comparing two lamps of the same wattage and different nominal voltages, the lamp with the higher current, and so with a lower nominal voltage, is more resistant to voltage fluctuations.

In order to characterize the behavior of a lamp with respect to voltage variations, a gain factor is defined that represents the amplification of the light output of the lamp when a voltage fluctuation occurs. The gain factor is obtained by placing the lamp in a test chamber and measuring relative changes in light level during voltage fluctuations. A typical gain factor for a 230 V, 60 W incandescent lamp is around 3 for the lower-fluctuation frequencies. This means that a 1 % voltage variation produces a 3 % light output change. The gain drops at higher frequencies because of the thermal inertia of the filament.

Compared to incandescent lamps, fluorescent lamps have a lower gain and, therefore, cause less flicker. On the contrary, the use of lamp dimmers substantially increases lamp susceptibility to voltage variations, so electronic dimmers increase the risk of flicker.

5.6 MITIGATION STRATEGIES

Flicker is caused by voltage fluctuations. Therefore, any strategy oriented to mitigate flicker must be based on a reduction of voltage fluctuations. Two kinds of approaches can be applied for this purpose:

1. Decreasing power variations (mainly reactive power variations) of the loads.
2. Increasing the short-circuit power level.

Flicker compensation devices are usually based on the first approach. The second tactic, in many cases, can be applied only at the design stage.

Very often, the solution can be provided not by a single method, but by a combination of several of them.

5.6.1 Devices Based on Decreasing Reactive Power Variations

5.6.1.1 Static Var Compensator

Static var compensators (SVCs) have been applied for flicker compensation since the end of the 1970s. At present, they constitute one of the most used methods for flicker compensation in arc furnace installations.

The SVC is a shunt device of the flexible a.c. transmission system (FACTS) family using power electronics to control reactive power flow. The term static is used to differentiate SVCs from rotating var compensators (synchronous motors or generators). The SVC regulates the voltage at its terminals by controlling the amount of reactive power injected into or absorbed from the power system. When the system voltage is low, the SVC generates reactive power (capacitive behavior). In doing so, the demand of reactive power of the load is provided by the SVC and the feeding lines are relieved from transporting it. As a result, the voltage drop decreases and the voltage at the load terminals increases. Similarly, when the system voltage is high, the SVC absorbs reactive power (inductive behavior).

SVCs usually consist of two parallel branches connected on the secondary side of a coupling transformer (Figure 5.9b). One of these branches consists of a thyristor-controlled

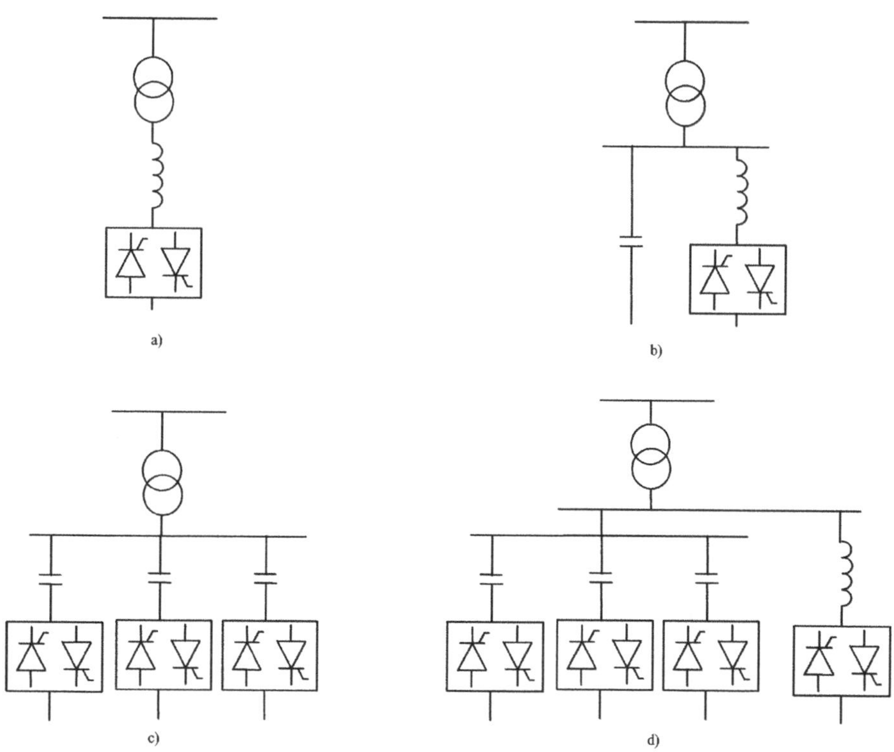

Figure 5.9 SVC configurations: (a) thyristor-controlled reactor; (b) thyristor-controlled reactor/fixed capacitor; (c) thyristor-switched capacitor; (d) thyristor-controlled reactor/thyristor-switched capacitor

reactor (TCR) (Figure 5.9a) and, in the other, a capacitor bank or a filter is placed. The variation of reactive power is achieved by controlling the thyristor's firing instants and, consequently, the current that flows by the reactance.

Another possibility is the use of a thyristor-switched capacitor (TSC) (Figure 5.9c). In this case, each capacitor bank is switched on and off by means of thyristors. With this, a discrete variation of the reactive power can be achieved, but never a continuous variation as in the TCR.

Finally, some structures combining the TCR and TSC can be also used (Figure 5.9d).

The control of the SVC is based on measuring the reactive component of the load current and deciding the firing angle in such a way that the SVC injects or absorbs the amount of reactive power required for compensation. This method of control is explained by means of Figure 5.10. At the instant of voltage zero crossing, the load current $i(t)$ is measured. At this instant, the active component of the load current is zero since the active current component is in phase with the voltage wave. Therefore, the current measured at the instant of voltage zero crossing is equal to the reactive component of the load current. In this way, the reactive current measured in a certain half cycle is used to determine the firing instant of the thyristors in the following half cycle. In doing so, the current through

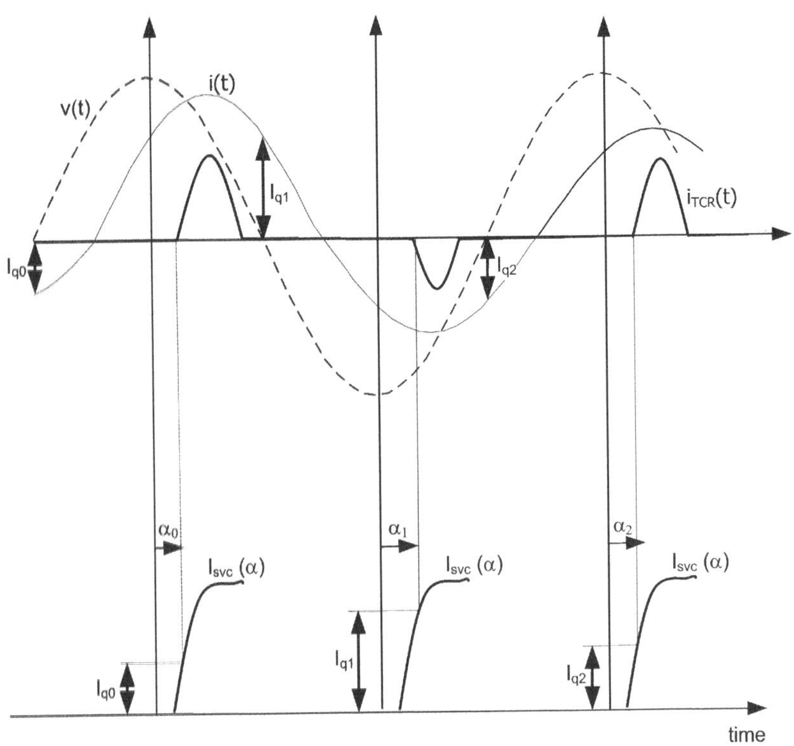

Figure 5.10 Reactive control in a TCR

the reactance (in the case of a TCR) is regulated and the total reactive current injected by the converter can also be controlled.

One of the main limitations of this kind of control to compensate rapid fluctuations is the inherent delay of its operation mode. There is a time interval between the instant of measuring the reactive component (in one half cycle) and the firing instant (in the following one). Once triggered, thyristor valves must remain conducting until natural commutation occurs [2]. On the other hand, it is important to take into account that the SVC can accurately control the current fundamental component, but it produces harmonic components. For this reason, generally, it is necessary to install filters in order to reduce the harmonic content introduced in the network by the compensation system.

As a first approximation, SVCs are considered cost effective for large installations above around 50 MVA [1]. The flicker attenuation factor that can usually be achieved by means of these compensators is between 1.5 and 2.

5.6.1.2 Voltage Source Converter

The rapid development of the technology of semiconductors in recent years has made possible their use in new applications.

The voltage source converter (VSC) is a power electronics device based on an inverter that obtains an alternating voltage at any frequency, magnitude and phase angle, from a direct voltage bus. The most common scheme used to switch on and off the semiconductors is a pulse-width-modulated pattern [15]. By regulating the output voltage of the inverter, the reactive power injected into or absorbed from the system can be controlled and, as in the SVC, in such a way that the reactive current variations of the system load compensator are kept as small as possible, mitigating voltage fluctuations.

One of the advantages of these devices is that, since the commutation frequency is not defined by the system frequency, as in the SVC, much higher commutation frequencies are used. With this, forced commutation compensators present a much lower inherent delay and, therefore, power variations can be more accurately compensated [2]. Another benefit in the use of this technology is that, besides power oscillations, it is possible to compensate also for the harmonics introduced by the load.

However, other important disadvantages must be also taken into account. Devices based on forced commutation are more complicated than SVCs and require semiconductors that are capable of supporting at the same time high current values and high commutation frequency. This fact considerably increases the cost of this equipment with respect to an SVC of the same power.

A family of controllers for transmission systems has been developed based on the VSC concept. These include the STATCOM (shunt-connected), SSSC (series-connected), UPFC (shunt and series units with common d.c. connection) and back-to-back STATCOM (two shunt units with common d.c. connection) [24]. Among these, the characteristics of the STATCOM, which is essentially a controlled a.c. voltage source connected to the power line through a suitable tie reactance, make it one of the most used for flicker mitigation [23].

5.6.1.3 Arc Furnaces: Insertion of a Series Inductance

In the case of arc furnaces, a quite common practice is the insertion of an inductance in series with the arc furnace. This method contributes reducing flicker since it increases the furnace's total impedance and, consequently, the relative variations of the furnace's impedance are lower [21].

Another beneficial effect of insertion of the inductance is that the phase shift between the current and voltage increases, allowing a faster reignition of the electric arc. However, this method also has important drawbacks since it reduces power supply capacity and, therefore, it decreases the arc furnace productivity.

5.6.1.4 Arc Furnaces: Insertion of a Saturable Inductance [1]

The earliest industrial flicker compensators installed in the late 1950s were based on the saturable reactor principle. This kind of reactor has a magnetizing characteristic that is deliberately saturated, often by means of a direct current flowing in a control winding. Once saturated, the inductance of the saturable reactor drops dramatically, so that great changes in the current only cause small variations in the magnitude of the voltage.

Saturable reactors up to 150 MVA have been made. However, nowadays they have been displaced by power electronics compensators which are cheaper and more flexible in use.

5.6.2 Methods Oriented to Increase the Short-Circuit Power

If the short-circuit power of the network is increased, the voltage variations are divided by the ratio

$$\frac{S_{sc\ after}}{S_{sc\ before}} \tag{5.20}$$

where $S_{sc\ after}$ indicates the short-circuit power level achieved after the intervention and $S_{sc\ before}$ indicates the original value of the short-circuit power.

Since flicker severity is a linear quantity with respect to the magnitude of the step change which caused it, the expected reduction in P_{st} will be of the same order than the reduction in voltage variations.

Often, the solutions oriented to increase short-circuit power are very expensive and not always applicable. In addition, sometimes these methods can force the re-evaluation of the whole electrical network, resulting in additional cost.

Measures that can be taken in order to increase the short-circuit power are:

- a supplementary line which can be constructed to reinforce the distribution line;
- connection to a network at a higher voltage level;
- parallel connection of the mains transformers. This solution increases the short-circuit level on the secondary side. However, the drawback of this method is that the disturbance is transmitted to the complete busbars.

5.6.3 Other Solutions for Flicker Mitigation

Finally, in certain cases, some other strategies can be helpful to reduce flicker levels, as follows:

- Separate, by isolating transformers, the supply terminal that feeds the fluctuating loads from the rest of the network, especially from the lighting power supply.
- In the case where fluctuating loads are single phase, connect the loads to a phase and feed the lighting systems from the non-disturbed phases.[4] This solution is applicable for some LV loads (such as large copying machines) or for MV loads, such as welding machines.
- Sometimes, it is possible to operate the perturbing loads at a time when they do not disturb the people nearby. In this case, flicker will not be avoided but it will be produced when people cannot experience it (e.g. operating the load by night if nobody works at this time).
- Sometimes, changes in the operating practice or equipment design can help to reduce voltage fluctuations. For instance, in the case of arc furnaces, restricting furnace transformer taps during the initial meltdown period (when flicker is more noticeable) reduces the operating voltage and can significantly decrease P_{st} during these intervals [1].
- In cases where the voltage variations are of the motor-start type, all the measures oriented to correct the inrush current can also be helpful in reducing voltage variations causing flicker
- An extreme solution consists of supplying the fluctuating loads from a decoupled source of the utility network (for instance, a diesel–electric group).

For a case study see web address

BIBLIOGRAPHY

[1] Ashmole P., Quality of supply – voltage fluctuations. Part 2. *Power Engineering Journal*, April, pp. 108–117, 2001.

[2] Garcia-Cerrada A., Garcia-Gonzalez P., Collantes R., Gomez T., Anzola J, Comparison of thyristor-controlled reactors and voltage-source inverters for compensation of flicker caused by arc furnaces. *IEEE Transactions on Power Delivery*, vol. 15, no. 4, pp. 1225–1231, 2000.

[3] Hanzelka Z. *et al.*, Comparative tests of flicker meters. Proceedings of CIRED. 17th International Conference on Electricity Distribution, Barcelona Spain, 12–15 May 2003. Technical Reports, Session 2, Paper no. 25, pp. 1–5.

[4] Hernandez A., Mayordomo J. G., Asensi R., Beites L. F., A new frequency domain approach for flicker evaluation of arc furnaces. *IEEE Transactions on Power Delivery*, vol. 18, no. 2, pp. 631–638, 2003.

[5] IEC 868, *Flickermeter. Functional and design specifications*, 1986.

[6] IEC 61000-4-15 (1997-11), *Electromagnetic compatibility (EMC) – Part 4: Testing and measurement techniques – Section 15: Flickermeter – Functional and design specifications.*

[7] IEC 61000-4-15-am1 (2003-01), *Amendment 1 – Electromagnetic compatibility (EMC) – Part 4: Testing and measurement techniques – Section 15: Flickermeter —- Functional and design specifications.*

[4] In this case, it is necessary to be careful with imbalance.

[8] IEC 60050 (161), *International Electrotechnical Vocabulary (IEV)* – Chapter: Electromagnetic compatibility.

[9] IEC 61000-3-7 (1996-11), *Electromagnetic Compatibility (EMC) – Part 3: Limits – Section 7: Assessment of emission limits for fluctuating loads in MV and HV power systems.*

[10] IEC 61400-21 (2001-12), *Part 21: Measurement and assessment of power quality characteristics of grid connected wind turbines.*

[11] IEC 61000-3-3 (1994-12), *Electromagnetic compatibility* (EMC) – *Part 3-3: Limits – Limitation of voltage changes, voltage fluctuations and flicker in public low-voltage supply systems, for equipment with rated current* $\leq$ *16 A per phase and not subject to conditional connection.*

[12] IEC 61000-3-5 (1994-12), *Electromagnetic compatibility (EMC) – Part 3: Limits – Section 5: Limitation of voltage fluctuations and flicker in low-voltage power supply systems for equipment with rated current greater than 16 A.*

[13] IEC 61000-3-11 Ed. 1 (2000-08), *Electromagnetic compatibility (EMC) – Part 3: Limits – Section 11: Limitation of voltage changes, voltage fluctuations and flicker in low-voltage power supply systems. Equipment with rated current* $\leq$ *75 A and subject to conditional connection.*

[14] IEC 61000-4-30 Ed. 1 (2003-02), *Electromagnetic compatibility (EMC) – Part 4-30: Testing and measurement techniques – Power quality measurement methods.*

[15] Larsson T. and Ekstrom Å., A PWM-operated voltage source converter for flicker mitigation. European Power Electronics Conference '97, Vol. 3, pp. 1016–1020, September 1997.

[16] De Lange H., Experiments on flicker and some calculations on an electric analogue of the foveal system. *Physica*, vol. 18, pp. 935–950, 1952.

[17] De Lange H., Eye's response at flicker fusion to square-wave modulation of a test field surrounded by a large steady field of equal mean luminance. *Journal of the Optical Society of America*, vol. 51, pp. 415–421, 1961.

[18] Mayordomo J. G., Prieto E., Hernández A., Beites L. F., Arc furnace characterization from an off-line analysis of measurements. Proceedings. Ninth International Conference on Harmonics and Quality of Power, 1–4 October 2000, Vol. 3, pp. 1073–1078, 2000.

[19] Mendis S. R., Bishop M. T., Witte J. F., Investigations of voltage flicker in electric arc furnace power systems. Industry Applications Society Annual Meeting, Denver, October, Vol. 3, pp. 2317–2325, 1994.

[20] Mombauer W., Calculating a new reference point for the IEC-Flickermeter. *European Transactions on Electrical Power*, vol. 8, no. 6, pp. 429–436, 1998.

[21] Montanari G. C., Loggini M., Pitti L., Tironi E., Zarinelli D., The effects of series inductors for flicker measurement in electric power systems supplying arc furnaces. Industry Applications Society Annual Meeting, 2–8 October 1993, Conference Record, Vol. 2, pp. 1496–1503, 1993.

[22] Rashbass C., The visibility of transient changes of luminance. *Journal of Physiology*, vol. 210, pp. 165–186, 1970.

[23] Reed G. F., Greaf J. E., Matsumoto T., Yonehata Y., Takeda M., Aritsuka T., Hamasaki Y., Ojima F., Sidell A. P., Chervus R. E., Nebecker C. K., Application of a 5 MVA, 4.16 kV D-STATCOM system for voltage flicker compensation at Seattle Iron and Metals. IEEE Power Engineering Society Summer Meeting, July, Vol. 3, pp. 1605–1611, 2000.

[24] Schauder C., STATCOM for compensation of large electric arc furnace installations. IEEE Power Engineering Society Summer Meeting, July, Vol. 2, pp. 1109–1112, 1999.

[25] Simons G., Das Flackern des Lichtes in elektrischen Beleuchtunganlages. *ETZ*, no. 37, pp. 453–455, 1917.

[26] UIE (Union Internationale d'Electrothermie), *Flicker Measurement and Evaluation*, Technical Report, Second Edition, 1991.

[27] UIE (Union Internationale d'Electrothermie), *Connection of Fluctuating Loads*, Technical Report, 1988.

6

Voltage and Current Unbalance

Irena Wasiak

Synchronous generators are the sources of a three-phase voltage in a power system. The voltages at generator phase terminals are equal in magnitude and displaced to each other by 120°. This vector system is called symmetrical [27]. If the components of the power system are linear and symmetrical, and individual phases of the system are equally loaded, the voltages measured at load buses remain symmetrical. Short-time unbalance can appear only under disturbance conditions, e.g. during an unsymmetrical short-circuit.

In practice, it is not possible to obtain the full symmetry at all nodes of the power system. The main sources of the long-term unbalance are unbalanced loads, primarily single-phase loads existing mainly in low-voltage and medium-voltage networks.

These loads cause non-equal phase currents to flow in the power system and unsymmetrical voltage drop on the system components. As a consequence, voltage balance at the nodes is lost.

Voltage unbalance can also result from different self- and mutual impedances of individual phases of transmission system components, and particularly those of overhead lines. In this case, different voltage losses are produced even when a load is symmetrical.

This chapter describes the voltage and current unbalance phenomenon in electric power networks. Considerations relate mainly to long-term unbalance occurring in normal steady-state conditions due to operation of load devices or asymmetry of network elements.

6.1 DESCRIPTION OF THE PHENOMENA

In three-phase systems voltage unbalance is defined as a condition in which the three phase voltages are not equal in magnitude and/or the displacement angles between them are different from 120° [30].

Handbook of Power Quality Edited by Angelo Baggini

An analogous definition is applied to currents.

For an analysis of unbalance the method of symmetrical components is commonly applied. This method was introduced in calculations of electric power systems at the beginning of the twentieth century. Its main idea consists in substitution of any three-phase unsymmetrical vector system of currents or voltages with the sum of three three-phase positive-, negative- and zero-sequence symmetrical systems.[1] Equations for the respective symmetrical sets of voltage may be written as follows:

- positive-sequence system (1):

$$\begin{aligned} \underline{U}_{1A} &= \underline{U}_{1A} \\ \underline{U}_{1B} &= a^2\underline{U}_{1A} \\ \underline{U}_{1C} &= a\underline{U}_{1A} \end{aligned} \tag{6.1}$$

- negative-sequence system (2):

$$\begin{aligned} \underline{U}_{2A} &= \underline{U}_{2A} \\ \underline{U}_{2B} &= a\underline{U}_{2A} \\ \underline{U}_{2C} &= a^2\underline{U}_{2A} \end{aligned} \tag{6.2}$$

- zero-sequence system (3):

$$\underline{U}_{0A} = \underline{U}_{0B} = \underline{U}_{0C} \tag{6.3}$$

where a is the rotational operator

$$a = e^{j\frac{2}{3}\pi} = -\frac{1}{2} + j\frac{\sqrt{3}}{2}$$

The respective phase voltage is the sum of the relevant components

$$\begin{aligned} \underline{U}_{A} &= \underline{U}_{1A} + \underline{U}_{2A} + \underline{U}_{0A} \\ \underline{U}_{B} &= \underline{U}_{1B} + \underline{U}_{2B} + \underline{U}_{0B} = a^2\underline{U}_{1A} + a\underline{U}_{2A} + \underline{U}_{0A} \\ \underline{U}_{C} &= \underline{U}_{1C} + \underline{U}_{2C} + \underline{U}_{0C} = a\underline{U}_{1A} + a^2\underline{U}_{2A} + \underline{U}_{0A} \end{aligned} \tag{6.4}$$

As an example, the unsymmetrical system of phase voltages in the form of the sum of symmetrical components is presented in Figure 6.1.

The symmetrical component method, when applied to describe a three-phase circuit, allows diagonalization of the impedance matrix which eliminates the couplings between phases and simplifies significantly the analysis of the circuit. Such a description and phase diagrams for the components are given in the next section.

[1] The terms *positive*, *negative* and *zero systems* are also used in the literature.

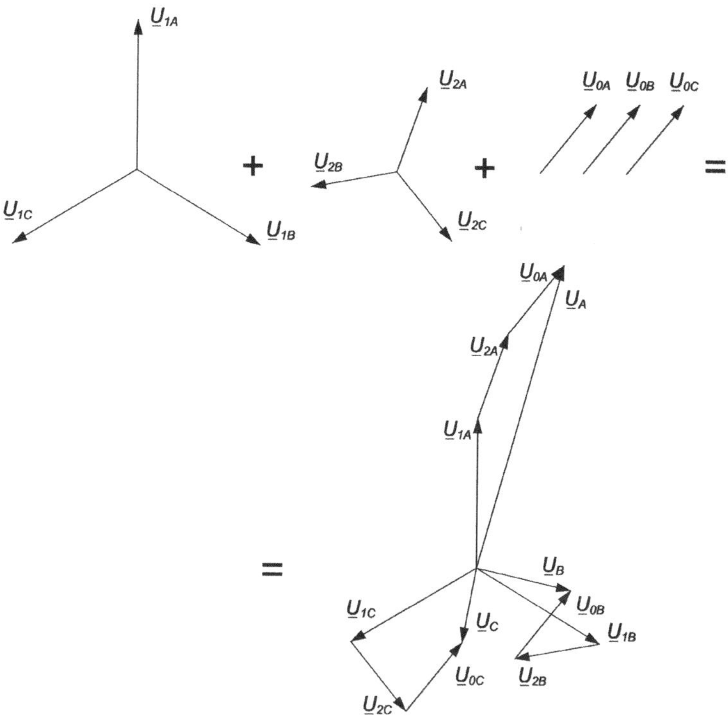

Figure 6.1 Unsymmetrical vector system and its symmetrical components

6.2 SYMMETRICAL COMPONENTS OF CURRENTS AND VOLTAGES

Let us consider the three-phase circuit presented in Figure 6.2. It is assumed that the components of the circuit are symmetrical whereas currents and voltages in general cases may form unbalanced systems.

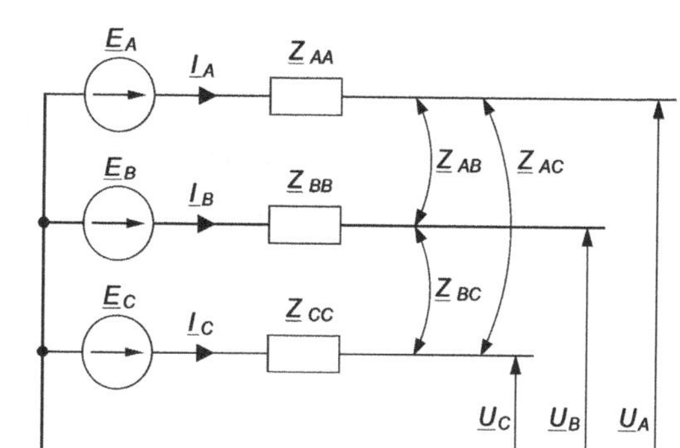

Figure 6.2 Diagram of a three-phase circuit

This circuit is described by the following equations

$$\begin{aligned} \underline{U}_A &= \underline{E}_A - (\underline{I}_A\underline{Z}_{AA} + \underline{I}_B\underline{Z}_{AB} + \underline{I}_C\underline{Z}_{AC}) \\ \underline{U}_B &= \underline{E}_B - (\underline{I}_A\underline{Z}_{BA} + \underline{I}_B\underline{Z}_{BB} + \underline{I}_C\underline{Z}_{BC}) \\ \underline{U}_C &= \underline{E}_C - (\underline{I}_A\underline{Z}_{CA} + \underline{I}_B\underline{Z}_{CB} + \underline{I}_C\underline{Z}_{CC}) \end{aligned} \tag{6.5}$$

Equations (6.5) may be transformed into the form

$$\begin{aligned} \underline{E}_A - \underline{U}_A &= \underline{I}_A\underline{Z}_{AA} + \underline{I}_B\underline{Z}_{AB} + \underline{I}_C\underline{Z}_{AC} \\ \underline{E}_B - \underline{U}_B &= \underline{I}_A\underline{Z}_{BA} + \underline{I}_B\underline{Z}_{BB} + \underline{I}_C\underline{Z}_{BC} \\ \underline{E}_C - \underline{U}_C &= \underline{I}_A\underline{Z}_{CA} + \underline{I}_B\underline{Z}_{CB} + \underline{I}_C\underline{Z}_{CC} \end{aligned} \tag{6.6}$$

or, in matrix notation,

$$\Delta\underline{U} = \underline{\underline{Z}}\,\underline{I} \tag{6.7}$$

where

$$\Delta\underline{U} = \begin{bmatrix} \underline{E}_A - \underline{U}_A \\ \underline{E}_B - \underline{U}_B \\ \underline{E}_C - \underline{U}_C \end{bmatrix} \quad \underline{\underline{Z}} = \begin{bmatrix} \underline{Z}_{AA} & \underline{Z}_{AB} & \underline{Z}_{AC} \\ \underline{Z}_{BA} & \underline{Z}_{BB} & \underline{Z}_{BC} \\ \underline{Z}_{CA} & \underline{Z}_{CB} & \underline{Z}_{CC} \end{bmatrix} \quad \underline{I} = \begin{bmatrix} \underline{I}_A \\ \underline{I}_B \\ \underline{I}_C \end{bmatrix}$$

Matrix $\underline{\underline{Z}}$ is the matrix of self- and mutual impedances of the component. Self-impedances $\underline{Z}_{ii}$ $(i = \text{A, B, C})$ are located on the main diagonal and mutual impedances outside it. For symmetrical components the self-impedances of individual phases are equal; this is

$$\underline{Z}_{AA} = \underline{Z}_{BB} = \underline{Z}_{CC} = \underline{Z}_s$$

Mutual impedances between individual phases depend on the type of components.

For a static component (line, transformer, reactor):

$$\underline{Z}_{AB} = \underline{Z}_{AC} = \underline{Z}_{BA} = \underline{Z}_{BC} = \underline{Z}_{CA} = \underline{Z}_{CB} = \underline{Z}_m$$

For a rotating component (electric motor) mutual impedances of a given phase in relation to the other phases are not equal because the windings of these phases are differently positioned in relation to the rotor:

$$\underline{Z}_{AB} \neq \underline{Z}_{AC} \quad \underline{Z}_{BC} \neq \underline{Z}_{BA} \quad \underline{Z}_{CA} \neq \underline{Z}_{CB}$$

However,

$$\begin{aligned} \underline{Z}_{AB} &= \underline{Z}_{BC} = \underline{Z}_{CA} = \underline{Z}_{m_1} \\ \underline{Z}_{AC} &= \underline{Z}_{CB} = \underline{Z}_{BA} = \underline{Z}_{m_2} \end{aligned}$$

Thus the matrices of impedances can be written in the following form:

- for a static component

$$\underline{Z} = \begin{bmatrix} \underline{Z}_s & \underline{Z}_m & \underline{Z}_m \\ \underline{Z}_m & \underline{Z}_s & \underline{Z}_m \\ \underline{Z}_m & \underline{Z}_m & \underline{Z}_s \end{bmatrix} \tag{6.8}$$

- for a rotating component

$$\underline{Z} = \begin{bmatrix} \underline{Z}_s & -\underline{Z}_{m_1} & -\underline{Z}_{m_2} \\ -\underline{Z}_{m_2} & \underline{Z}_s & -\underline{Z}_{m_1} \\ -\underline{Z}_{m_1} & -\underline{Z}_{m_2} & \underline{Z}_s \end{bmatrix} \tag{6.9}$$

A minus sign indicates that currents in the other phases induce in the given phase e.m.f.s having the opposite sense in relation to self-induction e.m.f.s, which results from the mutual location of stator phase windings.

This results from Equations (6.7), taking into account (6.8) and (6.9), i.e. the voltage loss in each phase of the component is dependent on all phase currents, and also on the self-impedance of this phase and the mutual impedances between this phase and the other phases. A transformation diagonalizing the matrix $\underline{Z}$ is introduced:

$$\underline{S}\,\Delta\underline{U} = \underline{S}\,\underline{Z}\,\underline{S}^{-1}\,\underline{S}\,\underline{I} \tag{6.10}$$

As a result of transformation

$$\Delta\underline{U}_t = \underline{S}\,\Delta\underline{U} \qquad \underline{I}_t = \underline{S}\,\underline{I} \qquad \underline{Z}_t = \underline{S}\,\underline{Z}\,\underline{S}^{-1}$$

Thus

$$\Delta\underline{U}_t = \underline{Z}_t\,\underline{I}_t \tag{6.11}$$

The form of Equations (6.7) has not changed after transformation. The matrix $\underline{S}$ is a third-order non-singular (det $\mathbf{S} \neq 0$) matrix whose elements are selected so that as a result of the transformation $\underline{S}\underline{Z}\underline{S}^{-1}$ the impedance matrix is diagonal. It may be demonstrated on the basis of the theory of linear transformations that a matrix of the following form satisfies this condition:

$$\underline{S} = \frac{1}{3}\begin{bmatrix} 1 & 1 & 1 \\ 1 & a & a^2 \\ 1 & a^2 & a \end{bmatrix} \qquad \underline{S}^{-1} = \begin{bmatrix} 1 & 1 & 1 \\ 1 & a^2 & a \\ 1 & a & a^2 \end{bmatrix} \tag{6.12}$$

Matrix $\underline{S}$ is called the operator matrix of symmetrical components and its element a is the complex operator.

Taking into account (6.12) the transformed matrices of currents and voltages are obtained:

$$\underline{I}_t = \underline{S}\,\underline{I} = \frac{1}{3}\begin{bmatrix} 1 & 1 & 1 \\ 1 & a & a^2 \\ 1 & a^2 & a \end{bmatrix} \cdot \begin{bmatrix} \underline{I}_A \\ \underline{I}_B \\ \underline{I}_C \end{bmatrix} = \frac{1}{3}\begin{bmatrix} \underline{I}_A + \underline{I}_B + \underline{I}_C \\ \underline{I}_A + a\underline{I}_B + a^2\underline{I}_C \\ \underline{I}_A + a^2\underline{I}_B + a\underline{I}_C \end{bmatrix} = \begin{bmatrix} \underline{I}_0 \\ \underline{I}_1 \\ \underline{I}_2 \end{bmatrix} \tag{6.13}$$

$$\Delta\underline{U}_t = \underline{S} \cdot \Delta\underline{U} = \frac{1}{3}\begin{bmatrix} 1 & 1 & 1 \\ 1 & a & a^2 \\ 1 & a^2 & a \end{bmatrix} \cdot \begin{bmatrix} \Delta\underline{U}_A \\ \Delta\underline{U}_B \\ \Delta\underline{U}_C \end{bmatrix}$$

$$= \frac{1}{3}\begin{bmatrix} \Delta\underline{U}_A + \Delta\underline{U}_B + \Delta\underline{U}_C \\ \Delta\underline{U}_A + a\Delta\underline{U}_B + a^2\Delta\underline{U}_C \\ \Delta\underline{U}_A + a^2\Delta\underline{U}_B + a\Delta\underline{U}_C \end{bmatrix} = \begin{bmatrix} \Delta\underline{U}_0 \\ \Delta\underline{U}_1 \\ \Delta\underline{U}_2 \end{bmatrix} \tag{6.14}$$

The transformed impedances are as follows:

- for a static component

$$\underline{Z}_t = \underline{S}\,\underline{Z}\,\underline{S}^{-1} = \frac{1}{3}\begin{bmatrix} 1 & 1 & 1 \\ 1 & a & a^2 \\ 1 & a^2 & a \end{bmatrix} \cdot \begin{bmatrix} \underline{Z}_s & \underline{Z}_m & \underline{Z}_m \\ \underline{Z}_m & \underline{Z}_s & \underline{Z}_m \\ \underline{Z}_m & \underline{Z}_m & \underline{Z}_s \end{bmatrix} \cdot \begin{bmatrix} 1 & 1 & 1 \\ 1 & a^2 & a \\ 1 & a & a^2 \end{bmatrix}$$

$$= \begin{bmatrix} \underline{Z}_s + 2\underline{Z}_m & 0 & 0 \\ 0 & \underline{Z}_s - \underline{Z}_m & 0 \\ 0 & 0 & \underline{Z}_s - \underline{Z}_m \end{bmatrix} = \begin{bmatrix} \underline{Z}_0 & 0 & 0 \\ 0 & \underline{Z}_1 & 0 \\ 0 & 0 & \underline{Z}_2 \end{bmatrix} \tag{6.15}$$

- for a rotating component

$$\underline{Z}_t = \underline{S}\,\underline{Z}\,\underline{S}^{-1} = \frac{1}{3}\begin{bmatrix} 1 & 1 & 1 \\ 1 & a & a^2 \\ 1 & a^2 & a \end{bmatrix} \cdot \begin{bmatrix} \underline{Z}_s & -\underline{Z}_{m_1} & -\underline{Z}_{m_2} \\ -\underline{Z}_{m_2} & \underline{Z}_s & -\underline{Z}_{m_1} \\ -\underline{Z}_{m_1} & -\underline{Z}_{m_2} & \underline{Z}_s \end{bmatrix} \cdot \begin{bmatrix} 1 & 1 & 1 \\ 1 & a^2 & a \\ 1 & a & a^2 \end{bmatrix}$$

$$= \begin{bmatrix} \underline{Z}_s - \underline{Z}_{m_1} - \underline{Z}_{m_2} & 0 & 0 \\ 0 & \underline{Z}_s - a^2\underline{Z}_{m_1} - a\underline{Z}_{m_2} & 0 \\ 0 & 0 & \underline{Z}_s - a\underline{Z}_{m_1} - a^2\underline{Z}_{m_2} \end{bmatrix} = \begin{bmatrix} \underline{Z}_0 & 0 & 0 \\ 0 & \underline{Z}_1 & 0 \\ 0 & 0 & \underline{Z}_2 \end{bmatrix} \tag{6.16}$$

Now Equation (6.11) may be written in the form

$$\begin{bmatrix} \Delta\underline{U}_0 \\ \Delta\underline{U}_1 \\ \Delta\underline{U}_2 \end{bmatrix} = \begin{bmatrix} \underline{Z}_0 & 0 & 0 \\ 0 & \underline{Z}_1 & 0 \\ 0 & 0 & \underline{Z}_2 \end{bmatrix} \cdot \begin{bmatrix} \underline{I}_0 \\ \underline{I}_1 \\ \underline{I}_2 \end{bmatrix} \tag{6.17}$$

or

$$\begin{bmatrix} \underline{E}_0 - \underline{U}_0 \\ \underline{E}_1 - \underline{U}_1 \\ \underline{E}_2 - \underline{U}_2 \end{bmatrix} = \begin{bmatrix} \underline{Z}_0 & 0 & 0 \\ 0 & \underline{Z}_1 & 0 \\ 0 & 0 & \underline{Z}_2 \end{bmatrix} \cdot \begin{bmatrix} \underline{I}_0 \\ \underline{I}_1 \\ \underline{I}_2 \end{bmatrix} \tag{6.18}$$

where $\Delta\underline{U}_t = \underline{S}\Delta\underline{U} = \underline{S}(\underline{E} - \underline{U}) = \underline{S}\,\underline{E} - \underline{S}\,\underline{U} = \underline{E}_t - \underline{U}_t$ has been taken into account.

The impedance matrix of symmetrical components is diagonal. The systems of symmetrical components are mutually independent. Ohm's law is kept independently in each system. So, as the result of transformation, the three-phase unbalanced system of Figure 6.2 is replaced by three three-phase symmetrical systems of zero-sequence, positive-sequence and negative-sequence as presented in Figure 6.3.

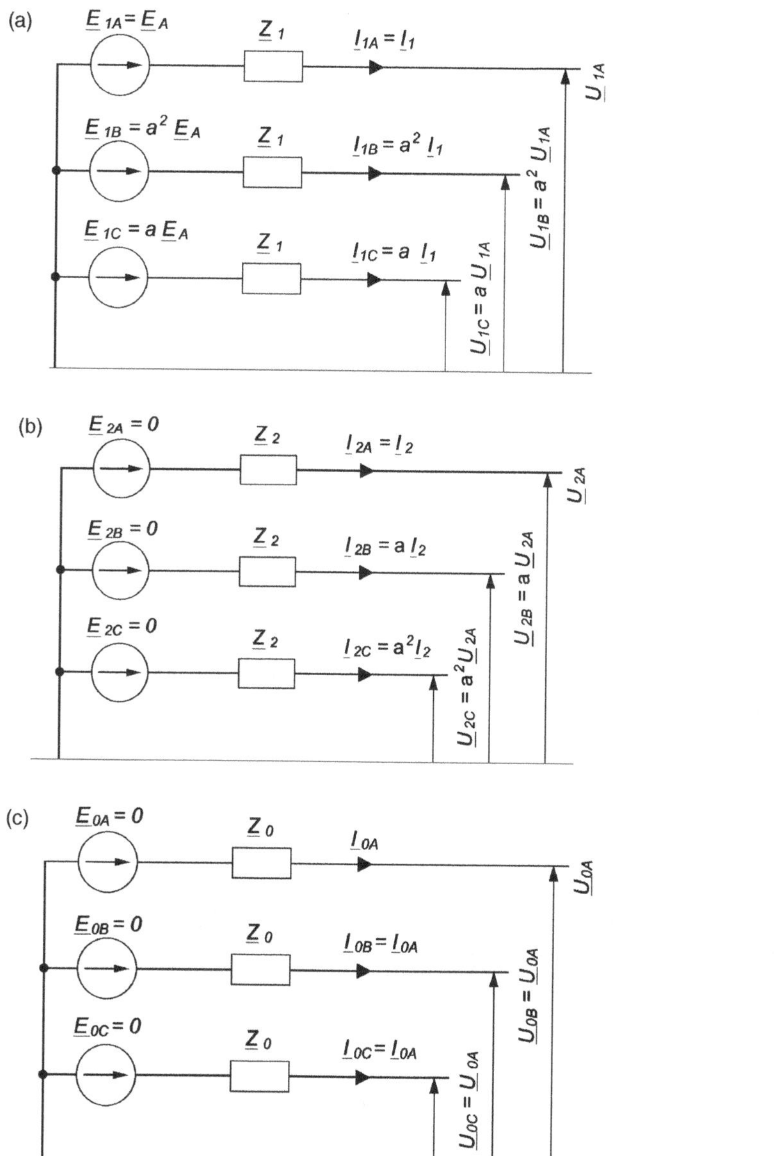

Figure 6.3 Systems of symmetrical components for the three-phase circuit of Figure 6.2 : (a) positive-sequence, (b) negative-sequence, (c) zero-sequence

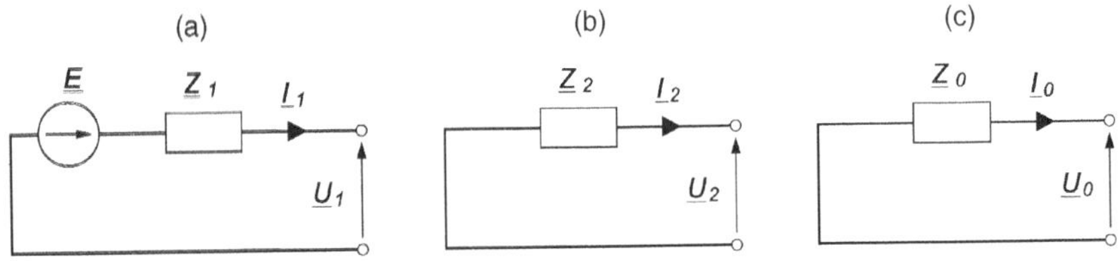

Figure 6.4 Single-phase equivalent diagrams of symmetrical components for the circuit of Figure 6.2: 9(a) positive-sequence, (b) negative-sequence, (c) zero-sequence

The source is symmetrical, thus an e.m.f. will be only in the positive-sequence diagram

$$\underline{E}_{1\text{A}} = \underline{E}_{\text{A}} = \underline{E} \qquad \underline{E}_{2\text{A}} = 0 \qquad \underline{E}_{0\text{A}} = 0$$

This can be written similarly for phases B and C.

On the basis of the diagrams in Figure 6.3 equations may be formulated that allow the determination of symmetrical components of voltages at the point of unbalance. These equations are usually written for phase A as a reference phase (index A has been omitted)

$$\begin{aligned}\underline{U}_1 &= \underline{E} - \underline{Z}_1\underline{I}_1 \\ \underline{U}_2 &= -\underline{Z}_2\underline{I}_2 \\ \underline{U}_0 &= -\underline{Z}_0\underline{I}_0\end{aligned} \tag{6.19}$$

Impedances $\underline{Z}_1$, $\underline{Z}_2$, $\underline{Z}_0$ are the impedances of circuits for the positive-sequence, negative-sequence and zero-sequence components. The single-phase circuits presented in Figure 6.4 correspond to Equations (6.19).

6.3 PARAMETERS

The voltage symmetrical components as a function of the phase voltages may be determined from Equations (6.4) (also transformation (6.14)):

$$\begin{aligned}\underline{U}_0 &= \frac{1}{3}(\underline{U}_\text{A} + \underline{U}_\text{B} + \underline{U}_\text{C}) \\ \underline{U}_1 &= \frac{1}{3}(\underline{U}_\text{A} + a\underline{U}_\text{B} + a^2\underline{U}_\text{C}) \\ \underline{U}_2 &= \frac{1}{3}(\underline{U}_\text{A} + a^2\underline{U}_\text{B} + a\underline{U}_\text{C})\end{aligned} \tag{6.20}$$

Similar relations may be given for currents.

As a measure of unbalance, the unbalance factor K is commonly taken; it is the ratio of the negative- and/or zero-sequence component to the positive-sequence component of voltage or current of any phase.[2]

For a voltage

$$K_{2U} = \frac{U_{2(1)}}{U_{1(1)}} \cdot 100\,\% \qquad K_{0U} = \frac{U_{0(1)}}{U_{1(1)}} \cdot 100\,\% \tag{6.21}$$

For a current

$$K_{2I} = \frac{I_{2(1)}}{I_{1(1)}} \cdot 100\,\% \qquad K_{0I} = \frac{I_{0(1)}}{I_{1(1)}} \cdot 100\,\% \tag{6.22}$$

The subscript (1) in the formulas above denotes that definitions are referred to the first harmonic.

The negative-sequence component generated by multi-phase loads is very important for the description of unbalance. It is transformed by a transformer irrespective of the windings connection, similarly as in the case of the positive-sequence component. The zero-sequence component is normally present only in low-voltage (LV) networks, and the delta-connected transformer prevents it from transferring to a network of higher voltage. The unbalance factor for the negative-sequence component is a function of phase voltages:

$$K_{2U} = \left| \frac{\left(\underline{U}_A + a^2 \underline{U}_B + a \underline{U}_C\right)}{\underline{U}_A + a \underline{U}_B + a^2 \underline{U}_C} \right| \cdot 100\,\% \tag{6.23}$$

Taking into account that $a + a^2 + 1 = 0$, the above formula can also be given as a function of line voltages:

$$K_{2U} = \left| \frac{\left(\underline{U}_{AB} - a \underline{U}_{BC}\right)}{\underline{U}_{AB} - a^2 \underline{U}_{BC}} \right| \cdot 100\,\% \tag{6.24}$$

Similar relations can be written for currents.

6.4 MEASUREMENTS

According to the definition, the amplitudes of positive- and negative-sequence voltage components must be known for the determination of the unbalance factor. In the literature [16], [19] the following relations are given, which allow calculation of the unbalance factor according to (6.20) and (6.21), making use of measurements of three phase voltages and two line voltages.

---

[2] In the literature two terms are encountered: *unbalance* or *unbalance*, defined by factors (6.21), and *asymmetry*, defined only by the ratio of the negative- to the positive-sequence component. This definition is mostly used in relation to the isolated neutral power network.

On the assumption that a voltage on phase A is put on the real axis, the complex values of phase voltages can be written in the form

$$\underline{U}_A = U_A$$
$$\underline{U}_B = U_B e^{-j\alpha} \tag{6.25}$$
$$\underline{U}_C = U_C e^{-j(\alpha+\beta)}$$

where α is the angle between voltages $\underline{U}_A$ and $\underline{U}_B$, and β is the angle between voltages $\underline{U}_B$ and $\underline{U}_C$. The angles α and β can be determined from the formulas

$$\alpha = \arccos\left(\frac{U_A^2 + U_B^2 - U_{AB}^2}{2U_A U_B}\right)$$
$$\beta = \arccos\left(\frac{U_B^2 + U_C^2 - U_{BC}^2}{2U_B U_C}\right) \tag{6.26}$$

In standardization documents the relations for determining the unbalance factor for practical use are given, being a function of the r.m.s. line or phase voltages. These are assembled in Table 6.1.

Table 6.1 Formulas for determination of the unbalance factor given in international standards and regulations.

Document	Formula for calculating the unbalance factor
IEC 61000-2-12 [13]	$K_{2U} = \sqrt{\frac{6(U_{AB}^2 + U_{BC}^2 + U_{CA}^2)}{(U_{AB} + U_{BC} + U_{CA})} - 2}$
IEC 61000-4-30 [15] IEEE P1159.1[16]	$K_{2U} = \sqrt{\frac{1-\sqrt{3-6\beta}}{1+\sqrt{3-6\beta}}} \cdot 100\,\%$ where $\beta = \frac{U_{AB}^4 + U_{BC}^4 + U_{CA}^4}{(U_{AB}^2 + U_{BC}^2 + U_{CA}^2)^2}$
Gost 13109-97 [7]	$U_1 = \sqrt{\frac{1}{12}\left[\left(\sqrt{3}U_{AB} + \sqrt{4U_{BC}^2 - \left(\frac{U_{BC}^2 - U_{CA}^2}{U_{AB}} + U_{AB}\right)^2}\right)^2 + \left(\frac{U_{BC}^2 - U_{CA}^2}{U_{AB}}\right)^2\right]}$
	$U_2 = \sqrt{\frac{1}{12}\left[\left(\sqrt{3}U_{AB} - \sqrt{4U_{BC}^2 - \left(\frac{U_{BC}^2 - U_{CA}^2}{U_{AB}} + U_{AB}\right)^2}\right)^2 + \left(\frac{U_{BC}^2 - U_{CA}^2}{U_{AB}}\right)^2\right]}$

Approximate formula:

$$U_2 = 0.62\left(U_{p\max} - U_{p\min}\right)$$

where $U_{p\max}$ and $U_{p\min}$ are the maximum and minimum values of line-to-line voltages

IEC 61000-2-1 [11], IEC 61000-4-27 [14], ANSI C84.1 [1], ER P29 [5]

$$K_{2U} = \frac{\Delta U_{\max}}{U_{av}} \cdot 100\,\%$$

where ΔU_{max} is the maximum deviation of any of the three phase voltages from the average phase voltage, and U_{av} is the average phase voltage.

ER P29 [5]

$$K_{2U} = \frac{\sqrt{3}\, I_2 U_p}{S_{sc}} \cdot 100\,\%$$

where I_2 is the negative-sequence component of the fundamental of a drawn current, U_p is a phase voltage and S_{sc} is the short-circuit power in the point of common coupling

For a single one-phase receiver with the power S:

$$K_{2U} = \frac{S}{S_{sc}} \cdot 100\,\%$$

6.5 SOURCES

Unbalanced operating conditions in an electric power system are caused mainly by the operation of unbalanced loads. Most low-voltage loads and certain medium-voltage ones, e.g. an electric traction motors, are single-phase appliances. Operation of such equipment in the three-phase system results in unbalanced load currents. Consequently, unsymmetrical voltage drops in individual phases of the supply system are produced, thus voltage at nodes of the network becomes unbalanced.

Three-phase loads which may introduce unbalance in the power system are arc furnaces. The disturbance results from different impedances of high-current paths of the furnace and not equal phase loads being the effect of the physical nature of the melting process, i.e. variations in the arc impedance.

As arc furnaces are the devices of relatively large power (tens or even hundreds of MVA), the furnace load unbalance may result in significant voltage unbalance in the supply system.

The sources of unbalance can also be three-phase components of the transmission system, in particular overhead lines. Due to different tower geometries the conductors of individual phases are not simultaneously at the same location as each other and to earth. Following this, the line has different values of phase parameters, and also the values of a voltage loss in individual phases are different.

6.6 EFFECTS

The negative-sequence and zero-sequence currents flowing in an electric power system result in

- additional losses of power and energy;
- additional heating, the consequence of which is the limitation of line transmission capability for positive-sequence currents;
- voltage unbalance at nodes of the network.

Voltage unbalance adversely affects the operation of loads. Asynchronous motors, synchronous generators and rectifiers are the most sensitive loads in this respect.

6.6.1 Asynchronous Motors

Asynchronous motors have their windings connected usually delta or star with an isolated star point. For this reason the operation of a motor is affected only by the positive- and negative-sequence components. The negative-sequence currents create a flux rotating in the direction opposite to the rotor. This flux causes:

- increased heating of the stator windings;
- additional losses of active power in the stator;
- additional torque operating in the opposite direction to the torque produced by the positive-sequence flux;
- inducting additional currents in the windings and rotor iron of a motor, and thereby additional power losses in the rotor.

Motor current under unbalanced conditions can be several times higher than the rated current. Long-lasting unbalance can cause the motor insulation to deteriorate more quickly and shorten its life. Motors may be equipped with protection which detects an overcurrent and switches them off.

Additional power losses due to the unbalance of a supply voltage reduce the maximum power of the motor to an extent which depends on the degree of unbalance, the type of motor and its construction.

A negative-sequence torque causes a reduction in the useful torque of the motor.

Moreover, in the case of unbalanced supply voltage, additional vibrations are produced in the motor resulting also in shortening its service life.

Laboratory tests have shown that most asynchronous motors are not affected by a supply voltage unbalance of 2 %.

6.6.2 Synchronous Generators

Load unbalance affects mainly the operation of generators in industrial heat and power stations supplying distribution grids. The generators of commercial power stations are located at some distance from unbalanced loads, thus in this case a load unbalance is of no importance.

Analysis of the effects of a load unbalance for synchronous generators can be limited to the negative-sequence component, because generators are connected to a system through transformers, in which the windings of one side are connected in delta preventing zero-sequence currents from entering the generator.

Negative-sequence currents generate in the machine a magnetic flux rotating in a direction opposite to the rotational direction of a flux generated by positive-sequence currents. Similarly to asynchronous motors, this flux affects the rotor and the stator of the generator; it induces eddy currents and increases heating and power losses. The negative-sequence flux produces also additional mechanical forces acting on the rotor and the stator of the generator, which are hazardous to the strength of structural components.

The fundamental criterion for evaluating the permissible operation of a generator under unbalanced conditions is the additional heating of the rotor.

In general, unbalanced loads are not a great problem for the operation of synchronous generators; unbalance may cause more severe hazards during disturbances, e.g. during unsymmetrical short-circuits.

6.6.3 Converters

Converter equipment in most cases is supplied from a three-phase, three-wire system, thus its operation can be affected only by the negative-sequence component of voltage. It generates:

- an additional variable component of a rectified voltage (current) whose amplitude depends on the unbalance factor;
- harmonics that are not characteristic for the given topology of a converter and interharmonics.

6.6.4 Other Loads

Unbalance can also affect the operation of other three-phase loads, changing the electric power, exploitative characteristics and their service life. Moreover, voltage unbalance associated with a change in voltage magnitude has an effect on the operation of single-phase loads. Some of them may be under the influence of a supply voltage that is too high or too low. Disturbances in the functioning of control systems can also occur, resulting in disturbing and even in interrupting the operation of equipment.

6.7 MITIGATION

The unbalance of power system components is eliminated by suitable design. In the case of overhead lines a transposition of conductors is applied for this purpose. The line is divided into sections, a number of which is divisible by 3, with three sections forming one transposition cycle. In each section the conductor of a given phase is routed at a different position to the other phases, which means the line, taken as a whole, can be considered as a symmetrical one.

Methods of unbalance mitigation concern primarily the load balancing.

The traditional approach to load balancing is to connect nominal loads evenly to each phase. Normally this is sufficient so that severe unbalance of voltage does not appear very frequently. Where significant load unbalance is unavoidable, particularly in the case of large single-phase loads, special balancing equipment to compensate for the disturbance should be applied. The purpose of its operation is usually the elimination or limitation of the negative- and zero-sequence components of load currents. This process is called balancing.

6.7.1 Principle of Balancing

In three-phase MV systems, which operate usually as isolated or compensated ones, unbalanced loads are connected to a line voltage. In this case, there is no zero-sequence component of currents, so the balancing resolves itself into the elimination or limitation of the negative-sequence component. LV grids are four-wire earthed neutral systems, so there the negative-sequence as well as the zero-sequence component is present.

Balancing equipment (BD), so-called baluns, is connected in parallel with the unbalanced load (UL) (Figure 6.5).

This equipment causes the currents $\underline{I}_{\mathrm{Ak}}$, $\underline{I}_{\mathrm{Bk}}$, $\underline{I}_{\mathrm{Ck}}$ to flow, which, added to the load currents $\underline{I}_{\mathrm{Al}}$, $\underline{I}_{\mathrm{Bl}}$, $\underline{I}_{\mathrm{Cl}}$, give as a result the symmetrical set of the source currents $\underline{I}_{\mathrm{A}}$, $\underline{I}_{\mathrm{B}}$, $\underline{I}_{\mathrm{C}}$. Then the following equation is satisfied:

$$\begin{aligned} \underline{I}_{\mathrm{A}} &= \underline{I}_{\mathrm{Al}} + \underline{I}_{\mathrm{Ak}} \\ \underline{I}_{\mathrm{B}} &= \underline{I}_{\mathrm{Bl}} + \underline{I}_{\mathrm{Bk}} = a^2 \underline{I}_{\mathrm{A}} \\ \underline{I}_{\mathrm{C}} &= \underline{I}_{\mathrm{Cl}} + \underline{I}_{\mathrm{Ck}} = a \underline{I}_{\mathrm{A}} \end{aligned} \tag{6.27}$$

The currents drawn from a network form a symmetrical set, so the negative- and zero-sequence components of these currents are equal to zero

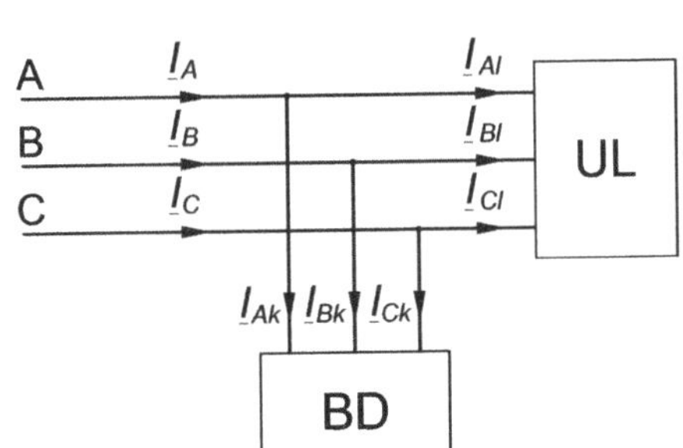

Figure 6.5 Diagram of unbalanced load with balancing device

$$\underline{I}_2 = \frac{1}{3}(\underline{I}_A + a\underline{I}_B + a^2\underline{I}_C) = 0$$
$$\underline{I}_0 = \frac{1}{3}(\underline{I}_A + \underline{I}_B + \underline{I}_C) = 0 \tag{6.28}$$

The above equations determine the overall conditions for balancing.

A load with a balun makes up an equivalent system whose phase admittances depend on the connection of these elements. Equivalent admittances for the elements connected in star are given by

$$\underline{Y}_A = \underline{Y}_{Al} + \underline{Y}_{Ak}$$
$$\underline{Y}_B = \underline{Y}_{Bl} + \underline{Y}_{Bk} \tag{6.29}$$
$$\underline{Y}_C = \underline{Y}_{Cl} + \underline{Y}_{Ck}$$

For the elements connected in delta

$$\underline{Y}_{AB} = \underline{Y}_{ABl} + \underline{Y}_{ABk}$$
$$\underline{Y}_{BC} = \underline{Y}_{BCl} + \underline{Y}_{BCk} \tag{6.30}$$
$$\underline{Y}_{CA} = \underline{Y}_{CAl} + \underline{Y}_{CAk}$$

Taking into account formulae (6.29) and (6.30), on the basis of Equations (6.28), the admittance balancing criteria can be easily introduced. For the elements of a load and a balun connected in star, it is obtained as

$$\underline{Y}_A + a\underline{Y}_B + a^2\underline{Y}_C = 0 \tag{6.31}$$

$$\underline{Y}_A + a^2\underline{Y}_B + a\underline{Y}_C = 0 \tag{6.32}$$

where equation (6.31) concerns the elimination of the negative-sequence component, and equation (6.32) concerns the elimination of the zero-sequence component. For delta connection, only an equation for the negative-sequence component is determined:

$$\underline{Y}_{AB} + a\underline{Y}_{BC} + a^2\underline{Y}_{CA} = 0 \tag{6.33}$$

The above equations are valid when a set of supply voltages is symmetrical.

There are a variety of systems for balancing load currents which differ in the number and type of elements and the art of connection. Comprehensive information on this matter can be found in the monograph [19]. Baluns accomplish their task at a constant value of loads. In the case of time-varying loads, such as electric traction motors or arc furnaces, follow-up compensation is necessary. This kind of compensation is implemented through thyristor-based controllers connected in shunt or series. The idea of compensation is to control currents or voltages in such a way as to minimize the negative- and zero- sequence components.

6.7.2 Static Compensators

6.7.2.1 Shunt Static Compensators

Conventional static var compensators (SVCs) are systems including thyristor-controlled reactors and fixed capacitors. Connected to a node of an electric power network, they may be considered as a controlled parallel susceptance. Balancing is one of the tasks which can be accomplished by these devices. They are usually applied also for the compensation of reactive power or the limitation of voltage fluctuations and a light flicker phenomenon.

The following elements may form a compensator circuit:

- thyristor-controlled reactors (TCRs)
- fixed capacitor (FC) banks
- thyristor-switched capacitors (TSCs)

In practice there are various configurations of circuits having different features. The compensator of FC/TCR type, composed of a capacitor bank with a constant power and a reactor branch with a controlled current connected in parallel, is numbered among the systems that are used most commonly. A diagram of the compensator is presented in Figure 6.6.

A reactor current is varied by changing the firing angle α of thyristors, measured from the instant of voltage zero crossing. The thyristors are fully turned on at $\alpha = \pi/2$. When the angle α increases ($\pi/2 \leq \alpha \leq \pi$) the fundamental harmonic of a reactor current decreases.

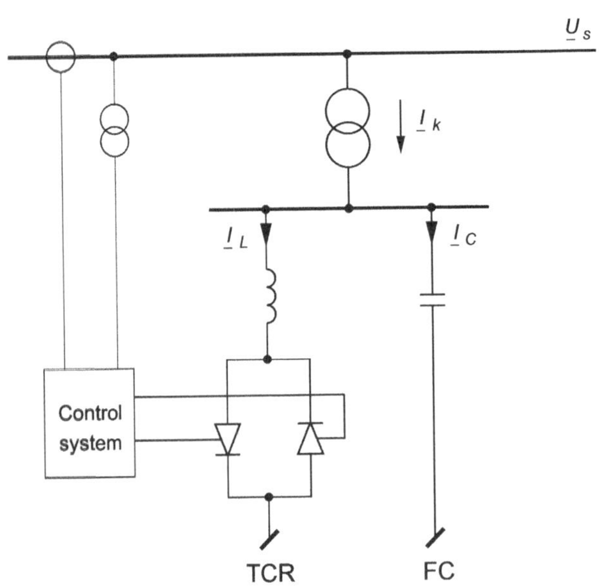

Figure 6.6 Schematic diagram of an FC/TCR static compensator

The resultant current of a compensator $i_k(t)$ is the sum of capacitor and reactor currents

$$i_k(t) = i_C(t) + i_L(t) \tag{6.34}$$

If a reactor branch current is equal to zero ($\alpha = \pi$) then the compensator gives out a reactive power to the network, and its current is capacitive. If the thyristors are fully turned on ($\alpha = \pi/2$), and the reactor power is higher than the capacitor power, then the compensator consumes the reactive power, and its current is inductive. A compensator current is continuously regulated within I_{Cmax} to I_{Lmax}. Operation of the SVC is aimed at the compensation of a load reactive power in each phase in such a way that the load and the compensator consume only an active power in the circuit of each phase. It can be demonstrated that such compensation also results in the balancing of active loads [23]. It is assumed that suitable filters eliminate higher harmonics of the current, and the compensation is related only to the first harmonic.

Newer solutions of the compensation systems use STATCOM-type devices based on the use of a.c./d.c. converters. STATCOM (STATic synchronous COMpensator) is a controlled voltage source connected to a node through an impedance which is usually the impedance of the coupling transformer. The basic operation of it is similar to that of a synchronous compensator. If the voltage generated by the compensator U_k is less than the supply voltage U_s then the compensator will act as an inductive load drawing reactive power from the network. Otherwise, if the compensator voltage is higher than the network voltage then STATCOM will act as a capacitor generating reactive power into the supply network. Schematic and phasor diagrams of the compensator are presented in Figure 6.7, under the assumption that system losses are neglected.

The STATCOM-type compensators applied in distribution systems for power quality improvement are called DSTATCOMs (Distribution STATCOMs). They can perform complex tasks simultaneously: reactive power compensation, load balancing and harmonic filtering. The basic element of the DSTATCOM system is the PWM voltage inverter in which fully controlled semiconductor devices are used, e.g. IGBT, IGCT. For load compensation the unit operates in current control mode. The control circuit is designed in such a way as to generate three independent compensator phase currents of a given shape so that the sum of the load and compensator currents gives the sinusoidal and positive-sequence currents flowing through the supplying network. A hysteresis controller is usually applied in this case and instantaneous real and reactive power theory is used for the determination of the reference currents. Practical structures of DSTATCOM devices that can be used in power distribution systems and control algorithms applied for compensation are discussed in details in [6].

The main features of the DSTATCOM compensators are the following [24]:

- the ability to realize different functions at the same time;
- the ability to follow rapid load changes;
- reactive components are not required – the size of these compensators is about four times smaller than that of corresponding SVC systems;
- adaptation to varying load condition in a power system is easy.

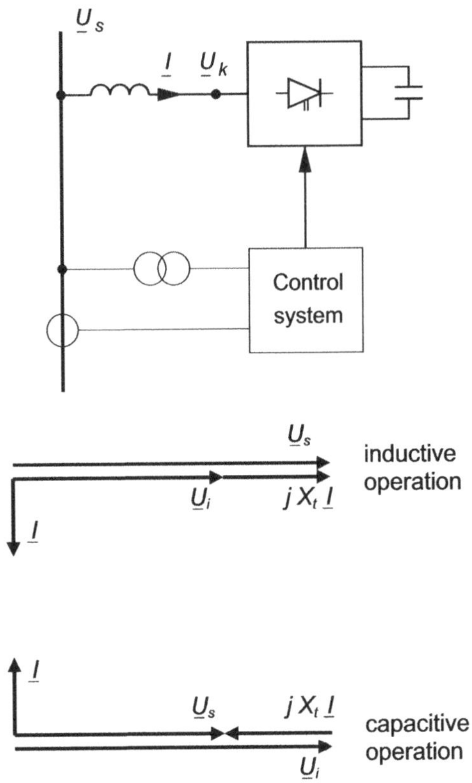

Figure 6.7 Schematic diagram of a STATCOM and phasor diagrams illustrating the principle of its operation

6.7.2.2 Series Static Compensators [9]

Compensation of load unbalance is usually accomplished by means of parallel compensators. However, units connected in series can be applied to mitigate the supply voltage unbalance.

The purpose of the system is to generate, for instance by means of a PWM converter, the boosting voltage $\underline{U}_b$ whose value and phase in relation to voltage $\underline{U}_s$/current $\underline{I}$ are regulated. The schematic diagram of the system is presented in Figure 6.8.

The converter can be considered as a voltage source represented on the $d-q$ plane by two orthogonal components with regulated values:

- component $\underline{U}_{bd}$ (in-phase or in opposite phase in relation to a line current) decides the exchange of active power between a supply system and a compensator;
- component $\underline{U}_{bq}$ (orthogonal to a line current) decides the value (also the character) of a series compensator reactive power.

If the fundamental harmonic of an output voltage, proportional and orthogonal to the line current, is lagging in relation to a line current, then a series compensator realizes the

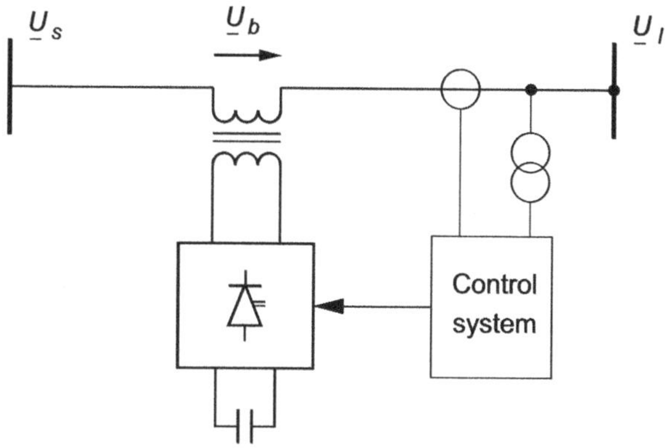

Figure 6.8 Schematic diagram of a series static compensator

compensation of a voltage drop or, the equivalent line reactance, as a series capacitor X_C. In this case, the output voltage of a converter is determined by the formula

$$\underline{U}_b = -jkX_s\underline{I} \tag{6.35}$$

where X_s is the phase reactance of a line, $\underline{I}$ is the line current, and the proportionality factor k is the degree of series compensation: $k = X_C X_s^{-1} = \mathrm{U}_b\,(\mathrm{I}\,X_s)^{-1}$.

The system can operate as a voltage stabilizer and compensator of voltage dips. In addition to the regulation of a load voltage, the circuit of a series compensator can also be equipped with the function of balancing. This function is performed by the three-phase circuit of boosting voltage components which are determined on the basis of measurement of the negative-sequence component of the load voltages. The following relation describes these voltages:

$$\frac{1}{U_{\mathrm{ref}}}\begin{bmatrix} \Delta\underline{U}_{\mathrm{bA}} \\ \Delta\underline{U}_{\mathrm{bB}} \\ \Delta\underline{U}_{\mathrm{bC}} \end{bmatrix} = -\frac{\underline{U}_2}{U_{\mathrm{ref}}}\begin{bmatrix} 1 \\ a \\ a^2 \end{bmatrix} \tag{6.36}$$

where U_{ref} is the predetermined reference value of a load voltage.

As a result of adding voltage components determined by relations (6.36) to the source voltages, the symmetrical voltage system at the load buses is obtained.

6.8 STANDARDS

6.8.1 Limits

Most international standards and documents are in agreement on the definition of the unbalance phenomenon and its parameters. Compatibility levels for LV and MV systems are 2 % and under special conditions 3 %. No separate specifications for HV networks are given in most analyzed documents, except for the document [5], in which a method of assessment of voltage unbalance is included for a network node to which an unbalanced load is connected, depending on the power of this load and the short-circuit power of the network.

When the limit values of the unbalance factor are determined, a possibility is assumed that the so-called *background* exists; this is a certain natural level of unbalance resulting from the asymmetric impedance of elements. Its value normally does not exceed 0.5 %, though sometimes it can reach 1 % for large loads or in the case of unbalanced operation of a network caused by switching-off operations.

The emission limit levels are not specified in the international standards. Only in the document [31] are the following values given: 0.7 % in the range of minutes and 1 % in the range of seconds. These values are related to individual customers and to all voltage levels.

When unbalance in a real network is evaluated, the measured values for the unbalance factor are compared to its permissible value. It is usually required that the permissible value is not exceeded for 95 % of an observation period. In mathematical statistics such a value is called the 95 % percentile. It means that the values which are present for 5 % of the observation period are not limited.

The values of the unbalance factor specified in various standard documents are presented in Table 6.2.

Table 6.2 Limit values of the unbalance factor in international standards and rules

Document	Values of unbalance factor	Remarks
IEC 61000-2-5 [12]	Two disturbance degrees are defined: Degree 1, $K_{2U} = 2\,\%$ Degree 2, $K_{2U} = 3\,\%$	Classification of phenomenon
IEC 61000-2-12 [13]	$K_{2U} = 2\,\%$ Under special conditions up to 3 %	Compatibility level
EN 50160 [25]	$K_{2U} = 2\,\%$ In some areas, unbalance up to 3 % may occur	Voltage characteristics, 95 % percentile
Gost 13109-97 [7]	$K_{2U} = 2\,\%$, referred to 95 % percentile $K_{2U} = 4\,\%$, referred to maximum value	Voltage characteristics
ANSI [1]	$K_{2U} = 3\,\%$	
ER P29 [5]	$K_{2U} = 2\,\%$ Under special conditions $K_{2U} = 1\,\%$ for systems with 33–132 kV voltages and $K_{2U} = 1.3\,\%$ for voltages up to 33 kV	Planning levels
UIE [30]	Compatibility levels for LV and MV systems: $K_{2U} = 2\,\%$	

	Planning levels for HV systems: $K_{2U} = 1\,\%$ Compatibility levels for industrial networks: For classes 1 and 2, $K_{2U} = 2\,\%$ For class 3, $K_{2U} = 3\,\%$ (95 % percentile) $K_{2U} = 4\,\%$ (maximum value)	
UNIPEDE [20]	$K_{2U} = 2\,\%$ In some areas, unbalance up to 3 % may occur	95 % percentile

6.8.2 Principles of Assessment

The basic document that describes the measurement technique and the evaluation of quality characteristic is IEC 61000-4-30 standard [15]. Two measurement classes, A and B, are defined in this document.

Class A is related to accurate measurements which are the basis for the assessment of quality level according to requirements of suitable standards. Class B covers measurements which can be carried out for statistical purposes, on solving practical problems, or in the other cases where high accuracy is not required. Detailed requirements concerned with the measurement of the unbalance factor, the method for aggregation of results and measuring instruments to be used are given for class A. Requirements for class B are to be set for each case of tests.

The unbalance factor shall be determined by the measurement of the first harmonic of line voltages according to the relation given in Table 6.1. The width of a measurement window at frequency 50 Hz shall be 10 cycles. Results obtained for successive windows are averaged over characteristic time intervals equal to 3 s (so-called very short interval), 10 min (so-called short interval) and 2 h. The following relations describe averaging:

$$K_{2U,3s} = \sqrt{\left(\sum_{k=1}^{N} K^2_{2U\%,k}\right)/N} \tag{6.37}$$

$$K_{2U,10min} = \sqrt{\left(\sum_{k=1}^{M} K^2_{2U,3s,k}\right)/M} \tag{6.38}$$

$$K_{2U,2h} = \sqrt{\left(\sum_{k=1}^{P} K^2_{2U,10min,k}\right)/P} \tag{6.39}$$

where $K_{2U,3s}$, $K_{2U,10min}$, $K_{2U,2h}$ are the values of the unbalance factor averaged over time intervals 3 s, 10 min and 2 h, respectively; $K_{2U\%,k}$ are the values of the unbalance factor determined for N successive measurement windows; M is the number of averaged values of $K_{2U,3s}$; P is the number of averaged values of $K_{2U,10min}$.

Unbalance shall be assessed over a minimum observation period of one day. The method of evaluation depends on the purpose of measurements. In the case of contractual applications, the unbalance factor averaged over a time interval of 10 min and 2 h is evaluated, and the total observation period shall include at least one week. The following methods of evaluation are proposed:

- determination of a number or a percentage of values which in the observation period exceed contractual values;
- comparison of maximum measured values to contractual ones (a longer observation period can be taken in this case, e.g. one year);
- comparison of statistical parameters of measured quantities to contractual values.

Statistical parameters that shall be compared to set limits are 95 % or 99 % percentiles of the averaged unbalance factor ($K_{2U,\ 0.95}$, $K_{2U,\ 0.99}$). The method or methods of evaluation are to be agreed between contracting parties.

The 10 min averaging time is also recommended by the 02 CIGRE–CIRED Working Group for comparison with planning levels and for evaluation of the unbalance effect on the operation of converter equipment. For testing the operation of electric motors the averaging time shall be equal to 2 h, because in that time the equipment reaches a steady-state temperature rise as a result of additional heating due to unbalance.

For a case study see web address

BIBLIOGRAPHY

[1] ANSI C84.1, 1995, *American national standard for electric power systems and equipment – voltage ratings.*

[2] Arrillaga J., Watson N. R., Chen S., *Power system quality assessment*, John Wiley & Sons, Ltd, Chichester, 2000.

[3] Australian Standard AS 1359.31 – 1997, *Rotating electrical machines – General requirements. Part 31: Three-phase induction motors – Operation on unbalanced voltages.*

[4] Dugan R. C., McGranaghan M. F., Beaty H. W., *Electrical power systems quality*, McGraw-Hill, New York, 1996.

[5] Engineering Recommendation P29, *Planning limits for voltage unbalance in the United Kingdom.* The Electricity Council (UK), London, 1989.

[6] Ghosh A., Ledwich G., *Power Quality Enhancement Using Custom Power Devices*, Kluwer Academic, Dordrecht, 2002.

[7] Gost 13109-97, Russian Standard, *Normy kaczestwa elektriczeskoj energii w sistemach elektrosnabzenia obszczego naznaczenia.*

[8] Häger M., Wahlström B., Friman E., Limits and responsibility sharing with regard to power quality for Swedish 220- and 400 kV transmission system. STRI Report S98-171.

[9] Hanzelka Z., *Szeregowe układy kompensacji*, Konferencja 'Energoelektronika w zastosowaniach przemysłowych', Warszawa-Międzylesie, 1995.

[10] IEC 60050-161, International Electrotechnical Vocabulary, Chapter 161. *Electromagnetic compatibility*, 1990.

[11] IEC 61000-2-1, *Electromagnetic compatibility – Part 2: Environment – Section 1: Description of the environment – Electromagnetic environment for low-frequency conducted disturbances and signalling in public power supply systems*, 1990.
[12] IEC 61000-2-5, *Electromagnetic compatibility – Part 2: Environment – Section 5: Classification of electromagnetic environments*, 1995.
[13] IEC 61000-2-12, *Electromagnetic compatibility – Part 2: Environment – Section 12: Compatibility levels for low-frequency conducted disturbances and signalling in public medium-voltage power systems*, 1995.
[14] IEC 61000-4-27, *Electromagnetic compatibility – Part 4–27: Testing and measurement techniques – Unbalance, immunity test*, 2000.
[15] IEC 61000-4-30, *Electromagnetic compatibility. Part 4–30: Testing and measurement techniques – Power Quality measurement methods*, 1999.
[16] IEEE P1159.1, *Guide for recorder and data acquisition requirements for characterization of power quality events.*
[17] International Telecommunication Union, Geneva, *Emerald Contract for the supply of electric at the Green Rate*. EdF – Electricite'de France, 1989.
[18] Kennedy B., *Power quality primer*. McGraw-Hill, New York, 2000.
[19] Kowalski Z., *Asymetria w układach elektroenergetycznych*, Warszawa, 1994.
[20] *Measurement guide for voltage characteristics electricity product characteristics and electromagnetic compatibility*. 230.02 NORMCOMP, UNIPEDE, 1995.
[21] *Measurements of quality in electric systems*. WG 39.04, Electra No. 185, 1999.
[22] P 1433/D5A, *A draft standard glossary of power quality terminology*. IEEE, 1999.
[23] Piróg S., *Negatywne oddziaływania układów energoelektronicznych na źródła energii i wybrane sposoby ich ograniczania*, Wydawnictwo AGH, Kraków, 1998.
[24] Plantive E., Courtois C., Poirrier J. P., Javerzac J. L., Application of a 20 MVA STATCOM for voltage balancing and power active filtering of a 25 kV AC single-phase railway substation connected to the 90 kV grid in France. CIGRE, 2000, 13/14/36-12.
[25] PN-EN 50160, *Parametry napięcia w publicznych zasilających sieciach elektrycznych*, 1995.
[26] PN-EN 61000-2-4, *Kompatybilność elektromagnetyczna. Środowisko. Poziomy kompatybilności dotyczące zaburzeń przewodzonych małej częstotliwości w sieciach zakładów przemysłowych*, 1997.
[27] PN–T-01030, *Kompatybilność elektromagnetyczna. Terminologia*, 1996.
[28] *Power quality workbook for utility and industrial applications*, EPRI, 1995.
[29] Robert A., Marquet J., Assessing voltage quality with relation to harmonics, flicker and unbalance. CIGRE, 1992.
[30] *UIE Guide to quality of electrical supply for industrial installations. Part 1: General introduction to electromagnetic compatibility (EMC), types of disturbances and relevant standards*, 1994.
[31] *UIE Guide to quality of electrical supply for industrial installations. Part 4: Voltage unbalance*, 1996.

7

Voltage and Current Harmonics

Angelo Baggini and Zbigniew Hanzelka

Harmonics have always been present in power systems. Recently, due to the widespread use of power electronic systems resulting in an increase in their magnitude, they have become a key issue in installations.

Harmonic disturbances come generally from equipment with a non-linear voltage/current characteristic. Nowadays a large part of industrial, commercial and domestic loads is non-linear, making the distortion level on the low-voltage (but not only) supply network a serious concern. Linear loads are comparatively rare today: the only examples which can be considered as common are undimmed filament bulbs and unregulated heaters. Non-linear loads represent a large percentage of the total loads. Under these conditions, total harmonic distortion (THD) may become very high and therefore dangerous for the system.

Harmonic distortion can be considered as a sort of pollution of the electric system which can cause problems if the sum of the harmonic currents exceeds certain limits. Knowledge of electromagnetic disturbances associated with this phenomenon is still developing; for this reason harmonics are currently an issue of great interest.

Interharmonics are not the subject of this chapter. They are discussed in detail in for example [16].

7.1 DESCRIPTION OF THE PHENOMENA

A harmonic is defined as a component with a frequency that is an integer multiple (the so-called order of harmonic n) of the fundamental frequency (Figure 7.1). The harmonic number indicates the harmonic frequency: the first harmonic is the fundamental frequency (50 or 60 Hz), the second harmonic is the component with frequency two times the fundamental (100 or 120 Hz), and so on.

Handbook of Power Quality Edited by Angelo Baggini

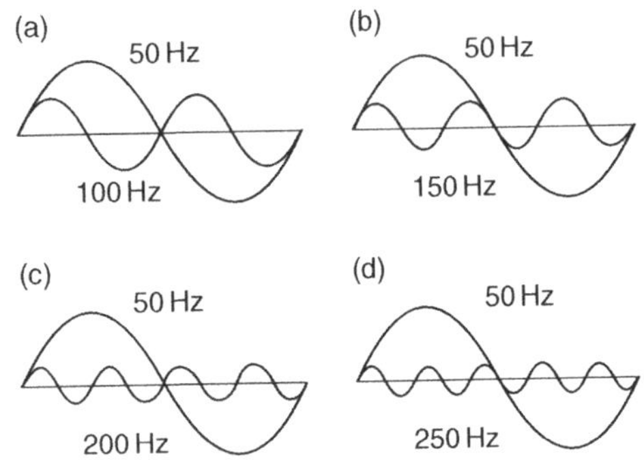

Figure 7.1 A sinusoidal waveform with fundamental frequency 50 Hz and its harmonics: (a) second (100 Hz); (b) third (150 Hz); (c) fourth (200 Hz); (d) fifth (250 Hz)

7.1.1 Composition of Distorted Waveform

A distorted waveform (i.e. different from the sinusoidal) of a chosen, arbitrary shape can be obtained by superposition of sinusoidal waveforms of various frequencies and amplitudes. For example, the rectangular waveform in Figure 7.3 is the result of the summation of an indefinite number of harmonics whose amplitudes are decreasing in inverse proportion to their order (n), and their frequencies $f_{(n)} = (2k + 1)50\,\text{Hz}$, $k = 0, 1, 2, \ldots$ are odd multiples of the fundamental harmonic $f_{(1)} = 50\,\text{Hz}$. Neglecting harmonics of small amplitude and taking into account a finite number of components *softens* the sharp top of the waveform and reduces the steepness of its edges.

The waveform shape depends not only on the frequencies and amplitudes of harmonic components but also on their mutual phase shift. Figure 7.2 illustrates the influence of harmonic amplitude and phase on the resultant waveform shape.

The waveforms in Figure 7.2(a) and (b) contain, apart from the fundamental component whose pu value is 1, also a cophasal fifth harmonic whose value is 0.15 (Figure 7.2a) and 0.30 (Figure 7.2b). The waveform in Figure 7.2(c), apart from the fundamental component, also contains the fifth harmonic with amplitude 0.15 shifted in phase by 180° with respect to it.

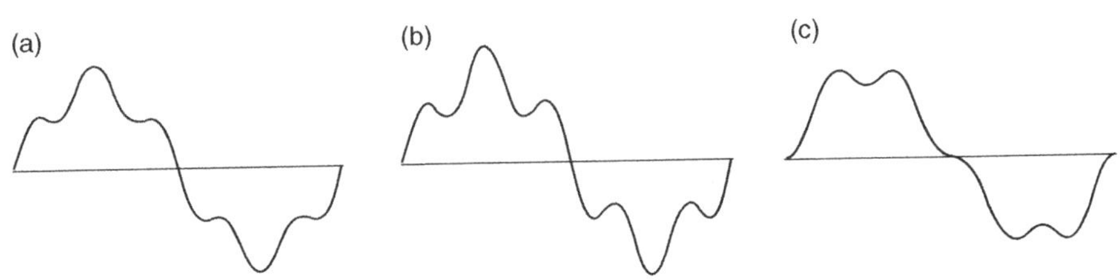

Figure 7.2 Waveforms differing in the amplitude and phase of a distorting harmonic: (a) 1-100 %, 5-15 %, $\varphi_{15} = 0$; (b) 1–100 %, 5–30 %, $\varphi_{15} = 0$; (c) 1–100 %, 5–15 %, $\varphi_{15} = 180°$ [28]

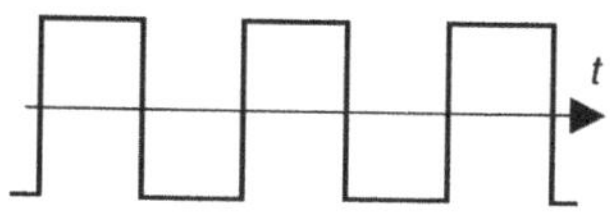

Harmonic order	Frequency [Hz]	Relative amplitude value
1	50	1.0
3	150	1/3
5	250	1/5
7	350	1/7
9	450	1/9
...	...	...
n	50*n*	1/*n*

Figure 7.3 Rectangular waveform as the superposition of the fundamental and odd harmonics

7.1.2 Decomposition of Distorted Waveform

As each distorted waveform[1] can be composed from harmonic components, so also can any periodic waveform be decomposed into harmonic components (Figure 7.4).

This analysis technique allows one to consider separately each component of the distorted waveform and, employing standard methods for circuit analysis, obtain a final result being the sum of partial results. The French mathematician Jean Baptiste Joseph Fourier (1768–1830), who is considered the author of this technique, demonstrated that any periodic waveform can be deconstructed into a sinusoid at the fundamental frequency with a number of sinusoids at harmonic frequencies. Depending on the kind of waveform, these coefficients may or may not exist. A d.c. component may complete these purely sinusoidal terms. This concept can be explained by the following equation:

$$
\begin{aligned}
f(x) = & \frac{a_0}{2} + a_{(1)} \cos x + b_{(1)} \sin x + a_{(2)} \cos 2x + b_{(2)} \sin 2x + \ldots + \\
& + a_{(n)} \cos\, nx + b_{(n)} \sin\, nx + \ldots \\
= & \frac{a_0}{2} + \sum_{n=1}^{\infty} a_{(n)} \cos nx + \sum_{n=1}^{\infty} b_{(n)} \sin\, nx = \frac{a_0}{2} + \sum_{n=1}^{\infty} c_{(n)} \sin\left(nx + \phi_{(n)}\right)
\end{aligned}
\tag{7.1}
$$

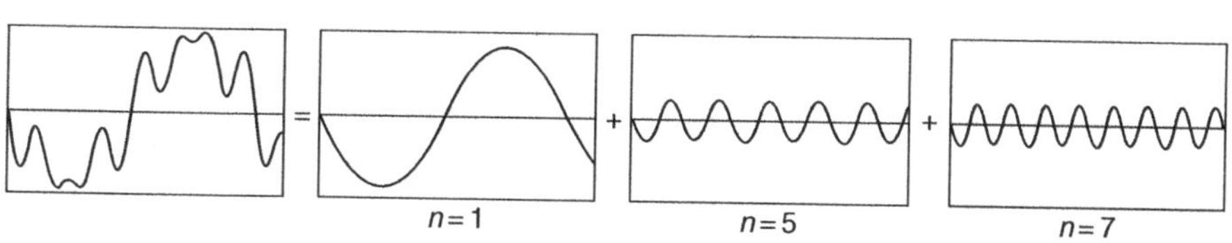

Figure 7.4 Decomposition of the distorted waveform into harmonic components: the fundamental (1) and the fifth and seventh

[1] Satisfying the so-called Dirichlet conditions.

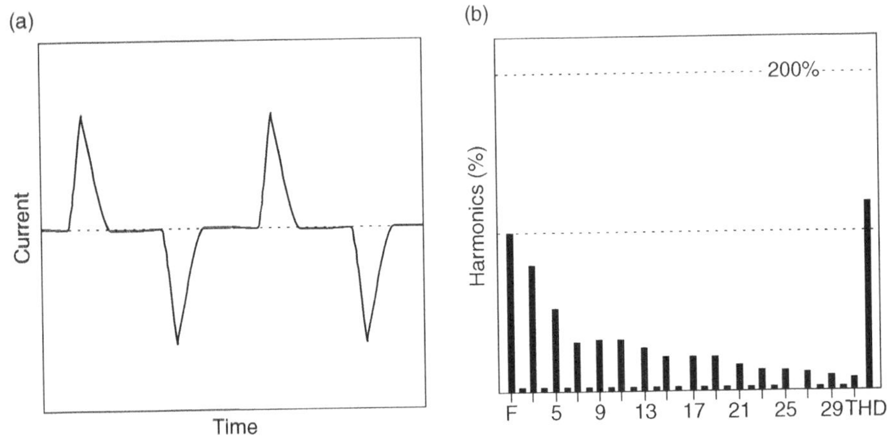

Figure 7.5 The current waveform of a compact fluorescent lamp (CFL) and its spectrum

where $f(x)$ is a generic periodic waveform; a_0 is the d.c. component, calculated as in (7.2); a_n, b_n are the coefficient of the series, calculated as in (7.2); n is an integer number between 1 and infinity; and $T = 2\pi$ is the period. That is,

$$a_0 = \frac{1}{\pi}\int_0^{2\pi} f(x)\,dx \quad a_{(n)} = \frac{1}{\pi}\int_0^{2\pi} f(x)\cos(nx)\,dx \quad b_{(n)} = \frac{1}{\pi}\int_0^{2\pi} f(x)\sin(nx)\,dx$$

$$c_{(n)} = \sqrt{a_{(n)}^2 + b_{(n)}^2} \quad \phi_{(n)} = \mathrm{arctg}\frac{a_{(n)}}{b_{(n)}} \tag{7.2}$$

Figure 7.5 shows the current waveform of an example non-linear load and its harmonic spectrum. It is possible to observe that current at fundamental frequency (F) is only a percentage of total current.

7.1.3 Harmonics and Symmetrical Components

Harmonics in a three-phase system, similar to the fundamental component, can be considered by applying the notion of symmetrical components. In a balanced three-phase system for the fundamental component and where each harmonic exists, there is a simple relation between the harmonic order and the corresponding phase sequence (Table 7.1).

Figure 7.6 shows the third-harmonic waveform as related to phase voltages or currents. It can be seen that in given phases these components are cophasal. Similarly, it can be demonstrated that the fifth-order harmonics in given phases are shifted with respect to each other by 120°, and their sequence is: L1 → L3 → L2.

7.1.4 Classification of Distortion Components

In terms of the type of the analyzed waveform, harmonics can be distinguished as voltage or current harmonics. A commonly used term is **triple harmonics**, which refers to the

Table 7.1 Relations between symmetrical components and harmonic orders

Symmetrical components	Positive sequence	Negative sequence	Zero sequence[2]
Harmonic order	1	2	3
	4	5	6
	7	8	9
	10	11	12
	...	...	...
	$3k+1$	$3k+2$	$3k+3$ for $k=0,1,2,3,\ldots$

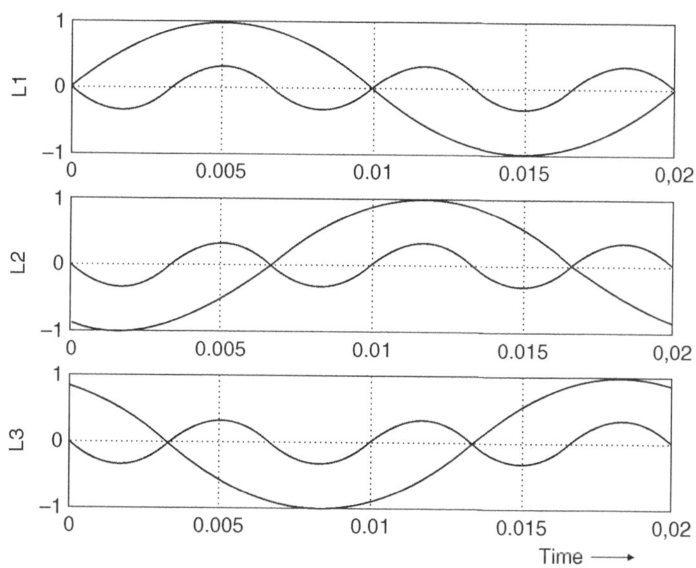

Figure 7.6 A three-phase balanced voltages or currents system with the third harmonic

components whose orders are multiples of three. Terms like even and odd harmonics do not need clarification.

In converter systems two additional groups of harmonics are distinguished: characteristic harmonics which occur in the converter currents under idealized operating conditions; and non-characteristic harmonics, occurring under actual operating conditions, whose presence is not related to the converter pulse number, e.g. the fifth harmonic in a 12-pulse converter (see Figure 7.42).

[2] Zero-sequence harmonics are also called homopolar harmonics. In a three-phase system homopolar currents are a sum in the neutral conductor as explained in Section 7.6.2.

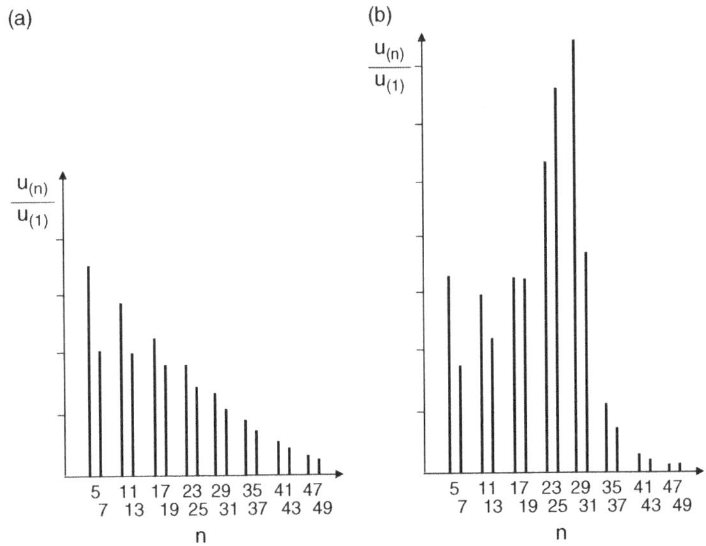

Figure 7.7 Example of harmonic spectra [28]

For a vast majority of non-linear loads the magnitude of the harmonic decreases with its increasing order. The typical form of harmonic spectrum shown in Figure 7.7(a) should be distinguished from an atypical, and rarely occurring, spectrum which can result from, for example, resonance amplification – Figure 7.7(b).

In terms of the relation between frequencies of the analyzed waveform components and the fundamental frequency, the so-called interharmonics can be distinguished, apart from harmonics. Interharmonics are voltages or currents with a frequency that is a non-integer multiple of the fundamental frequency. Another term is used too, namely subharmonic, which does not have any official definition. It is a particular case of an interharmonic with a frequency less than the fundamental frequency. However, the term has appeared in numerous references and is in general use in the professional community.

7.1.5 Quantities Describing Voltage and Current Distortion

Voltage/current distortion can be characterized in either the time or the frequency domain. Description in the time domain consists of finding the differences between the actual, distorted waveform values and the reference sinusoidal waveform values. The difficulty in determining these differences by means of measurement causes this method of description is seldom used. The distortion description in the frequency domain is commonly accepted. The most complete information is provided by the set of numbers determining the orders, amplitudes (r.m.s. values) and phases of individual harmonics. Standardization documents adopt various numerical values defined on this set (usually the harmonic phase is not taken into account). The quantities listed in Table 7.2 are the bases of power quality standardization. Their actual values (most often, the percentiles CP95) are compared to limit values set forth in standards and related regulations. On this basis, the technical conditions of connection or equipment certificates are issued.

Table 7.2 Voltage and current distortion factors

$D_{(n)} = \frac{U_{(n)}}{U_{(1)}}$; $D_{(n)} = \frac{U_{(n)}}{U}$	*The* n*th voltage harmonic ratio* (analogously for a current harmonic). Any harmonic component can be represented as a percentage of fundamental (%f) or as a percentage of r.m.s. value (%r,mostly nominal) of the total voltage (current – Figure 7.8a)
$D_{(1)} = \frac{U_{(1)}}{U}$	*Fundamental voltage harmonic ratio* (analogously for a current harmonic)
$\frac{U_{peak}}{U}$	*Peak factor* – the ratio of the peak and r.m.s. value of a periodic waveform, which for a sinusoidal wave is 1.41. The logical consequence of this is that any other value means a waveform distortion. This factor is curtailed for certain design tasks, e.g. selecting semiconductor devices according to the voltage/current peak value
$\frac{U}{U_{average}}$	*Crest factor* – the ratio of r.m.s. value to average, which for a sine wave is 1.11. A different value indicates the waveform distortion
$D_{hh} = \sqrt{\sum_{n \geq 2} D_{(n)}^2}$	*R.M.S. value of harmonic*
$TDC = \sqrt{Q^2 - Q_{(1)}^2}$	*Total distortion content*
$THD_U = \frac{\sqrt{\sum_{n=2}^{n_{limit}} U_{(n)}^2}}{U_{(1)}} 100\,\%$	*Total harmonic distortion factor* (THD_U, Figure 7.8b) – the ratio of the r.m.s. values of harmonic components to the r.m.s. value of fundamental component. Typically the upper limit of summation is taken $n_{limit} = 50$. If the risk of resonance for high-order harmonics is low the summation can be limited to the 25th order. The total harmonic current distortion factor (THD_I) takes a similar form
$TDR = \frac{\sqrt{U^2 - U_{(1)}^2}}{U_{(1)}}$	*Total distortion ratio*
$TDF = \sqrt{\sum_{n \geq 2}^{n \leq 50} \left(\frac{U_{(n)}}{U_N}\right)^2}$	*Total distortion factor*
$PWHD = \sqrt{\sum_{n \geq 14}^{n \leq 40} n \left(\frac{I_{(n)}}{I_{(1)}}\right)^2}$	*Partial weighted harmonic distortion*
$TDD_i = \sqrt{\sum_{n=2}^{40} \left(\frac{I_{(n)}}{I_{(1u)}}\right)^2} \cdot 100\,\%$	*Total demand distortion* for a load current

Table 7.2 (Continued)

$U_{(n)}$	the amplitude of current harmonic n
$U_{(1)}$	the amplitude of the fundamental component
U_N	nominal voltage
U_{peak}	peak value
$U_{average}$	average value
U	r.m.s. value
$I_{(1u)} = \frac{S_u}{\sqrt{3} \cdot U_N}$	current fundamental component, pertinent to the demand power S_u
S_u	demand power (MVA)

7.2 PARAMETERS

As previously stated, with Fourier's method it is possible to analyze every periodic signal through the sum of its harmonics content. Any component has a proper frequency which is an integer multiple of rated frequency. According to IEC 61000-2-1, harmonic frequency is a frequency which is an integer multiple of the fundamental frequency. The ratio of the harmonic frequency to the fundamental frequency is the harmonic order. According to the IEC recommendation, the order of a harmonic is denoted by the letter 'h' (IEC 61000-2-2).

Any harmonic component can then be represented as a percentage of the fundamental (%f) or a percentage of the r.m.s. value (%r) of the total current (see Figure 7.8a), with the following formula:

$$I_h = 100\frac{I_n}{I_1}$$

where I_n is the amplitude of current harmonic n and I_1 is the amplitude of fundamental current (or the r.m.s. value of the total current).

For harmonic voltages the approach is the same. The ratio of the r.m.s. value of the sum of all the harmonic components up to a specified order to the r.m.s. value of the fundamental component is called total harmonic distortion and can be represented as

$$THD = 100\sqrt{\sum_{n=2}^{\infty} I_h^2} = 100\sqrt{\sum_{n=2}^{\infty} \left(\frac{I_n}{I_1}\right)^2}$$

This parameter is used in low-voltage, medium-voltage or high-voltage systems.

Conventionally, current distortion parameters are suffixed with 'I', e.g. 35 % THD_I, and voltage distortion figures with 'V', e.g. 4 % THD_V.

More specific parameters are described in Table 7.2.

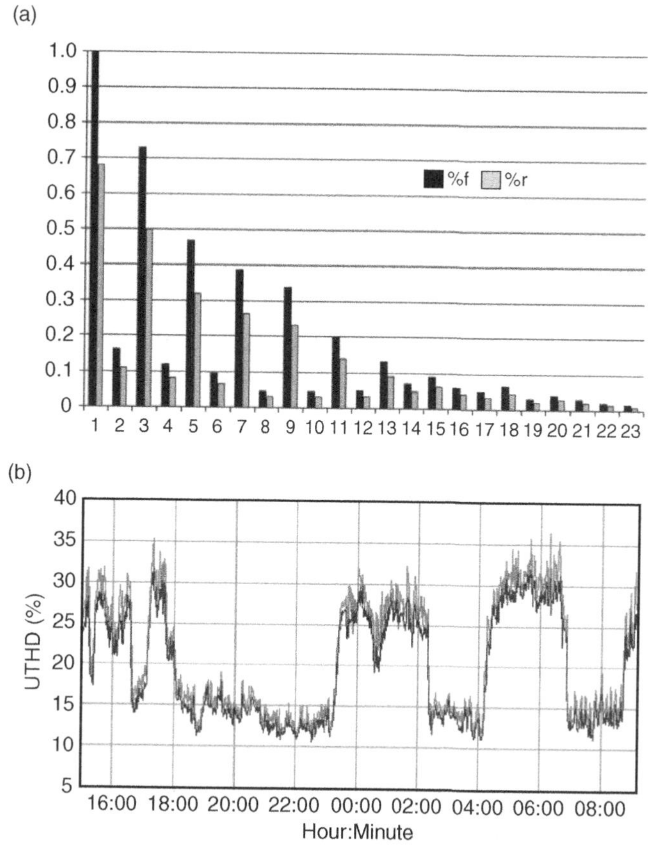

Figure 7.8 (a) Spectrum of a waveform represented in %f (fundamental) and %r (r.m.s. value); (b) example of THD characteristic in industrial network

7.3 MEASUREMENTS

The main analysis tool used by measurement devices is the Fourier transform (FT) and its modifications and improvements. This transform can be applied to an arbitrary function, both periodic and non-periodic. The result of this transform is the spectrum in the frequency domain which in the case of a non-periodic function is continuous. Its special case is a periodic function whose spectrum is discrete and its lines are components: the fundamental and harmonic.

The discrete Fourier transform (DFT) is a digital application of the classical Fourier transform. The analogue signal to be analyzed is sampled, A/D converted (sampled, quantized and then coded) and the results of sampling are stored. In practical applications the signal is analyzed in a limited time interval (measurement window of duration T_w) using a limited number of samples (M) of the actual signal. Results of the DFT depend on the choice of the T_w and M values. The inverse of T_w is known as the fundamental Fourier frequency (f_F) (DFT resolution). The DFT is applied to the measured signal within the measurement window. The signal outside the window is not processed and is assumed to be identical to the waveform inside the window, i.e. the signal is assumed to be M-samples periodic. In this way, the real signal is substituted with a virtual one, which is periodic with a period equal to the window width. To ensure that the results of the DFT applied to the functions which are considered periodic will be the same as those obtained from the Fourier analysis of an infinite length signal, the duration T_W of the time window should be an integer multiple of the fundamental period, i.e. $T_W = NT_1$; this requires the sampling rate f_S to be an integer multiple of the basic DFT frequency $f_s = Mf_b = M/(NT_1)$.

Sampling synchronization is of key importance. The loss of synchronization can alter the spectrum by adding extra components (lines) and changing the amplitudes of actual components. Before DFT processing, the samples obtained in the time window T_W are often weighted by multiplying them with a special symmetrical function (windowing function). However, for periodic signals and synchronous sampling, it is preferable to use a rectangular weighting window which multiplies each sample by unity.

In practical applications, when equipment and software limitations require the number of samples M to be no greater than a certain maximum number, the measurement time is limited. A measurement time different from the fundamental Fourier period leads to a discontinuity between the signal at the beginning and the end of the measuring window. This issue gives rise to errors in identification of the components known as spectrum leakage. A possible solution is the use of the *weighted* time window to a time-varying signal before DFT (or fast FT, FFT) analysis. In common practice two kinds of measuring windows are used: rectangular and Hanning windows.

The FFT is a special algorithm allowing computation times to be shortened. It requires that the number of signal samples (M) be an integer power of 2 ($M = 2^i$). In other words, it requires that the ratio of the sampling rate and basic DFT frequency be expressed by an integer power of 2. Considering the capabilities of state-of-the art digital signal processors, sine and cosine tables used in the DFT could be a helpful modification of the algorithm.

If the supply system voltage is analyzed, the component with fundamental frequency is that of the largest amplitude. It is not always the first line in the spectrum obtained from DFT processing of a time function. Where the current is analyzed, the fundamental frequency component is not necessarily the one with the largest amplitude.

Most instruments for measurements in the frequency domain work correctly when only integer harmonic components are present in the measured signal. These instruments feature a phase-locked loop aiming to synchronize measurements with the fundamental component and sample the signal during one or several cycles in order to analyze it using the FFT. Due to this phase-locked loop, the single-cycle samples provide an accurate representation of the waveform spectrum only when it does not contain interharmonics. If other non-harmonic frequencies (in relation to the measuring period) are present and/or the sampled waveform is not periodic in this time interval, the interpretation of results becomes difficult.

Because of the contemporaneous presence of both harmonic and interharmonic components, the Fourier frequency (the greatest common divisor of all component frequencies contained in the signal) is different from the supply voltage fundamental frequency and usually very small. This leads to two problems:

1. The minimum sampling time may be long and the number of samples large.
2. It is difficult to predict the fundamental Fourier frequency because not all the components of the signal are known.

This can be easily understood with the following examples.

The signal to be analyzed has a fundamental component (50 Hz) and an interharmonic (71.2 Hz) and harmonic (2500 Hz). The fundamental Fourier frequency is 0.2 Hz, much lower than the frequency of the fundamental. The corresponding period is 5 s making the permissible minimum sampling time equal to 5 s. If the sampling rate is 10 kHz, which is practically the minimum applicable value resulting from the Nyquist criterion, the minimum required number of samples M is 50 000. Without the interharmonic component, the minimum time measurement would be 20 ms and the number of samples would be 200.

Another example can be a remote control signal with frequency 175 Hz superimposed on the sinusoidal supply voltage signal frequency 50 Hz. The superposition yields a periodic voltage with a period of 40 ms and DFT resolution 25 Hz. Classical Fourier analysis of this voltage yields a 25 Hz fundamental component with zero amplitude and two components with non-zero amplitudes: the second harmonic (50 Hz) with amplitude equal to the supply voltage; and the seventh harmonic (175 Hz) with amplitude equal to that of the remote control signal.

The greatest difficulty associated with sampling a continuous signal is the problem of ambiguity. The essence of this problem is illustrated in Figure 7.9. It follows from the figure that the same set of sampled data may describe several waveforms, indistinguishable by measuring equipment. The principle of frequency analysis is the representation of an arbitrary waveform by the sum of a series of sinusoidal signals. Such a method of presentation allows the quantitative analysis of the problem of ambiguity. For this purpose, consider the waveform shown in Figure 7.10.

A signal $x(t)$ is sampled in equal intervals of time h, determining the instants of sampling, for which values of the measured signal are indicated in the figure. Assume that the function $x(t)$ is sinusoidal with frequency f_0. The same points could also represent

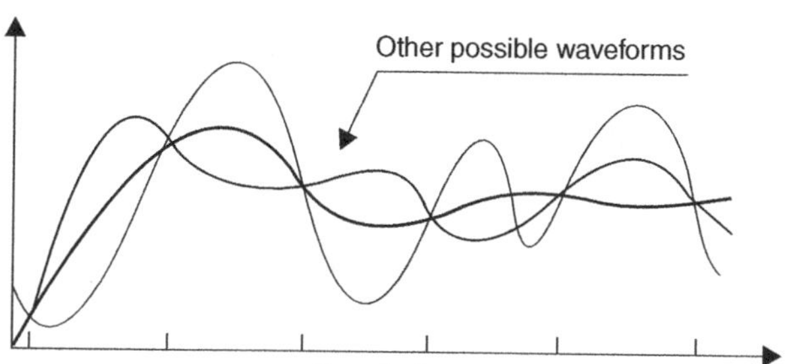

Figure 7.9 Ambiguity

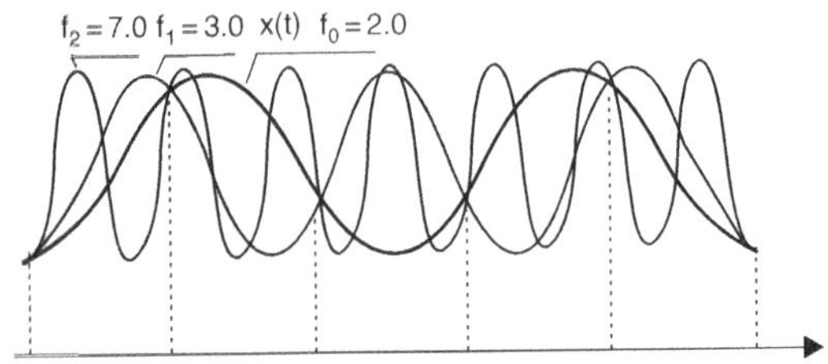

Figure 7.10 Analysis of ambiguity

sinusoids with frequencies f_1 or f_2. These various frequencies are obviously associated with the sampling period. The frequency f_0 is referred to as the fundamental frequency.

It could be stated, without presentation of the mathematical proof, that the range of frequencies for which the effect of ambiguity does not occur extends from $f_0 = 0$ to $f_0 = f_N$, where f_N, the maximum frequency, is referred to as the Nyquist frequency. It determines the limit frequency of data sampling, the so-called Shannon limit, beyond which a unique reconstruction of a continuous signal is not possible. Thus, if the signal being analyzed does not contain any component frequencies greater than f_N, then the minimum sampling rate necessary to allow the sampled signal to represent the real signal is given as $f_S \geq 2f_N$, or, because $f \geq 1/h$, then $f_N \geq 1/2h$.

This is the so-called sampling theorem. It follows that, for a given spectrum of frequencies, the components situated between $f_0 = 0$ and $f_0 = f_N$ can be considered separately. If the signal contains components of frequencies $f > f_N$, these components will not be distinguished. Therefore it is necessary to limit the bandwidth of the measured signal to reduce a direct consequence of the ambiguity during its sampling. That implies the need to filter the signal to be measured through a low-pass filter before sampling, in order to eliminate all frequencies greater than f_N.

The precise computing of harmonics is a difficult task and it often leads to a 'blurring' of the spectrum, even with synchronous sampling; for this reason the so-called grouping was introduced [22]. A given harmonic is then assessed using not only its spectral line, but also a group of lines around the harmonic being sought. The concept of grouping is particularly useful for the assessment of interharmonics. It allows only the spectral components to be determined in 5 Hz intervals, thus finding for example the 278 Hz interharmonic is not possible, since such a spectral line does not exists. The energy of this component will be distributed over several adjacent spectral lines. The approximate value of the sought interharmonic can be given by the value of this component group or subgroup.

The concept of grouping is illustrated in Figure 7.11 and Figure 7.12. Values of groups and subgroups are determined from the following relationships.

For harmonics, the group is

$$G_{g,n} = \sqrt{\frac{C_{k-5}^2}{2} + \sum_{i=-4}^{4} C_{k+i}^2 + \frac{C_{k+5}^2}{2}} \tag{7.3}$$

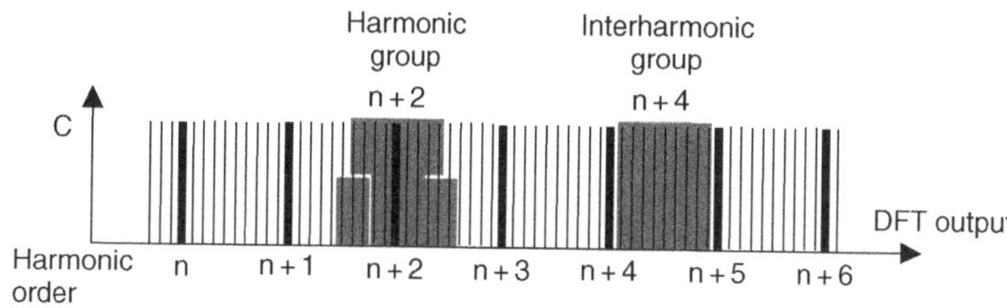

Figure 7.11 Illustration of harmonic and interharmonic groups [22] (Reproduced from current-carrying capacity and related overcurrent protection (Revision of section 523"), IEC TC 64 WG 2)

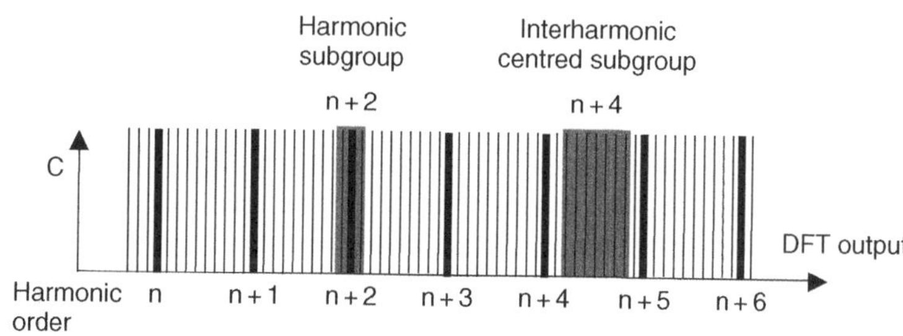

Figure 7.12 Illustration of a harmonic subgroup and an interharmonic centered subgroup [22] (Reproduced from current-carrying capacity and related overcurrent protection (Revision of section 523"), IEC TC 64 WG 2)

and the subgroup is

$$G_{sg,n} = \sqrt{\sum_{i=-1}^{1} C_{k+i}^2} \tag{7.4}$$

where $G_{\mathrm{g,n}}$, $G_{\mathrm{sg,n}}$ are the values of the harmonic group or subgroup, respectively, and C_{k+i} is the r.m.s. value of the ith line with respect to the harmonic of order n.

For interharmonics, the group is

$$G_{ig,n} = \sqrt{\sum_{i=1}^{9} C_{k+i}^2} \tag{7.5}$$

and the subgroup is

$$G_{isg,n} = \sqrt{\sum_{i=2}^{8} C_{k+i}^2} \tag{7.6}$$

where $G_{ig,n}$, $G_{isg,n}$ are the values of the interharmonic group or subgroup, respectively, and C_{k+i} is the r.m.s. value of the ith line with respect to the harmonic of order n.

Detailed information on harmonics measurements can be found in standard IEC 61000-4-7 [22].

7.4 SOURCES OF CURRENT HARMONICS

Among the sources of harmonic voltages and currents in power systems three groups of equipment can be distinguished:

- magnetic core equipment, like transformers, electric motors, generators, etc.
- arc furnaces, arc welders, high-pressure discharge lamps, etc.
- electronic and power electronic equipment.

7.4.1 Transformers

Transformers were historically the first sources of harmonics in a power system. The relationship between the primary voltage and current – shown in Figure 7.13 as a magnetization curve – is strongly non-linear and hence its location within the saturation region causes distortion of the magnetizing current (Figure 7.14).

The mechanism of harmonic generation is illustrated in Figure 7.15. For each time instant the shape of distorted magnetizing current can be reconstructed by finding the subsequent values of waveforms on the magnetization curve.

Transformers are designed so that the magnetizing current will not exceed 1–2 % of the nominal current. The nominal operating point is then located below the knee of the magnetizing curve, within its linear region. Consequently, even if a large number of transformers are operated in the power system, they are not a significant source of harmonics under normal operating conditions. This condition may change due to, for instance, a slight voltage increase. Within the saturation region even a small voltage increase above the nominal value results in a large increase in the magnetizing current. Also the harmonic content rises significantly.

For instance, at the voltage above the nominal U_N, as in Figure 7.14, the magnetizing current third-harmonic value may increase up to 50 %. This may occur under low-load conditions in the cable networks or as a consequence of switching on or off large reactive power loads, e.g. switching off shunt reactors or switching on a capacitor bank. The effects of switching are transients which propagate in the system and can cause transformer saturation, sometimes over a large area.

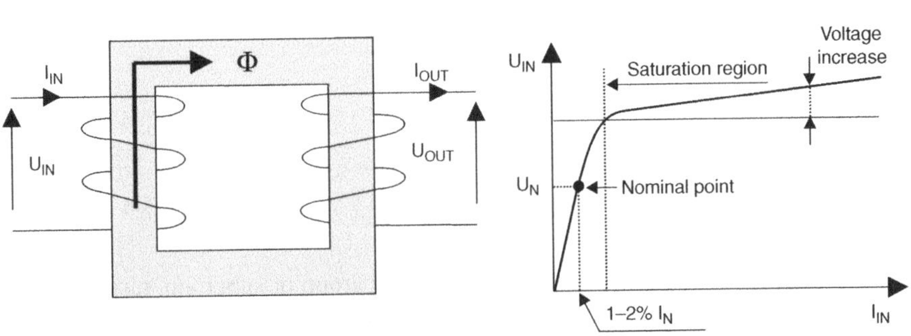

Figure 7.13 (a) Schematic diagram; (b) transformer magnetization curve (ϕ-flux)

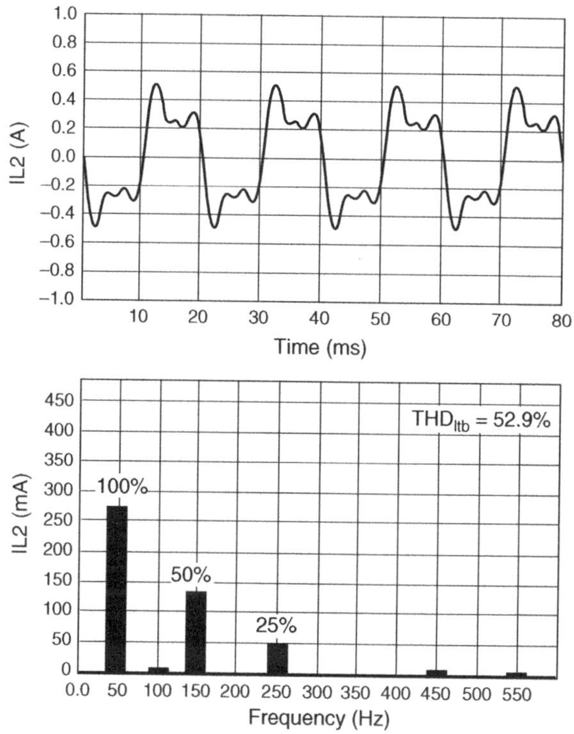

Figure 7.14 An example of a transformer-distorted magnetizing current and its harmonic spectrum

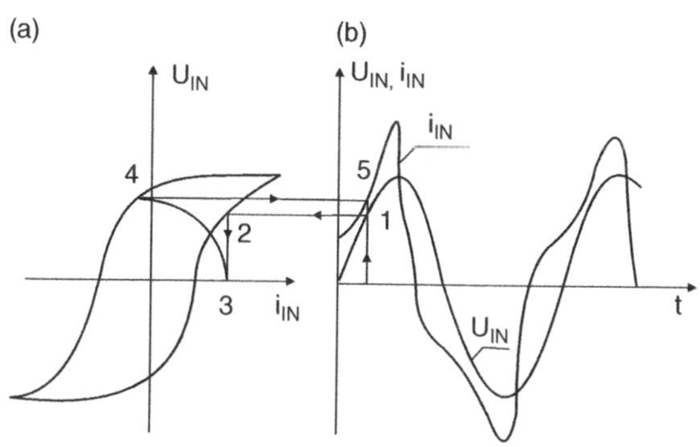

Figure 7.15 Generation of harmonics in the transformer magnetizing current

Considering a large number of transformers in the system and the fact that many of them are operated at low load, the effect can be a significant increase in voltage distortion. In transmission systems, Y_0/y_0-connected transformers are commonly used. With this winding connection a distorted magnetizing current may cause significant distortion of the transformer secondary voltage. On the contrary, the delta/y connection guarantees a low-impedance delta-connected circuit for the third harmonic. Therefore this harmonic does not distort the secondary voltage waveform. In distribution networks and plant networks, delta/y transformers are commonly used, which eliminates the problem of third-harmonic distortion (see Section 7.6.2).

7.4.2 Motors and Generators

Similar to transformers, motors can also generate harmonic currents in order to produce a magnetic field. Their contribution, however, is very small as the motor magnetizing characteristic, due to the presence of an air gap, is much more linear compared to the transformer magnetization characteristic.

The pitch of motor winding can also be a cause of harmonic currents. Typical motor windings have 5–7 slots per pole, which results in the generation of the fifth or seventh harmonic. In spite of the fact they are incomparably smaller than high harmonics in converter equipment, their presence is noticeable in the case of very large motors.

Harmonics (of very small magnitude) also occur in generator voltage, since for both practical and economic reasons a spatial distribution of the stator windings which could guarantee a purely sinusoidal voltage waveform is neither advisable nor possible. The induced voltages are therefore slightly distorted, and usually the third harmonic is the dominant one. It causes the third-harmonic current flow under generator load conditions.

7.4.3 Arc Furnaces

Distortion of arc furnace currents, and in consequence also of voltages, is an important issue because of their common use and large – in comparison to the short-circuit capacity at the point of connection – individual powers. Moreover, for technological reasons, arc furnaces are presently operated at a lower power factor than in the past. One of the consequences of this, as well as more stringent requirements regarding reactive power compensation, is the increasing rated power of the compensating capacitors. This results in lowering the resonant frequency. As the amplitudes of high harmonics are of significant value in this range of the spectrum, a magnification of the supply voltage harmonics may occur.

The form of the furnace voltage and current waveforms implies that representation of their distortion employing a discrete spectrum is only a certain, and commonly accepted, approximation of technical reality. These waveforms, having the nature of stochastic variables, are non-periodic functions of time (Figure 7.16). A continuous spectrum between the dominant harmonics has the nature of white noise of significant value.

Conditions for arc discharge change in subsequent phases of the heat. The highest level of current distortion occurs during the melting phase, whereas it is much lower in the other phases (air refining and refining). With the occurrence of a liquid metal surface a short arc occurs, the current fluctuations are smaller and the current waveform is closer to the sinusoidal one. A typical amplitude spectrum of the current – during the melting and refining – is shown in Figure 7.17. This spectrum exhibits the dominant harmonics with the largest amplitudes and of orders being both even and odd multiples of the fundamental frequency: 2, 3, ... It is regular for these amplitudes (determined as expected values)

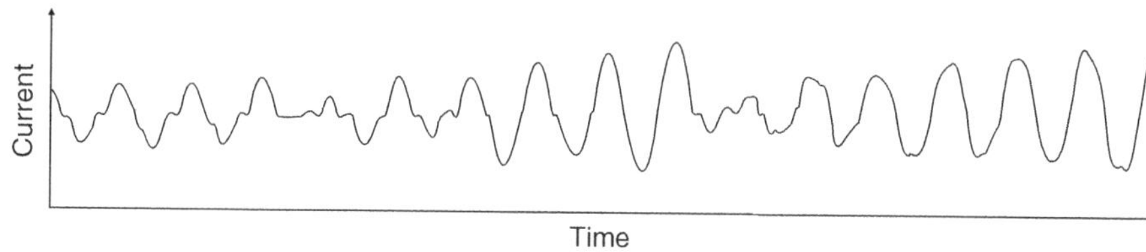

Figure 7.16 Example time graph of a furnace current during the starting phase of melting

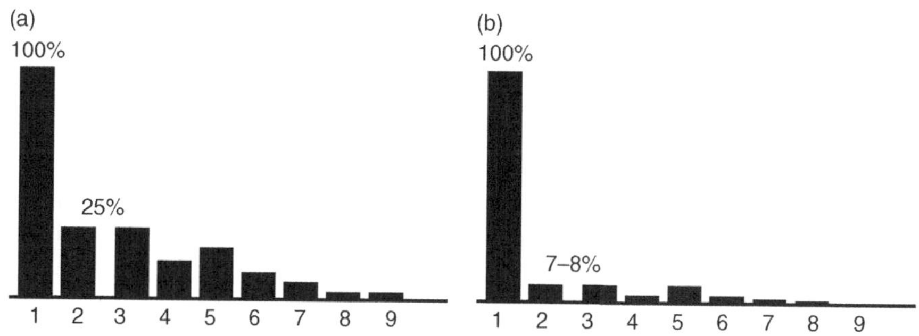

Figure 7.17 Typical harmonic spectrum of an arc furnace current (a) during melting and (b) during refining

to decrease quickly with the increase in the harmonic frequency. With increasing furnace power the voltage distortion increases, while the current distortion decreases.

Table 7.3 shows the results of high-harmonic measurements for arc furnaces of various charge capacity (as a percentage of the fundamental harmonic). Values of high harmonics of orders $n > 11$ are negligible. The presence of the third harmonic (generally of orders being integer multiples of 3) results from unbalanced operation of the furnace. Publications regarding high harmonics show a substantial differentiation of results, which is reasonable considering the measurement methods and instrumentation employed and the randomness of arc discharge phenomena and the types of arc furnaces and supplying networks.

Table 7.3 Example values of nth harmonic ratio in the arc furnace current, depending on the furnace transformer nominal power

Furnace transformer nominal power	Harmonic order								
	2	3	4	5	6	7	9	11	13–25
MVA	%	%	%	%	%	%	%	%	%
2.5	36	25	8	10	4	3	2	1	0
5	26	20	5	7	2	3	2	1	0
10	26	13	4	5	1	2	1	1	0
16	16	18	6	8	3	3	2	1	0
60	7	10	4	5	1	2	2	1	0

Harmonic emission from d.c. arc furnaces is analogous to that of multi-pulse rectifiers. Average values of harmonics are lower than the high-harmonic values of a converter operated in applications different from the arc furnace (e.g. an electric drive) and at the same time higher than those of an a.c. arc furnace.

7.4.4 Fluorescent Lighting

Electronic lighting ballasts have become popular recently because of their improved efficiency compared to magnetic ballasts, even if they are only a little more efficient than the best magnetic ballasts, and in fact most of the gain is attributable to the lamp which is more efficient when driven at high frequency rather than to the electronic ballast itself. Their main advantage is that the light level can be maintained over an extended lifetime by feedback control of the running current – a practice that reduces the overall lifetime efficiency. These devices have unfortunately a great disadvantage because of the harmonics they generate in the supply current. Today power-factor-corrected types are available in order to reduce the harmonic problems, but at a cost. In any case, smaller units are usually uncorrected.

Compact fluorescent lamps (CFLs) are now sold as replacements for tungsten filament bulbs. In these lamps, a small electronic ballast, installed in the connector casing, controls a folded 8 mm diameter fluorescent tube. CFLs rated at 11 W are sold as replacements for a 60 W filament lamp and have a life expectancy of nearly 8000 h. A typical harmonic current spectrum for these devices is shown in Figure 7.5. These types of lamps are widely used today to replace filament bulbs in domestic properties and especially in hotels where in fact serious harmonic problems have suddenly become common.

7.4.5 Electronic and Power Electronic Equipment

7.4.5.1 Switched Mode Power Supplies (SMPS)

The major part of modern electronic devices is fed by switched mode power supplies (SMPS) with single-phase rectifiers. The main difference from older units is in the lack of the traditional step-down transformer and rectifier: they are replaced by direct controlled rectification of the supply to charge a reservoir capacitor from which the direct current for the load is derived in order to obtain the output voltage and current required.

With this approach, the main advantage is that size, cost and weight have been reduced and the power unit can be made with practically any form factor. The disadvantage introduced is that now, instead of drawing continuous current from the supply, the unit draws pulses of current which contain large amounts of third- and higher-order harmonic components. Figure 7.18 shows the waveform and spectrum of a typical current for most of presently used electrical and electronic equipment (the current of a single-phase rectifier with a capacitive filter at the dc side). It can be clearly seen that the third-, fifth- and seventh-harmonic values are comparable to the fundamental component value.

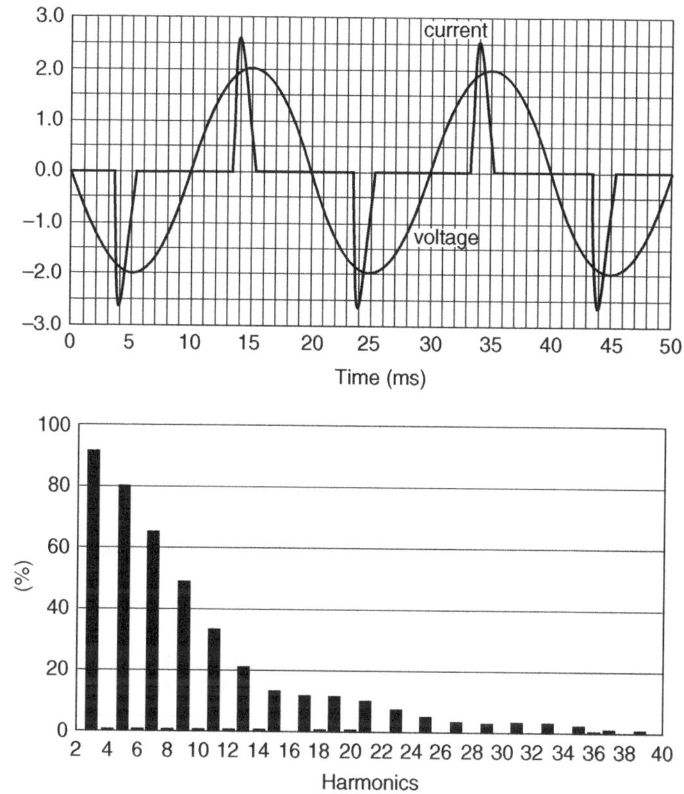

Figure 7.18 The typical waveform (shown against the voltage waveform) and spectrum of the current in a diode bridge with dc side capacitive filter ($THD_I = 130\,\%$)

7.4.5.2 Three-Phase Rectifier

All equipment containing static converters, as variable speed controllers, UPS units and a.c./d.c. converters in general, are based on a three-phase bridge, also known as a six-pulse bridge because there are six voltage pulses per cycle (one per half cycle per phase) on the d.c. output.

This bridge produces in supply networks current harmonics of order $6n \pm 1$, which means one more and one less than each multiple of six. In theory, the magnitude of each harmonic should be equal to the reciprocal of the harmonic number, so there would be 20 % of the 5th harmonic and 9 % of the 11th harmonic, etc. Figure 7.19 shows an example waveform of a thyristor bridge current against the phase voltage. Commutation notches are clearly visible in the voltage waveform (the source of high-frequency distorting components).

The magnitude of the harmonics is significantly reduced by the use of a 12-pulse converter (see Section 7.7.2). The harmonic spectrum of these two converters, according to [25], is shown in Table 7.4.

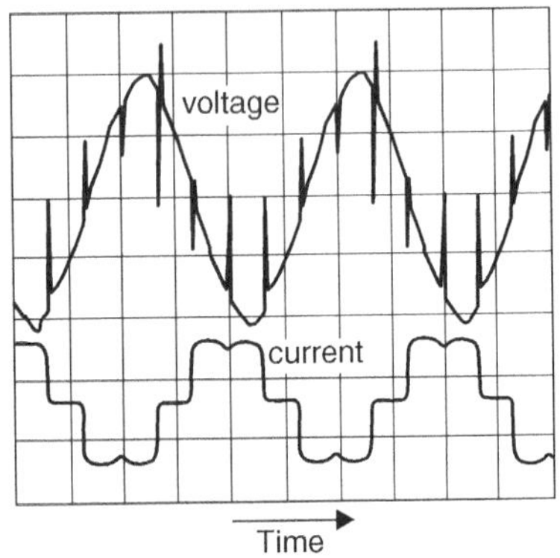

Figure 7.19 Example waveforms of the supply voltage and current of a six-pulse thyristor bridge with d.c. side reactor

Table 7.4 Example harmonic current pu values with respect to the fundamental harmonic value [25]

Harmonic order	5	7	11	13	17	19	23	25	29	31
6-pulse (pu)	6	0.132	0.073	0.057	0.035	0.027	0.02	0.016	0.014	0.012
12-pulse (pu)	0.0192	0.0132	0.073	0.057	0.0035	0.0027	0.02	0.016	0.0014	0.0012

7.4.5.3 Static Var Compensator – FC/TCR (see Chapters 12 and 15)

The schematic diagram of this circuit is shown in Figure 7.20. For the thyristor's control angle $\alpha = \pi/2$ (with respect to the positive zero crossing of the supply voltage) the phase reactor current is sinusoidal. Increasing the control angle ($\alpha > \pi/2$) not only reduces the current value but also causes the current discontinuity. Harmonics of odd order occur when control angles are identical for both thyristors in the switch T. Their r.m.s. values are determined by the relationship

$$I_{(n)} = \frac{4}{\pi}\frac{U}{X}\left[\frac{\sin(n+1)}{2(n+1)}\alpha + \frac{\sin(n-1)}{2(n-1)}\alpha - \cos\alpha\frac{\sin\alpha}{n}\right] \quad (7.7)$$

where $n = 3, 5, 7, \ldots$, U is the phase-to-phase r.m.s. voltage, X the total reactance in one reactor branch and α the control angle (Figure 20c).

The maximum values of harmonics (up to 25th) are listed in Table 7.5. It should be emphasized that they occur at different control angles. Under the reactors balanced operating conditions the percentage values given in the table apply to both the phase and line currents. Under balanced operation of the compensator, triplen harmonics (in parenthesis in Table 7.5) present in phase currents do not occur in line currents.

Table 7.8 shows also example current waveforms for different converter configurations.

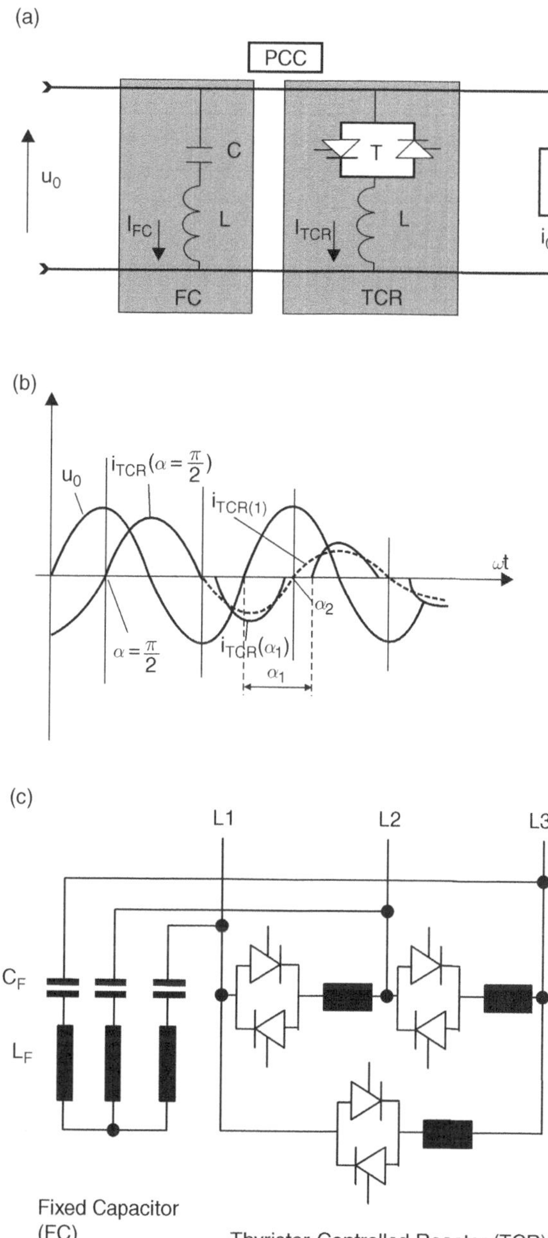

Figure 7.20 (a) Single-phase equivalent circuit of FC/TCR; (b) the voltage and current waveforms illustrating the principle of operation; (c) a three-phase system

Table 7.5 The maximum value of harmonics of FC/TCR reactor

Harmonic order	Max. value [%]	Dependence of harmonics on the control angle α
1	100	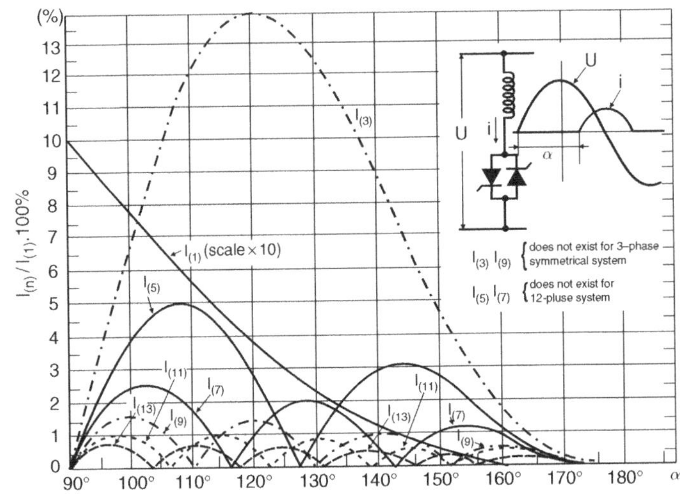
3	(13.78)	
5	5.05	
7	2.59	
9	(1.57)	
11	1.05	
13	0.75	
15	(0.57)	
17	0.44	
19	0.35	
21	(0.29)	
23	0.24	
25	0.2	

7.4.6 Harmonic Current Values/Magnitudes for Selected Loads

Magnitudes of harmonic currents for selected, commonly used loads are listed in Table 7.6 and Table 7.7. These values should be regarded as an example only, since they undergo substantial changes.

Table 7.6 Example magnitudes of harmonic currents of selected single-phase loads

Harmonic order	Welder [%]	PC [%]	Fluorescent lamp [%]
1	100	100	100
3	29.6	75	12.3
5	8.8	47.3	13.8
7	2.0	22.9	3.0
9	2.3	9.0	1.1
11	2.3	3.3	0.7
13	1.1	3.0	0.5
15	0.4	2.1	
17	0.9	1.9	

Table 7.7 Example magnitudes of harmonic currents of selected three-phase loads

Harmonic order	Induction furnace (with static frequency converter)	DC ASD[3]	DC arc furnace	ASD (PWM)[4]
1	100	100	100	100
5	20.9	37.1	18.9	25
7	12.7	1.1	10.3	11
11	7.8	8.6	5.4	7.5
13	7.2	2.5	3.9	5.0
17	4.3	4.7	1.8	4.4
19	4.9	2.3	1.3	3.2
23	2.6	3.1	0.6	2.6
25	3.6	2.1	0.5	2.0
29	1.7	2.2	0.5	1.7
31	2.7	1.9	0.5	1.3
35	1.2	1.7	0.4	1.0
37	2.0	1.8	0.4	0.8
41	0.8	1.4	0.3	0.6
43	1.4	1.6	0.3	0.5
47	0.5	1.1	0.2	0.4
49	1.0	1.3	0.2	0.3

7.5 VOLTAGE AND CURRENT HARMONICS

The component $I_{(n)}$ of a current supplying non-linear load produces across the supply network's equivalent impedance $Z_{s(n)}$ the voltage drop $\Delta U_{(n)} = Z_{s(n)} I_{(n)}$ (Figure 7.21). This voltage drop is the cause of voltage distortion at the point of connection of a harmonic sensitive load. Figure 7.22 illustrates the propagation of distortion along the supply line. The increase in the equivalent impedance – from the supply source (generator terminals) to the end user – results in an increased harmonic distortion level.

Figure 7.23 shows the waveform of the voltage supplying a typical office building. Although the absolute values of harmonic currents generated by personal computers and other office equipment are rather small, the resultant effect can be a significant voltage distortion. The voltage waveform's flat-topped characteristic for this type of load is clearly visible.

A typical change in the voltage THD during a working day over 24 hours is shown in Figure 7.24. Over a week the THD achieves its maximum value at the final phase of the weekend (Figure 7.25). Changes of this factor occur regularly on specific days of the week. For most of the time, these values are correlated with the current changes. From this figure it is evident that often the main sources of harmonics are not industrial loads but

[3] Phase-controlled converter with large value of control angle.

[4] Electric drive was supplied with a line reactor. Without the reactor the percentage content of the 5th, 7th, 11th and 13th harmonics will be much larger.

Table 7.8 The current waveforms for various types of converters (Reproduced from [18])

Converter type	Current waveform	Comments
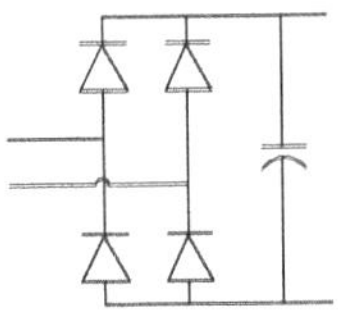	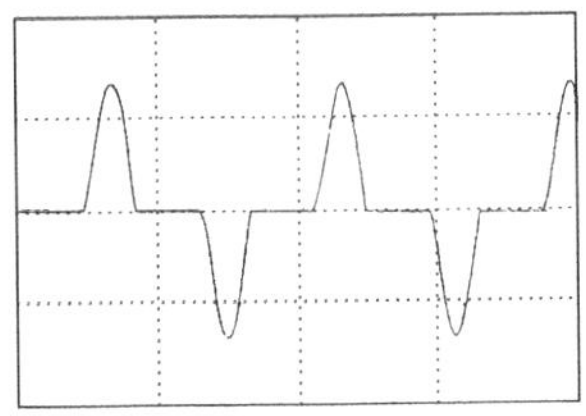	Single-phase rectifier $THD_I \approx 80\,\%$ dominant third harmonic
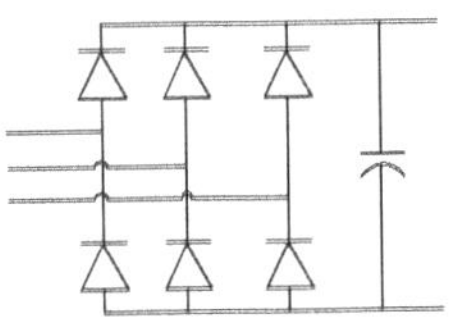	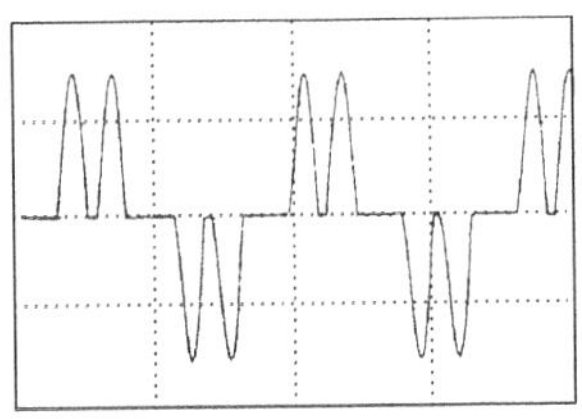	6-pulse rectifier with capacitor on the DC side $THD_I \approx 80\,\%$
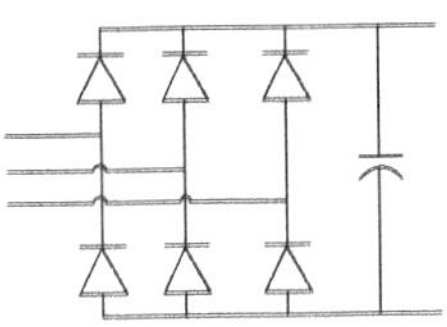	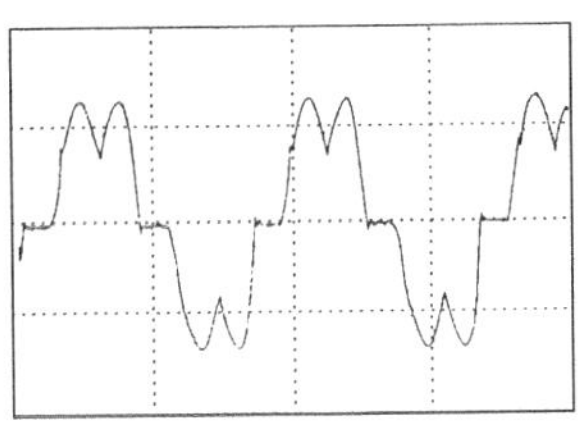	6-pulse rectifier with capacitor on the DC side and AC input reactors $> 3\,\%$, or a DC drive $THD_I \approx 40\,\%$
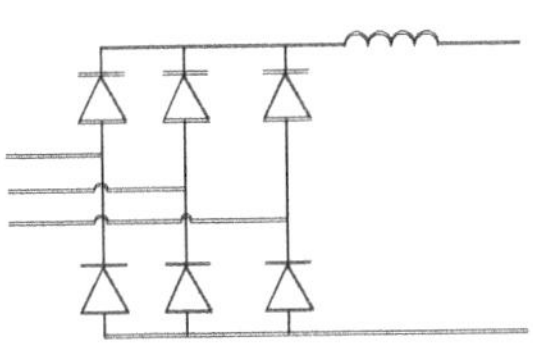	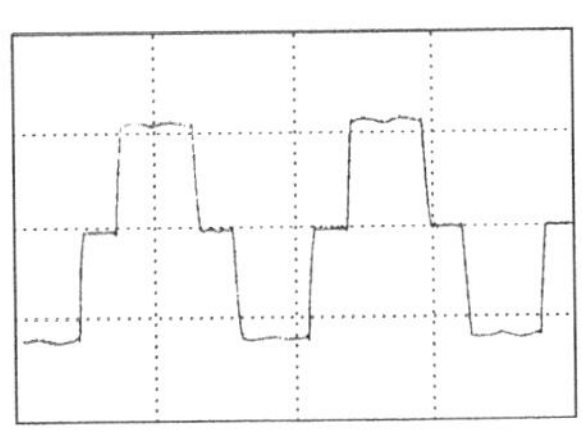	6-pulse rectifier with DC side reactor $THD_I \approx 28\,\%$
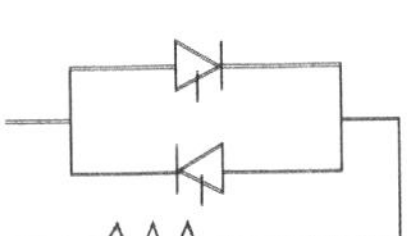	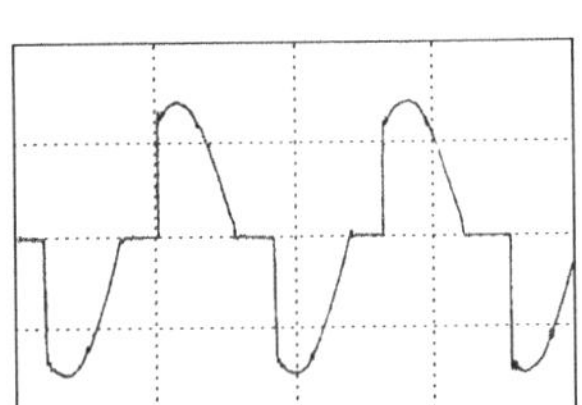	AC power controller (resistive load) THD_I variable with control angle

residential loads, particularly power suppliers of electronic and electrical home equipment. Their cumulative negative impact on the supply network becomes particularly visible during the evening peak load period. This kind of equipment is used both by households and by industry. It consists of: TVs, video recorders, computers, printers, microwave ovens,

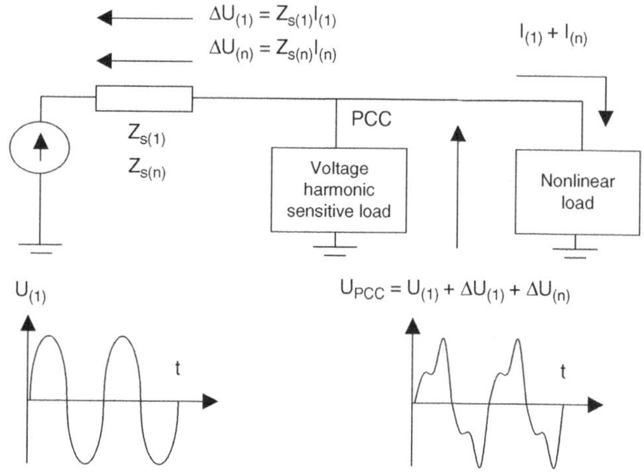

Figure 7.21 The effect of the voltage drop $\Delta U_{(n)}$ – voltage distortion at the point of common coupling (PCC)

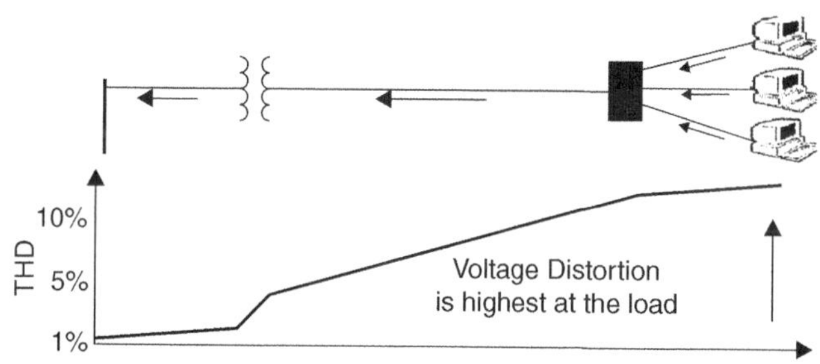

Figure 7.22 The increase in voltage distortion along the supply line

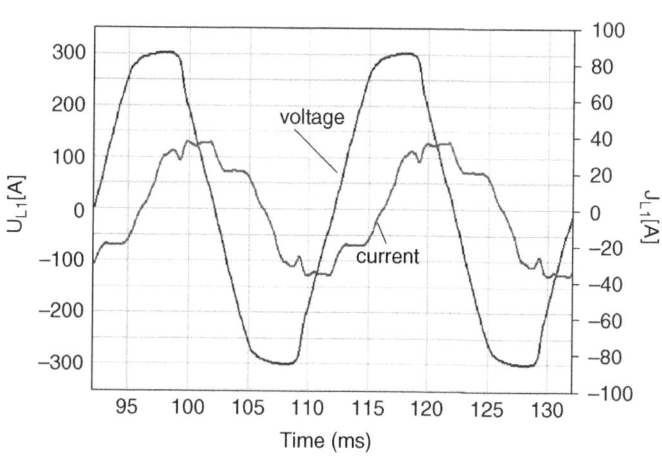

Figure 7.23 Example waveform of the office building supply voltage and its harmonic spectrum ($THD_U = 4\,\%$)

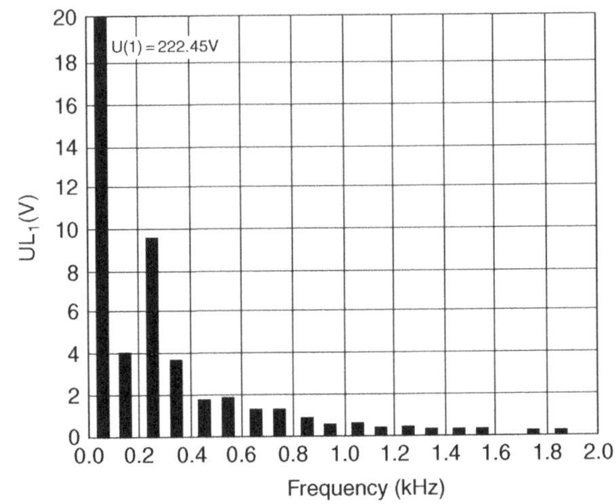

Figure 7.23 (Continued)

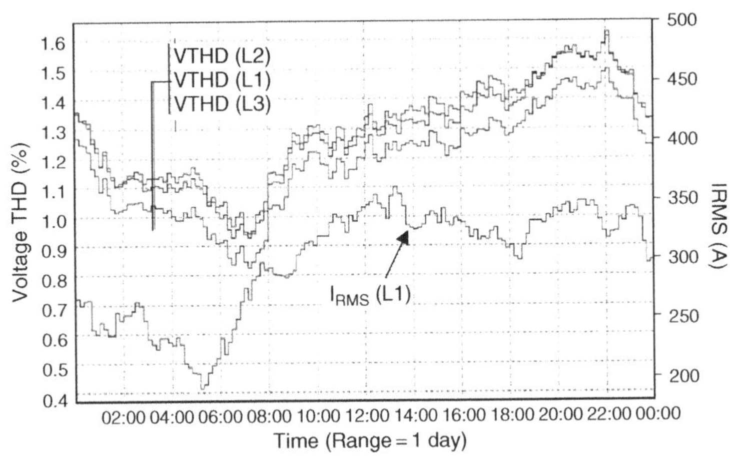

Figure 7.24 Change in the phase-to-neutral voltage distortion factor during the example workday 24 hours (110 kV network)

adjustable-speed drives (low power), HF fluorescent lighting, small UPS, etc. Their unit power is small, but their number is huge.

An increase in the voltage distortion factor due to the greater share of household loads (TV sets) during the World Cup Final football match, and reduction of the total load at the same time, is shown in Figure 7.26.

Figure 7.27 shows a typical harmonic spectrum. Odd harmonics are always dominant, the contribution of the third harmonic is insignificant and, practically, there are no even harmonics. Harmonic magnitudes are recorded over one week and averaged in 10-minute intervals. Their pu values with respect to the fundamental and THD factor are the basis

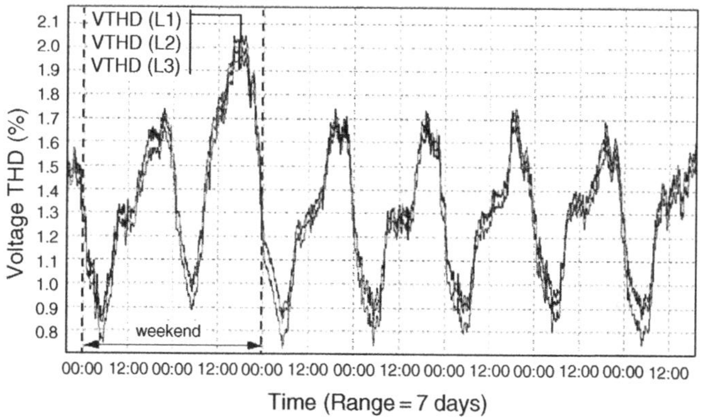

Figure 7.25 Week's variation of the phase-to-neutral voltage THD (110 kV network)

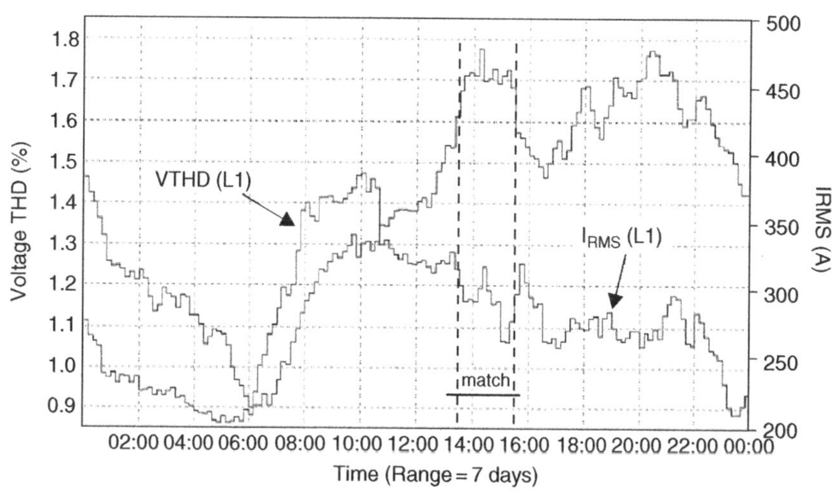

Figure 7.26 World Cup Final football match – the phase-to-neutral voltage and phase current THD (110 kV network)

for computing the percentiles (usually CP95, sometimes also CP99). Comparison of these values to the limits set forth in standards is the basis for power quality assessment at a given point of the supply network [12].

7.6 EFFECTS

The voltage/current distortion limit is determined by the sensitivity of loads (also of power sources) which are influenced by the distorted quantities. The least sensitive is heating equipment of any kind. The most sensitive kind of equipment is those electronic devices

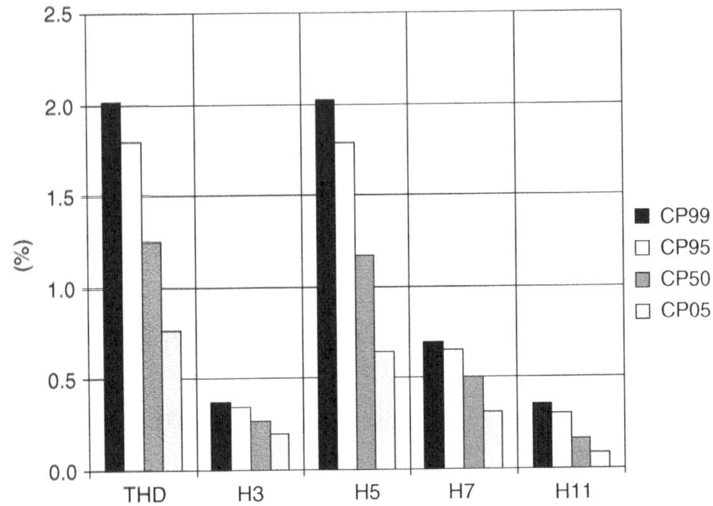

Figure 7.27 An example harmonic spectrum and percentile values

which have been designed assuming an ideal (almost) sinusoidal fundamental frequency voltage or current waveforms. Electric motors – the most popular loads – are situated between these two categories. Most of them tolerate significant distortion levels. An example classification of the effects of the presence of high harmonics is given in Table 7.9.

7.6.1 Power Factor

In the general case the voltage and current waveforms, as in Figure 7.28, are determined by relationships

$$u(t)=\sum_{n}\sqrt{2}U_{(n)}\sin\left(n\omega t+\alpha_{(n)}\right)+\sum_{m}\sqrt{2}U_{(m)}\sin\left(m\omega t+\alpha_{(m)}\right) \tag{7.8a}$$

$$i(t)=\sum_{n}\sqrt{2}I_{(n)}\sin\left(n\omega t+\alpha_{(n)}+\phi_{(n)}\right)+\sum_{p}\sqrt{2}I_{(p)}\sin\left(p\omega t+\alpha_{(p)}\right) \tag{7.8b}$$

where n, m, p are the orders of voltage and current harmonics; $U_{(n)}$, $U_{(m)}$, $I_{(n)}$, $I_{(p)}$ are harmonic voltages and currents of orders n, m, p (r.m.s. values); and $\alpha_{(n)}$, $\alpha_{(m)}$, $\phi_{(n)}$, $\alpha_{(p)}$ are the initial phases and phase shifts of individual harmonic voltages and currents.

The power factor (PF) of a load is determined by the relationship

$$PF=\frac{P}{S}=\frac{\frac{1}{T}\int_{0}^{T}u(t)\,i(t)\,dt}{UI}=\frac{P}{P_{(1)}}\frac{P_{(1)}}{S_{(1)}}\frac{S_{(1)}}{S}=[I]\cdot DPF\cdot[II] \tag{7.9}$$

where P, S are the active and apparent power; $P_{(1)}$, $S_{(1)}$ are the active and apparent power of the fundamental harmonic; and U, I are the voltage and current r.m.s. values.

Table 7.9 Classification of high harmonics effects

Classification criterion	Type of effect	Comment
Duration	Very short-term effects	These effects are associated with a failure, malfunction or inoperative state of equipment exposed to high harmonics, such as control and instrumentation equipment, electronic equipment, IT equipment, etc.
	Long-term effects	Chiefly of thermal nature. The thermal effect (causing accelerated ageing of insulation or, rarely, damage to equipment) is a function of many variables, of which the most important are the values and orders of harmonics
Physical nature of the distorted quantity	Current effects	Related to the instantaneous or time-averaged current value (overheating of electric machines, capacitor fuses blowing, increased losses in transmission lines, unwanted operation of relays, etc.). Harmonics in power supply systems are the main cause of temperature rise in the equipment and shortening of in-service time. This effect attains extremely high values under conditions of resonant amplification of harmonic currents
	Voltage effects	Associated with the peak, average or r.m.s. value of distorted voltage

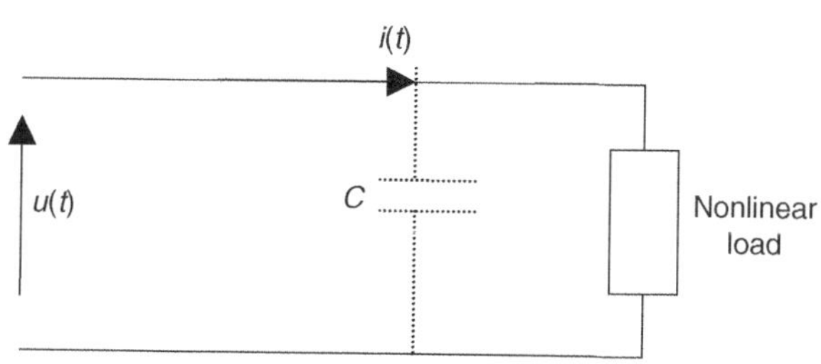

Figure 7.28 A single-phase non-linear load with compensating capacitor

Factor [I] is

$$[I] = \frac{P}{P_{(I)}} = \frac{\sum_n U_{(n)} I_{(n)} \cos\phi_{(n)}}{U_{(1)} I_{(1)} \cos\phi_{(1)}} \tag{7.10}$$

The fundamental harmonic power factor, also termed the displacement power factor (DPF), is given by

$$DPF = \frac{U_{(1)} I_{(1)} \cos\phi_{(1)}}{U_{(1)} I_{(1)}} = \cos\phi_{(1)} \tag{7.11}$$

Factor [II] is

$$\frac{S_{(1)}}{S}=\frac{1}{\sqrt{\left(1+THD_I^2\right)\left(1+THD_U^2\right)}} \tag{7.12}$$

The voltage and current r.m.s. values and total distortion factors are given by

$$\begin{aligned}U&=\sqrt{\sum_n U_{(n)}^2+\sum_m U_{(m)}^2}=U_{(1)}\sqrt{1+THD_U^2}\\ I&=\sqrt{\sum_n I_{(n)}^2+\sum_p I_{(p)}^2}=I_{(1)}\sqrt{1+THD_I^2}\end{aligned} \tag{7.13}$$

$$THD_U=\frac{\sqrt{U^2-U_{(1)}^2}}{U_{(1)}} \qquad THD_I=\frac{\sqrt{I^2-I_{(1)}^2}}{I_{(1)}} \tag{7.14}$$

If the voltage is sinusoidal then the relationship (7.9) takes the form

$$PF^*=DPF\frac{1}{\sqrt{1+THD_I^2}} \tag{7.15}$$

If the compensation is performed in a passive way, i.e. by means of capacitors, then the question arises: *What should be the value of capacitance C of the capacitor connected in parallel with the load (Figure 7.28) to be considered optimal, i.e. such that it minimizes the value of apparent power of the supply source, and therefore ensures the maximum value of the PF and the minimum value of the r.m.s. source current?*

The answer to the question formulated in such a way was presented by Shepherd and Zakikhani [30]. With the parallel capacitance C connected, the value of the source apparent power can be expressed by the relationship

$$S=\sqrt{\sum_n U_{(n)}^2+\sum_m U_{(m)}^2}\times\sqrt{\sum_n\left(I_{(n)}^2+U_{(n)}^2n^2\omega^2C^2+2U_{(n)}I_{(n)}n\omega C\sin\phi_{(n)}\right)+\sum_m U_{(m)}^2m^2\omega^2C^2+\sum_p I_{(p)}^2} \tag{7.16}$$

Optimal value of the capacitance C_{opt} can be found by comparing the apparent power derivative to zero:

$$\frac{dS}{dC}=0\Rightarrow C_{opt}=\frac{\frac{1}{\omega}\sum_n nU_{(n)}I_{(n)}\sin\phi_{(n)}}{\sum_n n^2U_{(n)}^2+\sum_m m^2U_{(m)}^2} \tag{7.17}$$

Consideration of power theory in electric circuits with periodic non-sinusoidal waveforms of voltages and currents is presented in Annex 3.

7.6.2 Overheating of Phase and Neutral Conductors

The effects of the presence of harmonics in the current can lead to overloading problems both on phase conductors and on the possible neutral conductor.

Under the conditions of current deformation, heat deformation inside the cable due to the Joule effect is evidently greater compared to ideal conditions and the line capacity is reduced, even without considering that neutral conductors, usually sized at most like phase conductors, can be overloaded without the neutral current exceeding the nominal phase current.

This issue is particularly important in low-voltage systems where harmonic pollution by single-phase loads is an increasingly serious problem. Triplen harmonic currents add arithmetically in the neutral conductor rather than summing to zero as do balanced fundamental and other harmonic currents. The result is neutral currents which are often significantly higher, typically up to 170 %, than the phase currents.

In a star-connected three-phase system, the current in the neutral conductor is the vector sum of the three line currents. The voltage waveforms of each phase referred to the neutral star point are displaced by 120°: when each phase is equally loaded, the combined current in the neutral conductor is zero, as shown in Figure 7.29. In other words, with a balanced sinusoidal three-phase system of currents, this sum is zero at any point in time and the neutral current is therefore zero.

If loads are not balanced a current flows in the neutral as a results of the vector sum of the three-phase currents which is now different from zero (Figure 7.30). It is important to point out that in a three-phase power system feeding linear single-phase loads, the current in the neutral conductor is rarely zero because the load on each phase is different. Typically the difference is small and is in any case far lower than the line currents.

In the past, the common practice of installers (with the approval of the standards authorities) has been to take advantage of this fact by installing *half-sized* neutral conductors. Where non-linear loads are being supplied, even when the load is well balanced across the phases, there is then likely to be substantial current in the neutral conductor. The supply current $i_i (i = L1, L2, L3)$ of such a load contains all odd harmonics, i.e. 1, 3, 5, 7, 9, . . . : $i_i = i_{i(1)} + i_{i(3)} + i_{i(5)} + i_{i(7)} + i_{i(9)} \ldots$ For a larger number of such loads connected to the

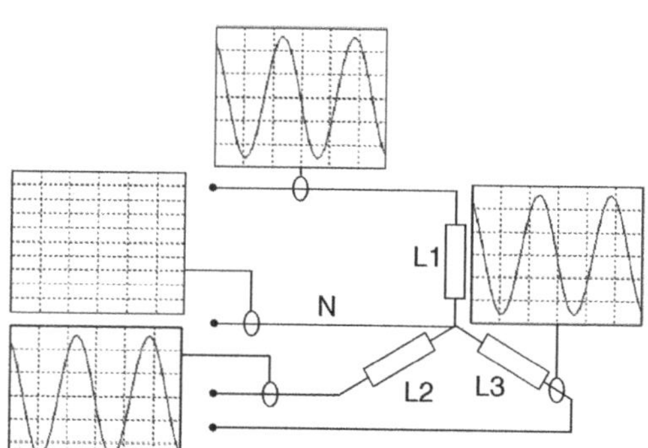

Figure 7.29 Currents in a three-phase system (linear and balanced load)

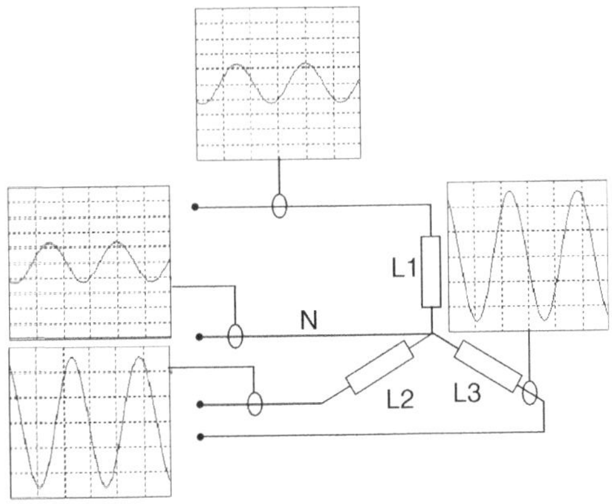

Figure 7.30 Currents in a three-phase system (linear and unbalanced load)

other phases in the supply network, the current in the neutral conductor is described by the relationship

$$\begin{aligned} i_N = i_{L1} + i_{L2} + i_{L3} &= \left[i_{L1(1)} + i_{L2(1)} + i_{L3(1)}\right] + \left[i_{L1(3)} + i_{L2(3)} + i_{L3(3)}\right] \\ &+ \left[i_{L1(5)} + i_{L2(5)} + i_{L3(5)}\right] + \left[i_{L1(7)} + i_{L2(7)} + i_{L3(7)}\right] \\ &+ \left[i_{L1(9)} + i_{L2(9)} + i_{L3(9)}\right] + \ldots \end{aligned} \tag{7.18}$$

Due to the equal powers of phase loads and balanced supply voltage, mainly triplen harmonics are present in the neutral conductor. In fact, the third-harmonic components (and all other harmonics where the order is a multiple of 3 – the sixth, ninth, etc.) of the line currents are all in phase with each other (i.e. they are homopolar components, Figure 7.6), so they sum arithmetically rather than cancelling by vector addition.

The effective third-harmonic neutral current is shown at the bottom of Figure 7.31. In this case, 70 % of the third-harmonic current in each phase results in 210 % current in the neutral conductor. Measurements performed in commercial buildings have shows neutral currents between 150 and 210 % of the phase currents, unfortunately flowing very often in a half-sized conductor.

An example waveform of the current in the neutral conductor of a student laboratory supply installation (about 20 PCs in different operating modes) is shown in Figure 7.32. The presence of harmonics that are not forming a homopolar system results from both the diversity of PCs and the unbalance of their supply phase currents.

For a case study see web address

7.6.3 Skin Effect

The skin effect is the tendency of alternating current to flow on the outer surface of a conductor. This effect is more pronounced at high frequencies; in fact it is normally ignored because it has very little effect at power supply frequencies, while above approximately

350 Hz (the seventh harmonic and above), it becomes significant, causing additional loss and heating. The a.c. resistance to d.c. resistance ratio is dependent on r/δ where r is the conductor radius and δ is the current penetration thickness, which can be expressed as

$$\delta = \sqrt{\frac{2\rho}{\omega\mu}} \tag{7.19}$$

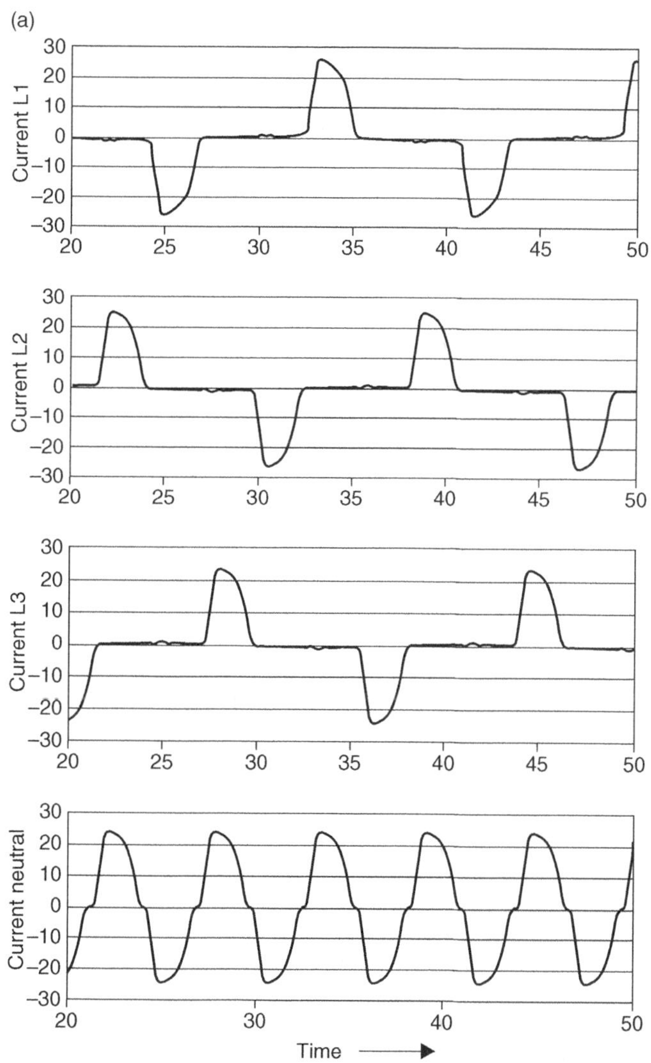

Figure 7.31 Sum of phase currents (see Figure 7.6) (a) and (b) third-harmonic current in neutral conductor [18] (Reproduced from Historical rationale for the limitation of power-frequency conducted harmonic current emissions from equipment in the frequency range up to 9 kHz, IEC 61000-1-4)

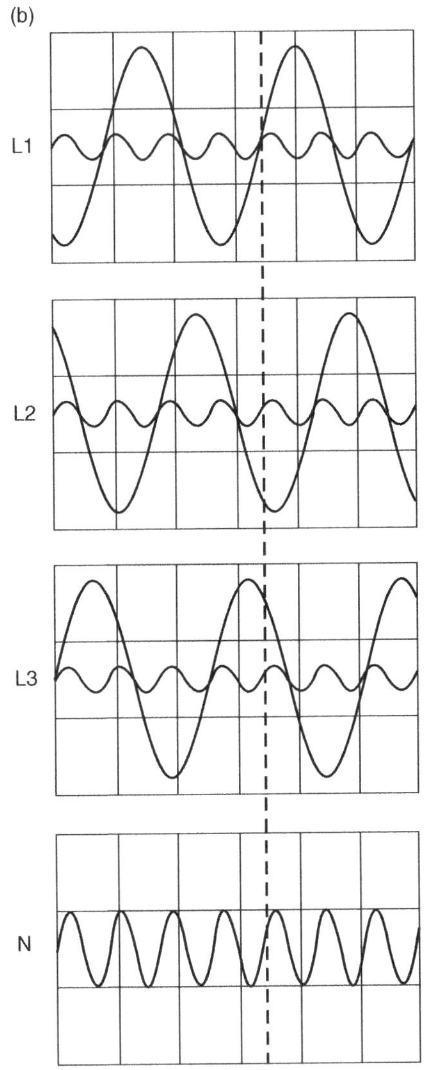

Figure 7.31 (Continued)

where μ is the magnetic permeability (H/m); ω the frequency (rad/s); and ρ the resistivity (Ω m/m^2).

It is evident that δ is dependent on the frequency; in particular it decreases as frequency increases. Figure 7.33 shows for round-section conductors the R_{ca}/R ratio as a function of conductor radius (r) and penetration thickness (δ). For a typical cylindrical copper conductor having a diameter equal to 20 mm, at 350 Hz, the R_{ca}/R ratio is equal to 1.6. For this reason, in harmonic-rich environments designers should take the skin effect into account and derate cables properly. A possible solution is also provided by multiple cable cores or

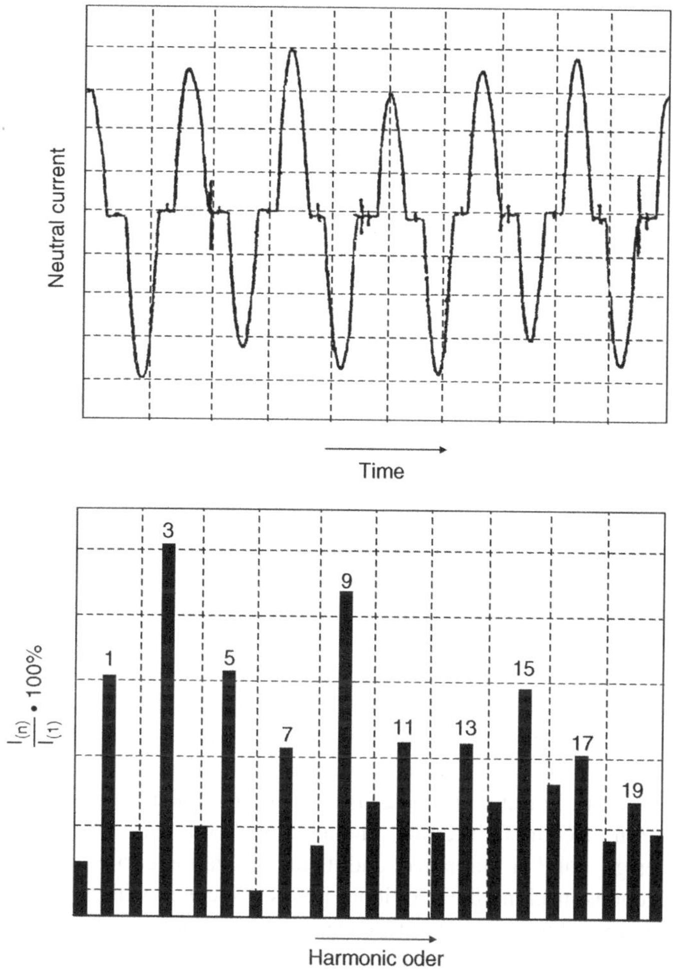

Figure 7.32 (a) Example waveform of the current in the neutral conductor of a student laboratory installation; (b) its harmonic

laminated busbars which can be used as an alternative way to overcome this problem. It is also important to pay attention to the fact that the mounting systems of busbars must be designed in order to avoid mechanical resonance at harmonic frequencies.

7.6.4 Motors and Generators

The main effect of the presence of voltage and current harmonics in electric rotating machines (both synchronous and asynchronous) is the operating temperature rise as a result of power losses. It is an additional stress on insulation which, if sustained, can shorten the in-service period. The losses occur mainly in the core and windings.

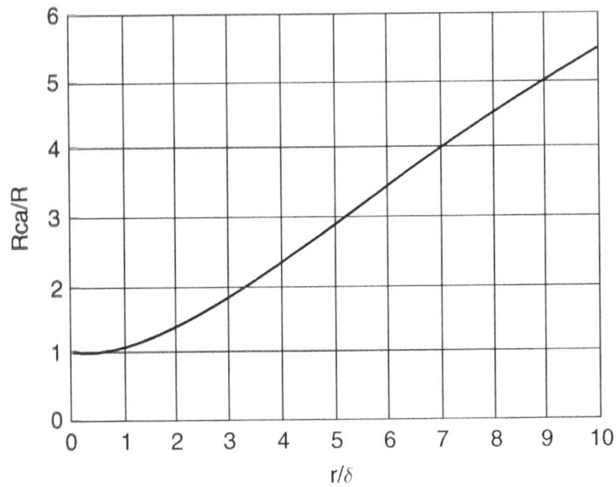

Figure 7.33 R_{ca}/R ratio as a function of conductor radius (r) and penetration thickness (δ)

The rise of stator and rotor winding losses results from both the increase in the r.m.s. current value, due to distortion, and the increase in the effective resistance of the windings, due to the skin effect.

In electric motors the increase of stator and rotor core losses (particularly three-phase induction motors) and even a small harmonic voltage distortion give rise to additional magnetic flux and hence additional currents in the rotor winding and core. These additional currents, which in the case of a strong voltage distortion can be of the same order of magnitude as the magnetizing current, are the cause of additional active power loss, temperature rise and increase in machine failure rate. The additional eddy current losses depend on the machine core design, solid or laminated, and its magnetization curve.

In a synchronous machine the additional losses associated with high harmonics occur mainly in the stator windings and damping cage. Generally, the stator losses are significantly smaller. Most significant are the harmonics which form a negative-sequence system, i.e. the 5th, 11th, 17th, 23rd, ...[5] In high-voltage induction motors the rotor and stator losses are approximately equal. The wound-rotor induction motors allow for a larger rotor loss than squirrel-cage motors.

Additional harmonic torques are the effect of interaction between the air gap flux (mainly the fundamental component) and fluxes produced by rotor harmonic currents. Their effect on the resultant, average motor torque is practically small. Moreover, they have a tendency to mutual cancellation. Positive-sequence harmonics produce a forward rotating field that adds to the torque and supports the machine rotation, whereas the other harmonics

[5] According to [37], eddy losses in synchronous machine with laminated rotor and stator usually do not exceed a single-digit percentage of the nominal loss even for a relatively high distortion of supply voltage – $THD_U =$ 10 to 15 % which, for a machine supplied directly from the network, is rarely the case. Additional losses in synchronous motors and compensators with solid poles are larger. According to the same source, the losses in rotating machines are of considerable significance for low-order harmonics, mostly the second and the third. The losses due to harmonics of orders above the 13th can be neglected.

(5th, 11th, 17th, 23rd, ...) have the converse effect. Harmonic torques influence the instantaneous value of the resultant torque and result in its fluctuation.

Mechanical oscillations of electric machines, supplied with distorted voltage, attain their maximum values when the frequency of the motor torque variations is equal or close to the mechanical resonance frequency of the motor and driven machine set. An important factor is the motor load torque. This phenomenon can also occur in a turbine–generator set.

The presence of harmonic currents in motor windings increases the acoustic noise emission compared to that for sinusoidal waveforms.

By affecting the air gap flux distribution, harmonics can hamper the soft start of a motor or increase its slip.

7.6.5 Transformers

The effects of harmonics on transformers manifest in two ways:

- Eddy current losses, which can be estimated to be normally around 10 % of the losses at full load, increase with the square of the harmonic order. For example, for a fully loaded transformer supplying a load comprising IT equipment, the total losses would be twice as high as for an equivalent linear load. These additional losses may result in a much higher operating temperature and a shorter life, therefore this effect must be taken into account at design stage.
- Triplen harmonic currents, when reflected back to a delta winding, are all in phase so they can circulate in the winding. These components are absorbed in the windings and do not propagate onto the supply; for this reason delta-wound transformers are useful as isolating transformers. Of course this circulating current has to be taken into account when rating the transformer.

The losses due to the load taking place in a transformer are a function of the current and are normally distinguished as main and additional. Referring to windings only, the main losses are the ones that can be calculated on the basis of the resistance (measured in d.c.) of the windings, and the additional losses are the ones that can be added to the main ones to give the global losses measured experimentally in a given condition. Additional losses are mainly due to the uneven distribution of the current in the sections of the windings.

7.6.5.1 Load Losses

The main characteristics of additional losses, particularly as concerns transformers made for feeding non-linear loads, concern the load current frequency. As a function of the winding type considered, additional losses have no univocal dependence from frequency, whilst ohmic losses are always independent. This is very important in non-sinusoidal current waves.

The problem can be analyzed using Fourier analysis, since it is possible to apply the principle of superimposition of the effects because the phenomena can be considered as linear. The total losses due to the load can therefore be calculated by adding the losses due to the single harmonics.

Table 7.10 Single loss components

Additional loss components

$$P_{dis} = P_{\Omega} \cdot \frac{5 \cdot m^2 - 1}{45} \cdot \xi^4$$

$$P_{circ} = N_2 \cdot \frac{\omega^2 \mu^2 \pi \cdot r_M \cdot s_l^3}{b \cdot \rho} \cdot (N_1 I_1)^2$$

$$P_{fr} = \frac{K_c \cdot K_e}{1 + K_\delta} \cdot \frac{f^2}{\rho \cdot \gamma} \cdot d^2 \cdot B_r^2 \cdot W_{av}$$

$$P_{fd} = \frac{\pi^2}{6} \cdot \frac{f^2}{\rho \cdot \gamma} \cdot a^2 \cdot B^2 \cdot W_{sb}$$

$$P_{ic} = \frac{\rho \ln_f \iint_s \left(N^2 \cdot \left(H + \sum_m D_m F_m \cosh(mky) \cos(mkx) \right)^2 \right) ds - P_{\Omega}}{P_{\Omega}} \cdot P_{\Omega}$$

P_{dis}	losses due to non-uniform density of current
P_{circ}	losses due to current circulation among parallel sections
P_{fr}	losses due to radial flux
P_{conn}	connection losses (proximity and skin effect)
P_{fd}	losses in connections inside leakage flux
P_{ic}	additional losses due to imperfect compensation of ampere-turns

A particular case is that of transformers with at least a sheet winding and a second winding made of an ensemble of sections connected in series. These happen to be far from one another, in an axial sense, with insulators interposed.

In load operating the current concentrates on the sheet winding according to height, mainly in the parts facing the sections of the other winding. In other words, the parts that are not facing are practically not crossed by the current.

In the winding sheet there is therefore a fictitious decrease in the useful section for the passage of the current with a consequent increase in losses. The phenomenon is quite complex and depends on several factors, such as the distance between windings and reels, the dimensions of spacers, etc.

Additional losses can often be a considerable part of total losses; single components are shown in Table 7.10.

In the presence of distorted currents, because the components vary with the frequency functionally in different ways, additional losses may be remarkably different from the case of sinusoidal currents. If the harmonic contents are known, additional losses can be calculated in an analytic way by applying the rules illustrated and using the method of effect superposition to calculate the contribute of the various harmonics. In principle the calculation procedure can be as follows:

1. Determination of all the components of additional losses in the sinusoidal current pointing out the frequency parameter.
2. Determination of the amplitude of the various harmonic components on the basis of the diagrams of the plant (particularly the conversion diagrams).

3. Calculation of the single contributions of the harmonics and their sum in order to determine the total additional loss value in the conditions considered.

Attention must be paid to the real amplitude of the harmonics and not the ideal one because, in this case, the losses due to the single harmonics do not tend to zero with the increase of the harmonic order (as is experimentally verifiable), but remain constant and total losses would tend to infinity.

7.6.5.2 No Load Losses

Non-sinusoidal waveforms lead to additional no-load losses, mainly localized in the magnetic circuit. No-load losses (P_o) can be divided into two parts: magnetic hysteresis losses and eddy current losses:

$$P_o = k_1 f B_{Mn} + k_2 f B_2 \qquad (7.20)$$

where $k_1 f B_{Mn}$ is the magnetic hysteresis losses; $k_2 f_2 B_2$ the eddy current losses; f the frequency; B the magnetic induction; and K_1, K_2 constants.

Magnetic induction maximum values are related to the voltage according to the relationship

$$B_M = K \int_0^{T/2} e\,dt \qquad (7.21)$$

It is then directly dependent on the voltage average value. In the case of a non-sinusoidal waveform, losses are the same as those that would occur with a sinusoidal waveform with the same average value.

For eddy current losses, when waveforms are not sinusoidal, it can be pointed out that the voltage r.m.s. value is directly dependent on the product fB and that for a non-sinusoidal waveform

$$E^2 \equiv \sum_{n=1}^{\infty} f_n^2 \cdot B_n^2 \qquad (7.22)$$

In this case, too, losses are still directly dependent on the voltage r.m.s. value. A formula to evaluate no-load losses in the case of a non-sinusoidal waveform is available from [4]

$$P_o = \frac{P_m}{P_1 + K\,P_2} \qquad (7.23)$$

where:

- P_o are the final no load losses;
- P_m are the no-load losses measured at rated voltage with an average value measurement device (with a measured value multiplied by a factor equal to 1.11);

P_1 is the ratio of hysteresis losses to total no-load losses ratio (typically 0.5);
P is the ratio of eddy current losses to total no-load losses (typically 0.5);
k is the ratio of the r.m.s. value of the measurement device reading to the average value of the measurement device reading ($k = 1$ in the case of a sinusoidal waveform)

Usually, when the waveform is not heavily distorted, the no-load loss variation is negligible.

A d.c. component may occur mainly in LV networks, due to the extensive use of electronic equipment in households and industry. However, its level is normally low but it may be sufficient to drive transformer cores into saturation (which can lead to transformer failure or at least to the generation of extra harmonics). It can also give rise to corrosion processes and have a detrimental effect on protective systems or other loads sensitive to current magnitude and distortion.

7.6.6 Capacitors

Capacitors are devices which experience to the largest extent the effects of electromagnetic environment distortion. They are exposed to voltage, current and power overloads. For this reason the permissible overload factors, provided by manufacturers and expressed in terms of the rated value ratio, are related to these quantities. They determine the non-destructive conditions for a capacitor bank, though operation under a long-duration overload significantly shortens the capacitors' life.

The increase in peak voltage value due to high harmonics is an additional dielectric stress. It may cause a partial discharge in the insulation, a foil short circuit and result in permanent damage to the capacitor. The permissible overvoltage factor of a capacitor normally does not exceed 110 % of rated value.

Most of the problems with capacitors, caused by harmonics, are *current related*. Voltage harmonics produce additional currents flowing through the capacitor which increase with the harmonic order (as a result of the reduction in capacitor equivalent impedance $Z_C \approx (n\omega C)^{-1}$), and can be of significant value. The relative value of a capacitor current for the nth harmonic, $I_{(n)}{}^*$, related to the fundamental harmonic, $I_{(n)}{}^* = I_{(n)}I_{(1)}{}^{-1}$, can be determined by the relationship $I_{(n)}{}^* = nU_{(n)}{}^*$ where $U_{(n)}{}^*$ is the nth harmonic relative value with respect to the fundamental $U_{(n)}{}^* = U_{(n)}U_{(1)}{}^{-1}$. For example, the seventh voltage harmonic of 15 % relative value will produce the seventh harmonic current with 105 % of the fundamental harmonic value. This may result in capacitor current overload. Generally, in presently manufactured capacitors the current overload factor typically does not exceed 1.6.

An excessive current through the capacitor bank results in additional losses and, consequently, adverse effects such as fuses blowing, physicochemical processes in the dielectric resulting in accelerated ageing of the insulation and reduced service life, permanent damage, etc. All these effects can be dramatically magnified by the series or parallel resonance.

Example waveforms of a capacitor current and voltage, and the current spectrum, are shown in Figure 7.34. In this case the fifth-harmonic value exceeds that of the fundamental.

Figure 7.35 shows a simplified schematic diagram and the equivalent circuit diagram for a non-linear load and supply network capacitor bank circuit. The load is represented by the ideal current source (i.e. the current value is independent of load impedance) for particular harmonics. For the sake of simplicity the resistances are neglected.

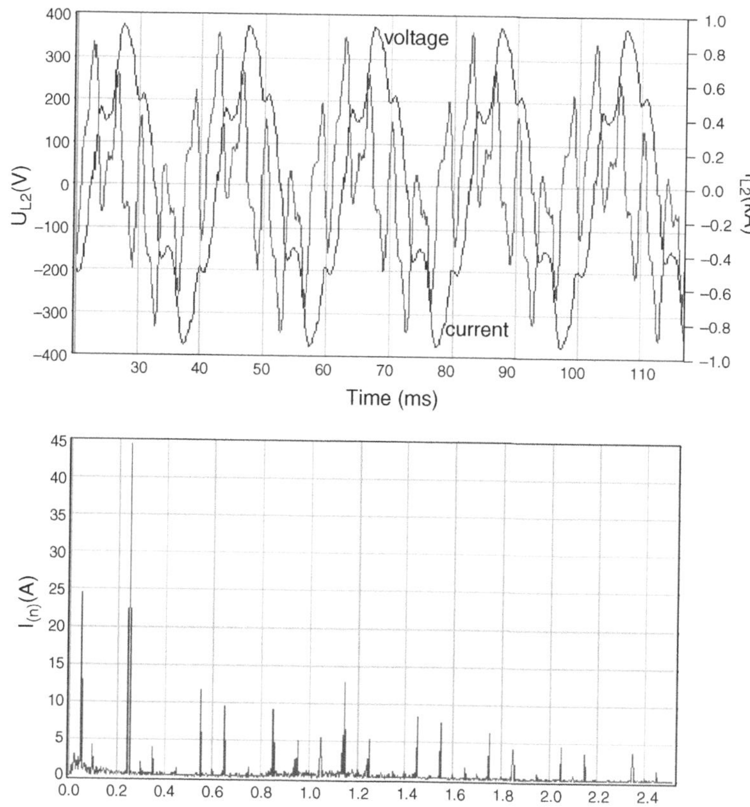

Figure 7.34 Example waveforms of a capacitor current and voltage, and the current spectrum

Connecting a capacitor bank to the bus supplying the non-linear load changes the amplitude voltage spectrum at PCC (Figure 7.35c) since the frequency–impedance characteristic, seen from the load terminals, has been changed. Voltage harmonic values $U_{(n)}$ of orders lower than the harmonic order $-n_0$ are increased, for which the impedance before and after switching the capacitor is the same, according to $U_{(n)} = Z_{\Sigma(n)} I_{(n)}$, where $Z_{\Sigma(n)}$ is the impedance for the nth harmonic seen from the load terminals, whereas the harmonic voltages of orders higher than n_0 are reduced (Figure 7.35c).

Table 7.11 shows example voltage and current waveforms and their spectra, both prior to and after the connection of a capacitor bank. The increase in magnitudes of certain harmonics, in effect of the increased equivalent impedance of the supply network and capacitor bank circuit for these frequencies, is evident.

7.6.7 Light Sources

The increase in the supply voltage peak value may shorten the service life of incandescent lamps. The discharge light sources, both fluorescent and high-pressure mercury lamps, have a series, current-limiting reactor which in connection with a commonly used input parallel

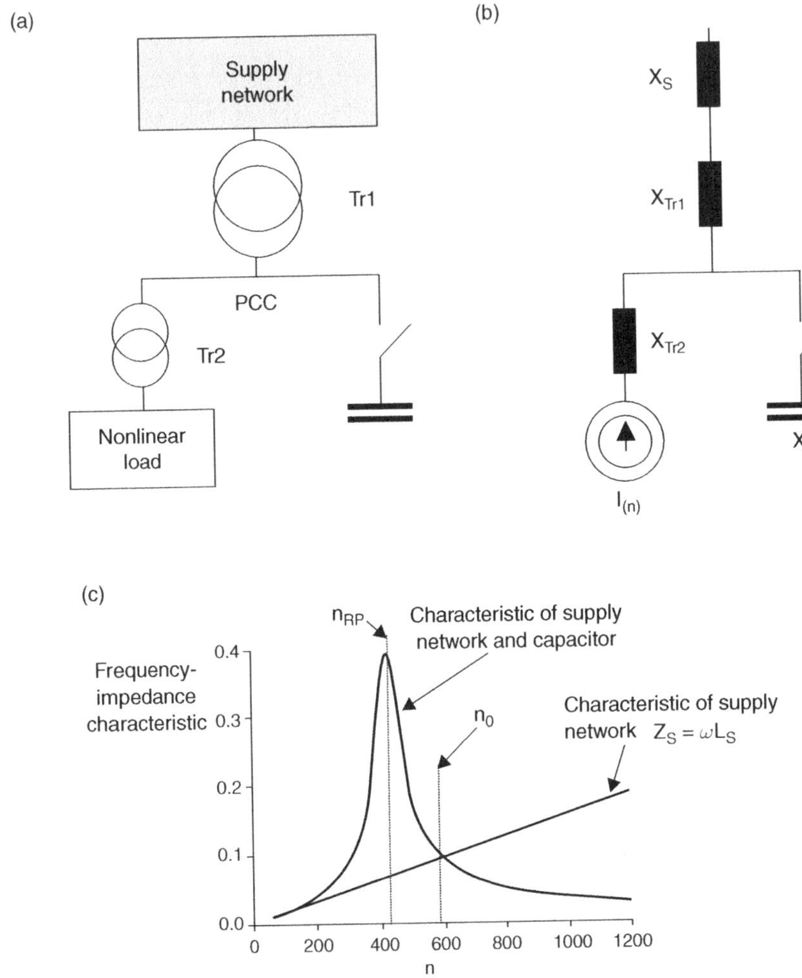

Figure 7.35 (a) The schematic diagram and (b) the equivalent circuit for the compensation circuit; (c) the frequency–impedance characteristics ($R \simeq 0$)

capacitor (for PF improvement) forms a resonant circuit. A close-to-resonance state is the source of additional losses.

7.6.8 Nuisance Tripping of Circuit-Breakers

Harmonic current distortion affects the switching capability of breakers only when breaking small currents, and has no effect on interrupting short-circuit currents. High harmonics may increase the current derivative *di/dt* value at zero crossing (compared to that of a sine wave), which hampers the current interruption process.

Residual current circuit-breakers (RCCBs) operate by summing the current in the phase and neutral conductors and then disconnecting the power from the load when the result is

Table 7.11 Example voltage and current waveforms and their spectra, before and after connection of capacitor bank

Voltage and current waveforms	Current spectrum	Voltage spectrum

Without capacitor

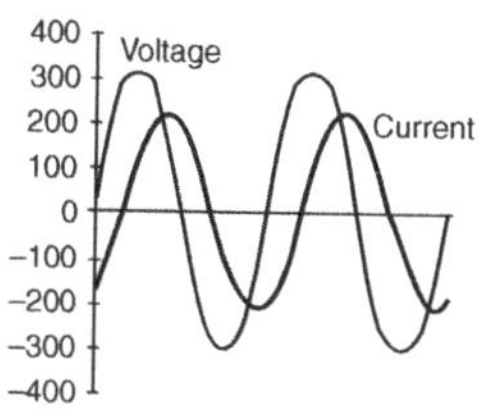

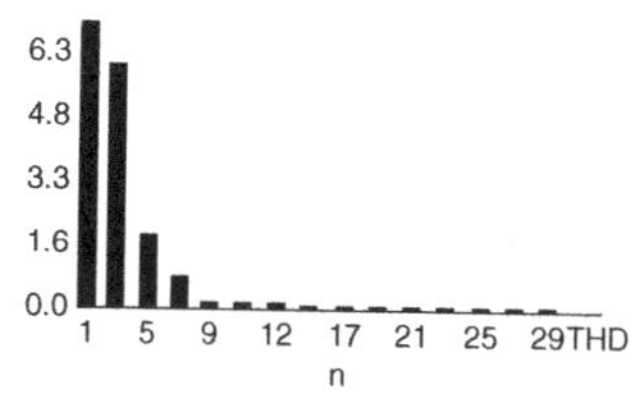

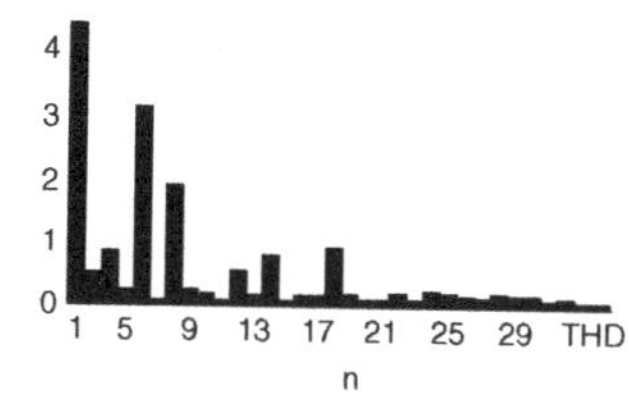

With capacitor

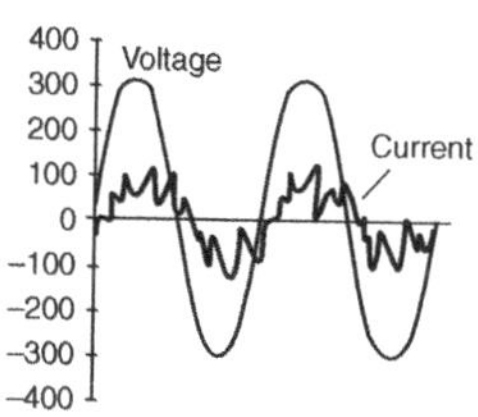

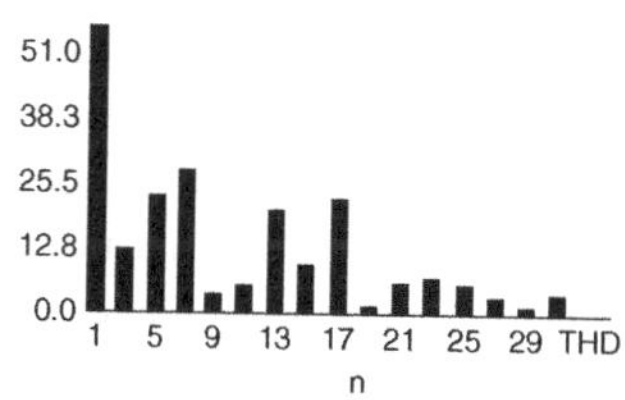

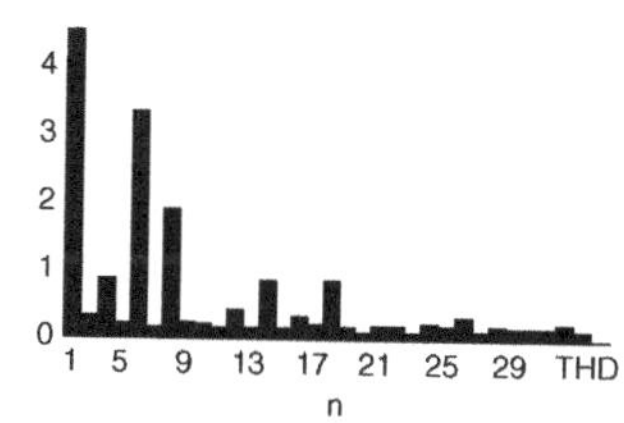

not within the defined limit. In harmonic-rich environments, the main issue that can occur is nuisance tripping, for two reasons. First of all, the RCCB, which is an electromechanical device, may not sum the higher-frequency components correctly and therefore may trip erroneously. The second reason is that the kind of equipment that generates harmonics, usually electronic equipment, also generates switching noise that must be filtered at the equipment's power connection. The filters which are normally used for this purpose have a capacitor installed between line and neutral and ground, leaking a small current to earth. This current value is limited by standards to less than 3.5 mA, but when equipment is connected to one circuit the total leakage current can reach a value sufficient to trip the RCCB.

This situation can be easily avoided by providing more circuits, each supplying fewer loads.

In the case of miniature circuit-breakers (MCBs) nuisance tripping is usually caused by the current flowing in the circuit being higher than that expected from calculation or simple measurement due to the presence of harmonic currents.

7.6.9 Earth-Fault Currents

In power supply systems, mainly MV ones, with an isolated or resonant-earthed neutral point, in which voltage is distorted, earth-fault currents may attain intolerably high values. This phenomenon is not observed in solidly earthed or impedance-earthed systems, because the short-circuit impedance has a resistive–inductive character and harmonic values are limited with frequency in a natural way. In isolated or compensated power systems the short-circuit impedance within the usual range of voltage harmonic frequencies has a resistive–capacitive character and it decreases with frequency. Short-circuit currents, determined by the capacitance to earth can, due to the voltage harmonics and interharmonics, attain the values which do not ensure self-extinction of earth-fault arcs.

7.6.10 Converters and Electronic Equipment

Converter systems are considered to be mainly a source of high harmonics, seldom as loads subjected to interference. However, this type of equipment, as well as most electronic equipment, is sensitive to various disturbances, including harmonics. The resulting irregularities in operation are associated with the following:

1. Zero-crossing noise – a lot of electronic controllers detect the point at which the supply voltage crosses zero volts in order to determine when loads should be turned on. The reason for using this technique is that switching reactive loads at zero voltage crossing avoids the generation of transients, and reduces electromagnetic interference (EMI) and stresses on semiconductor switching devices. When harmonics or transients are present on the supply waveforms, the rate of change of voltage at the crossing may become faster and then more difficult to identify, leading to improper operation, because there may in fact be several zero crossings per half cycle (Figure 7.36).
2. In line-commutated converter control systems, synchronized with the supply voltage zero crossing, the voltage distortion around zero may result in the inequality of the control angles of semiconductor devices. As a consequence, the converter generates non-characteristic harmonics, including even and triplen harmonics, interharmonics and, in particular cases, a d.c. component. Synchronization errors may also occur in the case of comparison of two waveforms. Improper switching of semiconductor devices into the on-state is particularly hazardous in the inverter mode of operation.

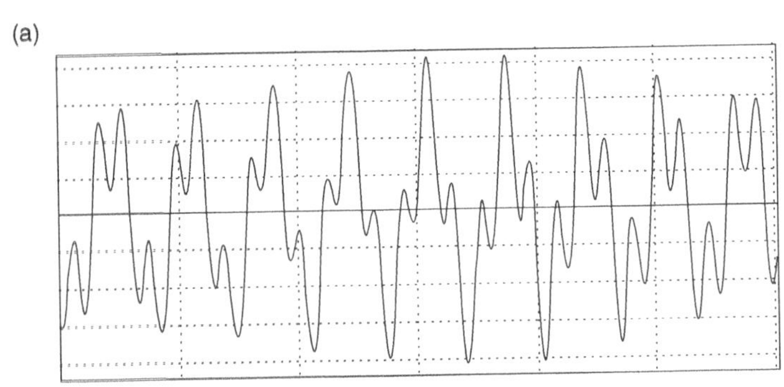

Figure 7.36 Examples of supply voltage with increasing number of zero crossings

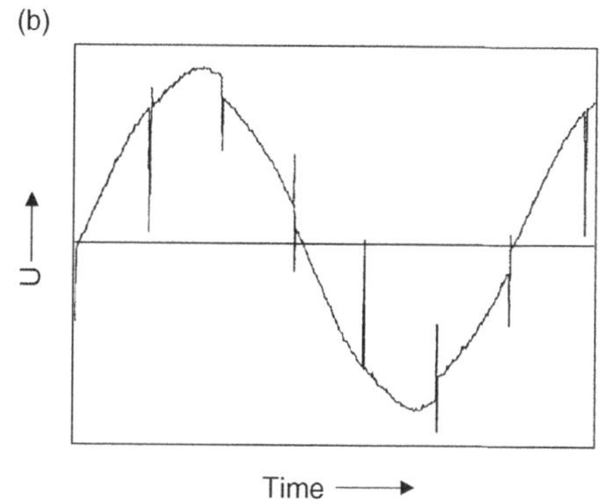

Figure 7.36 (Continued)

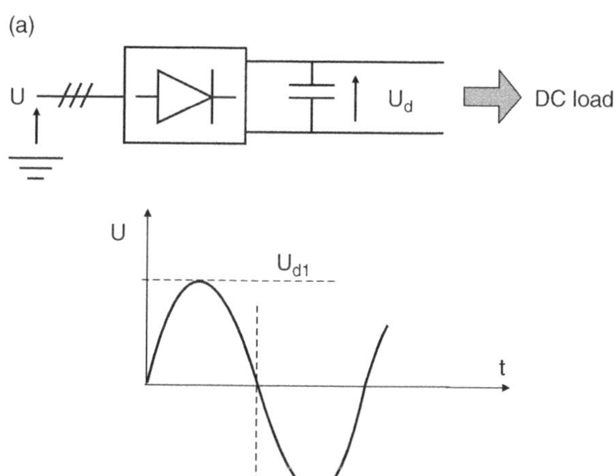

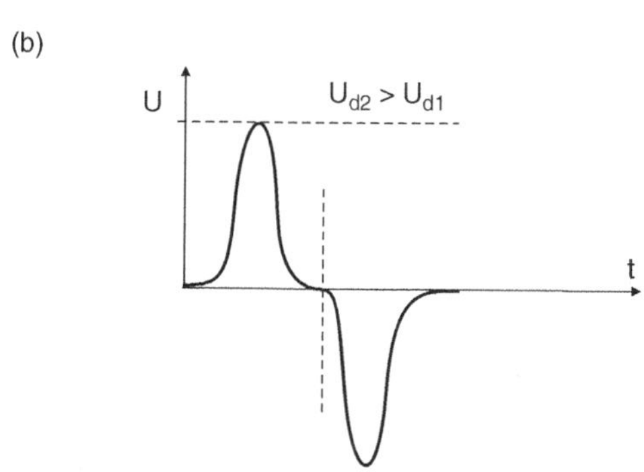

Figure 7.37 Effect of the shape of the supply voltage waveform on the rectifier d.c. output voltage value

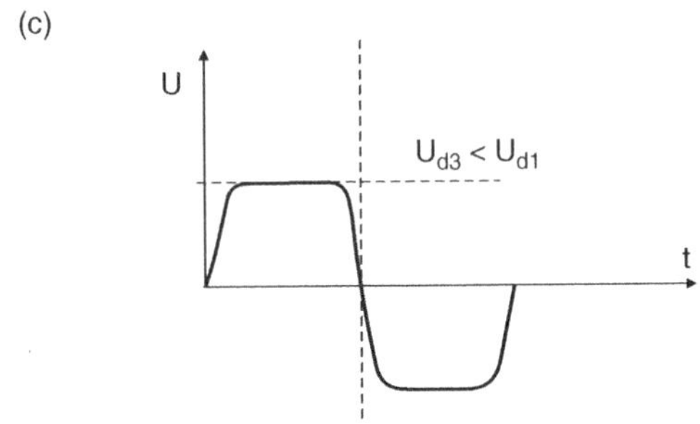

Figure 7.37 (Continued)

3. Component failures as a result of the increase in the maximum value of supply voltage due to harmonic distortion. Equipment power supplies commonly incorporate non-controlled rectifiers with a d.c. side capacitor. The supply voltage peak value can decrease or increase, depending on the harmonics' frequency and their phase shift with respect to the fundamental. In effect, for the same r.m.s. value of input voltage, the output d.c. voltage may differ from its nominal value (Figure 7.37).
4. Errors of input signal transducers in control systems (e.g. errors in SVC reactive power measurement, current feedback loops in electric drives, etc.) and disturbances in converters' digital control systems.
5. Disturbed operation of diagnostic and protective devices.
6. Adverse impact on capacitors in power electronics systems (as well as in other electronic equipment), in overvoltage protection circuits, EMC filters, etc.

IT equipment, as well as programmable logic controllers, requires the THD factor and relative value of each harmonic present in the supply network not to exceed specified limit values. Higher distortion levels cause incorrect operation, errors or data loss, characteristic 'humming' of disk drives, etc. This may lead to dangerous consequences, in particular for health services, banking, air transport, etc. For these reasons critical equipment is supplied through power-conditioner systems with improved power parameters. Most electronic equipment is also affected by radiated electromagnetic emissions from harmonic frequencies.

7.6.11 Measuring Instruments

Measuring instruments are most often calibrated for sinusoidal quantities, and their use under distorted conditions can be a source of errors. The error values, both positive and negative, depend on numerous factors, such as type of measurement, type of instrument involved, order, magnitude and phase of given harmonic, etc. Considering a wide diversity of measuring instruments, various methods of operation and different design techniques (analogue or digital), general opinions can hardly be formulated. The following remarks concern several, selected instruments used in measurements.

7.6.11.1 True R.M.S.

True r.m.s. measurement is essential in any installation where there is a significant number of non-linear loads. Due to harmonic pollution it is common to have in the system currents with values higher than the expected one: measurement campaigns aiming to assess real currents are therefore mandatory in order to avoid potentially catastrophic events. Unfortunately, measurement campaign performed with inappropriate devices can lead to a wrong estimation of current values.

Under-measurement occurs very frequently in fact, even if nowadays digital instruments are increasingly accurate and reliable. The problem is that many instruments are not suitable for measuring distorted currents, which is the most common case in modern installations.

Distorted waveforms are not sine waves, therefore all the usual sine wave measurement devices are no longer able to correctly assess them. For this reason, when analyzing the performance of a power system, it is essential to use the correct tools: in this case, tools that can deal with non-sinusoidal currents and voltages. Averaging reading meters will then provide an under-measurement of up to 40 %, which can result in potentially dangerous conditions: for example, circuit-breakers will be underrated with the consequent risk of failure and nuisance tripping.

In addition, since cable ratings are given for particular installation conditions (which determine how fast heat can be dissipated) and a maximum working temperature, and since harmonic-polluted currents have higher r.m.s. values than that measured by an averaging meter, cables may be underrated and consequently reach hotter temperatures than expected, resulting in degradation of the insulation, premature failure and risk of fire.

7.6.11.2 Apparent Power

Voltage and current distortion may result in significant errors in measuring the apparent power, defined for single-phase systems by the relationship

$$S_1 = UI \tag{7.24}$$

In certain types of instruments, the apparent power value is determined from measurement of the average values of voltage U_{AV} and current I_{AV} (obtained by means of rectification, for example). These instruments are calibrated for sinusoidal waveforms, assuming a constant factor of proportionality, according to the relationship

$$U = (1.11)\,U_{AV} \quad I = (1.11)\,I_{AV} \tag{7.25}$$

Hence

$$S_2 = (1.11)^2 U_{AV} I_{AV} \tag{7.26}$$

The current or voltage distortion yields a different value of the proportionality constant ($\neq 1.11$). It is the source of the measurement inaccuracy with respect to the result obtained according to the definition (7.24).

7.6.11.3 Energy Meter

A typical electromechanical energy meter, still used in some countries, by the principle of its operation, is a small electric motor with an aluminum rotor disk. A load current component, cophasal with the supply voltage, produces the torque which drives the disk at a speed proportional to the power. The number of revolutions is thus proportional to the energy supplied to the load. Voltage and current harmonics produce additional harmonic torques acting upon the disk. These torques may act in the same, or the opposite, direction as the main torque. The resulting measurement errors, of both positive and negative value, depend on many factors, including voltage and current distortion, harmonic orders, magnitudes and phase, direction of power flow, etc. The meter's ferromagnetic elements containing magnetic flux are non-linear with respect to the frequency and magnitude of harmonics and therefore the flux components, intended for instrument calibration (e.g. the friction torque compensation), change their values.

A noticeable measurement error occurs only for a strong voltage and current distortion (exceeding 20 %), which is rarely the case in real supply networks [1, 2]. A digital electric energy meter computes the energy by means of sampling the voltage and current waveforms. Similar to other instruments of this type, the bandwidth is limited by the sampling rate. Commercially available digital energy meters have generally a flat frequency characteristic up to an input frequency of 1000 Hz. The meter readings in this range are therefore correct. This, however, does not exclude errors due to the sampling technique and data processing.

7.6.12 Relay and Contactor Protective Systems

Most published research on relays and contactors operated under distorted conditions concerns their electromechanical structure and, to a lesser extent, electronic solutions. These publications (e.g. [4]) justify the following conclusions:

- Contactor/relay operation may differ significantly in the presence of harmonic interference. The response not only depends on the device type and manufacturer, but also varies with each piece of equipment tested, as well as with changes in the characteristic features of the spectrum.
- Contactor/relay sensitivity to current or voltage harmonics decreases with the increase of harmonic order.
- Most contactors/relays are insensitive to voltage distortion not exceeding 20 % [8]. Irregularities in their operation may occur above this value, under both regular and irregular operating conditions for the protected equipment.

Current relays are actuated at large, e.g. short-circuit, currents when the current value considerably exceeds the nominal current. In such cases the harmonics have no effect on the correctness of their operation as their contribution to the short-circuit current is small.[6]

In certain cases, under normal operating conditions of a power system, harmonic distortion may cause irregular operation of protective devices. This can occur in earth-fault

[6] Due to possible saturation of current instrument transformers, a strong distortion of secondary current may occur, particularly when the primary current d.c. component is large. In such a case, a correct reconstruction of the primary current wave, which is essential for operation of the relays, is difficult.

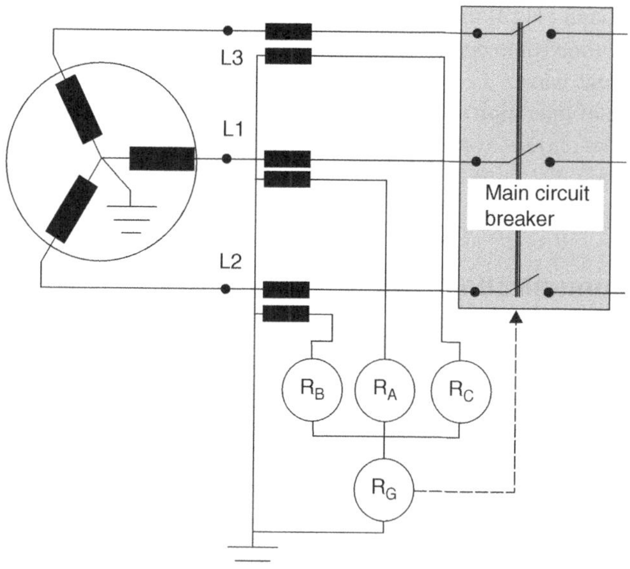

Figure 7.38 Generator overcurrent protection [28] (Reproduced from Power System Harmonics, Power Technologies, Inc.)

protection relays installed together with overcurrent protection systems. For instance, the generator in Figure 7.38 is provided with phase overcurrent relays R_A, R_B, R_C, for detecting short circuits in the system. Their settings are high in order to prevent the tripping action of the main circuit-breaker during normal switching operations, e.g. starting large motors.

The fourth relay R_G, which responds to the earth fault, is normally installed in the neutral conductor, as in Figure 7.38, and responds to zero-sequence currents associated with the impedance unbalance of loads. Because this current is usually small, the earth-fault relay setting can be much lower than the phase relay settings. A low setting of this relay enables the detection of earth faults with currents too small to be detected by phase relays. In case the load current contains triplen harmonics, mainly the third, of significant magnitude, they can cause a false actuation of the earth-fault relay and, as a consequence, trip the main circuit-breaker.

Another type of relay which can be affected by high harmonics is under-frequency relays. Their purpose is to disconnect loads in the event of a power shortfall which manifests itself as a decrease in the system frequency below a set value. In some solutions the frequency recording consists of counting the voltage zero crossings. More than two zero crossings during the period can cause an erroneous frequency reading and, as a consequence, the load will not be disconnected despite the actual decrease in the system frequency.

In LV networks the main circuit-breaker is sometimes supplied with voltage/current sensors whose operation is based on the peak value measurement. Harmonic distortion may increase the peak value and cause the breaker to trip even when the r.m.s. value remains within the safe range.

Also important is the effect of distortion on the impedance measurement in impedance relays. Harmonics may result in a significant measurement error with respect to the impedance value at the fundamental frequency.

It has been found [1],[8] that in many solutions of time-dependent overcurrent protection, the operating time (time delay) is a function of the amplitudes and phases of harmonics distorting the current wave.

The principle of operation of many solutions is the inverse proportionality of their time of operation and the current average value. The latter may vary significantly, depending on the type of harmonic distortion. The same studies have demonstrated that present solutions of electronic relays are less sensitive to harmonics.

7.6.13 Telecommunications Interference

This type of interference is one of the earliest identified and understood problems, associated with the presence of high harmonics in supply systems. In the 1920s, when static rectifiers were applied in industry for the first time, their supply cables were often placed close to telephone networks. The resulting noise in telecommunications circuits caused discomfort to users. A high level of noise severely hindered the transmission, sometimes leading to a total loss of communication. Numerous cases were recorded, particularly in the 1930s and 1940s, when switching on large rectifier loads caused disruption in telephone communication, sometimes over very large areas.

Also, currently, the improvement of transmission quality requires a continuous analysis of interferences occurring in telephone lines located in the vicinity of power systems; however, the risk of interference is small. There are three main factors causing interference to telecommunications lines:

- *The influence of electric power circuits.* The effect of power circuits on telecommunications circuits depends on the location of harmonic sources with respect to the telecommunications circuits, and amplitudes and frequencies of disturbing components.
- *The type and level of coupling in telecommunications circuits.* The mechanism of the influence of extraneous disturbing factors (harmonic voltages and/or currents) on telecommunications circuits can be that due to electromagnetic or electrostatic induction, or conduction.
- *Sensitivity of telecommunications circuits* to external disturbances.

The interference occurs as a result of the coincidence of these three factors. Although this type of interference is still present, it now poses a lesser problem. Commonly, harmonic disturbance has become less significant due to the use of new hardware and software techniques.

7.7 MITIGATION

7.7.1 Methods of Voltage Distortion Reduction

The interference source, receptor (load) and transmission path are three components involved in a mutual interaction in the electromagnetic environment (Figure 7.39):

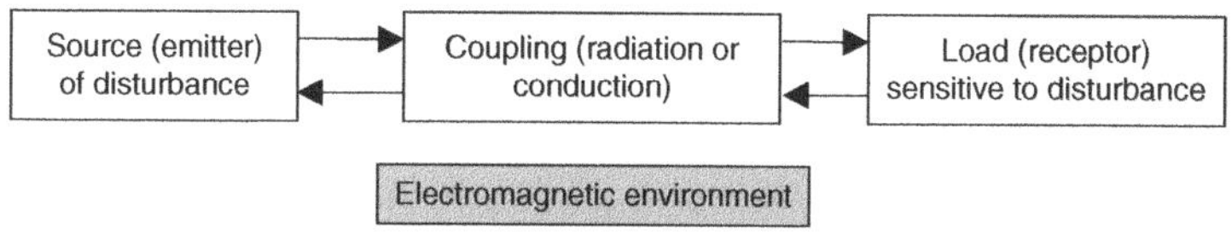

Figure 7.39 Mutual interaction of the electromagnetic environment elements

- the source of disturbance, in the case of harmonic interaction, is a non-linear load which is the source of harmonics;
- the load subjected to an external disturbance which may degrade its performance;
- the coupling, or transmission path, between the source and the load; for disturbances in general, mainly conducted ones, it is the supply network.

The reduction of the magnitude of voltage harmonics, and therefore of their effects, involves a comprehensive set of technical actions, concerning each of the above components of the electromagnetic environment. It is erroneous to hold solely the end users responsible for the poor quality of electric power and force them to limit the magnitude of generated harmonics. Also, the electricity supplier should, among other actions, continuously monitor the voltage distortion level to prevent magnification of the voltage resonance.

7.7.2 Reduction of Harmonic Emission

Three different solutions can be adopted in the reduction of the emission level: (i) reduction of harmonic emission from non-linear loads, by modifications to their structure; (ii) high harmonic filters (passive and active); and (iii) isolation and harmonic reduction transformers. Each solution has its advantages and disadvantages, so it is not possible to define which is the best one. In order to avoid spending a large amount of money on an inappropriate and ineffective solution it is necessary to carry out a preliminary analysis of the problem and choose the most effective solution.

7.7.2.1 Reduction of Harmonic Emission from a Non-Linear Load

The technical solutions taken to reduce the magnitude of generated harmonics depend on the load type. They concern modifications in its structure or technology. For instance, in the case of an arc furnace it is a comprehensive set of actions aimed at steadying its operation. Below only the most popular methods are discussed.

7.7.2.2 Line Reactors in Converter Circuits

The use of a.c. line reactors or d.c. side reactors significantly reduces the level of converter current distortion. Figure 7.40 shows the output current waveform and THD factor of a single-phase rectifier for different values of input reactor.

Figure 7.41(b) shows the current waveforms in a given phase of a non-controlled rectifier with constant d.c. load for different values of the inductance L and capcitance C.

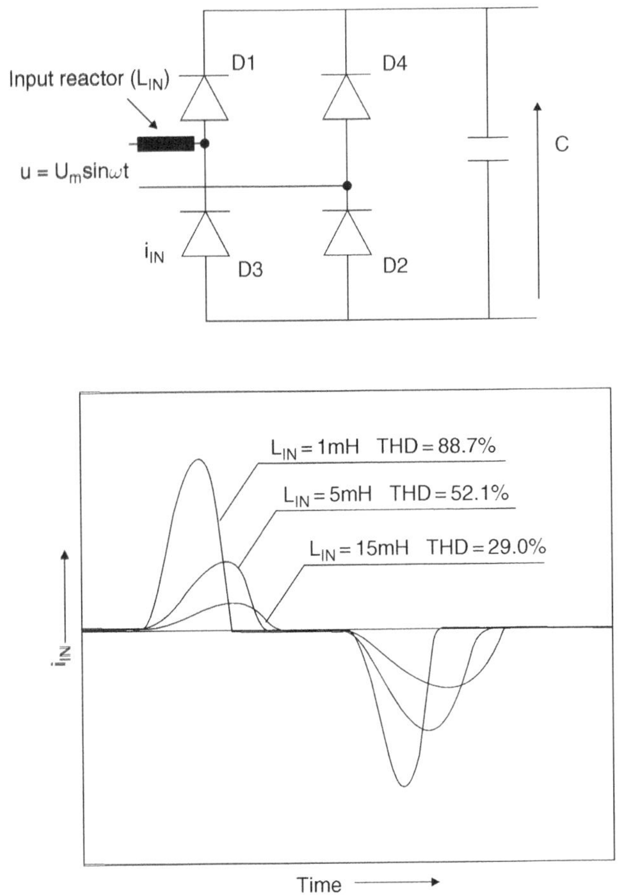

Figure 7.40 Effect of the input reactor on the waveform and THD factor of input current

In Figure 7.41(b) the current distortion factor THD is 30 %, whereas in Figure 7.41(e), due to the reactor's small inductance, the distortion factor reaches 180 %.

7.7.2.3 Multi-Pulse Converter Systems

Increasing the number of pulses in the converter system is the most common way to reduce the converter current distortion factor and therefore to mitigate the adverse effects of harmonics in a power system. An equivalent, multi-pulse mode of operation can be achieved by means of connecting in series, or in parallel, converters with a smaller number of pulses and ensuring an appropriate phase shift between the voltages supplying the bridges. Figure 7.42(a) illustrates an example 12-pulse converter configuration (the required phase shift is 30°), and Figure 7.42(c) shows a 24-pulse converter (the required phase shift is 15°).

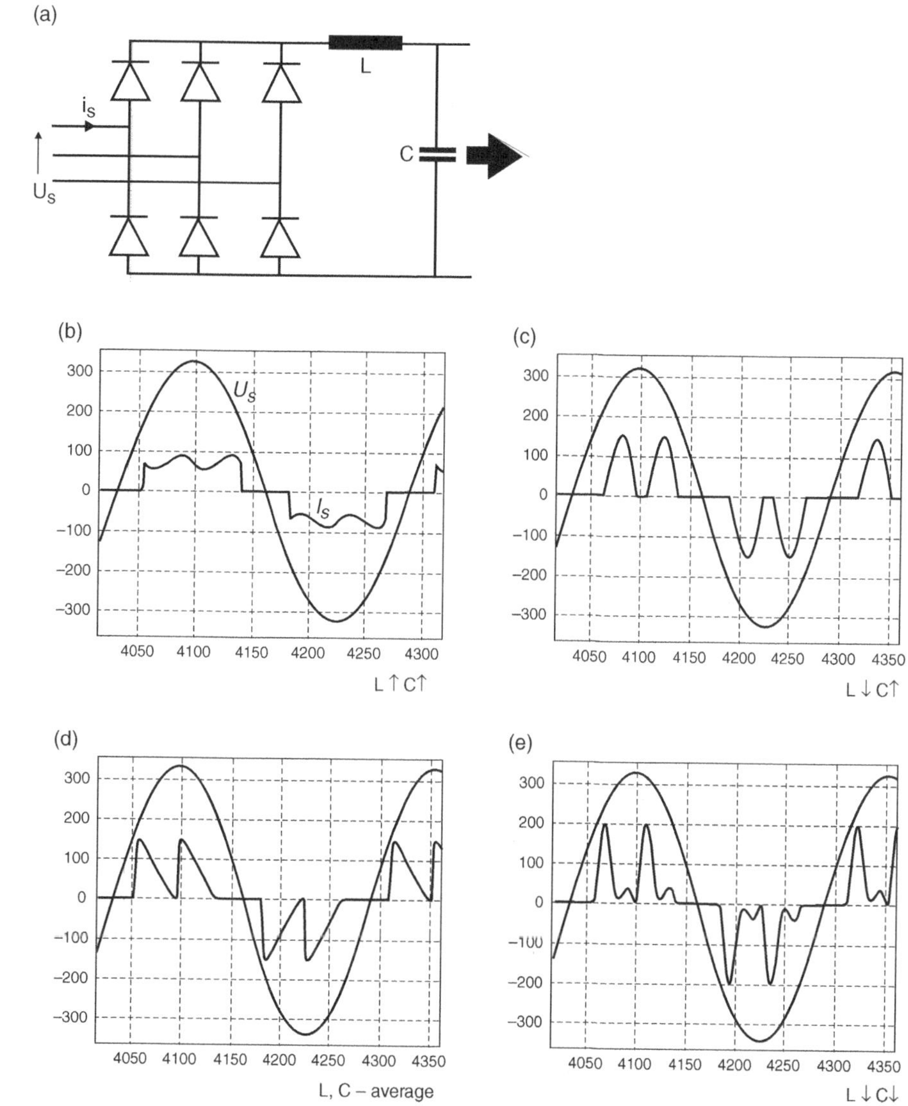

Figure 7.41 The effect of the d.c. side reactor and capacitor parameters on the rectifier input current distortion [28]

Taking into account the relationships which determine the orders of characteristic harmonics, $n = pk \pm 1$, $k = 1, 2, 3$, and their amplitudes, $I_{(n)} = 1/n$ (theoretically), the effect of the increased number of pulses, which results in the elimination (actually reduction in magnitude) of lower-order harmonics, is evident (see Table 7.4). Practically, the input current contains, apart from characteristic harmonics, also non-characteristic harmonics (e.g. 5th,

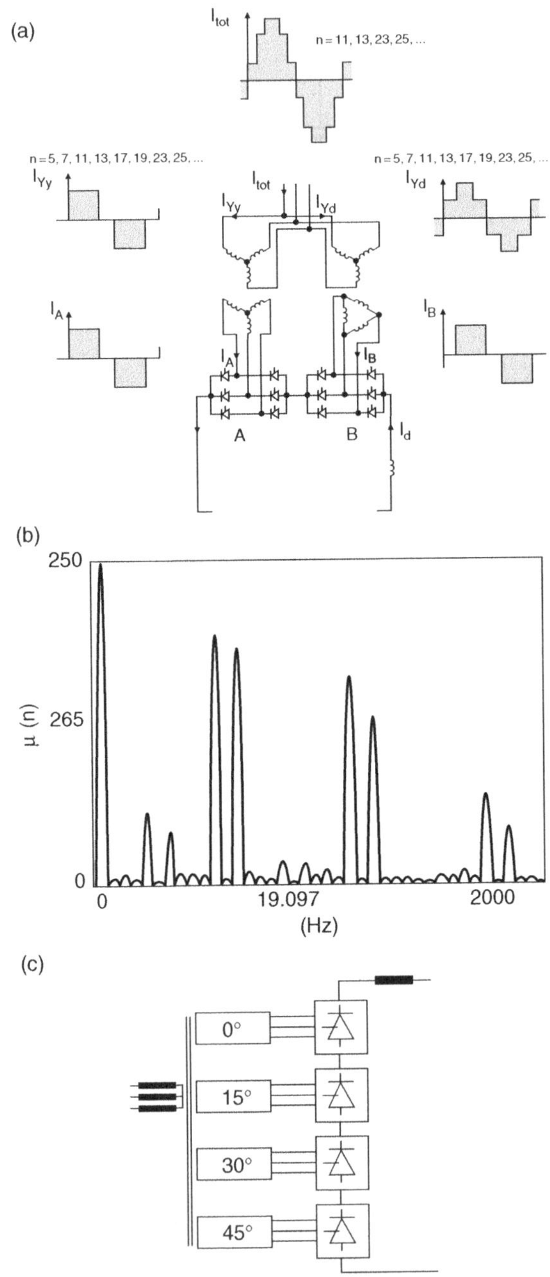

Figure 7.42 Reduction of the converter input current harmonics: (a) schematic diagram of a 12-pulse converter; (b) spectrum of the 12-pulse converter input current (non-characteristic harmonics can be seen); (c) schematic diagram of a 24-pulse converter

7th and 17th, 19th which appear in the 12-pulse converter – Figure 7.42b). This is due to the imprecise phase shifts of supply voltages, their unbalance and distortion, unbalanced equivalent impedances of the supply network, asymmetrical control of bridges and low d.c. inductance.

Of course this configuration leads to increased costs, making this type of controller useful only when absolutely necessary, e.g. in the case of limits imposed by the electricity supplier.

7.7.2.4 Active Shaping of a Converter Input Current

One of the most important goals of power electronics today is to design a.c./d.c. converters with reduced influence on the supply network. These are the so-called power-factor-corrected converters (PFC converters). Figure 7.43 illustrates one of the possible concepts of input current active shaping for a single-phase rectifier with a d.c. filter capacitor, commonly used in residential and office applications. With the switch S turned on, the rectifier input current builds up, and the energy stored in the magnetic field of reactor *L* increases: t_1 (Figure 7.43b). Turning the switch S off forces the current i_D to flow, due to the electromotive force of self-inductance, through the diode *D* and capacitor *C*. In the interval t_2 the current i_S decreases (Figure 7.43b). The instances of turning the switch S off and on are determined

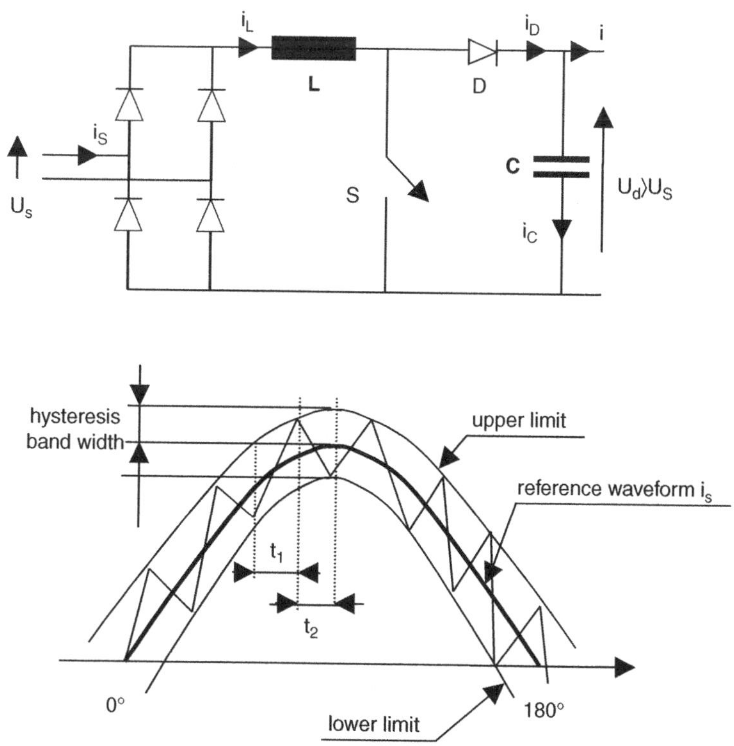

Figure 7.43 The principle of active input current shaping in a single-phase rectifier

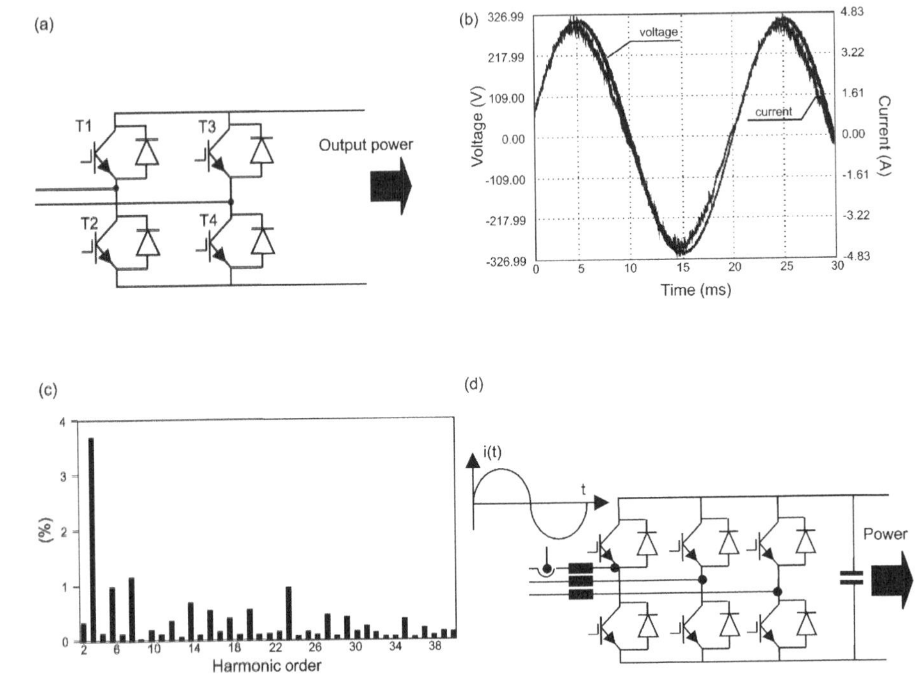

Figure 7.44 Typical current waveform (shown versus the voltage waveform) and its spectrum in a single-phase PFC converter (a, b, c); three-phase configuration (d)

by the source current attaining the upper or lower limit value, respectively. Narrowing the hysteresis band as in Figure 7.43b enables a more accurate reconstruction of the reference current, cophasal with the supply voltage.

Figure 7.44 shows a single-phase converter employing fully controlled semiconductor devices, which enables almost a sinusoidal input current (bipolar or unipolar control).

The use of the PWM technique and developments in large-power semiconductor devices (e.g. insulated gate bipolar and power transistors) with high permissible switching frequency allow for active input current shaping also in medium- and large-power converters (Figure 7.44d).

7.7.2.5 Parallel Filters

7.7.2.5.1 Passive Filters

Where the voltage distortion factor exceeds, or could exceed (e.g. in planned installations), the limit value, there is a need for connecting harmonic filters at the supply busbars. Currently they are mostly parallel filters with passive *LC* components selected so as to form a low-impedance branch shunting the supply network impedance. They play a double role: that is, eliminate harmonic currents and reduce the system load by reactive power

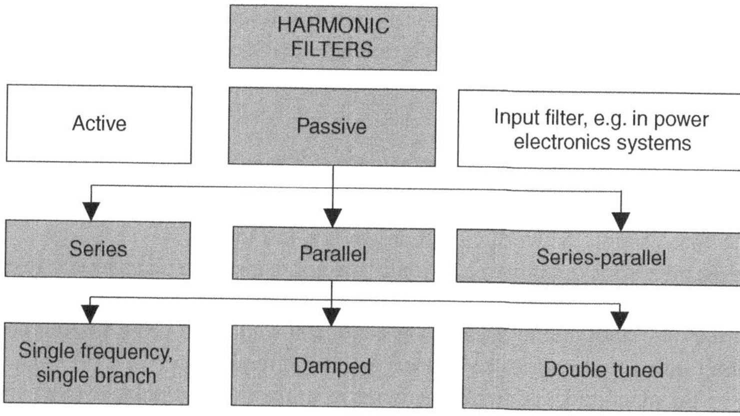

Figure 7.45 An example classification for various types of filters

compensation – for the fundamental harmonic. All filter configurations for this harmonic have a capacitive character.

The filter configuration is individually designed for a given point of supply so as to obtain the required frequency–impedance characteristic of the supply system. Figure 7.45 shows a classification diagram for various types of harmonic filters.

Resonant filters, both single frequency and double tuned, guarantee low impedance for selected series resonance frequencies, whereas damped filters have a low impedance over a broad frequency range. Hence they are referred to as broadband filters. The most often used configuration is single-frequency resonant filters plus a broadband filter. This, of course, does not exclude other solutions which are technically advantageous and economically viable for specific applications.

Figure 7.46 shows schematically the set of various, mutually dependent factors that have an effect on filter design.

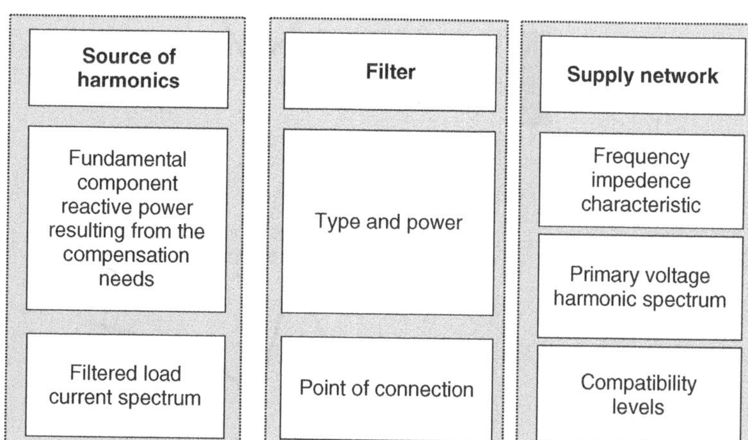

Figure 7.46 The set of data necessary for filter design

The filter branch is connected to a non-linear load terminal and tuned to the harmonic of order *n* generated by the load. In effect, under ideal conditions, the harmonic current flows only in the filter circuit and does not occur in the supply network.

7.7.2.5.2 Single Harmonic Resonant Filter (single-frequency filter)

The equivalent circuit diagram and typical impedance characteristics of a single-frequency filter and the filter–supply network circuit are shown in Figure 7.47.

The resistance *R* is mainly that of the reactor, since the capacitor's resistance is negligible.

Most filters are designed in such a way that each frequency to be filtered has its own filtering circuit tuned to the series resonance with this frequency.

Knowing the magnitudes of harmonic currents at the point of planned filter installation allows their elimination to be carried out starting from the harmonics of smallest magnitude and checking the voltage distortion factor. This pattern is repeated, step by step, until the desired value is reached. The effects of several filters are shown in Figure 7.48. With a six-pulse converter generally the fifth-harmonic filters are applied or, less frequently, mainly for large power converters, the fifth- and seventh-harmonic filters. In this case the essential issue is the distribution of the fundamental harmonic reactive power among the filters, which should be optimized in terms of the adopted criterion, e.g. minimum loss, cost, overall dimensions, etc. Figure 7.48 shows an example of filter installation with its equivalent circuit and impedance characteristic at PCC.

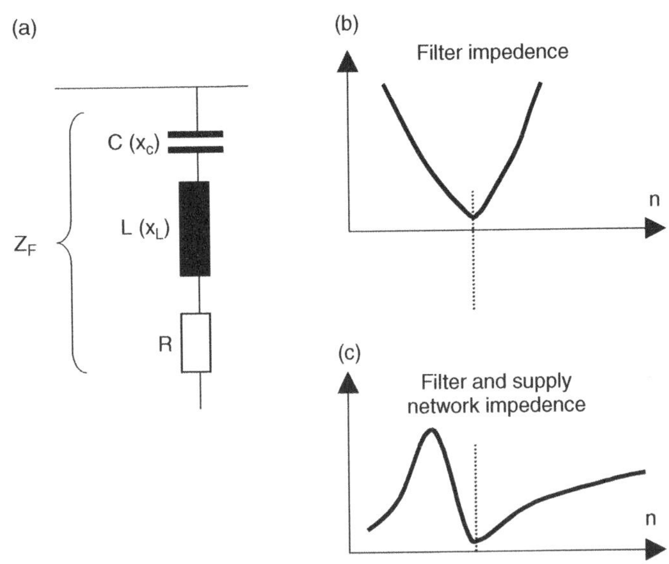

Figure 7.47 The equivalent circuit diagram of a single-frequency filter (a) and typical impedance characteristics of (b) the filter and (c) the filter with the supply network

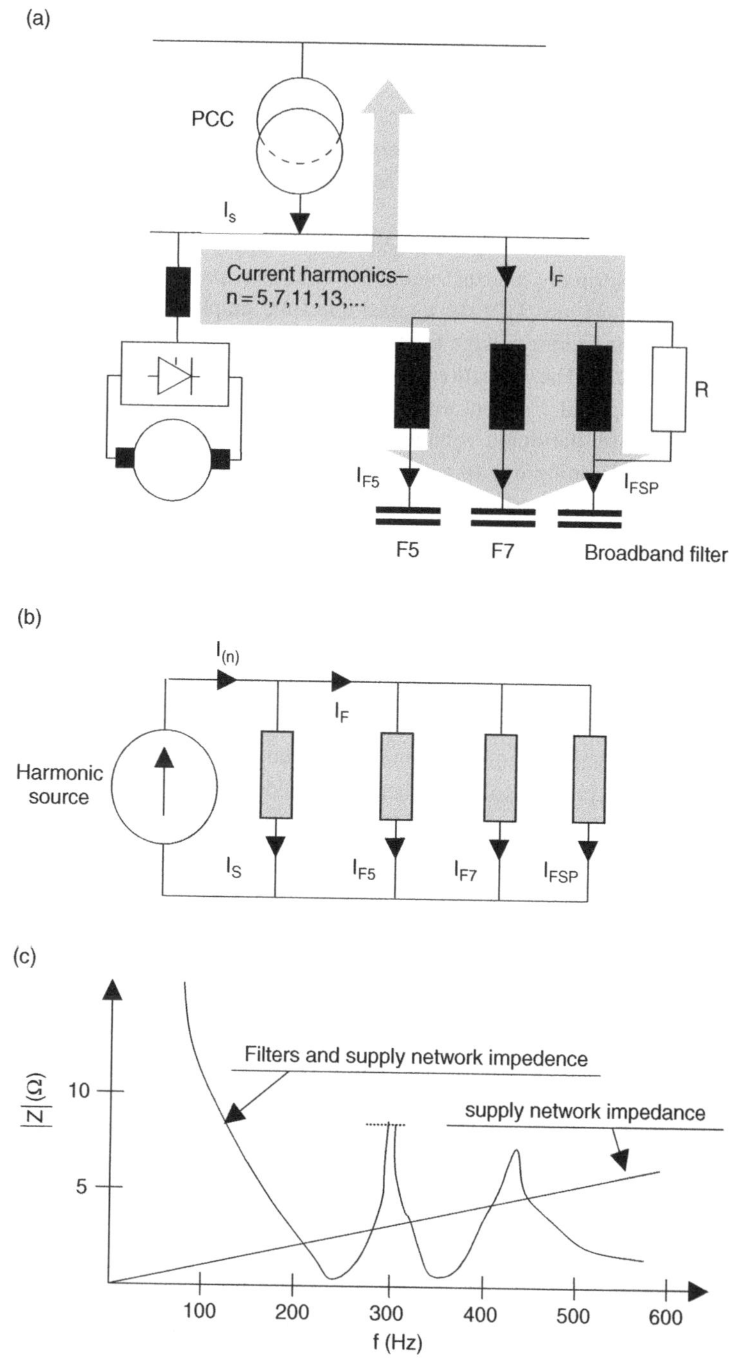

Figure 7.48 (a) An example filter installation for a converter d.c. drive; (b) the equivalent circuit diagram; (c) impedance characteristic at PCC

7.7.2.5.3 Drawbacks of Passive Filters

The drawbacks are as follows:

1. An electric power system with passive filters is a weakly damped *LCR* circuit which, in order to exclude resonance phenomena, requires a careful analysis of the frequency characteristics at the design stage.
2. The effectiveness of the filter strongly depends on the supply network impedance at the point of connection. Normally its exact value is not known and it varies with changes in network configuration. Figure 7.49 shows example time graphs of r.m.s. load and filter current and the fifth-harmonic voltage relative value, both prior to and after connecting the fifth-harmonic filter. The reduction of the current value (due to the effect of the fifth-harmonic filtering and compensation of the load current reactive component) and the reduction of the fifth-harmonic voltage magnitude are evident.
3. Filters are subject to detuning due to variations in the supply frequency and changes in *LC* component values (e.g. due to the effect of capacitor ageing). The adverse effect of detuning can be mitigated by an appropriate tuning or reducing the filter quality factor. The latter method, however, increases both the active power loss and the unfiltered harmonic content in the supply voltage.
4. The filter current also contains harmonic currents produced by the supply source voltage harmonics.
5. Only selected harmonics of dominant magnitude are filtered. The load non-characteristic harmonics, which may occur in the load supply current, are not filtered.
6. Passive filters are a large and expensive component of compensation systems. The number of single harmonic filters equals the number of filtered harmonics. Their use instead of high-order filters lowers the effectiveness of filtration, requires large power components and worsens the installation efficiency.

7.7.2.5.4 Parallel Active Power Filters (APFs)

Their principle of operation consists in eliminating those components in a non-linear load current which are not active currents, i.e. not sinusoidal and cophasal with corresponding phase voltages (Figure 7.50). A power electronics system generates a current waveform opposite in phase to the unwanted component in the load current. This device, connected in parallel to the disturbing load, unbalanced and non-linear, as seen in Figure 7.51, causes the supply currents to be near sinusoidal, balanced and cophasal with the corresponding supply voltages.

7.7.2.6 Series Filters

Series filters belong to those technical solutions which modify the impedance of supply networks. Their purpose is not to reduce the equivalent impedance of the source, but on the contrary to increase it for selected harmonic(s) (Figure 7.52). As a result, they prevent the injection of selected harmonic currents into the supply network. For instance, the effectiveness of a parallel passive filter connected to a non-linear load, represented by

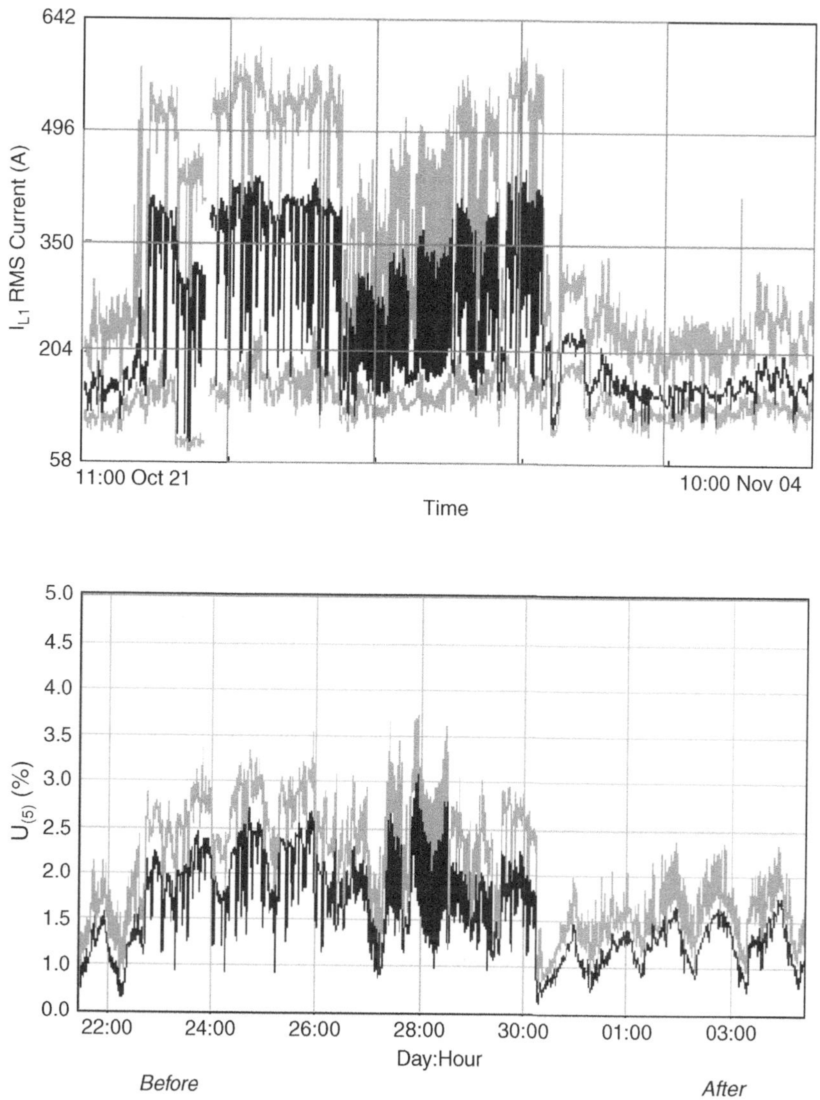

Figure 7.49 Time graphs of r.m.s. load current and the fifth-harmonic voltage in a feeder line supplying the non-linear load: before and after connecting the parallel passive filter of the fifth harmonic (recording at MV level)

the current source $I_{(n)}$ in Figure 7.52, is substantially improved. The value of the equivalent impedance Z_F of a series filter should be:

- close to zero for the fundamental harmonic, in order not to affect the energy exchange between the supply source and the load (in the domain of this harmonic);
- very large for the filtered (blocked) harmonic – in effect this harmonic current will not be present in the supply network.

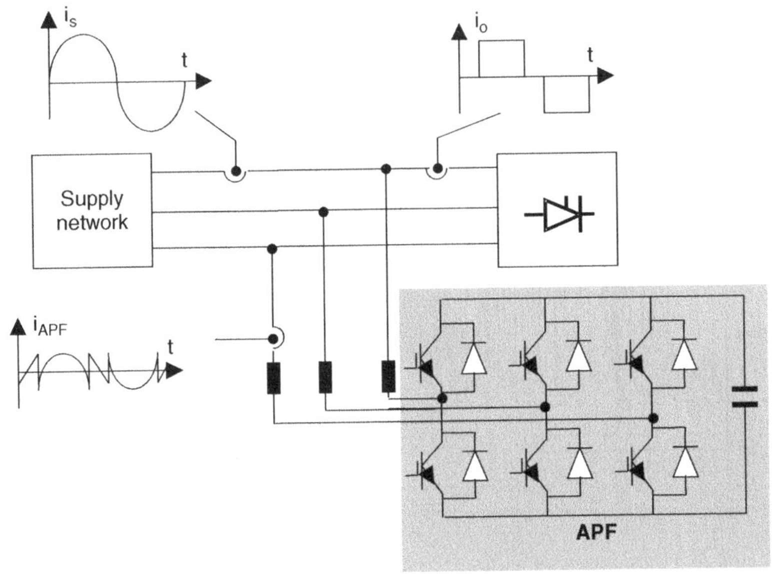

Figure 7.50 Shunt active power filter

The series filter can be implemented as a passive or active device. In the first case, in its simplest form, it is a two-terminal *LC* network, in which parallel resonance for the harmonic of order n_R occurs, connected in series between the supply source and the load. For this harmonic the filter impedance is of very large value, thus its current flow in the supply network is blocked.

The impedance Z_F of the series active filter is controlled by means of an appropriate power electronics circuit.

If the supply voltage contains, apart from the fundamental harmonic, also a high harmonic, the latter can be eliminated at the sensitive load terminals by means of the series filter, as in Figure 7.53. This filter generates the voltage which is opposite in phase to the unwanted distorting component and in this way guarantees that the supply voltage at PCC is sinusoidal.

7.7.2.7 Isolation Transformers

Third-order harmonic currents circulate in the delta windings of transformers. Although this is a problem for transformer manufacturers and specifiers due to the extra losses to be taken into account, it is useful for systems designers because it isolates these harmonic components from the supply.

The same benefit can be obtained using a zigzag wound transformer. Zigzag transformers are star configuration autotransformers with a specific phase relationship between the windings that are connected in shunt with the supply.

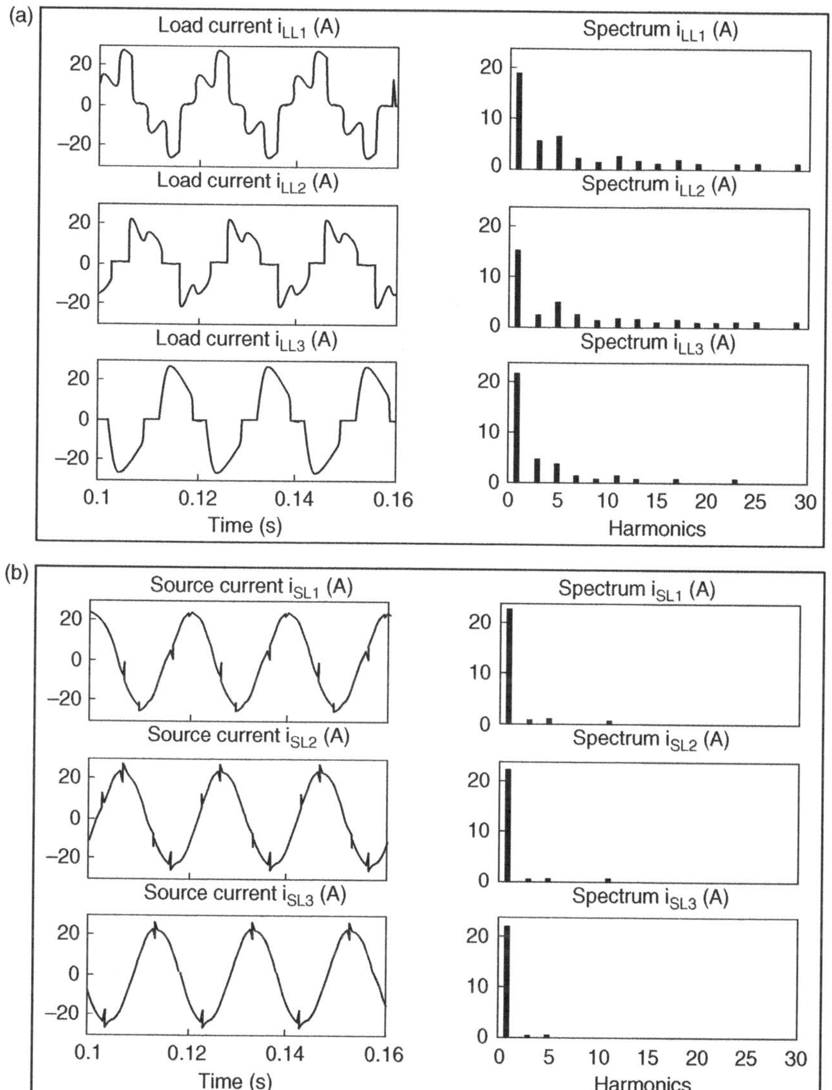

Figure 7.51 The time waveforms and frequency spectra of the supply current: prior to (a), and after (b) the active filter connection

Where a number of non-linear loads are supplied from transformers with different connection groups, a self-compensation of harmonic currents occurs because of their different phase shifts. Figure 7.54 illustrates the principle of mutual cancellation of the fifth and seventh harmonics in the case when non-linear loads are supplied from sources A and B whose voltages are shifted by 30°. The reduction of harmonics emitted from different loads, due to their vector summation, is strongly evident at the LV level.

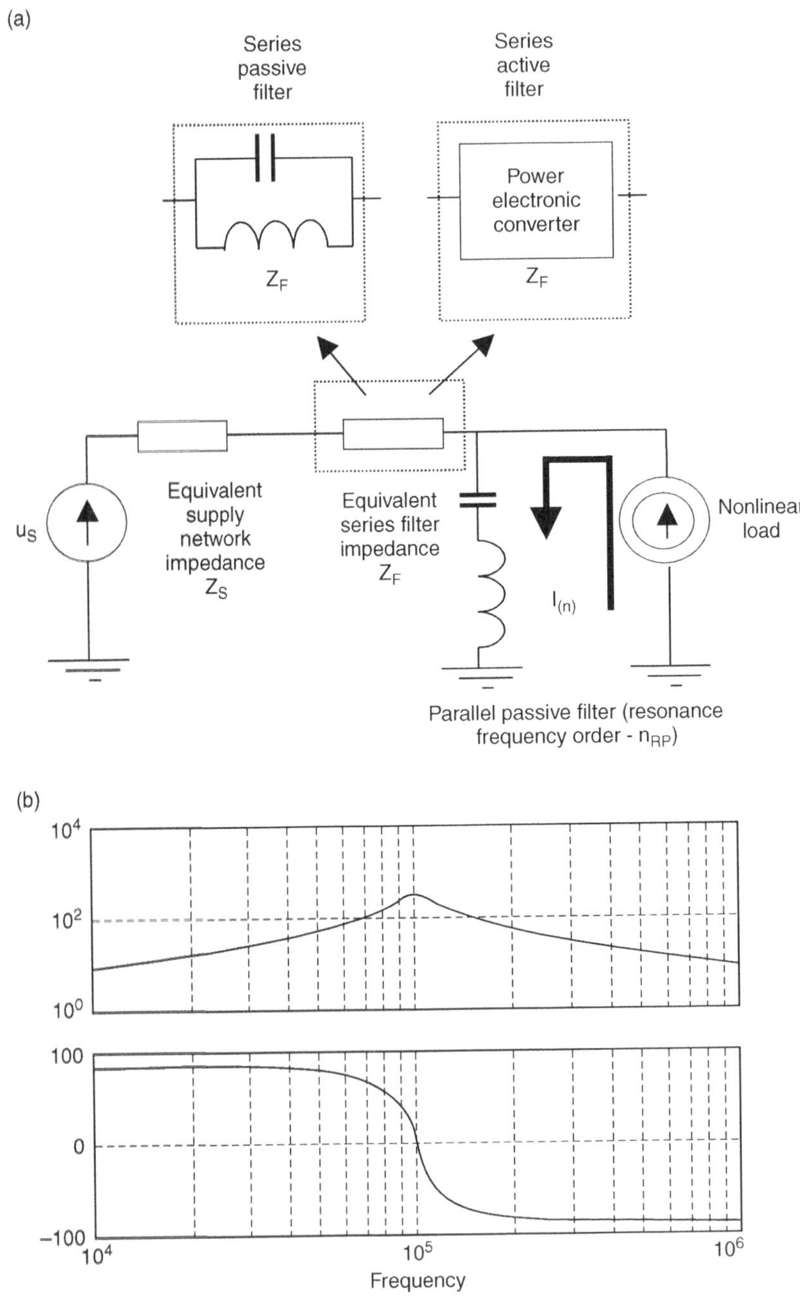

Figure 7.52 Series filter for supply voltage harmonic elimination (a) and Bode characteristics of passive series filter (b)

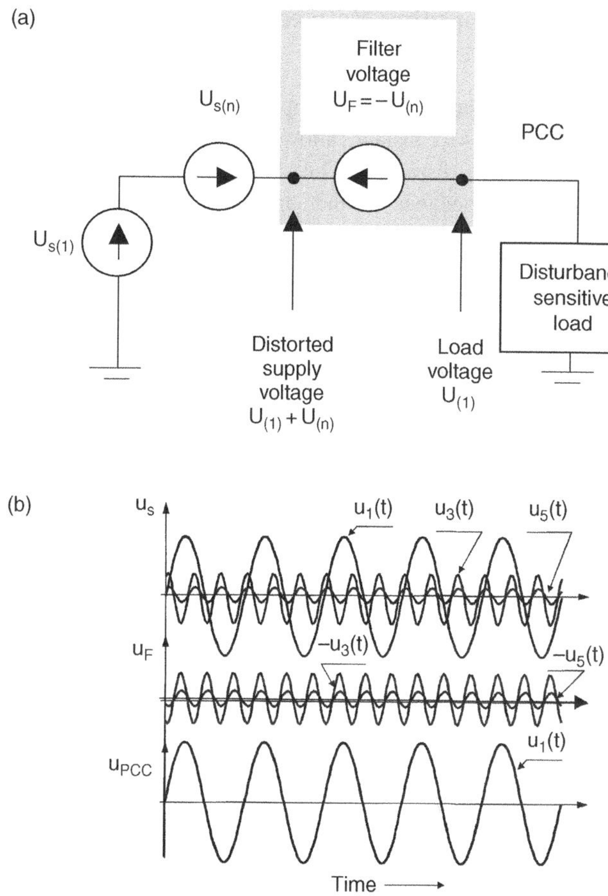

Figure 7.53 Active series filtering of supply voltage harmonics: (a) schematic diagram; (b) waveforms illustrating the principle of operation

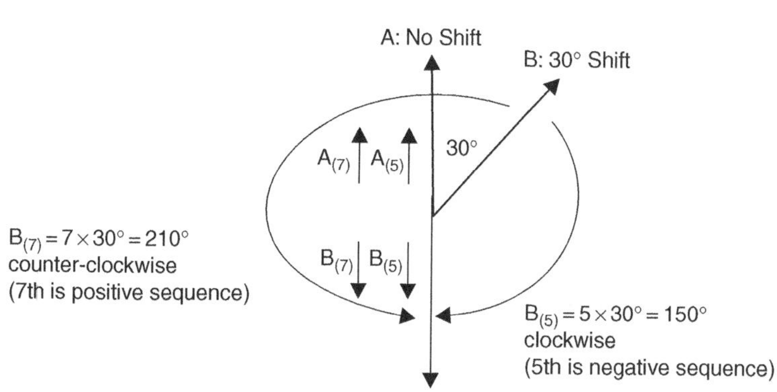

Figure 7.54 The principle of mutual cancellation of the fifth and seventh harmonics in the case when non-linear loads are supplied from sources A and B whose voltages are shifted by 30°

7.7.3 Reduction of the Coupling Between Sensitive Load and Harmonic Source

The reduction (or generally: shaping) of the supply network inpedance $Z_{s(n)}$, assuming constant load current value $I_{(n)} = \text{constant}$, allows a decrease in the voltage drop $\Delta U_{(n)}$ and thereby reduces the voltage distortion. The short-circuit capacity at the PCC can be increased, i.e. the equivalent impedance[7] of the supply line reduced, by means of:

- extension of the supply system;
- elimination of series reactors;
- parallel operation of supply lines and transformers;
- application of transformers of larger power and/or lower short-circuit voltage, etc.

These may also include compensation of the line impedance by means of series capacitors of fixed or, recently, variable capacitance.

Supplying large non-linear loads directly from an HV line is an advantageous solution. It guarantees a sufficiently large short-circuit capacity at the point of connection, and a larger voltage distortion can be accepted in such a line since harmonic-sensitive loads are not connected to it.

7.7.4 Reduction of Load Sensitivity to Disturbances

In the case of harmonic disturbances a process contrary to the intended one is observed: the sensitivity of present-day loads is increasing, with only a few exceptions whose prime example is TV sets.

Another available method to mitigate harmonic effects is oversizing the components and equipment. For example, for choosing transformers, among the available methods described in the following paragraphs it is possible to consider the *K* factor. This is an index for the amount of harmonic current in the system and can help in selecting transformers or motors. It may be used along with kVA to select a replacement transformer or motor to handle non-linear, harmonic-rich loads. The *K* factor is a number that quantifies potential losses in transformers due to harmonic currents. An example calculation for a transformer is shown in the case study.

7.7.4.1 Transformers Derating

7.7.4.1.1 *K* Factor

There are different approaches to considering additional losses when selecting a transformer. The first come from the USA, and is based on the calculation of the factor increase in eddy

[7] In practice the equivalent inductive reactance, since $R_S \cong 0$ and $Z_{s(n)} \cong n\omega L_{(s)}$, where R_s and L_s are the line equivalent resistance and inductance.

current loss and then in the specification of the transformer designed to cope. This factor is known as the 'K factor':

$$K \equiv \sum_{n=h}^{h\max} n^2 I_{(n)}^2 \tag{7.27}$$

where n is the harmonic order and $I_{(n)}$ the fraction of total r.m.s. load current with harmonic order n.

Today, many power quality meters read the K factor of the load current directly. Once the K factor of the load is known, designers just have to specify a transformer with a higher K rating within the standard range of 4, 9, 13, 20, 30, 40, 50.

Of course a purely linear load will have a K factor equal to unity. A higher K factor indicates that the eddy current loss in the transformer will be K times the value at the fundamental frequency. 'K-rated' transformers are therefore designed to have very low eddy current loss at fundamental frequency.

7.7.4.1.2 Factor K

The second method is used in Europe, and is based on the estimation of how much a standard transformer should be derated so that the total loss on harmonic load does not exceed the fundamental design loss: this is called factor K:

$$K = \left[1 + \frac{e}{1+e}\left(\frac{I_{(1)}}{I}\right)^2 \sum n^q \left(\frac{I_{(n)}}{I_{(1)}}\right)^2\right]^{\frac{1}{2}} \tag{7.28}$$

where:

- e is the ratio of fundamental frequency eddy current loss to ohmic loss, both at the reference temperature;
- I is the r.m.s. value of the current including all harmonics;
- $I_{(n)}$ is the magnitude of the nth harmonic;
- $I_{(1)}$ is the magnitude of the fundamental current;

Q is the exponential constant dependent on the type of winding and frequency. Typical values are 1.7 for transformers with round or rectangular cross-sectional conductors in both windings and 1.5 for those with foil LV windings.

7.7.4.1.3 The Additional Loss Factor

The third available method is based on the calculation of an additional loss factor. A resistance factor is defined as

$$K_{\Delta R}(f) = \frac{R_{ac}(f) - R_{dc}}{R_{ac}(f_1) - R_{dc}} \tag{7.29}$$

where R_{dc} is the equivalent series d.c. resistance and R_{ac} the series a.c. resistance. R_{ac} is dependent on frequency, due partly to current redistribution in the winding, and should be determined for each harmonic frequency. The type of construction and the placement of the windings have a major effect on the shape of the relationship between R_{ac} and frequency.

The total additional loss factor $K_{\Delta P}$ is then calculated as the sum of the frequency-dependent losses at each frequency arising from R_{ac}. This of course requires knowledge of the harmonic current spectrum of the load:

$$K_{\Delta P} = \sum_{f>f_1} K_{\Delta R}(f) \left(\frac{I_f}{I_R} \right)^2 \tag{7.30}$$

where $K_{\Delta P}$ is the additional loss factor; $K_{\Delta R}$ the resistance factor; I_f the current at harmonic frequency f; and I_R the rated current.

To determine this factor for a given transformer, prototype or computation model, the series resistances or short-circuit resistances have to be determined.

For a case study see web address

7.7.4.2 Motor Derating

To determine the motor derating factor, it is possible to use the curve shown in Figure 7.55, where the derating factor is shown as a function of the harmonic voltage factor (HVF). HVF is defined as

$$HVF = \sqrt{\sum_{n=5}^{n=\infty} \frac{U_{(n)}^2}{n}} \tag{7.31}$$

where $U_{(n)}$ is the r.m.s. value of the nth-harmonic component expressed as a percentage of the fundamental and n is the harmonic order, with the exclusion of third-order harmonics.

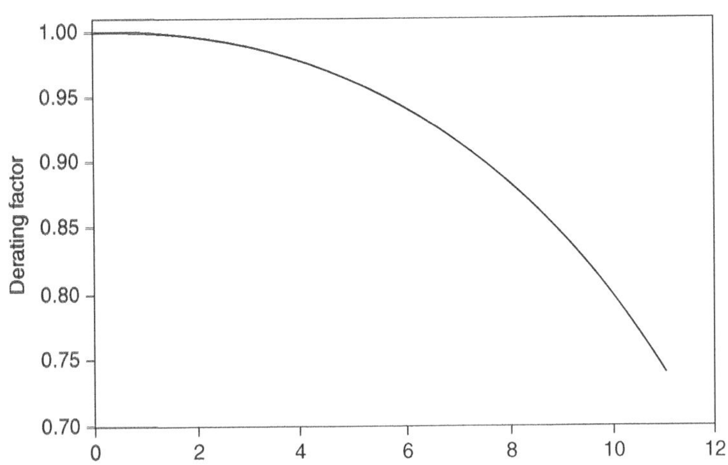

Figure 7.55 Derating factors for motors as a function of HVF

7.7.4.3 Cable Derating

The first and maybe easier solution, for single-core cables, to avoid problems related to harmonics, is to install a double-sized neutral conductor, either as two separate conductors or as one single large conductor. Unfortunately, when multi-core cables are used this method is not applicable.

The ratings of multi-core cables (e.g. as specified in IEC 60364–5-523 [17]) is based on the assumption of a balanced load and neutral conductor carrying no current: in this case only three of the four or five cores carry current and generate heat. Since the cable current-carrying capacity is defined by the amount of heat it can dissipate at the maximum permitted temperature, the consequence is that cables carrying triplen currents must be derated.

Although it is not possible to determine the neutral current in absolute terms, unless the real or theoretical waveform of the load currents is known, it is not unreasonable to refer [15] to a value that can be equal to 1.61 times the phase current in the case of loads made by computers, but that can reach the value of 1.73 times the phase current in the worst conditions with controlled rectifiers ($\alpha = 60°$).

The simplest approach for the solution of the problem is to apply appropriate corrective coefficients to the cable capacity.

Also, the document [27], without providing instruments, and a recent, more concrete IEC standard [23] state that the current flowing in the neutral has to be taken into consideration in the calculation of the cable capacity when no corresponding load reduction in the phase conductors is available.

In order to simplify the approach, it can be assumed that:

- the system is three-phase balanced;
- the only significant harmonic not being cancelled in the neutral is the third one;
- the cable is made of four or five cores and the neutral is of the same material and section as the phase conductors.

In a strict sense the calculations of the current harmonic effects should be made also as a function of the conductor sizes, but in a first approximation this dependence can be neglected.

Table 7.12 lists possible reduction factors, based on the hypothesis mentioned, that, when applied to the capacity of a cable having three active conductors, supply the capacity

Table 7.12 Reduction factors for current harmonics

Third-harmonic current (%)	Value selected on the basis of the line current	Value selected on the basis of the neutral current
0–15	1.00	—
15–33	0.86	—
33–45	—	0.86
45	—	1.00

of a cable with four conductors charged, where the current in the fourth conductor is due to harmonics.

In those cases where a neutral current greater than the phase current is expected, the cable sizes must be taken on the basis of the neutral current.

If the neutral current is not much higher than the phase current, then, considering cables with all cores made of the same material and section, it is necessary to reduce the capacities for the three conductors charged as shown in Table 7.12.

If the neutral current is greater than 135 % of the phase current then, under the same hypothesis, it is not necessary to apply any corrective factor since the three-phase conductors will be completely charged and the greater heat generated by the neutral will be compensated by a reduction in the heat generated in the former ones.

It seems obvious that the situation is harder when the load is no longer balanced and particularly when only two phases of the three are charged, because in this situation the neutral will carry a harmonic current in addition to the unbalanced one.

In the case where this unbalance among the phases is greater than 50 %, smaller reduction factors must be considered.

Since the data reduction factors have been calculated on the basis of only the current third harmonic, if the greater harmonics of multiples of 3 (the 9th, the 12th, etc.) are present in a size greater than 10 %, then still smaller reduction factors are to be applied. The situation described can be particularly critical when the neutral is common to more circuits.

On the contrary, referring to unipolar conductors, on the one hand, the choice of the neutral sections and the phase conductors becomes somehow independent and, on the other hand, the thermal interaction among them is analytically more difficult to model because of the respective variable positions.

The most direct method is to utilize the independent sizing of the neutral conductor, tending, however, toward the phenomena related to the mutual position of the conductor, versus the heating and the reactance.

Particular attention must be given to armored cables or those with a metal screen for which the contribution given by the current harmonic circulation in the same screen or armor may be not negligible. In conclusion, when a load current deformation is expected with a consistent component of the third harmonic, the neutral circuit section should never be chosen smaller than the corresponding phase conductor together with all the accessories of the same neutral circuit.

When, as can happen in standard electrical systems, an increase in elements of the neutral circuit beyond the corresponding phase components is difficult or not even feasible because of the commercial unavailability of suitable components to integrate correctly in the system, it is acceptable either to limit the load or to size the phase sections accordingly.

For terminal circuits, neutral circuits separated for each line and circuits separated for each deforming load should be foreseen (this also ensures the greatest possible electromagnetic independence among disturbing and susceptible elements).

The use of the best possible balance of the loads, as pointed out, avoids further contributions to the neutral current due to unbalance.

The considerations here are just as important for large-section cables, but are applicable to modest-section cables too and can be extended, of course, to at least an approach level, as in the case of busbars.

For a case study see web address

7.8 STANDARDS

The philosophy of developing harmonic limits in recommended practice is to limit the harmonic injection in the network from individual customers so that they will not cause unacceptable voltage distortion levels for normal system characteristics and to limit overall harmonic distortion of the system voltage supplied by the utility.

The harmonics standardization process can be considered as quite complete since this problem is relatively old. The same cannot be said about the interharmonics standardization process, which is in its infancy, with knowledge and measured data still being accumulated.

Standards prescriptions define the limits on voltage supply, measurement method and instruments, mitigation process, etc. The main IEC standards dealing with harmonics and interharmonics are presented in Table 7.13. Essential standards regarding the harmonics are EN 50160, IEEE 519-1992 and IEEE 1159-1995. A list of harmonics and interharmonics standards appears in Annex A2 on the website. For example, Table 7.14 shows the voltage distortion limit according to [12].

Table 7.15 shows the limits according to IEEE 519-1992 [24]

Currently defined distortion limits assume that there will be some diversity between the harmonic currents injected by different customers. This diversity can be in the form of different harmonic components being injected, differences in the phase angles of the individual harmonic currents, or differences in the harmonic injection versus time profiles. In recognition of this diversity, the current limits are developed so that the maximum individual frequency harmonic voltage caused by a single customer will not exceed the limits in Table 7.16 for systems that can be characterized by a short-circuit impedance.

Table 7.13 Main IEC harmonic standards

Subject	Standard
General	IEC 61000-1-4
Emission (description)	IEC 61000-2-2, 61000-2-3, 61000-2-4, 61000-2-6, 61000-2-12
Limits	IEC 61000-3-2, 61000-3-4, 61000-3-9, 61000-3-6, 61000-3-10, 61000-3-12
Testing and measurement techniques	IEC 61000-4-7, 61000-4-13, 61000-4-30, 61000-4-31

Table 7.14 Values of individual harmonic voltages at the supply terminals for orders up to 25, given as a percent of U_n[12][8] (Reproduced from voltage characteristics of electricity supplied by public distribution systems, EN 50160)

Odd harmonics				Even harmonics	
Not multiples of 3		Multiples of 3			
Order h	Relative voltage (%)	Order h	Relative voltage (%)	Order h	Relative voltage (%)
5	6	3	5	2	2
7	5	9	1.5	4	1
11	3.5	15	0.5	6–24	0.5
13	3	21	0.5		
17	2				
19	1.5				
23	1.5				
25	1.5				

Table 7.15 Voltage distortion limits [24][9]

Bus voltage at PCC	Individual voltage distortion (%)	Total voltage distortion THD (%)
69 kV and below	3.0	5.0
69.001 kV through 161 kV	1.5	2.5
161.001 kV and above	1.0	1.5

Table 7.16 Basis for harmonic current limits

SCR at PCC	Maximum individual frequency voltage harmonic (%)	Related Assumption
10	2.5–3.0	Dedicated system
20	2.0–2.5	1–2 large customers
50	1.0–1.5	A few relatively large customers
100	0.5–1.0	5–20 medium-size customers
1000	0.05–0.10	Many small customers

[8] No values are given for harmonics of order higher than 25, as they are usually small, but largely unpredictable due to resonance effects.

[9] High-voltage systems can have up to 2.0 % THD where the cause is an HV d.c. terminal that will attenuate by the time it is tapped for a user.

Table 7.17 Classification of equipment [21]

Class of equipment	Type of equipment	Note
A	Balanced three-phase equipment Household appliances (not those in Class D) Tools (not portable) Light dimmers Audio equipment Everything else not covered by B, C or D	
B	Portable tools Arc welding equipment (non-professional)	
C	Lighting equipment	
D	Personal computers and monitors TV sets	Power $\leq$ 600 W

If individual customers meet the current distortion limits, and there is not enough diversity between individual customer harmonic injections, then it may be necessary to implement some form of filtering on the utility system to limit voltage distortion levels. However, it is more likely that voltage distortion problems would be caused by system frequency response characteristics that result in the magnification of harmonic current at a particular harmonic frequency. This changing of the system impedances versus frequency characteristic is a result of the system's physical configuration. This situation has to be solved on the utility system either by changing capacitor locations or sizes, or by designing a harmonic filter.

Standard IEC 61000-3-2 [21] deals with the harmonic current emission limits of individual equipment connected to public networks. This standard imposes limits on the current harmonics drawn from the mains supply. The standard requires that electrical appliances be type tested to ensure that they meet the requirements of the standard. The standard defines four classes of waveform according to the different types of equipment (Table 7.17). For example, one of the classes (Class B) applies to portable tools, whereas Class D refers to PCs and TV sets. Each class has different harmonic limits up to the 40th, which must not be exceeded. Some classes have dynamic limits which are set according to the power drawn by the device.

BIBLIOGRAPHY

[1] Arrillaga J., Watson N. R., Chen S., *Power system quality assessment*, John Wiley & Sons, Ltd, Chichester, 2000.

[2] Arsenau R., Filipski P. S., Zelle J., A VA-meter-error analyzer. *IEEE Transactions on Power Delivery*, vol. 6, no. 4, 1991.

[3] Baggini A., Zanoli F., Progetto di trasformatori per l'alimentazione di azionamenti e carichi non lineari. VIII Seminario Interattivo su Azionamenti elettrici innovazioni tecnologiche e problematiche emergenti, Bressanone (BZ), 10–12 marzo 1997.

[4] CEI 14-4/1983, *Trasformatori di potenza.*

[5] CEI UNEL 35024/1, *Cavi elettrici isolati con materiale elastomerico o termoplastico per tensioni nominali non superiori a 1000 V in corrente alternata e 1500 V in corrente continua. Portate di corrente in regime permanente per posa in aria*, 1997.
[6] Chapman D., Harmonics – causes and effects. *Leonardo Power Quality Application Guide – Part 3.1*, 2001.
[7] Correggiari F., *Costruzione di macchine elettriche*, Cisalpino Goliardica, Milan.
[8] Datta S. K., Nafsi A., Distribution relay performance under harmonics conditions. PQA'92, Atlanta, Georgia, USA, 1992.
[9] Desmet J., Baggini A., Harmonics – neutral sizing in harmonic rich installations. *Leonardo Power Quality Application Guide – Part 3.5.1*, 2003.
[10] Desmet J., Delaere G., Harmonics – selection and rating of transformers. *Leonardo Power Quality Application Guide – Part 3.5.2*, 2005.
[11] Elmore W. A., Kramer C. A., Zocholl E., Effect of waveform distortion on protective relays. *IEEE Transactions on Industry Applications*, vol. 29, no. 2, 1993.
[12] EN 50160, *Voltage characteristics of electricity supplied by public distribution systems.*
[13] Fassbinder S., Harmonics – passive filters. *Leonardo Power Quality Application Guide – Part 3.3.1*, 2003.
[14] Girgis A. A., Nims J. W., Jacomino J., Dalton J. G., Bishop A., Effect of voltage harmonics on the operation of solid-state relays in industrial applications. *IEEE Transactions on Industry Applications*, vol. 28, vol. 5, 1992.
[15] Gruzs T. M., A survey of neutral currents in three-phase computer power systems. *IEEE Transaction on Industry Applications,* vol. 26, no. 4, 1990.
[16] Hanzelka Z., Bien A., Harmonics – interharmonics. *Leonardo Power Quality Application Guide – Part 3.3.1*, 2004.
[17] IEC 60364–5-523, *Electrical installations of buildings – Part 5-52: Selection and election of electrical equipment – Wiring systems.*
[18] IEC 61000-1-4, *Historical rationale for the limitation of power-frequency conducted harmonic current emissions from equipment in the frequency range up to 9 kHz*, Technical Report.
[19] IEC 61000-2-1, *Electromagnetic compatibility (EMC) Part 2-1: Environment – Description of the environment: Electromagnetic environment for low-frequency conducted disturbances and signalling in public power supply systems*, 1990.
[20] IEC 61000-2-2, *Electromagnetic Compatibility (EMC) – Part 2-2: Environment – Compatibility levels for low frequency conducted disturbances and signalling in public low-voltage power supply systems.*
[21] IEC 61000-3-2, *Limits for harmonic current emissions (equipment input current* $\leq$ *16 A per phase).*
[22] IEC 61000-4-7, *Electromagnetic compatibility (EMC) Part 4: Testing and measurement techniques Section 7: General guide on harmonics and interharmonics measurements and instrumentation for power supply systems and equipment connected thereto.*
[23] IEC TC 64 WG 2, *Current-carrying capacity and related overcurrent protection*, Revision of section 523", September 1996.
[24] IEEE 519-92, *IEEE Recommended Practices and Requirements for Harmonic Control in Electrical Power Systems*, 1992.
[25] IEEE 1159, *Recommended practice for monitoring electric power quality.*
[26] Karve S., Harmonics – active harmonic conditioners. *Leonardo Power Quality Application Guide – Part 3.3.3*, 2001.
[27] Norma CEI 64-8/5, *Impianti elettrici utilizzatori a tensione nominale non superiore a 1000 V in corrente alternata e a 1500 V in corrente continua. Parte 5: Scelta ed installazione dei componenti elettrici*, 1992.

[28] *Power System Harmonics*, Power Technologies, Inc., 1989.
[29] Purkayastha I., Savoce P. J., Effect of harmonics on power measurement. *IEEE Transactions on Industry Applications*, vol. 26, no. 5, 1990.
[30] Shepherd W., Zakikhani, P., *Energy flow and power factor in non-sinusoidal circuits*, Cambridge University Press, New York.
[31] Stade D., Shau H., Influence of voltage harmonics on single-phase earth fault currents. PQA'91.
[32] Tsukamoto M., Kouda I. N., Minowa Y., Nishimura S., Advanced method to identify harmonics characteristic between utility grid and harmonic current sources. 8th International Conference on Harmonics and Quality of Power, Athens, Greece, 14–16 October, 1998.
[33] West K., Harmonics – true RMS – the only true measurement. *Leonardo Power Quality Application Guide – Part 3.2.2*, 2001.
[34] Xu Wilsun, Liu Yilu, A method for determining customer and utility harmonic contributions at the point of common coupling. *IEEE Transactions on Power Delivery*, vol.15, no.2, 2000.
[35] Xu Wilsun, Liu Xian, Liu Yilu, An investigation on the validity of power-direction method for harmonic source determination. *IEEE Transactions on Power Delivery*, vol. 18, no. 1, 2003.
[36] Yacamini R., Chang S. C., Noise and vibration from induction machines fed from harmonic sources. Proceedings of IEEE ICHPS VI, Bologna, 21–23 September, 1994.
[37] Żeżelenko I. W., *Harmonics in power system supplying industrial loads*, Elektroatomizdat, Moscow, 1994 (in Russian).

8

Overvoltages

Franco Bua, Francesco Buratti and Alan Ascolari

IEC 60071-1 [5] defines:

- **Insulation and coordination** – the selection of the dielectric strength of equipment in relation to the voltages which can appear on the system for which the equipment is intended and taking into account the service environment and the characteristics of the available protective device.
- **Overvoltage** – any voltage between one phase conductor and earth or between phase conductors having a peak value exceeding the corresponding peak of the highest voltage for equipment.

Insulation coordination is then a discipline which aims at achieving the best possible technical and economic compromise in protection of persons and equipment in case of overvoltages occurring in electrical installations.

The importance of this discipline is emphasized by the fact that it applies to high-voltage networks. For insulation coordination it is therefore necessary that the level of the possible overvoltage must be known so that the right protective devices can be used when necessary and the correct overvoltage withstand level can be chosen for each network component.

The main issue is that overvoltages' random nature makes them hard to characterize: it is then possible to apply only a statistical approach to their duration, amplitude and effects.

Overvoltages are the most frequent cause of disturbances and faults, even if, in general, the worst consequence are not due to overvoltage energy, which usually is very small (the transient phenomenon is very fast), but to the power frequency energy at the point of

Handbook of Power Quality Edited by Angelo Baggini

defective equipment. The overvoltage is then only the starting point of the fault, through which the power of generators can give rise, in a few seconds, to disastrous consequences.

8.1 DESCRIPTION OF THE PHENOMENA

An overvoltage is defined as an average line-to-line voltage value greater than the maximum acceptable for the equipment installed.

With reference to shape and duration, overvoltage can be classified as:

- Continuous (power frequency)
- Temporary
- Transient
- Combined

8.1.1 Continuous (Power Frequency) Voltage

In insulation coordination overvoltages and impulse voltages are defined in terms of peak value (referred to earth). It is also convenient to consider the system voltage phase-to-earth peak value: that is, $\sqrt{2}/\sqrt{3} = 0.816$ times the usual r.m.s. phase-to-phase voltage.

Under normal operating conditions, the power frequency voltage is usually expected to vary in a certain way in magnitude and can be described in terms of the probability distribution of the average operating value. Of course the parameters of this distribution differ at the various points of the system. In insulation design and coordination the power frequency voltage must, however, be considered as constant and equal to the highest voltage for the equipment:

- in voltage range C it does not differ from the highest system voltage, with a phase-to-earth peak value of $U_m = \sqrt{2}/\sqrt{3}$;
- in range A and in range B up to 72.5 kV, the highest voltage for equipment may be higher than the highest system voltage, as indicated in the note under Clause 5 of IEC Publication 60071-1.[1]

It is, however, usually considered that equipment insulation will always be able to operate satisfactorily at the highest voltage for equipment immediately above, if not equal to, the highest system voltage.

8.1.2 Temporary Overvoltage

The term temporary overvoltage describes sustained overvoltages, or overvoltages with several successive peaks with decreased amplitude and then comparable to a sustained overvoltage at power frequency.

[1] Range A, above 1 kV and less than 52 kV; range B, from 52 kV to less than 300 kV; range C, 300 kV and above.

The severity of temporary overvoltages is mainly due to both their amplitude and duration. The importance of temporary overvoltages is related to:

- The characteristics of temporary overvoltage in corresponding to the surge arrester location: they are of great importance in surge arrester selection.
- Repetition of successive overvoltage peaks of opposite polarity, even if, at lower amplitude, they may affect the design of both the internal insulation of equipment and the external insulation (the surface exposed to contamination).

This kind of overvoltage generally comes from:

- earth faults;
- sudden changes of load;
- resonance and ferroresonance.

8.1.3 Transient Overvoltage

IEC Publication 60071-1 defines transient overvoltage as a short-duration overvoltage of a few milliseconds or less, oscillatory or non-oscillatory, usually highly damped.

Transient overvoltages (see Table 8.1) are divided into:

- **Slow front** – transient overvoltage, usually unidirectional, with time to peak $20\,\mu s < T_p \leq 5000\,\mu s$, and tail duration $T_2 \leq 20$ ms.
- **Fast front** – transient overvoltage, usually unidirectional, with time to peak $0.1\,\mu s < T_1 \leq 20\,\mu s$, and tail duration $T_2 \leq 300\,\mu s$.
- **Very fast front** – transient overvoltage, usually unidirectional, with time to peak $T_f \leq 0.1\,\mu s$, total duration < 3 ms, and with superimposed oscillations at frequency $30\,kHz < f < 100\,MHz$.

8.1.4 Combined Overvoltage

Combined (temporary, slow-front, fast-front, very fast-front) overvoltage consists of two voltage components simultaneously applied between each of the two phase terminals of a phase-to-phase (or longitudinal) insulation and earth. It is classified according to the component with higher peak value.

The standard combined switching impulse is a combined impulse voltage made of two components of equal peak value and opposite polarity:

- the positive component is a standard switching impulse;
- the negative one is a switching impulse with times to peak and half value no lower than those of the positive impulse.

Both impulses should reach their peak value at the same instant. The peak value of the combined voltage is then the sum of the peak values of the components.

Table 8.1 Representative overvoltage shapes and tests [5] (Reproduced from Insulation co-ordination – Part I: Definitions, principles and rules, IEC 60071-1, Eighth edition)

Class	Low frequency		Transient		
	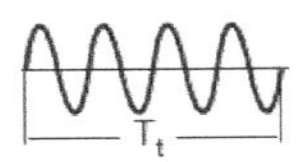	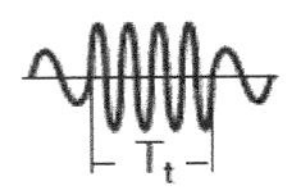	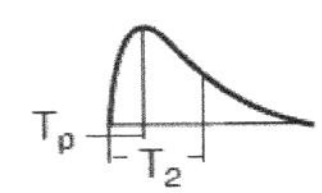	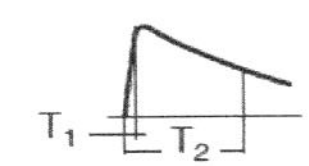	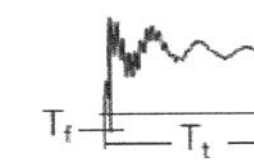
Shape	Permanent	Temporary	Slow front	Fast front	Very fast front
Shape range (frequency, rising front, term)	$f = 50$ or 60 Hz $T_t \geq 3600$ s	$10 < f < 500$ Hz $3600 \geq t_t \geq 0.03$ s	$5000 > T_p > 20$ ms $20\text{ ms} \geq T_2$	$20 > T_1 > 0.1$ ms $300\text{ ms} \geq T_2$	$100 > T_f > 3$ ns $0.3 > f_1 > 100$ MHz $30 > f_2 > 300$ kHz $3\text{ ms} \geq T_t$
Standardized shape	$f = 50$ or 60 Hz T_t^a	$48 \leq f \leq 62$ Hz $T_t = 60$ s	$T_p = 250$ ms $T_2 = 2500$ ms	$T_1 = 1.2$ ms $T_2 = 50$ ms	N/A[a]
Standardized withstand test	N/A[a]	Short-duration power frequency test	Switching impulse test	Lightning impulse test	N/A[a]

[a] To be specified by the relevant product committee.

8.1.5 Overvoltage Propagation

MV lines are generally more exposed to lightning flashes than LV lines: they have longer and higher structures than other structures located in their vicinity (houses, trees). The number of lightning flashes which can reach the lines depends on the keraunic[2] level of the local area.

The propagation of surges through the MV lines and their transfer rate to the LV system depend on the constructive characteristics of the system. Some important differences can also exist between different countries due to different design approaches.

The lightning surges in MV systems are caused by:

- direct flashes;
- nearby flashes causing induced surges;
- flashes striking earth wires or extraneous metal parts of structures or equipment, or striking the earth close to a line structure; in these cases back-flashovers can occur.

Surge propagation depends essentially on the MV system structure and on the surge-protective devices installed. Lightning surges are usually attenuated rapidly by their propagation on the line (losses and flashover across the line insulators). After a short distance, the magnitude of an overvoltage is reduced to the insulation levels of the line isolators; therefore it can be assumed that overvoltages in an MV system are limited by the insulation level of the line isolators: in a 20 kV system; this is about 150 to 180 kV. This excludes direct strokes to the MV/LV transformer or its vicinity and in wooden-pole lines without earthed cross-arms where much higher surges can occur.

Another limitation to the surge level is provided by the surge-protective devices located on the system, generally at the primary side of the MV/LV transformer, or at the entrance to an underground network. The residual overvoltage (e.g. in the range of 70 kV for a 20 kV system) depends on the rated value and earthing impedance of the devices.

If air gaps are present, the lightning surge is expected to be followed by a power frequency current generating a temporary overvoltage.

Propagation support can be easily modeled with values per unit length of inductance and resistance in a longitudinal direction and of capacitance and conductance in a transverse direction.

The inductance and capacitance of an electrical system influence successive and periodic transformations of electrostatic energy to magnetic energy (of the magnetic field).

The line impedance, for sinusoidal waveforms, can be expressed as

$$Z=\sqrt{\frac{L\omega+R}{C\omega+G}} \tag{8.1}$$

where L is the inductance (H); R is the resistance (Ω); C is the capacitance (F); and G is the conductance (S).

---

[2] Keraunic levels are a tentative evaluation of stormy activity, quantified by a number representing the monthly average of earthed lightning strikes.

At high frequency, or if we consider a lossless line, it can be considered that the impedance is formed only by inductance and capacitance, so the so-called 'characteristic impedance' can be expressed by

$$Z = \sqrt{\frac{L}{C}} \tag{8.2}$$

The magnitudes of the characteristic impedance are usually: EHV lines, 300 to 500 Ω; HVA lines, 1000 Ω.

The propagation speed can be estimated to be approximately the velocity of light (3×10^8 m/s). This speed is then equal to 300 meters per microsecond allowing an estimation of the distribution along the conductor of a very short-term wavefront.

According therefore to the theory of guided propagation, when a wave propagating along a conductor faces an impedance change, then a partial reflection and transmission occur.

If Z_c and Z_a are respectively the characteristic impedance of the first and the second conductor, the transmission and reflection coefficients are given by

$$T = 2Z_a/(Z_a + Z_c) \tag{8.3}$$

$$R = (Z_a - Z_c)/(Z_a + Z_c) \tag{8.4}$$

The limit values of these coefficients in some simple and frequent cases are:

- $Z_a = 0$ (occurring when the line is closed at the frame): this corresponds to no transmission and a reflected wave with a factor of -1.
- $Z_a = Z_c$ (the situation of a homogeneous conductor): transmission equal to 1 and no reflection.
- $Z_a = \infty$ (the line is open): the voltage at the reflection point then comprises the superimposition of the incident and the reflected wave with a factor of $+1$. Its maximum value will then be equal to twice the peak of the incident wave. There is no propagation in the medium Z_a, but the border value is given anyway by T which is also equal to 2. This leads to a high stress at the reflection point and its vicinity. This may lead to confusion and the assumption that the reflected wave is twice as large as the initial wave, but it is only at the reflection point that the maximum value observed is twice the value of the incident wave, because this is the only point where the incident and the reflected wave join their peaks.

8.2 PARAMETERS

Overvoltages can manifest in various waveforms. It is not possible describe and standardize all the waveforms, therefore standard IEC 60071-1 defines three types of overvoltages:

- **Standard short-duration power frequency voltage** – a sinusoidal voltage with a frequency between 48 and 62 Hz and a duration of 60 s.

- **Standard switching impulse** – an impulse voltage having a time to peak of 250 μs and a time to half value of 2500 μs.
- **Standard lighting impulse** – an impulse voltage having a front time of 1.2 μs and time to half value of 50 μs.

On the basis of the type to be studied, one of the waves described previously can be adopted. For example:

- Basic lightning impulse insulation level: A reference impulse insulation strength expressed in terms of crest value of the withstand voltage of a standard lightning impulse voltage wave. The standard lightning impulse wave rises to a crest value in 1.2 μs and drops to one-half crest value in 50 μs.
- Basic switching impulse insulation level: A reference switching impulse insulation strength expressed in terms of crest value of the withstand voltage of a standard switching impulse voltage wave; the standard switching impulse voltage wave rises to a crest value in 250 μs and drops to one-half of crest value in 2500 μs.

On the basis of the characteristics of wave shape, overvoltages can be classified (Table 8.2) with reference to shape and duration; and cause.

8.2.1 Lightning Impulse Protective Level of a Surge Arrester

This is characterized by the following voltages:

- sparkover voltage for a standard full lightning impulse;
- residual (discharge) voltage at the selected standard nominal current;
- front of wave sparkover voltage;
- protective level under lightning impulses, taken for insulation coordination purposes at the highest of the following values:
  - maximum sparkover voltage with 1.2/50 impulse;
  - maximum residual voltage at the specified current;
  - maximum front of wave sparkover voltage.

Table 8.2 Characteristics of the various overvoltage types

Classification	Overvoltage type (cause)	MV–HV overvoltage coefficient	Term	Steepness of frequency front	Damping
Internal	At power frequency (insulation fault)	$\leq \sqrt{3}$	Long > 1 s	Power frequency	Low
	Switching (short-circuit disconnection)	2 to 4	Short < 1 ms	Medium, 1 to 200 kHz	Medium
External	Atmospheric (direct lightning stroke)	> 4	Very short, 1 to 10 ms	Very high, 1000 kV/ms	High

8.2.2 Switching Impulse Protective Level of a Surge Arrester

This is characterized by the following voltages:

- maximum sparkover voltage for the standard wave shapes;
- total surge arrester voltage exhibited by the surge arrester when discharging switching surges.

Currently, since a standard for total surge arrester voltage has not yet been specified by IEC, reference should be made to the manufacturers.

For any overvoltage type, standards should define the type of test to which the protective device, electrical machine or lines must be subordinate.

It is necessary to define any type of different wave by describing it through its properties in order to characterize overvoltage protective devices for coordination insulation. Manufacturers will have then to realize their products according to regulatory prescriptions and designers will have the opportunity to choose between normalized products.

8.3 SOURCES

With reference to cause, overvoltages can be classified in relation to:

- **Internal origin** – produced from phenomena related to system operation and in general due to abrupt variations of system conditions. They can be generated mainly from unexpected load reduction, self-excited generators, resonance phenomena, line faults and switching operations. It is possible to include also in this category abnormal overvoltages due to different voltage-level circuit contacts (e.g. contacts between conductors with different voltage levels).
- **External origin** – due to atmospheric electricity. Their amplitude is independent of operating voltage, and they can be greater than it. They can originate from the electrostatic charge, electrostatic or electromagnetic induction of lines, and in the worst cases from direct lightning stroke.

8.3.1 Internal Overvoltages

This category (Table 8.3) includes all overvoltages with frequencies under 500 Hz (the usual network frequencies are 50, 60 and 400 Hz).

Internal overvoltages are power frequency or medium-frequency overvoltages (up to 20 kHz) with:

- a duration between a millisecond and some seconds;
- an amplitude up to dangerous values (even in modern devices like maximum voltage relays, fast voltage regulations, neutral earthing, etc.) which can be limited to 2–2.5 pu.

Table 8.3 Characteristics of internal overvoltages

Overvoltage origin	Maximum amplitude (pu)	Wave type	Duration
Unload HV line	≤ 1.15	Sinusoidal, 50 Hz	Permanent
Load shedding	≤ 1.4	Sinusoidal, 50 Hz	Many seconds
Alternators' overspeed	≤ 1.4	Sinusoidal, 50 Hz	Many seconds
Self-excitation generators	≤ 1.5–1.6	Sinusoidal, 50 Hz	Many seconds
Resonance	Indefinite	Sinusoidal ≥ 50 Hz	Permanent[a]
Earth fault	≤ 1.8 neutral earthing ≤ 2.8 neutral insulated	Oscillation $\gg 50$ Hz	0.1–0.5 s[b]
Switching operation	2.3–2.5	Damped oscillation	2–10 ms

[a] If protection does not operate.
[b] Operation time of equipment.

8.3.1.1 Earth Fault

Earth-fault overvoltages are abnormal voltages to earth arising in healthy phases in the case of phase-to-earth faults. These overvoltage can be either at power frequency (with, in this case, maximum value equal to system voltage) or at higher frequency, such as in neutral insulated networks in the case of earthing arcs.

At arc extinguishing, corresponding to current zero crossing, the neutral-to-earth voltage reaches its maximum value, equal to $1.5E_{\max}$ (where $E_{\max} = \sqrt{2}E$ is the phase voltage crest value).

The electric charge on the conductor when the arc is extinguishing can be expressed as

$$Q = \sum_{i=a,b,c} Q_i = \sum_{i=a,b,c} C_i U_{it} = C(0 + 1.5E_{\max} + 1.5E_{\max}) = 3CE_{\max} \tag{8.5}$$

At arc extinguishing, the grounding connection is eliminated, so the electric charge cannot discharge to ground restoring the null value it has in the absence of the fault. Therefore the system will assume unidirectional voltage to earth equal to the maximum phase voltage value.

Assuming that the three capacities to earth are equal, the residual charge will be divided into three equal parts. From (8.5)

$$Q_a = Q_b = Q_c = \frac{1}{3}Q = CE_{\max} \tag{8.6}$$

This residual charge will discharge toward earth through connections coming from line transversal conductances and conductances due to TVs with a terminal connected to earth. This is a typical case of capacitance discharge through a resistance or resistance and inductance. During the attempt to eliminate arcing, a new striking of the arc is possible due to the ionization of the pre-existing arc. This phenomenon is called an **interrupted arc**: that is, arcs which are eliminated at each current zero crossing and strike again immediately.

Abrupt variations of stored charge on conductors are the cause of overvoltages higher than $2.8E_{max}$. These earth-interrupted arcs generally start from atmospherically originated overvoltages whose values are bigger than the voltage discharge of the line insulators, or from temporary accidental contacts between a conductor and ground through foreign bodies (branches, birds, etc.) or from faulty insulators.

Protection from these overvoltages can be effectively obtained through neutral earthing, either direct or through an impedance.

Overvoltages usually due to a fault, in lines with neutral connected to earth, can reach values of approximately 1.8–1.9 pu.

8.3.1.1.1 Overvoltage Due to Faults Between MV and Earth

Depending on the configuration of the earthing system of MV and LV networks, MV fault current flows into one or more earth electrodes and generates a.c. overvoltages in LV systems by earth coupling.

The main parameters influencing the value and duration of overvoltages are:

- The configuration of earth electrodes of both MV and LV networks:
  - one, two or three distinct earth electrodes;
  - common earthing electrodes or separated;
  - earthing electrodes for MV and LV networks;
  - values and number of earth electrodes of LV distribution system.
- The type of earthing system of the MV network:
  - isolated;
  - resonant earthed;
  - earthed through an impedance;
  - solidly earthed.
- The method used to clear the MV fault:
  - long time for isolated resonant-earthed types;
  - short times (< 5 s) for resonant-earthed or impedance-earthed types;
  - shorter time for solidly earthed types.

Temporary overvoltages can occur in different places and in different ways:

- In MV/LV substations, overvoltage stresses the insulation of LV equipment between live parts and accessible conductive parts if there is no common MV/LV earthing.
- In LV electrical installations, overvoltage stresses the insulation of LV equipment, between live parts and accessible conductive parts if the neutral is not connected to the local earthing electrode.

8.3.1.1.2 Overvoltage due to LV Insulation Faults to Earth

A temporary overvoltage occurs after a transient situation during such a fault:

- In TN systems earth faults can have results comparable to those occurring in circuits between phase and neutral. The return path to the neutral of the transformer consists of a cross-section comparable to that of the phase conductors;
- In TT systems, earth-fault currents flow through the protected-earth (PE) conductors and two earth electrodes. These earth electrodes are separated or anyway not intentionally connected so the fault current remains relatively low. The fault is cleared generally by the residual current circuit-breakers. The corresponding overvoltage is usually lower than $\sqrt{3}U_o$.
- In IT systems, the earth-fault current is very low in the case of a first fault, the capacitive leakage current of insulated conductors of installation and of filters in electrical equipment. Therefore the first phase-to-earth fault will not make protective devices trip but will cause transients and establish an overvoltage condition approximately equal to the phase supply voltage.

8.3.1.2 Fault Between MV and LV Networks

This fault can originate, for example, from an HV conductor falling on an LV network, or from an insulation fault in a transformer etc. It is always an extremely dangerous fault because it very often has major consequences: destruction of LV plan, fires, etc.

A first protective measure which can be adopted is neutral earthing on the LV network. Contacts between the two networks cause the operation of instantaneous protection in the HV network.

In this case, the connection of a neutral conductor to earth is subjected to an HV single-phase current I_g, and, if R_e is the earth resistance, the neutral conductor will have a voltage to earth equal to $R_t I_g$.

This voltage can reach high values, but for a short time, corresponding to the protection trip time. Protection against this fault, if the LV neutral conductor is insulated from earth, can be obtained through dischargers.

8.3.1.3 Ferroresonance

Ferroresonance can be considered as a special kind of resonance occurring in circuits containing a capacitance and an iron-core inductance.

The ferroresonance phenomenon is different from resonance in linear system elements, where the resonance condition leads to high sinusoidal voltages and currents at the resonant frequency. Ferroresonance can generate high voltages and currents too, but the resulting waveforms are usually irregular and with chaotic shape.

The concept of ferroresonance can be explained by considering a simple series *LCR* circuit as shown in Figure 8.1. If the resistance *R* is not considered, the current flowing in the circuit can be expressed as

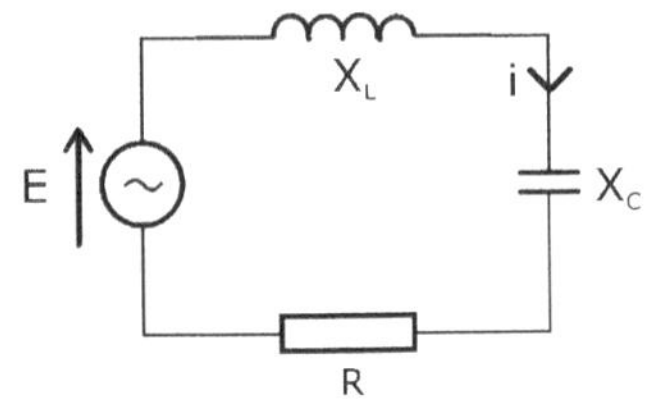

Figure 8.1 Series *LCR* circuit

$$I = \frac{E}{j(X_L - |X_C|)} \tag{8.7}$$

where E is the voltage, X_L the reactance of L and X_C the reactance of C.

When $X_L = |X_C|$, a series-resonant circuit is formed, and the consequence is an infinitely large current.

Another way to get a solution to the series *LCR* circuit is to write two equations defining the voltage variable V:

$$V = jX_L I \tag{8.8}$$

$$V = E + j|X_C|I \tag{8.9}$$

Figure 8.2 shows the graphical solution of this two-equation system for two different reactances, X_L and X_L', which represents the case of series resonance. At resonance frequency, in fact, the two lines will intersect at infinitely large voltage and therefore current since the X_C line is parallel to the X_L' line.

If it is considered that the inductive element has a non-linear reactance characteristic like that found in transformer magnetizing reactance, the graphical solution of the equations will be as shown in Figure 8.3.

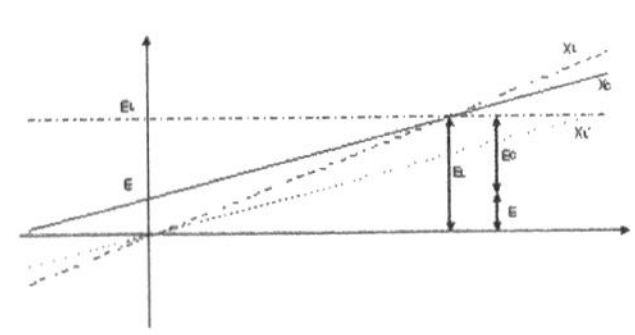

Figure 8.2 Graphical solution of the circuit

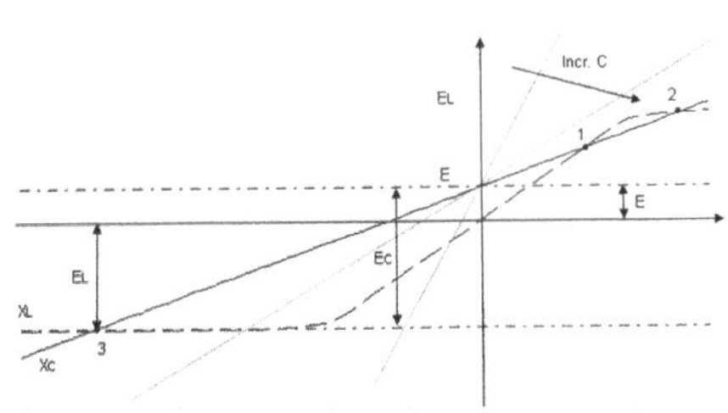

Figure 8.3 Graphical solution of the linear circuit

This figure is useful in helping to understand ferroresonance phenomena. There may be many intersections between the capacitive reactance line and the inductive reactance curve:

- Point 2 is an unstable solution, and this operating point gives rise to some of the chaotic behavior of ferroresonance.
- Point 1 is stable and will exist in the steady state.
- Point 3 is stable and will exist in the steady state too; it leads to high voltages and high currents.

When C increases and becomes very large, the X_C line will intersect the X_L line only at points 1 and 3. One operating state is of low voltage and lagging current (1), and the other is of high voltage and leading current (3). Depending on the applied voltage, operating points during ferroresonance can oscillate between intersection points 1 and 3. Often, the resistance in the circuit prevents operation at point 3 and no high voltages will occur.

Ferroresonance phenomena therefore can usually arise in the case of unloaded transformers isolated on underground cables within a certain range of lengths, while the capacitance of overhead distribution lines often is not sufficient to generate appropriate conditions for ferroresonance.

The minimum length of cable which can cause ferroresonance depends on the system voltage level: the capacitance of cables is approximately the same for all distribution voltage levels, varying from 40 to 100 nF per 300 m, depending on conductor size. However, the magnetizing reactance of a 35 kV distribution transformer is several times higher (then with a steeper curve) than a comparably sized 15 kV transformer, hence damaging ferroresonance has been more common at the higher voltages. For delta-connected transformers, ferroresonance can occur in less than 30 m of cable; many utilities thus avoid this connection on cable-fed transformers.

The most common actions which can lead to ferroresonance problems are:

- Manual switching of unloaded, cable-fed, three-phase transformers where only one phase is closed: ferroresonance may be noted when the first phase is closed upon energization or before the last phase is opened on de-energization.
- Manual switching of unloaded, cable-fed, three-phase transformers where one of the phases is open: in this situations ferroresonance may happen again during energization or de-energization.
- One or two riser-pole fuses may blow leaving a transformer with one or two phases open (single-phase reclosers may also cause this condition): many modern loads are equipped with controls to transfer the load to backup systems when this problem conditions.

Of course all the mentioned events do not necessarily lead to a noticeable ferroresonance phenomenon; it is frequently the case that underground cable systems operate for a long time without any ferroresonance issues. There are in fact some particular system conditions that help increase the possibility of ferroresonance:

- Switching of lightly loaded and unloaded transformers.
- Low-loss transformers.
- Ungrounded transformer primary connections.

- Very lengthy underground cable circuits.
- Higher distribution voltage levels, in particular 25 and 35 kV systems.
- Cable damage and manual switching during construction of underground cable systems.
- Weak systems, e.g. low short-circuit currents.
- Three-phase systems with single-phase switching devices.

It is important to remember that the occurrence of ferroresonance is possible at all distribution voltage levels, not just at the higher level, even if the proportion of losses, magnetizing reactance and capacitance at lower levels may limit the effects of ferroresonance. It can occur anyway, with several modes of physical and electrical manifestations as high voltages and currents, or simply voltages close to normal – there may or may not be failures or other evidence of the problem in the electrical components. For this reason, in some cases it may be difficult to notice if ferroresonance has occurred, unless power quality measurement instruments have recorded it or some staff have witnessed some of its effects.

Common indicators of ferroresonance are generally:

- **High overvoltages and surge arrester failure.** When ferroresonance leads to overvoltages, this could result in electrical damage to both the primary and secondary circuits, with surge arresters being common casualties. They are designed to intercept brief overvoltages and clamp them to an acceptable level. While they may be able to withstand several overvoltage events, there is a definite limit to their energy absorption capabilities. Failure of LV arresters in end-user facilities is sometimes the only indication that ferroresonance has occurred.
- **Overheating.** Transformer overheating can be a symptom of some ferroresonance problems, especially when the iron core is driven deep into saturation. When this situation occurs the magnetic flux find its way into parts of the transformer where the flux is not expected, such as the tank wall and other metallic parts. One of the indicators of this overheating is the presence of charring or bubbling of the paint on the top of the tank. This does not necessarily indicate that the unit is damaged, but damage can occur in this situation if ferroresonance has persisted sufficiently long to cause overheating, or in some of the larger internal connections, which may in turn damage solid insulation structures beyond repair. Transformers exhibiting signs of ferroresonance such as loud, chaotic noises do not show signs of appreciable heating.
- **Audible noise.** This noise is caused by the magnetostriction of the steel core of transformers being driven into saturation. The noise is distinctively different and louder than the normal noise generated by a transformer, and most electric system operating personnel are able to recognize it immediately.
- **Flicker.** During ferroresonance the voltage magnitude may fluctuate so end users at the secondary circuit may actually see its effect as flicker. Some electronic appliances may be very susceptible to such voltage excursions, and the main issue is related to the fact that prolonged exposure can shorten their expected life if not immediate failure. In facilities that transfer over to the UPS system in the event of utility-side disturbances, repeated and persistent sounding of alarms on the UPS may occur as indicator of this problem.

8.3.1.4 Switching Overvoltages

Sudden changes in electrical network structure caused by switching operations give rise to transient phenomena which usually result in the creation of overvoltages or high-frequency wave trains of a periodic or oscillating type.

8.3.1.4.1 Overvoltage on a Long Off-Load Line (Ferranti Effect)

When a long line is energized at one of its ends and not connected at the other, an overvoltage may occur, generated by a resonance resulting in a voltage wave increasing in linear fashion along the line.

Considering L and C, the line inductance and capacity respectively, and U_s and U_e voltages at the open end and at line entrance, the overvoltage factor can be expressed as

$$\frac{U_s}{U_e} = \frac{1}{1 - \dfrac{LC\omega^2}{2}} \tag{8.10}$$

This overvoltage factor is approximately 1.05 for a 300 km line and 1.16 for a 500 km line. This phenomenon is particularly frequent when a long line is suddenly discharged.

8.3.1.5 Overvoltages Due to Sudden Load Changes

Overvoltage due to sudden changes of load may start with a high switching surge followed by a temporary overvoltage. An abrupt power reduction can make overvoltage cause an internal voltage drop in system elements (alternators, transformers, lines). The worst case occurs for unexpected and complete load sheddings. In this case, supposing that speed does not increases, generator voltage could rise to the maximum value it could have in correspondence to the excitation current needed in case of full load. This kind of overvoltage can be easily mitigated with voltage regulators: experience shows that overvoltage amplitude can be limited to 15–20 % of their value.

In HV networks, in this situation, voltage drop is replaced by negative voltage drop due to the Ferranti effect.

8.3.1.5.1 Overvoltages Caused by Making and Breaking of Small Inductive Currents

This type of overvoltage can be a consequence of three events:

- current pinch-off;
- rearcing;
- prearcing.

A network supplying a load through a circuit-breaker is shown in Figure 8.4. The network contains:

- a sinusoidal voltage source u with an inductance L_A and a capacitance C_A;
- a circuit-breaker, S, with its stray elements L_{cb} and C_{cb};
- an inductive load, L_B, with distributed capacitance C_B;
- a line inductance, L_0, generally negligible.

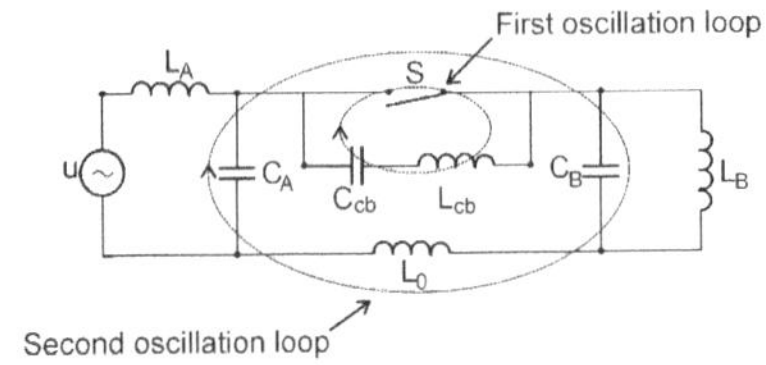

Figure 8.4 Circuit for the study of overvoltages caused by inductive current breaking

8.3.1.5.1.1 Current pinch-off

When breaking low currents, in particular those less than the circuit-breaker's rated current, the arc occurring takes up little space because of the cooling due to the circuit-breaker's capacity to break higher currents. The arc then becomes unstable and its voltage may present high relative variations even if its absolute value remains below the network voltage. These e.m.f. variations may generate oscillating currents of high frequency in the adjacent capacitances. The amplitude of these currents can become non-negligible when compared to a 50 Hz current reaching 10 % of its value. Through a superimposition of the 50 Hz current with this high-frequency current in the circuit-breaker the current will as a result tend to zero several times around the zero of the fundamental wave, as shown in Figure 8.5.

The circuit-breaker, only little affected by these low currents, is usually capable of breaking at the first current zero occurring, when the currents in the generator and load circuits are not zero. The energy stored in the circuit (Figure 8.4) depends on the impedances involved, mainly resistive and inductive ones. Small inductive currents present a load with a high inductance which, when the arc is extinguished, will have an energy given by

$$\frac{1}{2}L_2I^2 \tag{8.11}$$

The L_2C_2 circuit is now in the slightly damped, free oscillation state, and according to the energy conservation hypothesis the peak value V of the voltage occurring at the terminals of $C_2$is given by

$$\frac{1}{2}L_2I^2 = \frac{1}{2}C_2V^2 \tag{8.12}$$

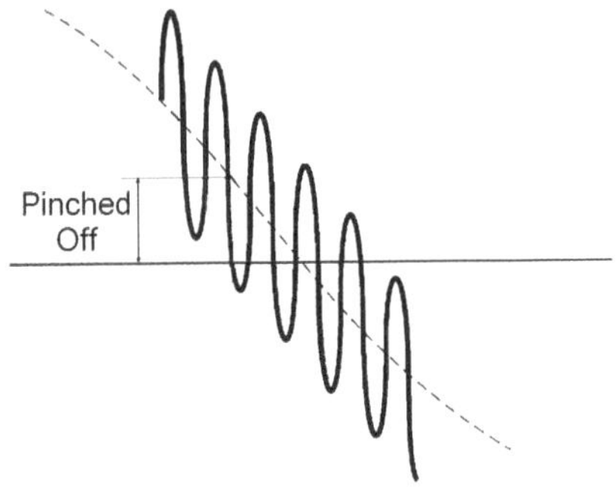

Figure 8.5 Superimposition of high-frequency oscillating current on fundamental (50 Hz). The value of the 50 Hz wave at the instant of arc extinguishing is called pinched-off current

The value of V could be dangerous for equipment insulation if C_2 is only made up of stray capacitances with respect to frames. The generator circuit has an equivalent behavior but its inductance is normally much smaller and then the voltages occurring at the terminals of C_1 are lower.

8.3.1.5.1.2 Prearcing

When a device such as a switch, contactor or circuit-breaker closes, there is a moment when the dielectric withstand between contacts is less than applied voltage. In the case of rapidly closing devices, the behavior depends on the phase angle during operation. An arc is then created between the contacts, and the circuit witnesses a voltage pulse due to the sudden cancellation of voltage at the device terminals, which may result in oscillation of existing parallel circuits (surging discharge of stray capacitances), reflections on impedance failures, and then in the appearance of high-frequency currents through the arc. If device operation is slow compared to this phenomenon, the arcing current may be made to move through zero by superimposition of the high-frequency current and the fundamental current. Extinguishing of the arc will then result in a behavior similar to that described for the phenomena above. However, since the dielectric withstand between contacts decreases with closing, the successive overvoltages decrease right up to complete closing.[3]

8.3.1.5.1.3 Rearcing

Rearcing occurs when the pinching-off phenomenon described above causes an input/output overvoltage at the terminals of the circuit-breaker unable to be withstood (Figure 8.4). After breaking and rearcing, three oscillating phenomena occur simultaneously at the respective frequencies Fp_1, Fp_2 and Fm:

1. Loop $D - Lp_1 - Cp_1$:

$$Fp_1 = \frac{1}{2\pi\sqrt{Lp_1 Cp_1}} \tag{8.13}$$

   (a few megahertz).
2. Loop $D - C_1 - L_0 - C_2$:

$$Fp_2 = \frac{1}{2\pi}\sqrt{\frac{C_1 + C_2}{L_0 C_1 C_2}} \tag{8.14}$$

   (from 100 to 500 kHz).
3. Throughout the circuit:

$$Fm = \frac{1}{2\pi}\sqrt{\frac{L_1 + L_2}{L_1 L_2 5(C_1 + C_2)}} \tag{8.15}$$

   (from 5 to 20 kHz).

[3] This type of overvoltage usually affects HV and MV off-load transformers during energization and motor starting.

Multiple rearcing then occurs (chopping) until it is stopped by increasing contact clearance. This rearcing is characterized by high-frequency wave trains of increasing amplitude which can present a considerable risk for equipment containing windings.

8.3.1.5.2 Overvoltage Caused by Switching on Capacitive Circuits

Breaking of capacitive circuits normally presents few difficulties because, as capacitances remain charged at the 50 Hz wave peak value after the arc is extinguished at current zero, voltage is resumed at the terminals without transients. However, at one alternation after breaking, the device is subjected to an input/output voltage twice the peak voltage. If it is unable to withstand this stress, reignition may occur, and this is followed by a voltage inversion at the capacitor terminals, raising them to a maximum load of three times peak voltage as shown in Figure 8.6. The current breaks yet again and a new reignition may take place with a value five times peak voltage at the next alternation. Such behavior may lead to a dangerous escalation and must be avoided through a correct choice of equipment to prevent reignition phenomena.

8.3.1.5.2.1 Characteristics of energizing an isolated capacitor bank
When a capacitor bank is energized, and this normally happens without initial load, in the case of slow operating devices arcing occurs between the contacts around the 50 Hz wave peak.

Transients have characteristics depending on a combination of the initiating mechanism and electric circuit characteristics. Circuit inductances and capacitances are responsible for the oscillatory nature of transients: they can be discrete components, such as shunt capacitance of power factor capacitor banks or inductance in transformer windings, but can also come from stray inductances or capacitances because of their proximity to other current-carrying conductors or voltages. Natural frequencies within the power system depend on the system voltage level, line lengths, cable lengths, system short-circuit capacity and the application of shunt capacitors.

Energizing a shunt capacitor bank from a predominantly inductive source will result in an oscillatory transient that can reach twice the normal system peak voltage (V_p).

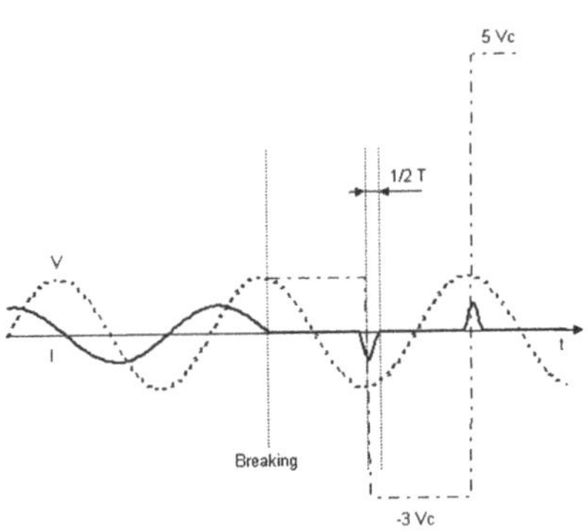

Figure 8.6 Voltage escalation on separation of a capacitor bank with a slow operating device

The characteristic frequency (f_s) of this transient is given by

$$f_s = \frac{1}{2\pi\sqrt{L_s C}} \approx f_{system}\sqrt{\frac{X_C}{X_S}} \approx f_{system}\sqrt{\frac{S_{SC}}{Q_r}} \approx f_{system}\sqrt{\frac{1}{\Delta V}} \tag{8.16}$$

$$I_p = \frac{V_p}{Z_S} \qquad Z_S = \sqrt{\left(\frac{L_S}{C}\right)} \tag{8.17}$$

where:

- f_s is the characteristic frequency (Hz);
- L_s is the positive sequence source inductance (H);
- C is the capacitance of the bank (F);
- f_{system} is the system frequency (50 or 60 Hz);
- X_s is the positive-sequence source impedance (Ω);
- X_c is the capacitive reactance of the bank (Ω);
- S_{sc} is the three-phase short-circuit capacity (MVA);
- Q_r is the three-phase capacitor bank rating (Mvar);
- ΔV is the steady-state voltage rise (pu);
- V_p is the peak line-to-ground bus voltage (V);
- Z_s is the surge impedance (Ω).

A simple and quick way to evaluate the expected frequency range for utility capacitor switching is based on the approach of relating the characteristic frequency of the capacitor energizing transient (f_s) to a steady-state voltage rise (ΔV) design range. For example, in a 60 Hz system with a design range of 1.0 to 2.5 % this corresponds, through this approach, to a characteristic frequency range of 380 to 600 Hz. When a shunt capacitor bank is installed on an HV bus, the transmission line capacitance and other nearby capacitor banks lead to an energizing transient with more than one natural frequency. In any case, Equation (8.16) can still be used to determine the dominant frequency in a reliable way.

When switching, capacitor voltage cannot of course change instantaneously, therefore the energizing process will result in an immediate drop in system voltage toward zero followed by an oscillating transient voltage superimposed on the fundamental power frequency waveform. The peak voltage value depends on the system voltage at the instant of energization, and can reach 2.0 times the normal system voltage (V_p in pu) in the worst-case conditions. The voltage surge has the same frequency as the inrush current (I_p) and will rapidly decay to the system voltage.

For safety reasons, capacitor banks are always featured with discharging resistors aiming to eliminate residual voltages with time constants of approximately one minute.

8.3.1.6 Overvoltages Due To Alternator Overspeed

When the prime mover of a generator, for any reason, gets into an overspeed condition, the voltage generated increases beyond the rated value. For example, for a Pelton turbine, whose overspeed is equal to $1.8v_r$ (rated speed), the voltage could theoretically increase to approximately $5.8V_r$ (rated voltage) even if saturation condition limits the voltage value to $2.5–3V_r$.

If, after the increase in prime-mover speed, a sudden change of load occurs, then the above-mentioned overvoltage must include the voltage variation of the now unloaded alternator; alternator voltages can reach values up to $3–3.5V_r$.

To limit the speed increase within tolerable limits, prime movers are manufactured with centrifugal relays in order to stop motor fluid (water, steam, etc.) when its speed exceeds 20–25 % of the rated value. These devices, if installed with a good-quality voltage regulator, allow the voltage increase to be limited to 30–40 % of rated voltage. To add further protection against this kind of overvoltage, it is possible to install, mainly to protect the alternator, a maximum voltage relay: when the voltage exceeds, for example, 150 % of the rated value, this device allows the alternator to be de-energized and separated from the network.

8.3.1.7 Overvoltages Due to Generator Self-Excitation

A capacitive load connected to the alternator absorbs a leading current which, since it circulates in the stator, is equivalent to an excitation current. If this leading current is higher than some specific value, the alternator can be self-excited and could generate on its terminals a voltage higher than its rated value, despite the exciting current also being null.

Self-excitation phenomena can become dangerous, especially when a very long HV network is fed, or when there is load shedding at the end of a long line, with the alternator suddenly connected with full excitation.

This kind of overvoltage depends on the characteristics of the alimentation machine, other than line capacitance. This overvoltage cannot reach large values due to the iron saturation phenomenon, but they are not transient and can produce serious disturbances mainly on system stability.

Measures to limit this inconvenience are the same as those used to improve transmission system stability. The best approach is to choose generators with a high short-circuit ratio, even if this increases machine cost and short-circuit current value.

8.3.1.8 Overvoltage Due to MV and LV Faults

These types of overvoltages usually originate from insulation faults or the loss of a supply conductor in MV or LV electrical installations. Another situation which could result in this kind of overvoltage could be the re-energizing after a power interruption of the system in a phase-by-phase sequence.

Other examples could be for an HV fall on an LV conductor, or as a consequence of the breakdown of the insulation of a transformer between the HV and LV windings. The latter has generally the most serious consequences for LV system destruction, with all its related effects. Product standards take these phenomena into account by prescribing appropriate insulation requirements and tests.

For example, in systems with MV and LV lines mounted on the same poles, or in systems with two different MV levels also mounted on the same poles, accidental commingling of the systems can occur, causing overvoltages on the LV system. If such exceptional events are to be considered, special surge protective devices (SPDs) need to be applied to deal with the event.

The most effective protection is LV direct neutral earthing. Contact between two networks determines when instantaneous HV network protection operates.

8.3.2 External Overvoltages

8.3.2.1 Lightning Overvoltages

Statistical data shows that an average of 1000 storms break out each day throughout the world. Of all the electric lines, overhead networks are the most affected by lightning overvoltages and overcurrents.

Lightning strikes are characterized by their polarization: they are generally negative even if approximately 10 % of them have reversed polarity, but these are the most violent. The rising front of lightning strikes defined by standards is 1.2 μs for voltage and 8 μs for current.

A distinction is often made between direct lightning strikes on a line and indirect lightning strikes, falling next to a line, or on the earth cable.

8.3.2.1.1 Direct Lightning Strikes

Direct lightning strikes result in the injection of a current wave of several dozens of kiloamperes in the line. This current wave, which may cause conductors to melt by propagating on either side of the point of impact, results in an increase of voltage U given by

$$U = Z_c \frac{i}{2} \tag{8.18}$$

where i is the injected current and Z_c the characteristic line zero-sequence impedance (usually from 300 to 1000 Ω).

U reaches values of several million volts, which no line can withstand, then at the first pylon which the wave meets, the voltage increases until clearance breakdown occurs (insulator string). According to whether or not arcing has occurred (depending on the value of the current injected into the line), the wave which continues to propagate beyond the pylon is said to be broken or full.

For various network voltages, arcing does not occur below the critical current indicated by the straight line in Figure 8.7. For networks with a voltage less than 400 kV, all direct lightning strikes result in arcing and an earth fault.

It is estimated that only 3 % of overvoltages exceed 70 kV and have an origin in direct lightning strikes. Moreover, as a result of attenuation of the voltage wave throughout its propagation along the line, maximum overvoltages at the entrance of a substation or building are estimated at 150 kV in MV systems.

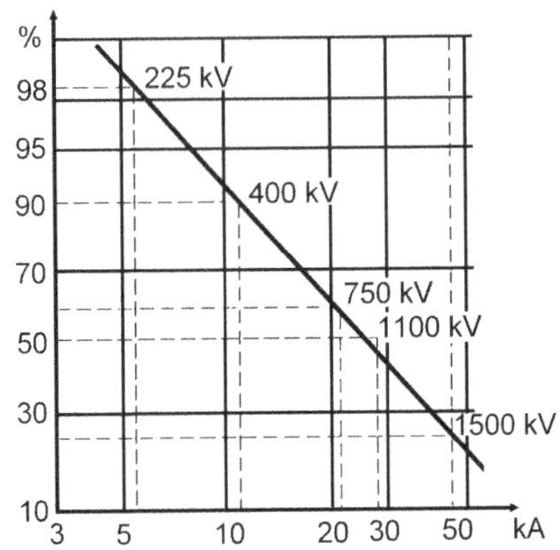

Figure 8.7 Strength of direct lightning strikes and minimum arcing strengths as a function of network voltage level

8.3.2.1.2 Indirect Lightning Strikes

When indirect strokes fall on a support or just next to a line, high overvoltages are then generated in the network. Indirect strokes are more frequent than direct ones and are proven to be also almost as dangerous.

If lightning hits the pylon or the earth cable, the current flowing off causes an increase in metal frame potential with respect to earth. The corresponding overvoltage U is

$$U = R\frac{i}{2} + \frac{L}{2}\frac{di}{dt} \tag{8.19}$$

where R is the earth connection steep wave resistance and L is the inductance of the pylon and/or the earthing conductor.

When this voltage reaches the value of arcing voltage of an insulator, an arcing return occurs between the metal structure and one or more of the conductors. When the voltage is greater than 150 kV, this arcing return is unlikely to happen. The quality of pylon earth connections plays an important role: from 750 kV upwards, there is virtually no risk of arcing return, thus justifying the installation of earth cables on EHV lines. In networks below 90 kV, these cables provide efficient protection if the pylon earth connection is excellent.

If lightning hits just near the line, the energy flowing off to the ground causes a very rapid variation in the electromagnetic field which induces waves on the line that are similar in shape and amplitude to those generated by a direct stroke. They are mainly characterized by their very steep front[4] and their very fast damping.

When the voltage wave resulting from a lightning stroke passes through an MV/LV transformer, transmission mainly occurs by capacitive coupling.

[4] Approximately 1 microsecond.

The amplitude of the overvoltage thus transmitted, observed on the secondary winding on the LV side, is less than 10 % of its value on the MV side (generally less than 70 kV). Therefore, on LV lines, induced overvoltages are generally less than 7 kV.

8.3.2.2 Electrostatic Overvoltages

Although a major part of induced overvoltages is of electromagnetic origin, some parts are electrostatic and occur in particular unearthed networks.

For example, in the minutes before a lightning stroke, when a cloud charged at a certain potential is fluctuating above a line, this line takes on a charge of opposite polarity.

Before the lightning strikes, thus discharging the cloud, an electric field exists between the line and the ground (it can reach 30 kV/m). The line–earth capacitor is then charged by this field to a potential up to 500 kV.

Unenergetic breakdown may then occur in the least well-insulated components of the network.

8.4 EFFECTS

8.4.1 Breakdown Consequences

A dielectric failure, breakdown or arcing, can lead to:

- tripping of protective devices;
- destruction of equipment;
- interruption of operation each time a failure occurs.

As a result of these faults, e.g. in an HV network, a power supply shortage can affect an entire town, a whole region or just an iron and steel plant, with potentially disastrous consequences, mainly as a:

- risk for people (e.g. in hospitals) and for computer data;
- risk of network destabilization;
- production loss for industrial consumers;
- loss of energy billed by the energy distributor.

For each new installation a detailed analysis must be performed in order to provide consistent and optimized risk protection to try to avoid such kinds of incidents.

A first approach can be to increase the installation insulation level by increasing clearances, even if this solution will result in considerable increases in cost because, by doubling clearances, volumes and costs can be approximately multiplied eight times.

For this reason, oversizing is therefore unacceptable in HV networks, while the only applicable solution is optimizing equipment.

In MV networks, the consequences of insulation faults are nearly the same, even if on a lesser scale, and therefore the consequences of the resulting electricity failures can also be serious for energy distributors (invoice losses), industrial consumers (production losses) and people (safety).

In LV systems the consequences of breakdown due to the low operating voltage are limited if considered in terms of power distribution. A non-negligible risk is due to the development of electronic equipment and systems, responsible for a large number of incidents coming from overvoltages. The reason is that the disturbance withstand level is not always specified or is not coordinated with the level corresponding to installation, even if these systems play an increasingly large role in the integrity of installations, production and management, and the economic consequences for the companies can be very serious.

Coordination of withstands is then fundamental, even in LV networks, and the use of arresters should be generalized. In fact, nowadays for LV consumers supplied by overhead lines, they are highly recommended.

8.4.2 Reduction of Overvoltage Risk and Level

Simple solutions to the various overvoltages can be foreseen at the installation design stage.

8.4.2.1 Overvoltage Due to Ferromagnetic Resonance

The only measure which can be adopted to completely avoid this issue is to make $1/(\omega C)$ greater than the slope at the origin of ωLi.

Other solutions can be considered too, in particular in MV networks, where an unbalance between the three phases can occur in the case of protection through a phase-by-phase controlled switch. The greatest simultaneity possible occurs on closing the three network phases (omnipole equipment); closing an off-load transformer may be the transient phenomenon causing ferromagnetic resonance. Capacitances must be reduced to prevent this situation by considering, for example, the transformer energizing equipment.

Connection of a load prior to energizing is useful since this load acts as a reducing resistance which can prevent resonance.

Neutral earthing is also a solution for phase/earth resonances.

8.4.2.2 Overvoltage Caused by Closing Off-load Lines

This is prevented in transmission networks by progressive energizing, obtained by adding insertion resistances to the circuit-breaker.

8.4.2.3 Overvoltage Caused by Capacitive Current Breaking

The solution here is to prevent successive reignitions by increasing the contact separation speed and using a good dielectric (vacuum or SF_6).

8.4.2.4 Overvoltage Caused by Lightning Stroke

There are three possibilities:

- good-quality earth connections;
- installation at vulnerable points (dischargers or arresters) of protective devices;
- installation of earth cables to prevent direct impulses.

8.5 MITIGATION

8.5.1 Principles of Protection

Overvoltage protection consists of limiting the entity in order to avoid damage to the system's electrical components. Protection against overvoltages can be divided into:

- **preventive** protection – aiming to limit internal or lightning impulse overvoltages (overhead protection cable, neutral earthing, regulators, protection relays, switching-impulse-limiting circuit-breaker);
- **repressive** protection – special equipment installed for draining the overvoltage to earth (dischargers, surge arresters).

The fundamental principles of overvoltage protection of electrical equipment are:

1. To limit the voltage for sensitive insulation.
2. To reduce, or prevent, surge current from flowing between grounds.
3. To drain the surge current away from the load.
4. To bond the ground and equipment.
5. To create a low-pass filter using limiting and blocking principles.

8.5.2 Insulation Coordination

Insulation coordination is a process in which the appropriate insulation levels for equipment and the corresponding overvoltage protection system (e.g. surge arresters) are coordinated with the expected overvoltages that can occur in a power system.

There are three basic elements to insulation coordination, which are:

- Determining the overvoltage stresses from the system.
- Knowing the strength of the insulation of specific equipment in the substation.
- Selecting surge arrester ratings and locations, or other mitigation equipment or operating restrictions, to ensure that the system-imposed overvoltages do not exceed the insulation strength of the equipment, including an appropriate protective margin.

Electrical equipment and protective devices are thus chosen according to the protection levels defined. Protection level is determined by three factors:

- installation;
- environment;
- equipment used.

Through a detailed analysis of these conditions it is possible to determine the overvoltage level to which the equipment could be exposed during use. This approach will ensure that this level will never be exceeded.

The expected stresses from the system can be determined by simulation; the insulation strength is an inherent characteristic of the equipment, leaving the output of an insulation coordination simulation to deal with the selection and placement (or verification) of properly sized surge arrester(s).

In power systems, there are four general types of overvoltage stresses that can be experienced by equipment. These are:

- maximum continuous operating overvoltage;
- temporary overvoltages (at or near the fundamental frequency);
- switching overvoltages;
- lightning overvoltages;

To face lightning overvoltages, it is necessary to adopt a compromise between insulation level, the protection level of arresters and acceptable risk of failure.

The final objective of insulation coordination is to ensure safe and optimized distribution of electric power, which means the best possible economic balance between the various costs, namely:

- insulation;
- protective devices;
- failures (operating loss and repairs) weighted with their probabilities.

The first action in removing the effects of overvoltages is to classify the phenomena generating them. Equipment switching overvoltages can be limited through several techniques, whereas it is impossible to have any effect on lightning: it is then necessary to locate the point of least dielectric withstand through which the current generated by the overvoltage will flow, to ensure that all equipment has a higher level of dielectric withstand capability.

8.5.2.1 Clearance and Voltage Withstand

8.5.2.1.1 Clearance

Clearance is defined for two cases, gas clearance (air, SF_6, etc.) and creepage distance of solid insulators (Figure 8.8), which do not have identical withstand:

- **gas clearance:** the shortest path between two conductive parts;
- **creepage distance:** the shortest path between two conductors following the outer surface of a solid insulator.

Voltage withstand is variable, and it varies in particular with the type of overvoltage applied (voltage level, rising front, frequency, time. etc.).

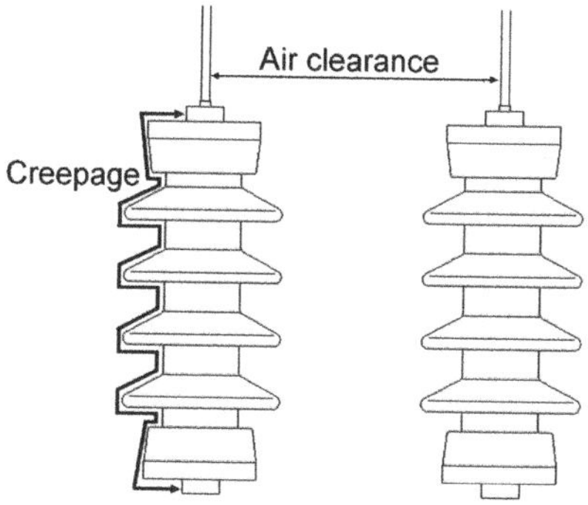

Figure 8.8 Air clearance and creepage distance

In addition, creepage distances may vary under particular conditions, specific to the insulating material, which cause deterioration of their characteristics:

- environmental conditions (humidity, pollution, UV radiation);
- age (deterioration of the material);
- permanent electrical stresses (local value of the electric field).

Creepage distances for busbar supports, transformer bushings and insulator strings are defined as distances to obtain a withstand similar to direct air clearance between two dry and clean end electrodes, even if rain and especially wet pollution considerably reduce their withstand voltage.

Gas clearance withstand depends on pressure:

- air pressure with altitude;
- device filling pressure.

Insulation withstand voltage in gases varies according to a highly non-linear function of clearance, but is practically unaffected by rain, a behavior arising from lack of uniformity of the electric field between electrodes of all shapes and not to intrinsic gas characteristics. For this reason it would not be observed in the case of flat electrodes of infinite size, corresponding to a uniform field. For example, in air, an r.m.s. voltage stress of 300 kV/m is acceptable under 1 m, but can be reduced to 200 kV/m between 1 and 4 m down to 150 kV/m between 4 and 8 m.

8.5.2.1.2 Power Frequency Withstand

Under normal operating conditions, the network voltage may present short-duration power frequency overvoltages with the duration depending on network protection and operating mode. Voltage withstand monitored through the standard one-minute dielectric tests is generally sufficient.

8.5.2.1.3 Switching Impulse Voltage Withstand

In the case of switching impulses, clearances can be characterized with the following properties:

- non-linear relation with voltage;
- unbalance, variation according to wave polarity;
- passage through a minimum curve value of the withstand voltage as a function of front time;
- dispersion withstand must be expressed in statistical terms.

When the gap between electrodes increases, this minimum value moves to increasingly higher front times. On average it is around 250 ms which accounts for the choice of standard test voltage rising front (standard tests of a wave of front time 250 ms and half-amplitude time 2500 ms).

8.5.2.1.4 Lightning Overvoltage Withstand

In the case of lightning, the withstand is characterized by far greater linearity than for the other stress types. Dispersion is present in this case also with a positive polarity withstand inferior to that of negative polarity.

Two formulas can be used to evaluate withstand to a 1.2 μs/50 μs positive-polarity impulse of an air gap for HV and MV networks:

$$V_{50} = \frac{d}{1.9} \tag{8.20}$$

where V_{50} is the voltage for which the breakdown probability is 50 %; and

$$V_0 = \frac{d}{2.1} \tag{8.21}$$

where V_0 is the withstand voltage and d is clearance in meters (V_{50} and V_0 are in MV).

A lot of experimental studies have provided tools to evaluate the relation between clearance and withstand voltage, taking into account a variety of factors such as front and tail times, environmental pollution and insulator type.

Figure 8.9 shows an example of the variations in voltage V_{50} as a function of clearance and tail time t for a positive peak-plane interval.

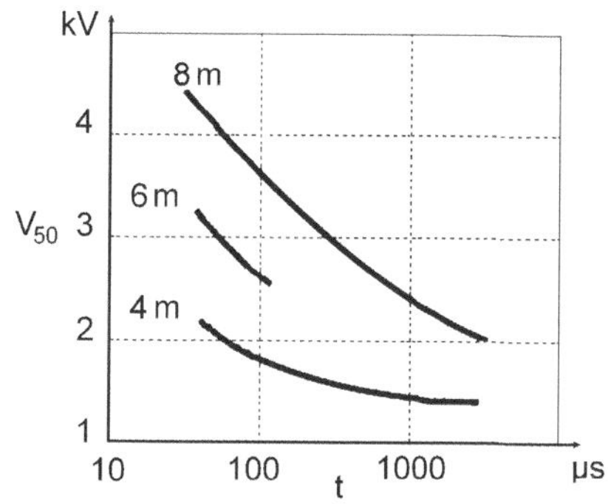

Figure 8.9 V_{50} to tail time relation at different clearances

8.5.3 Overvoltage Protective Devices

8.5.3.1 Preventive Protection

8.5.3.1.1 Neutral Earthing

Neutral earthing of an electric system has several advantages from the point of view of overvoltage protection:

- voltage to earth of all phases is stabilized and equal to phase voltage value;
- elimination of internal overvoltages due to interrupted earth arcs;
- static charges are continuously drained to earth so they cannot be accumulated on conductors.

The neutral earthing condition in a three-phase system is characterized by the earth-fault factor, the ratio between the biggest earth voltage r.m.s. value (at power frequency) of an operating phase during an earth fault and phase voltage r.m.s. value.

8.5.3.1.2 Shielding (LPS)

In overhead lines shielding is provided by overhead protection cables, which are steel armored conductors placed over line conductors with effective connection to earth through transmission towers. They essentially have the task to avoid line conductors being subjected to the electrostatic induction of charged clouds and attracting direct lightning strikes.

In order to reach an effective protective position, it is necessary that overhead protection cables are located in an elevated position with respect to line conductors. It is then necessary that the angle between the ideal vertical line passing through an overhead protection cable and the line linking it to the most external conductor does not exceed 30°.

Overhead protection cables can also mitigate overvoltages coming from direct lightning strikes on pylons, by introducing between the ground and pylon other parallel circuits that drain the discharge current to the pylon–earth resistance circuit and then reduce its phase-to-earth and phase-to-phase overvoltages.

In addition, a voltage impulse propagating along an overhead protection cable when subjected to lightning stroke induces an equal polarity impulse on conductors and then reduces the voltage difference on insulators.

A similar function is performed by transforming and switching substation shielding, which is also obtained through protection cables linked to the earth system.

This is an example of LPS (Lightning Protection Systems). In general they have three functions: intercepting lightning strokes, draining them to earth and dissipating them in the ground.

Interception devices such as lightning rods are available in different forms such as guard wires on HV overhead lines or Franklin antennas at the top of steeples. They are earthed in order to allow the lightning currents to flow to earth, by one or several conductors. The earth connection, which must be accurately realized, is usually formed by several, separately buried, copper conductors.

The choice of lightning rod is performed on the basis of the maximum acceptable lightning current for the installation and the area to be protected: having defined this maximum current (peak current of the first impulse), an electrogeometrical model is used to calculate the critical arcing distance. This distance is used as the radius of a fictitious sphere rolling along the ground and which fetches up against the buildings to be protected (Figure 8.10). Only the area under the sphere is protected against lightning currents greater than or equal to the reference value. All elements in contact with this sphere are exposed to direct lightning strikes.

8.5.3.2 Repressive Protection

Surge arresters are devices used to limit high-amplitude transient overvoltages. They are normally designed to be able to deal with lightning overvoltages.

Protective devices are installed in order to provide protection against the indirect effects of lightning and/or switching overvoltages and power frequency surges.

In fact these devices have two functions: to limit the impulse voltage (parallel protective devices) or to limit the power transmitted (serial protective devices).

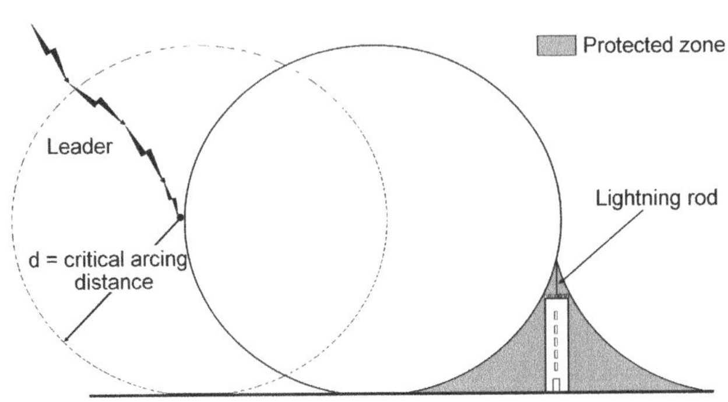

Figure 8.10 Principle of rolling sphere used to define the zone protected by lightning rod

8.5.3.2.1 Spark Gap Arresters

Spark gap arresters are used in MV and HV networks, usually placed in highly exposed network points and at the entrance to MV/LV substations. Their function is to be an intentional weak point in network insulation aiming to force any arcing to occur systematically there. In terms of age, the first protective device was the point discharger. It basically consisted of two points facing each other, the electrodes, connected respectively to the conductor to be protected and to earth.

The most common spark gap arresters used today are based on the same principle. In addition, they contain two horns to elongate the arc and then to simplify restoration of dielectric qualities by deionizing the arcing gap and, in certain cases, to extinguish it. Some models are also equipped with a rod placed between these two electrodes (Figure 8.11), designed to prevent unwanted short-circuiting by birds and their electrocution.

The gap between the two electrodes allows the modification of protection level. This device is very simple, efficient and economical, but has some issues:

- The arcing voltage is dispersed and depends on atmospheric conditions, which can lead to variations of more than 40 %.
- The arcing level depends on the amplitude of the overvoltage.
- The arcing delay increases as overvoltage decreases.

With these kinds of conditions an impulse may lead to arcing of a device with a withstand voltage greater than that of the discharger, just because this device has a smaller arcing delay (e.g. cables). In addition, after arcing, ionization between the electrodes maintains

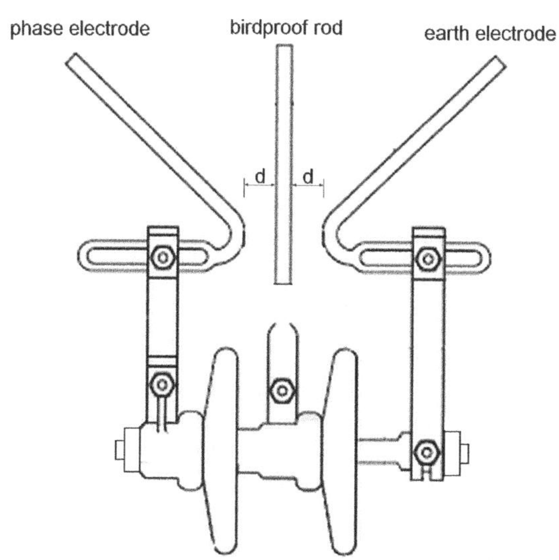

Figure 8.11 Example of spark gap arrester with birdproof rod[5]

[5] Distance *d* varies for different voltage levels.

the arc, which is then supplied by network voltage and may give rise (according to neutral earthing) to a power frequency retaining current. This current is a full earth-fault current and requires the intervention of the protective devices.

Arcing also produces a steep-front broken wave which could damage windings (transformers and motors) placed nearby. For all these reasons, dischargers today are increasingly replaced by surge arresters, even if they are still used in networks.

8.5.3.2.1.1 Non-linear resistance arresters and air gap arresters

These types of arresters connect in series air gap protectors and non-linear resistances (varistors) able to limit current in case of a surge. When the discharging current wave has flown off, the arrester is only subjected to the network voltage, which maintains an arc on the air gap protector, but with a corresponding current, the retaining current, which flows through the resistance which has a high value. Current is then sufficiently low not to damage the air gap protector and to be cleared when the current tends to zero for the first time (arc naturally extinguished). The non-linearity of resistances maintains a residual voltage which appears at the terminals of the device, close to the arcing level, since resistance decreases as current increases.

Various techniques have been used to produce varistor arresters and air gap protectors but the most common kind uses a silicon carbide (SiC) resistance.

Some devices are also equipped with arc blowing systems (magnets or coils for magnetic blowing) and voltage distribution systems (resistive or capacitive dividers). The main characteristics of these types of arresters are:

- impulse current evacuation capacity, or its energy dissipation capacity, generally given by withstand to rectangular current waves;
- extinction voltage or rated voltage, or the highest-power-frequency voltage under which the arrester can be spontaneously de-energized, which must be greater than the highest short-duration power frequency overvoltage which could occur on the network;
- arcing voltages, related to wave shape: power frequency, switching impulse, lightning impulse.

8.5.3.2.1.2 Zinc oxide (ZnO) arresters

These devices are made up only of varistors and are increasingly replacing air gap protectors and non-linear resistance arresters (Figure 8.12). Having no air gap, ZnO arresters are permanently conductive, but under protected network rated voltage, they have a very small earth leakage current (less than 10 mA).

The ZnO arrester operating principle is based on the highly non-linear characteristic of ZnO varistors, such that resistance decreases from $1.5\,\mathrm{M\Omega}$ to $15\,\Omega$ between the operating voltage and the voltage at rated discharging current. The advantage is in their increased limitation and reliability when compared to silicon carbide arresters (Figure 8.13).

Improvements have been recently added, in particular those aiming to add thermal and electrical stability on ageing.

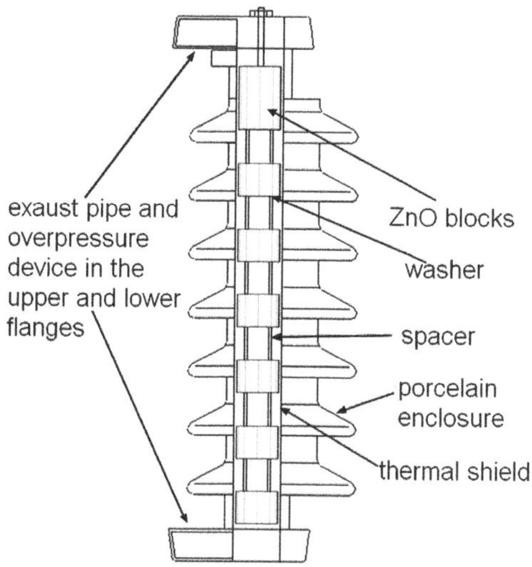

Figure 8.12 ZnO arrester in a porcelain enclosure

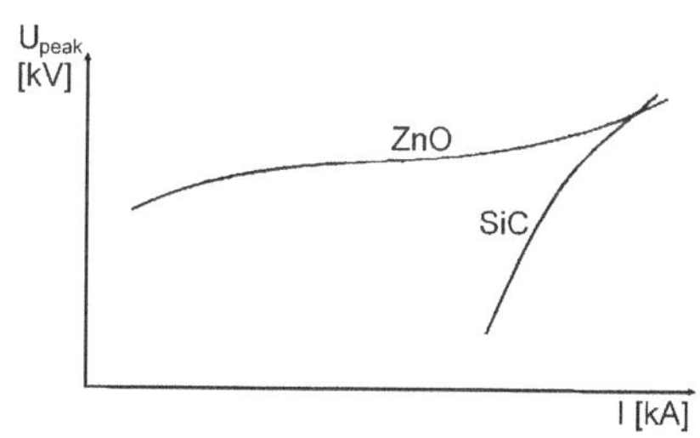

Figure 8.13 Typical characteristics of two arresters with the same level of protection

ZnO arresters (see Table 8.4) are available in porcelain enclosures[6] and in synthetic enclosures[7] (glass fiber plus resin). The second type is the more recent introduction and it is characterized by being lighter, less vulnerable to vandalism and with better live-part protection against humidity since it is completely compound filled (humidity is the main cause of failure for ZnO arresters).

The outer part of these arresters is usually made of silicon polymer in order to provide environmental resistance and sufficient creepage distances. Their internal composition and silicon enclosures mean that these arresters can be placed in far more positions with optimization of implementation (e.g. horizontal mounting).

These various arrester types are used for the protection of equipment, transformers and cables. In this case, practically all the arresters used are zinc oxide ones, gradually replacing

[6] Practically all operating voltages.
[7] Generally used in distribution networks.

Table 8.4 Example of characteristics of a ZnO arrester

Maximum permanent voltage	12.7 kV
Rated voltage	24 kV
Residual voltage at rated discharging current	< 75 kV
Rated discharging current (8/20 μs wave)	5 kA
Impulse current withstand (4/10 μs wave)	65 kA

horn gaps and silicon carbide arresters, with the purpose of increased accuracy of protection levels to guarantee insulation coordination to a higher degree.

For a case study see web address

8.6 STANDARDS

There are three levels of standardization: international, continental and national, respectively:

- The **IEC (International Electrotechnical Commission)** is the global organization that prepares and publishes international standards for all electrical, electronic and related technologies. IEC standards are used as a basis for national standardization and as references for international tenders and contracts.
- For example, **CENELEC (European Electrotechnical Standardization Committee)** produces the 'EN' standards and covers 29, mostly European, countries. Application of its standards is mandatory.
- **National standardization bodies**, such as CEI (Comitato Elettrotecnico Italiano), UTE (Union Technique de l'Electricité), BEC (British Electrotechnical Committee), etc.

8.6.1 Insulation Coordination

One of the objectives of the standards is to explain and define the various factors to consider to achieve withstand voltages through an approach aiming at the optimization and reduction of voltage withstand levels.

Standard IEC 60071 proposes conventional modeling of actual stresses by wave shapes reproducible in laboratories and having shown satisfactory equivalence. Two main concepts are outlined:

- longitudinal insulation (between the terminals of the same phase of an open device);
- consideration of altitude and of installation ageing.

Moreover, the standard defines internal insulation and external insulation:

- internal insulation, which covers everything not in ambient air such as liquid insulation for transformers;
- external insulation, which refers to air clearances.

It also distinguishes two voltage ranges:

- range I: from 1 kV to 245 kV;
- range II: above 245 kV.

The standard also provides a table of standardized rated withstand voltages for each range, which have been defined mostly on the basis of empirical experience.

8.6.1.1 Determination of Insulation Levels

The standard does not define invariable withstand voltages with validity for every case, but provides a list of steps to enable insulation coordination analysis to be carried out:

- Relationships between network type and choice of its insulations, aiming to establish the characteristics of the maximum possible permanent voltages and the foreseen temporary overvoltages as a function of:
  - network structure and its rated voltage;
  - neutral earthing connection diagram;
  - substations and rotating machines present on the line;
  - type and position of surge limitation devices, if any, and according to considerations common to all overvoltage classes defined by the standard (as shown in Table 8.1).
- Coordination of insulation: the withstand voltage must be determined for each overvoltage class considering the required performance and the acceptable insulation failure rate. The value obtained is of course specific to the studied network and its situation and is the lowest withstand voltage to the overvoltage that the network has in its operating conditions. For the choice of the components of a network, their specified withstand voltages must be defined. Determination of coordination withstand voltages consists of defining the minimum values of insulation withstand voltages satisfying performance criteria when insulation is subjected to the representative overvoltages in operating conditions.
- Determination of specified insulation withstand voltages: this consists of converting the coordination withstand voltages into appropriate standardized test conditions. This is done through multiplication of the coordination withstand voltages by factors compensating for the differences between actual insulation operation conditions and standardized withstand test conditions.
- Rated insulation level: chosen by selecting the most economical series of standardized insulation withstand voltages, so that all the specified withstand voltages are satisfied.
- Rated withstand voltage or insulation level: the specified withstand voltage for overvoltages which can be tested:
  - power frequency test;
  - switching impulse test;
  - lightning impulse test.
- Equivalence factors: as proposed by standard IEC 60071, these indicate that only two withstand voltages need to be specified among the three considered. In the case of

Table 8.5 Standardized insulation levels for r.m.s. voltage networks between 1 and 245 kV (a similar table is available for voltages greater than 245 kV)

Highest voltage for equipment U_m (kV r.m.s.)	Standardized short-duration withstand voltage at power frequency (kV r.m.s.)	Standardized withstand voltage to lightning impulses (kV r.m.s.)
3.6	10	20/40
7.2	20	40/60
12	28	60/75/95
17.5	38	75/95
24	50	95/125/145
36	70	145/170
52	95	250
72.5	145	325
123	(185)/ 230	450/550
145	185/230/275	(450)/550/650
170	(230)/275/325	(550)/650/750
245	(275)/(325)/360/395/460	(650)/(750)/850/950/1050

operating voltages under 245 kV, the power frequency test and lightning impulse test are normally chosen.

- Final choice: made from standardized levels from all the rated voltages (Table 8.5).

Several IEC publications give standards with the withstand and protective levels considered separately. The main publications on overvoltages and insulation coordination are reported in Annex A2 on the website.

BIBLIOGRAPHY

[1] De Metz-Nobla B., *Lightning and HV electrical installations*, Merlin Gerin, Cahier Technique No. 168, 1994.

[2] Dugan R. C. *et al.*, *Electrical Power Systems Quality*, McGraw-Hill, New York, 2002.

[3] Faletti N., Chizzolini P., *Trasmissione e distribuzione dell'energia elettrica*, Vol. 1, Patron editore, Milan, 1985.

[4] Fulchiron D., *Overvoltages and insulation coordination in MV and HV*, Merlin Gerin, Cahier Technique No. 151, 1995.

[5] IEC 60071-1, *Insulation co-ordination – Part 1: Definitions, principles and rules*, Eighth Edition, 2006.

[6] IEC 60071-2, *Insulation co-ordination – Part 2: Application Guide*, Third Edition, 1996.

[7] IEC 60947-1, *Low-voltage switchgear and controlgear – Part 1: General rules*, Fifth Edition, 2007.

[8] Seraudie C., *LV surges and surge arresters, LV insulation co-ordination*, Merlin Gerin, Cahier Technique No. 179, 1999.

[9] UIE (International Union for Electricity Applications), *Transient and temporary overvoltages and current*, Paris, 2001.

9

Analysis of Waveforms in Modern Power Systems

Johan Rens and Piet Swart

A comprehensive study of electric power quality phenomena in modern power systems more often than not requires the analysis of the non-sinusoidal but periodic behavior of voltage and current waveforms. Advances in instrumentation have increased the availability of digital waveform data that can now be pre-selected and captured to characterize specific power quality parameters of concern in sufficient detail for analysis through the capturing of pre- and post-events.

It is necessary to understand the practical limitations and the wide-ranging features offered by waveform data acquisition within the context of the analysis of harmonic power. This analysis of harmonic power forms an integral part of the search for sources of harmonic pollution in the network and in the devising of tariff structures to furnish incentives for consumers to reduce harmonic contributions. Foremost among these considerations are the extremely small levels of harmonic power when compared to that of the total system and the precautions of its most optimal computation. These aspects are discussed and demonstrated.

The Fortesque transform is a well-known tool for analyzing the contribution of zero-, positive- and negative-sequence components and a minor modification of the classic single-frequency expression is able to expand its application also to multi-frequency systems by redefining the $\boldsymbol{a}$-operator to be harmonic-order dependent. The availability of this tool furnishes additional insight into the behavior of harmonic systems and permits superfluous data related to asymmetry to be isolated for the purpose of optimizing specific types of computations.

Handbook of Power Quality Edited by Angelo Baggini

Because three-phase systems are self-contained entities it is possible to use single-phase equivalent models and power definitions under perfectly balanced and sinusoidal waveform conditions to represent these more complex systems. The Park transform is a powerful tool for the analysis of power phenomena in modern three-wire power systems, whilst a new transformation technique termed the Ferrero–Guiliani–Willems transform in this book is preferred to analyze four-wire power systems. Both are formulated and demonstrated through a case study.

To analyze power phenomena in modern power systems requires practical power definitions based on tangible, physically related principles. Several alternative power definitions have been made that emphasize alternative characteristics. The IEEE 1459 standard, based on one such approach, is examined, demonstrated and used in a case study.

Research has shown that the pinpointing of pollution-producing consumers through single-point measurements is impractical. Such measurements would have been ideal and would have enabled the direct implementation of structures that could have been incorporated with the normal tariffs. Alternative methods, based on agreed indices and synchronous multi-point measurements, are now under investigation. Aspects on the localization of sources of waveform distortion in power systems are investigated and the theoretical considerations are reviewed and demonstrated through a case study.

9.1 FREQUENCY ANALYSIS OF NON-SINUSOIDAL WAVEFORMS: PRACTICAL CONSIDERATIONS

Frequency analysis of non-sinusoidal waveforms is well known to the power system analyst. Modern power system waveform data is readily available in digital format and the discrete Fourier transform is normally used to transform time-domain data to frequency-domain data. Such frequency analysis is in general integrated into proprietary software that supports commercial instrumentation used to obtain power system waveform data.

The power system analyst who needs, for example, to study energy phenomena in higher-frequency components has to consider the impact of instrumentation performance on the reliability of such waveform analysis. Practical considerations are subsequently presented.

9.2 ANALOG-TO-DIGITAL CONVERSION

9.2.1 Signal-to-Noise Ratio

Assuming a 16-bit analog-to-digital converter (ADC) is used with a voltage input range of 0–10 V, the smallest detectable voltage change measurable will be $152.6\,\mu\text{V}(10\,\text{V}/2^{16})$. It is common to have an ADC deviation from linearity, or the non-linearity of the ADC is specified as ±0.5 of the LSB (Least Significant Bit). This number is an indication of how sensitive the analog-to-digital (A/D) conversion process is.

The smallest detectable change in analog input signal to be recognized with certainty by the ADC can be calculated. For example, if the input analog signal is slightly above

5000 μV, such as an increase to 5100 μV, it will not be recognized by the ADC. This voltage increase is within the 0.5 LSB ($0.5 \times 152.6\,\mu V$) uncertainty range. But, if the signal increases to 5200 μV, it will be recognized with certainty.

The point to note is that higher-frequency components that are relatively very small in magnitude compared to the magnitude of the fundamental frequency component could be misrepresented due to the signal-to-noise ratio (SNR) of a specific ADC. A resolution higher than 13 bits in the A/D conversion of waveforms in practical power systems will most probably be into the noise band.

9.2.2 Spectral Leakage and Aliasing

The sampling frequency of an A/D card should be adjustable to select a sampling rate which will minimize spectral leakage and aliasing. Practical power system investigations will rarely require frequency information on a waveform above the 50th-harmonic component. The implication is that in a 50 Hz power system, the Nyquist criterion dictates that the sampling frequency is at least 5 kHz.

The Fourier transform in general requires 2^n data points in the fundamental frequency component to avoid spectral leakage. Assuming an overall maximum sampling rate of 100 000 samples per second for a specific A/D card when only three voltage and three current channels are being used as input channels, a 12.8 kHz sampling frequency per channel will result in 256 data points per channel (2^8 data points in a 50 Hz fundamental frequency cycle), which also satisfies the Nyquist information criterion.

A sampling frequency that is not synchronized with the fundamental frequency component of the signal to be sampled is a cause of errors in the determination of the phase and amplitude of harmonic components. Such a sampling frequency can satisfy the Nyquist theorem, but because of the non-synchronization, spectral leakage can exist. Unavoidable truncation of a sampled signal to a finite length of samples is another cause of spectral leakage.

The literature in References [8], [9], [10], [11], [18] and [19] proposes several mathematical solutions such as interpolation algorithms and windowing functions. These approaches can reduce the measurement errors, but cannot completely remove them [10]. A solution to spectral leakage is to synchronize the sampling frequency with the signal to be measured and to obtain an integer number of samples.

A hardware solution to the synchronization problem is described in [9] and [10]. The hardware determines the fundamental frequency and multiplies it by an appropriate integer in order to generate a synchronized sampling frequency.

9.2.3 Spectral Leakage and Windowing

Certain windowing techniques deliver satisfactorily results in moving harmonic energy to the harmonic number where it should have been, in other words correcting the amplitude spectrum. Although the error in amplitude is reduced, the phase errors can be worse than before and affect the energy calculation. Ferrero and Ottoboni [8] have shown that for the usual non-symmetrical sequences employed in DFT algorithms, the phase errors are the dominant errors in 'short-range' leakage and cannot be reduced by data windows. Complex interpolation algorithms, which are computationally intensive, should be used provided the harmonic interference is negligible. An alternative DFT algorithm, which reduces this phase error significantly without an increase in the computational burden, is described in [10].

Windowing functions are integrated in some mathematical packages and can be used with relative ease, but may be unsatisfactorily when used to study power phenomena in non-sinusoidal power systems. Energy calculations require accurate phase information. If non-synchronized digitizing measurement systems are used, the alternative DFT algorithm in [8] should be used in analysis of the data.

9.2.4 Metrological Features of Measurement System

It is not easy to estimate the overall metrological characteristics of a digital measurement process. The error/uncertainty results in influencing the subsequent digital signal processing. The accuracy class of voltage and current transducers is one aspect influencing the certainty of the results obtained. Signal conditioning electronics such as anti-aliasing filters can introduce additional non-linearities as a function of frequency.

A standard instrument against which to evaluate the accuracy of the comprehensive measurement system is not readily obtainable. A very useful contribution to establish the overall accuracy of a digital measurement system is reported in [15]. A method to estimate the uncertainty associated with the signal conditioning elements that includes voltage and current transducers, filters, amplification, isolation and the A/D conversion process is presented. This method requires a practical investigation with specialized supporting instrumentation and it estimates the final uncertainty based on a Monte Carlo procedure.

9.2.5 Harmonic Power Measurement in Non-simultaneous Sampling

If it is not possible to sample the different input channels simultaneously as generally found in A/D cards and other power system instrumentation, a small time difference will result between the time when the data point of one waveform is obtained at a specific channel and the time when the corresponding data point is obtained at another input channel. Assume that a certain A/D card has a time interval of 10 μs between successive samples taken at adjacent channels. Further assume that two adjacent channels are used as the input of a pure sinusoidal voltage and current waveform which are harmonic frequency components at 1250 Hz,[1] each with a peak amplitude value of 10 V and a phase difference between the voltage and current waveforms of 40° (current lagging voltage).

[1] If a 50 Hz voltage signal is fed to a six-pulse rectifier, then the current withdrawn will have harmonic frequencies which are an integer number $h = 6k \pm 1$. With $k = 4, h =$ the harmonic order of the fundamental frequency signal, 1250 Hz is the one harmonic component obtained. The goal is to demonstrate the effect of the 10 μs mismatch between two adjacent A/D input channels on one of the higher-frequency components ('worst case') in the 'distorted' current and voltage digitized signals.

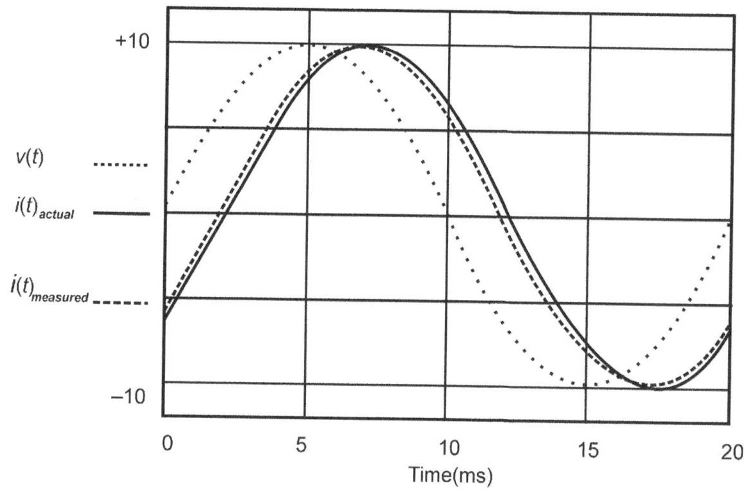

Figure 9.1 Voltage and current waveforms at 1250 Hz to demonstrate non-simultaneous digitizing, 10 μs between two channels (adjacent channels)

The phase error in percent due to the non-simultaneous sampling is then

$$\text{Phase error} = \frac{10 \cdot 10^{-6}}{800 \cdot 10^{-6}} \cdot 100 = 1.25\,\%$$

The waveforms $i(t)_{actual}$ and $i(t)_{measured}$ in Figure 9.1 respectively represent a simulated *error-free* signal and a non-simultaneously measured signal, referenced to the voltage signal $v(t)$.

With $v(t)$ as the reference signal, the phase angle error in Figure 9.1 between $i(t)_{actual}$ and $i(t)_{measured}$ can be written as

$$v(t) = 10 \sin(2 \cdot \pi \cdot 1250 \cdot t)$$
$$i_{actual}(t) = 10 \sin(2 \cdot \pi \cdot 1250 \cdot t - 40°)$$
$$i_{measured}(t) = 10 \sin(2 \cdot \pi \cdot 1250 \cdot t - 35.5°)$$

The phase angle between the current and voltage, if measured simultaneously, should have been −40° (current lagging voltage). In terms of degrees, the measurement error is 4.5°. The error caused by this multiplex time interval resulted in the erroneous phase angle of −35.5°.

The calculation of active power ($P_{measured}$) is in error when based on these measured values:

$$\begin{aligned} P_{measured} &= V_{1\phi} I_{1\phi} \cos(\alpha - \beta) \\ &= \frac{10}{\sqrt{2}} \cdot \frac{10}{\sqrt{2}} \cos[0 - (-35.5°)] \\ &= 40.706\,\text{W} \end{aligned}$$

where $V_{1\phi}$ and $I_{1\phi}$ are the r.m.s values of the single-phase voltage and current waveforms, respectively; and α and β are the phase angles of voltage and current, respectively.

Without a measurement error, the active power (P_{actual}) should have been

$$\begin{aligned} P_{actual} &= V_{1\phi} I_{1\phi} \cos(\alpha - \beta) \\ &= \frac{10}{\sqrt{2}} \cdot \frac{10}{\sqrt{2}} \cos[0 - (-40°)] \\ &= 38.302\,\text{W} \end{aligned}$$

The percentage error in the active power is

$$\begin{aligned} \%\,\text{Error} &= \frac{P_{measured} - P_{actual}}{P_{actual}} \cdot 100 \\ &= +6.275\,\% \end{aligned}$$

A significant error results at high harmonic frequencies but at the fundamental frequency this error is very small. Note that the assumption was made that the channels above are adjacent, which represents the smallest possible error. If three-phase measurements are done, the error can be worse.

Assume that the channels of an A/D card are wired in the following sequence: $v_a(t)$ (indicated as $v(t)$ in Figure 9.2) to channel 1, $v_b(t)$ to channel 2, $v_c(t)$ to channel 3, $i_a(t)$ (indicated as $i(t)_{measured}$ in Figure 9.2) to channel 4, $i_b(t)$ to channel 5, $i_c(t)$ to channel 6.

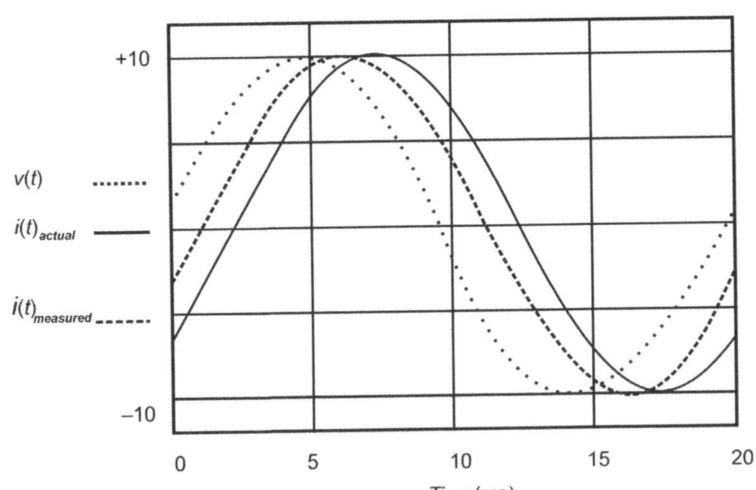

Figure 9.2 Voltage and current waveforms at 1250 Hz, non-simultaneous digitizing, 30 μs (three-channel separation)

The phase error in the 25th harmonic between the voltage and current signal in a phase is now based on a 30 μs difference in digitizing and demonstrated in Figure 9.2:

- $v_a(t)$: Phase a time-dependent voltage (referenced to an earth plane), similarly for phases b and c.
- $i_a(t)$: Phase a time-dependent current, similarly for phases b and c.

The *phase error* in Figure 9.2 is now 13.5° and the *active power error* is 16.825 %. The error is progressive as a function of frequency and of channel number. It is not negligible, even when two adjacent channels are considered.

9.2.5.1 Error in Apparent Power and Complex Power Measurement

When the concept of effective values [20], [21] for current and voltage is used in the calculation of apparent power, it will not be influenced by the phase angle error introduced by non-simultaneous sampling. The same is not true, however, when the single-phase *complex power* $(\boldsymbol{S}_{1\phi})^2$ is calculated per phase and per harmonic number (h):

$$\boldsymbol{S}_{1\phi} = \boldsymbol{V}_{1\phi} \times \boldsymbol{I}^*_{1\phi} = V_{1\phi}\underline{|\alpha}(I_{1\phi}\underline{|-\beta})^* = V_{1\phi}\underline{|\alpha}(I_{1\phi}\underline{|\beta}) = P_{1\phi} + jQ_{1\phi} \tag{9.1}$$

In the above equation it can be seen that the phase angle error in the phase angle difference $(\alpha - \beta)$ due to non-simultaneous sampling influences the values of single-phase active $(P_{1\phi})$ and imaginary power $(Q_{1\phi})$. $\boldsymbol{V}_{1\phi}$ and $\boldsymbol{I}_{1\phi}$ are the phasor notation for a single-phase fundamental frequency voltage and current waveform, respectively.

9.2.5.2 Frequency-Domain Compensation of Phase Error

A simple solution does not exist for the non-simultaneous sampling problem by compensating the time-domain data for the multiplex time error. The error in phase measurement can be compensated by a frequency-dependent (or harmonic-dependent) phase adjustment function. Such a *phase compensation function* must have the following characteristics:

- The amplitude must be unity and independent of frequency.
- The phase correction must be a function of frequency.

The compensation function below satisfies the above requirements:

$$\text{Phase correction function}(h) = 1 \cdot \mathrm{e}^{-\mathrm{j}\omega_1 h t_{error}} \tag{9.2}$$

In the above equation, $\omega_1 = 2\pi f_1$ with f_1 the fundamental frequency, whilst h is the harmonic order and t_{error} the digitizing time interval that causes the phase error. This phase correction function has the amplitude and phase dependency shown in Figure 9.3 on harmonic order.

[2] Italic font is used to indicate variables. Regular (roman) font indicates constants. Bold italic font is used to indicate complex numbers and phasor, vector and matrix quantities. Lower-case font is associated with time-dependent quantities, whilst capital letters are used to indicate r.m.s. quantities.

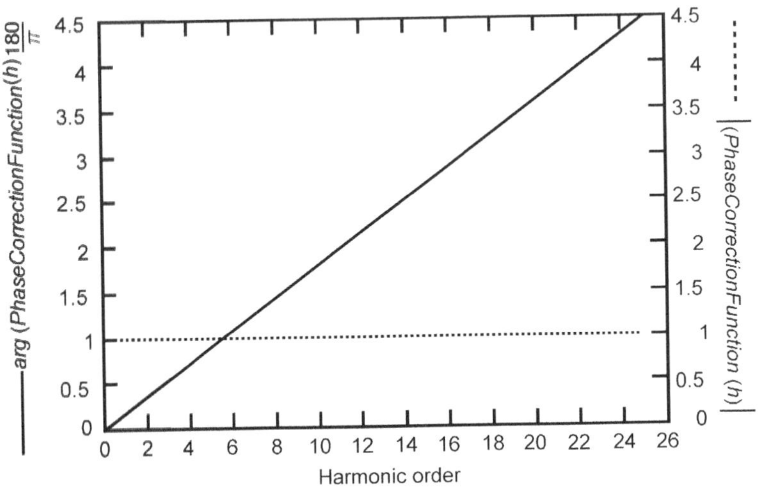

Figure 9.3 The phase and amplitude dependency on harmonic order of the phase compensation function

An additional requirement indicated in Figure 9.3 is that the channel number[3] should be included in the phase compensation function as redefined below:

$$\text{Phase correction}(h, c) = 1 \cdot e^{j \cdot \omega_1 \cdot c \cdot t_{error} \cdot h} \tag{9.3}$$

9.2.5.3 Practical Application of a Phase Compensation Function

Assume that the digitized data acquired by an A/D process is to be in row–column format and arranged according to the following matrix of data indexed against time:

$$\boldsymbol{Matrix}_{time} = \begin{bmatrix} t_0 & v_a(t_0) & v_b(t_0) & v_c(t_0) & i_a(t_0) & i_b(t_0) & i_c(t_0) \\ t_1 & v_a(t_1) & v_b(t_1) & v_c(t_1) & i_a(t_1) & i_b(t_1) & i_c(t_1) \\ t_2 & v_a(t_2) & v_b(t_2) & v_c(t_2) & i_a(t_2) & i_b(t_2) & i_c(t_2) \\ \cdot & \cdot & \cdot & \cdot & \cdot & \cdot & \cdot \\ \cdot & \cdot & \cdot & \cdot & \cdot & \cdot & \cdot \\ t_{end} & v_a(t_{end}) & v_b(t_{end}) & v_c(t_{end}) & i_a(t_{end}) & i_b(t_{end}) & i_c(t_{end}) \end{bmatrix} \tag{9.4}$$

$\boldsymbol{Matrix}_{time}$ is a matrix containing digitized time-domain data. Timing instances $t_0 \ldots t_{end}$ represent the A/D timing values assigned to the data points. For instance, $v_a(t_0)$ in the second column (column number $j = 2$ and row number $i = 1$) is the amplitude of voltage in phase a at time t_0. Element $v_b(t_0)$ is the amplitude of voltage in phase b taken at (t_0+ one multiplex time interval). Element $v_c(t_0)$ is the amplitude of voltage in phase c taken at (t_0+ two multiplex time intervals). Similarly, the amplitude of current in phase c, $i_c(t_0)$, is

[3] c: Channel number $c \in [0, 1, 2, \ldots, N-1]$. N = number of channels.

taken at real time ($t_0 + 5\times$ multiplex time intervals). After another multiplex time interval, the data row is completed for t_0 and the process is repeated in the next row.

The Fourier transformation of the time-domain data in $\boldsymbol{Matrix}_{time}$ results in

$$\boldsymbol{Matrix}_{FFT} = \begin{bmatrix} \boldsymbol{V}_a(1) & \boldsymbol{V}_b(1) & \boldsymbol{V}_c(1) & \boldsymbol{I}_a(1) & \boldsymbol{I}_b(1) & \boldsymbol{I}_c(1) \\ \boldsymbol{V}_a(2) & \boldsymbol{V}_b(2) & \boldsymbol{V}_c(2) & \boldsymbol{I}_a(2) & \boldsymbol{I}_b(2) & \boldsymbol{I}_c(2) \\ \boldsymbol{V}_a(3) & \boldsymbol{V}_b(3) & \boldsymbol{V}_c(3) & \boldsymbol{I}_a(3) & \boldsymbol{I}_b(3) & \boldsymbol{I}_c(3) \\ \cdot & \cdot & \cdot & \cdot & \cdot & \cdot \\ \cdot & \cdot & \cdot & \cdot & \cdot & \cdot \\ \boldsymbol{V}_a(N) & \boldsymbol{V}_b(N) & \boldsymbol{V}_c(N) & \boldsymbol{I}_a(N) & \boldsymbol{I}_b(N) & \boldsymbol{I}_c(N) \end{bmatrix} \tag{9.5}$$

where $\boldsymbol{Matrix}_{FFT}$ is a matrix containing the frequency-domain data, written as voltage harmonic phasors $\boldsymbol{V}_P(h) = |\boldsymbol{V}_P(h)|e^{j\alpha_h}$, with P the phase number (a, b and c), α_h the phase angle of the voltage harmonic phasor and h the harmonic number with N the highest harmonic order considered. Similarly, the harmonic current phasors are $\boldsymbol{I}_P(h) = |\boldsymbol{I}_P(h)|e^{j\beta_h}$.

Phase compensation requires application of the *phase compensation function* to the matrix of frequency data:

$$\boldsymbol{Matrix}_{FFT}(h, c) = \boldsymbol{Matrix}_{FFT}(h, c) \cdot \text{Phase correction}(h, c) \tag{9.6}$$

The *phase correction function* simulates a simultaneous sampling process as it has the required progressive influence on compensation of the phase error. It is important to realize that this compensation scheme is valid only when steady-state conditions prevail.

In an industrial power system, it is not necessary valid to assume steady-state conditions. The wavelet transform should then be used to evaluate the time dependency of harmonic components and hence the assumption of a steady state.

When digital measurement instrumentation is used to study power quality issues, simultaneous sampling is important to avoid errors in power calculations. Non-simultaneous sampling could cause erroneous interpretations of measurements.

9.3 SEQUENCE COMPONENT ANALYSIS

Fortesque formulated a mathematical transformation that enables the linear transformation of phase-domain components to a sequence domain containing a set of symmetrical components. Analysis of a three-phase circuit with non-sinusoidal waveforms requires a modification of the conventional Fortesque transformation derived in 1918 [17]. Note that it is valid only for those periodic signals showing harmonic distortion (harmonic number h can only be an integer).

9.3.1 Fortesque Transform Redefined for Non-sinusoidal Circuits

The complex operator $\boldsymbol{a}$ used in the classical Fortesque transform requires redefinition as voltage and current signals to be transformed are no longer pure sinusoidal waveforms. It is accomplished by defining a harmonic-dependent complex operator $\boldsymbol{a}(h)$:

$$\boldsymbol{a}(h) = 1 \cdot e^{j\frac{2\cdot\pi\cdot h}{3}} \tag{9.7}$$

The 3×3 Fortesque transformation matrix $\boldsymbol{A}(h)$ is then defined as

$$\boldsymbol{A}(h) = \begin{bmatrix} 1 & 1 & 1 \\ 1 & \boldsymbol{a}(h)^2 & \boldsymbol{a}(h) \\ 1 & \boldsymbol{a}(h) & \boldsymbol{a}(h)^2 \end{bmatrix} \tag{9.8}$$

Transformation of the three-phase voltage vector $\boldsymbol{V}(h)_{abc}$ to the sequence domain is defined as

$$\begin{bmatrix} \boldsymbol{V}_0(h) \\ \boldsymbol{V}_1(h) \\ \boldsymbol{V}_2(h) \end{bmatrix} = \frac{1}{3} \begin{bmatrix} 1 & 1 & 1 \\ 1 & \boldsymbol{a}(h) & \boldsymbol{a}(h)^2 \\ 1 & \boldsymbol{a}(h)^2 & \boldsymbol{a}(h) \end{bmatrix} \begin{bmatrix} \boldsymbol{V}_a(h) \\ \boldsymbol{V}_b(h) \\ \boldsymbol{V}_c(h) \end{bmatrix} \tag{9.9}$$

where:

$\boldsymbol{V}_a(h)$ is the vector of phase a voltage harmonic phasors listed as a function of harmonic number h, and similarly for phases b and c.

$\boldsymbol{V}_0(h)$ is the vector of zero-sequence voltage harmonic phasors listed as a function of the harmonic number h; the subscript '0' indicates the zero-sequence components – similarly the interpretation for subscript '1' which indicates the positive-sequence components and subscript '2' which indicates the negative-sequence components.

The compact notation of the above equation is written as

$$\boldsymbol{V}_s(h) = \boldsymbol{A}(h)^{-1}\boldsymbol{V}(h)_{abc} \tag{9.10}$$

Transformation of the three-phase current harmonic vector $\boldsymbol{I}(h)_{abc}$ to the sequence domain is defined as

$$\begin{bmatrix} \boldsymbol{I}_0(h) \\ \boldsymbol{I}_1(h) \\ \boldsymbol{I}_2(h) \end{bmatrix} = \frac{1}{3} \begin{bmatrix} 1 & 1 & 1 \\ 1 & \boldsymbol{a}(h) & \boldsymbol{a}(h)^2 \\ 1 & \boldsymbol{a}(h)^2 & \boldsymbol{a}(h) \end{bmatrix} \begin{bmatrix} \boldsymbol{I}_a(h) \\ \boldsymbol{I}_b(h) \\ \boldsymbol{I}_c(h) \end{bmatrix} \tag{9.11}$$

where:

$\boldsymbol{I}_a(h)$ is the vector of phase a current harmonic phasors listed as a function of harmonic number h, and similarly for phases b and c.

$\boldsymbol{I}_0(h)$ is the vector of zero-sequence current harmonic phasors listed as a function of the harmonic number h; the subscript '0' indicates the zero-sequence components, subscript '1' indicates the positive-sequence components and subscript '2' indicates the negative sequence components.

The compact notation of the above is written as

$$\boldsymbol{I}_s(h) = \boldsymbol{A}(h)^{-1}\boldsymbol{I}(h)_{abc} \tag{9.12}$$

The inverse Fortesque transforms follows as

$$\boldsymbol{V}(h)_{abc} = \boldsymbol{A}(h)\boldsymbol{V}_s(h) \quad \boldsymbol{I}(h)_{abc} = \boldsymbol{A}(h)\boldsymbol{I}_s(h) \tag{9.13}$$

For a case study see web address

9.3.2 Three-Wire Power System Analysis

The Park transform is a mathematical tool that transforms three-phase quantities to quadrature coordinates (d–q) which can completely describe a three-wire, three-phase power system in an equivalent synthetic domain. In the case of an unbalanced four-wire, three-phase power system, a third set of zero-sequence quantities results that requires separate consideration.

In a three-phase, three-wire power system it is convenient to express the phase-domain voltage and current vector as a two-dimensional orthogonal voltage and current vector that completely describe all three-phase phase-domain phenomena. The d–q theory is widely employed in analyzing the transient behavior of electrical machines. Park[4] formulated this transformation of three-phase quantities to orthogonal d–q quantities in 1922.

9.3.2.1 The Park Transform and Non-sinusoidal Waveforms

The Park transformation function converts the time-dependent three-phase, three-wire voltage vector $\boldsymbol{v}(t)_{abc}$ and the time-dependent three-phase line current vector $\boldsymbol{i}(t)_{abc}$ to time-dependent Park voltage vector $\boldsymbol{v}(t)_{\text{Park}}$ and Park current vectors $\boldsymbol{i}(t)_{\text{Park}}$, respectively. The Park transform is a special case of the Clarke transform. The d–q axis is stationary and also known as the p–q theory. It is a linear orthogonal transformation which makes use of an orthogonal matrix $\boldsymbol{T}_{\text{Park}}$:

$$\boldsymbol{T}_{\text{Park}} = \begin{bmatrix} \sqrt{2/3} & -\sqrt{1/6} & -\sqrt{1/6} \\ 0 & \sqrt{1/2} & -\sqrt{1/2} \\ \sqrt{1/3} & \sqrt{1/3} & \sqrt{1/3} \end{bmatrix} \quad |\boldsymbol{T}_{\text{Park}}| = 1 \tag{9.14}$$

The Park transformation of the voltage phase-domain vector is defined as

$$\begin{bmatrix} v_d(t) \\ v_q(t) \\ v_0(t) \end{bmatrix} = \begin{bmatrix} \sqrt{2/3} & -\sqrt{1/6} & -\sqrt{1/6} \\ 0 & \sqrt{1/2} & -\sqrt{1/2} \\ \sqrt{1/3} & \sqrt{1/3} & \sqrt{1/3} \end{bmatrix} \cdot \begin{bmatrix} v_a(t) \\ v_b(t) \\ v_c(t) \end{bmatrix} \tag{9.15}$$

with $v_a(t) = \sum_{h=1}^{N} v_a(t)_h$, and similarly for phases b and c, where $v_a(t)_h$ is the phase a time-dependent voltage at harmonic frequency $h\omega_1$ (referenced to earth plane), and similarly for phases b and c; and

$$v_d(t) = \sum_{h=1}^{N} v_d(t)_h \quad v_q(t) = \sum_{h=1}^{N} v_q(t)_h \quad v_0(t) = \sum_{h=1}^{N} v_0(t)_h$$

[4] Note that the Park transformation is a particular case of the Clarke transformation.

where $v_d(t)_h$ is the Park direct-axis time-dependent voltage at harmonic frequency $h\omega_1$, $v_q(t)_h$ the quadrature-axis component and $v_0(t)$ the additional zero-sequence component but not part of the Park voltage vector $\boldsymbol{v}(t)_{\text{Park}}$.

The compact notation of the above is

$$\boldsymbol{v}(t)_{\text{Park}} = \boldsymbol{T}_{\text{Park}}\boldsymbol{v}(t)_{abc} \tag{9.16}$$

The Park transformation of the current phase-domain vector is defined as

$$\begin{bmatrix} i_d(t) \\ i_q(t) \\ i_0(t) \end{bmatrix} = \begin{bmatrix} \sqrt{2/3} & -\sqrt{1/6} & -\sqrt{1/6} \\ 0 & \sqrt{1/2} & -\sqrt{1/2} \\ \sqrt{1/3} & \sqrt{1/3} & \sqrt{1/3} \end{bmatrix} \cdot \begin{bmatrix} i_a(t) \\ i_b(t) \\ i_c(t) \end{bmatrix} \tag{9.17}$$

with $i_a(t) = \sum_{h=1}^{N} i_a(t)_h$, and similarly for phases b and c, where $i_a(t)_h$ is the phase a time-dependent current at harmonic frequency $h\omega_1$, and similarly for $i_b(t)_h$ and $i_c(t)_h$; and

$$i_d(t) = \sum_{h=1}^{N} i_d(t)_h \quad i_q(t) = \sum_{h=1}^{N} i_q(t)_h \quad i_0(t) = \sum_{h=1}^{N} i_0(t)_h$$

where $i_d(t)_h$ is the Park direct-axis time-dependent current at harmonic frequency $h\omega_1$, $i_q(t)_h$ the quadrature-axis component and $i_0(t)$ the additional zero-sequence component but not part of the Park current vector $\boldsymbol{i}(t)_{\text{Park}}$.

The compact notation of the above is

$$\boldsymbol{i}(t)_{\text{Park}} = \boldsymbol{T}_{\text{Park}}\boldsymbol{i}(t)_{abc} \tag{9.18}$$

The time-dependent Park voltage $\boldsymbol{v}(t)_{\text{Park}}$ and current $\boldsymbol{i}(t)_{\text{Park}}$ vectors are complex and contain two orthogonal quantities:

$$\boldsymbol{v}(t)_{\text{Park}} = v(t)_d + \mathrm{j}v(t)_q \tag{9.19}$$

$$\boldsymbol{i}(t)_{\text{Park}} = i(t)_d + \mathrm{j}i(t)_q \tag{9.20}$$

The multi-frequency Park transformation also yields zero-sequence components $v_0(t)$ and $i_0(t)$ that are similar to the zero-sequence components yielded by its single-frequency counterpart. As in the single-frequency case, no zero-sequence components exist when the phase-domain voltage vectors have a common reference and current vectors do not have a common return (neutral).

The inverse multi-frequency Park transform for voltage is given by

$$\begin{bmatrix} v_a(t) \\ v_b(t) \\ v_c(t) \end{bmatrix} = \begin{bmatrix} \sqrt{2/3} & \sqrt{1/6} & -\sqrt{1/6} \\ \sqrt{0} & \sqrt{1/2} & -\sqrt{1/2} \\ \sqrt{1/3} & \sqrt{1/3} & \sqrt{1/3} \end{bmatrix}^{-1} \begin{bmatrix} v_d(t) \\ v_q(t) \\ v_0(t) \end{bmatrix} \tag{9.21}$$

$$\begin{bmatrix} i_a(t) \\ i_b(t) \\ i_c(t) \end{bmatrix} = \begin{bmatrix} \sqrt{2/3} & \sqrt{1/6} & -\sqrt{1/6} \\ \sqrt{2/3} & \sqrt{1/2} & -\sqrt{1/2} \\ \sqrt{1/3} & \sqrt{1/3} & \sqrt{1/3} \end{bmatrix}^{-1} \begin{bmatrix} i_d(t) \\ i_q(t) \\ i_0(t) \end{bmatrix} \tag{9.22}$$

Because $\boldsymbol{T}_{\text{Park}}$ is an orthogonal matrix, $\boldsymbol{T}_{\text{Park}}^{-1} = \boldsymbol{T}_{\text{Park}}^{T}$.

The sequence powers yielded by the multi-frequency Park transform follow analogously from that of its original single-frequency counterpart [4].

The relationship between the Fortesque sequence-domain components in the fundamental frequency and the Park components [11] is written below for a single frequency, as follows:[5]

$$\boldsymbol{v}(t)_{\text{Park}} = \boldsymbol{V}_1(1)\mathrm{e}^{\mathrm{j}\omega_1 t} + \boldsymbol{V}_2^*(1)\mathrm{e}^{-\mathrm{j}\omega_1 t} \tag{9.23}$$

$$v_0(t) = \sqrt{2}\cdot \mathrm{Re}[\boldsymbol{V}_0(1)\mathrm{e}^{\mathrm{j}\omega_1 t}] \tag{9.24}$$

$$\boldsymbol{i}(t)_{\text{Park}} = \boldsymbol{I}_1(1)\mathrm{e}^{\mathrm{j}\omega_1 t} + \boldsymbol{I}_2^*(1)\mathrm{e}^{-\mathrm{j}\omega_1 t} \tag{9.25}$$

$$i_0(t) = \sqrt{2}\cdot \mathrm{Re}[\boldsymbol{I}_0(1)\mathrm{e}^{\mathrm{j}\omega_1 t}] \tag{9.26}$$

The above equations show that the Park voltage and current vectors can be obtained from only the positive-sequence components in a balanced three-phase system with positive phase rotation as the negative- and zero-sequence components will not exist.

If the phase-domain voltage and current waveforms are non-sinusoidal and are transformed to the Park domain, then the Park voltage and current vectors can also be decomposed into a Fourier series [11]:

$$\boldsymbol{v}(t)_{\text{Park}} = \sum_{h=-\infty}^{\infty} \boldsymbol{V}_{\text{Park}}(h)\,\mathrm{e}^{\mathrm{j}h\omega_1 t} \tag{9.27}$$

$$\boldsymbol{i}(t)_{\text{Park}} = \sum_{h=-\infty}^{\infty} \boldsymbol{I}_{\text{Park}}(h)\,\mathrm{e}^{\mathrm{j}h\omega_1 t} \tag{9.28}$$

The quantities $\boldsymbol{V}_{\text{Park}}(h)$ and $\boldsymbol{I}_{\text{Park}}(h)$ are Park harmonic phasors of voltage and current, respectively. $\boldsymbol{V}_{\text{Park}}(h)\,\mathrm{e}^{\mathrm{j}h\omega_1 t}$ and $\boldsymbol{I}_{\text{Park}}(h)\,\mathrm{e}^{\mathrm{j}h\omega_1 t}$ are Park vectors with constant magnitudes $V_{\text{Park}}(h)$ and $I_{\text{Park}}(h)$ that rotate in a positive or a negative direction as a function of the sign of the harmonic index h and with the rotational speed proportional to the index number h. The Park voltage and current vectors are related to positive- and negative-sequence components, Park harmonic components with positive harmonic frequencies ($h > 0$) are related to positive-sequence components and Park harmonic components at negative harmonic frequencies ($h < 0$) represent negative-sequence symmetrical components:

$$\begin{aligned} \boldsymbol{V}_{\text{Park}}(h) &= \boldsymbol{V}_1(h) \text{ and } \boldsymbol{I}_{\text{Park}}(h) = \boldsymbol{I}_1(h) \text{ if } h > 0 \\ \boldsymbol{V}_{\text{Park}}(h) &= \boldsymbol{V}_2^*(h) \text{ and } \boldsymbol{I}_{\text{Park}}(h) = \boldsymbol{I}_2^*(h) \text{ if } h > 0 \end{aligned} \tag{9.29}$$

The r.m.s. value of the Park voltage vector (similar for the Park current vector) is expressed as

$$V_{\text{Park}} = \sqrt{\sum_{h=-\infty}^{+\infty} V_{\text{Park}}^2(h)} \tag{9.30}$$

[5] The frequency component assumed in this example is the fundamental frequency component of positive-sequence and negative-sequence phasors, $h = 1$ in $\boldsymbol{V}_1(h)$ and $\boldsymbol{V}_2(h)$. The superscript '*' indicates the complex conjugate.

The special consideration of separate zero-sequence quantities of the Park transform, when applied to unbalanced four-wire, three-phase power systems, is overcome by the transformation presented in Section 9.3.3.

9.3.2.2 Application in Power Systems

The Park voltage and current vectors can be studied in the frequency domain and classical power definitions can be applied to these quantities. The formal properties of the original and physical three-phase power system (three-wire) in which these quantities originate are not lost.

The application of the Park transform in power systems was demonstrated in principle when Ferrero and Superti-Furga [11] derived power definitions based on the Park voltage and current vector. It achieved practical status when Akagi and Nabae [1] developed a novel compensator design methodology using just that.

Energy analysis in the Park (or only termed *d–q*) domain makes use of conventional single-phase power definitions. The advantage is that the resulting power components in the *d–q* domain have been assigned a physical significance. Ferrero *et al.* [11], [12] explained the Park real/active power, whilst the Park imaginary power was explained through application of the Poynting vector [14].

9.3.2.3 Power Definitions in the Park Domain

Time-dependent power $p(t)_{\text{Park}}$ can be defined based on the Park vectors of voltages and currents, $\boldsymbol{v}(t)_{\text{Park}}$ and $\boldsymbol{i}(t)_{\text{Park}}$, respectively. This power is numerically the exact same quantity as the three-phase time-dependent power $p(t)_{3\phi}$ because the Park transform is a linear orthogonal transform:

$$\begin{aligned} p(t)_{\text{Park}} &= \boldsymbol{v}(t)_{\text{Park}}\boldsymbol{i}(t)_{\text{Park}} = [v_d(t) \quad v_q(t) \quad v_0(t)]\begin{bmatrix} i_d(t) \\ i_q(t) \\ i_0(t) \end{bmatrix} \\ &= \boldsymbol{v}(t)_{3\phi}\boldsymbol{i}(t)_{3\phi} \\ &= [v_a(t) \quad v_b(t) \quad v_c(t)]\begin{bmatrix} i_a(t) \\ i_b(t) \\ i_c(t) \end{bmatrix} \\ &= p(t)_{3\phi} \end{aligned} \tag{9.31}$$

The Park time-dependent complex power $\boldsymbol{a}(t)_{\text{Park}}$ is defined from the product of the Park time-dependent complex voltage and current:

$$\boldsymbol{a}(t)_{\text{Park}} = \boldsymbol{v}(t)_{\text{Park}} \bullet \boldsymbol{i}(t)^*_{\text{Park}} \tag{9.32}$$

The Park time-dependent real power $p(t)_{\mathrm{Park}}$ and the Park time-dependent non-active/imaginary power $q(t)_{\mathrm{Park}}$ is defined as

$$\begin{aligned} \boldsymbol{a}(t)_{\mathrm{Park}} &= p(t)_{\mathrm{Park}} + \mathrm{j}q(t)_{\mathrm{Park}} \\ &= \mathrm{Re}[\boldsymbol{a}(t)_{\mathrm{Park}}] + \mathrm{j}\,\mathrm{Im}[\boldsymbol{a}(t)_{\mathrm{Park}}] \\ &= \left[v(t)_d i(t)_d + v(t)_q i(t)_q\right] + \mathrm{j}\left[v_q(t) i(t)_d - v(t)_d i(t)_q\right] \end{aligned} \tag{9.33}$$

The three-phase time-dependent power $p(t)_{3\phi}$ relates to the Park time-dependent real power $p(t)_{\mathrm{Park}}$ and the zero-sequence power $p_0(t)$:

$$\begin{aligned} p_{3\phi}(t) &= p(t)_{\mathrm{Park}} + p_0(t) \\ p_0(t) &= v_0(t)\, i_0(t) \end{aligned} \tag{9.34}$$

Akagi and Nabae [1] showed that the Park imaginary power $q(t)_{\mathrm{Park}}$ does not involve time-dependent three-phase power. The Park imaginary power is a characteristic quantity of three-phase systems and is a result of a ratio between time-dependent line voltages and line currents that are not the same for each phase [11].

The average values of the Park time-dependent complex power are

$$\boldsymbol{A}_{\mathrm{Park}} = P_{\mathrm{Park}} + \mathrm{j}Q_{\mathrm{Park}} = \frac{1}{T}\int_T \boldsymbol{v}(t)_{\mathrm{Park}} \bullet \boldsymbol{i}(t)^*_{\mathrm{Park}}\,\mathrm{d}t \tag{9.35}$$

P_{Park} is equal to the average value of the Park *active power* when the zero-sequence time-dependent power $p_0(t)$ is zero. Q_{Park} is the average value of the Park imaginary power.

Because these power definitions are formulated in a single synthetic domain, accepted power theory principles can be applied. This is contrary to other approaches (e.g. that of Czarnecki [6]) that extended single-phase-domain power definitions to the three-phase domain. The following characteristics of the Park power quantities $\boldsymbol{a}(t)_{\mathrm{Park}}$, $p(t)_{\mathrm{Park}}$, P_{Park}, $q(t)_{\mathrm{Park}}$, Q_{Park} and $p_0(t)$ are important:

- The Park power definitions are quantities that can be regarded as actual powers and not apparent powers because their sign depends on the reference directions chosen for the voltages and currents.
- These powers satisfy the energy conservation principle: that is, the algebraic sum of powers associated with each element of an isolated network will sum to zero.

The Park real and imaginary powers are by definition the three-phase positive-sequence active and reactive powers when the three-phase system is sinusoidal, symmetrical, balanced and of positive phase rotation. The Park voltage and current vectors are then [11]

$$\begin{aligned} \boldsymbol{v}(t)_{\mathrm{Park}} &= \sqrt{3}V_e \mathrm{e}^{\mathrm{j}\omega_1 t} \\ \boldsymbol{i}(t)_{\mathrm{Park}} &= \sqrt{3}I_e \mathrm{e}^{\mathrm{j}\omega_1 t-\phi} \end{aligned} \tag{9.36}$$

where V_e and I_e are three-phase effective values of voltage and current according to the IEEE 1459 [20] definition and presented in Section 9.4.

The time-dependent Park *real* ($p(t)_{\text{Park}}$) and *imaginary* ($q(t)_{\text{Park}}$) powers relate to the average three-phase real/active power ($P_{3\phi}$) and the reactive power ($Q_{3\phi}$):

$$\begin{aligned} p(t)_{\text{Park}} &= P_{\text{Park}} = P_{3\phi} = 3V_e I_e \cos\phi \\ q(t)_{\text{Park}} &= Q_{\text{Park}} = Q_{3\phi} = 3V_e I_e \sin\phi \end{aligned} \tag{9.37}$$

When the balanced three-phase system is of negative phase sequence, only negative-sequence voltages and currents result. Again in terms of the effective three-phase values, the Park vectors of voltage and current are [11]

$$\begin{aligned} \boldsymbol{v}(t)_{Park} &= \sqrt{3}V_e \mathrm{e}^{-\mathrm{j}\omega t} \\ \boldsymbol{i}(t)_{Park} &= \sqrt{3}I_e \mathrm{e}^{-\mathrm{j}(\omega t-\phi)} \end{aligned} \tag{9.38}$$

The *real* and *imaginary* Park powers then are:

$$\begin{aligned} p(t)_{\text{Park}} &= P_{\text{Park}} = P_{3\phi} = 3V_e I_e \cos\phi \\ q(t)_{\text{Park}} &= Q_{\text{Park}} = -Q_{3\phi} = -3V_e I_e \sin\phi \end{aligned} \tag{9.39}$$

Application of the Park power definitions to non-sinusoidal and unbalanced but periodical three-phase systems requires the Fourier series components of the Park voltage and current vectors. This application is valid because the Park vectors are periodical too and the Park transform is a linear transform. The Park time-dependent complex power is then defined in terms of Park harmonic phasors of voltage $\boldsymbol{V}_{\text{Park}}(h)$ and current $\boldsymbol{I}_{\text{Park}}(h)$ [11]:

$$\boldsymbol{a}(t)_{\text{Park}} = \sum_{h=-\infty}^{+\infty} \boldsymbol{V}_{\text{Park}}(h)\boldsymbol{I}^*_{\text{Park}}(h) + \sum_{-\infty}^{+\infty}{}_{k\neq h} \boldsymbol{V}_{\text{Park}}(h)\boldsymbol{I}^*_{\text{Park}}(h)\mathrm{e}^{\mathrm{j}(h-k)\omega t} \tag{9.40}$$

The Park *real* and *imaginary* powers also follow from the average value of the Park time-dependent complex power:

$$\begin{aligned} \boldsymbol{A}_{\text{Park}} &= P_{\text{Park}} + \mathrm{j}Q_{\text{Park}} \\ &= \sum_{h=-\infty}^{+\infty} \boldsymbol{V}_{\text{Park}}(h)\boldsymbol{I}^*_{\text{Park}}(h) \\ &= \mathrm{Re}\left[\sum_{h=-\infty}^{+\infty} \boldsymbol{V}_{\text{Park}}(h)\boldsymbol{I}^*_{\text{Park}}(h)\right] + \mathrm{j}\left[\mathrm{Im}\left(\sum_{h=-\infty}^{+\infty} \boldsymbol{V}_{\text{Park}}(h)\boldsymbol{I}^*_{\text{Park}}(h)\right)\right] \\ &= \sum_{h=-\infty}^{+\infty} P_{\text{Park}}(h) + \mathrm{j}\left[\sum_{h=-\infty}^{+\infty} Q_{\text{Park}}(h)\right] \end{aligned} \tag{9.41}$$

The Park power components apply only to a three-phase power system and not to a single-phase system. The Park *real power* is an expression of all the active powers found in each of the harmonic components and its symmetrical components ($h > 0$ for the positive-sequence harmonic components and $h < 0$ for the negative-sequence harmonic components). It implies a summation of the real powers for the positive- and negative-sequence components.

The Park *imaginary power* has a different meaning and definition than the Budeanu reactive power, although at a first glance the mathematical formulation appears to be similar. The Park imaginary power is based on a summation of the reactive powers in the positive-sequence harmonic components ($Q_1(h)$) with the summation of the reactive powers in the negative-sequence harmonic components ($Q_2(h)$), subtracted:

$$Q_{\text{Park}} = \sum_{h=1}^{\infty} Q_1(h) - \sum_{h=1}^{\infty} Q_2(h) + Q_{\text{Park}}(0) \tag{9.42}$$

The component ($h = 0$) is shown separately in Equation (9.42) as the d.c. component of current can be spatially shifted with respect to voltage in the Park vectors [11].

Budeanu adds the reactive power in both the positive- and negative-sequence harmonic components as shown below (the Budeanu three-phase reactive power $Q_{B,3\phi}$ rewritten in terms of sequence components):

$$Q_{B,3\phi} = \sum_{h=1}^{\infty} Q_1(h) + \sum_{h=1}^{\infty} Q_2(h) \tag{9.43}$$

The characteristics of the Park imaginary power (Q_{Park}) are

- The Park imaginary power (Q_{Park}) does not exist in a single-phase power system.
- It does not yield information on possible time-dependent energy exchange between reactive elements in the load and the source.
- Q_{Park} is defined in the time domain, but $Q_{B,3\phi}$ is not.
- The physical meaning of Q_{Park} is therefore completely different from the reactive power definitions of Budeanu, Czarnecki and others.
- It can be used successfully to improve the power factor of the load [1].
- It can be measured easily, whereas Budeanu's reactive power cannot be easily measured.
- If the three-phase load is passive, linear and time invariant, Q_{Park} is associated only with the susceptance of the load.

9.3.2.4 The Park Power Factor

The power factor of a load, based on Park power components, is defined as

$$PF_{\text{Park}} = \frac{P_{\text{Park}}}{S_{3\phi}} \tag{9.44}$$

9.3.2.5 Compensation with Park Power Components

Power factor correction, based on the Park power definitions, requires complete compensation of the time-dependent values of $q(t)_{\text{Park}}$. It was first shown by Akagi and Nabae [1] that it is not sufficient to compensate only for $q(t)_{\text{Park}}$. Without energy storage elements, compensation of $q(t)_{\text{Park}}$ will represent the maximum power factor correction that is possible. Unity power factor requires a holistic consideration of the Park time-dependent power $p(t)_{\text{Park}}$.

For a case study see web address

9.3.2.6 Consideration on Park's Transform

Important features of the Park voltage, current and power components [11] are:

- Measurement thereof is easy.
- The Park active power P_{Park} describes the real power associated with useful energy in a three-phase power system. In the case of a balanced sinusoidal three-phase power system it is equal to the three-phase active power $P_{3\phi}$.
- The Park three-phase imaginary power is a power applicable only to a three-phase power system. It is not an extension of the concept of the single-phase imaginary/reactive power.
- The Park three-phase imaginary power is not an erroneous concept of the Budeanu reactive power.
- The Park three-phase imaginary power has a physical meaning explainable by field theory as it represents the variation in the ratio of energy exchange between the dielectric and magnetic energy of different phases.
- Compensation equipment can successfully be designed based on the Park voltage and current vectors as proved by Akagi and Nabae [1].

Application of the Park transform on unbalanced four-wire, three-phase power system quantities results in both the Park vectors *and* additional zero-sequence components. This problem is overcome with the transform of Ferrero, Giuliani and Willems[6] [13]. In the case of either balanced four-wire, three-phase systems or three-wire, three-phase systems, the transformed quantities of the FGW transform are the two-dimensional Park vectors of current and voltage. Only in the case of four-wire, three-phase quantities are the resultant voltage and current vectors three dimensional.

9.3.3 Four-Wire Power System Analysis

A *three-wire*, three-phase power system can be completely described by its Park voltage and current vectors as demonstrated in the previous section. When a *four-wire*, three-phase system is analyzed, the zero-sequence components represent an *extra* system and have to be considered separately. The transformation principle devised by Ferrero, Giuliani and Willems [13] enabled analysis of transformed quantities in a single new domain. A space-vector transformation was formulated that transforms voltage and current vectors of both three-wire and four-wire power systems into a single domain.

9.3.3.1 FGW Transformation Principles and Mathematical Definitions

A four-wire, three-phase power system is shown in Figure 9.4.

The FGW transformation transforms both three-wire and four-wire, three-phase power system quantities to three-dimensional *hypercomplex quantities*. The concept of

[6] This transform is abbreviated as the FGW transform in this book.

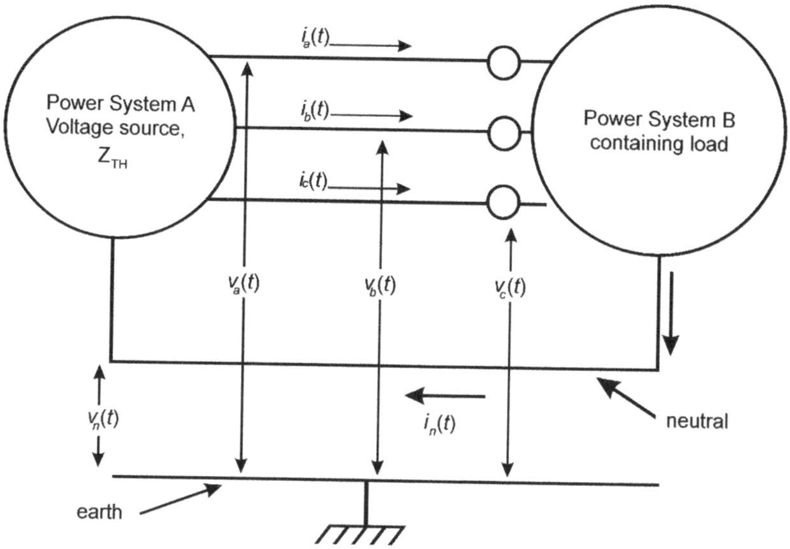

Figure 9.4 Three-phase, four-wire power system

a space vector defined in the d–q domain is extended by means of a space vector in a three-dimensional domain by a 4×4 transformation matrix that has the following characteristics:

- It is orthogonal.
- A diagonal impedance matrix results if the phase impedance matrix is symmetrical with equal off-diagonal elements.
- Three linearly independent voltage and current components in the FGW domain represent the four linearly dependent voltage and current components of a four-conductor, three-phase power system.
- The FGW transformation matrix transforms three-wire, three-phase quantities to Park quantities just like the Park transformation matrix.

The FGW transformation matrix is defined below:

$$\boldsymbol{T}_{FGW} = \sqrt{\frac{2}{3}} \begin{bmatrix} 1 & -\frac{1}{2} & -\frac{1}{2} & 0 \\ 0 & \frac{\sqrt{3}}{2} & -\frac{\sqrt{3}}{2} & 0 \\ \frac{1}{\sqrt{2}} & \frac{1}{\sqrt{2}} & \frac{1}{\sqrt{2}} & -\frac{3}{\sqrt{2}} \\ \frac{\sqrt{3}}{2\sqrt{2}} & \frac{\sqrt{3}}{2\sqrt{2}} & \frac{\sqrt{3}}{2\sqrt{2}} & -\frac{\sqrt{3}}{2\sqrt{2}} \end{bmatrix} \qquad |\boldsymbol{T}_{FGW}| = 1 \tag{9.45}$$

Transformation of the four-wire, three-phase quantities to the three-dimensional space vectors is defined as follows:

$$\begin{bmatrix} v_d(t) \\ v_q(t) \\ v_z(t) \\ v_0(t) \end{bmatrix} = \begin{bmatrix} 1 & -\frac{1}{2} & -\frac{1}{2} & 0 \\ 0 & \frac{\sqrt{3}}{2} & -\frac{\sqrt{3}}{2} & 0 \\ \frac{1}{2\sqrt{2}} & \frac{1}{2\sqrt{2}} & \frac{1}{2\sqrt{2}} & -\frac{3}{2\sqrt{2}} \\ \frac{3}{2\sqrt{2}} & \frac{3}{2\sqrt{2}} & \frac{3}{2\sqrt{2}} & -\frac{3}{2\sqrt{2}} \end{bmatrix} \begin{bmatrix} v_a(t) \\ v_b(t) \\ v_c(t) \\ v_n(t) \end{bmatrix} \tag{9.46}$$

$$\begin{bmatrix} i_d(t) \\ i_q(t) \\ i_z(t) \\ i_0(t) \end{bmatrix} = \begin{bmatrix} 1 & -\frac{1}{2} & -\frac{1}{2} & 0 \\ 0 & \frac{\sqrt{3}}{2} & -\frac{\sqrt{3}}{2} & 0 \\ \frac{1}{2\sqrt{2}} & \frac{1}{2\sqrt{2}} & \frac{1}{2\sqrt{2}} & -\frac{3}{2\sqrt{2}} \\ \frac{3}{2\sqrt{2}} & \frac{3}{2\sqrt{2}} & \frac{3}{2\sqrt{2}} & -\frac{3}{2\sqrt{2}} \end{bmatrix} \begin{bmatrix} i_a(t) \\ i_b(t) \\ i_c(t) \\ i_n(t) \end{bmatrix} \tag{9.47}$$

where $v_n(t)$ is the time-dependent voltage between the neutral conductor and earth plane and $i_n(t)$ the time-dependent current in the neutral conductor.

The compact notations for the space vectors of voltage and current are

$$\boldsymbol{v}(t)_{dqz0} = \begin{bmatrix} v_d(t) \\ v_q(t) \\ v_z(t) \\ v_0(t) \end{bmatrix} \tag{9.48}$$

$$\boldsymbol{i}(t)_{dqz0} = \begin{bmatrix} i_d(t) \\ i_q(t) \\ i_z(t) \\ i_0(t) \end{bmatrix} \tag{9.49}$$

where:

$v(t)_d$ is the voltage space vector component of the d-axis;
$v(t)_q$ is the voltage space vector component of the q-axis;
$v(t)_z$ is the voltage space vector component of the z-axis;
$v(t)_0$ is the voltage space-vector component of zero sequence, similar to the Fortesque zero-sequence component;
$i(t)_d$ is the voltage space current component of the d-axis;
$i(t)_q$ is the voltage space current component of the q-axis;

$i(t)_z$ is the voltage space current component of the z-axis;
$i(t)_0$ is the voltage space current component of zero sequence, similar to the Fortesque zero-sequence component.

The sum of the four line currents will always be zero [13], hence $i_0(t) = 0$. If the line voltages are measured with respect to a virtual star connection, it would also sum to zero, hence $v_0(t) = 0$. The space vectors of voltage and current then reduce to three-dimensional space vectors:

$$\boldsymbol{v}(t)_{dqz} = \begin{bmatrix} v_d(t) \\ v_q(t) \\ v_z(t) \end{bmatrix} \tag{9.50}$$

$$\boldsymbol{i}(t)_{dqz} = \begin{bmatrix} i_d(t) \\ i_q(t) \\ i_z(t) \end{bmatrix} \tag{9.51}$$

The four-wire, three-phase system is fully presentable in a three-dimensional space by the three linearly independently transformed voltages and currents defined in Equation (9.50) and Equation (9.51). The r.m.s. values of the space vectors relate to the r.m.s. values of phase vectors as follows:

$$\begin{aligned} V_{dqz} &= \sqrt{V_d^2 + V_q^2 + V_z^2} \\ &= \sqrt{V_a^2 + V_b^2 + V_c^2 + V_n^2} \\ &= V_{3\phi} \end{aligned} \tag{9.52}$$

$$\begin{aligned} I_{dqz} &= \sqrt{I_d^2 + I_q^2 + I_z^2} \\ &= \sqrt{I_a^2 + I_b^2 + I_c^2 + I_n^2} \\ &= I_{3\phi} \end{aligned} \tag{9.53}$$

where:

V_d is the r.m.s. value of space-vector voltage component $v_d(t)$;
V_q is the r.m.s. value of space-vector voltage component $v_q(t)$;
V_z is the r.m.s. value of space-vector voltage component $v_z(t)$;
I_q is the r.m.s. value of space-vector voltage component $v_q(t)$;
I_z is the r.m.s. value of space-vector voltage component $v_z(t)$;
I_d is the r.m.s. value of space-vector voltage component $v_d(t)$.

The space vectors can also be written in terms of hypercomplex quantities $\overline{\boldsymbol{v}}_x$, $\overline{\boldsymbol{v}}_y$, $\overline{\boldsymbol{v}}_z$. The voltage space vector can be written as

$$\overline{\boldsymbol{v}}(t)_{dqz} = v(t)_d \overline{\boldsymbol{v}}_x + v(t)_q \overline{\boldsymbol{v}}_y + v(t)_z \overline{\boldsymbol{v}}_z \tag{9.54}$$

The current space vector can be written as

$$\overline{\boldsymbol{i}}(t)_{dqz} = i(t)_d \overline{\boldsymbol{v}}_x + i(t)_q \overline{\boldsymbol{v}}_y + i(t)_z \overline{\boldsymbol{v}}_z \tag{9.55}$$

The latter equations will allow the use of hypercomplex algebra to formulate power definitions to analyze a four wire, three-phase power system under non-sinusoidal conditions.

9.3.3.2 Power Definitions of FGW Space-Vector Components

The time-dependent hypercomplex power $\overline{\boldsymbol{a}}(t)$ is based on the accepted complex power definition but mathematically written here in terms of hypercomplex algebra:

$$\begin{aligned}\overline{\boldsymbol{a}}(t) &= \overline{\boldsymbol{v}}(t)_{dqz} \cdot \overline{\boldsymbol{i}}(t)^*_{dqz} \\ &= a_s(t) + a_x(t)\overline{\boldsymbol{v}}_x + a_y(t)\overline{\boldsymbol{v}}_y + a_z(t)\overline{\boldsymbol{v}}_z\end{aligned} \tag{9.56}$$

The components of the voltage space vector $\overline{\boldsymbol{v}}(t)_{dqz}$ and current space vector $\overline{\boldsymbol{i}}(t)^*_{dqz}$ were defined in Section 9.3.3.1. The power components in the hypercomplex power $\overline{\boldsymbol{a}}(t)$ result from the application of hypercomplex algebraic rules:

$$a_s(t) = v_d(t)\, i_d(t) + v_q(t)\, i_q(t) + v_z(t)\, i_z(t) \tag{9.57}$$

$$a_d(t) = v_z(t)\, i_q(t) - v_q(t)\, i_z(t) \tag{9.58}$$

$$a_q(t) = v_d(t)\, i_z(t) - v_z(t)\, i_d(t) \tag{9.59}$$

$$a_z(t) = v_q(t)\, i_d(t) - v_d(t)\, i_q(t) \tag{9.60}$$

where:

$a_s(t)$ is the scalar component of the hypercomplex power $\overline{\boldsymbol{a}}(t)$;
$a_d(t)$ is the d-axis component of the hypercomplex power $\overline{\boldsymbol{a}}(t)$;
$a_q(t)$ is the q-axis component of the hypercomplex power $\overline{\boldsymbol{a}}(t)$;
$a_z(t)$ is the z-axis component of the hypercomplex power $\overline{\boldsymbol{a}}(t)$.

Power component $a_s(t)$ is a scalar equivalent to the time-dependent power transferred through a cross-section [13].

Compensation of the hypercomplex power component $\overline{\boldsymbol{a}}_i(t) = \overline{\boldsymbol{a}}(t) - a_s(t)$ will maximize time-dependent power transfer. In the case of a three-conductor power system the $a_z(t)$ component is the Park imaginary power component.

For a case study see web address

9.4 IEEE 1459: POWER DEFINITIONS FOR MODERN POWER SYSTEMS

The modern three-phase power system has to be considered as a self-contained entity when studying power phenomena. Formulations have to be carried out in this domain or in an equivalent transformed domain. Power definitions based on quantities formulated in a single-phase domain are only valid for three-phase systems under perfectly balanced three-phase conditions. Application of classical power theory in modern power systems containing distorted waveforms and asymmetry in supply or loading has created many pitfalls and unsatisfactory performance is widely demonstrated in the literature.

The importance of the inadequacies found in the classical power theory to describe the power phenomena of non-linear power system operation was highlighted in 1922 by Bucholz. The well-known Budeanu formulation followed in 1927. Fryze [16] reinterpreted Budeanu's definitions in 1932. Many other attempts followed thereafter when it became increasingly important for a power theory to describe power under non-sinusoidal conditions. The need to reach a conclusion on power definitions was again emphasized from the middle 1980s [5], [6], [7], [1]. For example, Czarnecki communicated this deficiency unambiguously in an IEEE Transactions paper entitled 'What is wrong with the Budeanu Concept of Reactive Power and Distortion Power and why it should be abandoned' [5]. Even today there is disagreement over several aspects and definitions in non-sinusoidal power theory. The need for universally accepted power definitions has come to the fore now, like never before.

The electrical engineer requires practical power definitions to analyze a power system. The IEEE published guidelines for the practical definition of electrical quantities and power definitions in non-sinusoidal power systems in 1996 [21]. These definitions were compiled by an IEEE Working Group on Non-sinusoidal Situations chaired by Professor Alexander Emanuel of Worcester Polytechnic.

These guidelines were later taken up in the IEEE Standard 1459-2000 [20]. It expanded on a number of principles and for example suggested the positive-sequence fundamental apparent power S_1^+; the positive-sequence fundamental reactive power Q_1^+; and the positive-sequence fundamental active power P_1^+ as the most important quantities when describing the conversion and utilization of three-phase electrical energy.

Digitized power system data is readily available in modern power systems and frequency analysis of data is made easy through developments in instrumentation and computing infrastructure. The adoption of IEEE 1459-2000 in the formulation of energy phenomena is straightforward and easy.

9.4.1 Voltage and Current Quantities under Non-sinusoidal Unbalanced Conditions

Customers expect utilities to generate and distribute perfectly sinusoidal voltage waveforms at the fundamental frequency. The power definitions that are proposed in [20] separate the ideal situation of only fundamental frequency components from the *polluting components* Only three-phase power systems with non-sinusoidal waveforms with unbalanced conditions (either because the supply voltage is asymmetrical or because the load elements are not balanced) are considered below. Single-phase, sinusoidal and balanced conditions are easily introduced with these definitions as a simplification of non-sinusoidal unbalanced three-phase power systems.

Assuming that the non-sinusoidal line–neutral voltages and line currents are defined as follows[7] (similarly for phases b and c):

$$v_a(t) = \sqrt{2}\sum_{h\neq 0}^{\infty} V_a(h)\sin(h\omega_1 t + \alpha_{a,h}) \tag{9.61}$$

$$i_a(t) = I_{a0} + \sqrt{2}\sum_{h\neq 0}^{\infty} I_a(h)\sin(h\omega t + \beta_{a,h}) \tag{9.62}$$

The d.c. components in the voltage, V_{a0}, V_{b0} and V_{c0}, should always be zero. The d.c. values in the line currents I_{a0}, I_{b0} and I_{c0} could be non-zero depending on the nature of the load.

The *r.m.s. line–neutral voltage* V_a and line current I_a (similarly for phases b and c) are related to the harmonic components by

$$V_a^2 = V_a^2(1) + \sum_{h\neq 2}^{\infty} V_a^2(h) = V_a^2(1) + V_{aH}^2 \quad \left(V_{aH}^2 = \sum_{h\neq 2}^{\infty} V_a^2(h)\right) \tag{9.63}$$

$$I_a^2 = I_a^2(1) + \sum_{h\neq 2}^{\infty} I_a^2(h) = I_a^2(1) + I_{aH}^2 \quad \left(I_{aH}^2 = \sum_{h\neq 2}^{\infty} I_a^2(h)\right) \tag{9.64}$$

The above formulations of voltage and current distinguish clearly between the fundamental frequency and the non-fundamental frequency (harmonic frequencies grouped) components. The *effective three-phase voltage* and *current* are written as

$$V_e^2 = V_{e1}^2 + V_{eH}^2 \tag{9.65}$$

$$I_e^2 = I_{e1}^2 + I_{eH}^2 \tag{9.66}$$

Because there is no neutral current in a three-wire power system, the expression for *effective three-phase voltage* and *current* is calculated as

$$V_e = \sqrt{\frac{V_a^2 + V_b^2 + V_c^2}{3}} \tag{9.67}$$

$$I_e = \sqrt{\frac{I_a^2 + I_b^2 + I_c^2}{3}} \tag{9.68}$$

If an artificial neutral point in a three-conductor, three-phase system is not used to find the line–neutral voltage values, the *effective three-phase voltage* can be calculated from the r.m.s. phase–phase voltage values as

$$V_e = \sqrt{\frac{V_{ab}^2 + V_{bc}^2 + V_{ca}^2}{9}} \tag{9.69}$$

[7] Formulas listed are a summary of Reference [20] and each formulation is not specifically referenced.

The fundamental frequency and non-fundamental frequency components of the *effective voltage* and *current* in a three-wire, three-phase power system are defined as

$$V_{e1} = \sqrt{\frac{V_{ab}^2(1) + V_{bc}^2(1) + V_{ca}^2(1)}{9}} \tag{9.70}$$

$$V_{eH} = \sqrt{\frac{V_{abH}^2 + V_{bcH}^2 + V_{caH}^2}{9}} \tag{9.71}$$

$$I_{e1} = \sqrt{\frac{I_a^2(1) + I_b^2(1) + I_c^2(1)}{3}} \tag{9.72}$$

$$I_{eH} = \sqrt{\frac{I_{aH}^2 + I_{bH}^2 + I_{cH}^2}{3}} \tag{9.73}$$

The *effective three-phase non-fundamental voltage* and *current* values are defined as

$$V_{eH} = \sqrt{\frac{\sum_{h\neq 1}^{\infty}\left(V_a^2(h) + V_b^2(h) + V_c^2(h)\right)}{3}} \tag{9.74}$$

$$I_{eH} = \sqrt{\frac{\sum_{h\neq 1}^{\infty}\left(I_a^2(h) + I_b^2(h) + I_c^2(h)\right)}{3}} \tag{9.75}$$

Unbalanced conditions in a four-wire, three-phase power system require the definitions of the *effective voltage* and *current* to be

$$V_e = \sqrt{\frac{1}{18}\left[3\left(V_a^2 + V_b^2 + V_c^2\right) + V_{ab}^2 + V_{bc}^2 + V_{ca}^2\right]} \tag{9.76}$$

$$I_e = \sqrt{\frac{I_a^2 + I_b^2 + I_c^2 + I_n^2}{3}} \tag{9.77}$$

where

$$V_{e1} = \sqrt{\frac{1}{18}\left[3\left(V_a^2(1) + V_b^2(1) + V_c^2(1)\right) + V_{ab}^2(1) + V_{bc}^2(1) + V_{ca}^2(1)\right]} \tag{9.78}$$

$$V_{eH} = \sqrt{\frac{1}{18}\left[3\left(V_{aH}^2 + V_{bH}^2 + V_{cH}^2\right) + V_{abH}^2 + V_{bcH}^2 + V_{caH}^2\right]} \tag{9.79}$$

$$I_{e1} = \sqrt{\frac{I_a^2(1) + I_b^2(1) + I_c^2(1) + I_n^2(1)}{3}} \tag{9.80}$$

$$I_{eH} = \sqrt{\frac{I_{aH}^2 + I_{bH}^2 + I_{cH}^2 + I_{nH}^2}{3}} \tag{9.81}$$

9.4.2 Apparent Power Definitions

Apparent power definitions, widely used, can be recognized to be either the arithmetic or vector apparent power approach. IEEE 1459-2000 [20] formulates and then demonstrates that the concept of *effective* (or *system*) *apparent power* is the preferred approach. Power definitions used in industrial metering equipment do not in general recognize the importance of the latter definition. Most metering equipment even wrongfully uses the Budeanu definition of reactive power and some report the existence of a distortion power that cannot be physically explained. The different apparent power approaches are summarized below.

9.4.3 Arithmetic Apparent Power

The Budeanu power definition for single-phase power systems is applied on a per phase basis:

$$\begin{aligned} S_a &= \sqrt{P_a^2 + Q_{Ba}^2 + D_{Ba}^2} \\ S_b &= \sqrt{P_b^2 + Q_{Bb}^2 + D_{Bb}^2} \\ S_c &= \sqrt{P_c^2 + Q_{Bc}^2 + D_{Bc}^2} \end{aligned} \tag{9.82}$$

where S_a is the apparent power in phase a (similarly for phases b and c); Q_{Ba} is Budeanu's reactive power, i.e.

$$Q_{Ba} = \sum_{h=1}^{\infty} V_a(h) I_a(h) \sin\left(\alpha_{a,h} - \beta_{a,h}\right) = \sum_{h=1}^{\infty} Q_a(h)$$

(similarly for phases b and c); and D_{Ba} is Budeanu's 'distortion' power

$$D_{Ba} \equiv \sqrt{S_a^2 - P_a^2 - Q_{Ba}^2}$$

(similarly for phases b and c).

The three-phase arithmetic apparent power (S_A) follows:

$$S_A = S_a + S_b + S_c \tag{9.83}$$

9.4.4 The Vector Apparent Power

The active, Budeanu's reactive and distortion power over all three phases can be calculated:[8]

$$\begin{aligned} P_{\Sigma,3\phi} &= P_a + P_b + P_c \\ Q_{B,3\phi} &= Q_{Ba} + Q_{Bb} + Q_{Bc} \\ D_{B,3\phi} &= D_{Ba} + D_{Bb} + D_{Bc} \end{aligned} \tag{9.84}$$

[8] The active power per phase ($P_a P_b P_c$) includes the harmonic components. The three-phase joint active power in this book is listed as $P_{\sum,3\phi}$. It is similar to the P used in IEEE 1459-2000 for three-phase definitions (the same symbol is used for single phase and three phase).

where $P_{\Sigma,3\phi}$ is the *three-phase total* (or *joint*) *active power*, including the fundamental frequency component.

The three-phase vector apparent power (S_V) follows:

$$S_V = \sqrt{P_{\Sigma,3\phi}^2 + Q_{B,3\phi}^2 + D_{B,3\phi}^2} \tag{9.85}$$

9.4.5 The Effective Apparent Power

The *three-phase* or *system effective apparent power* S_e can be written in terms of the contribution of the fundamental and non-fundamental frequency voltage and current components:

$$S_e^2 = (V_e I_e)^2 = (V_{e1} I_{e1})^2 + (V_{e1} I_{eH})^2 + (V_{eH} I_{e1})^2 + (V_{eH} I_{eH})^2 \tag{9.86}$$

The *three-phase effective apparent power* contains a *fundamental frequency apparent power* S_{e1} and a *non-fundamental frequency apparent power* S_{eN}:

$$S_e^2 = S_{e1}^2 + S_{eN}^2 \tag{9.87}$$

The *non-fundamental frequency apparent power* is defined so as to consist of three distorting components:[9]

$$\begin{aligned} S_{eN}^2 &= (V_{e1} I_{eH})^2 + (V_{eH} I_{e1})^2 + (V_{eH} I_{eH})^2 \\ &= D_{eI}^2 + D_{eV}^2 + D_{eH}^2 \end{aligned} \tag{9.88}$$

The three components above are termed:

- $V_{e1} I_{eH}$: The *current distortion power*, D_{eI}
- $V_{eH} I_{e1}$: The *voltage distortion power*, D_{eV}
- $V_{eH} I_{eH}$: The *harmonic distortion power*, D_{eH}

The *effective harmonic apparent power* S_{eH} relates to the *harmonic distortion power* D_{eH} and the *effective* (or *joint*) *harmonic active power* $P_{H,3\phi}$:

$$S_{eH}^2 = P_{H,3\phi}^2 + D_{eH}^2 \tag{9.89}$$

The level of distortion in a three-phase power system is respectively defined by the *voltage total harmonic distortion* factor THD_{eV} and the *current total harmonic distortion* factor THD_{eI}:

$$THD_{eV} = \frac{V_{eH}}{V_{e,1}} \tag{9.90}$$

[9] IEEE 1459-2000 used the symbol '*e*' to indicate that these distortion powers are based on 'effective' three-phase values.

$$THD_{eI} = \frac{I_{e,H}}{I_{e,1}} \tag{9.91}$$

The equivalent total harmonic distortion factors can be used to find S_{eN} (non-fundamental frequency apparent power), D_{eI}, D_{eV}, D_{eH}:

$$S_{eN} = S_{e1}\sqrt{THD_{eV} + THD_{eI} + (THD_{eV}THD_{eI})^2} \tag{9.92}$$

$$D_{eI} = S_{e1}THD_{eI} \tag{9.93}$$

$$D_{eV} = S_{e1}THD_{eV} \tag{9.94}$$

$$D_{eH} = S_{e1}THD_{eI}THD_{eV} \tag{9.95}$$

9.4.6 Harmonic Pollution and Unbalance

The normalized *non-fundamental apparent power* S_{eN}/S_{e1} relates to the *total harmonic distortion factors*:

$$\left(\frac{S_{eN}}{S_{e1}}\right)^2 = (THD_{eI})^2 + (THD_{eV})^2 + (THD_{eI}THD_{eV})^2 \tag{9.96}$$

Unbalanced loading can cause the three-phase fundamental frequency apparent power to increase without an increase in the transfer of fundamental frequency active power and is isolated by IEEE 1459-2000 as follows:

$$S_{u1} = \sqrt{S_{e1}^2 - [S_1(1)]^2} \tag{9.97}$$

$$S_1(1) = 3V_1(1)I_1(1) \tag{9.98}$$

The component $S_1(1)$ is the *positive-sequence fundamental apparent power* calculated from the r.m.s. values of the *fundamental frequency positive-sequence voltage* $V_1(1)$ and *fundamental frequency positive-sequence current* $I_1(1)$. The term S_{u1} is the unbalanced contribution termed the *unbalanced fundamental frequency apparent power*:

- A quantitative measure of the level of harmonic pollution is contained in Equation (9.96). A zero value indicates that no harmonic pollution is present. The ratio S_{eN}/S_{e1} progressively relates to the level of harmonic pollution.
- Similarly, the ratio S_{u1}/S_{e1} furnishes a progressive measure of the level of unbalance (it includes both the effect of voltage asymmetry and unbalance in loading).

9.4.7 Fundamental Frequency Active and Reactive Power

The fundamental frequency *apparent power*, *active power* and the *fundamental reactive power* for a three-phase non-sinusoidal unbalanced power system should be based on positive-sequence quantities [20]:

$$\begin{aligned} S_1(1)^2 &= P_1(1)^2 + Q_1(1)^2 \\ P_1(1) &= 3V_1(1)I_1(1)\cos[\theta_1(1)] \\ Q_1(1) &= 3V_1(1)I_1(1)\sin[\theta_1(1)] \end{aligned} \tag{9.99}$$

where $S_1(1)$ is the fundamental frequency *apparent power* in the positive-sequence quantities; $P_1(1)$ is the fundamental frequency *active power* in the positive-sequence quantities; and $Q_1(1)$ is the fundamental frequency *reactive power* in the positive-sequence quantities

9.4.8 Power Factor Definitions

The *power factor*[10] to be used is defined as

$$PF_e = \frac{P_{\Sigma,3\phi}}{S_e} = \frac{P_{3\phi}(1) + P_{H,3\phi}}{S_e} \tag{9.100}$$

The utilization of energy in fundamental frequency components is defined by the positive-sequence components as

$$PF_1 = \frac{P_1(1)}{S_1(1)} \tag{9.101}$$

where PF_1 is the power factor formulation based on fundamental frequency positive-sequence components.

The *three-phase total* (or *joint*) *active power*[11] $P_{\Sigma,3\phi}$ contains both the *three-phase fundamental active power* $P_{3\phi}(1)$ and the *three-phase total harmonic active power* $P_{3\phi,H}$:

$$P_{\Sigma,3\phi} = \sum_{h=0}^{\infty} P_{3\phi}(h) = P_{3\phi}(1) + \sum_{h\neq 1}^{\infty} P_{3\phi}(h) = P_{3\phi}(1) + P_{3\phi,H} \tag{9.102}$$

The arithmetic or vector apparent power should not be used in the power factor formulation when the three-phase power system is non-sinusoidal and unbalanced:

$$PF_A = \frac{P_{\Sigma,3\phi}}{S_A} \quad \text{(arithmetic power factor)} \tag{9.103}$$

$$PF_V = \frac{P_{\Sigma,3\phi}}{S_V} \quad \text{(vector power factor)} \tag{9.104}$$

[10] The subscript '*e*' is used to indicate that this power factor formulation makes use of effective apparent power.
[11] The three-phase *joint active power* is exactly the same concept as the three-phase *joint real power*.

When the power system is balanced and sinusoidal, $PF_A = PF_V = PF_e$. IEEE 1459-2000 and Section 9.3.4.9 demonstrate that if the power system is unbalanced and non-sinusoidal, $PF_e < PF_A < PF_V$, and propose the *effective* power factor formulation to be used.

The above formulations are easy to implement when waveform data is available. A case study is to be found at www.wiley.com/go/powerquality

9.5 LOCALIZATION OF SOURCES OF WAVEFORM DISTORTION IN A MODERN POWER SYSTEM

Various contributions to measurement techniques that are aimed at localizing a specific source of distortion in a power system are found in the literature and mostly based on the erroneous assumption that the direction of harmonic active power can be used to determine the localization of a distortion source through single-point measurements such as [24]. However, Swart *et al.* [25] showed that it is not possible to localize distortion sources in a network in which more than one distortion source is present. This finding is confirmed in [23].

Customers of electrical utilities that experience high ambient levels of voltage harmonics at their point of common coupling (PCC) invariably have to contend with a range of detrimental effects in their equipment that range from lightning equipment malfunctioning to overheating and the necessity for derating rotating equipment and transformers. The utilities themselves are seldom responsible for generating the distortion but their networks are responsible for transmitting it from other guilty consumers to them. These innocent customers look on the utility for the low power quality, and not on the guilty parties, as the primary source of their problems. It is therefore in the interest of the utility to dissuade active and potential harmonic-producing customers through proper management and tariff penalties, but knowledge is required as to the localization and relative contribution of a specific customer.

Network analysis in the frequency domain shows that harmonic-producing sources can be modeled reasonably accurately by current sources at the relevant discrete frequencies and that these currents produce voltages across the system immittances with phase angles corresponding to the angles of these immittances. In frequency-domain analysis, networks and their elements allow harmonic superposition to be carried out and individual frequencies may be modeled separately. A logical consequence of the interaction of these voltages with the given immittances is that the power angles at those harmonic frequencies will also correspond to the immittance angles. It follows, therefore, that the harmonic current sources will export real power only to the network immittances with real components and that these harmonic real power component magnitudes will be commensurate with the aggregate magnitude of the system real immittances.

When harmonic power measurements are carried out at the PCC of a given consumer, an outflow of harmonic real power furnishes proof that the harmonics originate behind the PCC and that they are caused by that consumer. Unfortunately, in practical systems, harmonic power can be sunk or generated.

Phase-controlled, line-commutated, a.c./d.c. multi-phase converters exhibit the ability to export and import harmonic real powers at the individual harmonic frequencies; the direction of harmonic active power is not absolute. In addition, these magnitudes and angles change with relative changes in the firing angles of the converters, rendering a set of observed data relating to one operating condition of the network unsuitable when one of these conditions changes.

All modern electric power networks contain harmonic-producing sources, which are distributed all over the interconnected electric power system. Therefore:

- It is not possible to localize a specific distortion-producing source in a practical three-phase power system through single point measurements [23].
- It is only possible to localize distortion-producing sources in a practical power system by measuring harmonic active power flow in all the terminals of all the consumers simultaneously and synchronously [25].

For a case study see web address

BIBLIOGRAPHY

[1] Akagi Y., Nabae. A., Instantaneous reactive power compensator comprising switching devices without energy storage components. *IEEE Transactions on Industrial Applications*, vol. IA-20, pp. 625–630, 1984.

[2] Andria G., Savino M., Windows and interpolation algorithms to improve electrical measurement accuracy. *IEEE Transactions on Instrumentation and Measurement*, vol. 38, no. 4, pp. 856–863, 1989.

[3] Budeanu G., *Reactive and fictitious powers*, Romanian National Institute (Bucharest), Publication No. 2, 1927; *The different options and conceptions regarding active power in non-sinusoidal systems*, Romanian National Institute, Publication No. 4, 1927.

[4] Cristaldi L., Ferrero A., A digital method for the identification of the source of distortion in electric power systems. *IEEE Transactions on Instrumentation and Measurement*, vol. 44, no. 1, pp. 14–18, 1994.

[5] Czarnecki L. S., What is wrong with the Budeanu Concept of Reactive Power and Distortion Power and why it should be abandoned. *IEEE Transactions on Instrumentation and Measurement*, vol. IM 36, no. 3, pp. 834–837, 1987.

[6] Czarnecki L. S., Powers in nonsinusoidal networks: their interpretation, analysis, and measurement. *IEEE Transactions on Instrumentation and Measurement*, vol. 39, no. 2, pp. 340–345, 1990.

[7] Czarnecki L. S., Misinterpretations of power properties in electric circuits. *IEEE Transactions on Power Delivery*, vol. 9, no. 4, pp. 1760–1769, 1994.

[8] Ferrero A., Ottoboni R., A new approach to the Fourier analysis of periodic signals for the minimization of the phase errors. *IEEE Transactions on Instrumentation and Measurement*, vol. 40, no. 3, pp. 694–698, 1991.

[9] Ferrero A., Ottoboni R., A low-cost frequency multiplier for synchronous sampling of periodic signals. *IEEE Transactions on Instrumentation and Measurement*, vol. 41, no. 2, pp. 203–207, 1992.

[10] Ferrero A., Ottoboni R., High-accuracy Fourier analysis based on synchronous sampling techniques. *IEEE Transactions on Instrumentation and Measurement*, vol. 41, no. 6, pp. 780–785, 1992.

[11] Ferrero A., Superti-Furga G., A new approach to the definition of power components in three-phase systems under nonsinusoidal conditions. *IEEE Transactions on Instrumentation and Measurement*, vol. 40, no. 3, pp. 568–577, 1991.

[12] Ferrero A., Morando A. P., Ottoboni R., Superti-Furga G., On the meaning of the Park power components in three-phase systems under non-sinusoidal conditions. *European Transactions on Electrical Power*, vol. 3, no. 1, pp. 33–43, 1993.

[13] Ferrero A., Giuliani G., Willems L., A new space-vector transformation for four-conductor systems. *European Transactions on Electrical Power*, vol. 10, no. 3, pp. 139–144, 2000.

[14] Ferrero A., Leva S., Morando A. P., An approach to the non-active power concept in terms of the Poynting-Park vector. *European Transactions on Electrical Power*, vol. 11, no. 5, pp. 291–299, 2001.

[15] Ferrero A., Lazzaroni M., Salicone S., A calibration procedure for a digital instrument for electric power quality measurement. *IEEE Transactions on Instrumentation and Measurement*, vol. 51, no. 4, pp. 716–722, 2002.

[16] Fryze S., Wirk-, Blind-, und Scheinleistung in Elektrisch Stromkreisen mit nichtsinusoidal-formingen Verlauf von Strom und Spannung. *ETZ*, vol. 53, no. 25, pp. 596–599, 625–627, 700–702, 1932.

[17] Glover J. D., Sarma M. S., *Power System Analysis*, Third Edition, Brooks/Cole, Pacific Grove, CA, 2002.

[18] Grandke T., Interpolation algorithms for discrete Fourier transforms of weighted signals. *IEEE Transactions on Instrumentation and Measurement*, vol. IM-32, no. 2, pp. 350–355, 1983.

[19] Hidalgo M. H., Fernandez J. G., Rivera R. R., Larrondo H. A., A simple adjustable window algorithm to improve FFT measurements. *IEEE Transactions on Instrumentation and Measurement*, vol. 51, no. 1, pp. 31–36, 2002.

[20] IEEE Standard 1459-2000, *Standard Definitions for the Measurement of Electric Power Quantities under Sinusoidal, Nonsinusoidal, Balanced or Unbalanced Conditions*, 2000.

[21] IEEE Working Group on Non-sinusoidal Situations, Effects on meter performance and definitions of power; practical definitions for power in systems with nonsinusoidal waveforms and unbalanced loads: a discussion. *IEEE Transactions on Power Delivery*, vol. 11, no. 1, pp. 79–101, 1996.

[22] Nabae A., Tanaka T., A new definition of instantaneous active-reactive current and power based on instantaneous space vectors on polar coordinates in three-phase circuits. *IEEE Transactions on Power Delivery*, vol. 11, no. 3, pp. 1238–1243, 1996.

[23] Rens A. P. J., Swart P. H., On techniques for the localisation of multiple distortion sources in three-phase networks: time domain verification. *European Transactions on Electrical Power*, vol. 11, no. 5, pp. 317–322, 2001.

[24] Srinivasan K., On separating customer and supply side harmonic contributions. *IEEE Transactions on Power Delivery*, vol. 11, no. 2, pp. 1003–1012, 1996.

[25] Swart P. H., Case M. J., van Wyk J. D., On techniques for localization of sources producing distortion in three phase networks. *European Transactions on Electrical Power*, vol. 6, no. 6, pp. 391–396, 1996.

[26] Webster J. G. (ed.), *Wiley Encyclopaedia of Electrical and Electronics Engineering*, Supplement 1, Harmonics and Power Phenomena, by L. S. Czarnecki, pp. 195–217, 2000.

10

Earthing

Franco Bua, Francesco Buratti and Antoni Klajn

This chapter deals with problems related to the basic properties and designing of earthing arrangements, which are also called in the literature earthing systems or grounding arrangements. A good earthing arrangement is required for the correct operation of electrical networks and is important for the quality of electricity supplied to consumers. One can distinguish the following main functions of an earthing system:

- safety of human and animal life by limiting touch and step voltages to safe values (protective earthing, earthing of work);
- correct operation of the electricity supply network and to ensure good power quality (power system earthing);
- achievement of a given electromagnetic compatibility (EMC) level, i.e. limitation of electromagnetic disturbances;
- protection of buildings and installations against lightning.

All these functions are provided by a single earthing arrangement that has to be designed to fulfill all the requirements. Some elements of an earthing system may be provided to fulfill a specific purpose, but are nevertheless part of one single grounding arrangement. Standards require all earthing measures within an installation to be bonded together, forming one arrangement.

10.1 TYPICAL EARTHING SYSTEM

An earthing system is the total of all means and measures by which electrically conductive parts (point of electric circuit, exposed conductive parts or extraneous conductive parts) are connected to earth.

Handbook of Power Quality Edited by Angelo Baggini

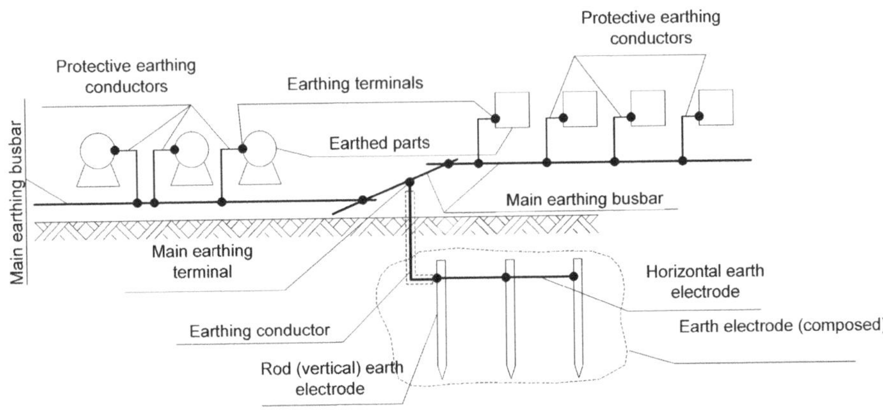

Figure 10.1 Typical elements of an earthing arrangement

Typical elements of the earthing arrangement are shown in Figure 10.1. A detailed explanation of these elements is provided in Annex 1 on the website and in the following introductory sections.

10.2 ELECTRIC RESISTIVITY OF SOIL

The electric resistivity of the soil ρ (soil resistivity) is the resistance of one cubic metre of earth, between two adjacent cubic faces (Figure 10.2). It is the main parameter which determines the electrical properties of the ground. In spite of this relatively simple definition of ρ, the determination of its value is often a complicated task for two main reasons:

- the soil does not have a homogeneous structure, but is formed of layers of different materials;
- the specific resistance of a given type of soil varies widely (Table 10.1) and is very dependent on water content and temperature.

The soil resistivity ρ is the fundamental parameter necessary in calculations or simulations of the resistance to earth. Thus, the large variation in the value of ρ is a problem. In many practical situations, a homogeneous ground structure will be assumed with an average value of ρ, which must be estimated on the basis of soil analysis or by measurement. One important point is that the current distribution in the soil layers during measurement should simulate the prospective distribution of the final installation. Consequently, measurements must always be interpreted carefully. Where no information is available about the value of ρ it is usually assumed $\rho = 100\,\Omega\,\text{m}$. However, as Table 10.1 indicates, the real value can be very different, so acceptance testing of the final installation, together with an assessment of likely variations due to weather conditions and over the lifetime, must be undertaken.

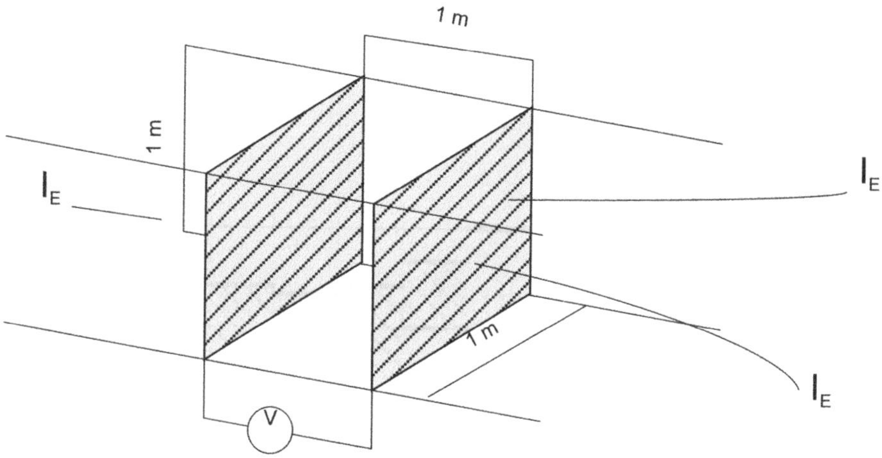

Figure 10.2 Illustration of the physical sense of electric resistivity of soil ρ

Table 10.1 Ground resistivity ρ for various kinds of soil and concrete [2],[7],[18],[21]

Type of ground	Ground resistivity, ρ (Ω m)	
	Range of values	Average value
Boggy ground	2–50	30
Adobe clay	2–200	40
Silt and sand–clay ground, humus	20–260	100
Sand and sandy ground	50–3000	200 (moist)
Peat	> 200	200
Gravel (moist)	50–3000	1000 (moist)
Stone and rock ground	100–8000	2000
Concrete: 1 part cement + 3 parts sand	50–300	150
1 part cement + 5 parts gravel		400
1 part cement + 7 parts gravel		500
Sea water	0.1–1	0.5
Ice	10 000–100 000	50 000

The soil resistivity depends significantly on the *ground temperature* and *water content (humidity)*. The temperature coefficient α of the soil resistance, inversely as in metallic conductors, has a negative value, which changes in the range $\alpha = (-0.023; -0.037)$ [18]. Thus the temperature-dependent electric resistivity ρ_ϑ of soil, i.e. resistivity at a given temperature ϑ (in °C), can be calculated:

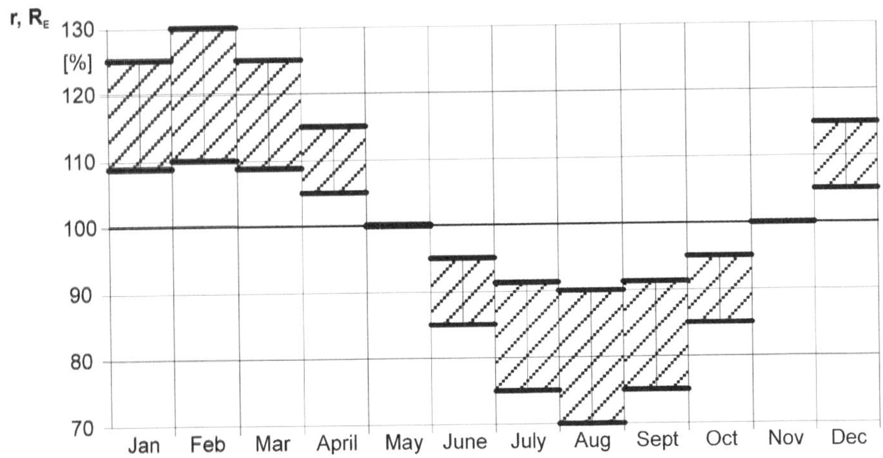

Figure 10.3 Changes of the soil resistivity ρ and resistance to earth R_E with season in Europe during the year, in relation to mean values in May and November (based on measurements performed in Switzerland [18]) (Reproduced from EMV nach VDE 0100, VDE Verlag)

$$\rho_\vartheta = \rho_0(1 + \alpha\vartheta) \tag{10.1}$$

Thus, in the winter the soil resistivity grows, and during the summer it is lower than the mean values (Figure 10.3). Frozen ground has a very high value of resistivity (see Table 10.1, ice), and because of this earth electrodes should be mounted beneath the freezing line. The depth of this line varies, depending on geographic position and local conditions (e.g. mountains). But apart from temperature, the next factor of significant importance is the water content of the soil. The moisture content in the soil can change over a wide range, depending on geographic location and weather conditions, from a low percentage for desert regions up to about 80 % for swampy regions. It also changes of course with the seasons, and has an influence on characteristics as shown in Figure 10.3.

In practice, in order to minimize the influence of the ground's humidity on the resistance to earth, the earth electrodes are mounted if possible deep in the ground. The preferable construction in this case is the vertical rod electrode, which commonly can reach parts of the ground at depths from 10 m up to 30 m. In such solutions, the dependence of the resistance from the soil temperature and humidity is reduced to minimal values. An illustration of changes in the soil resistivity for clay ground is shown in Figure 10.4. One can see here that, for humidity values higher than 30 %, changes in ρ are very slow and not significant. However, when the ground is dry, i.e. values of h lower than 20 %, the resistivity grows very rapidly. It must be remembered that the effect of drying is similar to that of freezing – the resistivity increases significantly.

The above-mentioned reasons of soil resistivity changes (layer structure of the ground, temperature, water content) mean that calculations of resistance to earth and the planning of electrodes can be performed with a limited level of accuracy.

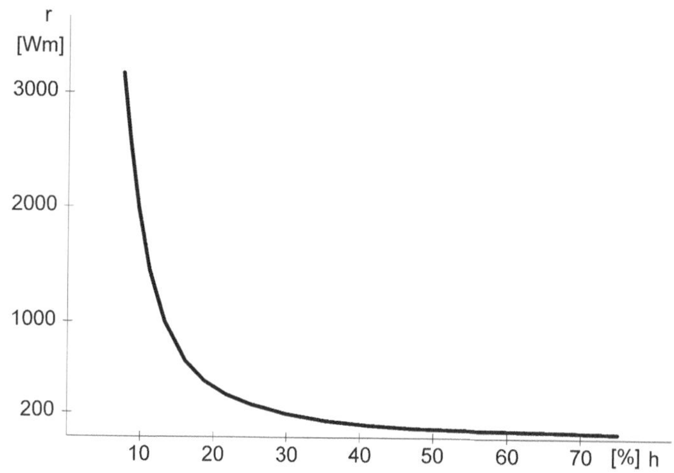

Figure 10.4 An example of soil resistivity ρ of clay versus soil humidity h

10.3 ELECTRICAL PROPERTIES OF THE EARTH ELECTRODES

The electrical properties of earth electrodes are characterized by the following parameters:

- resistance to earth or impedance to earth;
- voltage-to-earth distribution during an earth fault.

These parameters are explained on a simplified hemispherical earth-electrode model, as illustrated in Figure 10.5.

10.3.1 Resistance and Impedance to Earth

The resistance to earth is determined by the relation between the voltage (potential) to earth V_E and the earth current value I_E:

$$R_E = \frac{V_E}{I_E} \tag{10.2}$$

The earthing resistance has two components (Figure 10.6):

- resistance of the metallic parts of the earth electrode and of the earthing conductor R_L;
- dissipation resistance R_D, which is the soil resistance between the range of the earth electrode and the reference earth:

$$R_E = R_L + R_D \tag{10.3}$$

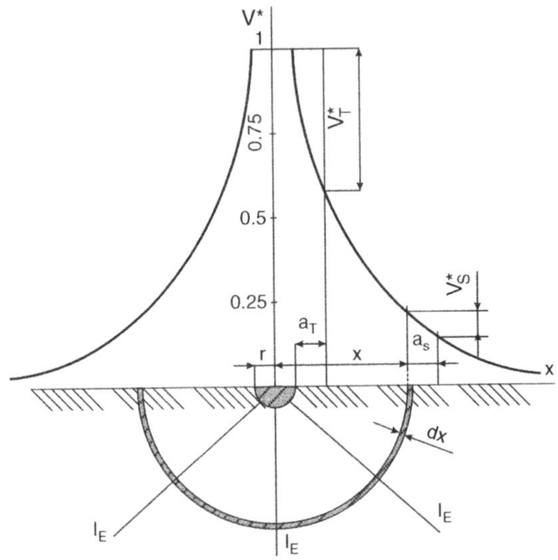

Figure 10.5 Illustration of the basic hemispherical earth-electrode model, showing the parameters required to calculate the resistance to earth and voltage (potential) to earth distribution on the surface of the ground (with $\rho = \text{constant}$): r, electrode radius; x, distance from the electrode center; a_T, a_S, touch and step distances, respectively; $V^* = V/V_E$, relative value of the voltage (potential) distribution; V_E, voltage (potential) to earth; V^*_T, V^*_S, prospective touch and step voltages, respectively

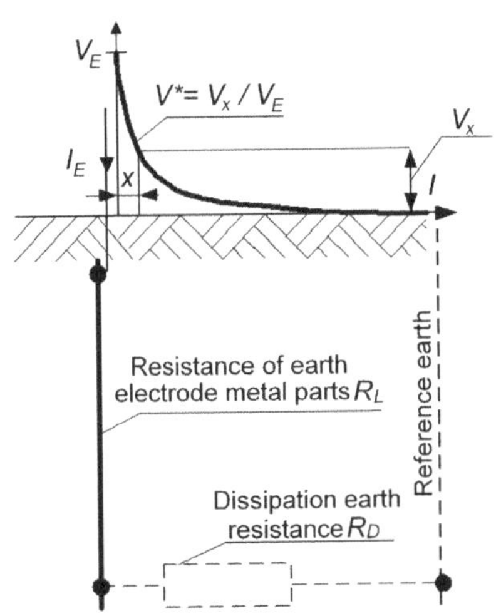

Figure 10.6 Illustration of components of the resistance to earth (R_L, resistance of the earth-electrode metallic parts; R_D, dissipation earth resistance) and relative value of the voltage-to-earth distribution V^*

The resistance R_L is usually much smaller than the dissipation resistance R_D. Thus, normally the resistance to earth is estimated to be equal to the dissipation resistance:

$$R_E \approx R_D \tag{10.4}$$

In the literature the qualification *resistance to earth* usually refers to the dissipation resistance.

In a.c. circuits one must consider essentially the impedance to earth Z_E, which is the impedance between the earthing arrangement and the reference earth at a given operating frequency:

$$Z_E = \frac{V_E}{I_E} \tag{10.5}$$

The reactance to earth is the actual reactance of the metallic parts of the earth electrode and the earthing conductor. At relatively low frequencies, such as the network frequency, the reactance is usually negligible in comparison to earthing resistance. However, it can reach significantly higher values at higher frequencies, such as lightning transients. Impedance to earth at lightning currents is described at the end of this section. For the network frequency the impedance to earth Z_E is assumed to be equal to the resistance to earth R_E and, according to the formula (10.4), equal to the dissipation resistance R_D:

$$Z_E \approx R_D \approx R_E \tag{10.6}$$

The resistance to earth R_E of an earth electrode depends on the soil resistivity ρ as well as on the electrode's geometry. In order to achieve low values of R_E the current density flowing from the electrode metal to the earth should be low, i.e. the volume of earth through which the current flows is as large as possible. Once the current flows from metal to earth it spreads out, reducing the current density. If the electrode is physically small, e.g. a point, this effect is large, but is very much reduced for a plate where spreading is only effective at the edges. This means that rod, pipe or wire electrodes have a much lower dissipation resistance than, for example, a plate electrode with the same surface area.

The calculation of resistance to earth is usually performed under the assumption that the ground is boundless and of uniform structure with a given value of soil resistivity ρ. It is possible to determine exact equations for resistance to earth but, in practice, their usefulness is very limited, especially in the case of complex and meshed earth electrodes where the mathematical relations become so complicated as to be little use. For that reason, in the case of extended and meshed earth electrodes approximations of resistance to earth are used. Another reason to justify the approximations here is that even a small inaccuracy in the soil resistivity ρ has a significant influence on the calculated value of the resistance to earth. Consequently, exact theoretical equations of resistance to earth are usually used in the design phase of the earth electrodes in order to estimate the prospective resistance to earth. However, these values should later be proved by measurement. For that reason, in the literature equations are given concerning resistance to earth and potential distribution at different levels of approximation. Furthermore, the equations of resistance to earth illustrate well the relationship between the earth voltage, earth potential distribution and earth current, especially for simple earth-electrode structures.

A basic model of the earth-electrode configuration, used for illustrating the fundamental electrical properties, is a hemisphere embedded in the ground surface (Figure 10.5). The earth current flowing to such an electrode is assumed to flow radially to the earth. The surface of the hemisphere, as well as all hemisphere cross-sections dx of the ground, are assumed to be equipotential, and the current lines are therefore perpendicular to these surfaces. Under these conditions the resistance of the hemisphere ground sheath of thickness dx and the radius x is expressed as follows (with ρ assumed constant):

$$dR_E = \frac{\rho}{2\pi \cdot x^2} dx \tag{10.7}$$

The resistance of the hemisphere earth electrode is

$$R_E = \frac{\rho}{2\pi} \int_r^{\infty} \frac{dx}{x^2} = \frac{\rho}{2\pi r} \tag{10.8}$$

The earth resistance depends significantly on the depth to which the electrode is embedded in the ground. This is because the moisture content is higher and more stable for deeper ground layers than for shallow ones. Layers near the surface are influenced more by seasonal and short-term weather variations and are subject to freezing. This problem is illustrated in Figure 10.7, for a rod earth electrode, where one can see the considerable reduction of earthing resistance as the depth increases. However, it is not always possible to place electrodes at the preferred depth for geological reasons; for example, where there are rocks

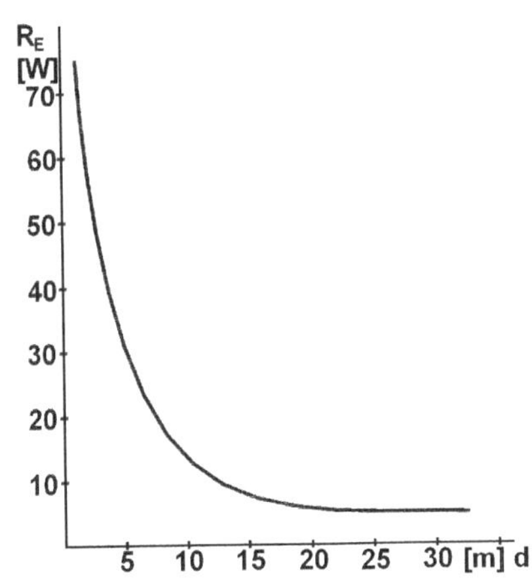

Figure 10.7 Example of resistance to earth R_E of a progressively longer rod earth electrode as a function of its length l, in a uniform soil with electric resistivity $\rho = 100\,\Omega\,\text{m}$

or obstructions close to the surface or where the electrode system covers a large area. For that reason, one can distinguish two basic types of earth electrodes:

- **Surface earth electrodes** – electrodes which are generally embedded at shallow depths, usually up to about 1 m. They consist of strips or wire arranged as radial, ring or meshed electrodes, or a combination of these.
- **Deep earth electrodes** – usually rod electrodes sunk vertically in the ground in order to reach greater depth, from about 3 m to 30 m or even more. They can be pipes or rods, sectioned and prepared to be embedded in the earth. Many rod electrodes placed vertically can also be connected to a wire or band at shallow depth in order to form one common deep electrode.

10.3.2 Voltage-to-Earth and Surface Potential Distribution

The *voltage (potential) to earth during an earth fault* is the voltage between a given point of the network and the reference earth. In particular, one can say about the voltage to earth of the main earthing busbar (Figure 10.1) during an earth fault that it is the voltage between that busbar and the reference earth. In Figure 10.6 it is represented by the voltage V_E, and in Figure 10.5 it is the relative value of $V^* = 1$.

However, in practice the main importance, especially in protection against electric shock, is the potential distribution on the earth surface between the main earthing busbar or earthed parts (Figure 10.1) and reference earth. The curve of potential distribution is illustrated in Figure 10.5 and Figure 10.6. It enables the determination of the voltage (potential) to earth of any point located at distance x from the middle of the earth electrode, into which earth current I_E flows. It can be formulated with the following equation:

$$V_x = \frac{\rho I_E}{2\pi x} \tag{10.9}$$

and its relative value

$$V_x^* = \frac{V_x}{V_E} \tag{10.10}$$

where V_E is the voltage to earth of the main earthing busbar or earthing conductor. In all these calculations it is assumed that the potential of the reference earth is equal to zero. The potential to earth can be described as follows:

$$V_E = I_E R_E = \frac{\rho I_E}{2\pi r} \tag{10.11}$$

The voltage to earth, as well as the distribution of the earth surface potential during the earth current flow, are important parameters for protection against electric shock. The basic relations are shown on the earth model presented in Figure 10.5.

The potential difference between two points, the first at distance x and the second at distance $x + a_S$, where a_S is assumed to be equal to 1 m, illustrates the prospective step potential V_S, i.e. the earth surface potential existing between the two feet of a person standing at that position on the earth surface:

$$V_S = \frac{\rho I_E}{2\pi}\left(\frac{1}{x} - \frac{1}{x+a_S}\right) \tag{10.12}$$

and its relative value

$$V_S^* = \frac{V_S}{V_E} \tag{10.13}$$

where $x \geq r$.

A similar relationship can be described for any other distances x and a. Particularly for $x = r$ and $a = a_T = 1$ m, the formula (10.12) enables calculation of the prospective touch voltage, i.e. the voltage between the palm and foot of a person who is just touching the earth electrode or metal parts connected to it:

$$V_T = \frac{\rho I_E}{2\pi}\left(\frac{1}{r} - \frac{1}{r+a_T}\right) \tag{10.14}$$

and its relative value

$$V_T^* = \frac{V_T}{V_E} \tag{10.15}$$

The touch and step voltages are more detailed subjects of Section 10.5.

As mentioned above, rod, vertical earth electrodes are recommended in order to ensure stable and low resistance to earth (Figure 10.7). On the other hand, rod electrodes have an extremely unfavorable potential distribution on the earth surface. The control of the field shape and the resultant touch and step voltages are important factors in the design of an earth electrode. Thus, resistance to earth and surface potential distribution are two significant parameters which should be considered when choosing the earth electrode construction.

10.3.3 Properties of Earth Electrodes at Lightning Currents

So far, the characteristics of earthing systems have been discussed assuming moderate current flow under steady-state conditions at the network frequency. However, the earthing arrangement must sometimes conduct the charge of lightning to earth. In this case one must consider the pulsed, or high-frequency current, flowing through the earth electrode. Lightning currents differ from currents at the network frequency, even short-circuit currents, in two main parameters:

- very high current values – the mean value of peak current is estimated to be about 30 kA, but it can reach even 200 kA;
- very fast current rise times – the mean value is estimated to be about 20 kA/μs, but the highest values are at a level of 200 kA/μs.

Lightning current causes the followings effects in the earth electrode and in the soil, which are schematically illustrated in Figure 10.8.

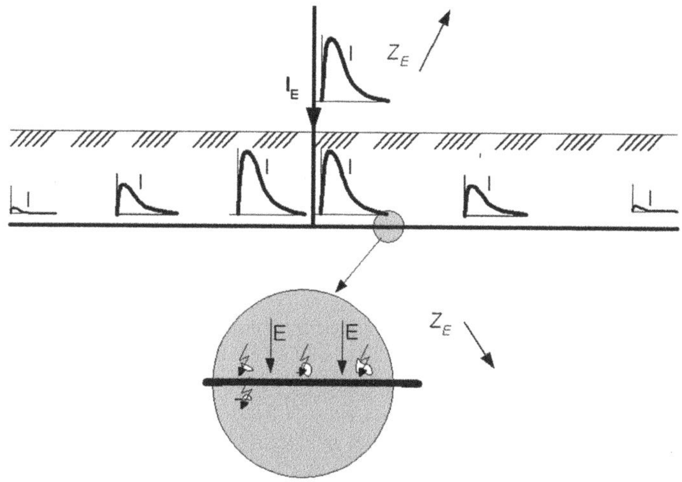

Figure 10.8 Schematic diagram illustrating two different effects of earthing current flow in the earth electrode and the soil; Z_E, with arrow directed down or up, indicates growing or diminishing impedance to earth

The first of them is that extremely high current density in the soil increases the electric field strength up to values which cause electrical discharges in small gaseous voids (lower picture in Figure 10.8). This phenomenon occurs mainly near the earth electrode, where the current density is highest and the influence is most significant. The intensity of it is especially high when the soil is dry or of high resistivity. It causes a decrease in the soil resistivity and consequently the earthing resistance.

The second effect of lightning current is damping of the current wave due to the reactance of the metallic earth-electrode parts (upper picture in Figure 10.8). The inductance of earth-electrode components can be estimated as equal to 1 μH/m. It is usually neglected when considering earth impedance at the network frequency. However, inductance becomes an important parameter when the rate of current increase is high, in the range of hundreds of kiloamps per microsecond, or more. During lightning strikes the inductive voltage drop $(L \cdot di/dt)$ reaches very high values. As a result, remote parts of the earth electrode play a reduced role in conducting current to earth, while the main part of the lightning current is conducted through those parts in the vicinity of the earthing conductor.

Thus, during lightning strikes both the phenomena described above have an effect, but operate in opposite directions. The first one diminishes the earth resistance, while the second one increases the impedance to earth. For that reason it is difficult to say, generally, if the effective impedance to earth during lightning strikes will grow or decrease. Usually, in practical considerations, the second factor is treated as dominant, and it is assumed that the impedance to earth for pulse currents increases in comparison to the impedance for static conditions. Thus, increasing the length of earth electrodes over the so-called critical length (Figure 10.9) does not cause any reduction in the impedance of the earth to transients. One solution in the design of meshed electrodes, considering lightning currents, is the denser grid of meshes in the vicinity of a terminal, where the earthing conductor is connected to the earth electrode.

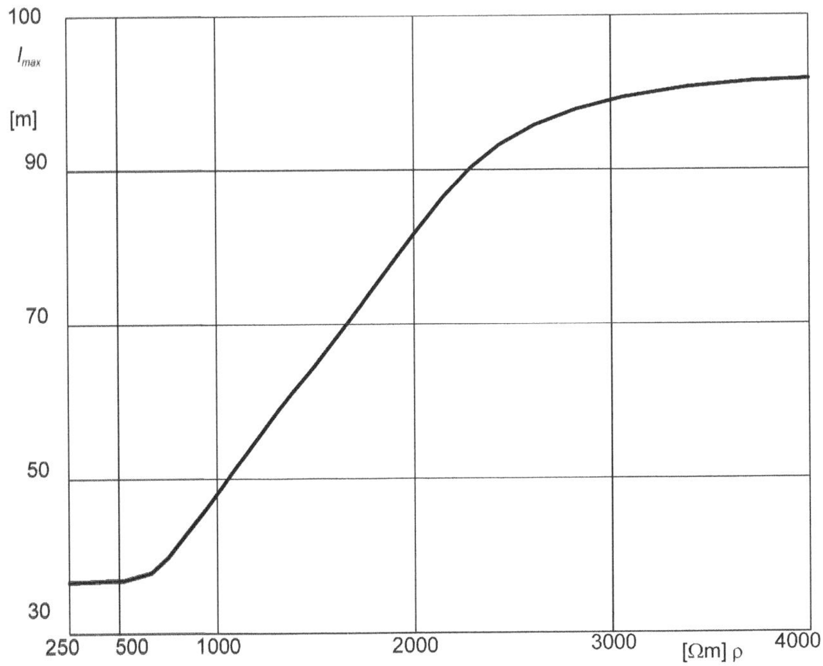

Figure 10.9 Maximum length l_{max} of earth electrodes in lightning earthings versus earth resistivity ρ

10.4 TYPICAL EARTH-ELECTRODE CONSTRUCTIONS

The properties of the earth-electrode arrangement significantly depend on the construction of the earth electrode. The conclusion from the previous section was that the earthing system designer is normally faced with two tasks:

- to achieve a required earth resistance (impedance) value;
- to ensure a satisfactory surface potential distribution; touch and step voltages especially should be at satisfactory levels.

Moreover, the construction of an earth electrode depends on other factors, such as the value of the expected fault current and the condition of the ground at the location where the earth electrode is planned.

In practice one can distinguish the following typical earth-electrode constructions:

- simple horizontal earth electrodes;
- simple rod electrodes;
- meshed electrodes;
- foundation earth electrodes;
- cable with exposed metal sheath.

A detailed description and properties of each of these electrodes are discussed next.

10.4.1 Simple Horizontal Earth Electrodes

Simple horizontal earth electrodes are metal rods, strips or pipes placed horizontally under the ground surface at a given depth t, which is usually in the range from 0.6 m to 1 m (Figure 10.10). In regions where the ground freezes deeper than 1 m, the horizontal electrodes should be mounted beneath the freezing line. Usually the length of the electrode elements, l, is much larger than t. Given this assumption, the resistance to earth can be calculated with the following formula [20]:

$$R_E = \frac{\rho}{2\pi l}\ln\left(\frac{l^2}{td}\right) \tag{10.16}$$

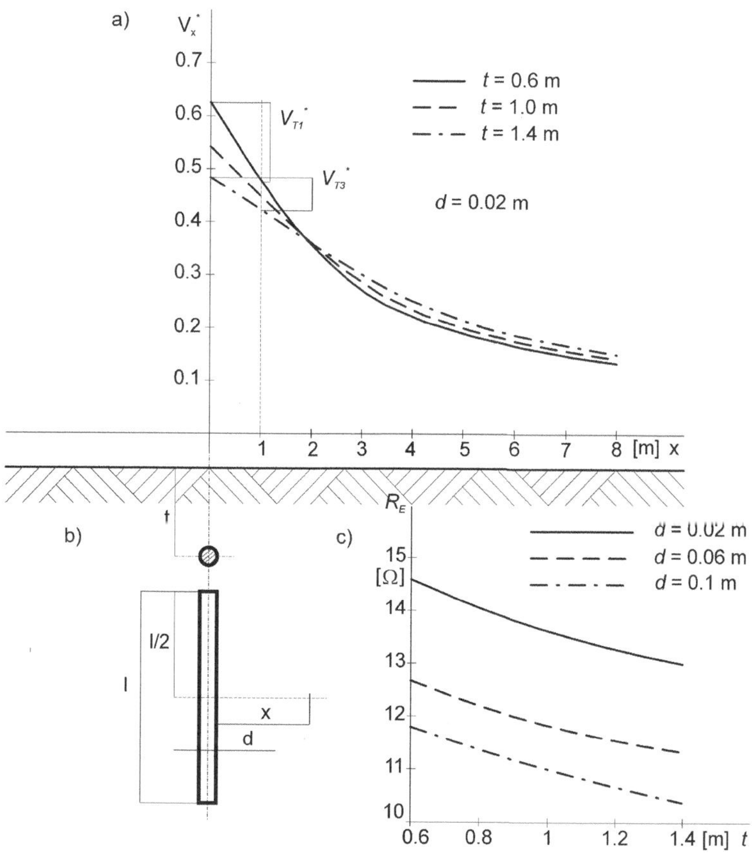

Figure 10.10 Illustration of the relative potential distribution V^* versus distance x from the horizontal electrode (a), for different buried depths t; (b) explanation of the earth electrode dimensions, where the length $l = 10\,\text{m}$; (c) R_E versus t for three electrode diameters d. V_{T1}^*, V_{T3}^* are exemplary relative values of touch voltages for $t = 0.6\,\text{m}$ and $t = 1.4\,\text{m}$, respectively; all calculations were made for a uniform soil with resistivity $\rho = 100\,\Omega\,\text{m}$

The surface potential distribution (10.9) during an earth current I_E, in a direction x perpendicular to the length l (Figure 10.10), is described by the following formula:

$$V_x = \frac{\rho I_E}{2\pi l} \ln \left(\frac{\sqrt{l^2 + 4t^2 + 4x^2} + l}{\sqrt{l^2 + 4t^2 + 4x^2} - l} \right) \tag{10.17}$$

and its relative value V_x^* can be calculated using Equation (10.10), where the full voltage to earth of the earth electrode is

$$V_E = R_E I_E \tag{10.18}$$

Horizontal earth electrodes are often made from a strip with a rectangular cross-section, usually 30–40 mm wide (b) and 4–5 mm thick (c). In this case the effective equivalent diameter d can be calculated by

$$d_e = \frac{2b}{\pi} \tag{10.19}$$

and $d = d_e$ substituted in formula (10.16). In some publications [1] it is suggested that $d_e = b/2$.

The results of example calculations, using formulas (10.16)–(10.18), for a horizontal earth electrode made from a strip of length $l = 10$ m, are presented in Figure 10.10. The analysis of these curves led to some general remarks:

- Increasing the buried depth of the horizontal earth electrode produces the smaller slope of the curve $V^* = f(x)$. For that reason, the touch and step voltages decrease with increasing buried depth.
- Increasing the buried depth causes a slight decrease in the resistance to earth. However, there is little to be gained, because increasing t from 0.6 m up to 1.4 m (more than two times) results in decreasing R_E by only about 10 %.
- The resistance to earth R_E decreases with increasing electrode diameter d. Increasing d from 0.02 m up to 0.1 m (five times) results in a decrease of about 18 % of R_E. There is little to be gained by extending the diameter of the earth electrode. However, extending the diameter is sometimes necessary in order to fulfill the mechanical and corrosion requirements.

In practice the simple, horizontally placed earth electrodes can have different, typical forms, besides that presented in Figure 10.10. The resistance can be calculated using the following formula:

$$R_E = \frac{\rho}{2\pi l_\Sigma} \ln \left(\frac{Bl^2}{td_e} \right) \tag{10.20}$$

where B is a factor depending on the electrode construction (given in Table 10.2), l_Σ is the sum of the lengths of all electrode elements, and d_e is given in (10.19).

Table 10.2 Values of the factor B (10.20) for various geometrical forms of surface electrodes

Earth electrode			Factor B in formula (10.20)
Name	Horizontal projection		
Line	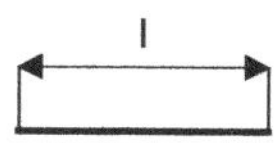		1
Square, two-arm	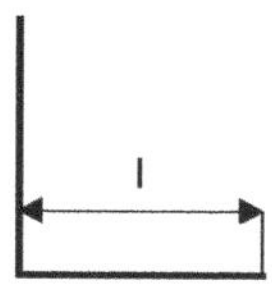		1.46
Three-arm, symmetrical	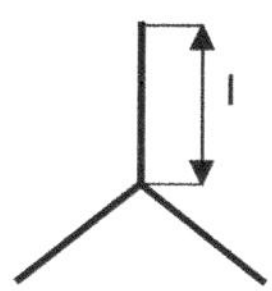		2.38
Four-arm, symmetrical	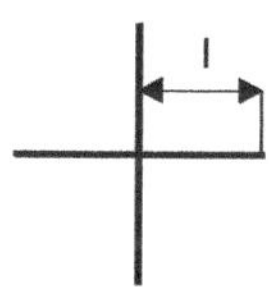		8.45
Six-arm, symmetrical	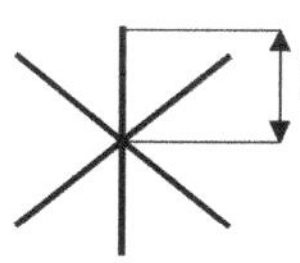		192
Two-arm, parallel	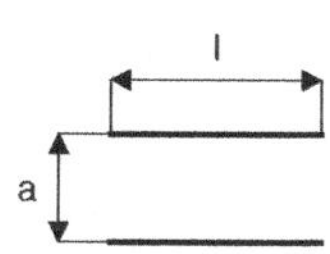		$\frac{l^2}{4a^2}$
Square	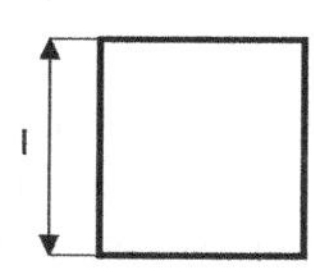		5.53
Rectangle, with various relations l_1/l_2 (1.5; 2; 3; 4)	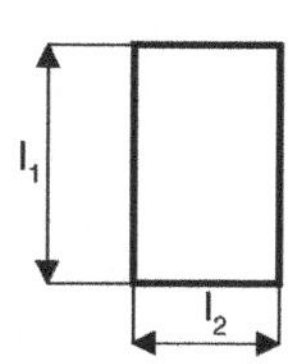	1.5 2 3 4	5.81 6.42 8.17 10.4

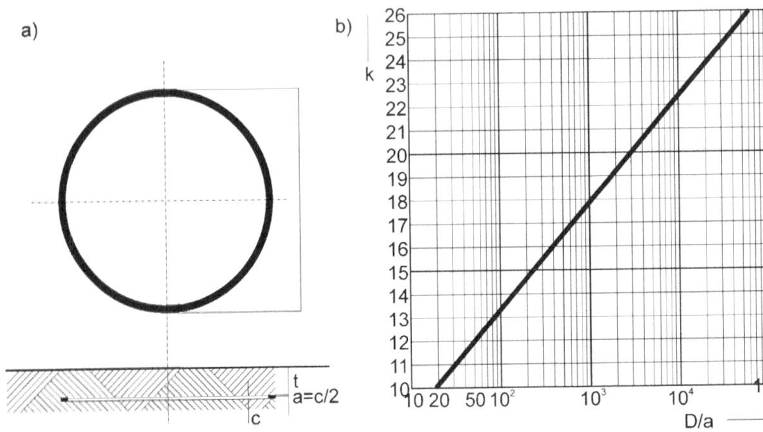

Figure 10.11 A simple ring earth electrode (a), according to Equation (10.21); (b) diagram of factor $k = f(D/a)$ which is useful in Equation (10.21)

The other typical construction of the simple horizontal earth electrode is a ring buried at a given depth t. The resistance of such an earthing electrode, with ring diameter D, made from a band with a thickness c (Figure 10.11), placed at a depth $t = 1$ m, can be calculated using the following formula [2]:

$$R_E = \frac{\rho}{2\pi^2 D} k \tag{10.21}$$

where k is the factor shown in Figure 10.11(b).

10.4.2 Simple Rod, Vertical Earth Electrodes

Simple rod, vertical earth electrodes are metal rods or pipes placed vertically in the ground. The typical length l of rods is in the range from 3 m up to 30 m or even more. They are usually mounted as a set of rods with lengths of about 1.5 m, placed successively one over the other during burying with a mechanical hammer.

As mentioned previously, such a construction has considerable advantages. Firstly, that the rod embedded into the ground passes through soil layers to the deepest ones, which contain more water and have a more stable level of temperature than the layers placed higher under the ground surface. Thus, the resistance to earth of a rod electrode is more stable in comparison with horizontal ones, because of the acute sensitivity to weather and ground conditions. The next advantage is that very little space on the ground surface is required to install the rod electrode. Thus, rods are recommended especially in areas of dense building, or where the surface is covered with asphalt or concrete. Further, vertical earth electrodes are often used in addition to horizontal ones in order to design a composite electrode and improve the resistance of existing horizontal electrodes. However, a significant disadvantage of rod electrodes is their unfavorable surface potential distribution, which can be calculated with the following formula for a given earth current I_E:

$$V_x = \frac{\rho I_E}{4\pi l} \ln\left(\frac{\sqrt{x^2+l^2}+l}{\sqrt{x^2+l^2}-l}\right) \tag{10.22}$$

where the symbols are as explained in Figure 10.12.

An example of the relative surface potential distribution $V_x^* = f(x)$, for certain electrode dimensions, is presented in Figure 10.12. A comparison of the characteristics in Figure 10.10 and Figure 10.12 shows that the potential gradients on the earth surface are considerably higher for a rod electrode and the touch voltages (V_{T1}^*, V_{T3}^*) are unfavorable in comparison to those for horizontal electrodes.

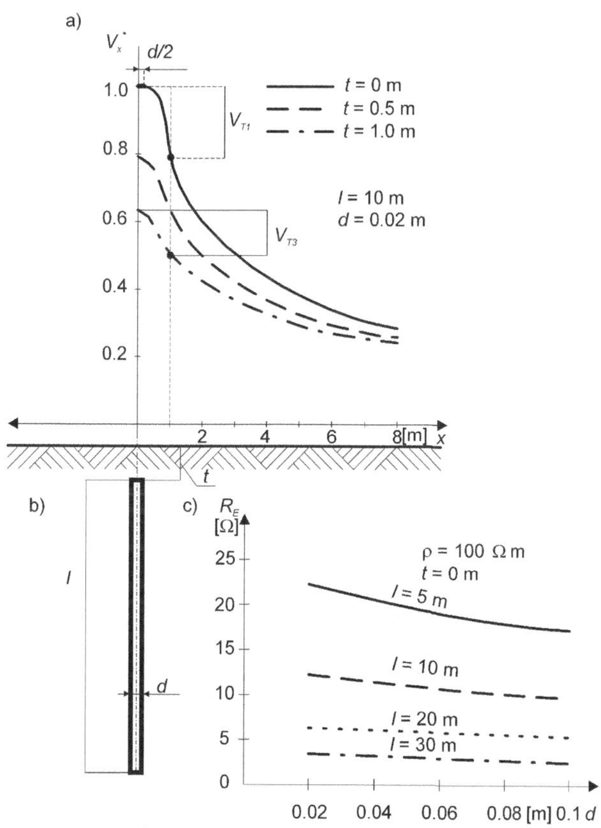

Figure 10.12 The relative potential distribution V^* versus distance x from the rod electrode (a), for different buried depths t; (b) explanation of the earth-electrode dimensions; (c) R_E versus electrode diameter d for different electrode lengths l. $V_{T1}{}^*$, $V_{T3}{}^*$ are exemplary relative values of touch voltages for $t = 0$ m and $t = 1$ m, respectively; parts (a) and (c) of the figure give the respective values of parameters l, d, ρ and t for which the curves were calculated

The unfavorable potential distribution of the rod electrodes can be improved by changing the burial depth (t, Figure 10.12b) of the electrode. In practice there are two solutions:

- the top of the electrode is buried at a given depth under the ground surface, as illustrated in Figure 10.12(b);
- the upper part of the rod is insulated from the soil.

Both solutions result in smaller gradients of the potential to earth, especially in the direct vicinity of the electrode. This is illustrated in Figure 10.12(a), where the curves $V^* = f(x)$ are shown for three different buried depths t.

The resistance to earth of the rod, vertical electrode is

$$R_E = \frac{\rho}{2\pi l} \ln\left(\frac{4l}{d}\right) \tag{10.23}$$

In Figure 10.12(c) changes in the resistance to earth versus electrode diameter d and length l are illustrated. The earth resistance changes considerably with changing electrode length, especially for lengths up to 20 m. However, the benefit which can be achieved by increasing the radius of the rod is not significant.

The important factor, which can change the resistance to earth, is the resistivity of the soil. In the curves shown in Figure 10.12(c) a uniform soil structure with resistivity

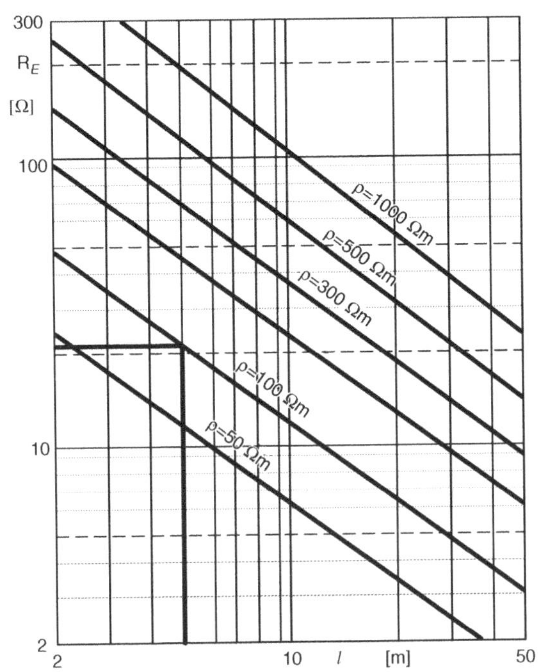

Figure 10.13 Earth resistance of a rod electrode of length l and diameter 0.02 m, in homogeneous ground with resistivity ρ [18],[19] (Reproduced from EMV nach VDE 0100, VDE Verlag and from [19] Switchgear Manual. ABB Pocket Book, Cornelsen Verlag)

$\rho = 100\,\Omega\,\text{m}$ was assumed. Figure 10.13 illustrates changes in the earth resistance versus electrode length l, for different values of the soil resistivity ρ. Such diagrams are used in the literature for prompt estimation of the resistance to earth [18],[19].

10.4.3 Simple, Combined Earth Electrodes

The simple earth electrodes described above are in practice rather rarely used as a single arrangement. When designing a single simple earth electrode there it is difficult to fulfill the requirements concerning resistance to earth and surface potential distribution. Much better effects are gained using a combination of both: horizontal and rod electrodes. Moreover, the rod electrodes are also used as an additional electrode element in cases when, during inspection of the existing earthing arrangement, the measured resistance to earth is too high.

In designing a combined electrode, the proximity effect of a number of rods must be considered. In the case of n vertical parallel electrodes (Figure 10.14) with individual resistances R_{E1}, R_{E2}, R_{E3}, ..., R_{En}, placed adjacent to each other in the soil a distance a apart, the effective earth resistance of n rods R_{Er} is as follows [2]:

$$\frac{1}{R_{Er}} = \left(\sum_{i=1}^{n} \frac{1}{R_{Ei}}\right) k \tag{10.24}$$

where k is the so-called *filling* or *duty* factor, and $k \geq 1$. The value of k is greater than 1 because of the mutual influence of electric fields produced by the adjacent rods in the soil. In effect the symmetry of current flow from each individual electrode is deformed and the current density in the soil is changed. Given in the literature [21] are exact values for the factor k for various configurations of parallel rod electrodes. In a simple configuration as shown in Figure 10.14, the value of k can be assumed as follows:

$$k \approx 1.25 \text{ for } a \geq 2l \quad \text{and} \quad k \approx 1 \text{ for } a \geq 4l \tag{10.25}$$

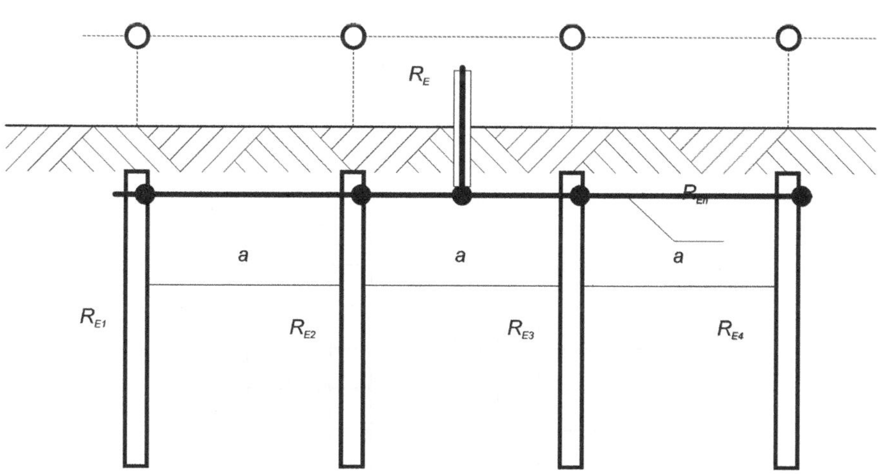

Figure 10.14 Earth electrode designed as n rods combined with a horizontal electrode

The main component of resistance to earth R_E of the combined electrode arrangement (Figure 10.14) is the effective resistance of rod electrodes R_{Er}. The resistance of the horizontal electrode R_{Eh}, which is connected in parallel to the effective resistance of rods R_{Er}, and additionally minimized by the influence of the rods, can be neglected in the calculation of the effective resistance of the whole electrode arrangement. Thus, in practical considerations it is assumed that

$$R_E \approx R_{Er} \tag{10.26}$$

10.4.4 Meshed Earth Electrodes

Meshed electrodes are designed as a grid of strips or rods placed horizontally in the ground. Strips are normally preferred as they have a larger surface for a given cross-sectional area and are considered to have a superior performance at higher frequencies, due to a slightly higher capacitance when installed in the soil. Meshed electrodes are used mainly in the earthing arrangements of a number of devices installed in a certain object of a larger area. In power systems a typical example is the power substation. Meshed electrodes usually have the form of a rectangle (Figure 10.15), and are usually constructed so that the dimensions of the rectangle correspond to the dimensions of the object area. In this way the earthing arrangement ensures a favorable, approximately uniform, surface earth potential distribution in the whole object area.

The resistance to earth of meshed electrodes is approximately equal to the resistance of a plate electrode buried in the soil. This approximation improves as the dimension of a single mesh a becomes smaller. The general formula for calculating the resistance to earth of the meshed electrode, from Figure 10.15, is as follows [2]:

$$R_E = \frac{\rho}{4r_e} k_m = \frac{\rho}{4\sqrt{\frac{lw}{\pi}}} k_m \tag{10.27}$$

where r_e is the equivalent radius of a ring, with the same area as the meshed electrodes; and k_m is the factor dependent on the single mesh dimensions:

$$k_m = 1.3 \text{ for } a \le 0.1l \quad \text{and} \quad k_m = 1.2 \text{ for } a \le 0.05l \tag{10.28}$$

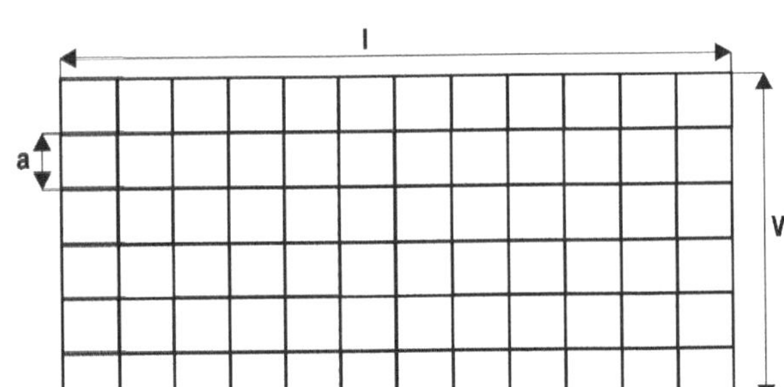

Figure 10.15 Typical diagram of the meshed earth electrode

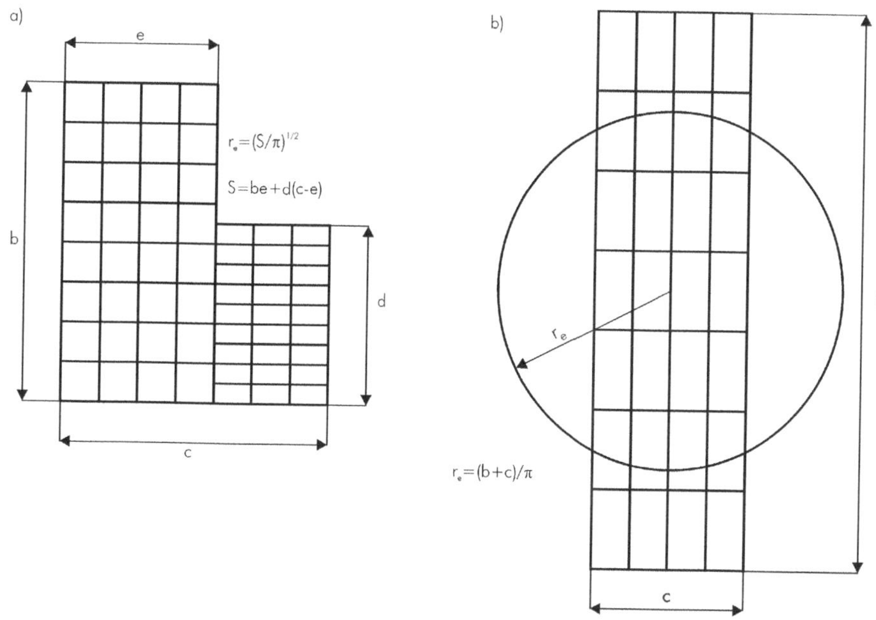

Figure 10.16 Examples of meshed earth electrodes with irregular forms, explaining the manner of calculation of the equivalent radius r_e in Equation (10.29), for two forms of the earth electrode: (a) similar to a square; and (b) an elongated rectangle, but with meshes that are not square

For meshed electrodes in which the meshes are not square, or if the electrodes are composed from a few rectangles (Figure 10.16), one can use the following simplified equation [21]:

$$R_E = \frac{\rho}{4r_e} + \frac{\rho}{l_\Sigma} \tag{10.29}$$

where l_Σ is the sum of the lengths of all conductors inside the grid.

Examples of the calculation of the equivalent radius r_e are presented in Figure 10.16.

One significant advantage of meshed electrodes is a favorable potential distribution in the earth surface. However, during an earth fault meshed electrodes increase the surface area that experiences a voltage rise as a result of current flow to the earth electrode (Figure 10.17). Over the area of the mesh an equipotential exists, but at the periphery of the electrode there is a potential gradient as shown in Figure 10.17(a). Although there is no touch potential – because the mesh extends beyond any metal structure by more than 1 m – dangerous step voltages can occur. This situation can arise, for example, in the earthing system of a substation. In order to avoid this phenomenon, the outer elements of the meshed earth electrode should be placed at a greater depth than the rest of the grid (Figure 10.17c).

One of the disadvantages of meshed earth electrodes is that, due to the large area covered by the grid system, it is not practical to bury them deeply, so they are more susceptible to changes in soil moisture content. Improved stability of resistance can be achieved by including a number of long vertical rods in the mesh.

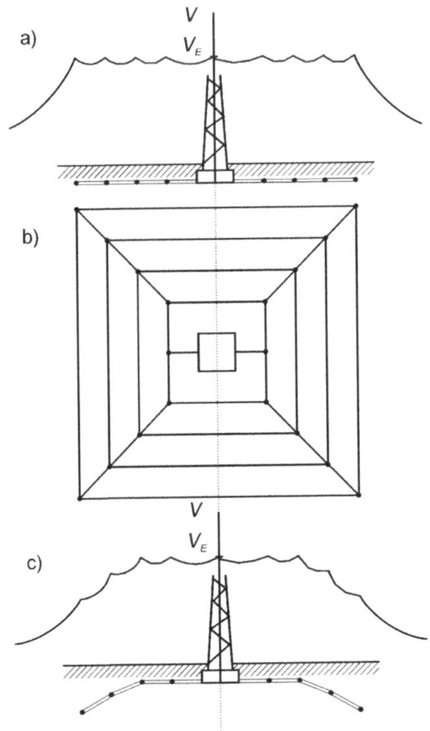

Figure 10.17 The phenomenon of potential carryover and the earth surface potential distribution for two meshed earth electrodes: (a) meshed plane electrode with almost full potential outside of the last envelope; (b) plan of electrode; (c) electrode with two end elements placed deeper

10.4.5 Foundation Earth Electrodes

Foundation earth electrodes are nowadays recommended as a very practical solution in building earthing [5]. They are constructed from conductive metal parts embedded in the concrete of a building's foundation. Concrete buried directly in the ground has a natural moisture content and can be considered as conductive matter, with a conductivity similar to that of the earth. Because of the large area of this type of electrode, low resistance can be achieved. Furthermore, the concrete protects the metal parts against corrosion and steel electrode elements embedded in the concrete do not need any additional corrosive protection.

In practice there are two basic foundation earth electrode constructions:

- in a foundation without concrete reinforcement (Figure 10.18);
- in a foundation with concrete reinforcement (Figure 10.19).

In both cases the earth electrode is made from:

- steel strip, usually with a rectangular cross-section not less than 30 mm × 3.5 mm;
- steel bar, usually with a round cross-section not less than 10 mm in diameter.

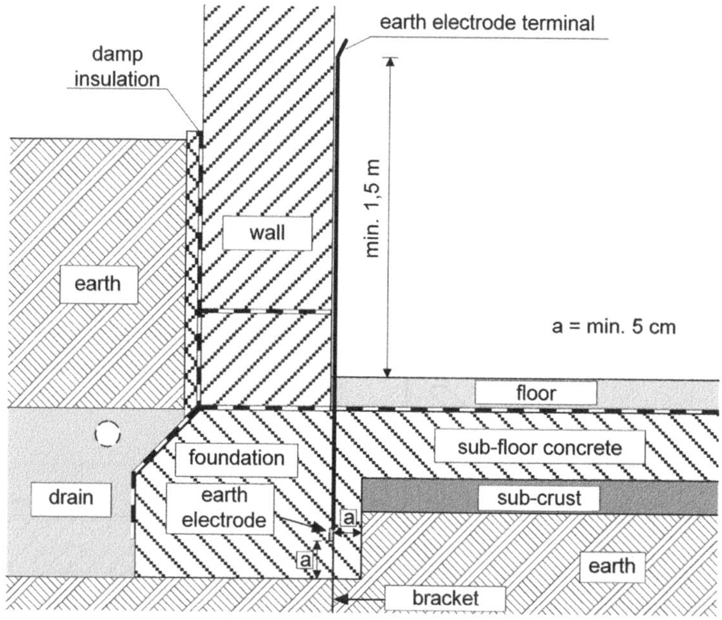

Figure 10.18 The placement of the foundation earth electrode in a foundation without concrete reinforcement

The steel elements can be galvanized (i.e. coated with zinc), but this is not necessary if the layer of concrete covering the electrode is greater than 50 mm [5], because the concrete ensures sufficient protection against corrosion, as shown in Figure 10.18.

In a foundation without concrete reinforcement (Figure 10.18), the electrode usually follows the contours of the building's foundation, i.e. it is placed under the main walls. In buildings with extensive foundations, the electrode is usually made in the form of loops, covering the parts of foundation outlines, and connected to each other.

In a foundation with concrete reinforcement, the earth electrode is placed over the lowest layer of wire mesh reinforcement (Figure 10.19) thus ensuring adequate corrosion protection for the electrode. The electrode should be fastened to the reinforcement mesh with wire strands at intervals of not more than 2 m over the length of the electrode. It is not necessary to make a sound electrical connection at each point because the main electrical connection is via the concrete. If the foundation is constructed as separate panels connected to each other with expansion joints, the earth electrodes of each panel should be galvanically connected to each other. These connections must be flexible and must be located so that they remain accessible for measurement and maintenance purposes [5].

The foundation earth resistance can be calculate using the following simplified equation:

$$R_E = 0.2\frac{\rho}{\sqrt[3]{V}} \tag{10.30}$$

where R is in Ω and V is the volume of the foundation in m^3.

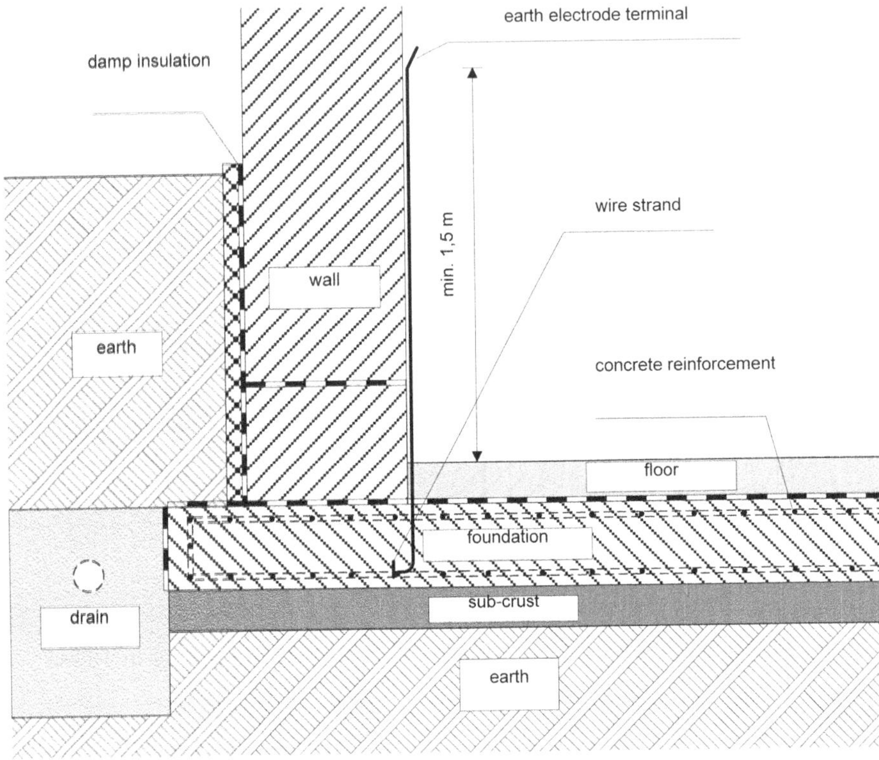

Figure 10.19 The placement of the foundation earth electrode in a foundation with concrete reinforcement

The terminal of the foundation earth electrode should have a minimum length of 150 cm above floor level (Figure 10.18 and Figure 10.19). It should be placed as close as possible to the main earthing terminal of the building installation. The terminals of the foundation earth electrode, which are connected to the lightning protective installation, should be placed outside the building.

10.4.6 Cable with Earthing Effect

In many cases, the cable whose metal shield, sheath or armor provides leakage to earth of a magnitude similar to that of a strip earth electrode can be used as an earth electrode. Metallic pipes (e.g. water pipes) embedded in the ground have a similar effect. Such earth electrodes, and also the foundation earth electrodes, are called 'natural' earth electrode systems.

Cables with earthing effect are mainly used as earth electrodes in power electrical substations supplied with cable lines [2]. Additionally, they can be connected to the foundation earth electrode of the substation building. In this way an earth electrode arrangement is formed, which is normally sufficient for substations supplied with the cable network. It is roughly estimated that such an earthing system, based on the cable earthing effect, fulfills

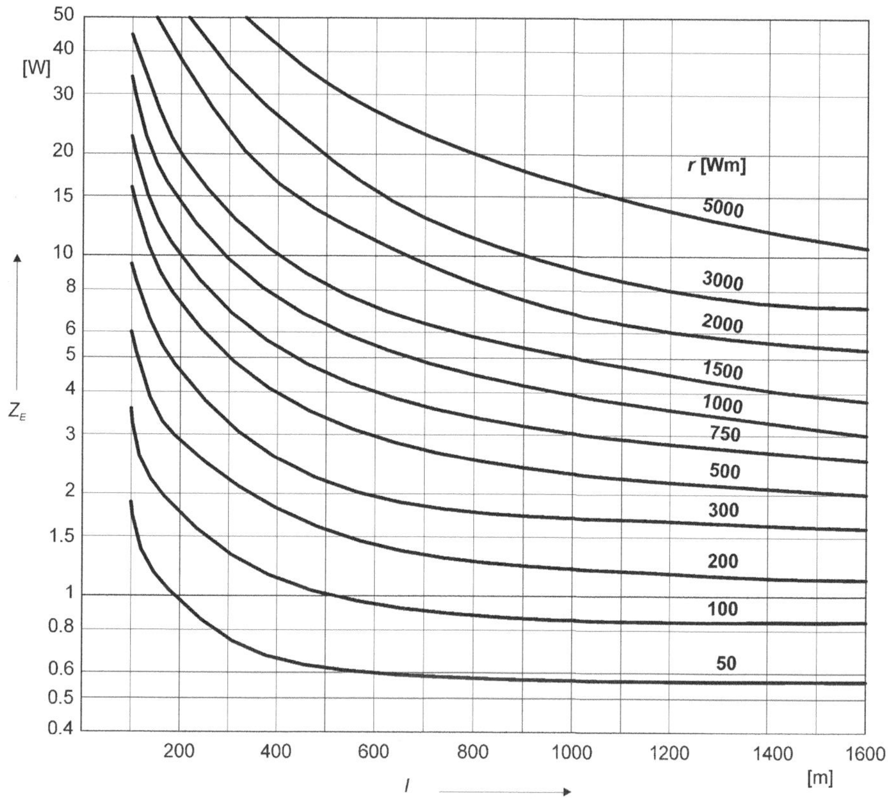

Figure 10.20 Impedance to earth of cables with earthing effect versus cable length, for different soil resistivities [2]

requirements if the fault current does not exceed 1500 A [2]. In other cases, an additional earth electrode system must be installed. The impedance to earth of a single cable line depends on the cable length and the electrical resistivity of the soil (Figure 10.20).

10.5 EARTHING ARRANGEMENTS IN PROTECTION AGAINST ELECTRIC SHOCK

The earthing arrangement is an important element in the protection against electric shock. As mentioned previously, the earthing system designer is faced with two tasks:

- to achieve a required value for the impedance to earth;
- to achieve touch and step voltages that are lower than given boundary values.

These tasks have different requirements in low-voltage (LV) and high-voltage (HV) networks.

So far, touch and step voltages have been described in previous sections. The voltages illustrated in Figure 10.5 are essentially the prospective values. However, if the person is affected by the touch or step voltage, the current of the shock changes the potential distribution in the vicinity of where the person is standing. This is the reason why in the literature [12],[21] there is a distinction between the prospective and effective touch and step voltages. A practical illustration of these parameters, for two different earth electrode constructions, is shown in Figure 10.21. Persons A and B are subject to the touch potential while person C is subject to the step potential. Voltages V_T and V_S are the pure, prospective values resulting from the potential distribution, whereas V_{TS} and V_{SS} are the effective values, i.e. they consider the small changes in potential distribution caused by the shocking current flowing through the affected person. The effective touch and step voltages are smaller than the prospective ones. On the other hand, in practice the difference between V_S and V_{SS} or V_T and V_{TS} is usually small. Considering the prospective values means that more rigorous conditions will be assumed. For that reason, in practice the same values for the respective potentials are often assumed: $V_S \approx V_{SS}$ and $V_T \approx V_{TS}$.

The left-hand side of Figure 10.21 shows the situation for a rod electrode while the right-hand side shows that for a meshed electrode. The rod electrode (1) has a low resistance but the most unfavorable potential distribution, while the meshed electrode (2) has a much flatter earth potential profile. The touch potential (person A) is considerably larger for the

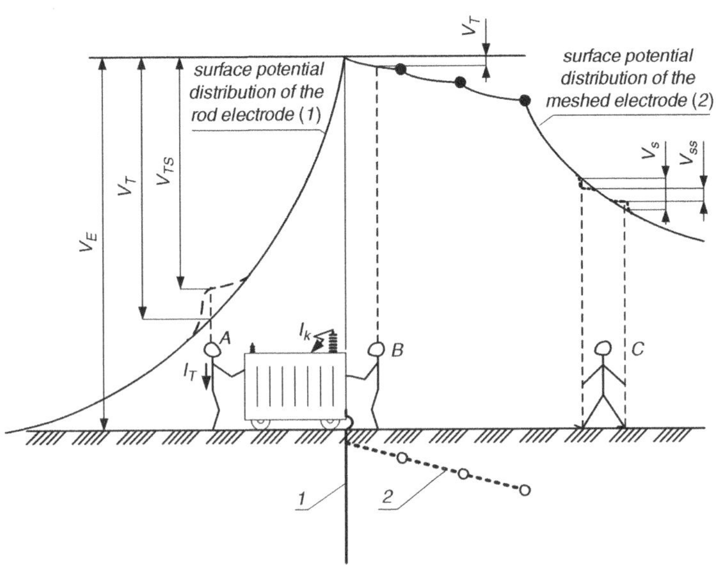

Figure 10.21 Comparison of earth surface potential distribution during current flow in the earthing system, for two earth electrode constructions: a rod electrode (1) and a meshed electrode (2). V_E is the earthing voltage; V_T, V_{TS} are the touch voltage and shocking touch voltage, respectively; V_S, V_{SS} are the step voltage and shocking step voltage, respectively; I_T is the shocking touch current; I_K is the short-circuit current equal to the current flowing to the earthing system; A, B, C are persons at various earth surface potentials

rod electrode (1) than for the meshed one (2) (person B). Step potentials (person C) are also less dangerous in the meshed electrode.

The resistance to earth determines the value of earthing voltage (10.9) and (10.10), whereas the configuration of the earth electrode has a significant influence on the potential distribution of the earth surface, (10.12)–(10.15). Naturally, the configuration also influences the earthing resistance – a meshed electrode has a larger area of contact with the soil than a single rod electrode – so both resistance and configuration need to be considered together.

As mentioned at the outset, in electrical networks, in a given object (home installation, substation), there is usually one earthing arrangement installed, common for all purposes, which offers protection against electric shock, functional earthing and lightning. Thus, under a protective earthing system it is understood here that there is one, common, earthing arrangement which has the protective task, among others.

10.5.1 Earthing Arrangements as Protection Elements in LV Networks

In LV networks the protective functions of the earthing arrangements can be distinguished between:

- the earthing and protective bonding system in electrical installations in buildings (residential, public and industrial);
- earthing in electrical lines and substations (public and industrial).

The electrical installation of a building is equipped with an earth-electrode system (preferentially a foundation earth electrode, described in the previous section) and an earthing conductor connected to the main earthing terminal, to which the following are connected [10]:

- protective main bonding conductors of all metallic installations coming into the building;
- depending of the network system, the following conductors:
  - in the TN-C system, the PEN conductor;
  - in the TN-S system, the neutral (N) and the protective conductor (PE);
  - in the TT system, protective earthing (E) and neutral (N) conductors;
  - in the IT system, protective earthing (E) conductors.

A detailed description of such systems is not the subject of this chapter, so only some general requirements concerning earthing arrangements are presented here.

The protective task of the earthing system in electrical installations of buildings is to provide the interconnection or bonding of all metallic parts (exposed and extraneous conductive parts) that a person or an animal could touch. Under normal, fault-free, circumstances there is no relative potential on these parts, but under fault conditions a dangerous potential may arise as fault current flows. The function of the earthing system and the respective bonding conductors is to protect lives against electric shock. The fundamental requirement is that the earthing potential, V_E, at a prospective short-circuit current, I_E, does not exceed the touch voltage limit, V_{TL}:

$$V_E \leq V_{TL} \tag{10.31}$$

In the TN network, where the fault-current loop is formed by PE or PEN conductors and phase conductors, the purpose of earthing is generally to ensure the earth potential on the main earthing terminal. The requirements concerning the value of the prospective current refer essentially to the impedance of these conductors and the parameters of fault current protection, but not the earth electrode. Because of this there are no specific requirements concerning earthing impedance and potential distribution in that kind of installation in buildings.

In the TT network the fault current flows through the earthing system to the exposed and extraneous conductive parts. For that reason the main requirement here is

$$(R_E + R_{EC})I_a \leq V_{TL} = 50\,\text{V} \tag{10.32}$$

where R_E is the resistance to earth of the earth electrode; R_{EC} is the resistance of the protective earthing conductor and exposed or extraneous conductive part; and I_a is the threshold value of the fault current protective device. In practice the resistance R_E dominates the sum $R_E + R_{EC}$. However, it is so difficult to achieve a low resistance to earth R_E that the values of current I_a are sufficient for common overcurrent protective devices, like fuses or LV power circuit-breakers. For that reason, it is recommended that circuits in TT networks are protected with a residual current devices (RCD), in order to fulfill easier the condition (10.32). In this case, the current I_a is equal to the rated residual current $I_{\Delta N}$ of the RCD. The touch voltage limit V_{TL} in LV networks is equal to 50 V.

In IT networks the single-phase short-circuit current can be sustained for a longer time. In resistance to earth must fulfill similar conditions to (10.32), with the difference that the current I_a in this case is the single-phase short-circuit current. In cases in IT networks when the single-phase fault current should be switched off immediately, only protection with RCDs is useful, because of the very low values of this current, which is a leakage one.

In substations from medium to low voltage, earthing systems of the LV and HV parts are usually a common arrangement. It is in effect one functional and protective earthing system.

10.5.2 Earthing Arrangements as Protection Elements in HV Networks

In HV power electric systems there is a distinction between [8],[19]:

- the basic protection (insulation, enclosures, protective enclosures and barriers);
- the fault protection against electric shock.

The earthing arrangement is the fundamental measure of fault protection in HV networks. In the event of a short-circuit current to earth, the earth carries at least a part of the fault current. The fault current flowing through the earthing system causes a certain potential difference on the ground surface, which can be shunted by humans. Thus, the main problem is here to ensure that the touch and step potentials are satisfactory during a fault. For that reason, the surface potential distribution is an important consideration for the earthing system designer. The design of an earthing system must be based on the fault current flowing through the earthing arrangement when a fault occurs in the network. Furthermore, the fault protection system in HV objects consists of connecting to an earth arrangement all metallic accessible parts in order to equalize the potentials in all those places where humans may be present. It concerns, for example, all the metallic parts of constructions in substations, barriers, armoring and screening of cables.

The protection criteria consist of describing permissible values of touch voltages, which can occur during the earth fault inside a substation or on HV lines [8],[19]. These permissible voltage values depend on the maximum duration of the fault and also on other circumstances which can diminish the effective touch voltage (Figure 10.22). Such positive circumstances considered here are all measures that can improve the insulation of a person from direct contact with the earth, e.g. shoes, the insulating effect of asphalt or concrete on roads, in parking places and similar areas. For that reason, the standard [8] gives a few characteristics of the permissible touch voltages, which can occur in the case of an earth fault in HV overhead lines, depending on the additional circumstances mentioned. The most critical circumstances represent in Figure 10.22 the *curve* $U_{D1.}$ at which no additional resistance to the earth is assumed. It concerns those places where people can have direct contact with the ground, like camping areas, swimming pools and recreational sites. *Curve* U_{D2}

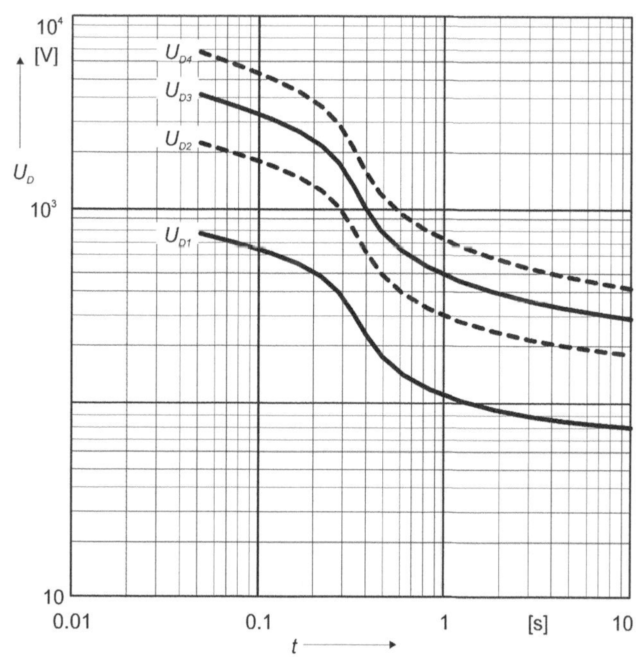

Figure 10.22 Examples of touch voltage limits U_D as a function of duration of fault current t in HV overhead lines [8] (Reproduced from Overhead electrical lines exceeding AC 45 kV, Part 6. Earthing systems, EN 50341: 2001)

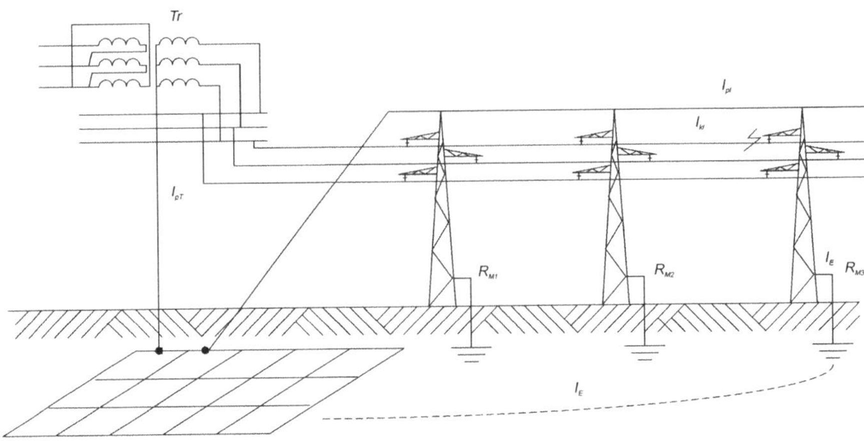

Figure 10.23 An illustration of the earth-fault current as the partial current during a short circuit in the HV line in a system with low-resistance neutral earthing: *Tr* is the transformer in the substation; R_{M1}, R_{M2}, R_{M3} are the earthing arrangements of the line posts; I_{kl} is the single-phase short-circuit current; I_{pl} is the partial short-circuit current flowing in the overhead earth wire of the line; I_E is the partial short-circuit current flowing through the earthing system; I_{pT} is the short-circuit current flowing to the neutral point of the transformer

concerns those locations where it can be assumed that people are wearing shoes, such as pavements, public roads, etc. The additional resistance to earth is assumed to be 1750 Ω. *Curves* U_{D3} and U_{D4} are representative of locations where people are wearing shoes and the soil resistivity is high – 2000 Ω m and 4000 Ω m, respectively. The touch voltage limits are considerably higher in such cases, assuming the additional resistance to be 4000 Ω and 7000 Ω, respectively.

During faults in HV lines, usually only a part of the short-circuit current flows through the earth (Figure 10.23). Another part of the short-circuit current (I_{pl}, Figure 10.23) flows through the overhead earth wires or screens in the cable lines. For that reason, the requirements concerning resistance to earth of the earthing arrangements are not so rigorous, as in the case when the whole short-circuit current I_{kl} is equal to the earthing current I_E (Figure 10.23). The partial fault current flowing through the earthing system I_E depends on the parameters of the overhead earth wire:

$$I_E = I_{kl} r \tag{10.33}$$

where factor r is as follows:

- $r = (0.9–0.95)$ in lines with a single steel overhead earth wire;
- $r = (0.87–0.9)$ in lines with two steel overhead earth wires;
- $r = (0.69–0.73)$ in lines with a single Al–Fe overhead earth wire;
- $r = (0.53–0.64)$ in lines with two Al–Fe overhead earth wires.

In cable lines the factor r is in the range (0.15–0.25), thus only 0.15–0.25 % of the fault current flows through the earthing system.

The fault current and lightning current values in HV networks usually reach high values and it is sometimes difficult to ensure full safety in each possible case. For that reason, in considering the protection of these systems the probability of certain dangerous situations must be taken into account. This concerns, for example, the earthing arrangements of posts in HV overhead lines, in regions where the density of population is very low, e.g. forests or mountains. The probability of the occurrence of an earth fault and the presence of a person in such a place, together with other extraordinary circumstances, is very small. These problems are presently under discussion, and it will likely be concluded that the requirements for earthing arrangements in HV lines will be differentiated according to the probability of danger in some locations.

10.6 ROLE OF EARTHING IN ELECTRONIC AND TELECOMMUNICATION SYSTEMS

The definition of an earthing system, according to IEC 60050-195, is: *functional earthing and protective earthing of a point or points in an electric power system.*

The growing proliferation of electronic equipment generates the need for new requirements and bonds when designing an earthing system, which is already a key element of electrical installations in terms of safety requirements.

Usually, electronic devices need to have a connection to earth for safety reasons (*protective earthing*), but also for operational reasons (*functional earthing*).

These different requirements create an EMC problem. The contemporaneous need of a connection to earth for safety and operational reasons, on the one hand, makes electronic devices liable to disturbances that usually they should not be vulnerable to, and, on the other hand, produces safety problems related to the permanent leakage currents that they generate.

An earthing system should be constructed to cover both safety and operational requirements, without prejudicing the safety level of the installation and, at the same time, equipment functionality.

This section deals with the problem of the correct earthing of electronic devices, in terms of such requirements. The considerations are also applicable to earthing systems where no electronic equipment, but devices with similar behavior, are connected to:

- high leakage currents;
- leakage currents with high frequency;
- permanent leakage currents;
- low immunity to disturbances.

The IEC EN 60617-2 standard defines different earthing system symbols (see Figure 10.24) as follows:

- On the left, there is protective earthing which aims to conduct fault and lightning currents to earth.
- On the right, there is functional earthing, which aims to conduct leakage currents to earth, and to set the neutral point voltage.

Figure 10.24 Graphic symbols: 02-15-03, protective earthing system; 02-15-03, functional earthing system, according to IEC EN 60617-2

10.6.1 Protective and Functional Earthing

The primary aim of the earthing system, as above, is the safety of electrical installations (indirect contact protection). In these situations, the earthing system has three objectives:

1. To ensure a predetermined path for fault currents in order to allow protective devices to detect them and break the fault circuit.
2. To limit step and touch voltage levels to safe values.
3. To ensure the equipotential of earth and extraneous conductive parts.

An example of functional earthing is the connection to earth of a neutral conductor. The earthing system function, in this case, is essentially:

- to set the phase voltage to a determined value;
- to limit ground overvoltages and to ensure coordination of rational insulation.

For electronic devices, the operational requirements which need a connection to earth are:

- to ensure a common reference voltage to circuits;
- to suppress disturbances.

10.6.2 Combined Earthing System

As previously stated, electronic equipment introduces new requirements and (safety) bonds into the design of an earthing system. In all electronic devices, for example, there are filters with the purpose of reducing radio-frequency disturbances. Usually the filters are made of one or more phase–earth connected capacitors. Under ordinary operation, the filter cleans the disturbances through an earth current. This solution can lead to the generation of leakage currents flowing towards earth, sometimes with high amplitude.

The presence of electronic equipment or of a generic device with a similar behavior sets, when making an earthing system, new:

- *requirements*: functional earthing with low noise level;
- *bonds*: permanent leakage currents with high amplitude and frequency (e.g. Figure 10.25).

These two aspects will be examined separately for a better understanding.

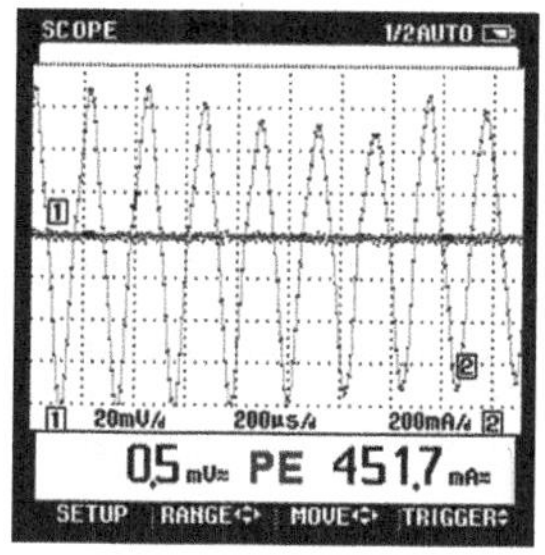

Figure 10.25 Example of leakage current towards earth for a 20 kVA device (amplitude 451.7 mA, frequency 5 kHz)

10.6.3 Safety Aspects

The presence in almost all electronic devices of filters with the purpose of reducing radio-frequency disturbances leads, under ordinary operation, to the generation of leakage currents flowing towards earth. To touch the exposed conductive part does not run the risk of electrocution, but if the connection to earth breaks, a dangerous condition occurs: if someone were to touch the exposed conductive part, the filter current would flow towards earth through the person's body. In addition, a key factor is represented by the fact that the fault circuit is the same as a normal indirect contact due to the exposed conductive part, but with an ever present dangerous condition, while the exposed conductive part becomes live only under fault conditions.

Actually, this problem occurs in the presence of any device generating current towards earth. The situation above is more dangerous because of the high current value. While leakage currents from non-electronic equipment are small, currents generated by electronic devices can have amplitude values from 3.5 to 10 mA and, sometimes, greater.[1]

For these reasons, the earthing requirements for the installation of data processing equipment have been included in the IEC 364-7-707 standard. This standard applies to devices with a leakage current > 3.5 mA (EN 60950), and also to non-data processing equipment.[2]

The PCs we use every day, for example, if complying with the EN standard, do not require to be connected to an earthing system under with prescriptions of Sec. 707 of the standard. Of course, 10–100 devices having currents < 3.5 mA and connected to the same earthing system will lead to a leakage current > 3.5 mA.

The standard contains requirements on safety and low noise. Safety requirements essentially aim to reduce the chance of an accidental interruption of a protective conductor. They are classified as a function of the neutral conductor state and of the leakage current (Figure 10.26, Figure 10.27): the reference is the TN system, while for IT and TT systems additional requirements have been added.

[1] Current conventionally chosen as the tetanization threshold is equal to 10 mA (duration > 2 s).

[2] Even though its section title refers only to data processing equipment, the standard seems also to consider, in Art. 2, that Sec. 707 applies to all devices, non-data processing equipment too, which have the same issues (e.g. VSDs).

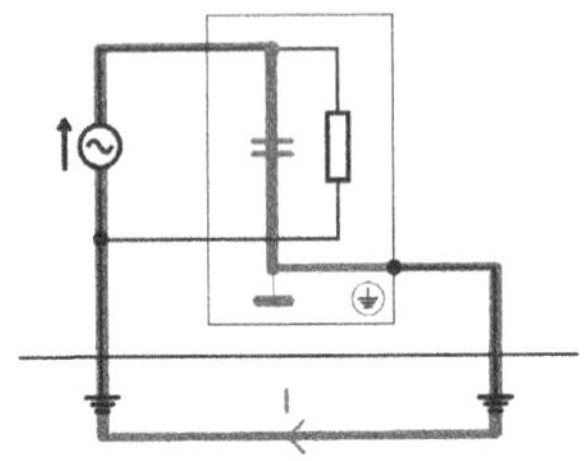

Figure 10.26 Leakage current path when a connection to earth is available

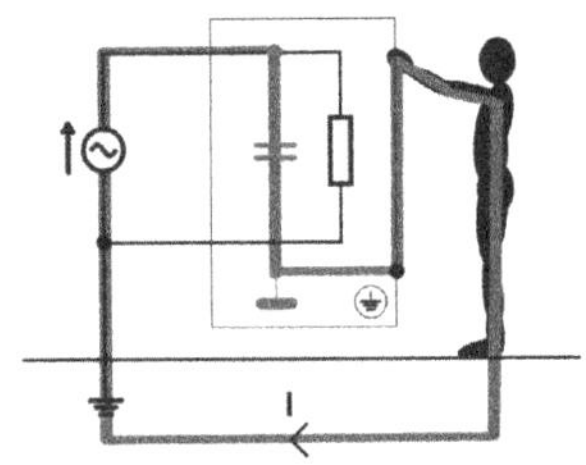

Figure 10.27 Leakage current path when a connection to earth is not available

If the leakage current is $< 10\,\text{mA}$ the prescriptions of Art. 707.471.3.2 suffice. That is, equipment shall be:

- stationary;
- either permanently connected to the building wiring installation or connected via industrial plugs and sockets (IEC 309-1).[3]

If the leakage current is $> 10\,\text{mA}$, Art. 707.471.3.3 defines two possibilities, the first of which with two alternatives:

- High-integrity protective circuits:
  - High-integrity connection.
  - Earth continuity monitoring.
- Use of transformers.

When the system is TT, in addition to previous prescriptions the standard introduces some limits[4] related to the rated residual current of the RCD:

$$I_1 \leq \frac{I_{\Delta N}}{2} \leq \frac{V_{TL}}{2R_A} \tag{10.34}$$

[3] Note that previous releases of the same standard did not allow use of a plug and the only possibility was a fixed connection to the appliance.
[4] Art. 707.471.4.

where I_l is the leakage current; $I_{\Delta N}$ is the rated residual current of the RCD; R_A is the earthing system resistance; and V_{TL} is the touch voltage limit.

This prescriptions can be classified from two points of view:

- one related to power quality (the left side of the inequality);
- one related to safety (the right side of the inequality).

Let us start from the first one: this is the usual inequality to be fulfilled when choosing RCDs, but to avoid nuisance tripping, and considering that the characteristics of the protective device and leakage currents are permanent, it is better to introduce a factor of 2. This is indeed a power quality design criterion not related to safety in the case of indirect contacts.

The second one is the usual inequality to be fulfilled to guarantee safety in the case of indirect contacts, but again there is a factor of 2.[5]

When the system is IT, instead of this last prescription and in addition to general ones, in Art. 707.471.5 the standard introduces some other extra prescriptions.

They are not real prescriptions but should be considered as suggestions, because, after the first fault to ground in these kinds of systems, it is very hard to maintain the touching voltage within admissible levels. Further, considering that an IT system is used in order to have the possibility of continuing to work even after the first earth fault, the standard suggests deriving a TN system by way of a transformer to connect appliances having high leakage currents and to refer to the prescriptions for TN systems.

10.6.3.1 High-Integrity Protective Circuits[6]

The aim of this prescription is to make reliable protective earthing connections. This requirement, according to the standard, can be satisfied by:

- use of one conductor with a cross-sectional area of not less than $10\,mm^2$;
- use of two conductors with independent terminations, each having a cross-sectional area of not less than $4\ mm^2$;
- use of a multi-core cable: the sum total cross-sectional area of all the conductors shall not be less than $10\ mm^2$, so that the protective conductor cross-sectional area will not be less than $2.5\,mm^2$.

10.6.3.2 Earth Continuity Monitoring[7]

To monitor earth conductor continuity, standard IEC 364-7 prescribes that a device or devices shall be provided which will disconnect the equipment in the event of a discontinuity occurring in the protective conductor, according to the requirements related to indirect contact protection.

---

[5] These factors should be considered as safety factors, not really widely used in electrical engineering, but very common in other technical fields, and introduced bearing in mind the importance of the phenomena being dealt with.

[6] Art. 707.471.3.3.1.

[7] Art. 707.471.3.3.2.

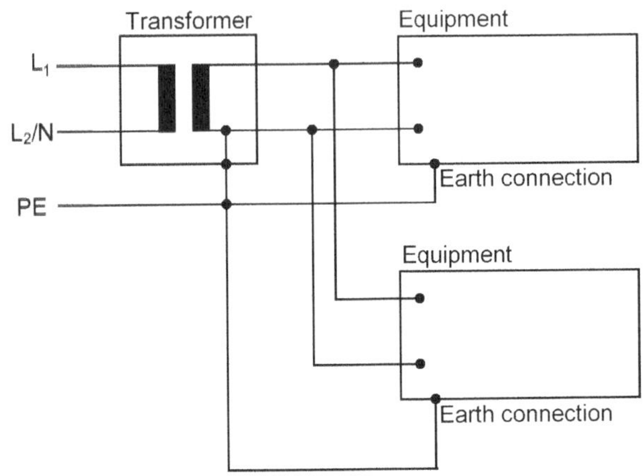

Figure 10.28 Use of transformers: the electronic appliance with leakage current greater than 10 mA is fed by way a transformer generating a TN-S system at the secondary side

10.6.3.3 Use of Transformers[8]

A last measure suggested by the standard to limit leakage currents consists of using transformers. If the equipment is powered through a transformer, and if the secondary circuit is connected as a TN system, leakage currents flow through the transformer secondary without involving earth conductor.

In this case it is possible to reduce the adoption of the above-mentioned requirements to improve earth conductor reliability to a reduced portion of the installation (i.e. the secondary side).

The use of special transformers is not necessary (safety transformers); a standard transformer with separate windings can be used and, in particular, autotransformers are not allowed. This solution is an economical option: designers have to choose to buy a transformer and then limit other safety systems to one portion of the installation, depending on the dimensions of the installation (Figure 10.28).

10.6.4 Functionality Aspects

Functional connection to the earthing system can lead frequently to the propagation of disturbances from the electrical installation to the equipment. These disturbances will usually degrade device performances.

Describing the problem in a schematic way, functional earthing establishes *de facto* a *coupling path* between the *disturbance source* (electrical installation) and the *receiver* (electronic device). The need for a protective and functional connection to earth then creates a problem of EMC that could be solved adopting suitable measures (*low-noise equipotential bonding*).

[8] Art. 707.471.3.3.3.

The electromagnetic environment in an electrical installation is characterized also by the presence of radiated disturbances which can produce interference phenomena in electronic devices. These disturbances can be mitigated through apposite solutions when designing the earthing system.

10.6.5 Disturbances and Coupling Mechanisms

Usually, disturbances in electronic devices appear as potential differences between device conductors and earth (*common mode voltages*) or as potential differences between couples of conductors (*differential mode voltages*). There are three coupling mechanisms by which disturbances transfer from source to receiver: *direct* (conducted disturbances), *inductive* and *capacitive* (radiated disturbances).

10.6.5.1 Direct Coupling

Disturbances are generated by the currents circulating in the various parts of the earthing system (earth electrodes, protective conductors). These currents are usually earth-fault currents, with power frequency, but they can be leakage currents generated by the devices themselves.

These kinds of disturbances are common mode voltages whose value depends on the current amplitude and on the earth electrode or connection to earth impedance value. These impedances are a function of the frequency and then the disturbance entity will depend on the nature of the currents.

In the case of power frequency earth-fault currents, equipment connected to the earthing system is usually subject to a common mode voltage with a value equal to the total earth voltage or a part of it.

10.6.5.2 Inductive Coupling

Even in the presence of inductive coupling, disturbances are due to currents circulating in the circuit. These currents generate a magnetic field whose flux can be linked with the turns formed by the installation's conductors (Figure 10.29). Usually, this kind of radiated disturbances can appear either as common mode voltages or differential mode voltages. To assess the disturbances is rather difficult, unless the circuit configuration is particularly simple and regular. The parameters which should be considered are:

- magnetic flux;
- turns characteristics:
  - dimensions;
  - geometry;
  - position and orientation related to the flux lines.

These kinds of disturbances can be found, for example, in the presence of earth faults, when fault currents circulate through different paths between the forward way (line conductor) and backward way (earth conductor and electrode) and when circulation paths are long. Particularly intense disturbances are generated by lightning currents whose interference is always considerably severe.

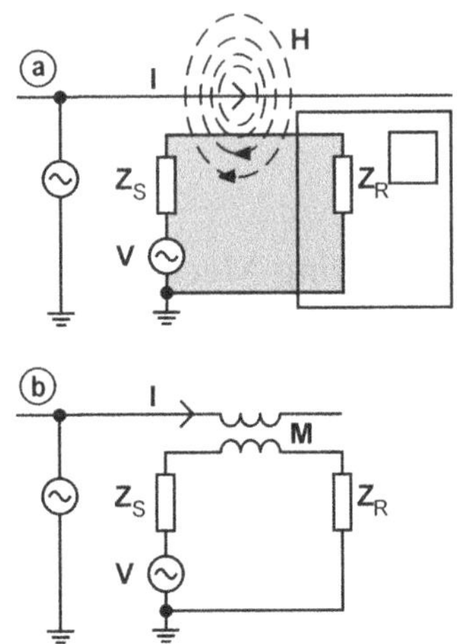

Figure 10.29 Inductive coupling: scheme showing the phenomenon (a) and equivalent circuit (b). A large current flowing into the ground system (e.g. a fault current, a lightning current or a leakage current) generates a magnetic field in the space around the linkage signal lines, creating common mode or differential voltages, i.e. disturbances

10.6.5.3 Capacitive Coupling

Capacitive coupling is always present when there is a circuit with conductors separated by a dielectric. Voltage fluctuations can induce in capacitively coupled circuits currents that are proportional to voltage amplitude variation (Figure 10.30). Devices usually subject to this kind of disturbance can be connection cables, because of the capacitances present between the conductors and between the conductors and earth.

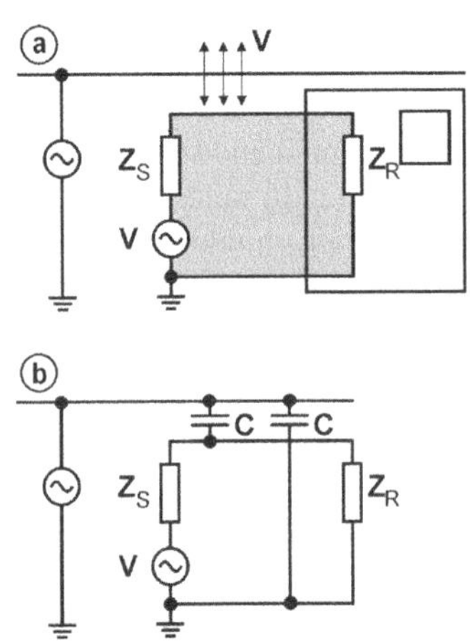

Figure 10.30 Capacitive coupling: scheme of the phenomenon (a) and equivalent circuit (b). Due to eddy capacitances between signal lines and earth, earth voltage fluctuations could become disturbances for signal lines

10.6.6 Low-Noise Equipotential Bonding

The solution to EMC problems requires a systematic approach which cannot be limited only to the design of a correct functional earthing connection. However, an accurate design of the earthing system can contribute in a decisive way to reducing disturbances. This subsection contains an overview of the prescriptions of IEC 364-7 related to combined earthing systems and to the main measures to create a low-noise functional earthing connection (Figure 10.31).

When functional earthing is required, conductive parts of electronic devices must be connected, directly or through intermediate terminals, to the main earthing terminal. The adoption of separate earthing systems is allowed only under the condition that the extraneous conductive parts connected to the two independent earthing systems are not simultaneously accessible.[9,10] However, the standard prescribes that for a combined earthing system (protective earthing system and functional earthing system), requirements related to protective measures must prevail.[11]

Functional earthing should be made with differentiated conductors. These conductors will then be connected directly, or through intermediate terminals, to the main earthing terminal, where protective conductors will also be connected. Eventually, to the conductors connected to the functional earthing system can be added additional turns aiming to introduce a high-frequency impedance.

Conductors used exclusively for functional earthing connection are not subject to protective conductor requirements. Their characteristics will be chosen on the basis of the EMC requirements of the installation. In the field of the same circuit, the functional earthing connection must be designed with the aim of avoiding multiple connections to earth (intentional or not).

Functional and protective earthing connections to the same earthing system allow common mode voltages occurring during earth faults to be cleaned. If it is not possible to reduce the disturbance transferred to the earthing system by a resistive way, it will be necessary to find alternative ways, such as increasing the equipment's immunity level.

Measures to be adopted to reduce disturbances propagating in an inductive way consist, first of all, of reducing the magnetic field by moving sensitive circuits away from the source of disturbances and avoiding parallelisms between functional earthing conductors, protective conductors and data and power cables. Alternatively, disturbances can be reduced by installing, for data transmission, twisted cables[12] or shielded cables. Shielding can be obtained by installing cables inside pipes or steel conduits, or directly by using shielded cables.

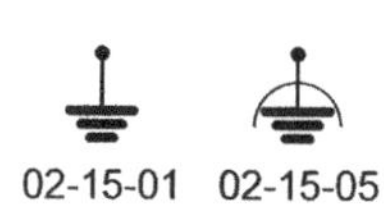

Figure 10.31 Symbols: 02-15-01, earthing system; generic symbol, 02-15-05 low-noise earthing system

[9] Art. 707.545.2.
[10] Art. 413.1.1.2.
[11] Art. 546.1.
[12] In this type of cable, electromotive forces induced in turns in one direction are compensated by electromotive forces induced in turns in the opposite direction.

These measures reduce differential-mode-generated voltages. If shielding is connected to earth at both sides, it creates, with the earthing system, a short-circuited turn, and then common mode generated voltages will be reduced. An appropriate solution, when either an inductive or a capacitive coupling is present, consists of installing double-shielded cables. The internal shield must be connected to earth only at one side, and the external one at both. An external shield may also be formed by the metallic armor of an armored cable or of a metallic pipe.

For a case study see web address

10.7 LIFETIME ASPECTS OF THE EARTHING ARRANGEMENTS

The metallic parts of an earth electrode are in direct contact with the soil and for this reason are affected by different negative measures limiting its life. The earth-electrode conductors must conform to certain minimum dimensions, in order to have:

- mechanical strength during installation and operation;
- adequate current-carrying capacity during fault or lightning currents;
- the capability to withstand corrosive attack in the soil.

All these parameters are interdependent by choosing the electrode material and its cross-section. This section focuses on problems related to the life or durability problems of earth electrodes. For the majority of power installations, the lifetime can exceed 25 years and, for power lines, 35–50 years. The earthing system should be included in repair and maintenance cycles.

Corrosion is the main problem when considering the lifetime of metallic elements buried in the soil. One can distinguish the following kinds of corrosion affecting earth electrodes:

- chemical corrosion, caused by chemical compounds and moisture in the soil;
- corrosion caused by stray (d.c.) currents flowing in the soil;
- galvanic (electrochemical) corrosion.

10.7.1 Chemical Corrosion

Chemical corrosion is caused by chemical reactions between the metal of the electrode and chemical compounds in the soil. The corrosive properties of the soil are characterized by the pH number, which indicates if the soil is acidic, neutral or alkaline. The pH number of neutral soil is equal to 7; smaller numbers show greater acidity, higher ones greater alkalinity. Protection against it consists of choosing:

- the proper material, to result in a longer life of the electrode;
- a proper cross-section of the electrode metal.

Table 10.3 Minimum dimensions of earth-electrode materials ensuring mechanical strength and corrosion resistance [8]

Material		Type of electrode	Minimum size				
			Core			Coating/sheath	
			Diameter (mm)	Cross-section (mm^2)	Thickness (mm)	Single values (μm)	Average values (μm)
Steel	Hot galvanized	Strip[b]		90	3	63	70
		Profile (incl. plates)		90	3	63	70
		Pipe	25		2	47	55
		Round bar for earth rod	16			63	70
		Round wire for horizontal earth electrode	10				50
	With lead sheath[a]	Round wire for horizontal, surface earth electrode	8			1000	
	With extruded copper sheath	Round bar for earth rod	15			2000	
	With electrolytic copper sheath	Round bar for earth rod	14.2			90	100
Copper	Bars	Strip		50	2		
		Round wire for horizontal, surface earth electrode		25[c]			
		Stranded cable	1.8[d]	25			
		Pipe	20		2		
	Tinned	Stranded cable	1.8[d]	25		1	5
	Galvanized	Strip[b]		50	2	20	40
	With lead sheath[a]	Stranded cable	1.8[d]	25		1000	
		Round wire		25		1000	

[a] Not suitable for direct embedding in concrete.
[b] Strip, rolled or cut with rounded edges.
[c] In extreme conditions, where experience shows that the risk of corrosion and mechanical damage is extremely low, 16 mm^2 may be used.
[d] Diameter of single wire.

There are some typical metals recommended for earth electrodes [2],[7],[16],[18],[19],[21]:

- steel (not coated steel or reinforced steel as in foundation earth electrodes, where the concrete protects it against corrosive attack);
- hot galvanized steel;
- steel coated with copper;
- zinc;
- copper;
- copper coated with tin or zinc;
- copper with lead sheath.

Standards give requirements concerning the minimum cross-section of different electrode materials used in different types of profiles. Such minimum cross-sections should ensure adequate mechanical strength and resistance against chemical corrosion. An example of such requirements for HV overhead lines is given in Table 10.3.

10.7.2 Corrosion Caused by Stray (Direct) Currents

Stray currents flow in the soil mainly in the vicinity of d.c. networks. The typical example here is substations supplying d.c. railway or tramway lines. Thus, this problem concerns essentially certain locations, and is not a general one. It consists of flowing part of the return current from the rails to the supplying substation through the ground (Figure 10.32). All metallic elements embedded in the ground (e.g. water pipes), which are in the way of such a current, form an electrode in that circuit and have a positive potential in comparison

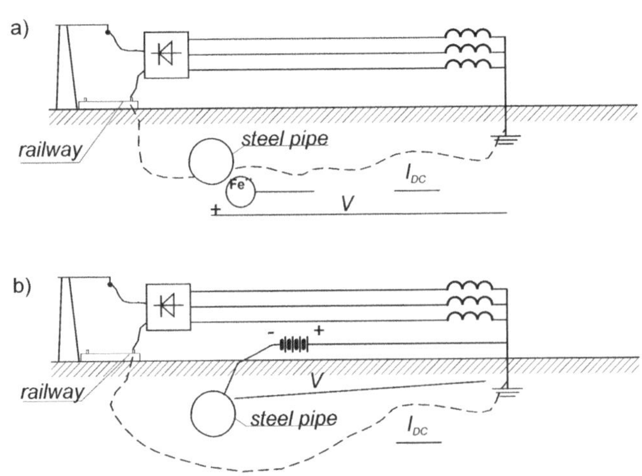

Figure 10.32 Illustration of the corrosion caused by stray (d.c.) currents (a) and the idea of cathodic protection (b)

to the potential of the earthing arrangement of the supplying substation. The ions from such an electrode begin to form part of the stray current and in this way the metallic part loses its material (Figure 10.32a). Protection against this corrosion consists of giving a potential to the protected metallic parts, which is negative with respect to the rails and the grounding point of the supplying substation (Figure 10.32b). This manner of protection is called *cathodic protection*. It concerns in practice metallic pipe systems first of all, but it is not a real problem of earthing electrodes in electric power networks.

10.7.3 Galvanic (Electrochemical) Corrosion

The soil, which contains a certain amount of water, has properties similar to an electrolyte. Metallic parts embedded in the electrolyte have its typical galvanic potential (Table 10.4), and form a galvanic current source, like a battery (Figure 10.33).

Table 10.4 The galvanic potential of some metals in an electrolyte [18]

Metal	Galvanic potential
Aluminum (Al)	−1.71
Zinc (Zn)	−0.9 ± 1.1
Galvanized steel	−0.7 ± 1.1
Steel (Fe)	−0.5 ± 0.8
Steel embedded in concrete	−0.1 ± 0.3
Tin (Sn)	−0.14
Lead (Pb)	−0.13
Copper	0.0 ± 0.1

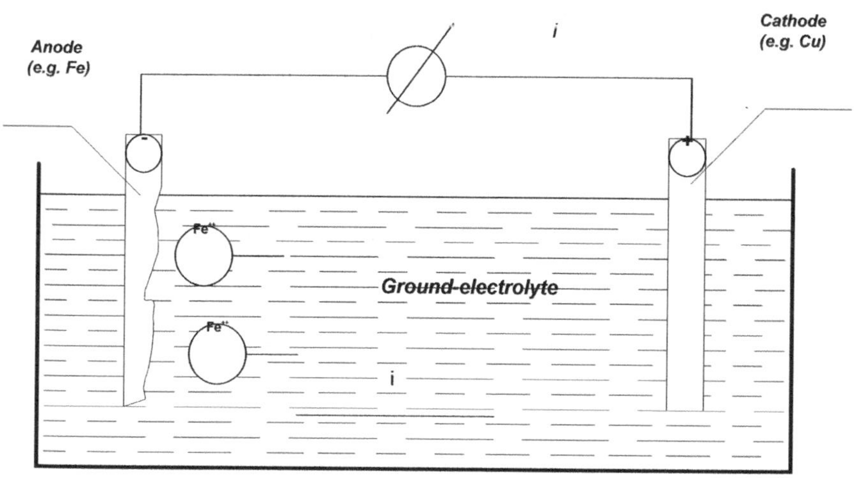

Figure 10.33 Illustration of galvanic corrosion

When two such metallic parts are connected with a wire, the galvanic battery supplies the circuit with a voltage, which is the difference in the galvanic potentials of both metals (Table 10.4). For example, electrodes made from steel and copper have a galvanic potential of about 0.5 V, where copper electrode has positive polarity with respect to steel. In the external part of the circuit (let's say in the wire connecting both electrodes) the current direction is from the positive to the negative electrode. However, inside the electrolyte, let's say in the moisture of the soil, the direction of the current is from the electrode with a negative potential to that with the positive. Thus, for the current inside the soil an *anode* is formed by the negative electrode, and a *cathode* by the positive, i.e. in an opposite manner just like the circuit outside the electrolyte (Figure 10.33).

The current flowing inside the electrolyte from the more negative to the more positive electrode transports ions, which leave the negative electrode. In such a way this electrode loses material, i.e. it is affected by the galvanic corrosion. The metallic parts with more negative potential diminish, while those with positive potential remain without loss.

This process is more intensive in soil with a greater amount of water than in a dry one, i.e. the phenomenon depends on the resistivity of the soil. A soil with a good resistivity, lower than 25 Ω m, is treated as aggressive to electrochemical corrosion, while ground with a resistivity over 100 Ω m is treated as nearly neutral for it. Thus, good soil resistivity also results as a disadvantage, from this point of view.

Protection against galvanic corrosion consists of choosing materials in such a manner that the designer can control these processes. This can be done in different ways. Firstly, an example is to use copper as the material for earth electrodes. Copper has a high potential in comparison to steel, zinc and other commonly used metals, and can be treated as a metal resistant to galvanic corrosion. Another good solution is steel embedded in concrete, as in foundation earth electrodes, which has a positive potential with respect to steel, or galvanized steel buried directly in the soil. On the other hand, one must also consider that other metals in the vicinity of copper and connected to it are affected by the electrochemical corrosion. This is for example the problem in buildings with foundation earth electrodes which are supplied by water pipes made from galvanized steel. They are connected together in the main earthing terminal of the building, but the galvanized steel has the lowest galvanic potential with respect to the steel in the concrete. For that reason the water pipes can be affected by the electrochemical corrosion. Possible protection in such case can be to use a so-called 'victim electrode' (Figure 10.34). It is an additional electrode with a potential lower than both that of the pipe and that of the steel in concrete, in order to design it to resist corrosion and protect the pipe. The idea of this method is similar to that of cathodic protection (Figure 10.32). However, the need for such solutions must be considered carefully.

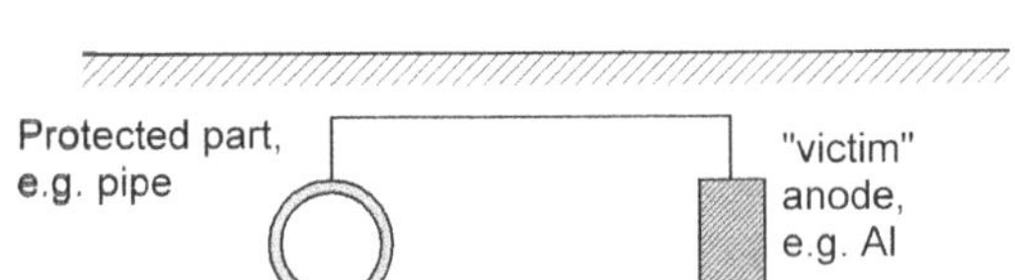

Figure 10.34 Illustration of protection against galvanic corrosion

Table 10.5 Suitable connections of different earth electrode materials; ratio of large area/small area $\geq 100/1$ [19]

Material of small area	Material of large area							
	Steel galvanized	Steel	Steel in concrete	Steel galvanized in concrete	Copper	Copper plated with tin	Copper galvanized	Copper lead-clad
Steel galvanized	+	+ zinc loss	–	+ zinc loss	–	–	+	+ zinc loss
Steel	+	+	–	+	–	–	+	+
Steel in concrete	+	+	+	+	+	+	+	+
Steel copper-clad	+	+	+	+	+	+	+	+
Copper	+	+	+	+	+	+	+	+
Copper plated with tin	+	+	+	+	+	+	+	+
Copper galvanized	+	+ zinc loss	+ zinc loss	+ zinc loss	+ zinc loss	+ zinc loss	+	+ zinc loss
Copper lead-clad	+	+	+ lead loss	+	+ lead loss	+	+	+

+ good for joining; – must not be joined.

Another solution in the protection against galvanic corrosion of earth electrodes consists of connecting different metals, but the metals affected by the corrosion are of large dimensions, and losses caused by corrosion can be neglected. In this case the protected metal can be used in relatively small amounts. In [19] there are recommendations for such possible connections (see Table 10.5).

10.8 MEASUREMENTS OF EARTHING ARRANGEMENTS

Measurements of earth-electrode resistance should allow for determination of its properties in the presence of currents of the greatest prospective value, usually the fault currents. Therefore, the most credible measurements are those carried out under conditions close to actual power system conditions. At present, earth-electrode systems are tested with specialized, earth-resistance testers; ammeter–voltmeter techniques, employing the alternating current of 10–100 A, are seldom used. Tests of earthing arrangements in HV substations by means of short-circuit currents are normally carried out only for research purposes.

The operational techniques for measuring the earthing resistance should meet the requirements formulated as follows:

- the measurement error should not exceed 15–20 %;
- the measurement method should allow determination of the resistance to earth $R_E \geq 0.1\,\Omega$;
- the results of measurement should be displayed directly on the instrument, or be determined using the simplest calculations;
- measuring instruments should be easy to operate and ensure safety;
- no significant disturbances in the power system operation should arise from the measurements;
- the measurement technique should provide elimination of stray currents in earth, both d.c. and a.c., mainly of the network frequency.

10.8.1 Measurement of Soil Resistivity

The measurement of the soil resistivity should be performed as accurately as possible, since the value of the resistance to earth of the earth electrode is directly proportional to the soil resistivity. Incorrect measurement can cause differences between the actual resistance to earth and the planned one.

The test is carried out using a four-terminal earth tester. Four spikes are driven into the ground as shown in the diagram in Figure 10.35. The spikes are located in a line on the earth surface, spaced the same distance *a* (measured in meters) apart from each other. The depth to which each spike is driven should not exceed *a* divided by 20, and is not normally greater than 0.3 meters. The outer two spikes C_1 and C_2 should be connected to the current terminals, the inner spikes to the potential terminals P_1 and P_2.

In the method it is assumed that the spike distance *a* corresponds to depth, for which the soil resistivity is measured. This is important for the designer, and the measured depth should be greater than the depth of the planned earth electrode. On the other hand, a series of measurements made for successively greater spacing *a* can give information about layering of the ground and about the soil resistivity of succeeding layers.

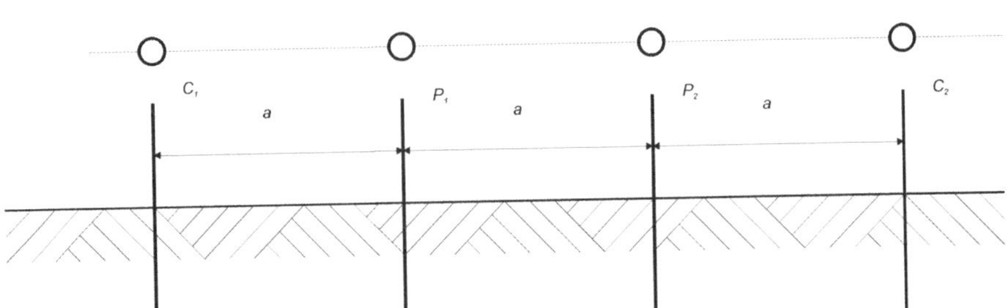

Figure 10.35 Diagram of placement of spikes for measurement of soil resistivity: C1, C2, current spikes; P1, P2, potential spikes

The soil resistivity can be read directly on the tester or calculated using the following formula:

$$\rho = 2\pi \frac{V}{I} a \tag{10.35}$$

where V is the potential difference between spikes P_1 and P_2 (V); I is the current flowing between spikes C_1 and C_2 (A); and a is the spike spacing in meters. The result will be obtained in $\Omega\,m$.

It is important to ensure that the line, in which the spikes are placed (Figure 10.35), is not inserted in line with buried metal parts, like water pipes or cables. Such a situation can be the reason for significant errors in measurement.

10.8.2 Measurement of Resistance to Earth of an Earth Electrode

In order to determine the resistance to earth of an earth electrode, its voltage has to be measured with respect to the reference earth at a specified current. Commonly employed measuring methods can be divided into two groups depending on the source of measuring current and measuring instruments used. Specialized earth-resistance testers, or voltmeters and ammeters in the so-called technical methods, are used for testing.

In earth-resistance testers the source of alternating current of a frequency different from that of the network (usually 75 Hz) is used. It is usually an alternator or a battery-fed frequency converter. As a rule, the measuring current is small, normally of several tens of milliamps, which can significantly influence the accuracy of measurement.

Measurements of the resistance to earth by means of earth-resistance meters should be restricted to closely located earth electrodes (earth electrodes of overhead transmission line supports, small substations, foundation and lightning protection earth electrodes). The measurements of resistance of extensive earth-electrode arrangements, which comprise meshed earth electrodes or long natural earth electrodes, e.g. earthing systems of large HV substations, carried out with earth-resistance meters may contain significant errors resulting from various sources. Such earth-electrode systems should be tested with other methods.

The tests of earth-electrode systems using earth-resistance meters can be carried out in various ways, depending on local conditions, mainly the accessibility of other earth electrodes not connected with the tested earth electrode and in an LV network (installation). The tests can be performed by:

- using an auxiliary spike electrode H and voltage spike S (Figure 10.36a);
- using the neutral conductor (N) or PEN conductor of the LV network and the voltage spike S (Figure 10.36b);
- using the functional earth electrode R_B of a substation and the neutral conductor (N) or PEN conductor (Figure 10.36c).

In these tests, the current circuit is closed through the tested earth electrode R_E and adequately distanced auxiliary spike electrode H or R_B. The voltage drop across the tested earth electrode at the measuring current I_M is measured with a voltmeter.

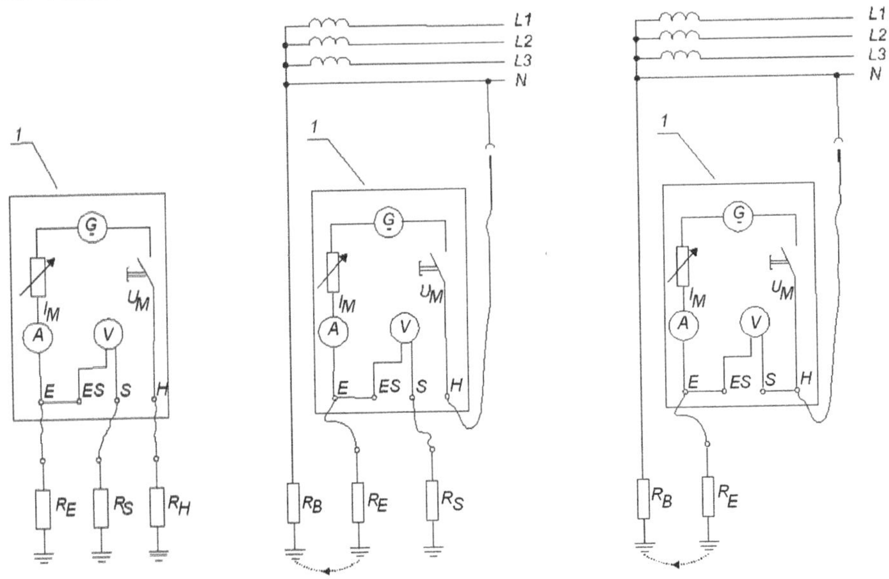

Figure 10.36 Different ways of using the earth-resistance tester (1) to measure the earth-electrode resistance R_e: I_m, U_m, measuring current and voltage, respectively; R_B, resistance of a functional earth electrode in the power system, e.g. of the neutral transformer point; H, auxiliary spike electrode and its resistance R_h; S, voltage spike electrode and its resistance R_s

The important question, which can have an influence on the results obtained, is the potential distribution on the earth surface in relation to the placement of spikes during measurement. This problem is illustrated in Figure 10.37. When testing a simple earth-electrode system, the zone of reference of earth potential is sufficiently extended, so that finding a place to locate the voltage spike electrode S is not difficult. When testing an extended earth-electrode system, this zone may occur in a narrow strip between the tested earth electrode R_E and auxiliary current spike electrode H. It may be very difficult, or even impossible, to determine this zone if the distance between the electrodes R_E and H is too small (Figure 10.37). Locating the voltage spike electrode S within the measuring current flow area, too close to the auxiliary current spike electrode H or to the tested earth electrode R_E, results in significant measurement error. In consequence, too large or too small a value of the earthing resistance is obtained. Table 10.6 contains the requirements for the spacing arrangements of the spike electrodes H and S with respect to the tested earth electrode, determined to minimize the measurement errors which result from inadequate location of the measuring spikes.

In the tests carried out by means of earth-resistance testers, the result of measurement is displayed on the instrument, or, using testers of older design, it is read from a potentiometer scale and multiplied by 1, 10 or 100, depending on the instrument range. Figure 10.37 also presents a simple method of measurement of the potential distribution on the earth surface, by changing the placement of the voltage spike electrode S.

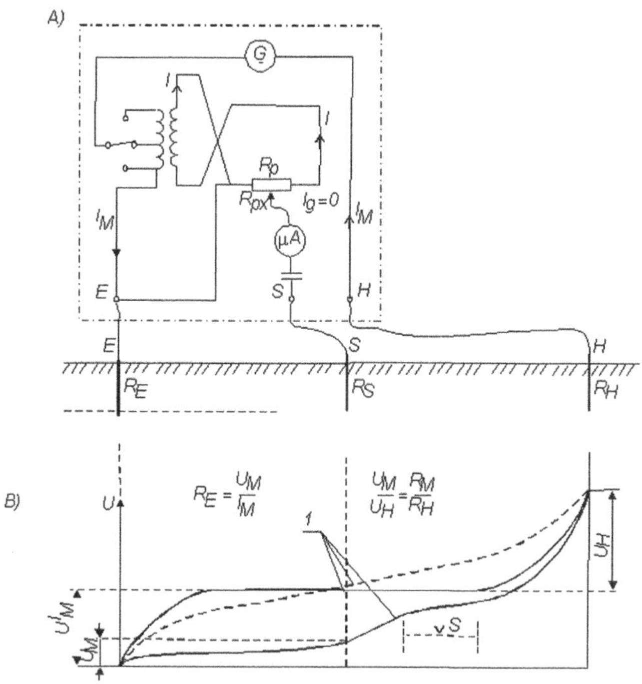

Figure 10.37 Measurement of earth-electrode resistance using an earth-resistance tester: (a) a simplified scheme of the measuring instrument and manner of connection of the electrodes; (b) voltage distribution (curve 1) on the earth surface at measuring current I_m, for various configurations of the tested earth electrode. R_e, R_s, R_h, resistance of the tested earth electrode, voltage spike electrode and current spike electrode, respectively; U_m, measuring voltage; U_h, potential of the voltage spike electrode

Table 10.6 Requested distances between the earth electrodes: tested earth electrode E, auxiliary voltage spike electrode S and current spike electrode H respectively (Figure 10.37) at measurement resistance to earth

Construction of the tested earth electrode E	Minimum sizes (m) or relative sizes	
	Spacing of electrodes along a line	Spacing of electrodes at vertexes of a triangle
A vertical rod electrode with length l_E	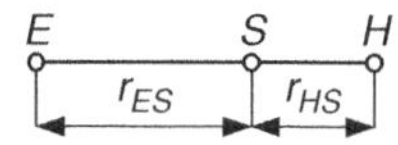 $r_{ES} \geq 1.8l_E$ $r_{HS} = (0.4–1.4)r_{ES}$ $r_{HS} \geq 20\,\text{m}$	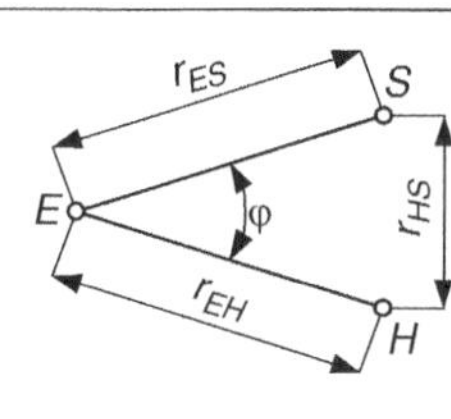 $r_{ES} = r_{EH}$; $r_{ES} = 2l_E$ $r_{HS} = (0.4–0.8)r_{ES}$ $r_{HS} \geq 20\,\text{m}$; $\varphi = (22°–45°)$

Table 10.6 (Continued)

Construction of the tested earth electrode E	Minimum sizes (m) or relative sizes	
	Spacing of electrodes along a line	Spacing of electrodes at vertexes of a triangle
A horizontal earth electrode with length l_E	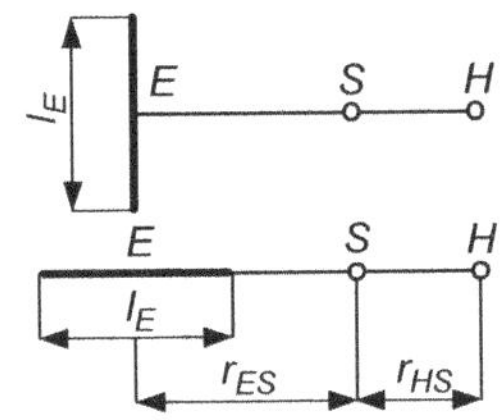 $r_{ES} \geq 1.2 l_E$ $r_{HS} = (0.4–1.5) r_{ES}$ $r_{HS} \geq 20\,\text{m}$	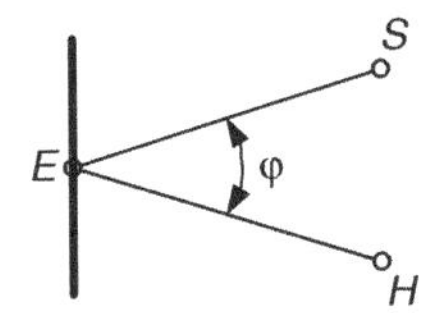 $r_{ES} = r_{EH}$; $r_{ES} = 1.2 l_E$ $r_{HS} = (0.3–1.0) r_{ES}$ $r_{HS} \geq 20\,\text{m}$; $\varphi = (22°–45°)$ (all symbols as in the figure above)
An extended, meshed earth electrode	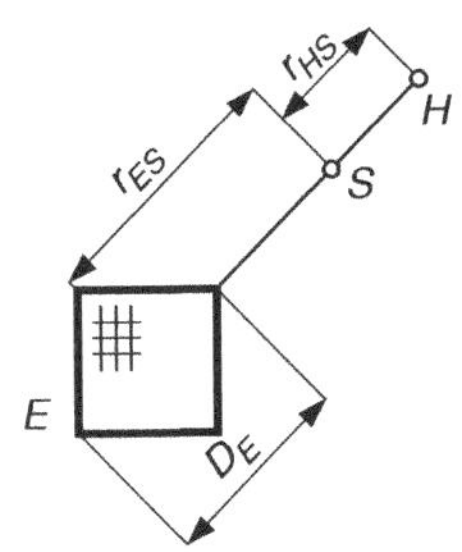 $r_{ES} \geq 1.5 D_E$ $r_{HS} = (0.53–0.74) r_{ES}$ $r_{HS} \geq 20\,\text{m}$	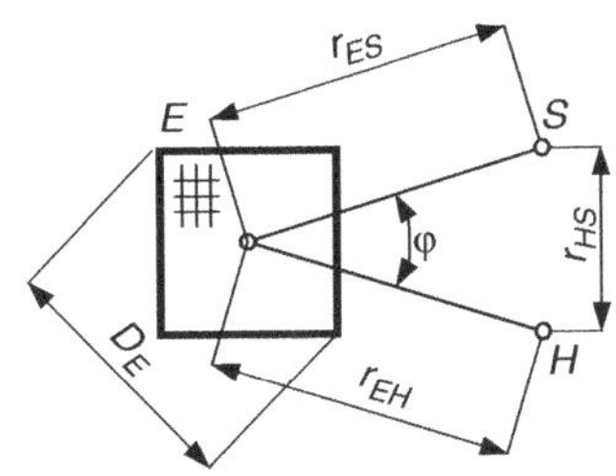 $r_{ES} = r_{EH}$; $r_{ES} = 1.5 D_E$ $r_{HS} = (0.3–1.0) r_{ES}$ $r_{HS} \geq 20\,\text{m}$; $\varphi = (22°–33°)$
Optimal test conditions	$r_{HS} = 0.62 r_{ES}$	$r_{HS} = 0.5 r_{ES}$; $\varphi = 30°$

BIBLIOGRAPHY

[1] Baggini A., Bua F., *La compatibilità elettromagnetica*, Manuale degli impianti elettrici, Tecniche Nuove, 2000.

[2] Bätz H., Loyda A., Müller B., Röllig H.-W., *Elektroenergieanlagen*, VEB Verlag Technik, Berlin, 1989.

[3] Carrescia V., G. De Bernardo, *Impianti di messa a terra*, Edizioni ENPI, Turin, 1974.

[4] Clayton R. P., *Compatibilità elettromagnetica*, Hoepli, Milano, 1995.

[5] DIN 18014, *Fundamenterder*, Beuth Verlag, Berlin, 1994.

[6] Directive EMC 89/336/EEC on the approximation of the laws of the Member States relating to electromagnetic compatibility.

[7] *Earthing Practice*, CDA Publication 119, 1997.

[8] EN 50341:2001, *Overhead electrical lines exceeding AC 45 kV, Part 6. Earthing systems.*

[9] IEC 364-1, *Electrical installations of buildings – Part 1: Fundamental principles, assessment of general characteristics, definitions*, 1989.

[10] IEC 364-5-54, 1980; A1, 1982, *Electrical installations of buildings. Part 5: Selection and erection of electrical equipment*, Chapter 54: Earthing arrangements and protective conductors.

[11] IEC 364-7-707, *Electrical installation of buildings – Part 7: Requirements for special installations or locations*, Section 707: Earthing requirements for the installation of data processing equipment.

[12] IEC 60050-195, *International Electrotechnical Vocabulary*, Chapter 195: Earthing and protection against electric shock, 1998.

[13] IEC 60364-7-707:1984, *Electrical installation of buildings – Part 7: Requirements for special installations or locations – Section 707 – Earthing requirements for the installation of data processing equipment.*

[14] IEC 60617-2, *Graphical symbols for diagrams – Part 2: Symbol elements, qualifying symbols and other symbols having general application.*

[15] Markiewicz H., Klajn A., *Earthing Systems – Fundamentals of Calculation and Design*, LPQI Application Guide No. 6.3.1. 2003 (available at www.lpqi.org).

[16] Markiewicz H., Klajn A., *Earthing Systems – Basic Constructional Aspects*, LPQI Application Guide No. 6.5.1, 2004 (available at www.lpqi.org).

[17] Norma CEI 110-26, *Guida alle norme generiche*, EMC, 1995.

[18] Rudolph W., Winter O., *EMV nach VDE 0100*, VDE Verlag, Berlin–Offenbach, 1995.

[19] *Switchgear Manual. ABB Pocket Book*, Cornelsen Verlag, Berlin, 1999.

[20] Venhuisen R., Earthing & EMC: a systems approach to earthing. APN LPQI 2002 (available at www.lpqi.org).

[21] Wolkowinski K., *Uziemienia urzadzen elektroenergetycznych* (Earthing of power electrical devices), WNT, Warsaw, 1967 (in Polish).

11

Reliability of Electricity Supply: Structure

Angelo Baggini, David Chapman and Francesco Buratti

To achieve power quality, the choice of schemes for electrical plants is one of the keys of primary importance in the electrical project which requires a full knowledge of the loads and the characteristics of supplies.

For a very modest plant, a simple examination of the map showing the point of common coupling (PCC) and the displacements of the loads defines an adequate scheme. But for a more complex plant, e.g. with more supply possibilities and more voltage levels, the final choices are not apparent.

This chapter discusses the following themes:

- The most common basic schemes of electrical grids, pointing out their ability to satisfy different service conditions regarding the exigencies of different characteristics of the subject and from a technical and an economic point of view.
- General criteria, for scheme choice, i.e. a crossed comparison among the indispensable requirements for users and the features of the possible supplies.
- Different aspects useful for the choice of the scheme, examined singularly: the availability and continuity of supplies are discussed, distinguishing even critically possible non-coinciding exigencies under the corresponding points of view.
- Redundancies of components and circuits, also introducing other important and correlated features such as infallibility, independence, bottlenecks capable of annulling redundancies.

The criteria shown are generally applicable to high-, medium- and low-voltage systems as well as to different types of user plants such as hospitals, computer centers, continuous process industries, telecommunication centers, etc.

Handbook of Power Quality Edited by Angelo Baggini

The evaluations proposed for comparison are not based on statistical data of reliability or rate of failure, but on comparisons between intuitive and simple conditions and situations. Such an approach leads to meaningful results that are no different from the ones obtainable by far more complex methods of analysis. The reason is that the random elements are varied and hard to value, so that experience and common sense are still effective aids.

Also the more complex grids, as long as they are rationally structured, can always be reduced to simpler equivalent grids where partial grids are considered as concentrated loads.

11.1 BASIC SCHEMES OF ELECTRICAL GRIDS

The scheme of an electrical grid is a graph representing the connections among feeder points (power sources) and utilization points (loads). The connections can be radial, ring shaped or other, passing from the smallest structure (the lowest number of branches to connect all the nodes) to a more and more complex structure that is redundant with alternative ways among the nodes.

The definition of the scheme is not just a geometric problem, but rather a basic problem in the electrical project second only to the analysis of load exigencies and the characteristics of the sources.

Once the type and architecture of the scheme suitable for the service exigencies has been found, it is necessary to analyze, even from a quantitative point of view, the rated and transient conditions of the grid, in order to compare the alternatives and make a final choice on the scheme. Calculations of the power load flow and short circuit enable optimization of the choice of voltages (both for systems supplied directly by public grids and for those originating inside the plant), of the power and the characteristics of the machinery.

An important problem concerns the method of connecting (where and at what voltage) generators and big loads [3].

11.1.1 Simple Radial System

At first sight, with a map of the plant in mind, a simple consideration of the feeder point and the load position leads to a definition of the grid scheme. This is true only for a very small plant having just a few loads that can be connected directly to a feeder switchboard, and the resulting scheme is radial.

The simple electrical installation of a flat or apartment points out immediately the problem of choosing a distribution with a pure radial or shunted or mixed scheme. Staying within the limits of a grid with only one voltage level, as soon as the power and load number exceed a certain limit, it is no longer possible to connect all lines to only one node (switchboard) but necessary to pass to a radial structure, i.e. with secondary switchboards at more levels. The grid study becomes complicated, the first electrical problems arise – voltage drop, protection selectivity – in addition to geometric problems (switchboard location, line ways, etc.) and economic ones (switchboards and structure costs). This is the typical case of an office building with floor and zone switchboards. The first step to be made is the location

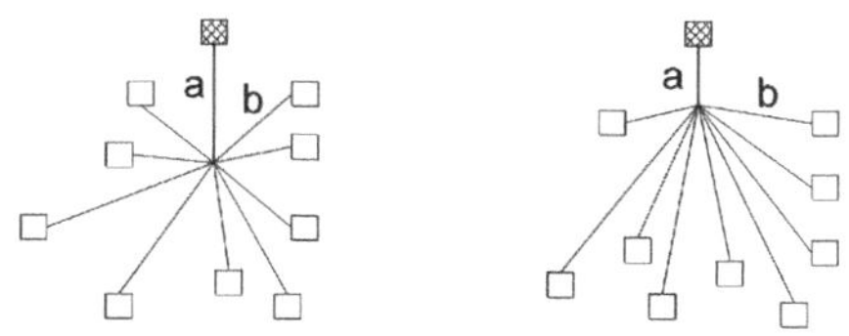

Figure 11.1 A coordinated solution, considering the voltage drop and cable cost problems, optimizes the position of the single nodes (under distribution switchboards) and this position determines the length of the feeder main line and of the derivations and the line sections

and connection of the switchboard as well as its load assignment (Figure 11.1). In order to avoid oversized lines, the biggest loads (particularly motors for which the voltage drop during starting is important) must be derived as much as possible at the bottom, or directly from the main switchboard. Switching and protection problems depending on the grouping of loads under the same protective device, in other words selectivity problems, can suggest modifications and adjustments to a first proposal of the grid structure, remembering the functional relations between single or associated loads.

Defining the grid structure can be harder than finding the load barycenter. Because of voltage drops and losses, over a certain length of the lines the biggest section of conductors will be required.

11.1.2 Ring Scheme

The ring scheme is characterized by at least one more branch than the least necessary to connect loads to the feeder node.[1] Electrically this corresponds to having at least one alternative way to feed one or more loads (Figure 11.2).

A ring scheme grid can be used with an open ring or closed ring (Figure 11.3). In the former case, in order to have an alternative way, it is necessary to remove the faulty part and restore supply. In the latter case it is necessary to remove the faulty part because the alternative connection is already activated (switch closed). In both cases the criteria of use of protection and switching devices will be different: simple disconnectors or switches for the derived loads can be used, making the scheme suitable to ensure the availability of supply or even continuity (see Section 11.2.3.1).

The alternative scheme (b) in Figure 11.3 enables the continuity of the ring to be found even when one or more secondary busbars are not available because these busbars are not necessary for continuity of the ring. The ring scheme can be in a double or triple ring redundant shape and used at all voltage levels.

Ring schemes are not used frequently in continuous cycle plants, where double radial schemes are preferred.

[1] If n is the total number of nodes in the grid, this least number of sides corresponding to the ring scheme is equal to $l = n - 1$.

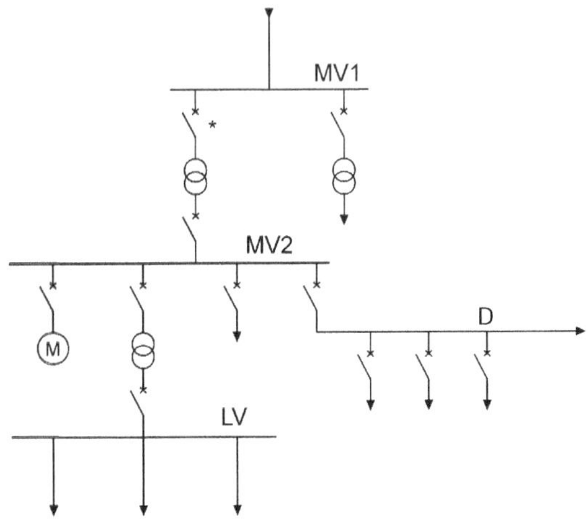

Figure 11.2 Generally speaking, the ring scheme for a grid with more voltage levels has a tree structure, with a possible spine supplying loads distributed along the way

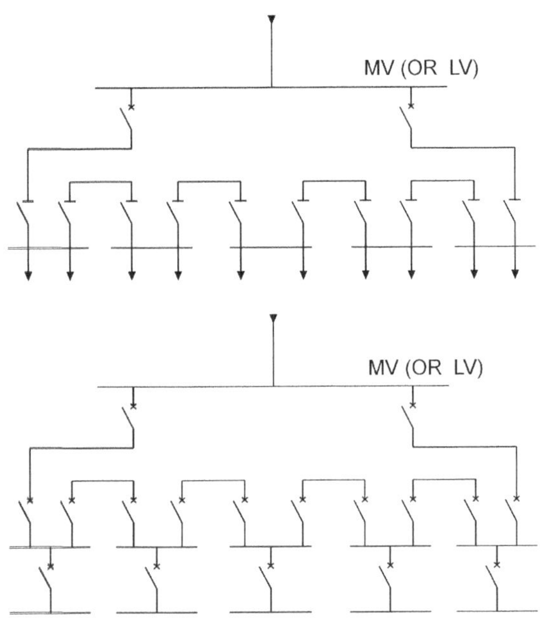

Figure 11.3 A ring scheme grid can be used with an open ring or closed ring: (a) a ring with disconnectors – the nodes are the load busbars; (b) a ring with switches – the nodes are independent of the load busbar

11.1.3 Double Radial Scheme

The benefit of this scheme is having two equal alternative ways, made of a doubled basic radial scheme. The duplication of the scheme can be extended to a single server, or more frequently to one or more distribution nodes (busbars) (Figure 11.4a).

At busbar level a double radial scheme can be used with an open or closed tie breaker. Management with a closed disconnector requires the two feeding lines to be derived from the same feeder, limiting greatly the functioning of a complex grid with more supplies (e.g. grid and self-production) that are not necessarily always in parallel. The functioning of the two half busbars in parallel requires the plant to be able to bear the sum of the contributions to the short circuit from the two supplies and from the rotating machinery. However, if

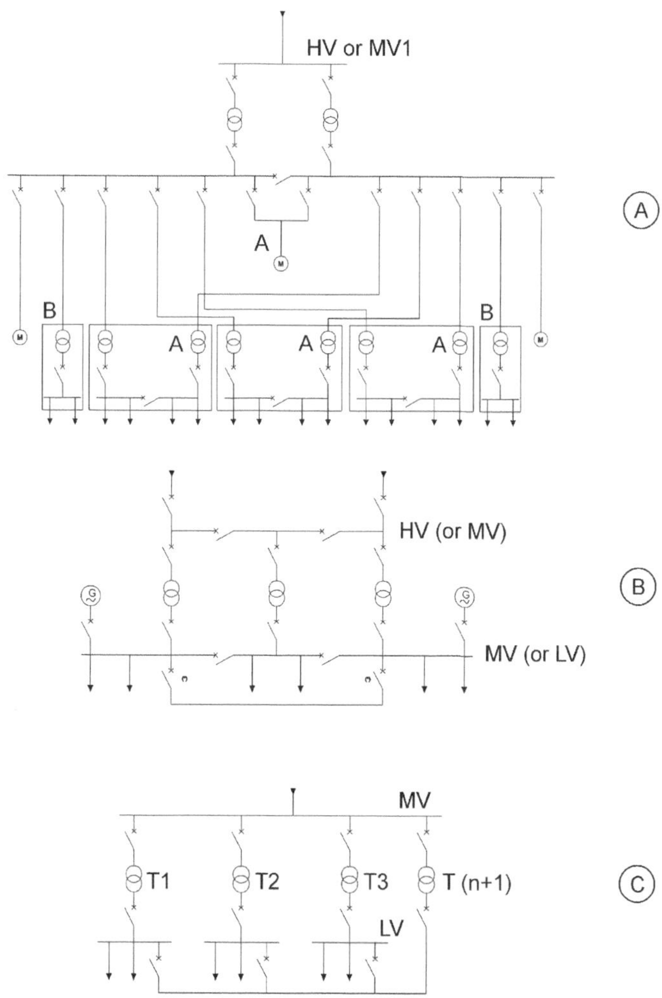

Figure 11.4 Multiple radial schemes

the functioning has an open disconnector and the switching among sources is possible only with short parallel after manual starting by the operator, the plant can be dimensioned for the contribution to the short circuit of one feeder source only.

Schemes of radial type are also used particularly to solve more economic problems of reserve for lines and transformers: only one branch is reserved for two adjacent branches connecting the nodes themselves (Figure 11.4b).

Scheme $n+1$ can be considered as derived from the triple radial scheme. It is the simple and economic solution to ensure, with one branch only (feeder or transformer), the reserve at more nodes (Figure 11.4c).

11.1.4 Meshed Scheme

This scheme is characterized by several connections among the grid nodes, to allow alternative supply ways for some of them, and able not only to establish a reserve connection, but also to improve subdivision of the load in various branches and among different feeder sources.

Figure 11.5 shows an example of this scheme characterized by a transformer named TT connecting the different voltage systems. The transverse connection allows transfer from a load busbar to the other part of the active and reactive power of self-production, improving service conditions [13].

11.1.5 Compound Scheme

In some rather complex plants the elementary schemes described can be combined with one another. It is therefore possible to encounter simple radial systems, double and triple systems, combined with simple or double ring systems.

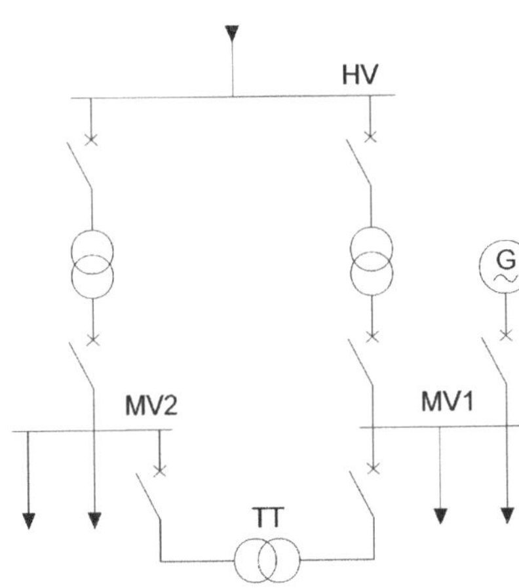

Figure 11.5 Meshed scheme

11.2 GENERAL CRITERIA FOR THE STUDY AND CHOICE OF THE SCHEMES

11.2.1 Parameters and Basic Conditions

In a complex plant the scheme structure depends first of all on the aspects that, in relation to the location of the feeder and load, determine the choice of the voltage levels.

The short-circuit aspects depend on outside supplies and on the machinery (transformers, generators and motors).

The voltage and power levels of the transformers determine the value of the rated currents of the single circuits.

Short-circuit currents and rated currents must not exceed their natural characteristic values (i.e. those well used in other plants and under comparable conditions to those of the plant being studied) for machinery and switchboards. With higher values than these, it is possible to have not only unacceptable stress for current reducers, cables and terminals, but also surprises that are difficult to foresee at design time.

With the data considered above it is possible to trace a simplified scheme of the grid (simple radial scheme) that can be completed only on the basis of further analysis of the characteristics of supplies, of service exigencies of the loads [17].

11.2.2 Scheme of the Grid as a Link Between Supplies and Loads

There are several possible choices concerning a plant scheme. First of all it is necessary to be free from secondary considerations and guided only by the analysis of the supply and load characteristics.

The basic analysis must evaluate the real exigencies of the users and of the service they have to do, as to:

- the availability and continuity of the supply required;
- the aptitude to tolerate various long planned or casual stops, because of faults or breakdowns, for maintenance or modification.

The toleration of stops may be very different, even with the same production or service unity.

The second analysis must evaluate the supply characteristics in order to evaluate whether they are suitable or have to be supplemented (by other outside supplies from self-production, reserves, emergency, safety, etc.).

Generally the supply from a public grid can be improved by making a connection with a better grid, or passing to a higher voltage level, if possible.

The third analysis concerns the distribution structure: it includes the choice of the scheme, the machinery and sizing, and the devices, lines and controls.

It is neither easy nor simple to find links (typical correlations) among the three elements considered. However, for small plants in which loads do not require a particular supply quality, only one supply is generally sufficient.

Important plants in relation to the functions assigned and to the dimensions and, further, industrial poles need multiple supplies, from the public grid and from self-production, and multiple schemes that are more or less widely redundant.

Regarding the importance and the dimensions of the plant, the grid study must find one or more nodes where supplies from sources converge and from which supplies depart to the load centers or to single loads. These nodes can be stations, medium- or low-voltage switchboards, with one or more busbar systems. On these nodes, if important, the voltage is kept constant acting on the tap load changer of the main transformers and on the static compensator. It is important for the function of these main nodes to be quite clear in the scheme structure and for the continuity and availability of their supplies.

Particularly when renovating and modifying existing plants, it is easy to mix up old and new parts with even remarkable power fluxes that may change direction. Such a situation leads to uncertainties, chiefly for the choice and calibration of protection and for the choice of the transfer transformer ratio. Operators may have the same uncertainties (including automatic systems).

11.2.3 Characteristics of the Installation

All the characteristics and the behavior dealt with so far can generally be estimated only in relative terms of comparison. As an example, action to be taken to reduce the supply unavailability to an hour a year, when for the system supplied it is expected with weekly frequency, means stops for maintenance or breakdowns can be judged unjustified and must be reconsidered.

Let's now define and analyze supply properties, referring, according to the context, both to the real power sources (external or internal) and, regarding terminal loads (single or associated), to their supply terminals – even if the prerogatives of an adequate supply must extend to the plugs of the single user.

A particular meaning is given to the terms *continuity*, *availability*, *redundancy*, *independence*, etc., used to describe the properties of the supplies and, more generally, of the installations.

11.2.3.1 Availability and Continuity

In this subsection, for availability of a supply (of a circuit or a component), it is intended that the aim is to become operative again, after a fault or a breakdown, within a certain time limit (delay) considered acceptable and compatible with reference to the service exigencies (see IEC 191-11-03) [1].

Continuity of a supply is the aim to be operative without interruptions, or to be restored after interruptions of a limited length of time, considered compatible with the service required (see IEC 604-01-32) [1].

The concept of *uninterruptible* supply joins the concept of availability for reasons of safety or preventing damage. The attributes of availability and continuity, easy to mistake, are not equivalent. They often have different exigencies and chiefly require different actions to be taken.

Availability does not always mean quickness in restoring the supply, but its certainty. Continuity suggests keeping a supply or its restoration within determined limits of time however short.

A particular availability and continuity level can be necessary for a single or groups of loads, almost never for a whole plant. This means that dedicated action can be taken, not necessarily extended to the whole plant.

Obviously this characteristic, originally only a technical one, has economic consequences and has to be considered as a sparse requirement.

As concerns availability, supply and connection between supply points and loads (alternative ways), multiplicity and independence are essential elements. Also, the possibility of rapidly replacing a faulty element with a spare part can warrant availability.

Besides the necessary action for availability, for continuity alternative supply systems are fundamental (double supplies in parallel, simultaneously; or double supplies with automatic commutation, more or less rapidly or instantaneously).

Availability is not always the necessary condition for continuity. More precisely, it can be said that continuity is an instantaneous availability, not necessarily of a long length of time and associated to availability, meant as a warrant for restoring the supply and persisting.

For instance, for a robot continuity of supply is necessary, but for a lack of it a stop is acceptable. Another example of continuity without availability can be the one of a data center that cannot bear interruptions, but can stop after a certain time and then without inconvenience, as long as the controlled stop is warranted. In this case continuity is ensured by the UPS. Generally, for availability motor generators are necessary.

The problem of safely stopping a process plant is the same. The dangerous situation is the instantaneous uncontrolled stop, not the stop itself.

The opposite case is given by a plant for which a stop is not detrimental, but can be seriously damaging if the supply is not restored at a given moment. Classical examples are an electric oven and a mixer of substances that can solidify if not kept agitated.

Availability is chiefly tied to safety functions, continuity to the functioning requirement (service, production).

These distinctions are not always recognizable in plants in reality. However, they are useful in order to clarify the basic concepts and find out the necessary actions to ensure the functioning conditions required.

Such actions are generally taken at the users' or the users' aggregate level, and cannot be taken case by case by the commercial electrical enterprise. The addition of a second (or third) supply line, connected to different nodes (or, as said earlier, to a higher voltage) of the public grid, can make a difference.

It is important to repeat that the actions indicated by an accurate analysis are generally sparse and dedicated and allow one to obtain sensible savings and answers more suited to the single exigencies.

Another remark concerns the specificity of the actions that, case by case, can satisfy these availability and continuity exigencies. This specificity is the higher, the more distributed are the actions taken. Sparsity allows a remarkable power reduction in the devices adopted, but it requires, in comparison to centralized devices, a more careful monitoring of their condition.

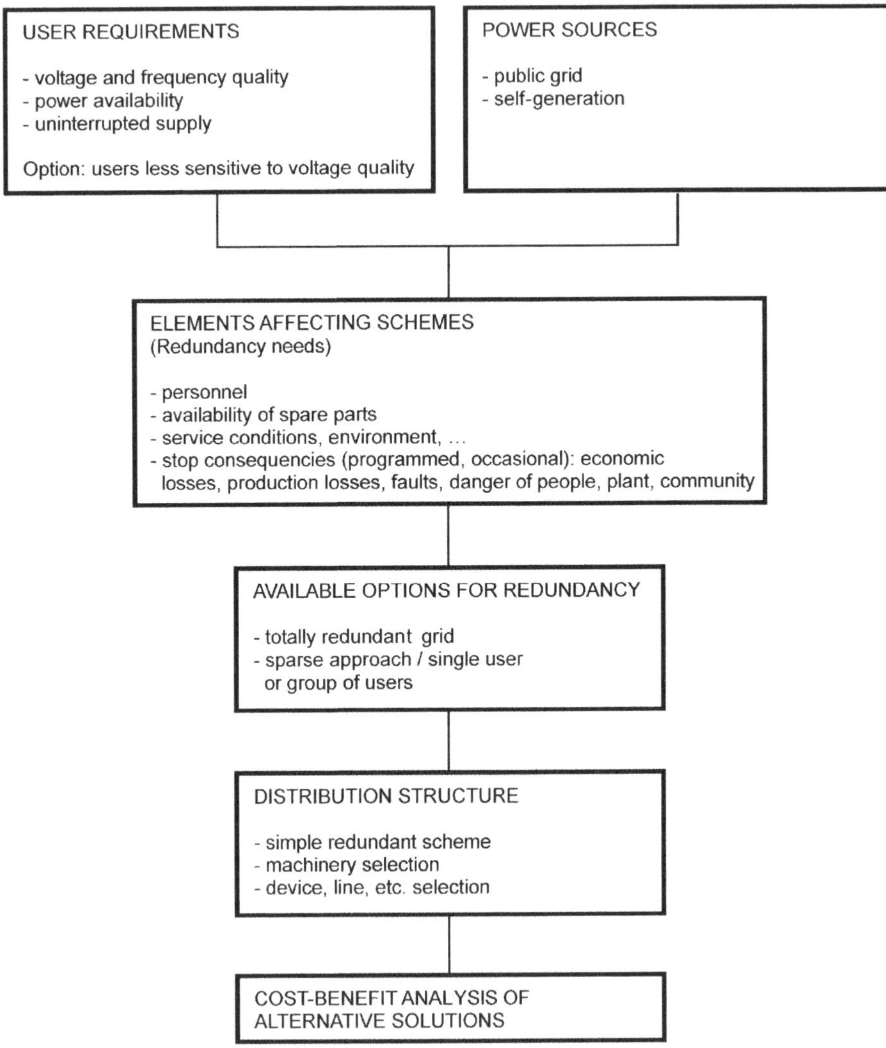

Figure 11.6 Scheme flowchart

Shown in the flowchart of Figure 11.6 is the analytical procedure to ascertain the characteristics of a power center, with reference to the requirement of availability and continuity, and explicitly introducing maintenance.

11.2.3.2 Reliability

The reliability (see IEC 191-12-01 [1]) and therefore the availability of a component is its own characteristic and depends on the project, on the manufacture of the component and on the test carried out. The maintenance of its original characteristics depends on the way it is used, i.e. on the correct choice, the dimensioning, the installation, the operation and the maintenance. For instance, with reference to switches, it is not sufficient for rated performances to be taken as adequate as originally, so, to be kept in time, they must be checked regularly.

The possibility to carry out control and inspections, preventive and occasional maintenance depends on the fact that, at some moments and for determined times, or casually in the case of faults or breakdowns, the service can be suspended. If the service conditions of the plant, or a part of it, do not permit suspensions or interruptions, it is necessary to adopt a wider redundant solution.

11.2.3.3 Redundancy

The concept of *redundancy* in technical terms is tied to the statistical one of a reduction in the probability of unavailability (see IEC 191-15-01 [1]).

Redundancy means the existence of one or more components, of one or more circuits, being able, in replacement of homologous parts of a system, to assume their functions totally or partially. When this function is consistent, or the function is assumed only in particular conditions, it is preferable to speak in terms of reserve or help.

Redundant components or circuits have generally equal importance on the screen, i.e. the qualifications of primary or secondary, normal or preferential, normal or reserve do not apply to them.

After recognizing, on the basis of the analysis already described, that determined functions of the grid must have particular requirements, it is necessary to ascertain where redundancy is essential and where it is not and can therefore be reasonably avoided.

Let's consider a double radial system supplied by the public grid through one medium-voltage line only. The question is whether the main switchboard must be made of two sections or if one section only is sufficient. The answer requires an evaluation of the reliability of the switchboard compared to the supply availability: how many times a year can the switchboard be unavailable due to faults, maintenance, modifications? If the answer is mostly favorable for the supply availability and therefore negative for the switchboard, the subdivision of its busbars into two separate sections can be justified.

In a different case the switchboard can be defined as infallible compared to supply and the type with a simple busbar can be correctly chosen.

So the concept has been introduced of an infallible element, a term which must not induce errors, but is particularly useful in this type of evaluation. Infallibility is not to be taken in an absolute sense, but in a relative sense, in order to define and characterize an element the reliability of which is significantly better compared to one of the other components of the same plant. Only the non-infallible components, if necessary, must be redundant, or made infallible, too. A service component or one in the warehouse is defined as infallible when it is subject to become faulty so that it reduces the actual safety of the circuit. The attribute infallible is therefore tied up, in the widest sense of the term, to the service conditions, and therefore has a relative value of comparison.

Where redundancy and, necessarily, the complexity that it involves are inevitable, complexity has to be controlled. In other words, it is necessary to evaluate the benefit that redundancy can assure in comparison to the risk of inconvenience (complexity, increased components and consequent fault conditions) [11].

It is important to remember that up to now reference has been made implicitly to alternative activated functions, i.e. those that can be made operative by a normal manual or an automatic action.

The concept of redundancy can also include the availability of spare or reserve components that are not activated, but that can be made readily available (replaced) to homologous damaged elements.

The result of the non-activated redundancy is measured in terms of availability and of restoration time, not of service continuity.

11.2.3.4 Independence

In the concept of redundancy it is fundamental for the redundant parts to be 'independent' of one another. The degree of independence cannot be defined in general, but has to be evaluated, case by case, with regard to the function of the part under examination and to the service conditions.

The following examples aim to clarify the concept of independence:

- An MV switchboard with a double system of busbars, with a degree of independence according to the following situations:
  - The two busbar systems are disposed adjacently and side by side within a protected switchboard without any segregation. They are independent only in relation to ordinary operation, for the possibility to effect different services on each in turn, but not in relation to faults on the busbars or on the disconnecting switches against the busbars, at maintenance or on repair (Figure 11.7a).

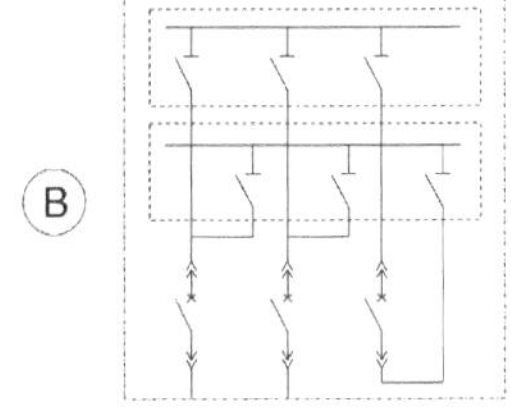

Figure 11.7 Switchboard with double system of bus bars: (a) busbar system without any segregation; (b) busbar system within compartments segregated metallically

- The two busbar systems, adjacent and side by side as above, are disposed, with their switches, within compartments segregated metallically and made short-circuit tight (Figure 11.7b). They can be considered as independent for determined maintenance and repair operations and for faults concerning one system of busbars only, but not for faults that concern disconnecting switches against the busbar or that, because of the emission of corrosive smoke, can deteriorate contacts.
- The two busbar systems are set out within the switchboard as a simple busbar (Figure 11.8) with the two sections set in line and with a tie breaker being the only common part to the switchboard; the independence of the two sections depends on the measures taken to prevent propagation of a section fault into the other through the tie-breaker unit.
- The two busbar systems are in two completely separate switchboards (double switchboard, scheme with two switches, Figure 11.9) put at a safe distance in case of fire, or even in different rooms (fireproof compartments); independence is extended to all the conditions considered above.

- Two lines in cables with separate ways and such that the fault of one line cannot compromise the functionality of the other.

An important element in evaluating the independence of common causes of faults is the container of the part being examined. Two switches in one box share the same environment and are immediately contaminated. In the examples shown the common container is the

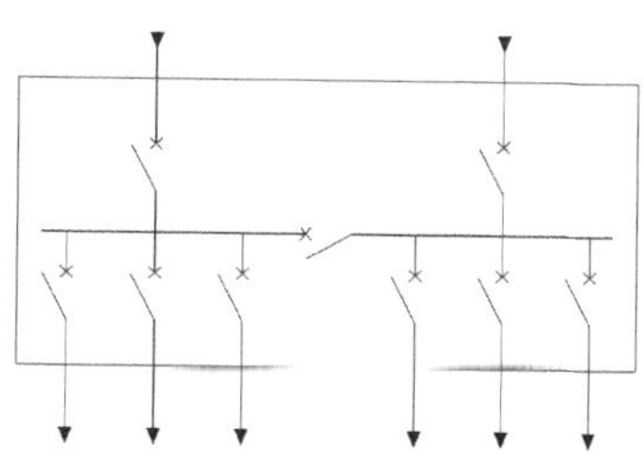

Figure 11.8 Switchboard with two sections set out in line and with a tie breaker

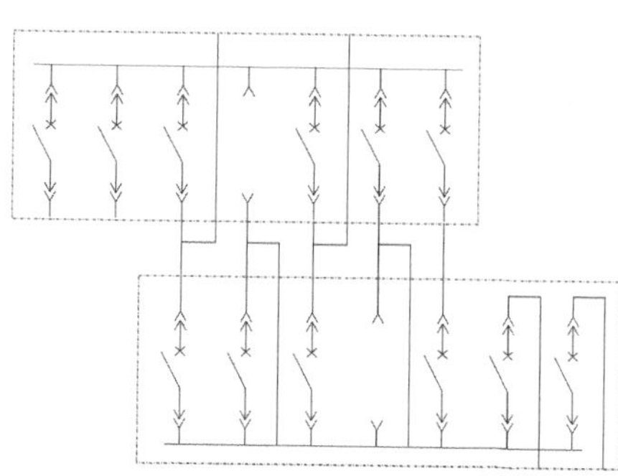

Figure 11.9 Two completely separate switchboards

switchboard with more or less effective insulation, but also the room is a container for more switchboards and, in the case of fire, independence can be compromised.

The analysis of independence becomes complicated when passing from consideration of the power circuits to protection and control systems, i.e. sources and supply circuits, circuits, devices.

The independence of alternative parts comprising a redundant system can never be absolute and total, unless they are completely separate and autonomous from supply up to the final user, including the respective control systems. In all other cases there are contact points, both for power circuits and for protection and control systems. For instance, in a double radial system independence can be compromised by the correspondence of the tie breaker to the busbars, a fault in which can be the common cause of disorganization of the two branches of the system, independently of each other for all the rest. This is the reason why in the schemes with a double switchboard and double switch (Figure 11.9) the disconnector is also doubled. The study of the control systems of the tie breaker requires particular attention to keep the independence level as high as possible.

From what has been said it is obvious that the concepts of independence and infallibility cannot be defined quantitatively (in terms of probability), but only in relative terms of comparison to the availability of other parts making up the portion of the plant under consideration.

11.2.3.5 Bottlenecks

In order to illustrate the considerations on redundancy in a railway example, it is possible to say that the advantages of the double rail cancel out one another when the rail becomes single and makes a bottleneck for circulation. In an ideal double radial system, supposed to obtain supply redundancy for the system below, the unique element could be (Figure 11.10):

- a feeder;
- the main bus bar;
- a terminal busbar;

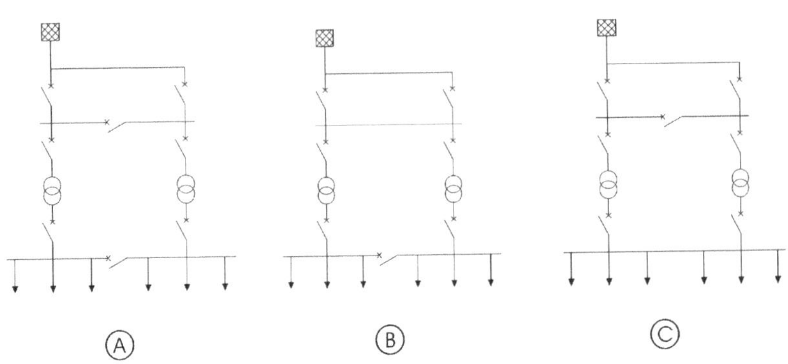

Figure 11.10 Examples of bottlenecks in a double radial scheme: (a) only one supply; (b) only one main busbar; (c) only one terminal busbar

- the end-user system;
- the control system.

The cases described correspond to real situations that may be found in most different structures (industrial plants, the tertiary sector, services). For each particular case the scheme can be either suitable if the unique element does not represent a bottleneck, or wrong if the only element cancels the result obtained by the redundancy of other parts. It is necessary to evaluate if this (not redundant) element can be considered as infallible, or such as not to be the determining cause of the unavailability of the service required.

For instance, in the case of Figure 11.10(a) the single supply cannot be a bottleneck if its availability is comparable to the one ensured by the double radial system below to the final user.

In case (c) only the last switchboard is a bottleneck if it can be made infallible by adopting constructive actions. In this type of analysis the evaluation of the final users is just as important. Installing an infallible last switchboard is meaningless when loads are actually susceptible to frequent disorganization. In a well-coordinated and harmonic project the concept of redundancy (or of infallibility) must generally extend also below the electric system. For instance, for the auxiliary services of a thermoelectric power station, steam circuits also have to be redundant (water, compressed air), by adopting double radial or ring schemes.

11.2.3.6 Uniform Availability and Limit of Tightness of the Components

The highest availability of a component system is obtainable with the lowest cost when these components have a uniform degree of availability, are without waste or lack, taking into account installation, use and maintenance conditions.

In particular cases a component can be chosen intentionally with reduced characteristics because it is required to be a *weak point* of the plant, destined to yield before others, as preventive protection of other more precious elements, or in the case of exceptional stress.

11.2.3.7 Redundancy of Protection

Faults or breakdowns are irregular situations during the functioning of a component and more generally of a plant that is correctly projected, built and run. Protection has the task of preventing serious damage to or the destruction of components, but chiefly to prevent direct or indirect damage to the plant itself and to people and the surroundings.

If it is well chosen, and coordinated, and regularly checked, protection has a high probability of accomplishing its task. In fact the breakdown of both the components and the protective device is rare, so the incorrect intervention of protection, coinciding with a fault in the plant, can be considered a most improbable event. However, this applies only if the fault in the plant is not itself the cause of the problem and thus of the corresponding protection, i.e. that the two damaging events are independent. For this reason (second-order event) no technical rule orders the redundancy of the protective devices [14].

Notwithstanding what has been said, good technical practice considers several situations where the late or non-intervention of a protective device is covered by the intervention of a support device (backup protection).

Some more significant examples are:

- large power transformers and rotating machines protected against internal short circuits (in order to obtain the lowest intervention times, without intentional delays) generally have backup protection for such short circuits;
- high- and medium-voltage busbar or line systems having differential protection have additional protection as a second step.

11.2.3.8 Resilience and Flexibility – Capability of Replacement

Up to now the various parts of an electrical plant have been examined regarding their ordinary functioning (basic functioning conditions supposed). In order to define the scheme, flexibility and elasticity exigencies are just as important.

With reference to the exigencies supposed to intervene for maintenance or fault repair, for modification and enlargement, it may be necessary to make other parts redundant that would not usually require it (parts considered infallible in service). This case is an opportunity for redundancy for reasons unconnected with consideration of the inadequate availability of components or circuits and which therefore does not involve the same necessary actions for (primary) redundancy intended to ensure high availability and continuity. These latest considerations lead to the definition of schemes for which, instead of redundancy, it would be correct to speak of 'capability of replacement'. An example is the double radial scheme with interconnections shown in Figure 11.11.

Another problem is that of the resilience of the grid. If the grid is very large and important, for instance, it is not acceptable for the whole power available (supply from the

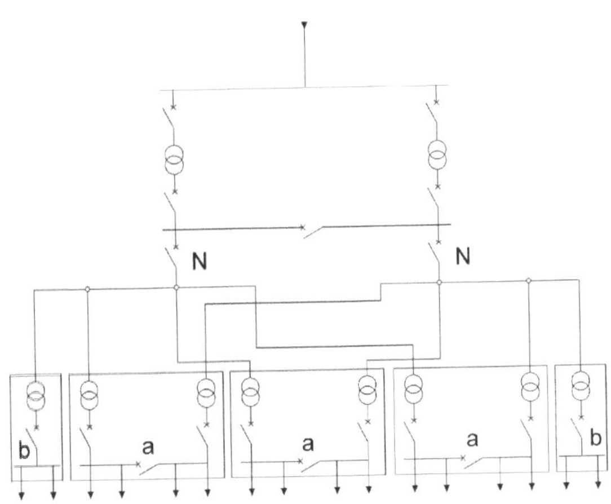

Figure 11.11 Double radial scheme with interconnections

grid and self-production) and all the lines to the secondary station to be directed to one principal switchboard only. If the switchboard were to break down, the whole plant would be brought to a stop. It is sufficient to subdivide the busbars and install a tie breaker in order to confirm the service of at least half a busbar, for instance supplied by one of the generators of the self-production.

11.2.3.9 Reserve Power

For any element of a grid capable of carrying out the function of another element (as a replacement for it) the problem arises of finding out where it can carry out the same functions and for what length of time. The value of its *reserve power* has to be ascertained. In order to analyze this problem it is necessary to distinguish two cases:

- the reserve element (redundant function) is sized for reduced power operation compared to the main element (e.g. a cable, a transformer, a reserve generator);
- the two elements, destined to cooperate in equal manner for the same function, are sized in the same way, and for neither of them is the function of normal or reserve element (double radial scheme) assigned a priori.

The first case needs no clarification – only the entity of the service required by the reserve element has to be ascertained, namely available power and duration.

The second case requires us to formulate the answer in an articulated manner:

- 100 % reserve: the two elements are sized for the full load and no load reduction is supposed for the unavailability of one of them;
- 50 % reserve (the least one): each of them is sized for half a load and load reduction is unavoidable;
- reserve for an intermediate value: included between 50 % and 100 % of the load.

With these sizing criteria, under regular conditions, each element works at a reduced load, and therefore with a temporary overloading margin due to the previous limited heating conditions. On this basis it is possible to define a length of time for which it is admissible to keep the total load invariant or almost, in order to pass subsequently to functioning at reduced load, equal to the rated power of the only element in service.

For HV and MV transformers, in order to satisfy conveniently the different load capacity, different ways of cooling are frequently adopted, e.g. ONAN–ONAF[2] [13].

These simple considerations, in many cases, for instance when a partial reduction of the load is acceptable, have the advantages of a redundant system, though still containing the oversizing of determined parts of the plant.

[2] Natural Oil–Natural Air, Natural Oil–Forced Air.

11.2.4 The System Supply Section and End Section

Similar considerations to the ones made for the grid components can be made for the supply section and the end section (users, loads).

Regular exercise of a supply (source) or of a user (e.g. motor) always depends on the existing conditions of the supply section and the end section of these elements. A careful examination of these conditions allows one to find out if, for an improvement in the scheme of the electrical grid, a global improvement in terms of greater availability of the final service can result. This analysis can ascertain if it is preferable, convenient, indispensable to intervene in the characteristics of the source, particularly if it is supplied by self-production, or of the final user, for instance foreseeing a reserve.

In plants a useless (complicated and expensive) overabundance is often remarked in some components of the supply chain, and a paucity in other parts, in the supply section or the end section or in between. An analysis of these situations leads to the opportunity of deriving a supply line from a node closer to the supply section (in the case of the public grid, directly from the station or room busbars, instead of a dorsal line) or of making it possible off grid, i.e. fed by an internal generator and disconnected from the public grid.

As far as loads are concerned, it is evident that the redundancy of the electric system does not correspond to a similar redundancy or intrinsic reliability of the end sections (e.g. double air compressors, double fireproof water pumps for cooling water, double fans for pressurizations, double data processing systems).

11.2.5 The Standard and Preferential Functions

In the systems where any redundancy is foreseen, it is important to find out what circuit, element, busbar is to be considered as standard, alternative, preferential, of help, emergency, safety. In the double radial systems as a rule the two branches are sized in the same way and their function is twofold: neither of them is destined a priori to be the *standard* or the *preferential* (reserve) service. The loads connected to both sides can also be twofold. In a different way it can be said that the two supplies are available or unavailable with the same probabilities and that one busbar's staying available during an irregular situation becomes preferential when the other is missing. The function stays assigned in a dynamic way (depending on casual events such as faults, irregularities, maneuverings, at the level of supplies, grid and load asset). This situation is nearer to reality, the more similar characteristics the two supplies have, in terms of availability.

In the case where one of the two available supplies can be considered more reliable than the other, the preferential function can be given to it, and the more important loads, or the loads necessitating supply continuity, are connected to it. The assignment is established beforehand, in a static manner, i.e. with connections modifiable only out of service. Remarkable engineering actions may correspond to this choice, such as the predisposition of a preferential grid or of devices that, when any trouble occurs, open the tie breaker between the two grids, standard and preferential. Each grid passes in a separate march, with its own loads, granted that it has stayed in service after the transient.

Such an assignment takes place even in cases where the second supply, considered available, is of reduced power compared to the standard one and can be defined as *emergency* or *safety*. This supply and the related circuits are different from the standard ones that they must replace. A load detachment can follow their intervention.

For a case study see web address

BIBLIOGRAPHY

[1] Baggini A., Bossi A., *Componenti e carichi suscettibili ai disturbi*, Corso 'Interazioni elettromagnetiche tra componenti e sistemi in ambito industriale: compatibilità elettromagnetica in bassa frequenza', Dipartimento di Elettrotecnica del Politecnico di Milano, 21–25 febbraio 1994.

[2] Bossi A., Bussei V., Tommazzolli F., *Trasformatori in olio e secco per gli impianti industriali. Le norme CEI nel progetto degli impianti*, Università di Pavia, Dipartimento di Ingegneria Elettrica, Giornate di studio 'Progetto degli impianti elettrici di energia', Pavia, 1989 – Atti, pp. 103–147 (*L'Elettrotecnica*, vol. LXXVI, no. 11, pp. 981–993, 1989).

[3] Caldon R., *Suscettibilità ed esigenze dei carichi*, Corso d-aggiornamento su Apparecchi, macchine e impianti elettrici: 'Il progetto degli impianti elettrici di energia. Le norme e la regola dell'arte', Dipartimento di Ingegneria Elettrica dell'Università di Pavia, 1991.

[4] Chapman D., Harmonics – causes and effects. *Leonardo Power Quality Application Guide – Part 3.1*, 2001.

[5] Chizzolini P., Noferi P. L., *Ottimizzazione degli interventi sulla rete di distribuzione mirati al miglioramento della continuità del servizio elettrico*, LXXXVII Riunione AEI, Firenze, 1986.

[6] De Cet C., Taglialatela: *Criteri di progetto dell'impianto elettrico in centro di telecomunicazioni: il centro RAI di grottarossa. L'Elettrotecnica*, vol. LXXVIIT., no. 4, pp. 247–115, 1991.

[7] Desmet J., Baggini A., Harmonics – neutral sizing in harmonic rich installations. *Leonardo Power Quality Application Guide – Part 3.5.1*, 2003.

[8] Desmet J., Delaere G., Harmonics – selection and rating of transformers. *Leonardo Power Quality Application Guide – Part 3.5.2*, 2005.

[9] Fava R., Lionetto P. L., *L'affidabilità, le anomalie e la verifica in esercizio degli apparecchi di manovra, di protezione e ausiliari.* Corso di aggiornamento, *Il miglioramento e l'innovazione nell'esercizio degli apparecchi, delle macchine e degli impianti elettrici*, Dipartimento di Ingegneria Elettrica dell'Università di. Pavia, 1990. Atti , vol. II, pp. 18/1–34.

[10] Gambirasio D., Lionetto P. F., Tommazzolli F., *Il progetto coordinato di una rete industriale e del suo sistema di protezione.* Università di Pavia - Dipartimento di Ingegneria Elettrica Giornate di studio 'Progetto degli impianti elettrici di energia', Pavia, 1989. Atti, pp. 231–266 (*L'Elettrotecnica*, vol. LXXVII, no. 5, pp. 433–444, 1990).

[11] Gruzs T. M., A survey of neutral currents in three-phase computer power systems. *IEEE Transaction on Industry Applications*, vol. 26, no. 4, 1990.

[12] *Guida di carico per trasformatori immersi in olio*, CEI, Supplemento S. 564, 1980.

[13] Hofstad P., Iorun E., Rolfseng I., Distribution system planning including the cost of non delivered energy. CIRED – 11th International Conference on Electricity Distribution 1991, Report. 6.3, pp. 1–5.

[14] IEC 364-5-523, *Electrical installations of buildings – Part 5-52: Selection and erection of electrical equipment – Wiring systems.*

[15] IEC 60050, *IEC International Electrotechnical Vocabulary.*

[16] Morsa R., Gli schemi dei quadri di alcune grandi centrali americane. *L'Elettrotecnica*, vol. I, pp. 4–10, 1914.

[17] Picciolo G., Lo Forti S., Camensig C., Scalcino S., Reliability based optimum choice of a MV electric power distribution system in a large petrochemical factory. CIRED – 11th International Conference on Electricity Distribution 1991, Report 5.13, pp. 1–6.
[18] Silvestri A., Tommazzolli F., *Schemi per gli impianti di energia: semplicità, affidabilità, risparmio, ridondanza dove e come*, Corso 'Il progetto degli impianti elettrici di energia. Le norme e la regola dell'arte', Dipartimento di Ingegneria Elettrica dell'Università degli Studi di Pavia, AEI, CNR, Pavia, 10–13 giugno 1991.
[19] Tomazzolli F., *Trasformatori per gli impianti industriali. Scelta delle caratteristiche e unificazione*. Politecnico di Milano. Corso di aggiornamento promosso dal dipartimento di Elettrotecnica, 'Trasformatori da distribuzione, da forno e per la conversione statica dell'energia elettrica', Milano 1986, pp. 3.1–3.25.
[20] Ward J., Power Quality – two different perspectives. IEEE 90WM 053–9 PWRD, 1990.

12

Reliability of Electricity Supply: Appliances and Equipment

Roberto Villafáfila-Robles and Joan Bergas-Jané

A discussion about the relation between reliability and power quality problems some years ago would have probably concluded that outages were the one and only main effects. This is understandable because the consequences were evident: lights, motors, belts, robots, computers and so on stopped working. There were likely other problems, but they were not a worry because they had no effect on most equipment connected to the mains. So, reliability was mainly related with continuity of electric supply – that is, availability.

Nowadays, the increasing use of electronic devices has led to concerns about the quoted relation because a lack of supply may imply a hazard to life and a significant loss of money. One must think what would happen in a hospital or in a telecommunications data center. But reliability can be related not only with a facility that is down. Power quality must also be considered as a part of the reliability of electricity supply. Voltage, current or frequency deviations may result in the failure or malfunction of equipment, such as the restart of a bank's servers by some sag or the tripping of thermal cut-off protection by harmonics.

Of course, no business is willing to take such risks. Nevertheless, a lot of consumers have not yet realized that systems are no more reliable than the power that operates their facilities. If continuity of operation is critical, some upgrading must be carried out throughout the power supply, namely power conditioning, emergency and standby power systems. Ensuring the power quality demanded by the load is not just the responsibility of the utility. It also involves extra investment, not always well analyzed.

This chapter will describe current power conditioning, emergency and standby power appliances for low-voltage facilities and equipment in order to help choose a cost-effective solution that fits better with a facility's capability to increase its reliability and then the availability of its power supply.

Handbook of Power Quality Edited by Angelo Baggini

12.1 POWER QUALITY, RELIABILITY AND AVAILABILITY

The utility is most commonly perceived by the user as the source of power quality and reliability problems. This is understandable because the effects of events in the mains, like lightning or breaker clearing, are fairly noticeable in a facility's performance. However, these events represent a small percentage of total events, although they may be true in some specific areas such as weak grids. The reality is that a large percentage of problems are generated within a user's own facility.

Even though it may seem unbelievable, the main source of power disturbances in a facility is its own activity. Normal utilization of the facility and normal operation of its own equipment, including turn-on and turn-off, may give rise to undesirable events which affect the sensitive equipment within the facility if it has not been taken into account in the design process.

The increasing spread of information technology and automation to practically every business and activity is mainly responsible. The increasing application of electronics to all kind of appliances makes possible real-time communication and/or continuous operation. This improves facilities' performance levels through effective data and energy management. But electronic devices are non-linear loads and they introduce more disturbances into the same facility they are fed by. Therefore, facilities and sensible equipment are becoming more vulnerable and dependent on both the quality and continuity of the electric power supply because of their operation. Typical examples of power quality problems are overloaded circuits and transformers, and unwanted operation of protection caused by the presence of harmonics (see Chapter 7).

Therefore, power quality, reliability and availability must be considered as a complete set from the design stage because they are closely related. Equipment is designed for normal operation within a rated range of power supply values. Small deviations from these values can result in increased losses, poor efficiency and unpredictable operation. Large deviations can cause protective devices to trip or the failure of a component. In addition, components always break down from time to time because they have a limited lifetime. Thus, both power quality and reliability influence the availability of any facility.

Availability is easier to estimate and quantify than reliability, so users are more concerned about the former. Nevertheless, the way to increase the availability is through power quality and reliability improvements. But high availability means much more than high power quality and reliability. At first, it requires a proper design to reduce failures or disturbances in the power supply. The reliability levels wanted or needed for the equipment of a business or an activity are known at the outset. So, if the reliability level must be higher than the one provided by the utility according to power quality and availability standards, additional investment is needed in order to achieve the expected power quality and availability for the facility. Here, the economics of improvement in reliability plays a major role in the decision-making process. The extra expense determines the final decision, which must be kept within an acceptable range indicated by the cost–benefit relation. In any case, customer cost associated with power disturbances always decreases as reliability increases. Moreover, any upgrade after a facility has come into operation may unleash higher expenses and fateful consequences as higher economic losses.

There are a number of appliances and different equipment configurations that may improve the power supply reliability for both critical facilities and loads. Nevertheless, each facility is unique and requires an individual risk and economic assessment for each option. The user's choice must be made as a function of several parameters like cost, type of usual disturbance, the electrical distribution on site, characteristics of the equipment to be protected, criticality of the application to be protected, etc. [23]. The aim of this chapter, as has been mentioned above, is to help users with this choice, so the following sections will cover the available technical solutions. Of course, it is advisable to seek advice from experts if an in-depth analysis is necessary.

Before going on to describe current reliability appliances, it should be noted that there another fact that has an influence on facility reliability and availability must be considered. This is the human factor, which is most of the time underestimated or forgotten, although it has an important role. It is always present because almost every power unavailability event implies human intervention, whether it is before or after the event. So, it must be taken into account during the design stage, although it is very difficult to cope with because it basically depends on the background, experience and training of the personnel involved. However, the wrong human operation may be quantified by the effects of its action.

The main tool for averting human errors is education and technical training. If education seems expensive, try ignorance. These tools provide a better understanding of power reliability problems and technologies, and of the operation of a facility. A resilient system philosophy can also be incorporated into the conceptualization of a facility to try to reduce to a minimum the rates of possible human effects. This can be achieved by either failure tolerance (elimination of single points of failure) or redundancy (alternative circuits or equipment) strategies [27]. Stocking spares and diversification of equipment manufacturers can help, too. Of course, most of these approaches mean significant extra investment.

Furthermore, it should be compulsory rather than advisable to set up emergency action and maintenance protocols at the same time as the facility design is developing [27]. An emergency action protocol sets operations that operators must follow to return the facility to a safe state. And as equipment and its components have a determined operational lifetime, maintenance procedures make possible the reduction of risk of the wrong operation or failure. Also included at the design stage should be a control monitoring system of the power supply in order to achieve high reliability levels [8].

12.2 GENERAL ASPECTS OF RELIABILITY APPLIANCES

The design of a suitable power supply system is a compromise between the investment levels and operating costs that utility companies reach to cope with the necessity of assuring power quality standards and the reliability and quality of supply that users want or need [2]. Most disturbances are totally beyond the control of the utilities because of the large number of stakeholders that form a complex environment. Further, on power quality standards, users are responsible for assuring the availability of power for the correct operation of their facilities. But a high availability level does not happen by chance. It is the result of careful design and maintenance protocols, as far as the cost–benefit ratio allows.

The importance of the reliability of appliances originates from present equipment that is much more sensitive to power quality issues than in the past. The reason is the extended use of electronics in this equipment, which may affect some elements of a user's facility and/or may be affected in turn, according to the electromagnetic compatibility (EMC) characteristics of each one. Some equipment includes some kind of built-in protection, but this is only against the most frequently encountered disturbances.

Therefore it is clear that users must care about the final power quality and the reliability of their facility. There are a wide variety of appliances available to solve power quality and reliability problems at a facility, and then to improve availability. The appliances chosen must match the requirements needed regarding the quality and reliability of power supply. The logical decision-making process about how to proceed to deal with this issue can only be done with accurate and documented data on the distribution network and typical disturbances that could be found, and the characteristics and criticalities of the equipment within the facility. Besides the technical factors, the cost will decide the chosen method [14].

12.3 POWER SYSTEM PROTECTION ALTERNATIVES

The susceptibility of equipment to disturbances in the power supply has to be known at the design stage in order to properly study the upgrades that facilities need. Equipment manufacturers usually provide specifications with rated values and the deviations that their products are designed to operate with. Then there are also several recommendations promoted by them, like the well-known voltage tolerance curve for computer equipment provided by the Information Technology Industry Council, ITIC (see Figures 4.8, 13.2 or 13.3).

This kind of curve illustrates the needs that facility designers must take into account, as design goals, when there are sensitive devices. It can help to select a suitable appliance for improving power quality and reliability. Power supply requirements may mean simple protection to avoid destruction or malfunction, or equipment that may enable the continuation of normal use of these sensitive devices. Thus, it is possible to distinguish two types of appliances for low-voltage facilities. The first type is power conditioners (Figure 12.1),

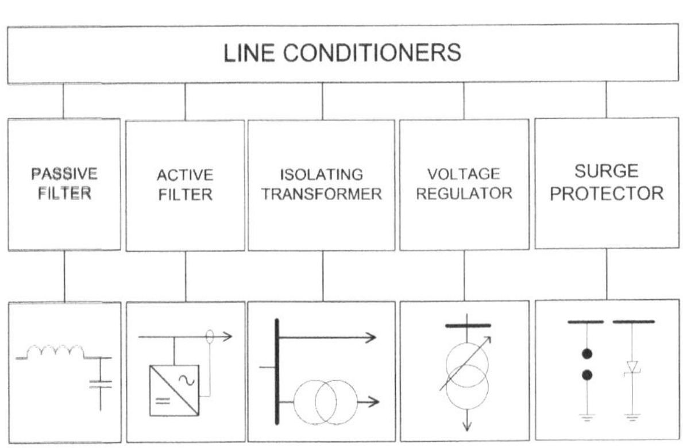

Figure 12.1 Low-voltage power conditioning appliances

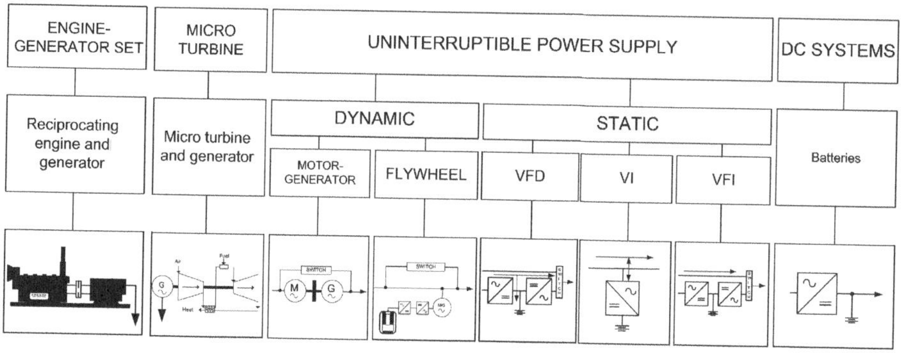

Figure 12.2 Low-voltage emergency and standby power supply appliances

whose main function is to preserve power parameters within the limits required for the load, although each appliance is designed to solve one specific power quality issue. On the other hand, the second type, emergency and standby power supply systems (Figure 12.2), can deal with both power quality and continuity issues because they have their own energy source that replaces the utility supply. So, their higher complexity also offers a higher performance. The former are simpler and cheaper, and the latest provide a higher reliability to the facility, but also at higher cost.

The final choice between the wide range of technologies shown in Figure 12.1 and Figure 12.2 should depend on both technical and economical factors. Emergency and standby power supply systems offer a higher level of protection against disturbances, but the costs of such systems are higher too. Most of the available power conditioners can provide adequate protection at a fraction of the cost. The following sections will describe each appliance.

Users can deal with the reliability and quality of supply at medium and high voltage too [7], although the cost is most of the time harder to justify. At this voltage level, the problem is that the utilities are usually the owners of the network and it is necessary to meet the economic and technical interests of both parties. The most common approach is to have a second feeder from a different substation at one's disposal (in a city if possible) or a flexible a.c. transmission system (FACTS) device like a dynamic voltage restorer (DVR) [16], [29].

There are also some other appliances available, such as software that can be used as a backup to save data, as in the air transportation sector, and other ways of storing energy different from batteries or flywheels, namely superconducting magnetic energy storage (SMES), ultracapacitors, pumped water storage, compressed-air storage (CAES) and thermal energy storage [4], [12]. Of course, some of these technologies need more than an electrical assessment and this is beyond the scope of this chapter.

12.3.1 Power Conditioners

Low-voltage power conditioners (Figure 12.1) are introduced here because they can deal with some power quality problems. They have no storage capacity, but they are sometimes combined in a single appliance to cope with more power quality problems, as for example in an isolating transformer with voltage regulation and overvoltage protection that provides output regulation plus load isolation and electrical noise attenuation. Such appliances are normally called power line conditioners.

This section lists appliances used to improve power quality, and then reliability, but that do not have energy storage associated with them so they cannot preserve continuity of operation of equipment. However, it is necessary to determine their voltage and size during the selection or design process, since they can be single-phase or three-phase and their power rating must match the proposed objectives.

12.3.1.1 Filters

Filters reduce harmonic distortion on power networks and then protect sensitive loads from polluting loads. It is possible to differentiate two groups:

- Passive filters have been in use for a long time and are a simple, reliable and cheap solution. They are based on combinations of resistors (R), inductors (L) and capacitors (C): capacitors block low-frequency signals and conduct high-frequency signals, while inductors do the reverse and resistors do not block any frequency, but, combined with L and C, determine the filter tuned frequency.
- Passive filters are designed to accommodate one harmonic order in a particular load condition. In another way, passive filters are an electric circuit tuned to one undesired frequency in order to eliminate it providing a path to earth. If their design conditions change, they could be detuned. It is possible to distinguish two types of passive filters:

  – EMI/RFI: suppress radiated or conducted electromagnetic and radio-frequency interference, i.e. high frequencies and noise;
  – RL, RC, LC and RLC: reduce low-frequency conducted disturbances, i.e. power supply harmonics.

- They are usually associated with a surge protector to increase protection against transients.
- Active filters are a newer but more complex and expensive solution than passive filters. They are power electronics devices able to eliminate different frequency disturbances at the same time, although initial conditions vary. They can be used alone or dedicated to significant polluting equipment like rectifiers of static uninterruptible power supplies.
- They work by sensing the harmonic content of the electrical supply and then injecting the same spectrum, but with opposite sign (Figure 12.3). This guarantees a lower harmonic content in the upstream power supply. Therefore, active filters provide a selective and instantaneous harmonic compensation reducing electrical losses. Moreover, they improve the power factor and even balance some power unbalances. They are commercially available and it is possible to parallelize several units up to hundreds of kVA.

12.3.1.2 Isolating Transformers

An isolating transformer is usually a 1:1 transformer with separate windings equipped with an electrostatic screen that enables the reduction of high-frequency interference in

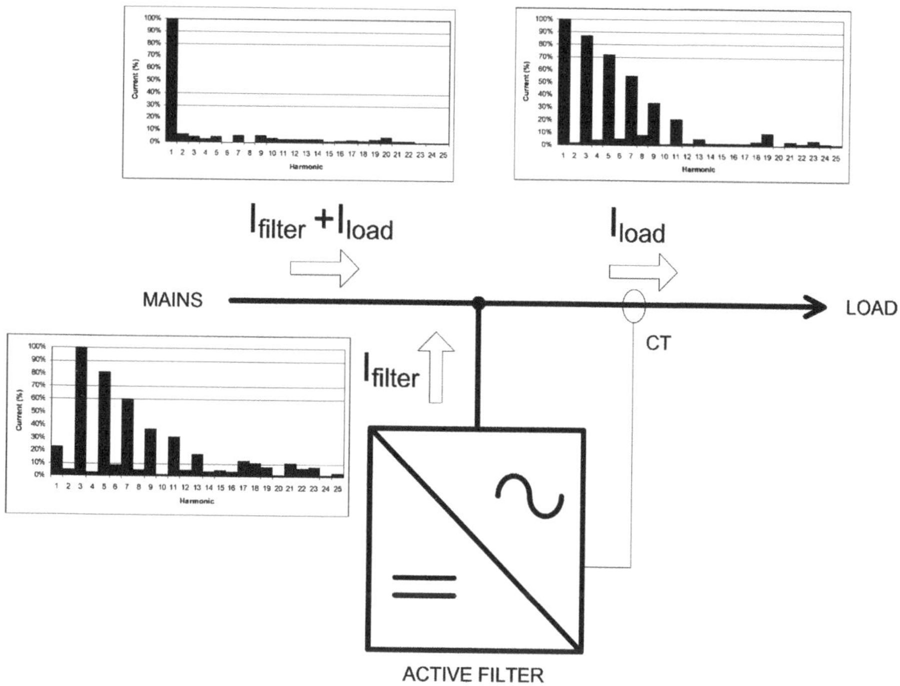

Figure 12.3 Active filter

common and cross-connected mode because it stops these disturbances transferring from the primary to secondary windings. It also allows earth leakage currents to be reduced since the secondary wiring permits an effective grounding to be built.

Moreover, certain winding arrangements in three-phase transformers allow certain harmonic currents to be reduced in the primary (third harmonic and its odd multiples) and to balance the current of unbalanced loads. Also, a transformer with a double secondary winding with a different arrangement makes it possible to diminish fifth and seventh harmonics. However, design and the manufacturing process are very important for the protection level achieved. And apart from decoupling circuits in order to block harmonics, interference and from offering an efficient grounding, no other protection is provided.

12.3.1.3 Voltage Regulators

A voltage regulator maintains a constant output voltage in spite of variations in its input voltage. Regulators are able to solve problems regarding voltage variations like sags, surges and spikes, but they are ineffective against noise transients and frequency variations. Then, they are usually combined with an isolating transformer. They are also completely ineffective against outages and frequency variations: only systems with emergency and standby power can solve this kind of problem. The criteria to be considered when evaluating the performance of regulators are the regulating range, the load variation response and the speed and flexibility of regulation.

There are mainly three types of regulators, namely ferroresonant, electromechanical and electronic. Thus the features of power line conditioners depend on voltage regulation technologies as well.

A ferroresonant voltage regulator is composed of a resonant winding and a capacitor to produce a constant output when the input varies. It works in the saturation region of the transformer design and it has a ride-through up to 1 cycle power outage. It also offers good short-circuit protection. However, it is heavier and less efficient than electromechanical regulators.

An electromechanical voltage regulator operates by selecting the proper tap on a multi-tap transformer. When the output voltage is too high, the tap-changer shifts to produce a lower voltage and vice versa. It is generally small and has a fast response that can draw some harmonics or interharmonics.

An electronic voltage regulator regulates line voltage to within tolerances of ±1 % by adding and subtracting windings driven by a precision electronic control system. However, it does not provide a conditioned power as does the ferroresonant regulator and it is larger and heavier than the electromechanical regulator.

12.3.1.4 Overvoltage Protection Devices

Overvoltage protection devices or surge protectors protect facilities and equipment against transient voltage surges produced by lightning (μs), electrostatic discharges (ns), switching (ms) or long voltage deviations at power frequency (s) [22]. They limit or eliminate voltage surges up to acceptable limits, blocking or shorting them to earth. Thus, earthing is a key issue in any facility (see Chapter 10).

There are different technologies commercially available: silicon components (Zener diode, triac, etc.), metal oxide varistors (known as MOVs) and spark-gap types (the gas arrestor type is the most common). They are sometimes embedded in electronic devices. They differ in two important facts: clearing time and dissipation of energy. They are listed like faster clearing time devices. In the reverse order, they are listed for higher conducted energy capability. A combination of two or even three of these technologies is the best solution. Moreover, spark-gaps are the only type that can withstand several lightning strikes. The others are usually destroyed.

12.4 EMERGENCY AND STANDBY POWER SYSTEMS

Improving the reliability and quality of power supply for critical loads to the required targets needs a proper design and implementation of emergency and standby power supply systems. These auxiliary power sources are independent from the ordinary power system, usually the utility grid. They are mainly used to supply backup energy in planned outages, during maintenance procedures of normal power, and when the mains fails or is beyond its standard limits.

Emergency and standby power supply systems are a reserve of power supply and their main aim is to ensure continuity of operation. However, it is possible to distinguish some different features between emergency and standby power systems since their scopes differ, too.

An emergency power system (normally legally required) feeds equipment or facilities whose failure, because of a power outage or some deviations from power quality of supply, may present a safety hazard to people or high economic disruption. It allows continuity of operation without disruption for a limited time, from milliseconds to several minutes, until a standby power system starts to feed the equipment or the facility. Emergency power systems permit controlled shutdown, too.

A standby power system allows operation of equipment or the facility in the event of a long normal power outage, e.g. a power failure or maintenance procedures, or continued low power quality of supply. It can provide continuity of operation from a few seconds to days, but it may imply some power disruption for the load if its use is not combined with an emergency power system.

Both are widely used in industry, from petrochemical to automotive plants, and in the service sector, from banks to hospitals, even in buildings for offices or residential purposes. These auxiliary power supply systems rely on engine-driven or microturbine-driven generators, uninterruptible power supplies (UPS) or d.c. systems based on batteries. The first could be considered a standby power system while the last two could be regarded as emergency systems. However, this division depends mainly on technological development. Upgrades in engine- and microturbine-driven generators for faster start-up, e.g. via compressed-air energy storage strategies, might avoid the likelihood of disruption. It is the same for uninterruptible power supplies and batteries if their energy storage capacity is increased in order to be able to give them a longer autonomy, e.g. by the use of fuel cells.

12.4.1 Engine-Driven Generators

An engine-driven generator consists of a reciprocating engine (internal combustion engine, Figure 12.4, 1) and an electric generator (Figure 12.4, 3) mounted together on the same

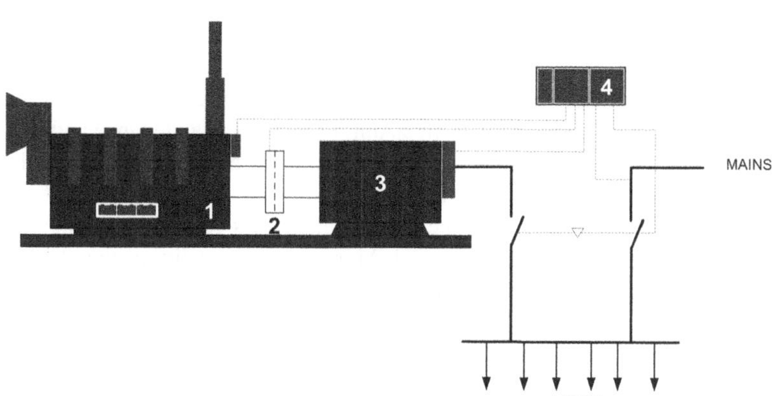

Figure 12.4 Engine-driven generator

shaft, usually mechanically coupled through an electromagnetic clutch (Figure 12.4, 2) to form one single equipment.

The reciprocating engine can be a two- or four-stroke cycle compression (diesel) or spark ignition (petrol, gas) engine. It is supplied by fuel from the fuel tank associated to the gen-set. The engine drives the generator, either single-phase or three-phase, that supplies energy to the critical loads. This combination is usually called either gen-set (an abbreviation of engine–generator set), or simply generator, since the engine is taken for granted in the combined unit.

In addition, it generally includes a control unit to provide a suitable power supply in terms of voltage and frequency, such as a speed regulator for the engine and a voltage regulator for the generator (Figure 12.4, 4). This control unit can usually start the gen-set automatically, to synchronize it with the power supply and to manage the transfer switch unit to disconnect the load from the utility power source and connect it to the generator without a power interruption to the load. This can be done manually, too. Most of the gen-sets are equipped with a battery and an electric starter to start up the engine.

Engine-driven generators are widely used as a standby power source to supply power in installations when utility power is not available or in situations where power is needed only temporarily because of outages, maintenance procedures or short-term high demand needs.

Engine-driven generators are commercially available in a wide range of power ratings, from small and portable units of several hundred watts to stationary or trailer-mounted units that can supply from several kilowatts to a few megawatts. The smaller units are usually petrol fuelled, and the larger ones can use various fuel types, including diesel, natural gas and propane. The engine, and mostly its ignition type, sets the efficiency of the gen-set. Diesel-fuelled gen-sets can reach an efficiency up to 45 %, petrol- or gas-fuelled ones up to 30 %.

Portable units are commonly single-phase generators, while stationary units are mostly three-phase generators. Gen-sets up to a few kilowatts are generally used in places where utility power is not available and in situations where power is needed only temporarily, e.g. on construction sites or in isolated cottages. They are especially popular in Third World countries to supplement grid power, which is often unreliable.

Gen-sets over some kilowatts are usually permanently installed and kept ready to supply power to critical loads during temporary interruptions of the utility power supply as standby power generators. Trailer-mounted generators are used to feed areas where grid power has been temporarily disrupted, such as during network maintenance and in disasters. Moreover, in some of these situations, large gen-sets can be considered as a distributed generator and support the grid operation.

Gen-sets are selected according to the proper generator's voltage, frequency and power ratings to suit the load that will be connected. Moreover, the quality of the electrical wave at the generator's output needs significant attention. Non-linear loads draw a high harmonic level to the upstream facility. If harmonic content is high enough, it could lead to an unsuccessful attempt of the gen-set to provide the power required. In this case, the gen-set might need to be oversized to avoid lost of supply.

Although gen-sets can be automatically switched on, they cannot take up the load immediately. There is a delay between a power disturbance and load transfer to the gen-set.

Then, they are generally installed in combination with some uninterruptible power supply system. This feeds the load till the gen-set reaches steady state and can take it up. The transfer time is usually between 15 and 30 s for medium-size units and up to 15 min for large units of some megawatts.

However, there are different strategies for upgrading gen-sets in order to handle the reduction of this transfer time. They are based on a flywheel, a rotating mass coupled to the gen-set shaft that stores mechanical energy:

- A motor connected to the mains drives a flywheel and a generator while the power is available. The flywheel accumulates energy like a kinetic battery. When there is a short power disturbance, the flywheel provides the energy to the generator to eliminate it. If the power disturbance is longer, the flywheel delivers its energy to the engine for a faster start-up. With this set-up, the effects of power disturbances are largely reduced.
- A motor mechanically coupled with an engine-driven generator shaft, which includes a flywheel, is continuously connected to the mains. While the normal power is available, this motor drives the generator that supplies the loads. When there is a power disturbance, the flywheel provides the energy to start the engine in order to reduce transfer time to zero.

Engine-driven generator systems are a mature and highly reliable technology. Gen-sets are flexible regarding fuels and they have a quicker start-up than turbines. Nevertheless, gen-sets are large appliances with considerable noise (above 70 dB) and gas emission levels (although the exhausted gases represent a low waste heat in comparison to turbines) that need large fuel storage. Therefore they require a specific conditioned place. Furthermore, they require frequent maintenance due to the need to change filters and oil. And their output power and efficiency could also be derated in case of high temperatures and altitude.

12.4.2 Microturbine-Driven Generators

Microturbines are becoming widely used in applications in hospitals, large office buildings and some industries like drying processes, where there are huge heating requirements and, at the same time, a standby power system is required because the mains has a lower reliability and/or power quality than desired. Moreover, advances in electronics allow unattended operation, parallel and redundant operation, and interfacing with the network.

A microturbine basically consists of a compressor and a turbine, which can be coupled with a power generator. If there is a recuperator, exhaust gases can be used to heat water (see Figure 12.5). In this case, they are known as combined heat and power (CHP) or cogeneration units, and their efficiency can reach a value of 85 %. However, if they are only used for power generation purposes, the efficiency diminishes by up to 30 % if a recuperator of exhausted gases is used to boost the temperature of combustion, or up to 15 % if there is no recuperator.

Commercial microturbines range from handheld units producing less than a kilowatt to commercial-sized systems that produce tens or hundreds of kilowatts. If a higher power rating is needed, conventional turbines can be used. Microturbines can be fuelled with

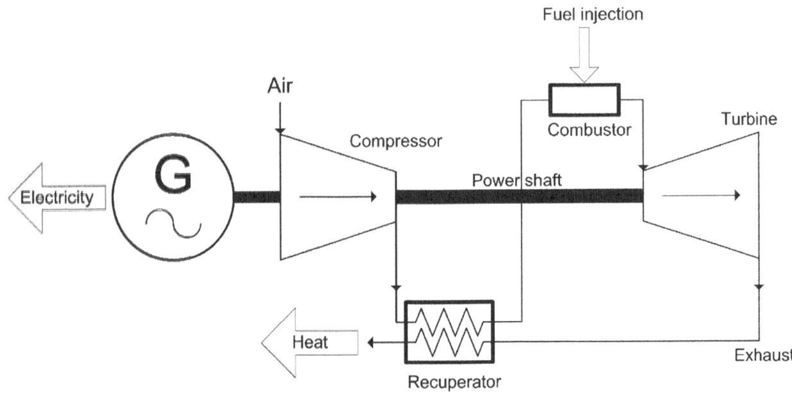

Figure 12.5 Microturbine-driven generator

commercial fossil fuels, such as natural gas or propane, and biofuels such as biogas from sewage plants.

Microturbine systems have many advantages over engine-driven generators, such as a small number of moving parts, extremely low emissions, higher power density (with respect to weight) and efficiency (in cogeneration units). Further, some microturbines can operate without oil or coolants, so having longer maintenance intervals. However, as engine-driven generators, their output power and efficiency reduce with high temperatures and altitude. Like gen-sets, microturbines can use a wide range of fuels, although engine-driven generators are quicker to respond to changes in output power requirement.

CHP units can be considered as distributed generation because they have the chance of injecting excess energy into the grid and, therefore, obtaining some extra revenue. Further developments are trigeneration applications. They generate electricity and heat as cogeneration units, besides cooling by absorption chillers, which can create cold air for air-conditioning from heat energy instead of electric energy. Thereby, the total efficiency increases even more.

12.4.3 Uninterruptible Power Systems

UPS systems are an interface between mains power supply and critical loads, which simultaneously provide uninterrupted power and the necessary power conditioning for loads sensitive to power disturbances. Therefore, UPS improve reliability and quality of power supply at the same time.

However, not all the current technologies of UPS (Figure 12.2) ensure reliability and quality of supply in the same way. Ideally, UPS should [1] always provide an excellent power supply, independent from the input power or the load, balanced or unbalanced, linear or non-linear. Thus the desired features should be high efficiency and reliability, low THD output voltage, low THD input current, low electromagnetic interferences and reduced acoustic noise. Unity input power factor, electric isolation and zero switching time from any change in operation mode are also desired features. Apart from the electrical issues,

UPS systems should have low weight, low size and, most important, low cost, including purchase, installation and maintenance.

On the other hand, UPS systems can provide power up to several minutes to allow a secure shutdown or the continuity of operation during their autonomy. In consequence, a standby power unit is needed for longer outages. This is usually an engine-driven or a turbine-driven generator. The UPS system supplies power to the critical load from the start-up of the standby power unit till it takes up the critical load. This combination requires that both auxiliary power supply systems have a control unit that permits communication between the systems to coordinate these actions.

Current UPS systems can be classified into three general types according mainly to their energy storage method: dynamic, static and hybrid static/rotary.

12.4.4 Dynamic UPS

Dynamic UPS systems are based on mechanical energy storage in flywheels. This concept has been used since the first UPS systems were developed some time ago [20]. They basically consist of a mechanically coupled motor/generator set that generates the output sine wave with a flywheel in the common shaft that stores energy as kinetic energy to provide a ride-through capability of several seconds to the generator. Thus, dynamic UPS are also called rotating or rotary UPS [5], [20].

The flywheel's stored energy from the motor during the normal mode of operation, when the mains is within power standard limits, is used to supply the generator for a few seconds in the stored energy mode, when there is some power excursion. There are a number of types of dynamic UPS. This technology has evolved from a simple coupled motor/generator with a flywheel in the common shaft to a development of the motor/generator within the same machine and modern flywheels based on new materials [5]. However, a generator always supplies the load.

Traditional dynamic UPS solutions (Figure 12.6) are not broadly used and they tend to be substituted by modern flywheels. They are only requested for some high-power critical applications up to several hundred kilowatts, because they are more reliable than static UPS systems and provide galvanic isolation. But they might be noisy, up to 95 dBA; they have considerable installation costs because of their significant size and weight that might require civil construction; and their maintenance requirements could imply a long downtime, e.g. for bearing replacement. Nevertheless, they can provide a high short-circuit current that

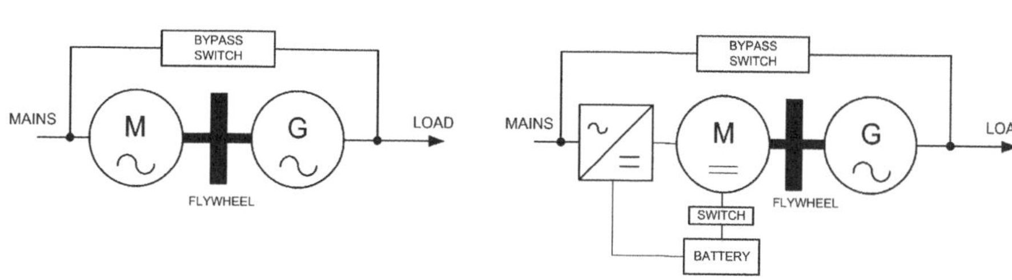

Figure 12.6 Traditional dynamic UPS

allows a transient overload capability up to six times full power for fault clearing. And at the same time, they provide stable electrical wave parameters and also present a good tolerance to non-linear loads because of their low internal impedance. Moreover, they do not draw either harmonics upstream (input THD_i is typically 3 % or less) or electromagnetic interference (EMI). Their efficiency is pretty high too, around 85 %, with a suitable power factor for the upstream facility.

Nowadays, current interest in dynamic UPS mainly concerns the flywheel itself [9], [19]. Recent advancements in materials, power electronics and control have led to commercial flywheels up to the megawatt range, with less noisy and space specifications. Newly developed dynamic UPS are essentially focused on the flywheel. New flywheels can be used in developments with a motor/generator set as the traditional philosophy, as well as energy storage for hybrid dynamic/static UPS or d.c. systems. Section 12.5 goes deeper into the description of flywheel technology.

12.4.5 Static UPS

Static UPS systems employ power electronics equipment that stores energy in electrochemical batteries. A simplified configuration of static UPS consists of four basic elements (Figure 12.7):

1. A rectifier, which converts input a.c. power to d.c. power.
2. An Inverter, which converts d.c. power to a.c. power for the critical load.
3. Batteries, which store energy during availability of normal supply and provide d.c. power to the inverter in the event of input a.c. disturbance. Thus, they are a key element of UPS. They are generally secondary batteries to allow charging/discharging cycles.
4. A static bypass switch, which transfers the critical load to the bypass source in the event of a failure in some of the elements.

Some static UPS have a rectifier and an inverter integrated in a single bidirectional power converter.

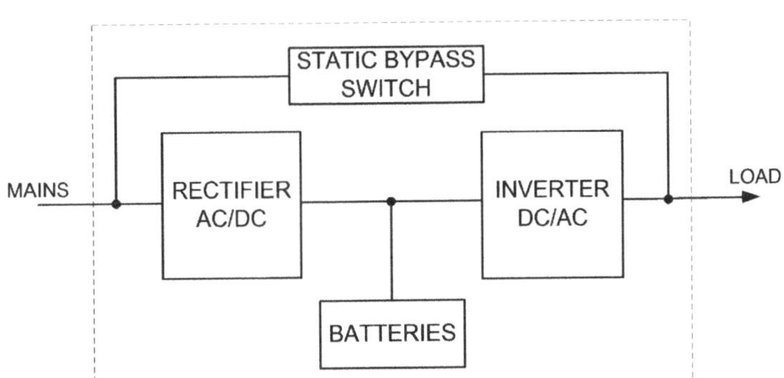

Figure 12.7 Static UPS

There are some other elements that UPS units could include: a manual bypass switch, which allows maintenance of power electronics components; an output transformer; and an input filter.

The first UPS were mainly dynamic because motor/generator sets were more reliable and efficient than power electronics devices. However, in the last few decades advances in power electronics have resulted in a wide variety of technological solutions for static UPS systems and have increased their efficiency. They can be classified in three topologies: double conversion, standby and line interactive [10]. For example, the use of IGBT has allowed bidirectional power converters and then a rectifier and an inverter have been joined in only one element, as in line interactive topology.

Static UPS are applied to a wide power range of systems, from low-power appliances such as a few watts for PCs to high-power facilities such as large data centers up to megawatts with several units operating in parallel. Furthermore, they come in many voltage ratings, both single- and three-phase. Low-power units are essentially single-phase, while medium- and high-power units are generally three-phase.

Their main advantages are high efficiency, high reliability and are almost 'plug & play' and expandable appliances because of their modular and flexible configuration. They also require less space and are ambient friendly and easy to integrate within communication and data management systems. In addition, pulse width modulation (PWM) control for inverters generates a low output THD_u and a good supply for non-linear loads. Thus static UPS systems are the most commonly used UPS system.

On the other hand, there are inherent problems related to static UPS systems: they draw some disturbances to the upstream facility, the transient overload capability for static UPS systems is limited typically to 150 % for a short-term and high cost for achieving very high reliability.

Section 12.6 contains a deeper description of static UPS technology.

12.4.6 Hybrid Static/Dynamic UPS

Hybrid static/dynamic UPS systems are a combination of both static and rotary UPS systems in order to achieve advantages compared to the single systems [28]. It is possible to identify different philosophies and their respective operation modes, as follows.

12.4.6.1 Motor/Generator Set with Battery Storage (Figure 12.8a)

This consists of an a.c. motor mechanically tied to an a.c. generator, a bidirectional a.c./d.c. converter, a battery bank and a static bypass switch. It has three modes of operation:

- **Normal mode.** The motor is fed from the normal supply, the utility power, and drives the generator that supplies the load. The bidirectional converter behaves as a rectifier and charges the batteries. It should be noted that when starting up the system, the a.c. generator is fed by normal supply to avoid overloads and allows the inverter to be rated for the normal operation. Later, when the generator is on, the normal supply is disconnected and the generator is supplied by the inverter. This is relatively easy because of the large inertia of the generator.

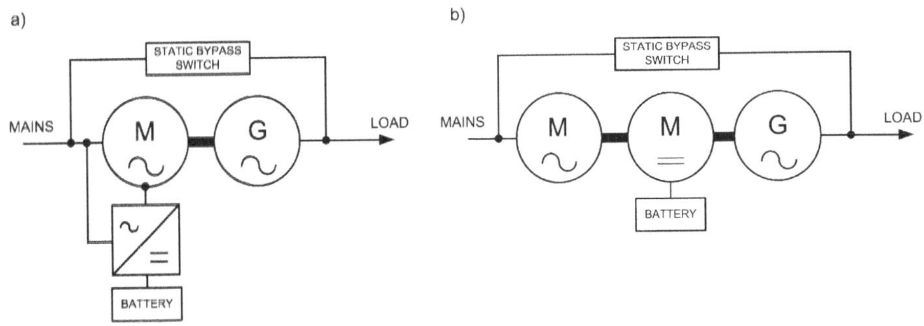

Figure 12.8 Hybrid static/dynamic UPS systems based on battery storage (Reproduced by permission of Jorge Sanchez Losada; ENDESA)

- **Stored-energy mode.** When the a.c. input voltage is outside the preset tolerances, the batteries supply the energy to the bidirectional converter that behaves as an inverter and feeds the a.c. motor that drives the generator supplying the load. As the power converter is always on, the transfer from the normal supply to the inverter takes place under controlled conditions and with close to zero switching to the stored-energy mode of operation.
- **Failure mode.** When an internal malfunction in the UPS system occurs, the static bypass switch turns on and the load is supplied directly from the normal supply. However, as the normal supply and the output voltage are not synchronized, the transition is not transient-free.

The main advantages of this hybrid UPS over static UPS include lower output impedance, lower THD with non-linear loads, higher reliability and better isolation. Moreover, its main advantages over dynamic UPS consist of the lack of mechanical elements: that means lower maintenance costs. Furthermore, this hybrid UPS can be used in applications with hundreds of kilowatts. However, it is more a specific design than a commercial solution.

12.4.6.2 Motor/Generator Set with D.C. Machine and Battery Storage (Figure 12.8b)

This is basically a modification of the previous one, adding a d.c. machine mechanically coupled to the a.c. machine's shaft. It also has three operating modes:

- **Normal mode.** During the normal mode of operation, the normal supply feeds the a.c. motor, which drives the d.c. machine. In its turn, the d.c. machine drives the a.c. generator, which supplies the load. Meanwhile, the battery bank is charging through a rectifier.
- **Stored-energy mode.** During the stored-energy mode of operation, the battery bank supplies the d.c. machine, which in turn drives the a.c. generator. The a.c. generator supplies the load.
- **Failure mode.** When an internal malfunction in the UPS system occurs, the static bypass switch turns on and the load is supplied directly from the normal supply. However, as the normal supply and the output voltage are not synchronized, the transition is not transient-free.

This variation of the previous system presents more disadvantages than advantages. It needs more maintenance and more space for machines, because of the d.c. machine's presence, although the supply to the a.c. generator can be regulated.

12.4.6.3 Rectifier/Inverter with a Flywheel (Figure 12.9a)

This consists of a rectifier which converts a.c. to d.c. power, a d.c. bus with a flywheel connected, an inverter which converts d.c. to a.c. power, and a static bypass switch. It is basically a static UPS system with its three modes of operation where a flywheel replaces batteries:

- **Normal mode.** The rectifier converts a.c. power from the normal supply, the utility power, to d.c. power, which feeds the inverter and the flywheel. Then, the flywheel stores energy as mechanical battery and the inverter supplies a suitable power to the load.
- **Stored-energy mode.** In front of some a.c. input disturbances, the flywheel supplies the energy to the inverter to feed the load. As in static converters, the transfer from the normal supply to the flywheel is close to zero switching.
- **Failure mode.** When an internal malfunction in the UPS system occurs, the static bypass switch turns on and the load is supplied directly from the normal supply.

The main advantage of this hybrid UPS system over static UPS is that it is battery-free, and thus it needs less maintenance, and environmental issues related to batteries are avoided. Furthermore, it is more reliable, its lifetime is longer, its recharging is faster and its operating time range is always known. This solution is commercially available in the range of hundreds of kilowatts, with a ride-through capability of up to 1 minute. Several flywheels can run in parallel to ensure desired operating time and for redundancy. However, it retains the disadvantages of drawing harmonics to the upstream facility. The main advantages over dynamic systems are that its footprint and noise are lower, and its downtime is also less for a bearing replacement.

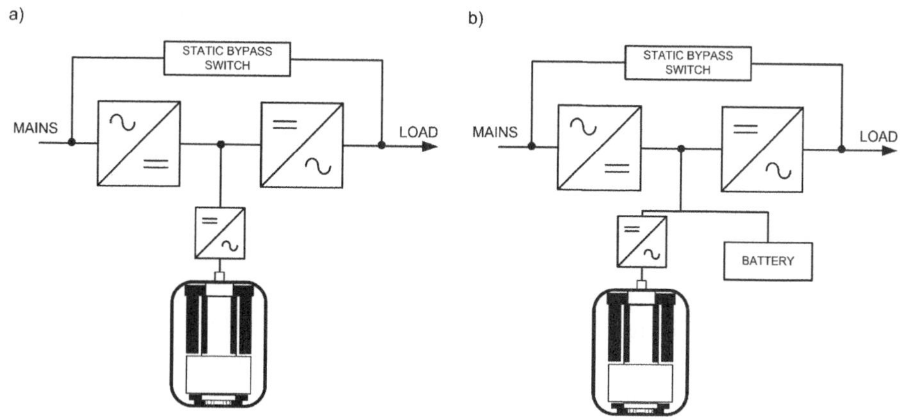

Figure 12.9 Hybrid static/dynamic UPS systems based on flywheel

12.4.6.4 Rectifier/Inverter with a Flywheel and Batteries (Figure 12.9b)

This is a modification of the previous one where a battery bank is connected to the d.c. bus in parallel with the flywheel. It also has three operating modes:

- **Normal mode.** The rectifier converts a.c. power from the normal supply, the utility power, to d.c. power, which feeds the inverter, the batteries and the flywheel. Then both batteries and flywheel store energy, and the inverter supplies a suitable power to the load.
- **Stored-energy mode.** In front of some a.c. input disturbances, the flywheel supplies energy to the inverter to feed the load. As in static converters, the transfer from the normal supply to the flywheel is close to zero switching. If the flywheel operating time range is exceeded, the batteries supply the energy.
- **Failure mode.** When an internal malfunction in the UPS system occurs, the static bypass switch turns on and the load is supplied directly from the normal supply.

The main advantages of this adaptation, also commercially available, related to the previously described ones, are that it provides more than 1 minute of backup power, extends battery lifetime and eliminates false starts-up of standby power system. Thus the reliability of this system is higher. Nevertheless, more maintenance, as well as more space and conditioning, are required.

12.4.7 D.C. Power Supply Systems

Although most devices with electronics involve an a.c. supply because they have their own embedded power converter, there are some that require a d.c. supply, like telecommunication transmitters and receivers for air traffic control or TV broadcasting. In such situations a d.c. power supply system is needed, as well as an auxiliary d.c. power supply in the event of a malfunction in the normal d.c. supply. Therefore, these kinds of devices have two different d.c. inputs in order to be supplied from two different d.c. systems. There are also other devices like emergency lighting and switchgear in substations for secure operation that can be fed both alternating and direct current. Here, d.c. systems operate as an emergency power system.

D.C. power supply systems are composed of two components: a rectifier/charger that converts a.c. to d.c. and secondary batteries that store the energy. There are two main configurations [13], although other structures are possible. The first and most common type is built with one full-rated rectifier that supplies the load and charges the batteries when the normal supply is available (Figure 12.10a). In this case, the batteries are continuously in parallel with the load and are then the real energy source for the load, so there is no transfer time. Therefore, this configuration is mainly used for uninterruptible loads, as in air traffic control or TV telecommunication devices.

The second type (Figure 12.10b) consists of two separate a.c./d.c. power converters: a full-rated rectifier that supplies the loads directly from the normal supply, and a non-full-rated rectifier that charges the batteries when the normal supply is available. In the event of a power disturbance, the load is transferred automatically to the batteries by a switch

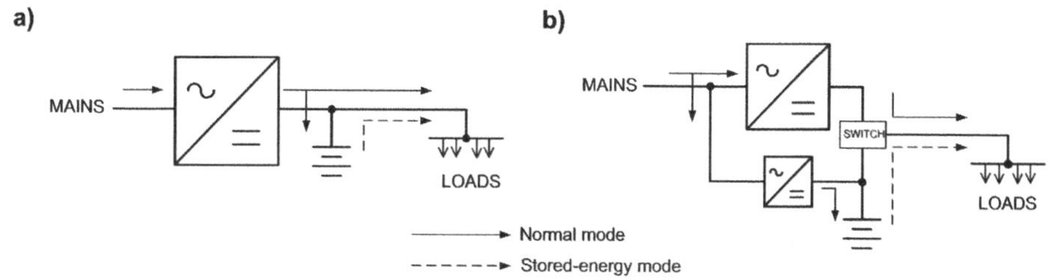

Figure 12.10 D.C. power supply systems

within a few milliseconds. This small delay means that this type mainly fits with emergency lighting requirements. However, its reliability is higher than the first type in the event of a failure in the main rectifier, since batteries have their own rectifier that will continue charging the load.

D.C. system capacity is large, over a few hours, in order to ensure load operation till normal supply is again available or controlled shutdown is completed. The d.c. system capacity is generally described in terms of current, from few amps to some kiloamps. Moreover, the charging time exceeds the discharge time, so charging/discharging cycles are low. Parallel operation of a rectifier/charger is possible to reach high current ratings. Both single- and three-phase units are commercially available, and the voltage ratings vary according to the application. The standard voltages are 24 V (e.g. control and emergency systems), 48 V (e.g. telecommunications), 110 V and 220 V (e.g. operation switchgear in substations), though other customized options are also available.

Valve-regulated lead–acid (VRLA) and Ni–Cd batteries with high-capacity range (up to several thousands ampere hours) are used in d.c. systems in general. They allow operating time ratings from some hours to a few days. They require low maintenance and their lifetime varies typically between 15 and 20 years. However, batteries are sensitive to temperatures, above all VRLA. The design temperature is normally between 20 and 25 °C. Higher temperatures reduce their operating time rating and lower temperatures affect their capacity. Furthermore, rectifiers are usually designed for a temperature range between 0 and 40 °C. Therefore, it is necessary to place them in an air-conditioned room.

Rectifiers for d.c. systems are usually composed of a transformer, to reduce the voltage level, a switching bridge to convert a.c. to d.c., and input and output filters to reduce harmonics and smooth the d.c. supply respectively. The switching bridge can be made with a thyristor or IGBT and operate in parallel charging. As in all electronic devices, d.c. systems draw harmonics, so it is necessary to assess the effect of their harmonic content on the upstream facility. Then, a power factor correction (PFC) may be implemented. Additionally, a rectifier power factor also influences the upstream facility. Moreover, both parameter values, as well as efficiency, get worse when the appliance is beyond its rated value.

12.4.8 Technological Comparison

Designing and installing an auxiliary power supply system to ensure the reliability and quality of power supply require a careful assessment in order to select suitable technology. This evaluation must start by analyzing the electric power supply required by the critical loads, which should mainly include the general power parameters, voltage and current, their maximum and minimum deviations and their related requirements like power factor, drawing harmonics, frequency margins and allowed change rate per second. At the same time, the normal power (usually utility supply) must also be studied in order to know if its reliability and quality parameters meet the desired requirements for the critical load.

These specifications vary widely depending on both the kind of load and its configuration, and the network that supplies the normal power. Then, the next step is to find the emergency or standby power technology that best matches these requirements. Therefore, a technical comparison between these auxiliary power technologies and specifications for supplying critical loads has to be carried out. The main factors of emergency and standby power systems that must be considered are [2], [26]:

- Input power parameters, including voltage, current, power factor and harmonic distortion.
- Output power parameters, including voltage, current, frequency and harmonic distortion, and their regulation in front of load steps and unbalanced load.
- Operating efficiency.
- Electromagnetic compatibility and noise.
- Overload and inrush power capacity and bypass capability.
- Energy storage technology and maximum power reserve time (backup).
- Transfer time for changes in operational mode.
- Neutral scheme (transformer configuration) upstream and downstream.
- System reliability and maintainability.
- Potential for future expansion.

The specifications of the facility's environmental conditions should also be included for an accurate selection and design because these might introduce some limitations. The environment in which an auxiliary power supply system operates has a significant effect on its own reliability and therefore on overall reliability. Altitude, humidity, extreme temperatures can reduce lifetime components or derated output power. Most emergency and standby power supply systems are rated for operation from 0 to 40 °C and for operating without any derating factor for altitudes below 1500 m.

Finally, an economic evaluation must be done. This compulsory analysis will be easier if a previous analysis has been done properly because it allows a balance between technical and economic points of view when making a decision to choose the cost-effective solution. Economic evaluation must include the loss of revenue for expected and unexpected power disturbances, both internal and external, not just the costs of purchasing and maintaining an auxiliary power system, as if it were an insurance policy.

Economic matters should not be the main factor or the only one. An emergency and standby power supply system that is not properly selected can often result in disastrous consequences because it does not perform as desired and its reliability may be lower than expected.

Table 12.1 summarizes and compares the emergency and standby power supply systems previously commented on in this section and their main applications. As has been said, usually the solution for a reliable and quality power supply is a combination of some of these systems since there is no ideal device: infinite power capacity and infinite stored energy

Table 12.1 Comparison of emergency and standby power technologies

Technology / Characteristic	Engine-driven generator	Microturbine-driven generator	Uninterruptible power supplies				D.C. power supply
			Dynamic		Static	Hybrid	
			Traditional	Flywheel			
Power disturbance protection	Long-term transients Voltage deviations Outages	Long-term transients Voltage deviations Outages	Transients Voltage deviations Frequency excursions		Transients Voltage deviations Frequency excursions Outages	Transients Voltage deviations Frequency excursions Outages	Voltage deviations Outages (for d.c. loads)
Power range	From kW to several MW	From kW to several MW	From a few kW to hundreds of kW	From hundreds of kW	From W to hundreds of kW	From hundreds of kW	From W to several kW
Energy storage capacity	Very high	Very high	Low	Low	Moderate	Moderate	High
Transfer time	Moderate to high	High	Almost zero	Almost zero	Almost zero	Almost zero	Almost zero
Operating time range	Very high (depends on fuel tank)	Very high (depends on fuel tank)	Low	Low	Moderate (depends on batteries)	Moderate (depends on storage technology)	Moderate (depends on batteries)
Efficiency (at design point)	Up to 45 % (diesel fuelled) Up to 30 % (petrol fuelled)	Up to 30 % (if only producing electricity) Up to 85 % (CHP)	Up to 85 %	Up to 90 %	From 80 to 95 %	Up to 95 %	>90 %

Table 12.1 (Continued)

Technology / Characteristic	Engine-driven generator	Microturbine-driven generator	Uninterruptible power supplies				D.C. power supply
			Dynamic		Static	Hybrid	
			Traditional	Flywheel			
Installation cost	Very high	Very high	Very high	High	High	High	Moderate to high
Maintenance cost	High	Very high	High	Moderate	Moderate	Moderate	Moderate
Size and space requirements	Very high	Very high	Very high	High	Moderate	High	Moderate
Environmental impact	Exhausted gases Oil Noise	Exhausted gases Noise	Noise	Friendly	Batteries	Batteries (if present)	Batteries
Application	Standby power for public buildings and ICT systems Grid reinforcement Isolated facilities	Hospitals Large office buildings Distributed generation Isolated facilities	Life-support systems ICT systems On-line management systems Industrial continuous processing				ICT systems Safety systems Control systems

with zero transfer time and low cost. The choice most commonly used is a combination of static UPS conditioning and uninterrupted power against short-term disturbances plus an engine-driven generator for a long-term disturbance or outage.

12.5 DYNAMIC UPS SYSTEMS

Traditional dynamic UPS systems have been introduced in Section 12.4.4. Generally, a steel flywheel mechanically stores the energy to give a ride-through capability to the rotating machine set. But this ride-through capability can be very limited up to few seconds because, when the flywheel slows down in the stored-energy mode, the generator output frequency drops.

Advances in power converters led static UPS systems to replace this motor/generator technology. But an a.c. generator will always be the best a.c. supply for any load. However, ongoing efforts have regenerated interest in dynamic UPS. On the other hand, this vigor has been concentrated on the flywheel itself [9], [19], in order to achieve higher power and energy densities because most of the power disturbances have a duration of less than 1 minute. Nowadays, the market has modern flywheels available in the megawatt range thanks to advancements in materials science, power electronics and control. This chapter is focused on these new appliances.

12.5.1 Modern Flywheels

New flywheels are the modern concept for dynamic UPS systems. They can be installed as standalone or combined with a motor/generator set. In any of these two cases, a flywheel should be considered as a battery that stores mechanical energy in the form of kinetic energy.

12.5.2 Components

Current flywheels are composed of different elements placed in a containment housing (Figure 12.11). One of their key components is the rotor, the rotating mass, i.e. the energy storage. The amount of kinetic energy is proportional to both the moment of inertia and the square of the rotational speed. Thus, flywheel kinetic energy can rise, increasing either the mass of the rotor or the rotational speed. Therefore, the improvements in former flywheels were concentrated on balancing mass and rotational speed to obtain the maximum energy.

Thus, at the beginning, metals were used for rotors because of their high density. However, they were limited in power. As centrifugal force is proportional to the mass and to the square of the rotational speed, the higher the density of the material, the greater the centrifugal force. So this centrifugal force can break the rotor if the rotational speed exceeds the tensile strength of the material [26].

Materials science has provided a solution. It has developed new, high-strength, lightweight fiber composites. This combination of strength and weight makes it possible for flywheels to spin at higher rotational speeds than conventional metal-based flywheels.

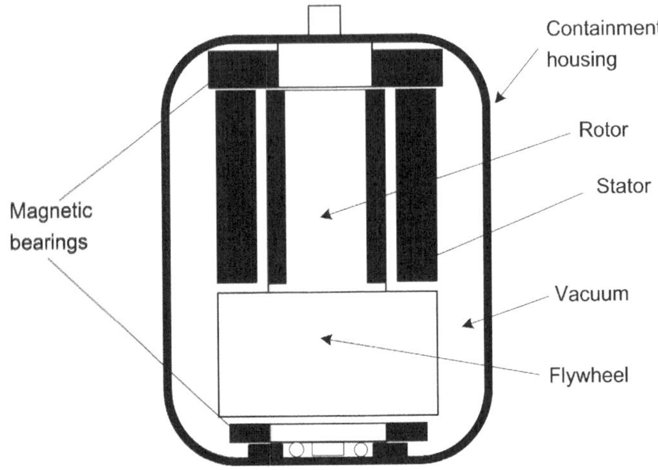

Figure 12.11 Modern flywheel

Then magnetic bearings were applied to be able to deal with such high speeds. Mechanical bearings cannot support such speeds because of frictional forces and subsequent wear. The magnetic bearings mean that the rotor levitates, reducing the frictional forces, although this implies a complex control. Moreover, the rotor is placed in a low-vacuum atmosphere, which reduces frictional losses even more and maximizes efficiency.

One last element of new flywheels is the motor/generator. It is integrated into the system included within the containment housing. It is mounted onto the stator and is usually of rotating field design. The magnetic field can be supplied by permanent magnets or rare earth permanent magnets, which can be placed either in the core or in the outer radius of the rotor, or by a conventional excitation.

A power electronics converter makes available the interaction with the power supply where they are connected, providing a ride-through capability with excellent reliability and recharging rates. These modern flywheels require little maintenance, even for the vacuum system. They also have other outstanding characteristics like long life and efficiency above 95 %. Furthermore, they are environmentally friendly and not noisy.

12.5.3 Classification

The design objective of flywheels is to reach higher power and energy density. As mentioned earlier, the method for maximizing stored energy is through increasing either the mass of the rotor or the rotational speed. The former is more often used, although research in new materials continues. Nevertheless, the speed is limited by the material properties, since inertial forces stress the rotor. Moreover, other facts like system cost, weight and size and the needs of the market also have an influence over flywheel design.

Flywheels can be classified into two groups based on the rotor speed, which basically depends on the rotor material: namely, low-speed and high-speed flywheels [5]. It should be noted that total energy capacity depends on the speed range of the flywheels since at very low speeds they cannot provide their rated power.

12.5.3.1 Low-Speed Flywheel Systems

Low-speed flywheels are usually made of high-strength engineering steel, which allows a rotating speed up to 6000 rpm. The rotor can be in a partial vacuum environment because the frictional losses are not significant at this speed rating. Both permanent magnets and conventional excitation can be considered for the motor/generator, with a generator output frequency between 100 and 200 Hz. This component actually limits the energy exchange ratings.

This sort of flywheel is mainly considered for a short operating time range. They are also known as power-wheels. And they are generally installed for stationary applications as UPS. The reason is their simplicity and their cost, in the order of five times lower than high-speed flywheels. On the other hand, their weight and size are nearly double high-speed flywheels. In conclusion, these flywheels provide the maximum energy storage per unit cost.

12.5.3.2 High-Speed Flywheel Systems

High-speed flywheels are made of high-tech fiber composites with a higher strength and lower moment of inertia than engineering steel, which allow rotating speeds above 10 000 rpm up to 100 000 rpm. Such high speeds are possible because the rotor spins in a vacuum environment and it levitates thanks to the magnetic bearings. These speeds also mean that the rotor can only be excited by permanent magnets. The stator is generally ironless gas ceramic, in order to limit hysteresis losses during the standby mode. Furthermore, the generator output is unregulated for a few kilohertz. And in any case, the power electronics converter is designed for both maximum voltage at high speed and maximum current at low speed.

This kind of flywheel can store a larger amount of energy to operate in a medium time range. They are also known as energy-wheels. Although the technology is developing, they are used in vehicular and aerospace applications mainly because of their high density and low weight and small size. Nevertheless, their cost is very large, the highest of all for the rotor advanced materials. Finally, the objective is to get the maximum energy storage per unit volume.

12.5.4 Application

High reliability is not the only main point that customers desire in the power supply. A high power quality is also strongly required, especially the inhibition of power disturbances to less than 1 minute. Flywheels represent a great solution to this power quality issue and it is their main application. Even dynamic batteries are a better solution than electrochemical batteries in this time range. Thus:

- Flywheels are more reliable and have a higher efficiency.
- Flywheels can reach high-cycle processes for load-following applications while batteries are slow, although they are good for load leveling.

- Flywheels can be extremely discharged and avoid the oversizing that batteries need in order to avoid reducing their lifetime.
- Flywheels have a fast recharge rate and a low sensitivity to temperature.
- Flywheels have a reliable state monitoring while for batteries this is difficult.

However, their installation is only justified for power ratings over some hundreds of kilowatts because of their cost and space requirements. It is possible to parallelize several units to increase the power rating or achieve redundancy. For a longer power supply time rating, flywheels require a standby power system that starts up within less than 1 minute – mostly a diesel gen-set.

Flywheel applications have been pointed out above. High-speed flywheels are still being developed, although they are focused on transportation applications [5]. For stationary applications, as power supplies, low-speed flywheels are generally used. In such a case two options are commercially available, either a horizontal or vertical shaft flywheel. Flywheels can be part of a hybrid UPS system as energy storage, whether a motor/generator set or static UPS system (see Section 12.4.6).

Flywheels can also perform as a standalone appliance although there are more elements involved than the flywheel itself. There are two main options depicted in Figure 12.12. Both are connected to the normal supply through a choke coil to smoother the interaction.

The simplest option (Figure 12.12a) consists of a flywheel connected directly through its bidirectional power converter to the normal supply. It has two modes of operation:

- **Normal mode.** The normal power is within power quality parameters and the flywheel speeds up and stores energy.
- **Stored-energy mode.** The normal supply parameters undergo some excursion and the flywheel supplies the energy in order to condition the power and it slows down.

The other option is more complex (Figure 12.12b). It includes a motor/generator, a flywheel and two bidirectional converters. There are two bidirectional converters in order to eliminate the speed dependency between the flywheel and the motor/generator. If there

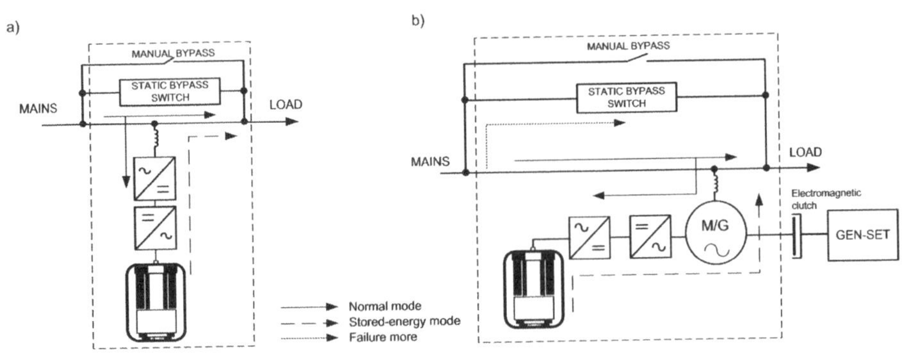

Figure 12.12 Flywheel-based dynamic UPS

is some speed dependency, the system might not supply all its energy. The motor/generator can be driven by a gen-set when the flywheel operating time is exceeded through an electromagnetic clutch.

It has three modes of operation:

- **Normal mode.** The motor spins the flywheel through the converters and it stores energy. In such a case, the generator could act as a dynamic filter for power factor correction.
- **Stored-energy mode.** In front of some a.c. input disturbances, the flywheel supplies the energy to the generator to feed the load.
- **Failure mode.** When an internal malfunction occurs in the UPS system, the static bypass switch turns on and the load is supplied directly from the normal supply.

12.5.5 Future Developments and Applications

Flywheels are basically high-power UPS appliances for power disturbances of less than 1 minute. Further developments aspire to make accessible solutions for a time range up to several hours at low commercial cost. This might be an ideal d.c. power supply system for telecommunications. This technology could also be used as energy storage for distributed generation based on renewable sources.

12.6 STATIC UPS SYSTEM

Static UPS systems have been briefly described in Section 12.4.5. They are composed of power electronics elements for processing energy, i.e. rectifier, inverter and static bypass switch, and electrochemical batteries to store energy. Most of the time, an external element, a manual bypass, is integrated to permit maintenance procedures.

There are a wide range of static UPS systems in terms of size, voltages and power ratings, in both single- and three-phase. The higher demand for reliability and quality of the power supply for critical loads, in combination with advances in power electronics, has meant that a lot of static UPS topologies have appeared in the last few decades, from a few watts to hundreds of kilowatts, or even some megawatts if they operate in parallel.

12.6.1 Components

12.6.1.1 Rectifiers

Rectifiers convert a.c. power from the normal supply to d.c. power.

Low-power rectifiers are generally made of diodes or thyristors. In order to improve their input power factor, a passive power factor correction (PFC) strategy is usually incorporated, which consists of passive filters to avoid drawing harmonics in the normal supply.

Depending on the type of silicon switches, high-power rectifiers can be classified into two sets:

- **Passive rectifier.** Passive rectifiers are not totally controlled. They can be made of diodes or thyristors. There are different commercial configurations of switches or bridges available. The most widely used are 6-pulse and 12-pulse thyristors. The former have a poorer performance regarding input power factor and harmonic content. The latter are built connecting two 6-pulse rectifiers in parallel through a transformer to shift their angles by 30°. The rectifier's performance is better, but its size and weight are larger and its efficiency lower. It is possible to connect more than two 6-pulse rectifiers to obtain 24 or 36 pulses, but this means more shifted transformers. In order to improve 6-pulse bridge behavior, whether it is made of thyristors or diodes, PFC strategy is implemented. PFC improves rectifier performance getting better values for both input power factor and harmonic content. It consists of a d.c./d.c. circuit that requires the rectifier to draw a sinusoidal wave in phase with the input voltage.
- **Active rectifiers.** Active rectifiers are totally controlled and permit a bidirectional energy flow. Their bridge is normally made of six IGBT switches PWM controlled, which permit UPS to be considered as a linear load. Then, the input power factor can be close to the unity and harmonic content is very small.

12.6.1.2 Inverters

Inverters convert d.c. power to a.c. power to supply the load. All current three-phase inverters are made of six IGBT switches, although the switching frequency for three-phase inverters varies up to 20 kHz, as well as the control method that can be through PWM switching or space vector PWM switching. However, single-phase inverters use either four IGBT switches or four thyristor switches since some single-phase applications only need a pseudo sine wave.

12.6.1.3 Battery Bank

Batteries are the most important element in static UPS systems, even more than the power electronics. They store the energy that the inverter needs to supply the load when there are power disturbances in the normal supply. Therefore, a static UPS system is useless if the batteries do not perform properly.

Batteries are the weakest part and this fact is often missed. Furthermore, batteries usually determine where the UPS systems are placed because they have a considerable weight and could need a large specially conditioned space. Thus, a battery bank represents a significant part of the cost of a static UPS system. Further, they require periodic maintenance.

Batteries are particularly sensitive to temperature, charging/discharging processes and d.c. ripple. Lower temperatures can decrease their capacity while higher temperatures can decrease their lifetime. Charging/discharging processes affect battery temperature too. And deep discharges and overcharging reduce battery lifetime as well. Meanwhile d.c. ripple increases their internal temperature [15].

Batteries can experience three types of failures [25]:

- Deterioration of capacity because of deep discharge, cycling usage, high temperature or dry-out.
- Low impedance due to a short circuit between the plates.
- High impedance due to corrosion of the plates, a loose contact between the active material on the plates, or low specific gravity of acid.

Static UPS systems, principally high-power units, include temperature compensation and charging/discharging control to optimize battery use for each situation. A bidirectional d.c./d.c. converter is used to perform these controls: constant current, constant voltage or even a combination of both. It allows implementation of suitable charging/discharging strategies and also maintaining the appropriate floating voltage as well as monitoring to estimate the battery condition.

Another function of this converter is to connect the battery bank to the d.c. bus. Then the battery can have low-voltage ratings while the d.c. bus has a high-voltage rate. Therefore, the number of batteries connected in series would be less than if the battery bank were connected directly to the d.c. bus. Thus space and maintenance requirements are reduced, as well as costs, while reliability increases.

As mentioned previously, batteries can also be connected directly to the d.c. bus in parallel with the d.c. bus capacitors. The rectifier charges the batteries, maintaining an appropriate d.c. bus voltage. This is a simple option and it is mainly used in single-phase low-power UPS units. But this means a high number of batteries connected in series, and this represents more space and maintenance, and then costs, while reliability decreases.

Lead–acid batteries are the most popular storage alternative for static UPS systems; in particular valve-regulated lead–acid (VRLA) batteries [6]. Lead–acid batteries can also be flooded (wet), but the VRLA type has replaced them because they reduce maintenance. There are basically two VRLA battery technologies: absorbed glass mat (AGM) and gel VRLA. They differ in the form of the electrolyte, which for AGM batteries is a liquid trapped in the separator and for gel batteries a gelled mass.

Both classes are capable of providing short-term high current. They come in 6 or 12 V modules that are connected in series to reach the voltage ratings required to be connected to the d.c. bus. However, the AGM type is employed more in static UPS and gel type in d.c. power supply systems.

These VRLA batteries are colloquially called sealed, but it should be noted that they are valve regulated and so some gases can be expelled. Then, a specific place with good ventilation is best for these batteries. Moreover, their ideal environment operational temperature range is between 20 and 25 °C.

Ni–Cd batteries are sometimes used in static UPS systems. However, their use is limited to applications with high energy density and small recharging time rating requirements because they are expensive.

12.6.1.4 Static Bypass Switch

This consists of a set of two elements, each one made of two anti-parallel-connected thyristors. These sets automatically change from normal to bypass mode when needed with practically zero transfer time.

12.6.1.5 Other Components

Apart from these four basic components, static UPS systems could include the following elements, depending on the UPS technology:

1. **Input filter.** The rectifier could either include or need an input filter in order to reduce the harmonics it draws from the upstream facility. This can be either a passive filter or an active filter like those described in Section 12.3.1.1.
2. **Output transformer.** Some static UPS units could include a transformer on their output. There are two possible causes. Firstly, it can increase the inverter voltage level to reach standard values and, at the same time, it could provide galvanic isolation. Secondly, it can be used as a part of the output filter, apart from the choice of providing galvanic isolation. The first option deals with low frequency, while the second deals with high frequencies. Then, transformer size for the first function is larger and has an inferior response to changes in the load and input voltage. Regarding the output transformer, it should be mentioned that it can provide electrical isolation between the normal power and the load. This takes place when the static bypass connection is upstream of the transformer. If the connection point is downstream, this isolation is lost in bypass operation mode.
3. **Output filter.** An LC filter is used at the output of the UPS or inverter to improve the quality of the output voltage. Nevertheless it can affect output power when the existing power factor of the load is different from the one selected at design stage or the UPS systems are performing away from the rated power.

12.6.2 Classification

The large variety of existing static UPS systems and their complexity require an advanced background to fully understand the technology. Additionally, widespread terms like on-line and off-line can mislead. The former means continuous power conditioning during normal operation while the latter only provides power conditioning during the stored-energy mode. Nevertheless, a simple interpretation of both words could be that the unit is normally operational or normally turned off, respectively. But if a unit is turned off, it is rather difficult for it to start up when it is required. Moreover, an operational unit does not always mean total power conditioning. Therefore, customers are easily confused and it is advisable to use other terms [11].

Since February 1999, the standard IEC 62040-3 [10] has provided a guide for the best understanding between manufacturers and costumers. It defines static UPS function and terms, and describes the methods of specifying the performance and the test requirements.

Moreover, it is applied to the whole of the UPS power rating. On the other hand, nowadays this standard is more a recommendation, and it does have other parts: IEC 62040-1-1/2 regarding general and safety requirements and IEC 62040-2 regarding electromagnetic compatibility.

12.6.2.1 Classification Code

IEC 62040-3 proposes an eight-digit classification code with three steps in order to describe the quality of output voltages under the different operating mode conditions according to load:

V F I – S S – 1 1 1

- Output dependency from input supply. The first three digits, which may be:

  - VFI: voltage and frequency independent.
  - VI: voltage independent.
  - VFD: voltage and frequency dependent.

- Output voltage waveform. Next two digits describe UPS generated voltage for normal mode and storage energy mode respectively. Each of these digits has three possible options:

  - S: sinusoidal under all kinds of loads.
  - X: non-sinusoidal under non-linear reference.
  - Y: non-sinusoidal at all.

- Dynamic tolerance curves describing output voltages limits. The last three digits explain UPS output voltage when there is: a change of operating mode performance, a step with linear load performance in normal/stored-energy mode (worst case) and a step with non-linear performance in normal/stored-energy mode (worst case), in that order. There are three possible classifications: 1, 2 and 3 (see Figure 12.13).

This codification can seem really intricate at first, but it is very helpful. It shows the quality of the output voltage under the different operating mode conditions according to load. Therefore, it is a first approach to know which UPS system is better suited to the critical load power requirements. For example, a static UPS unit with codification VFI–SS–111 provides the best performance to protect critical loads, because its output voltage is never zero when there is disturbance in the input voltage. Of course, there are other technical facts to take into account when selecting static UPS [24], and price as well.

12.6.3 UPS Technology

The basic function of UPS systems is to supply continuous and conditioned power to the loads. This can be done with different strategies. IEC 62040-3 defines the following three technologies of static UPS based on the input/output power path during the normal mode.

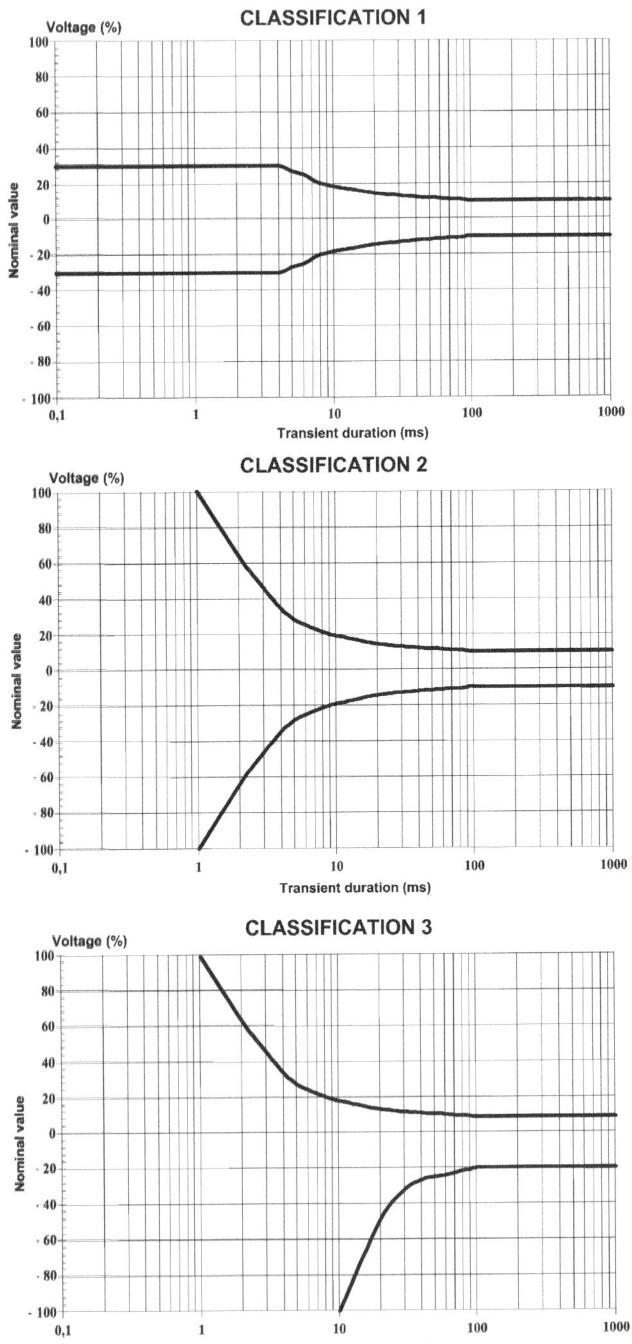

Figure 12.13 UPS dynamic performance classification [10] (Reproduced from July 2001 Uninterruptible Power Systems [UPS] Part 3: Method of Specifying the Performance and Test Requirements, IEC 62040-3)

12.6.3.1 Double Conversion

Its name comes from the double conversion, from a.c. to d.c. and from d.c. to a.c., that occurs during the normal mode of operation. It consists of a rectifier/charger (a.c./d.c.), a battery bank, an inverter (d.c./a.c.) and a static bypass switch (Figure 12.14a). A manual bypass switch can be included for maintenance.

The rectifier and the inverter are full rated because they supply energy to the load, as well as to the battery, during the normal mode. The battery bank is rated in order to have the desired backup time according to the application. It should be mentioned that a phase-locked loop (PLL) control is needed to assure that input and output voltages are in phase, and then it is possible to use the static bypass switch if needed.

There are three operating modes related to this topology:

- **Normal mode.** The load is continuously supplied with high-quality power via the rectifier and inverter. Then the output voltage and frequency are independent of normal supply. At the same time, the rectifier or a dedicated smaller power converter charges the batteries.
- **Stored-energy mode.** If the normal power is outside predetermined values, the inverter is supplied by the battery to maintain continuity of power to the load until the normal power returns within preset values or battery backup time is exceeded. The transfer time is virtually non-existent.
- **Static bypass mode.** The load is fed through the static bypass. UPS operate in this mode in the event of an internal malfunction and for fault clearing and overloads.

This UPS technology is commercially available from a few watts to hundreds of kilowatts. Figure 12.15 shows the normal configuration.

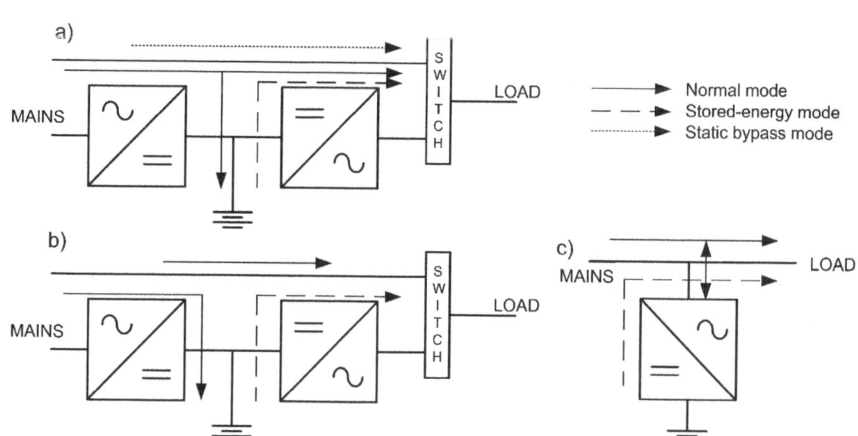

Figure 12.14 UPS technologies: (a) double conversion; (b) passive standby; (c) line interactive [10] (Reproduced from Economic Evaluation of Power Quality, Power Engineering Review IEEE, IEEE)

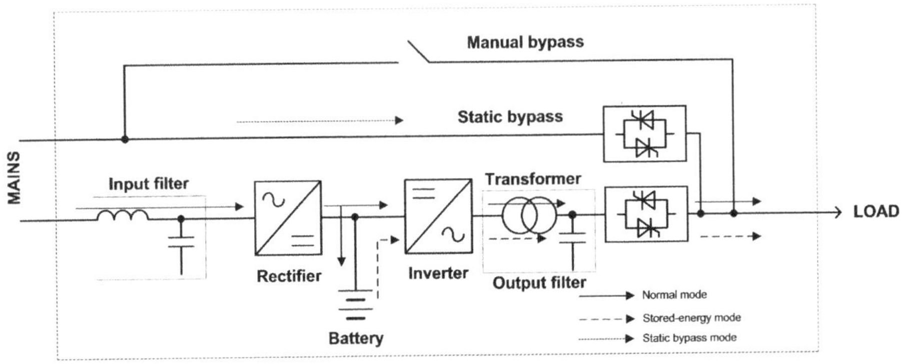

Figure 12.15 Double conversion UPS

12.6.3.2 Passive Standby

Its name comes from the fact that it supplies the load in the event of disturbances in the normal power. It consists of the same elements as double conversion: rectifier/charger, battery bank, inverter and a static bypass (Figure 12.14b). However, the load is supplied through the static bypass during normal operation. As in double conversion, a manual bypass switch can be included.

The inverter is full rated because it has to supply the whole load in stored-energy mode. On the other hand, the rectifier can be smaller as it has to charge the battery and is not necessary to meet the load power. It should be noted that the load is not electrically isolated from the normal power since the unit is connected in parallel and the inverter is off. Therefore, power conditioning during the normal mode can only be provided by adding a power conditioner.

It has two modes of operation:

- **Normal mode.** The normal power supplies the load through the static bypass. Then, the output voltage and frequency are dependent on normal supply. At the same time, the rectifier charges the batteries.
- **Stored-energy mode.** If the normal power is beyond predetermined values, the battery delivers its energy to the inverter and this supplies the load until normal power returns within preset values or battery backup time is exceeded. There is a significant transfer time compared to double conversion. Its duration depends on the starting time of the inverter after sensing the disturbance in the normal supply.

This UPS technology is commercially available from a few watts to hundreds of kilowatts. There are few high-power static UPS units that have been designed as passive standby, although all double conversion units can perform such a passive standby.

12.6.3.3 Line Interactive

This consists of a bidirectional converter, a battery bank and a static switch (Figure 12.14c). A manual bypass switch can be included. The bidirectional converter is full rated to meet the demand of the load. It is connected in parallel and it can charge the batteries while it regulates the voltage or improves the power factor. It should be pointed out that if the inverter has to balance some disturbance in the normal supply, it cannot perform the power factor correction.

There are two operating modes for it:

- **Normal mode.** The load is supplied by the normal power, as in the passive standby. The converter charges the batteries and ensures a stabilized and sinusoidal output voltage for the load at once. Then the output voltage is independent of the normal supply. Moreover, it also performs as an active filter, improving the power factor and suppressing the harmonics.
- **Stored-energy mode.** When the normal supply is beyond the preset values, the batteries deliver their energy to the converter, which now performs just like an inverter, until normal power returns within preset values or battery backup time is exceeded. The transfer time is very small and depends on the time needed to make that the batteries deliver energy to the inverter.

This UPS technology is commercially available for low-power ratings. For high-power ratings there is a new UPS solution called delta conversion [21]. It is an evolution of the simple line interactive concept in order to be capable of simultaneously performing a voltage regulation and a power factor correction. The delta conversion configuration consists of two bidirectional converters, a battery bank, a static switch and a series transformer (Figure 12.16).

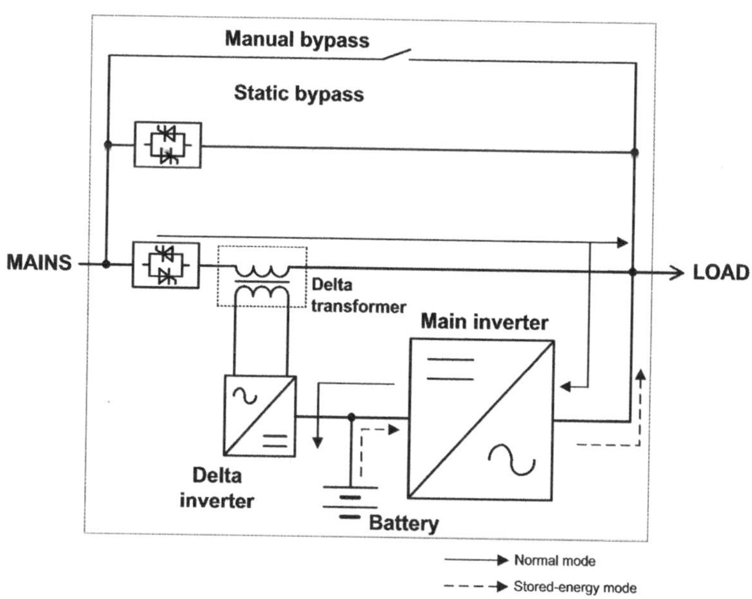

Figure 12.16 Delta conversion UPS

The main inverter is the usual inverter of line interactive UPS. It is full rated and connected in parallel with the load. It uses PWM control to regulate the output voltage. On the other hand, the delta inverter is connected with the normal supply by a series transformer and is rated at 20 % of the UPS power. Its function is to compensate for any voltage difference between the input and the output. Furthermore, it controls the charging of the batteries and the input power factor.

The load is fed by the normal supply whenever it is within the preset values. The power conditioning is limited to 15 % of the total power to make the input power factor unity. Figure 12.17 depicts different situations of the power flow in front of normal power supply deviations. As this figure illustrates, control is rather complex and the load is not electrically isolated from the normal supply. Nevertheless, their efficiency is high because a small part of the energy flows through the converters.

12.6.4 Application

Static UPS systems are widely used because they provide both uninterrupted supply and power conditioning for a wide range of power ratings, from watts to hundreds of kilowatts, and even megawatts if some units operate in parallel. The three technologies previously commented on are commercially available for this power range. However, their performance and structure, as well as costs, determine the choice.

A performance comparison between the three UPS technologies is given in Table 12.2. Figure 12.18 shows the relation between power disturbances and the static UPS technologies.

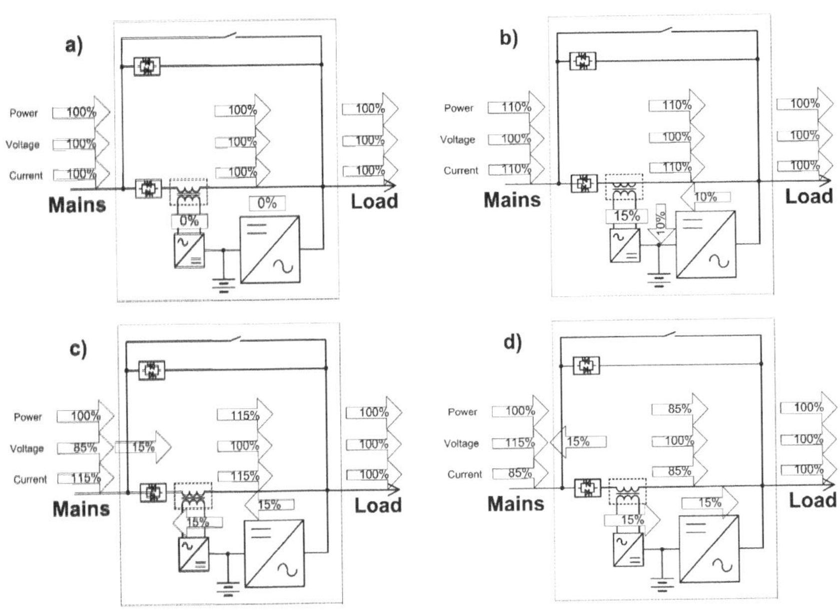

Figure 12.17 Delta conversion UPS power flow [21] (Reproduced from New generation UPS technology, the delta conversion principle, 31st Industry Applications Conference, IEEE Record)

Table 12.2 Comparison between UPS technologies

UPS technology	Double conversion	Passive standby	Line interactive
IEC 62040-3	VFI	VFD	VI
Voltage regulation	Yes	No	Limited
Frequency regulation	Yes	No	No
Drawn disturbances	High	Few	Moderate
Transfer time	Zero	Short	Very short
Efficiency	Low	High	Medium
Operation in normal mode	Continuous	Occasional	Intermittent
Power supply in normal mode	Inverter	Mains	Mains
Power supply when mains absent	Batteries (without interruption for loads)	Batteries (commutation <10 ms)	Batteries (commutation <4 ms)
Maintenance	Medium	Low	Medium
Cost	High	Low	Medium

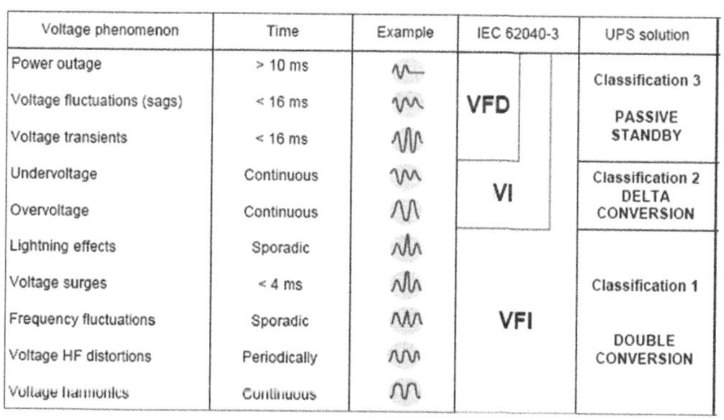

Figure 12.18 Normal supply disturbances and UPS solutions [2] (Reproduced from UPS European Guide, CEMEP)

None of these static UPS technologies matches exactly the features of the ideal UPS listed in Section 12.4.3. However, despite some disadvantages like efficiency and induced disturbances, double conversion is the most preferred UPS technology in terms of performance, power conditioning and load protection. This is the reason why double conversion is broadly installed in high-power applications. On the other hand, for low-power applications, all three technologies can be found [11].

Apart from the performance factors pointed out in Table 12.2, there is another important feature to be considered. This is the availability of the UPS system and then the supply of the load. It is also related with their reliability and their maintenance. One UPS system can be installed to protect the load. However, this means a high initial inversion in the equipment and the facility; and a single point of failure, since if this unit fails or needs maintenance, load is no longer protected. In order to achieve a higher availability, UPS units can be connected in parallel.

Parallel operation permits a redundant structure to be built. Redundancy is called $N+X$, where N is the number of units necessary to meet the load power and X is the number of additional units that are in reserve. When an operating unit fails or needs maintenance, the additional unit takes its place while the former is being fixed or replaced. Moreover, parallel operation allows an easy and cheap update regarding load enlargement by adding additional units. UPS parallel operation can be performed by different control strategies: distributed, master–slave and wireless independent.

UPS performance is mainly evaluated in terms of the output voltage quality, as IEC 62040-3 notes. However, other factors should be included in order to obtain a proper assessment [24]. Redundancy is good in terms of availability of power. But it means that the UPS units are running below half of their capacity and thus far from their design point. In this situation, UPS features such as input power factor, input current THD_i, efficiency and output voltage THD_u get worse, even more so with non-linear loads. This issue has to be considered in the final choice in order to include it at design stage.

12.6.5 Future Developments and Applications

Static UPS systems are mostly used as emergency and standby power supply systems because they can provide uninterrupted power for several minutes and power conditioning. However, in future there will be even more solutions within each UPS technology, due to the market pressures on reducing costs and improving performance. Moreover, new developments in static UPS systems will increase their application areas. Among the different possible ways for decreasing the costs of static UPS systems, there are two that should be pointed out.

The first is to reduce the cost of power electronics converters. The main option is to play with switch technology by reducing the number of switches, replacing controlled switches (IGBT and thyristors) with diodes, or employing switches with less silicon. Apart from reducing costs, these might mean enhancing efficiency.

The second is to make use of emerging storage technologies for replacing lead–acid batteries. These batteries are a mature technology and are prevalent in static UPS systems. But they represent a significant cost and are the weakest element, too. Distributed generation and renewable energy have increased interest in alternative energy storage technologies for energy management. Nowadays, some of them are under development for UPS systems. The most promising are described following [4], [13], and Figure 12.19 shows the power rating and discharge time of these technologies.

- **Advanced batteries.** Some mobile applications, like aircraft, have difficultly coping with the weight and space requirements that involve the use of lead–acid batteries for UPS applications. Here, high-energy-density storage batteries with a high performance are required. Advanced batteries, Ni–hybrid metal and Li-ion, meet these requisites of high energy density. However, their cost is still high for stationary applications with a demand of several kilowatts, as in static UPS.

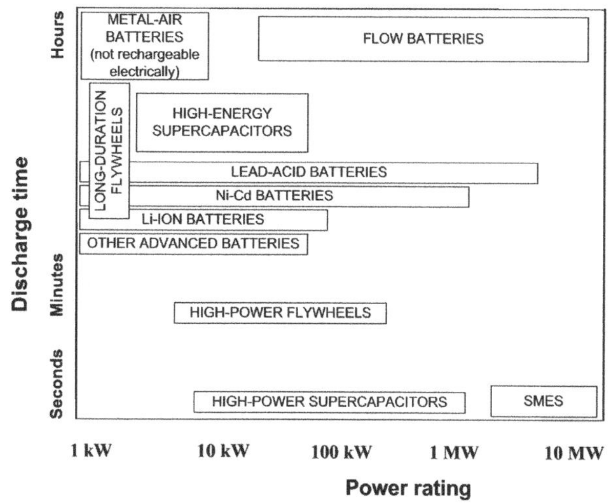

Figure 12.19 Range of storage technologies for power quality and reliability appliances [4] (Reproduced from Economic Evaluation of Power Quality, Power Engineering Review IEEE, IEEE)

- **Flywheels.** As mentioned in Section 12.5, recent developments in flywheel technology have allowed the development of hybrid UPS systems where they can replace batteries in static UPS systems or connect both of them in parallel. However, they mainly fit with high-power ratings for disturbances of less than 1 minute.
- **Supercapacitors.** Supercapacitors are electrochemical capacitors with a capacitance and energy density thousands of times larger than in electrolytic capacitors. Their electrodes are often made of porous carbon material. And the electrolyte can be aqueous or organic. In comparison to lead–acid batteries, supercapacitors have a lower energy density although they can achieve thousands of charge/discharge cycles. Moreover, they have a fast charge/discharge capability, and in that way they are more powerful than batteries. By combining supercapacitors and lead–acid batteries in a UPS system, the former will provide the energy for very short disturbances and the latter will do so for longer ones. Therefore, the charge/discharge battery cycle is reduced and battery lifetime will then be extended. However, only low-power supercapacitors for small electronic devices are well developed.
- **Superconducting magnetic energy storage (SMES).** SMES systems store energy in the magnetic field created by the flow of d.c. in a superconducting coil that has been cooled to superconducting temperature. SMES can be classified as low-temperature SMES (cooled by helium) and high-temperature SMES (cooled by liquid nitrogen). SMES systems consist of a superconducting coil, a bidirectional power converter and a cooler. The power converter controls the charging/discharging processes of the coil. This performs as a current source due to its high inductance. The losses in the superconducting coil are insignificant. Once the SMES is charged, then the magnetic energy can be stored indefinitely and the current will not decay whenever the cooler maintains the temperature. The global efficiency for SMES is 95 %. SMES systems are mainly used for short-duration energy storage of up to 1 second because of their maximum energy and

refrigeration requirements. Their charging/discharging time range is very short and they can provide instantaneously power, up to 100 MW, for this brief period of time. There are several SMES units commercially available, although for low power. High-power units are in development, although their high cost is limiting their widespread use.

- **Fuel cells.** There is a growing interest in replacing gen-sets as standby power systems because of their conditioning and maintenance requirements. Fuel-cell-based UPS are the promising technology that can do that. Moreover, some fuel-cell-based UPS solutions have been developed, although the power of existing units is around tens of kilowatts [18]. Fuel cells provide d.c. power from an electrochemical reaction between oxygen and hydrogen. Hydrogen can be obtained from hydrogen deposits or synthesized from conventional fuels. It is not a combustion process and the main product is water. Therefore, there are no exhausted gases. Also, fuel cells are not noisy and their power density is higher than gen-sets, so they need less space. Fuel cells' backup power depends on the hydrogen deposit. This deposit determines the total size of the system [3]. There are actually two main technologies of fuel cells: high- and low-temperature [17]. The former do not match UPS requirements since their start-up can take several tens of minutes. On the other hand, low-temperature fuel cells can start up and take up the load in approximately 2 minutes. Therefore, as gen-sets, fuel cells need a short-term power supply associated to them, such as lead–acid batteries, advanced batteries or supercapacitors. Although their start-up time is longer than in gen-sets, fuel cells are a scalable technology and have a higher efficiency and reliability than gen-sets.

12.7 GOOD ENGINEERING PRACTICE

The previous sections have described the appliances and equipment used to improve reliability of supply. There are a wide range of possibilities to achieve a goal like availability for twenty-four hours seven days per weeks. However, this target is cumbersome and involves an infinite cost.

The best option for maintaining high availability is a resilient system, which means fault tolerance [27]. This requires careful design to identify the best disposition of reliable appliances in a redundant scheme and good maintenance management in order to meet the cost-effective solution. Moreover, contingency plans for dealing with emergency situations are strongly recommended to avoid a possible lack of understanding of how the system works that could lead to disastrous situations. Periodic training and reinforcing are essential.

This fault tolerance can be achieved through power distribution flexibility and redundancy. If there is more than one path for supplying the load, the critical elements can be easily changed from one power source to another as required for load balancing, system renovation or maintenance or equipment failure.

Furthermore, the following issues are highlighted:

- It is advisable to separate physically redundant emergency and standby power supply appliances.
- Static transfer switches (STS) shift automatically from one source to another without any interruption of the load. There are also automatic or manual switches, but their transfer time is higher.

- Hot-standby redundancy permits the same quality of supply to be retained during UPS failures or maintenance. It consists of a UPS unit that supplies the load and a second UPS unit that is connected to bypasses of the former.
- Critical loads and non-critical loads split in different boards mean that non-critical loads can be turned off in the event of power failure while critical loads can be connected to an emergency and standby power system.
- Each static UPS downstream branch should be rated to 10 % of UPS power as a maximum order to avoid overloading the unit; this can offer protection in front of a fault without shifting to bypass.
- TN-S is advisable for facilities where most of the loads are electronic (non-linear loads). Moreover, grounding is a key element for reliable operation of electronic devices and it requires regular inspection, above all at interconnection points.
- Ground-fault protection should be installed at appropriate points in the power distribution system.
- Switchgear requires regular inspection and cleaning.

For a case study see web address

BIBLIOGRAPHY

[1] Bekiarov S. B. and Emadi A., Uninterruptible power supplies: classification, operation, dynamics, and control. 17th IEEE Applied Power Electronics Conference Proceedings, Dallas, TX, March 2002, pp. 597–604.

[2] CEMEP, European Committee of Manufacturers and Power Electronics, *UPS European Guide*, Paris, 1999.

[3] Choi W., Enjeti P., Howze J. W., Fuel cell powered UPS systems: design considerations. Power Electronics Specialist Conference, 15–19 June 2003, IEEE 34th Annual Meeting, pp. 385–390.

[4] Electric Storage Association (ESA), www.electricitystorage.org.

[5] Emadi A., Nasiri A., Bekiarov S. B., *Uninterruptible Power Supplies and Active Filters*, CRC Press, Boca Raton, FL, 2004.

[6] Essig B., Braun G., Selecting batteries for uninterruptible power supply systems. Applied Power Electronics Conference and Exposition, APEC'91, Conference Proceedings, Sixth Annual Meeting, 10–15 Mar 1991, pp. 642–646.

[7] da Costa Teixeira C., Power quality solutions for low and medium voltage critical loads. Transmission and Distribution Conference and Exposition: Latin America, 2004 IEEE/PES, 8–11 November 2004, pp. 326–331.

[8] Gatine G., *High availability electrical power distribution*, Cahier technique Merlin Gerin No. 148, October 1991.

[9] Hebner R., Beno J. and Walls A., Flywheel batteries come around again. *IEEE Spectrum*, vol. 39, no. 4, pp. 46–51, 2002.

[10] IEC 62040-3, *Uninterruptible power systems (UPS) Part 3: Method of specifying the performance and test requirements*, July 2001.

[11] Karve S., Three of a kind [UPS topologies, IEC standard]. *IEE Review*, vol. 46, no. 2, pp. 27–31, 2000.

[12] Kusko A., DeDad J., Short-term, long-term, energy storage methods for standby electric power systems. Industry Applications Conference, 40th IAS Annual Meeting, Conference Record of 2005, Vol. 4, 2–6 October 2005, pp. 2672–2678.

[13] Markiewicz H., Klajn A., Resilience. Improving reliability with standby power supplies. *Power Quality Application Guide – LPQI*, June 2003.
[14] McGranaghan M., Roettger B., Economic evaluation of power quality. *Power Engineering Review IEEE*, vol. 22, no. 2, pp. 8–12, 2002.
[15] Nelson R. F. and Kepros M. A., AC ripple effects on VRLA batteries in float applications. 14th IEEE Battery Conference on Applications and Advances Proceedings, January 1999, pp. 281–289.
[16] Nielsen J. G., Blaabjerg F., A detailed comparison of system topologies for dynamic voltage restorers. *IEEE Transactions on Industry Applications*, vol. 41, no. 5, pp. 1272–1280, 2005.
[17] Perry M. L., Kotso S., A back-up power solution with no batteries. 26th Annual International Telecommunications Energy Conference, INTELEC 2004, 19–23 September 2004, pp. 210–217.
[18] Perry M. L., Strayer E., Fuel-cell based back-up power for telecommunication applications: developing a reliable and cost-effective solution. 28th Annual International Telecommunications Energy Conference, INTELEC'06, September 2006, pp. 1–8.
[19] Plater B. B., Andrews J. A., Advances in flywheel energy storage systems. Power Quality '97, Intertec International, Ventura, CA, 1997, pp. 460–469.
[20] Platts J., Aubyn J. St., *Uninterruptible Power Supplies*, IEE Power Series, 1992.
[21] Rathmann S., Warner H. A., New generation UPS technology, the delta conversion principle. Industry Applications Conference, 31st IAS Annual Meeting, IAS'96, Conference Record of 1996, IEEE, Vol. 4, 6–10 October 1996, pp. 2389–2395.
[22] Séraudie C., *LV surge and surge arresters. LV insulation co-ordination*, Cahier technique Schneider Electric No. 179, Collection Technique, March 1999.
[23] Stebbins W. L., Power line disturbances: a user's perspective on the selection and application of mitigation equipment and techniques. Textile Industry Technical Conference, IEEE, 1989, pp. 4/1–4/7. 3–4 May 1989.
[24] Sudrià A., Jaureguialzo E., Sumper A., Villafáfila R., Rull J., High-power UPS selection methodology and installation guideline for high reliability power supply. International Conference on Renewable Energy and Power Quality (ICREPQ), Zaragoza, Spain, 16–18 March 2005.
[25] Waltari P., Suntio T., Survey and evaluation of battery monitoring methods and results from user's viewpoint. 21st IEEE International Telecommunication Energy Conference Proceedings, 1999.
[26] Whitaker J. C., Power systems protections alternatives. *AC Power Systems Handbook*, Second Edition, CRC Press, Boca Raton, FL, 1999.
[27] Whitaker J. C., Standby power systems. *AC Power Systems Handbook*, Second Edition, CRC Press, Boca Raton, FL, 1999.
[28] Windhorn A., A hybrid static/rotary UPS system. *IEEE Transactions on Industry Applications*, vol. 28, no. 3, pp. 541–545, 1992.
[29] Woodley N. H., Morgan L., Sundaram A., Experience with an inverter-based dynamic voltage restorer. *IEEE Transactions on Power Delivery*, vol. 14, no. 3, pp. 1181–1186, 1999.

13

Monitoring Power Quality

Andreas Sumper and Samuel Galceran-Arellano

Typically, power quality phenomena are physical phenomena that, in many cases, are appearing and disappearing arbitrarily. Therefore, recording them is more than a simple measurement of an electrical parameter; it is necessary to record them over a certain time interval. In order to reduce the huge amount of data by recording and analyzing several electrical parameters over a long period of time, recording limits are set. If these limits are exceeded, the monitoring instruments record the essential data of the event. For analyzing the tendency, aggregation algorithms are used to reduce the amount of data without losing the trend information. It is necessary to define power quality by establishing limits for measurable power quality magnitudes. If the measured values are in the established interval, this can be assumed as good power quality. So, once these limits are defined, measurements can be compared and the level of power quality in the grid can be determined. Those limits can be found in standards, e.g. in the EN 50160 standard, which shows the most important parameters of voltage quality in distribution networks or taken from experience or from site characteristics.

Monitoring alone is not the solution for power quality problems. In order to solve such problems, something more than the simple installation of power quality monitors at the site is needed. This chapter attempts to dispel any doubts of engineers about organizing a monitoring program to solve power quality problems. It can be organized in three separate programs: an overall power quality program, the power quality survey and the immunization program. In all three, monitoring plays a decisive role. Furthermore, an overview of the main monitoring features will be given.

13.1 MONITORING OBJECTIVES

When the assembly line stops, or the computer network crashes for no apparent reason, very often electric power quality is involved. Voltage dips, harmonics, interruptions, high-frequency noise, etc., are the most important power quality problems that we find in

Handbook of Power Quality Edited by Angelo Baggini

industrial and commercial installations. Troubleshooting these problems requires measuring and analyzing power quality and that leads us to the importance of monitoring instruments in order to localize the problems and find solutions.

Power quality problems are not just solved by the simple installation of a power quality monitor, there are other aspects to consider, either technological or non-technological aspects. The former are all very well known and discussed in engineering societies. Conversely, the latter, like managing power quality projects and the economic impact of power quality problems, are still unknown. The awareness of these non-technological aspects of power quality helps to apply the most effective solution to the problem, which in some cases can differ from the most appropriate technical solution. In many projects related to finding a solution for power quality problems, monitoring plays a decisive role, and, therefore, managing monitoring properly helps to minimize the cost of solving problems [9].

It is useful to know the relationships behind power quality in order to solve the related problem. Firstly, quality of electric supply depends on both the utility power supply and the loads. Secondly, it must be recognized that electric power is similar to one of the raw materials in the production process: a better quality will increase the quality of the end product. Better quality of the end product helps to justify the costs in power quality solutions. An economic study of the impact of poor power quality is recommended for every project related with power quality solutions. Thirdly, technology is good, but knowledge is better. Solving power quality problems depends not only on the technology applied to solve the problem, but also on a profound knowledge of the power quality phenomena, the applied solution and the electrical installation needed to find the most effective solution. Troubleshooting and simple fixes are short-term solutions; knowledge is the only way to find long-term solutions. Finally, the personal relationship between the maintenance personnel and the company offering power quality solutions is sometimes decisive as to what kind of solution is applied. In many cases, maintenance personnel are skeptical of new power quality solutions offered in the market, because they are not sufficiently informed about the different alternatives available.

13.1.1 Benefits of Power Quality Monitoring

There are several reasons for monitoring power quality. The most important reason is the economic damage produced by electromagnetic phenomena in critical process loads. Effects on equipment and process operations can include malfunctions, damage, process disruption and other anomalies [4].

Monitoring requires an investment in equipment, time and education. In many cases management, production and plant engineers must be sufficiently convinced of the benefits of monitoring. It is an essential analytical tool used in order to improve the availability of power. The investment in monitoring can be justified by its increased availability due to the following:

- Preventive and predictive maintenance.
- Determining the need for mitigation equipment.
- Ensuring equipment performance.
- Sensitivity assessment of process equipment to disturbances.

Monitoring can help to identify power quality problems and minimize losses in the production process and increase plant productivity. Monitoring is an essential component of the customer care process for his business.

13.2 MEASUREMENT ISSUES

The recorded measurements can vary from the way some instruments record the disturbance levels and how the signals are interpreted. This can lead to non-existent errors and recording disturbances. Consequently, users are not able to interpret the importance of the disturbance on equipment; this can lead to incorrect conclusions and costly decisions.

Therefore, users need to revise the measurement instrument specification of the monitoring instruments. The most important are listed below [4]:

- sampling rate;
- accuracy;
- precision;
- resolution;
- bandwidth amplitude/frequency;
- differential mode argument and amplitude accuracy;
- anti-aliasing filter;
- sampling window width;
- number of windows analyzed per second;
- type of weighted window used;
- synchronization technique used;
- accuracy of the synchronized technique;
- common mode rejection ratio;
- flag when the phase-locked loop is not synchronized;
- flag when hardware or software error occurs;
- flag when some frequency components present in the signal are not recorded;
- instrument immunity to disturbances in the supply voltage;
- environment of operation.

13.2.1 Measurement Chain

The measurement chain stated by IEC 61000-4-30 [3] is shown in Figure 13.1. There are three elements of power quality measurement, namely the measurement transducer, the measurement unit and the evaluation unit.

The first is used to adapt the electrical input signal for the measurement unit: for example, to step down the voltage, to isolate the input circuits or to transmit the signals over a distance. Two important considerations when using transducers are the signal level in relation to the full scale of the monitor and the frequency and phase response.

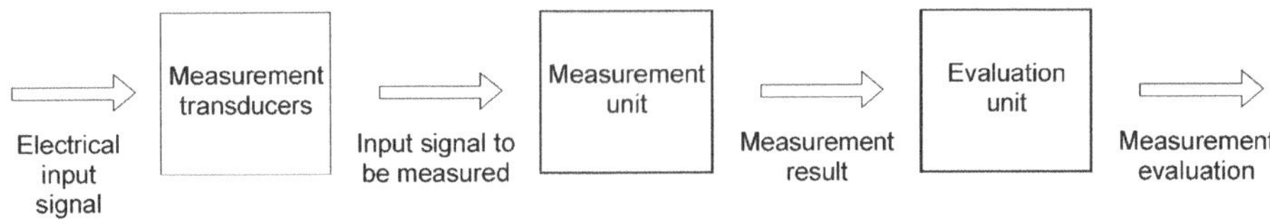

Figure 13.1 Measurement chain [3] (Reproduced from IEEE 1159: Recommended practice for monitoring electric power quality, 1995, IEEE)

The measuring unit realizes the measurement of the input signal and provides the measurement results to the evaluation unit, which processes them to obtain the measurement evaluation. A measurement evaluation can be, for example, the FFT for the harmonic or interharmonic analysis.

The first step of measurement is to obtain samples of the event waveforms, typically for three phases. It is necessary that the sampling rate and the resolution of the signal are adequate to record the disturbance. Due to the fact that the monitor is also used to obtain the harmonic spectrum, typical samples per cycle are 128 or 256 [2], [5].

13.2.2 R.M.S. Measurement

In order to calculate the magnitude of the supply voltage or current, the r.m.s. value is calculated based on the sampled values over a specified period. For example, for the measurement of a voltage dip and swell $U_{rms(1/2)}$ is to be used on each measurement channel, where $U_{rms(1/2)}$ is defined as *the value of the r.m.s. voltage measured over one cycle and refreshed each half cycle* [3]. The sampled voltages are squared and averaged over a window of one cycle duration [2], [5]:

$$U_{rms(1/2)} = \sqrt{\frac{1}{N} \sum_{i=1+k\frac{N}{2}}^{(k+1)\frac{N}{2}} u^2(i)} \tag{13.1}$$

where N is the number of samples per cycle and $u(i)$ the recorded (sampled) voltage waveform at $k = 1, 2, 3$, etc.

The first value is obtained over a cycle (from sample 1 to sample N), the next from sample $^1/_2N+ 1$ to sample $N + {}^1/_2N$, and so on.

13.2.3 Transients Measurement

Transients are by nature rapidly varying signals and have a wide range of waveforms, amplitudes and durations. Transients can be classified into two categories, namely impulsive transients and oscillatory transients. These terms reflect the wave shape of a current or voltage transient. Obtaining their signatures allows their classification into a few typical waveforms. The requirements for the data acquisition of transients are higher than for the r.m.s. measurement. The frequency spectrum of transients of, for example, a.c. main transients can be up to 10 MHz (100 ns) and for large band amplitude up to 1 MHz. The sampling rate must be at least twice the maximum frequency, but practically it should be 10 times higher in order to record the original waveform.

13.2.4 Sampling Rate

The sampled signal is a sequence of values of a signal taken at discrete instants. The resulting signal is only intermittently observed and therefore it represents the measured parameter as a series of values for subsequent use, such as for modulation, coding and quantization.

Methods that are used to interpolate a sampled signal in the case when the signal has components which are faster than monitoring's capability to collect sufficient sample points are called linear interpolation, which connects sample points with straight lines, and sine interpolation, which connects samples with curves.

13.2.5 Bandwidth Amplitude/Frequency

The bandwidth of a measurement instrument is understood as the width of a frequency band over which a given characteristic of this instrument is in accordance with certain limits or ratios.

13.2.6 Accuracy

Accuracy is defined as the degree of conformity of a measured or calculated value to its actual or specified value. In other words, accuracy is the degree of agreement between a measured value and the true value. Generally, it is expressed as percent inaccuracy. For example, instrument accuracy is expressed in terms of its uncertainty, the degree of deviation from a known value. An instrument with an uncertainty of 0.5 % is 99.5 % accurate during the full normal life of the calibration recommended by the manufacturer.[1]

13.2.7 Precision

The degree of mutual agreement among a series of individual measurements, values or results is often, but not necessarily, expressed by the standard deviation. With respect to a single device, put into operation repeatedly without adjustment, precision is the ability to produce the same value or result, given the same input conditions and operating in the same environment. In other words [4], precision is the quantity of coherence or repeatability of measurement data, customarily expressed in terms of the standard deviation, of the extended

[1] Accuracy is not related to the repeatability, nor to the reproducibility of the measurement, in contrast to the term precision. On the other hand, measurement uncertainty is defined as 'A parameter associated with the result of a measurement that characterizes the dispersion of the values that could reasonably be attributed to the measurand.'

set of measurement results from a well-defined (adequately specified) measurement process in a state of statistical control. The standard deviation of the conceptual population is approximated by the standard deviation of an extended set of actual measurements. When not specified, the precision specification is assumed to be repeatable during one standard deviation interval that is 68.27 % of the measurements.

13.2.8 Resolution

Resolution is the ability of the measurement system to detect and consistently indicate small changes in the characteristic of the measurement result. In other words, resolution is defined as the smallest change in the measured or supplied quantity to which a numerical value can be assigned without interpolation. It can be seen in the level of detail of the measurement.

13.2.9 Measurement Aggregation Algorithm

The measurement of supply voltage, harmonics, interharmonics and unbalance are carried out in every cycle. In order to evaluate the measurement results during a longer time interval, it is necessary to aggregate these cycle measurements. IEC 61000-4-30 proposes four different aggregation intervals. Aggregations are performed using the square root of the arithmetic mean of the squared input values. The basic measurement time interval is a 10-cycle time interval for 50 Hz power systems or a 12-cycle time interval for 60 Hz power systems:

$$U_{rms\_200\mathrm{ms}} = \sqrt{\frac{1}{200\ \mathrm{ms}} \int_{200\mathrm{ms}} U^2(t)\, dt} \tag{13.2}$$

The obtained 200 ms measurements are again aggregated by the *very short-term interval aggregation* in 3-second intervals (150/180-cycle time interval). This is made from 15 basic measurement time intervals of 200 ms (10/12-cycle):

$$U_{rms\_3\mathrm{s}} = \sqrt{\frac{1}{15} \sum_{i=1}^{15} U^2_{rms\_200\mathrm{ms}}} \tag{13.3}$$

Short-term interval aggregation (10 min): 200 very short-term intervals of 3 seconds are aggregated to one short-term interval. The resulting 10 min measurement value should be tagged with the absolute time (e.g. 01H10.00), which indicates the time by the end of the 10 min aggregation:

$$U_{rms\_10\mathrm{min}} = \sqrt{\frac{1}{200} \sum_{i=1}^{200} U^2_{rms\_3\mathrm{s}}} \tag{13.4}$$

Long-term intervals (2 h): 12 short-term intervals (10 min) are aggregated to a long-term interval:

$$U_{rms\_2h} = \sqrt{\frac{1}{12} \sum_{i=1}^{12} U_{rms\_10min}^2} \tag{13.5}$$

13.3 SELECTION OF MONITORING INSTRUMENTS

There are different types of monitoring instruments available on the market, but sometimes their characteristics are not easy to compare. In order to choose the right instrument, it is necessary to analyze the monitoring needs first, e.g. type of disturbance to be monitored, monitoring period, requested accuracy. As a function of these needs, the selection of monitoring can be realized more objectively.

Standard IEC 61000-4-30 is aimed at defining measurement methods which make the comparison of the monitoring results available, no matter which instrument is used. The first version of this standard was published in 2003 [3].

Measuring parameters defined in this standard are power frequency, nominal voltage, flicker, voltage changes, voltage dips and swells, voltage transients, unbalance, harmonics, interharmonics and signaling voltages. It also specifies measurement uncertainties for voltage, current and frequency as well as for derived values, even though uncertainties of connected current and voltage transducers are not considered.

One of the most important parts of the standard is the specification of measurement intervals and their aggregation as well as the time-clock uncertainty.

Two different classes of measurement performance are defined in this standard:

1. **Class A performance.** This class of measuring instrument is recommended for contractual measurements between network providers and customers to verify compliance with standards or resolving disputes.
2. **Class B performance.** This class of measuring instrument should be used for statistical surveys, troubleshooting applications, etc., where high accuracy is not necessary.

This standard has two different product lines for monitor manufacturers on the market: high-end class A and low-end class B instruments. Table 13.1 shows a comparison of the technical requirements of both classes. It is easily seen that the requirements of class A measurement devices are very high, the accuracy of voltage and current is 0.1 % and the aggregation of the measurements results requested is very strict. Therefore, class A instruments are definitely more expensive than class B instruments and there are few manufacturers offering instruments which comply with all specifications. Moreover, there is no authorized test laboratory for this standard; mostly the certificates are given by firms and laboratories on their own self-reliance. For troubleshooting applications, where the accuracy is less important than the type of disturbance and the location of its source, class B instruments or existing measuring instruments may be adequate. Also, for measurements according to the EN 50160 standard, class B devices are adequate.

Table 13.1 Comparison between class A and class B monitoring instruments from [3] (Reproduced from EC 61000-4-30, Electromagnetic Compatibility, 2003 IEEE)

		Class A		Class B
Parameter	Magnitude	Meter interval	Accuracy	Measurement procedure
Voltage	*V*rms	10 cycles	±0.1 %	The manufacturer shall indicate which process is used
Dips, overvoltages, interruptions	*V*rms, *t*, *T*	*V*rms 1/2 cycle (10 ms in 50 Hz; 12 ms in 60 Hz)	±0.2 %	*V*rms 1/2 cycle
Unbalance	% unbalance	Symmetric component method		The manufacturer shall indicate which process is used
Harmonics and interharmonics	THD, harmonics and interharmonics	IEC 61000-4-7		The manufacturer shall indicate which process is used
Flicker	Plt	IEC 61000-4-15		The manufacturer shall indicate which process is used
Frequency	Hz	10 s	±10 mHz	The manufacturer shall indicate which process is used
Signaling	*V*rms	Interharmonic measurement (for $f > 3$ kHz see IEC 61000-3-8)		The manufacturer shall indicate which process is used
Flagging during dip, swell or interruption		The measurement of power frequency, voltage magnitude, flicker, supply voltage unbalance, voltage harmonics, voltage interharmonics, mains signaling and measurement of underdeviation and overdeviation parameters will be flagged		
Time synchronization		Via GPS, external clock, etc.		The manufacturer shall indicate which process is used

13.3.1 General Features of Monitoring Instruments

The general features of power quality monitoring instruments [6], [12] are:

- **Enclosure options.** Handheld, portable and fixed. The choice of the enclosure option depends on the user's requirement. The handheld and portable options are more specific for engineering and troubleshooting applications. The fixed installation option is more often used by utilities, industrial plants and equipment that are integrated in a power quality monitoring system.

- **Enclosure protection.** The environmental limits for the power quality monitors are usually specified by the manufacturers. The IP (Ingress Protection) rating also must be specified by the manufacturer; instruments could have a similar NEMA rating as well.
- **Power supply.** The power supply of a power quality monitor is also an important consideration. Supply voltage and frequency, battery backup during power failure or a separate supply should be checked.
- **Memory.** The memory options for the recorded events can be hard disks, floppy disks, internal RAM and PCMCIA memory cards.
- **User interface.** Instrument–user communication is usually realized by built-in displays, external viewing devices or personal computers. User–instrument communication is done by keypads, keyboards or by a personal computer. The personal computer options require a connection interface from the instrument to the personal computer. Many instruments allow remote monitoring operation and real-time display of the signals.
- **Software and data analysis tools.** The software and data analysis tools supplied with most power monitors have a variety of functions and data manipulating abilities.
- **Printer.** Printers may be installed internally, with a direct connection or connection by a personal computer.
- **Accessories.** Leads, probes, sensors, current clamps, frames, handles and carrying cases are typical accessories supplied by the manufacturer.
- **Warranty.** The manufacturer or supplier may provide a warranty for a year or more.
- **Update ability.** The update ability of software and hardware (optional modules or cards) is an important deciding factor in the purchase of the monitoring instrument.
- **Maintenance and calibration.** A power quality monitor requires periodic maintenance and calibration. This is an important factor for the lifetime costs of monitoring and should be considered.
- **Accuracy.** The accuracy of a power quality monitoring is specified by the manufacturer.
- **Resolution.** The resolution of an instrument is a measure of the detail of the digital sampled data after the analogue-to-digital conversion process, and it is represented in bits. The larger the number of bits, the finer the resolution with which the sampled data is captured.
- **Sampling rate.** This defines the rate at which the input channels are sampled and should be stated in samples per cycle. For detecting transients, high sampling rates in MHz are necessary.
- **Voltage withstand.** Manufacturers may specify the voltage withstand of the monitor and the complied standards.

13.3.2 Signal Input/Output

The following list describes the signal input and output interface of power quality monitoring instruments:

- ***Input channels.*** The input channels are the main analogue inputs to which the voltage/current transducers are connected.

- **Analogue inputs/outputs**. Analogue inputs can, for example, monitor additional parameters while analogue outputs are used, e.g. as signals to other monitors.
- **Digital inputs/outputs.** These types of inputs and outputs are mainly used to trigger other monitors.
- **Communication and networks.** Internal and external modems via RS232, Ethernet and direct PC connection (USB, RS232, RS485 and Infrared) are provided with the monitoring instrument. Many instrument manufacturers allow the user to download information or operate the monitor via the Internet.

13.3.3 Functions

The following list describes the functions of power quality monitoring instruments:

- **Data capture by present thresholds.** Parameters to be measured are usually captured when the disturbance exceeds a present threshold (event logging) or at repeated set time intervals. The thresholds and time intervals are set up by the user. By event logging, the captured waveform is usually logged one or more cycles before and after the event to provide a full picture of the event.
- **Data capture by self-adjusting thresholds.** The monitor can set its own thresholds by an established steady-state norm. This method allows the detection of small deviations and trends.
- **Externally triggered data capture.** Most monitors provide this feature.
- **Manual data capture.** For a snapshot of the present situation, most monitors provide a manual trigger function.
- **Data logging and time interval recording.** With data logging, the parameters are continuously monitored and can be captured at set time intervals established by the user.
- **Waveform capture.** Some power quality monitors have the ability to capture waveforms (mainly voltage and current). These captured waveforms can be viewed by built-in displays or downloaded to a PC. Often, the instruments provide functions such as harmonic analysis and waveform analysis.
- **Time synchronization.** Some power quality monitors have the option of time synchronization by an external time signal, GPS or a radio signal.
- **Firmware.** Some monitor manufacturers periodically provide new releases of monitor firmware. Firmware releases, typically provided free of charge by manufacturers, are sometimes used to correct errors in the metering algorithms or to enhance the existing feature set of the meter without requiring the purchase of new hardware.

13.4 SUCCESSFUL POWER QUALITY MONITORING

13.4.1 Power Quality Program

When power quality problems occur in industry, monitoring may be useful to determine the cause of the lack of power quality, but it should be a part of the whole power quality program. Monitoring is an important part of the program, but it should go with a power quality program. The following list summarizes its principal parts [7], [12], [10]:

1. Secure a commitment. The first task is to obtain internal support for the program. Engineering, production and management must be convinced of the benefits of this program.
2. Assemble a team. A team of people representing different plant divisions should be assembled.
3. Obtain the participation and cooperation of the electricity utility. An important part is to obtain the participation of the electricity utility. Usually, the utility is interested in collaborating in order to improve customer satisfaction.
4. Establish a power quality specification. This is needed to specify the lack of power quality, defining limits for the measured magnitudes. The limits refer to the process, which suffers the consequences of power quality problems. For example, the CBEMA and ITIC curves (Figure 13.2 and Figure 13.3) can be taken as bases for the voltage dip analysis.
5. Establish a communication process with the electricity utility. Events in the power system that lead to interruptions etc. should be expressed to the customer by the utility through a defined communication process. For example, restarting a production process should be delayed when more grid-side disturbances are expected.
6. Establish a monitoring project and execute a power quality survey. This part deals with monitoring power quality and recording event data. The organization of this survey is described in a later section.
7. Establish an immunization program. In this part a process for event analysis, root cause analysis and corrective action should be realized. A principal description of an immunization program is detailed in a later section.

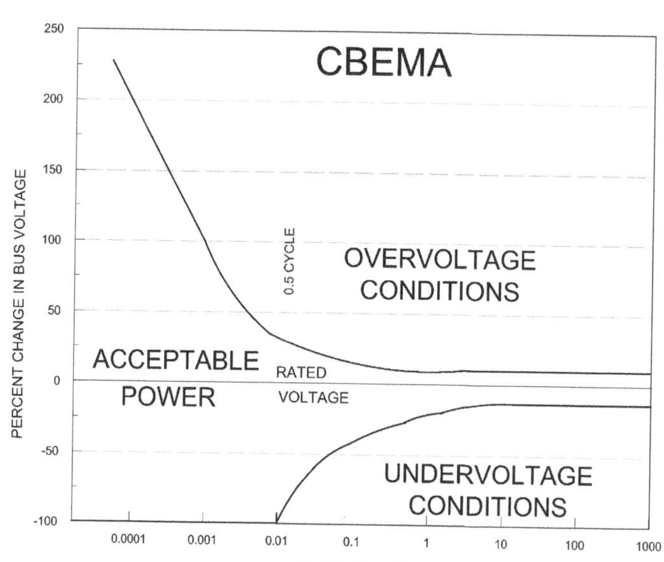

Figure 13.2 CBEMA curve is used to establish criteria for acceptable power quality

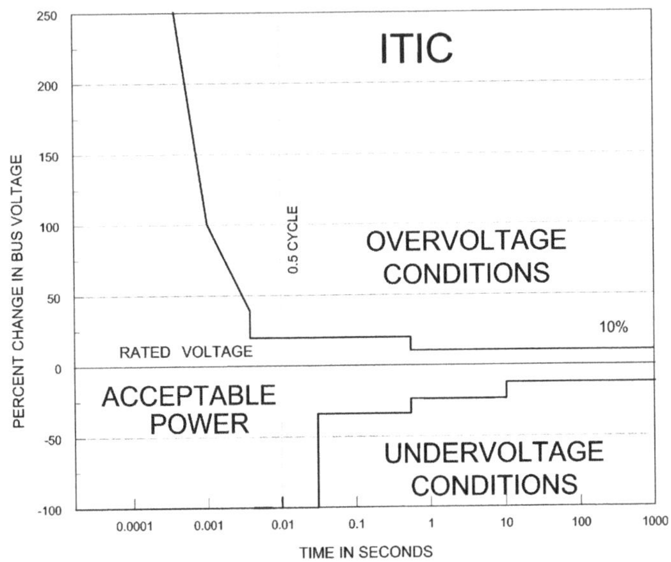

Figure 13.3 ITIC curve is also used in order to establish criteria for acceptable power quality in industrial plants

8. Set improvement targets. Business losses can be reduced by improving power quality or immunizing processes. An evaluation of long-term improvement actions, taking into account these business losses, should be done.
9. Establish a process to evaluate the power quality program. It is necessary to evaluate the effectiveness of this program periodically and, if required, continue the program, redefine targets or terminate the program.

13.4.2 Managing Monitoring Projects

When managing a monitoring project, significant questions will arise depending on the targets established previously [12]:

1. *Why measure?* This question clarifies the monitoring objectives as they determine the choice of measurement equipment, the triggering thresholds, the methods for collecting data, the data storage and analysis requirements, and the overall level of effort required.
2. *What kind of power quality parameter do we want to measure?* Power quality includes a wide variety of conditions on the power system. Important disturbances can vary in duration from very high-frequency impulses (lightning strokes) to long-term overvoltages and interruptions. Standards, like IEC 61000 and EN 50160, and grid codes define the power quality parameter to be measured. IEC 61000-4-30 defines the methods for measurement and interpretation of results for power quality parameters in 50 Hz systems and is part of the standard IEC 61000 of electromagnetic compatibility.

3. *Where should the measurement equipment be located?* Power quality monitoring can be very expensive due to the number of possible monitor locations. It is very important, therefore, to carefully select the monitoring locations based on the monitoring objectives to minimize the involved costs. For example, for troubleshooting applications, the monitor should be placed as near to the sensitive load as possible. On the other hand, for overall power quality monitoring, the monitor is located at the power entrance. Often, monitor placement is limited by access to the power lead, especially for current metering.
4. *How should the measurement be carried out?* The physical organization of the measurement should be carried out carefully; the number of monitors needed should be defined along with the kind of current clamps required. Also, the accessibility of the measurement point should be determined.
5. *What kind of equipment should be used for the measurement?* The instruments can be separated into two main types: power quality monitor and power quality analyzer. Power quality monitors are instruments equipped with memory and the ability to record power quality parameters over some period of time by triggers. Modern monitors can self-adjust thresholds to capture the most relevant events and also ignore non-relevant events. On the other hand, power quality analyzers are instruments that measure and analyze real-time data, sometimes only harmonics. These instruments have limited data storage, so the data and the analyzed results may not be recorded.

   Monitors and analyzers on the market enclose a wide range of features, and it is not easy for customers to choose the right instrument for the required application. In Section 13.5, an overview of the most important features is given.
6. *How long should we measure for?* Often it is important to define the period of monitoring depending on the event expected to take place and the available budget. In critical processes, permanent monitoring is also applied.

After clarifying these six questions, the survey is developed by the following steps:

1. Planning the survey. All persons involved in the monitoring survey should participate in planning the survey and agree on a schedule for installation of the monitor, the monitoring period, the communication process and de-installation of the monitor.
2. Preparing for the survey. Appropriate monitoring instruments and current clamps should be available. Staff using the monitoring equipment should have received sufficient training on the instrument in order to avoid human errors.
3. Inspecting the site. The optimal monitoring location may not be easily accessible and therefore the possible location should be decided by a site inspection. Modifications to the electrical cabinet or programmed power interruption may be necessary. Furthermore, during the inspection, the most significant electrical and non-electrical data of the installation should be collected for further analysis.
4. Installing the monitor. The installation should be done according to the security code, in particular when the installation is done without interruption. Isolating gloves, protective goggles and a helmet are indispensable in order to prevent accidents during the installation.

5. Monitoring the power. In this part, the monitor is installed for the agreed monitoring period. It may be necessary to discharge the monitored data from the instrument periodically, if a lot of power quality events are expected and the monitor's memory is limited. Furthermore, the trigger levels should be chosen with care, because very high levels could fill the memory with unimportant events without significance. It is recommended that the memory is checked one hour after installation of the monitor and the trigger levels readjusted.
6. After the monitoring period, de-installation of the instrument is done, respecting the security code. If the event expected to take place is not recorded, a decision to prolong the monitoring or abort the survey must be determined.
7. Analyzing monitoring and inspection data. The most difficult part is analyzing and interpreting the data collected during the survey. Several kinds of analysis can be carried out with this data: r.m.s. analysis, waveform analysis, trend analysis, transient analysis, harmonic analysis, etc. Monitor manufacturers offer special software in order to help the user visualize the data, and some software is also able to generate reports automatically. Nevertheless, correct interpretation of the results of these tools can only be carried out by an experienced engineer.
8. Defining corrective solutions. The next logical step is to perform an immunization program, following the steps described in the next subsection.

13.4.3 Immunization Program

Basically, there are two families of immunization techniques: to stop the plant with a controlled stop and try to restart after the dip, or to keep the plant working during the dip.

The steps to characterize the problems and, when possible, solve them are as follows [10]:

1. Analyze the process. This activity has technical and management components. The technical component is the basis for identifying the origin and streaming of dip disturbances due to monitoring as well as analysis of the production process. The management component is the basis for evaluating the economic impact of the disturbance. Sometimes, it is useful to monitor the power supplies and to compare the results to the overall production plant monitor.
2. Identify the critical parts. As a result of the previous step, it is possible to identify the critical parts. These are the parts that make a major contribution to production losses, or that indicate the origin of disturbance, or both. Due to the high level of plant complexity, the critical parts should ranked according to their importance.
3. Choose the immunization technique. Basically, there are two families of immunization techniques: to stop the plant with a controlled stop and try to restart after the dip, or to keep the plant working during the dip. In both cases, the process controls of the plant must be on, so UPS or other techniques must be used to ensure this. To implement any one of the two immunization techniques, it is necessary to combine a set of the following immunization tools into a coherent form:

   (a) Timed undervoltage relays.
   (b) UPS to supply the process controls.

(c) Time and level protection sets.
(d) Specific programming of process controls.
(e) Ride-through features in static converters.

Figure 13.4 shows a current inrush peak of a power supply after a voltage dip. These inrush peaks will activate the protection of the power supply and disconnection will happen. 'Ride-through' features of static converters are disconnecting the static converter during voltage dips and reconnecting them after the re-establishment of the voltage.

4. Estimate the attainable theoretical level of immunity. In the case when the parameters for a good plant characterization are known, the calculation of the attainable immunity level is possible.
5. Simulate and/or test the proposed actions. If a correct simulation is possible, in addition to the attainable immunity level, other useful information can be obtained, such as relay protection set points, speed changes, temperature changes, peak torques, etc.
6. Project the concrete case. As a result of the previous steps, a project plan for every case can be established. It includes changes in the wiring scheme, or the new programs and the new set points, installation of immunization equipment, etc.
7. Estimate costs. Execution cost, operation cost and maintenance cost must be estimated.
8. Make the decision. The plant management, after considering cost estimation and future benefits due to fewer production losses, must decide if the immunization projects will be executed. The final decision to immunize the plant and the strategy to be applied is always taken in accordance with economic parameters such as payback time.

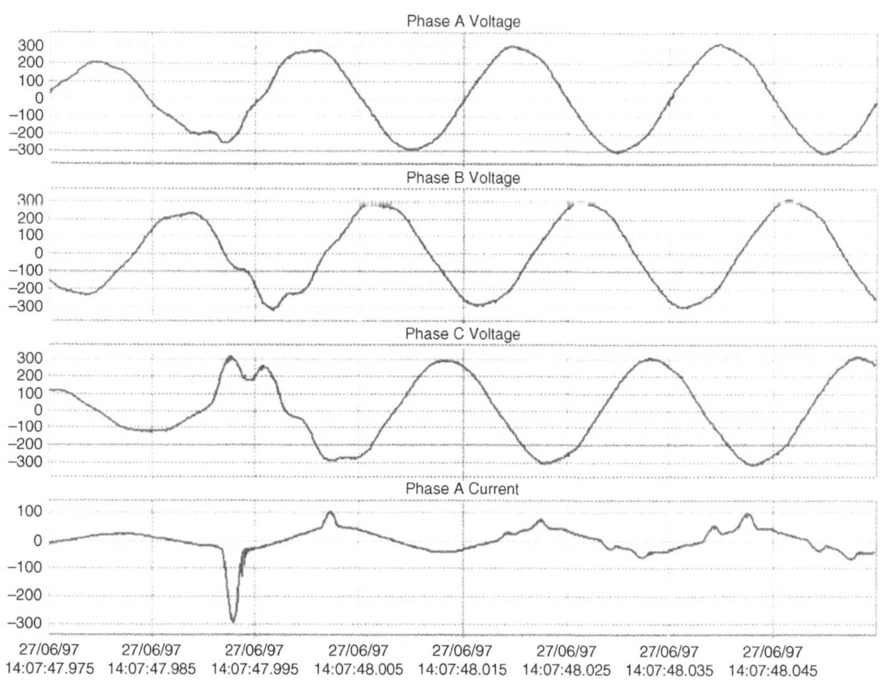

Figure 13.4 Inrush peak current after voltage dip of a power supply

13.5 POSTPROCESSING MONITORING RESULTS

13.5.1 Interpreting Monitoring Results

Analyzing power quality data is a complex process and may be the most critical part of the process of power monitoring. The researcher should have enough knowledge and skills to produce a solution from the available data. Though there is no standard solution for interpreting monitoring results, the following paragraph should be a brief example of a methodology for interpreting monitoring results.

Interpreting power quality reports starts with the interpretation of data summaries, which provide the information of what data needs to be examined more closely. It should be possible to identify the critical events, which can contain more than one disturbance; for example, during a short interruption the monitor can also report transients, dips or swells. Once the critical data events have been determined, the events should be verified. During a dip, swell or interruption, the measurement algorithm for other parameters can produce inaccurate values or misinterpret. Therefore, during this kind of events, the measurements are flagged to indicate to the user that an aggregated value might be unreliable.

Subsequently, the interpreting of the critical events can be started. The possible causes can be deduced by the analysis of signature and wave shape, high-frequency analysis, harmonic analysis and dip/swell analysis. There is a plentiful literature on this topic that provides solutions for the most common power quality problems [4], [8], [13].

13.5.2 Data Collection and Monitoring Systems

Instruments used for permanent monitoring are able to communicate through network cards or modems with centralized monitoring systems, which control the monitoring instruments and download the recorded data from them [1]. These systems are normally used for fixed installed monitors from utilities in order to process the vast amount of data proceeding from monitoring.

The core of any centralized monitoring system is a database, which stores the acquired power quality data. A communication process collects the data from the monitors and stores them in the database. The user interface is carried out by a reporting system, which can automatically generate disturbance reports, inform the user (customer or utility) by means of Web-based Internet applications or transfer data to other information systems (SCADA, ERP, etc.).

13.5.3 Software and Data Analysis Tools

Most power quality monitor manufacturers provide a software and data analysis tool. The following list gives an overview of these functions:

- Download of captured data. This is one of the most important features. Due to the limited memory of the monitors, the data should be downloaded periodically to a personal computer to conserve this data over a long period.
- Setting up the instrument for monitoring.

- Display of data and waveforms (some instruments display real-time values).
- Correlating and categorizing (e.g. event type, magnitude and duration are automatically classified according to EN 50160 or IEEE 1159 standards).
- Detection of trends for predictive maintenance.
- Plotting the captured data against ANSI/CBEMA/ITIC plots (ITIC replaced CBEMA plots, older monitors can have the CBEMA plot option).
- Providing customized alarms and triggers.
- Producing automatic reports.
- Printing the data/waveforms.
- Postprocessing the captured data.
- Analyzing threshold levels.

13.5.4 Power Quality Data Interchange Format (PQDIF)

In many cases, new monitoring instruments are placed alongside legacy equipment. There is a strong interest in data formats that enable different types of instruments to work together. Currently, most software programs for downloading and analyzing data are incompatible with other instruments, hence making comprehensive analysis difficult or impossible. In response to this, EPRI developed the PQDIF [11], which allows engineers to incorporate a wider range of instruments into their overall systems. Nowadays, several manufacturers have adopted the PQDIF.

For a case study see web address

BIBLIOGRAPHY

[1] Bergas J. *et al.*, Centralised supervision of energy quality using analogical cellular telephony. PQA'97 Europe, Stockholm, 1997.
[2] Bollen M. H. J., Ongoing standard work on statistical presentation of voltage dips. Probabilistic Methods Applied to Power Systems, Naples, Italy, September 2002.
[3] IEC 61000-4-30, *Electromagnetic compatibility (EMC) – Part 4-30: Testing and measurement techniques— Power quality measurement methods*, 2003.
[4] IEEE 1159: 1995, *Recommended practice for monitoring electric power quality*, 1995.
[5] IEEE P1564, *Voltage sag indices draft 2, working document for IEEE project group 1564*, 2001.
[6] Lam D., A review of power quality monitoring instruments. ERA Report 2002-0213, ERA Technology Ltd, Leatherhead.
[7] Lang R. G., A 10-step program for improved power quality. *Plant Engineering*, vol. 53, no. 4, pp. 84–87, 1999.
[8] McEachern A., *Handbook of Power Signatures*, BMI, Foster City, CA, 1989.
[9] McGranaghan M., Monitoring morphs into customer service quid pro quo. *Electric Light & Power*, vol. 79, no. 6, pp. 15–16, 2001.
[10] Rull J. *et al.*, Voltage dips in industrial plants: identification methodology, effects and solutions. Immunization cases. CIRED 2003, Barcelona, 2003.

[11] Sabin D., McGranaghan M., Sundaram A., *A Systems Approach to Power Quality Monitoring for Performance Assessment*, www.dranetz-bmi.com/pdf/systemsApproach.pdf, 2007.

[12] Sumper A. *et al.*, Successful power quality monitoring. *Electrical Power Quality and Utilization*, vol. I, no. 1, pp. 37–43, 2005.

[13] *The Dranetz Field Handbook for Power Quality Analysis*, Dranetz Technologies, Edison State, NJ, 1991.

14

Static Converters and Power Quality

Mircea Chindris and Antoni Sudrià-Andreu

The past couple of decades have witnessed a dramatic increase in the number of electronic power converter loads (adjustable-speed motor drives, computers and other office equipment, electronic power supplies, d.c. motor drives, battery chargers, electronic ballast and many other rectifier/inverter applications). According to the EPRI (Electric Power Research Institute in Palo Alto, CA), about 50–60 % of the electric power is flowing through some kind of power electronics equipment, and eventually over 90 % is likely [55]. However, it is difficult to sustain this allegation as the number and the power of purely resistive loads (resistance heater or furnaces, for instance) are also in a continuous expansion.

That equipment, with its aptitude for producing harmonic currents, now constitutes the most important class of non-linear loads in the power system, and draw distorted, and often fluctuating, line current; they also generate high-frequency noise due to the sharp edges of the waveforms, representative of the switching power devices existing in them. As a result of the finite grid impedance, the distorted line currents cause voltage distortion, increase the distribution losses and produce other disagreeable effects like the flicker of the emitted light of lamps. These types of noise interfere with radio and TV reception, communication via cellular telephones, and data transmission; the result is a gradually deteriorating electromagnetic environment with an influence on different domains [30].

Electromagnetic pollution introduced by line-connected power electronics equipment can propagate by conduction, and by near-field or far-field radiation. The line-borne pollution includes harmonic distortion (caused by non-linear loads, mostly rectifiers), interharmonics (caused by asynchronous active loads, such as cycloconverters, or by indirect frequency converters), high-frequency noise (brought about by the fast edges of switching power

Handbook of Power Quality Edited by Angelo Baggini

converters), notches (caused by the momentary short circuit of the line during a commutation interval) and voltage fluctuation or flicker (a series of voltage changes, generated by periodically or continuously changing loads) [6],[8],[55].

On other hand, static converters are very sensitive to the quality of the electromagnetic environment and the different disturbances existing in power supply can dangerously affect their operation and performance.

14.1 IMPACT OF STATIC CONVERTERS ON THE SUPPLY NETWORK

14.1.1 Industrial Equipment

The main industrial power electronics equipment is rectifiers (diode or thyristor), drives, inverters and switching power supplies.

14.1.1.1 Uncontrolled Rectifier (Diode Rectifier)

In almost all electronic equipment, the devices directly connected with the power network are rectifiers; their characteristics determine the harmonic behavior of the complete system and the impact on power supply depends on the rectifier topology and the type of power devices employed [18].

The diode Graetz bridge represents the most used configuration – Figure 14.1 [58]; it requires a rectangular pulsed alternating current when the load is highly inductive, or tips when the bridge is followed by a capacitor. The characteristic harmonic components of the current pulses supplying rectifiers have harmonic orders n, such as $n = k \cdot p \pm 1$, where $k = 1, 2, 3, 4, 5, \ldots$ (integer) and p is the number of rectifier arms (pulse number).

It can be seen that they are all odd numbers and their r.m.s. values, under theoretical conditions (ideal commutation process), respect approximately the relation

$$I_n = \frac{I_1}{n} \tag{14.1}$$

where I_1 is the amplitude of the fundamental current. In fact, the current spectrum is slightly different, and IEC 60146-1-2 [36] proposes the following formula:

$$I_n = \frac{I_1}{\left(\frac{n-5}{n}\right)^{1.2}} \tag{14.2}$$

14.1.1.2 Controlled Rectifier (Thyristor Rectifier)

Thyristor rectifiers have the advantage of a relatively simple control system and can be found in d.c. drives or other many applications. The harmonics generated by a phase-controlled rectifier on the a.c. side may be calculated as for uncontrolled rectifiers, depending on pulse number; as an example, for six-pulse rectifiers, the main harmonic components are the

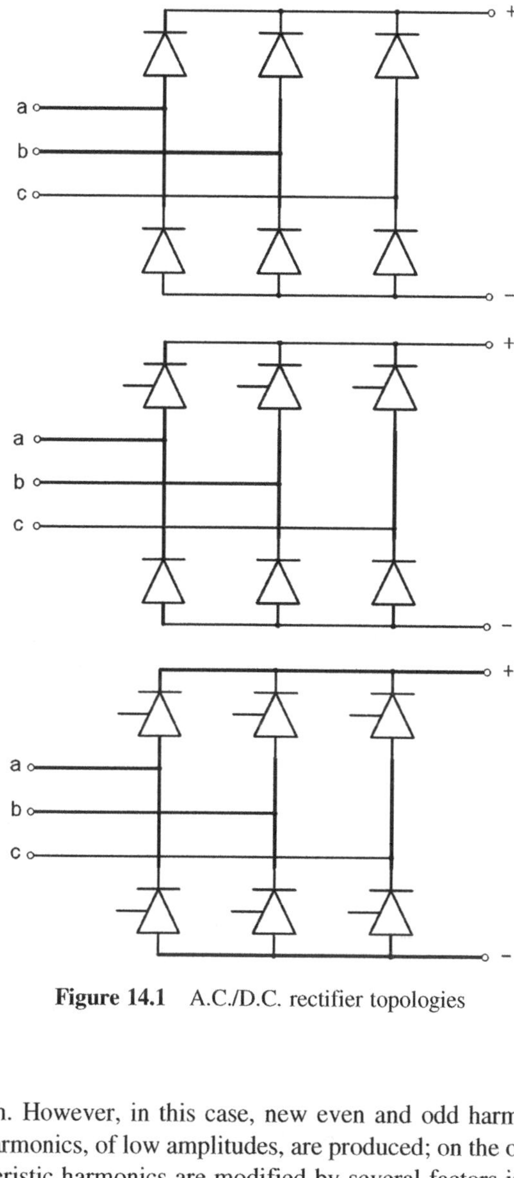

Figure 14.1 A.C./D.C. rectifier topologies

fifth and the seventh. However, in this case, new even and odd harmonics, referred to as non-characteristic harmonics, of low amplitudes, are produced; on the other hand, the amplitudes of the characteristic harmonics are modified by several factors including asymmetry, inaccuracy in thyristor firing times, switching times, imperfect filtering. A displacement of the harmonics as a function of the thyristor phase angle may also be observed.

If overlap (or commutation) and delay angles (μ, respectively α) are considered, and assuming ripple-free direct current, the equation for line current harmonics is [40]

$$I_n = I_d \left\{ \sqrt{\frac{6}{\pi}} \cdot \frac{\sqrt{A^2 + B^2 - 2 \cdot A \cdot B \cdot \cos(2\alpha + \mu)}}{n \cdot [\cos\alpha - \cos(\alpha + \mu)]} \right\} \tag{14.3}$$

where

$$A = \frac{\sin\left[(n-1)\cdot\frac{\mu}{2}\right]}{n-1} \tag{14.4}$$

$$B = \frac{\sin\left[(n+1)\cdot\frac{\mu}{2}\right]}{n+1} \tag{14.5}$$

n is the order of the harmonic and I_d is the average d.c. load current of the rectifier.

Mixed thyristor – diode bridges generate even harmonics whose order is given by $n = 3k \pm 1$. These even harmonics, particularly the second, produce serious disturbances (exceedingly disruptive to power electronics devices) and are very difficult to eliminate; that is why mixed rectifiers are used only at low ratings.

14.1.1.3 Adjustable Speed Drives

An adjustable speed drive (ASD) is equipment designed to control the speed of an induction motor, generating sinusoidal voltages and currents with the necessary frequency and magnitude. ASDs are extensively used in both industrial and commercial applications as the most versatile and efficient means of achieving motion control (with a precision up to 1 % or more). They have become the primary choice for most new and retrofit precision motor control applications based on the following advantages: improved energy efficiency, reduced noise levels, minimal space requirement, good reliability, etc.

The common ASD structure for medium- and low-power equipment is an indirect converter. This device first converts the power supply a.c. voltage, of fixed magnitude and frequency, into d.c. voltage by a rectifier; that voltage is then converted by the inverter in three-phase adjustable a.c. voltage with variable frequency and magnitude – Figure 14.2.

Practically, the rectifier output is inverted to produce a variable frequency a.c. voltage for the motor and voltage source inverters (VSIs) or current source inverters (CSIs) are used for this purpose. For the first configuration, a capacitor or an *LC* filter is placed in the d.c. link to obtain a constant d.c. (i.e. low-ripple) voltage input to the inverter stage. In the latter case, only a series inductor is placed in the d.c. link for a constant current input.

Basically, any ASD equipment contains an input converter *P1* (rectifier) on the supply network side and an output converter *P2* (usually operating as an inverter) on the load

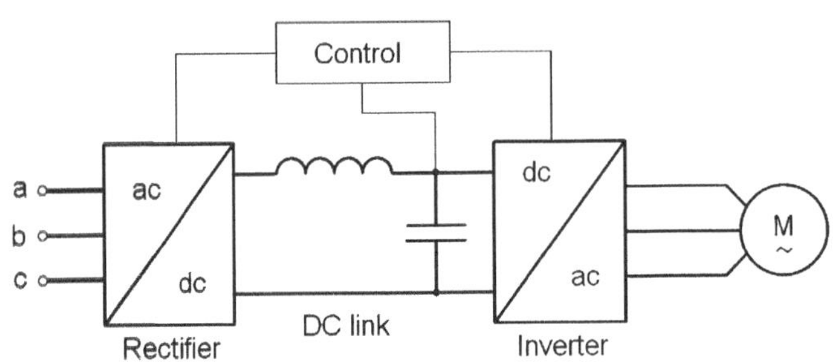

Figure 14.2 Basic topology of an ASD

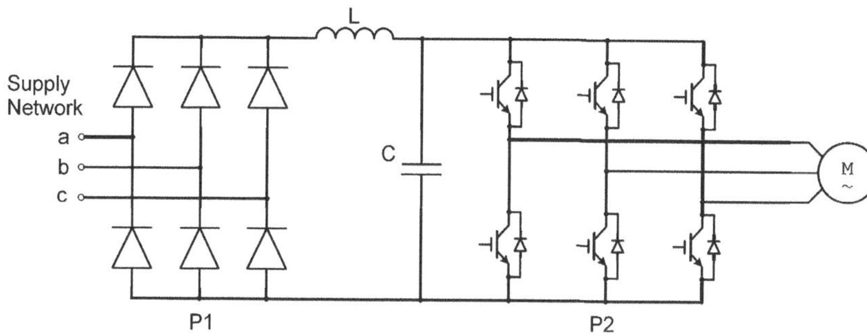

Figure 14.3 Typical frequency converter topology

side, interconnected by a d.c. link – Figure 14.3 [33]. A popular a.c. drive configuration uses VSI employing PWM techniques to synthesize an a.c. waveform as a train of variable width d.c. pulses; the inverter uses either silicon-controlled rectifiers (SCRs), gate turn-off (GTO) thyristors, or power transistors for this purpose. For the last period, insulated gate bipolar transistors (IGBTs) are the most used power switches in the inverter as they allow high switching frequencies resulting in a high dynamic control of the current. In the control systems, total digital control, implemented by high-performance digital signal processors (DSPs), has replaced the previous analog control systems.

Under balanced line excitation conditions, the current waveforms drawn by the rectifier are symmetrical in all the three phases, although they may be rich in harmonics if a controlled rectifier is implemented. The line current waveform changes significantly, depending on motor speed and torque, and consequently the values of harmonic distortion in ASDs are not constant. Despite the fact that the waveform at a reduced speed is much more distorted proportionately, the drive injects considerably higher-magnitude harmonic currents at rated speed; this aspect could be a limiting design factor more important than the THD value.

In both a current or voltage configuration, the d.c. link contains a filter which decouples the current or the voltage of the supply and load systems; at least theoretically, the two fundamental (the supply and the load) frequencies are mutually decoupled. However, ideal filtering does not exist, and there is always a certain degree of coupling. As a result, current components associated with the load are present in the d.c. link, and consequently, in the supply side. These components are subharmonic and interharmonic with respect to the power system frequency; their magnitude depends on the topology of the power electronics and the degree of coupling and filtering between the rectifier and inverter sections.

Due to above-mentioned reasons, the supply network current will contain harmonic components of frequencies

$$f_{k(1)} = (k \cdot p_1 \pm 1) \cdot f_1 \pm n \cdot p_2 \cdot f_2 \tag{14.6}$$

where p_1, p_2 are pulse numbers, respectively, of the converters *P1* and *P2*; f_1 is the fundamental frequency of the system 1 (supply network), in Hz; f_2 is the fundamental frequency of the system 2 (load), in Hz; $k = 0, 1, 2, \ldots$ (integer); and $n = 0, 1, 2, \ldots$ (integer), k and n not simultaneously equal to 0.

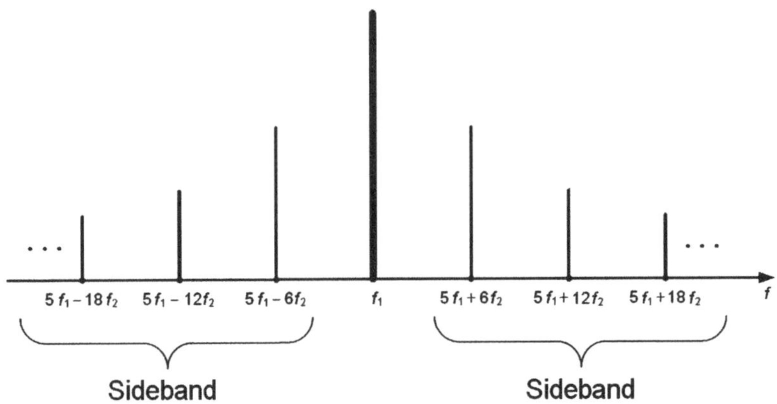

Figure 14.4 Sidebands adjacent to the characteristic fifth harmonic of a six-pulse converter *P1* and *P2*

For $n = 0$, orders of characteristic harmonics for a given configuration of the converter *P1* may be obtained. Components determined for $k =$ constant and $n \neq 0$ are the sidebands adjacent to the inverter characteristic frequencies; each characteristic harmonic has its own sidebands, as shown in Figure 14.4 for the fifth harmonic of a six-pulse bridge ($p_2 = 6$). The first pair of interharmonics, occurring in the neighborhood of the fundamental component, i.e. with frequencies $f_1 \pm p_2 \cdot f_2$, has the largest amplitude [33].

A special issue is represented by induction motors with wound rotor using subsynchronous converter cascades and other doubly fed configurations; in this case, the stator is coupled directly to the utility supply, while the rotor is connected to a three-phase diode bridge (converter 1) with its d.c. bus fed from the grid through a three-phase thyristor bridge (converter 2). The interharmonics transferred to the utility are of two kinds [41]:

1. Interharmonics related to the rotor slip frequency, transferred through the thyristor bridge with frequencies:

$$f_{k(1)} = (k \cdot p_2 \pm 1) \cdot f_1 \pm n \cdot p_1 \cdot s \cdot f_1 \qquad k = 0, 1, 2, \ldots; \; n = 1, 2, \ldots \tag{14.7}$$

   where p_2 is the pulse number of the converter 2 and s is the motor slip with respect to the synchronous speed.
2. Interharmonics related to the converter 1, circulating in the rotor windings and coupled through the air gap to the stator, with frequency given by

$$f_{k(2)} = (p_1 \cdot s \cdot k \pm 1) \cdot f_1 \qquad k = 0, 1, 2, \ldots \tag{14.8}$$

   where p_1 is the pulse number of the converter 1 (diode bridge).

Numerical determination of the supply current harmonic and interharmonic values represents a difficult task and supposes the precise analysis of a particular frequency converter including the load, or information from the manufacturer. For instance, voltage source frequency converters with a PWM input rectifier emit current components at the semiconductor device's switching frequency and its harmonics, which are not synchronized with the line frequency. Normally they are within the range from several hundred hertz to several tens of kilohertz.

14.1.2 Control and Informatics Equipment

Control and informatics equipment belongs to a large class of single-phase equipment together with TV receivers, hi-fi sets, automotive and telecommunication equipment, and so on, and is commonly referred to as RCD (Resistance, Capacitors, Diodes) types of loads. The percentage of load which contains electronic power supplies is increasing at a dramatic pace, with the increased utilization of personal computers in every commercial sector, and some authors estimate that power supply cost will soon reach 50 % of the total cost of a typical electronic product [54].

Two major types of single-phase power supplies are common. Older technologies use a.c. side voltage control methods, such as transformers, to reduce voltages to the level required for the d.c. bus. The inductance of the transformer provides a beneficial side effect by smoothing the input current waveform, so reducing harmonic content. Newer technology, such as switched mode power supplies (SMPS) – Figure 14.5, uses d.c./d.c. conversion techniques to achieve a smooth d.c. output with small, lightweight components [34].

The input diode bridge is directly connected to the a.c. line, eliminating the transformer. This results in a coarsely regulated d.c. voltage on the capacitor that is then converted back to a.c. at a very high frequency by the switcher and subsequently rectified again – Figure 14.6 [29].

The key advantages of SMPS are light weight, compact size, efficient operation and lack of need for a transformer. Also, they can usually tolerate large variations in input voltage.

Because there is no large a.c. side inductance, the input current to the power supply comes in very short pulses as the capacitor regains its charge on each half cycle. Consequently, the supply current is highly distorted compared to a perfect sine wave and, in addition, slightly out of phase with respect to the supply voltage, depending on the source impedance. The harmonic distortion is much higher than those caused by current source converters [11].

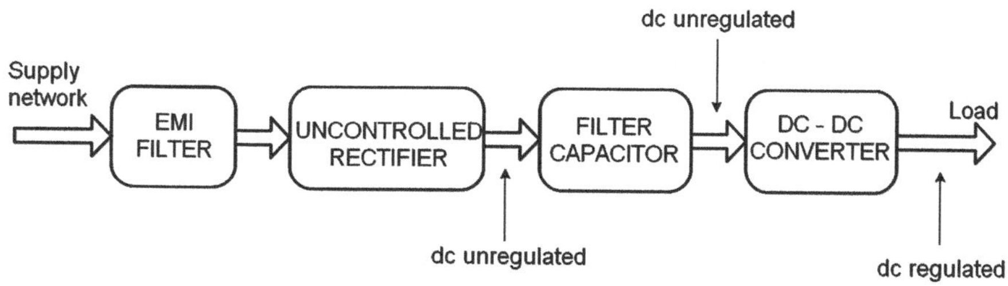

Figure 14.5 General structure of a switched mode power supply

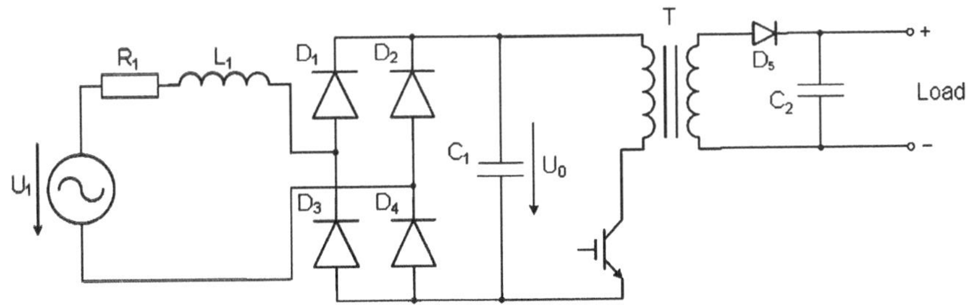

Figure 14.6 Basic diagram of SMPS for RCD load

Different characteristic parameters related to current and voltage at the rectifier input are strongly influenced by source impedances; when the latter increase, the power factor improves, whereas the distortion rate of the voltage in the input of the user installation increases [29].

A representative characteristic of SMPS is a very high third-harmonic content in the line current. The increasing application of this equipment may cause problems, especially in older buildings where an undersized neutral may have been installed. This apprehension is produced by the fact that the third-harmonic current components add in the neutral of three-phase, four-wire LV electrical systems, and consequently they could determine the overloading of neutral conductors. Concern for transformer heating is also important when the load includes a significant amount of SMPS. These aspects should demand the attention of power engineers and designers of industrial or domestic electrical installations.

In order to reduce the size and weight of transformers and to achieve silent operation, the switching frequency is always high and in any case in excess of 20 kHz.

14.1.3 Solutions

Practically, there are three possible ways of suppressing or at least reducing the influence of harmonics:

- reducing generated harmonic currents;
- adding filters to siphon the harmonic currents off the system, block the currents from entering the system, or supply the harmonic currents locally;
- altering the frequency response of the system by filters, inductors and capacitors.

14.1.3.1 Classic Solutions

The following technical measures are referred to as classic solutions aiming to reduce the harmonic pollution: line chokes, harmonic isolation, multi-pulse methods and passive filtering.

14.1.3.1.1 Line Choke

In this case, a three-phase or single-phase choke is connected in series with the power supply (or integrated into the d.c. bus for frequency inverters). The inductor slows the rate at which the capacitor on the d.c. bus can be charged and forces the load to draw current over a longer time period; thus, it reduces the line current harmonics (especially high-number harmonics) and therefore both the r.m.s. value of the current consumption and the distortion at the converter connection point.

The method is particularly effective for ASDs based on PWM techniques; in these cases, the mounted chokes do not change the converter structure and they may be used for several drives. The net effect is a lower-magnitude current with much less harmonic content while still delivering the same energy. Chokes also help reduce nuisance drive tripping due to capacitor-switching transients and voltage unbalance. An addition of 2–5 % of line reactors is usually suggested [9],[17],[27],[51],[66].

14.1.3.1.2 Harmonic Isolation

By using suitable coupling transformers it is possible to limit the circulation of harmonics currents to a small part of the installation. For instance, a Δ/Y transformer does not allow $3k$-order harmonic currents generated at LV to flow into the MV network; moreover, for a Y/Z transformer, the same harmonic currents do not flow at the transformer primary, and the transformer impedance Z_s depends only on the secondary windings. In this case, the inductive part of the impedance is very low ($U_{\mathrm{CCX}} \approx 1\,\%$), and resistance is practically halved compared to a Δ/Y transformer of identical power [5].

14.1.3.1.3 Multi-pulse Methods

These involve multiple converters connected so that the harmonics generated by one converter are cancelled by harmonics produced by other converters. By these means, certain harmonics, related to the number of converters, are eliminated from the power source. Multi-pulse systems have two major advantages that are achieved simultaneously: reduction of a.c. input line current harmonics and reduction of d.c. output voltage ripple. Phase-shifting transformers are an essential ingredient and provide the mechanism for cancellation of harmonic current pairs, for example the fifth and seventh harmonics, or the 11th and 13th, and so on [51].

The most common is the so-called 12-phase rectifier where two six-pulse rectifiers feed the same d.c. load; one converter is supplied through a Δ/Y transformer that produces a three-phase set of secondary voltages shifted by 30° with respect to the primary voltage. The other one is connected to a Δ/Δ transformer, which has no phase shift. As a result, the fifth and seventh harmonics are cancelled and the system will see the equivalent of a 12-pulse converter; the effect is important as these harmonics often cause the most disturbances because of their large amplitude. The $6k$ harmonics are theoretically removed, but, in practice, the amount of reduction depends on the matching of the converters and is typically by a factor between 20 and 50. The $12k$ harmonics remain unchanged.

This solution presents two advantages: the total harmonic current is reduced, and the remaining harmonic components are of a higher order, making the design of the filter much easier. A further increase in the number of pulses to 24, achieved by using two parallel 12-pulse units with a phase shift of 15°, reduces the total harmonic current to about 4.5 %. The extra sophistication increases cost, of course, so this type of controller would be used only when absolutely necessary to comply with limits imposed by electricity suppliers [18].

On the other hand, although the d.c. load can be connected directly to the rectifiers, in practice an interphase transformer is included in order to enable the converters to operate independently of each other.

14.1.3.1.4 Passive Filtering

This consists of connecting supplementary passive components (inductor, capacitor and resistor) in order to modify the natural impedance of different parts belonging to the electrical system; this will allow limiting harmonic components present on a network to a specified low value.

Generally speaking about harmonic filtering techniques, two approaches are commonly used: shunt filters and series filters [17],[18],[24].

The shunt filter works by short-circuiting the harmonic currents as close to the source of distortion as practically possible. This is the most common type of filtering because it is most convenient from an economic point of view and keeps the currents out of the supply system; the solution removes the harmonic currents and tends to smooth the load voltage as well.

Another approach is to apply a series filter that blocks the harmonic currents to flow between parts of the system by tuning the elements to create a resonance at a selected harmonic frequency. This is a parallel-tuned circuit that offers high impedance to the particular harmonic current. The filter is not used very frequently because it is difficult to insulate it and the load voltage is much distorted; on the other hand, all load current flows through the filter that should be rated to the full power. One common application is in the neutral of a grounded-wye capacitor to block the flow of triplen harmonics while still retaining a good ground at fundamental frequency.

Passive filters are relatively inexpensive compared to other means for eliminating harmonic distortion, but they have the disadvantage of potential adverse interactions with the power system [52].

Shunt filters are implemented as single-tuned or high-pass configurations – Figure 14.7.

The most common type of passive filter is the single-tuned notch filter; it presents the most economical structure and is frequently sufficient for practical situations. The notch filter is series tuned to present low impedance to a particular harmonic frequency depending on its components (capacitance C and inductance L); this frequency is determined, as known, by

$$f_r = \frac{1}{2 \cdot \pi \cdot \sqrt{L \cdot C}} \tag{14.9}$$

The selected structure is connected in shunt with the power system, usually at the same point with the non-linear receiver – Figure 14.8; as a result, harmonic currents are diverted from their normal flow path on the line into the filter. Due to the capacitive component,

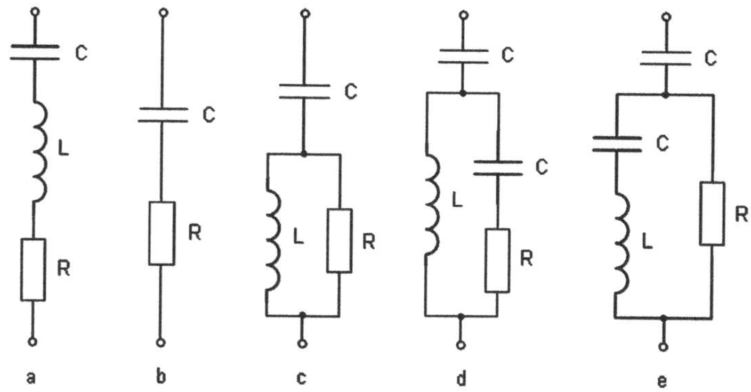

Figure 14.7 Passive shunt filters: (a) single-tuned; (b) first-order high-pass; (c) second-order high-pass; (d) third-order high-pass; (e) C-type high-order

tuned filters can also provide power factor correction (on the fundamental wave) in addition to harmonic suppression.

A common delta-connected LV capacitor bank can be converted into a filter by adding an inductance in series. In this case, the resonant harmonic order is

$$n_r = \sqrt{\frac{X_c}{3 \cdot X}} \tag{14.10}$$

where X_c is the reactance of one leg of the delta and X is the reactance of the series inductance.

One important side effect of adding a filter is that it creates a sharp parallel resonance point at a frequency below the notch frequency – Figure 14.8(c). This resonant frequency must be safely away from any significant harmonic and can be calculated by

$$f_{ar} = \frac{1}{2 \cdot \pi \cdot \sqrt{(L + L_{sc}) \cdot C}} \tag{14.11}$$

where L_{sc} represents the short-circuit inductance in the point of installation.

Filters are commonly tuned slightly lower than the harmonic to be filtered (usually around 3–8 %) to provide a margin of safety in case there is some change in system parameters. If they were tuned exactly to the harmonic, capacitance and inductance manufacturing tolerances, changes in either capacitance or inductance with temperature or normally expected failure of some elements in the capacitor might shift the parallel resonance higher, too close to the harmonic. This could present a situation worse than without a filter because the resonance is generally very sharp.

For this reason, filters are added to the system starting with the lowest harmonic that has to be eliminated. Passive filters should always be placed on a bus where X_{sc} (short-circuit reactance) can be expected to remain constant. While the notch frequency will remain fixed, the parallel resonance will move with system impedance. Also, filters must be designed taking into account the capacity of the bus as the temptation is to size the current-carrying

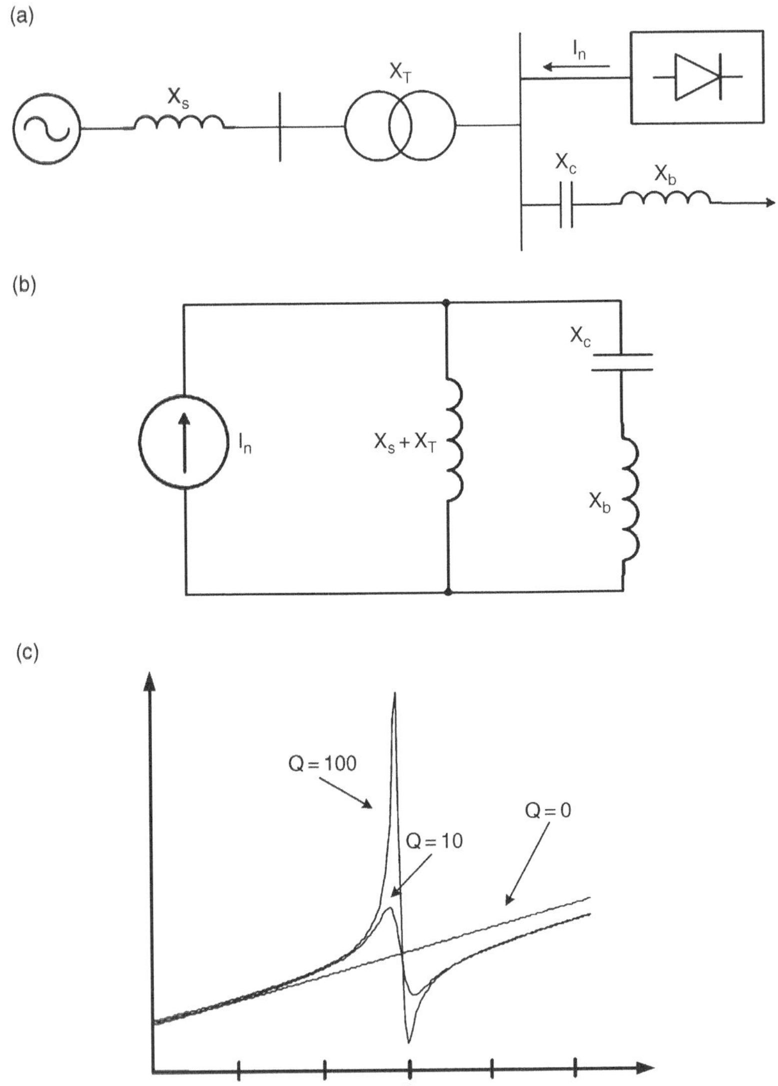

Figure 14.8 Tuned-filter arrangement: (a) basic diagram; (b) equivalent diagram; (c) series and parallel resonance for different quality factors

capability based solely on the load that is producing the harmonic. However, a small amount of background voltage distortion on a very strong bus may impose excessive duty on the filter.

The high-pass filter provides attenuation over a broad range of harmonic frequencies. In fact, to higher frequencies, it appears to be the resistance R tied directly to ground. The high-pass filter is not as effective in attenuating as the notch filter but the associated parallel resonance is much less severe than those associated with the notch filters.

The main disadvantage of these filters is that fundamental frequency current flows through the resistor. This fact increases the cost of the resistor, and the monthly costs of the consumed energy can be significant. The solution is *C filter* where the capacitor C_2 is sized so that its capacitive reactance is equal to the inductive reactance at fundamental frequency [17],[26]: the series branch of C_2 and L prevents fundamental current from flowing in R, reducing its initial cost, and practically eliminating the cost of kilowatt losses in the filter.

14.1.3.2 Power Factor Controller

Power electronics loads, especially those with front-end rectifiers, are sources of high harmonic components drawn out from the electric power utility ($THD_i > 100\,\%$) and have very low power factor ($PF \approx 0.5$–0.65). Consequently, limitation of the high harmonic contents in the line currents and improvement of the input power factor of the loads supplied by power electronics converters became one of the most important problems of power electronics.

Diode rectifiers using a power factor controller can fulfill these demands and give power factor correction (PFC) rectification in a simpler way than PWM converter systems. In this case, the power factor controller is a d.c./d.c. converter used as an intermediate stage, just after the rectifier and before the load-supplying converter; it shapes the input alternating current in an appropriate manner for decreasing the harmonic content and improving the power factor.

By suitable command signals applied to drivers of controllable switches in the d.c./d.c. converter, the average output voltage is modified as requested, and the waveform of the supply current can be forced to follow a sinusoidal reference, with the subsequent advantages:

- high value of power factor ($PF \approx 0.95$–0.99);
- mitigation of harmonic pollution ($THD_i \leq 3\,\%$ if necessary);
- larger range of supply voltage (90–270 V);
- filter capacitor is smaller and cheaper.

Different solutions, classified as clean power converters, have been proposed. For low powers, flyback, S^2IP^2 (Single-Stage Isolated Power-factor-corrected Power supplies), BIFRED (Boost Integrated with Flyback Rectifier/Energy storage/D.c./d.c. converter) and BIBRED (Boost Integrated with Buck Rectifier/Energy storage/D.c./d.c. converter) topologies are most common. With medium and high power levels, step-up (boost), step-down (buck), SEPIC (Single Ended Primary Inductance Converter), Cuk and Vienna rectifier are the basic structures. The boost converter is especially popular in PCF applications – Figure 14.9 [5].

In this case, the control system receives a sinusoidal-type current reference with double wave rectification and commands the transistor T according to this reference; i will follow this pattern and, as a result, i_1 is a sinusoidal wave, in phase with v_1 (from the supply

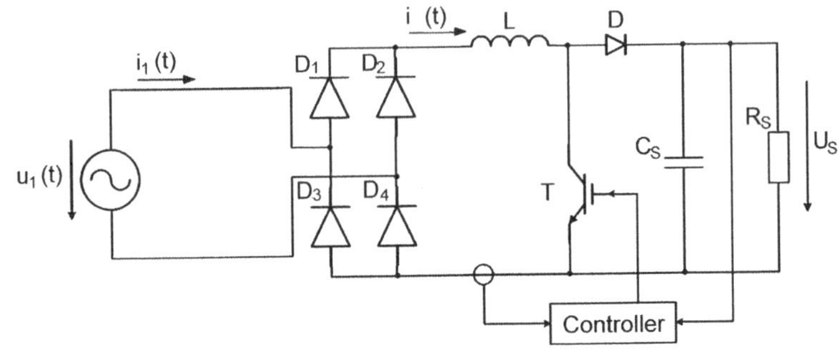

Figure 14.9 Single-phase boost PFC rectifier

network point of view, the converter represents a resistance). The correspondence between the absorbed current and the reference sinusoidal wave depends on the control frequency: the higher is the switching frequency, the smaller the errors will be. The harmonic distortion of the supply network is also highly attenuated compared to common solutions, obeying international power quality standards requirements [37]. Practically, by using PWM techniques, the frequency spectra of the input waveforms can be shaped and harmonic components moved to a higher frequency; this substantially reduces the size of filter components.

The input current also contains low residual harmonic currents whose frequencies are the frequency of modulation (normally tens of kilohertz) and its multiples; filtering of these high-frequency currents is easy and inexpensive.

The Vienna rectifier is a more complex structure, i.e. a three-level voltage boost converter. Its main advantages are: three phases, three-level input voltage generation, controlling only three electronic switches, less voltage stress of the electronic switches than in the case of the PWM converter. Unfortunately, the Vienna rectifier is a unidirectional system and can convey the electrical energy only from the a.c. circuit to the d.c. circuit [53].

An important issue is the design of input electromagnetic interference filters for PFC circuits; the use of standard input filter design procedures is inappropriate as the obtained devices determine large reactive currents through input filter capacitors, excessive power dissipation in damping resistors, or converter instability due to filter–converter impedance interaction. To avoid these disadvantages, new techniques are proposed in the literature [60].

A PFC circuit input filter has to meet the following three main requirements:

- required switching noise attenuation;
- low input displacement angle between filter input voltage and current;
- overall system stability.

Meeting all these requests supposes a careful design and use of new passive and active structures providing acceptable damping without affecting the filter high-frequency attenuation characteristic.

14.1.3.3 Active Rectifier

Power quality standards establish emission limits for different classes of non-linear loads; the implementation of these constraints was the motivation for the development of active methods to improve the quality of the input current and, consequently, the power factor.

The continuous evolution of the semiconductor power component performances (especially IGBTs) made possible the industrial development of power converters able to guarantee no disturbance at the point of common coupling. The line current harmonics generated by rectifier bridges can be considerably reduced using PWM strategies: in this case, instead of diodes, the device uses transistors switched with a high frequency, around 30 kHz – Figure 14.10 [34]. The line current becomes more sinusoidal but contains a high-frequency ripple (up to 100 kHz).

The application area of the PWM rectifiers includes inverter drives, small-scale power generation, d.c. transmission, solar energy, arc welding, etc. Besides the reduced distortion level and unity power factor, this solution provides the possibility for bidirectional power flow, so saving lots of energy in some applications [50].

Filtering requirements for PWM rectifiers are usually satisfied through the use of small line side inductors or low-pass *LC* input filters. Their design involves the positioning of the resonant frequency to meet the harmonic attenuation requirements, and introducing damping at the resonant frequency to avoid amplification of residual harmonics.

14.1.4 Electromagnetic Compatibility Issues

Electromagnetic compatibility represents a very important issue in practical applications of power electronics. For instance, conducted EM perturbations generated by inverter-fed induction motor drive systems with PWM represent a difficult current technical problem that limits power electronics drive's evolution [47].

14.1.4.1 Voltage Notches Due to Semiconductor Commutation

Voltage notches due to semiconductor commutation belong to so-called synchronous wave shape phenomena and occur in power electronics load currents in which the switching process is synchronized to the power frequency [24],[35].

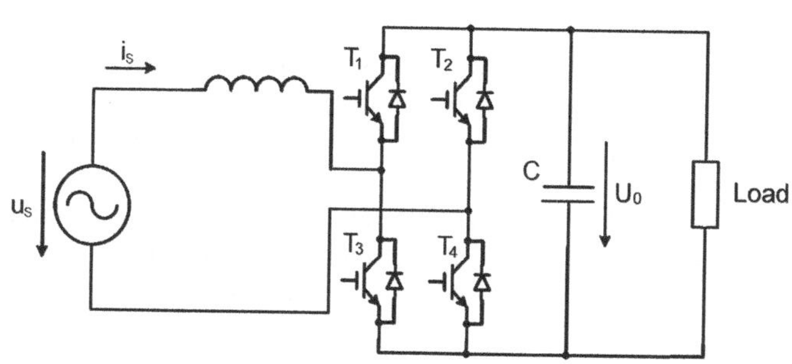

Figure 14.10 Single-phase PWM rectifier

Notching is a periodic voltage disturbance caused by the normal operation of power electronics devices when current is commutated from one phase to another; during this period, there is a momentary short circuit between two phases pulling the voltage as close to zero as permitted by system impedances. This disturbance is associated with high-frequency oscillations that affect the insulation coordination of the plant and can give rise to radiated interference. The annoying effect can be reduced by the introduction of damper circuits (snubbers) across the switching devices.

Notches can disturb electronic equipment and damage inductive components by their high level of voltage rise and the additional zero crossing of the mains voltage. However, most problems caused by notches are confined to the customer's own installation as, practically, the high-frequency content of notches is filtered by the utility's power transformer at the service entrance.

The depth of the notch at points nearer to the power source is proportional to the system impedance up to that point, while the width of the notch is the commutation angle [40]:

$$\mu = \cos^{-1}[\cos\alpha - (X_s + X_t)\cdot I_d] - \alpha \tag{14.12}$$

where α is the delay angle; X_s is the system reactance in per unit on converter base; X_t is the converter transformer reactance in per unit on converter base; and I_d is the direct in per unit on converter base.

14.1.4.2 High-Frequency Disturbances

High-frequency disturbances are mainly produced by the switching of control devices. In power electronics, the major sources of disturbance tend to be voltage rather than current transients. The voltages can vary by hundreds of volts in a matter of a few nanoseconds giving *dv/dt* in excess of 10^9 V/s. PWM, for example, works with voltage changes from 0 to U_{dc} (660 V for rectified three-phase) occurring in a very short time, nano- to microseconds depending on the technology used.

The high-frequency commutations do generate high-frequency perturbations that are propagated by conduction and radiation and give rise to the presence of parasitic interference along the line upstream of the switching device, respectively in the mains.

14.1.4.3 Common Mode Disturbances

Rapid voltage changes are the source of various disturbance phenomena; practical experience has highlighted that the most problematic of them consists of the generation of high-frequency currents due to the existence of high levels of *dv/dt* in the output voltage [19]. These high-frequency disturbance currents will flow through any stray capacitances and the zero reference conductor, and can produce different disagreeable effects: modifying signals (data or commands), disturbing sensitive measurements and perturbing other equipment by injecting the disturbance back into the public distribution network.

The currents can flow between power lines (differential mode components) or between power lines and ground (common mode components), but the common mode components are of major importance. All converters generate common mode voltages relative to the

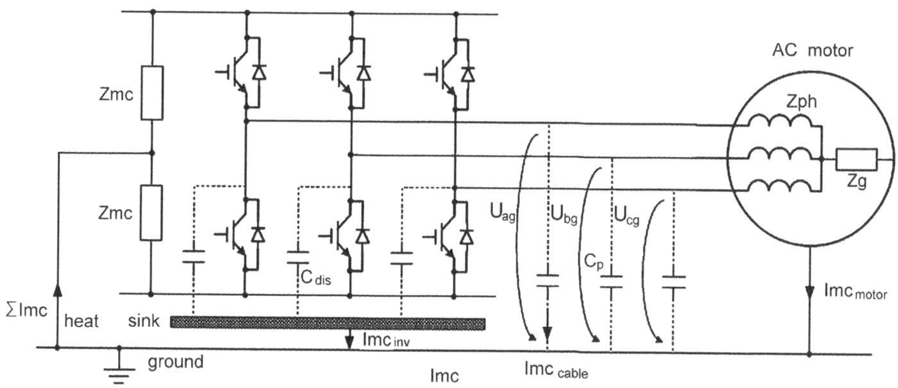

Figure 14.11 Common mode current propagation path

power source ground that cause coupling currents through the parasitic capacitances inside the supply system and power electronics. The disturbance strongly depends on the high-frequency characteristics of the load connected to the converter since the whole equipment is the generator of perturbations in this case [47].

For ASDs, the major source of noise is the voltage sources in the middle point of the switching cells generating high-frequency currents. Figure 14.11 illustrates the different propagation ways of the common mode current in the system inverter/cables/motor [59]; impedance Z_{mc} symbolizes the return paths of the common mode currents.

The main propagation paths are:

- the capacitive path between the semiconductor module and the thermal heat sink connected to the ground, C_{dis};
- capacitive couplings within the feeding cable, C_{cable};
- capacitive couplings between the stator windings of the motor and the iron core connected to the ground, C_{motor}.

In order to limit the circulation of these high-frequency currents, constructors of electronic equipment may use the following solutions – Figure 14.12 [59]:

- Mitigating perturbations (common mode transformer coupled to a resistor, passive filtering).
- Changing the propagation paths by offering a low-impedance preferential way where damage is reduced or even avoided.
- Increasing the impedance of the propagation paths in order to reduce the common mode currents.
- Balancing the perturbations by passive or active methods (common mode transformer and/or active compensation circuit).

For control and IT equipment, a widespread solution consists of inserting filters upstream of the SMPS unit; a typical circuit is shown in Figure 14.13 [29].

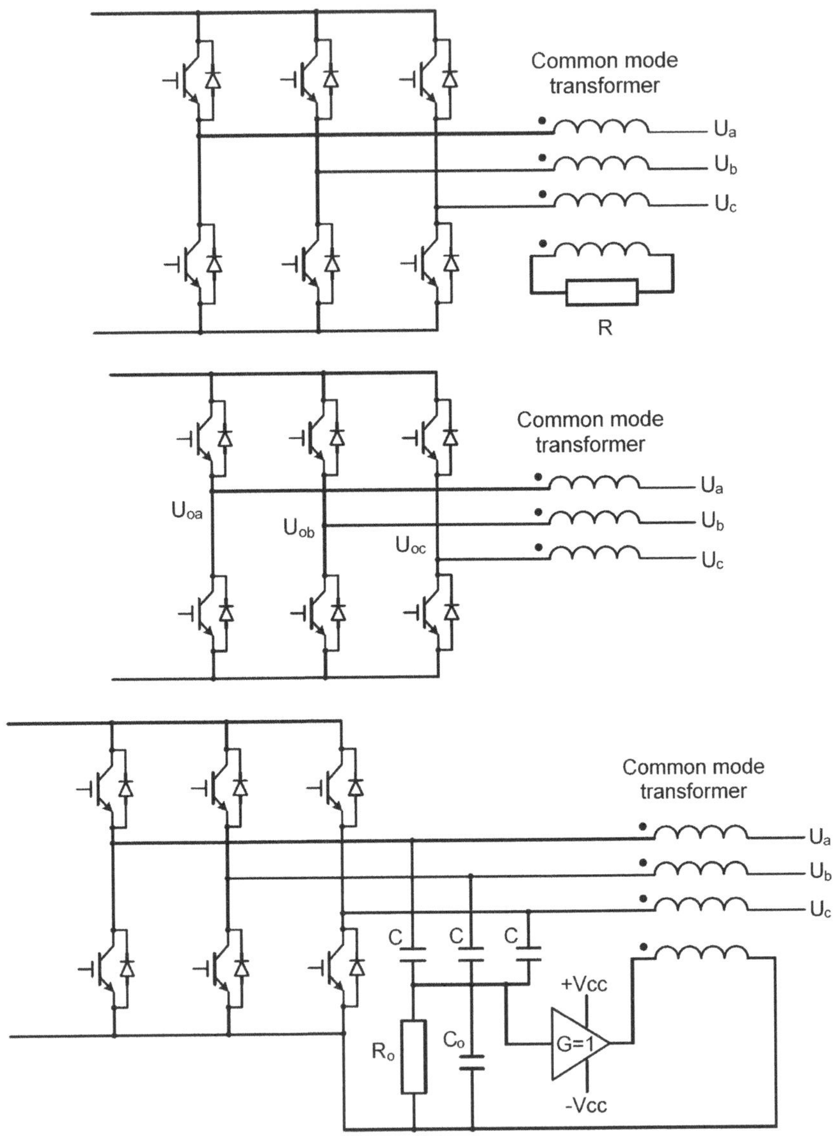

Figure 14.12 Methods to mitigate common mode currents

Inductance L opposes high impedance to common mode components, blocking them from penetrating into the supply network; at the same time, common mode currents are conducted to earth by the group of capacitors C_1–C_4. For differential mode components, inductance L has almost zero impedance (its windings are wound in opposition), while all capacitors short-circuit them.

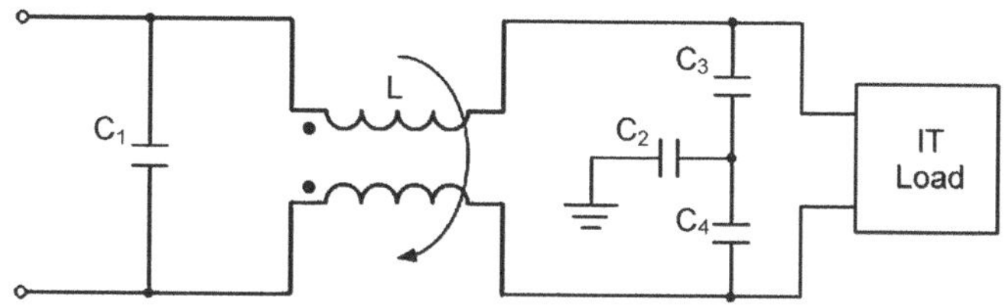

Figure 14.13 Basic circuit diagram of an anti-parasitic interference filter

The filter inserted between the a.c. mains and the SMPS ensures a second function: it protects the supply from impulse-type overvoltages and from high-frequency interference of the differential and common modes which are present in the mains.

14.1.5 Impact on Loads Supplied by Power Converters

14.1.5.1 Inverter Output Characteristics

The converter output characteristics depend on converter structure, type of electronic devices and control method. Besides the connection circuit, they impose the proprieties of output voltage and consequently the global impact of the power converter on load supplied.

For instance, in adjustable-speed drives a frequency converter produces a series of pulses in a specific pattern between the three output phases, and the stator is supplied by these pulses. The r.m.s. value of the output voltage depends on the actual output frequency, i.e. on the duration of the pulses: longer pulses equal a higher voltage.

The non-sinusoidal voltage generated by the converter produces a circulating harmonic current on the stator winding; its magnitude depends on the stator winding configuration, harmonic reactance and rotor damping effect. Each harmonic voltage will produce a corresponding harmonic current; the flow of these supplementary currents will generate additional heating in the stator winding that will add to the temperature rise caused by the fundamental flow of current [32].

On the other hand, the stator harmonics generate space harmonic magnetomotive forces (MMFs) in the air gap and, as a result, harmonic currents in the rotor. The order of the space harmonics existing in the air gap can be determined by the relationship

$$n = 2 \cdot k \cdot m \pm 1 \tag{14.13}$$

where n represents the order of space harmonic; m is the number of stator winding phases; and k is any integer.

According to their harmonic order, the rotation of space harmonic MMFs will be either forward or backward with respect to the rotor rotation. The resulting pulsating torque can have undesirable effects including vibration and audible noise; it can also excite the natural frequencies of the mechanical drive train, degrading system performance. The flow of harmonic currents will produce additional heating of the rotor. Mitigation of these unpleasant effects may be obtained by increasing the number of the stator winding phases.

An important factor influencing the load supplied by power converters is the rate of voltage variation in time; indeed, the output voltage change per time unit is determined by the components used in the frequency converter. There is a general trend to minimize the transition time, because it minimizes the losses inside the frequency converter; as a result, an increase of *dv/dt* can be expected in the future due to the fact that there is a constant demand for smaller-sized frequency converters. A high value of *dv/dt* has a negative impact on the insulation system used in the motor; however, this influence decreases with increasing cable length as the cable introduces some additional inductance in the circuit. Still, a cable that is too long between the frequency converter and the motor can raise other harmful problems.

14.1.5.2 Bearing Currents

Bearing currents were discovered about a century ago; at that time, the source of some problems in electric machines was found to be a net flux enclosing the motor shaft. This flux, produced by the magnetic dissymmetry in the equipment, induces a back e.m.f. and causes a current to circulate in the conductive loop formed by the stator case, motor shaft and bearings; the current is frequently referred to as a bearing current. Voltage discharge associated with this flow can produce damage to the bearings or even their failure if not properly insulated. For a long time, shaft voltages were quite a common event only for medium and large motors (more than 250 kW); nowadays, the increased use of variable frequency drives has resulted in shaft voltages in much smaller motors. The technical literature has reported such incidents in motors of 75–250 kW and sometimes even smaller.

This phenomenon has two explanations: (i) for large motors, the recent design regulations and technological achievements have practically eliminated the asymmetry of the motor; (ii) for smaller motors, the extended use of power electronics, associated with the rapid switching of the static devices, may generate high-frequency current pulses through the bearings. Rotor voltages can be generated by harmonics, asymmetries in the inverter voltage, incorrect cable length and grounding between inverter and motor, and they produce current leakage in the motor bearings.

For all types of electric motors, the constant passage of bearing current causes fluting in the outer and inner ring and this will accelerate the wear of the bearing, i.e. a reduced life [1].

The bearing damage in inverter-driven motors is mainly caused by the shaft voltage and bearing currents created by the common mode voltage and its sharp edges [12],[63],[64]. All inverters generate common mode voltages relative to the power source ground, and through the parasitic capacitances existing in the motor these voltages produce coupling currents. Bearing damage of motors driven by inverters has been reported mainly in cases where the shaft is not grounded.

The main culprit is the *dv/dt* deviation associated with common mode voltages. Taking into account the numerous electrical paths that can materialize in the drive, different types of bearing currents can be produced by different processes. According to their generation practice, the following four types of bearing currents are more frequent [64]:

- *Capacitive bearing currents.* Fast deviations of common mode voltage in the stator windings generate currents through the stray capacitance between the stator winding and the rotor surface. From the whole value of these capacitive currents, the part that flows through the bearing is dependent on the speed and mechanical load on the shaft. Although the amplitude of this current is small compared to the total common mode current, it can overheat the lubricating medium.

- *Electric discharge machining (EDM) bearing currents.* The common mode voltage produces an electric charge in the capacitance formed across the rotor body and the grounded stator frame. The voltage of this capacitor can reach a value high enough to produce the breakdown of the lubricating layer; as a result, the capacitor charge will discharge through the insulating film of the bearing. This current is also referred to as EDM bearing current, and frequently the stored energy can produce bearing damage. This incident may be avoided if the motor shaft is grounded or the rotating speed is low. EDM current flows only if the dielectric breakdown of the insulation film decreases under an imposed value.

  Recently, a rising number of EDM-type bearing failures relatively soon after start-up (within one to six months) have been reported. The incidents appeared in a.c. drive systems and their frequency depends on the a.c. drive system architecture and the installation techniques used.
- *Common mode current flow through shaft due to poor grounding.* This case occurs when the motor frame is poorly grounded and the motor shaft is connected to a mechanical load with a better ground connection; under these circumstances, the common mode current flowing through the stator to the rotor capacitor can now flow to the external ground through the shaft. This typically appears when the shaft is grounded by means of an external grounding brush kit: in this situation, the current flows to ground through the shaft or the load structure connected to the shaft.
- *Circulating bearing currents.* The shaft voltage is produced by the asymmetry in the magnetic field of long axial machines and this phenomenon, based on electromagnetic induction, is typical for medium and large motors (greater than 110 kW). This voltage has a very low frequency and produces currents flowing in an electric circuit consisting of the rotor, the stator frame and the bearings located at the two ends of the motor. For smaller a.c. machines, this current may be neglected.

The size of the motor, the technique used to ground the motor frame and/or the shaft and the quality of this grounding (its impedance) are the most important factors that decide which of the above mechanisms is more active. An important role is played by the electrical installation, i.e. the selection of the connection cable, proper bonding of the protective conductors and the electrical shield. If the distance between the inverter and motor is too long, the abrupt voltage transient and distributed inductance–capacitance combination of the cable impose higher voltages at the motor terminals; this aspect will be analyzed in detail in the next section. High voltage appearing across the motor terminals may damage the insulation material of the windings.

Normally, small motors need to have insulated bearings in both the drive end and non-drive end or ceramic bearings to cut off the small flow of bearing current. Larger motors, however, have to be fitted with one single insulated bearing or one single ceramic bearing to cut off the large flow of bearing current.

For high-frequency converters, special coatings made of special materials are used on the rings and balls of the bearings, but applying these coatings is an expensive and time-consuming process. The latest bearing types on the market make use of the spin-off effect from the aviation industry, where the following types of bearings are used:

- hybrid bearings;
- full ceramic bearings;
- ceramic-coated bearings.

There are some approaches that prevent bearing current damage of a.c. machines. They are as follows [1],[63],[64],[65]:

- *External passive/active common mode filters.* Common mode noise filters represent a quite frequent solution to mitigate or even to cancel the common mode noise; unfortunately they are expensive and reduce efficiency. Additionally, these filters are massive and voltage oscillations may appear if the passive components are not tuned properly.
- *Motor shaft ground brushes or insulated bearings.* The most effective method to prevent bearing current problems consists of using insulated bearings (full ceramic or ceramic-coated) and such a solution is presented in newer motors; for older equipment, the replacement of the existing bearings is expensive and time consuming. In these cases, an effective way to deviate the current from flowing through the motor bearing consists of the grounding of the motor shaft; for this purpose, a brush is connected between the motor shaft and the motor frame. The solution supposes regular maintenance as the brush has a limited lifetime.
- *Multi-level inverter technologies.* As the common mode voltage represents the source of bearing currents, the most effective method to prevent the problems consists of reducing its amplitude and transition step. The multi-level inverter topology, recently proposed by different scientists, assures this goal; for instance, a three-level inverter [42],[61],[64] was newly introduced on the general purpose inverter market. This new structure has more favorable features: capability to handle high voltage, lower line-to-line and common mode voltage steps, more frequent voltage steps in one carrier cycle, and lower ripple component in the output current for the same carrier frequency.

14.1.5.3 High-Voltage Spikes at Motor Terminals

Another problem for electric motors consists of voltage spikes produced by the rapid rise time of pulses out of ASDs. These voltage spikes are on top of the d.c. pulses traveling out to the motor. Figure 14.4 presents waveforms that indicate that, when a step voltage is applied to an LC resonant circuit, the voltage can oscillate up to twice the input voltage. As a result, for an original voltage E, the peak value V_{peak} at the motor terminals will reach the value $2E$.

Any induction motor represents a series LC load, and practical measurements have indicated a resonance frequency ranging between 30 and 100 kHz; this value mainly depends on the electric motor's characteristics (winding inductive reactance and parasitic capacitance). In all cases, the corresponding harmonics of motor voltage will be amplified, the level of

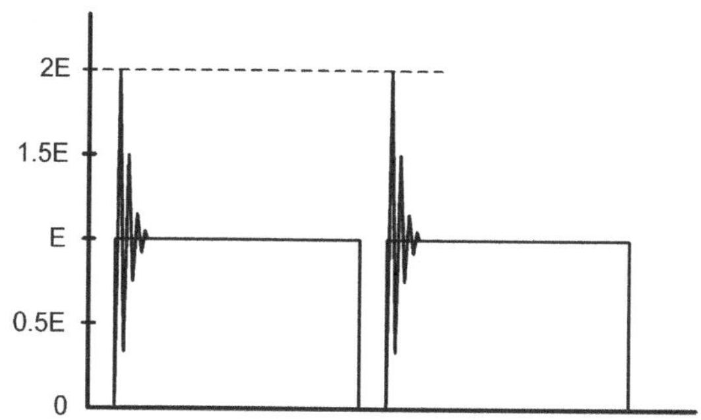

Figure 14.14 Voltage overshoot at the motor terminals

the overvoltages at the motor terminals depending on the quality factor of the resonant circuit. This fact is very harmful to the machine as it can produce (suddenly or in time) insulation failure of the motor, especially of the first few turns of the stator windings. For frequencies higher than the resonance frequency the motor behaves as a capacitive load and its impedance decreases.

Figure 14.4 shows a very high gradient of the voltage rise (5–10 kV/μs), its value depending on parameters such as carrier frequency, voltage boost, load frequency, U/f characteristics, belonging to both inverter and load. According to the control algorithm, these transients may occur 20–100 times per period, and as much as 85 % of their peak value is found across the first turn of the first coil of the motor windings [21].

When the modulation frequency of the inverter happens to be close to the resonance frequency of the fed motor, the above-mentioned phenomena are more prominent. The impedance characteristic of the induction motor is considerably influenced by the electric line between the converter and the motor as the parasitic capacitance of this cable moves the resonance frequencies toward lower ranges. V_{peak} also depends on the length of the cable, because the capacitance in the cable increases with increasing cable length. The experimental measurements have proved that the V_{peak} voltage does not create problems if the cable connecting the motor to the converter has a length less than 15–20 m. Nevertheless, special attention must be paid to the ASD applications with long motor cable (e.g. over 100 m) [47].

In order to avoid these problems, the connecting cable must have a length less than a critical value that can be determined by the relationship [21]

$$l_c = \frac{v \cdot t_r}{2} \qquad (14.14)$$

where v is the pulse velocity given by [22]

$$v = \frac{1}{\sqrt{L_c \cdot C_c}} \qquad (14.15)$$

L_c, C_c are cable parameters and t_r is the rise time of the voltage pulse.

The expected overvoltages for a given structure (inverter, cable and motor) are expressed by the relationship

$$\frac{V_{max}}{V_{dc}} = (1+\Gamma_L) \cdot \left[1 + \sum_{k=1}^{n} (-1)^k \cdot \Gamma_L^k \cdot \left(1 - 2 \cdot k \cdot \frac{t_t}{t_r} \right) \right] \tag{14.16}$$

where

$$\Gamma_L = \frac{R_L - Z_0}{R_L + Z_0} \tag{14.17}$$

is the load reflection coefficient; R_L is the load resistance;

$$Z_0 = \sqrt{L_c / C_c} \tag{14.18}$$

is the characteristic impedance of the cable; and

$$t_t = \frac{l_c}{v} \tag{14.19}$$

is the travel time of a PWM pulse from the inverter to the motor.

To mitigate these overvoltages with steep slopes, the following solutions can be implemented:

- Use of motors provided with an improved winding insulation.
- Use of a cable with a length less than the critical value.
- *LC*-type filter (sine filter) at the output terminals of the inverter (the filter's resonant frequency must be higher than the inverter carrier frequency in order to avoid filter or inverter damage).
- Reactors placed at the inverter terminals.
- *RC* snubber network at motor terminals (the disadvantage of this solution is high dissipation losses).

The first two methods are common sense and derive from the above-mentioned phenomena. In contrast, a filter at the output of the frequency converter reduces the values of dv/dt and V_{peak} at the motor terminal and consequently the stress on its insulation. The typical solution consists of connecting a reactor in series with the motor windings (additional components, such as capacitors, can be added to get a better filter performance); the risk of overshoot is minimized, because the charging and discharging of the cable is slower.

However, introducing a filter at the output has some disagreeable implications. One of them is represented by the power losses in the filter: since their value depends on the switching frequency of the power converter, it is common to reduce this frequency when an output filter is connected. Nevertheless, an output filter will always affect the overall efficiency of the system.

The output filter also produces a voltage loss and the motor will by supplied by a lower voltage than the rated value. Although in many cases the voltage losses are insignificant, the following troubles may become visible: current drawn by the motor from the power converter is a little higher, the slip in the motor increases, and the performance of mechanical load drops.

Because output filters for frequency converters transform a square wave into a more sine-shaped pulse, some noise is detectable occasionally; the reason is the forces that act upon the reactors when the voltages changes in square waves. In order to reduce the noise, some equipment available on the market is encapsulated, also providing for a better thermal performance.

If the filter contains capacitors connected to the ground, the leakage current may increase, affecting the protective circuit-breaker that is used in the application. The problem must be tackled and solved during the design of the motor–converter system.

14.2 IMPACT OF SUPPLY NETWORK DISTURBANCES ON STATIC CONVERTERS

The main supply network disturbances are voltage sags and interruption, harmonics, overvoltages, unbalance and voltage fluctuations. The impact of these disturbances on static converters is generally presented as in Table 14.1.

14.2.1 Impact of Voltage Disturbances on Static Converters

14.2.1.1 Harmonic Voltage Distortion

In the case of single-phase voltage source converters, if the supply system voltage is distorted by the presence of harmonics, the phenomenon of discontinuous conduction can be observed (more conduction time intervals during a half period of the fundamental). The following conclusions have been highlighted by analyses performed in [11]:

- The discontinuous current condition can appear in various working circumstances.
- The condition with two conduction periods is the most frequent situation.
- The discontinuous current conditions are more frequent for reduced values of absorbed power and system impedance ratio and produce harmonics of higher values than in continuous current conditions.

Table 14.1 Sensitivity to disturbances of power converter

Equipment	Sensitivity to disturbance				
	Voltage sags	Interruptions	Overvoltages	Harmonics	Unbalance
Adjustable-speed drive	×	×	×	×	×
Control and informatics equipment	×	×	×	×	×
Rectifiers	×	×	—	—	×

14.2.1.2 Voltage Unbalance

Unbalanced supply voltages produce non-characteristic harmonics of significant amplitudes in a.c./d.c. converters, especially in the case of higher d.c. ripple [3]; this is the case of high dynamic drive applications, where the size of a.c. and d.c. smoothing reactances must be limited to moderate values. High values of non-characteristic harmonics can also be generated if resonance phenomena emerge in the supply network or in a front-end rectifier of VSI. Different research reports have indicated the existence of uncharacteristic triplen harmonics (third, ninth, etc.) and the rise of input current THD. On the other hand, test results presented in the literature indicate that raising the voltage unbalance from 0.6 to 2.4 % causes the input current unbalance to increase from a nominal 13 % to a maximum 52 %.

Steady-state voltage unbalance conditions in the three-phase input line voltages can additionally cause the transition of the rectifier stage into single-phase rectifier operation. In this case, significant amounts of harmonic voltage having double the line frequency appear on the d.c. bus. If the rectifier is a component part of an adjustable-speed drive, this link voltage ripple at the inverter input terminals affects the output voltage waveforms, generating into the motor low-frequency harmonic currents which create undesired torque pulsations. These pulsations are dominated by the second harmonic of the input line frequency as the existence of the second line frequency ($2\omega_i$) harmonic voltage component in the d.c. bus voltage generates low-frequency harmonic components in the inverter output voltages at the sum and difference frequencies with the inverter output frequency (ω_o). The circuit analysis performed in [41] undoubtedly shows how input line voltage unbalance gives rise to d.c. bus voltage ripple and torque pulsations that fall principally at the second harmonic of the input line frequency. In addition, this analysis also shows that harmonic current components having frequencies of $[(\omega_o + 2\omega_i)$ and $(\omega_o - 2\omega_i)]$ appear in the machine stator currents as a result of the input line unbalance conditions.

14.2.1.3 Voltage Sags

Adjustable-speed drives probably are the type of equipment most sensitive to voltage sags, as far as operation of the equipment can be interrupted; it is also generally assumed that it is difficult to make them more tolerant against sags. Since ASDs are often applied in critical process control environments, nuisance tripping can be very disruptive with potentially high downtime cost implications [4],[7],[48].

When the power supply voltage decreases below the d.c. bus voltage, only the d.c. bus capacitors will supply the load. Therefore, the d.c. bus voltage will decrease to a voltage level at which the ASD inverter will be disconnected. The d.c. voltage drop depends on the characteristics of the disturbance, its magnitude, duration, unbalances and phase jump, on the capacity available at the d.c. bus and on the power consumed by the load. If the d.c. voltage drop is not limited this can cause the following problems on the ASD [58]:

- The power switches of the inverter can be damaged (depending on the control methodology employed) if the power supplied to the load remains constant, since the voltage decrease will make the current increase.
- To provide a maximum torque across the speed range, and avoid motor saturation, the ratio of voltage to frequency must remain constant. If a voltage drop is detected, output

frequency has to decrease proportionally, and consequently so does load speed. In some processes this drop in the speed may be unacceptable.

- In many cases, control electronics and IGBT drivers can be powered off, since they are supplied from the d.c. bus through a buck d.c./d.c. converter.
- Rectifier power switches can be damaged or the fuses blown, due to charge overcurrents of capacitors on the a.c. line voltage recovery.

On the other hand, under unbalanced type C and D voltage sag conditions [6],[7], ASDs can easily transfer into the single-phase excitation conditions, even with a small amount of voltage unbalance. During type C and D voltage sags, the input line voltages have different amplitudes. The diodes in the front-end rectifier bridge only conduct to charge the d.c. bus capacitor bank when the input line-to-line voltage amplitude is higher than the d.c. bus voltage. As a result, the ASD enters single-phase operation when only four out of the six diodes conduct current during each cycle [43]. This behavior corresponds to phase-to-phase faults or to phase-to-ground faults if the load is Δ connected.

To avoid these problems all modern ASDs have an undervoltage protection that disconnects the inverter when d.c. bus voltage drops below a preset value. When nominal a.c. voltage is recovered, some ASDs restart immediately, others restart after a certain time, and still others only admit a manual restart operation. Automatic restart is not necessary in all cases but it is very interesting in those processes where a limited drop in the speed is admitted. In those cases, the problem consists of the synchronization of the inverter output with the actual motor speed, to avoid important electrical and dynamic transients that can damage the equipment.

Different solutions are available depending on the drop of speed admissible by the process. Naturally the solution chosen will have an effect on the cost of the equipment. One solution would be to disconnect the inverter for 1 or 2 s, to assure that no induced voltage remains in the stator phase, and, after this time, restarting the motor from zero speed. This solution will suppose no extra cost to the equipment. At the other extreme, voltage transducers can be used to sense the induced voltage in the stator, to generate a synchronized voltage with the inverter. This solution needs more transducers and thus ASD cost will be much higher.

Another method, only adding one voltage transducer to monitor d.c. bus voltage, consists of using the mechanical energy stored in the load inertia to maintain constant the d.c. bus voltage. Thus the rectifier is protected from a.c. line recovery, since no charge currents will occur, and there is no loss of synchronization between the inverter and the motor speed.

Voltage sags can also produce link voltage ripple at the inverter input terminals and undesired torque pulsations dominated by the second harmonic of the input line frequency, as was presented in Section 14.2.1.2.

At the same time, problems caused by overvoltage transients are more limited. All ASDs incorporate overvoltage suppressors, snubbers or other filters that, with a suitable design, will protect the equipment against these disturbances. If no protection exists, the rectifier will be the most affected: d.c. bus voltage will remain practically constant, due to the large capacitors, but a sudden voltage rise on the a.c. side will produce overcurrents that can blow the fuses, or, worse, can cause the destruction of the power switches if the maximum blocking voltage is exceeded.

14.2.2 Voltage Sag Susceptibility

Sensitivity of ASDs to voltage sags is usually expressed as a voltage tolerance curve, in terms of only one pair of sag magnitude/duration values. These two values are denoted as the threshold values – if the voltage sag is longer than the specific duration threshold and deeper than the specific voltage magnitude threshold, the ASD will malfunction/trip. For ASDs, reported threshold values vary from 50–60 % to 80–90 % of rated voltage for magnitude, and from a cycle (or even less) up to 5–6 cycles for the duration. The use of magnitude/duration threshold values is straightforward for single-phase equipment and balanced polyphase sags, but it cannot be applied for assessment of ASD sensitivity to polyphase unbalanced voltage sags. The ASDs are three-phase equipment and voltage sags with different combinations of phase voltages will have different effects on their operation. These effects can be assessed only if the sensitivity of ASDs and voltage sag characteristics are expressed considering the three-phase nature of the power supply and drive itself.

The following specific conclusions can be drawn from the test performed and presented in [23]:

- Drive sensitivity to voltage sag characteristic (magnitude and duration) varies in wide ranges, depending on the perturbation type and drive/motor operating condition.
- Generally, if speed is lower than rated, ride-through capabilities improve.
- Point on wave of sag initiation and phase shift during the sag do not influence drive sensitivity.
- Variations in supply voltage magnitude and different harmonic contents influence both vertical and horizontal parts of voltage–tolerance curves. The overall drive sensitivity decreases if pre-sag and post-sag voltage is overrated (i.e. 110 % of rated) and increases if they are underrated (90 % of rated).
- Most tests indicated that the response of the drive is mainly determined by the sag type, not by its shape.

14.2.3 Immunization Techniques

Any power system is characterized by different intrinsic disturbances, and consequently all equipment connected to the power supply system is subjected to these disturbances. The challenge nowadays is that modern electronic equipment is more susceptible to the electrical environment; as the use of this equipment increases the problem increases too [58].

There are two solutions to solve the problem: improve the power supply or improve the connected equipment. The attempt to solve these aspects led to electromagnetic compatibility (EMC): all equipment must be compatible with the power supply where it is connected. According to this concept, electromagnetic emission and immunity limits are fixed to electronic equipment to prevent, on one hand, an excessive emission of disturbances and, on the other hand, to provide equipment with a suitable immunity level. IEC 61000 (Parts 6-1 and 6-2) [38],[39] specifies that all equipment must ride through a voltage sag of a residual voltage of 70 % for 10 ms, and must not be damaged for a voltage sag of a residual voltage of 40 % for 100 ms.

In many cases, this will be enough since the equipment will not be damaged. But critical processes, such as a continuous process in which a shutdown is not admissible, will need mitigation devices to increase the immunity level of susceptible equipment like ASD.

14.2.3.1 Ride-through

ASDs' susceptibility to different electric power disturbances depends on the capability of the d.c. bus capacitor to store energy. As generally this capacitor has a relatively small amount of stored energy, when the mains voltage decreases under the imposed limits, the d.c. link energy is absorbed by the induction motor load within a few milliseconds and nuisance tripping will occur.

There are many possibilities to achieve a higher immunity; selection of the right one will depend on the following criteria: the type of disturbance, the cost of the mitigation device and the cost of damage produced by the disturbance [9]. Some of them are of global nature, such as the well-known DSTATCOM (Distribution Static Synchronous Compensator), DVR (Dynamic Voltage Regulator) or UPQC (Unified Power Quality Conditioner), but all of them lead to high investments.

However, the problem of voltage sags and short interruptions is energetic and most of the solutions consist of increasing the stored energy [25]. A variety of energy storage technologies are candidates for providing the needed full-power ASD ride-through under short-term power interruptions and sags. Battery backup systems, supercapacitors, motor–generator sets, flywheel energy storage systems, superconducting magnetic energy storage (SMES) and fuel cells are some examples of these technologies. Already many customers with sensitive loads have installed uninterruptible power supplies (UPS) to provide the required ride-through capability; however, this solution is often inappropriate from both economic and technical points of view as it is expensive, assures a quite reduced stored energy and requires intensive maintenance.

There are also other methods, such as active rectifier and boost or flyback converters on the d.c. side:

- **Active rectifier (active front end).** Replacing the diode rectifier by an active rectifier has three main advantages. The first is that the d.c. bus voltage is controlled so it can remain constant in the presence of voltage sags and voltage transients. The second is that the input current has a sinusoidal waveform, without harmonics and with unitary power factor. Finally this converter works in the four quadrants, thus regenerative braking can be made. The main disadvantages are the lower efficiency of the system, the higher cost and the produced EMC noise.

  The ability of active rectifiers to ride through voltage sags at full load will depend on the nominal current of the power switches. In addition, it must be considered that three inductors in series are needed to filter the current input, and that the d.c. bus voltage is higher than with a diode rectifier, so higher *dv/dt* and switching losses are produced.

On other hand, under sag conditions the PFC functions, an important attribute of the active rectifier, are inhibited, maximizing the available ride-through current.

- **Boost converter in the d.c. link.** A d.c./d.c. boost converter is a device that maintains the d.c. bus voltage constant during voltage sag. It is connected at the output of the rectifier and before the capacitor filter. The converter can be integrated in the ASD supplied by the manufacturer, or it can be an add-on module in ASD with the accessible d.c. bus. In the first case, the rectifier must be sized properly since the current will be doubled when the voltage sag has a depth of 50 %; therefore the maximum depth that the ASD can ride through with this system is limited (usually 50 % voltage sag with a duration up to 2 s). In the second case the ASD needs no modification since the add-on boost converter can be supplied by a different rectifier or even by an energy storage system such as those mentioned before.
- **Flyback converter powered by supercapacitors.** The solution, proposed in [25], represents an approach to achieve ride-through under short-time power interruptions (STPIs). A block of supercapacitors is connected to a flyback d.c./d.c. converter with a view to supplying energy to the d.c. link of the ASD in the event of an STPI. The main advantages of this solution are:

  – a 5 s ride-through of an ASD for voltage sag and/or an STPI;
  – supercapacitors have a long life and assure fast recharge rates;
  – requires minimal maintenance;
  – the structure is modular and additional supercapacitor and d.c./d.c. converter blocks can be added for higher-power ASDs or for longer ride-through times.

For the first two methods, during sag the rectifier unit draws significant discontinuous currents producing an important voltage drop in the LV distribution system. As a result, the available supply voltage supplementary decreases while the protective devices installed in the distribution system are highly stressed.

14.2.3.2 Converter Design

For a typical three-phase voltage source inverter PWM ASD, the vulnerability of the drive to transient overvoltages depends on three important parameters: (i) the size of the d.c. bus capacitor; (ii) the presence of a d.c. bus choke or an a.c. line choke; and (iii) the overvoltage trip setting of the drive. Special attention has to be paid during the converter design process in order to ensure correct operation of the drive [4]:

(i) Therefore, in order to improve the transient overvoltage ride-through capability the d.c. bus capacitance value should be as large as economical possible; this value is also designed so as to minimize the ripple current and to provide acceptable momentary power loss or voltage sag ride-through capability. For most cases, a three-cycle or 50 ms power loss ride-through capability is common. Commercially available equipment usually presents d.c. bus capacitance values ranging from 75 to 367 pF per kW; capacitance values in excess of 136 pF per kW are most typical.

(ii) Another goal in converter design is to reduce the harmonic pollution of the supply; consequently, the d.c. bus choke must be designed to ensure continuous current conduction. The relationships indicated in the literature lead to a minimum inductance value of the d.c. bus choke of approximately 1.8 % (based on drive kVA rating). If the d.c. bus does not include any inductor, an a.c. line choke with effective reactance at supply frequency of 3–5 % is normally recommended (especially in situations where resonance phenomena at frequencies close to harmonics produced by the converter can appear due to PFC capacitors connected to the load side of the transformer supplying the drive) [45],[49]. As they limit the slope of the transient inrush current to the d.c. bus capacitor, both the d.c. bus choke and a.c. line choke improve the transient overvoltage ride-through capability of the converter. However, since the drive responds to line-to-line voltages, the a.c. line choke is much more effective than the d.c. bus choke; unfortunately, the presence of a.c. line inductors produces voltage drops that may alter the drive performance under steady-state conditions. For ASDs not including a d.c. bus choke, the use of a.c. line inductors is unavoidable for improved transient overvoltage mitigation, otherwise any transient overvoltage at the drive input will be transferred directly to the d.c. bus. Nevertheless, in some cases this solution may not be sufficient to avoid overvoltage tripping, even in the absence of voltage magnification.

(iii) Normally, for 460 V drives, electronic devices in the power circuit of the ASD inverter are rated at 1000–1200 V. In order to avoid damage to these transistors, drives typically have a d.c. bus overvoltage trip setting of about 760–820 V (1.17–1.26 per unit), with trip settings exceeding 780 V (1.2 per unit) being most common. For a d.c. bus voltage of 650 V, the overvoltage margin thus ranges from about 110 to 170 V; this margin is further reduced to 80–140 V when utility practice allows an a.c. supply voltage 5 % above nominal. Theoretical analyses and *in situ* tests have proved that ASDs not equipped with an a.c. line or d.c. bus choke may trip due to line-to-line overvoltages of as little as 1.17 per unit. In usual distribution systems, overvoltages in the range from 1.2 to 1.6 per unit line-to-line can quite frequently emerge at the utility's bus if a capacitor bank existing to that bus is switched without transient overvoltage control. In the worst case, nuisance tripping can occur during this normal operation in a supply system due to low overvoltage trip settings on ASDs.

14.2.3.3 System Design/Re-engineering

Other types of solutions to improve the behavior of power converters during supply system perturbations are of a local nature, and can be implemented directly on the problematic ASD. One of them consists of reducing the severity of different perturbations; other solutions refer to enlarging the inertia of the drive or making a design oriented to a low-speed or low-load operating point.

As already seen, numerous trippings of voltage source inverter ASDs, particularly those of PWM drives, are the result of a transient overvoltage on the d.c. bus of the ASD; this phenomenon is often caused by the transient perturbation associated with the switching of shunt capacitor banks placed at the utility's substations near the customer coupling point. This perturbation may be amplified by the resonant circuit created by the total inductance at the LV customer bar and the capacitance of the PFC capacitors connected to the same bus;

obviously, the level of magnification depends on how closely the natural frequency of the transient matches that of the resonant circuit, on the instant of switching and on the quality factor of the resonant circuit.

To mitigate these phenomena, three switching devices with transient overvoltage control are available [4]:

1. *Circuit-breakers with pre-insertion resistors.* In this case, suitable resistors are inserted into the capacitor bank supply circuit for 10 to 15 ms prior to the closing of the main contacts.
2. *Circuit-switchers with pre-insertion inductors.* This time, suitable inductors are inserted into the capacitor bank supply circuit for 7 to 12 cycles of the supply frequency during the closing of the high-speed disconnect blade; devices with high resistance are recommended in order to limit the effects of voltage magnification.
3. *Controlled closing devices.* The goal of this solution is to close the individual poles of high-speed vacuum switches or SF_6 circuit-breakers near voltage zero to minimize voltage and inrush current transients.

The results presented in [4] show that all three methods are very effective when ASDs are equipped with 3 % a.c. line chokes. However, the relative performance of the different methods of control for a specific installation is best determined by simulation based on the actual system parameters.

Another approach consists of the implementation of kinetic buffering, a method aiming to use the kinetic energy stored in the rotating parts of machinery for keeping the d.c. bus voltage at an imposed value during sag. It is activated by a special d.c. undervoltage protection and consequently motors operate like generators during the sag supplying energy back into the d.c. bus. The method is more valuable in common d.c. bus applications: here, the energy stored by non-critical drives can be used to keep the d.c. voltage up and maintaining the normal operation of critical ones. Usually, a ride-through of at least 0.5 s is provided, for very severe sags or even interruptions.

For a case study see web address

BIBLIOGRAPHY

[1] ABB, *Bearing Currents in Modern AC Drive Systems*, Technical Guide No. 5, 1999.
[2] Arrillaga J. *et al.*, *Power System Quality Assessment*, John Wiley & Sons, Ltd, Chichester, 2000.
[3] Bauta M., Grotzbach M., Harmonics in AC/DC converters under unbalanced voltage supply. International Conference on Power Conversion, Intelligent Motion and Power Quality, 2000.
[4] Bellei T. A. *et al.*, Evaluating capacitor-switching devices for preventing nuisance tripping of adjustable-speed drives due to voltage magnification. *IEEE Transactions on Power Delivery*, vol. 11, no. 3, pp. 1373–1378, 1996.
[5] Bettega E., Fiorina J. N., *Active harmonic conditioners and unity power factor rectifiers*, Cahier Technique Merlin Gerin No. 183, 1999.
[6] Bollen M., *Understanding Power Quality Problems. Voltage Sags and Interruptions*, IEEE Press, Piscataway, NJ, 2000.

[7] Bollen M., Zhang L. D., Analysis of voltage tolerance of ac adjustable-speed drives for three-phase balanced and unbalanced sags. *IEEE Transactions on Industry Applications*, vol. 36, no. 3, pp. 904–910, 2000.

[8] Bollen M. *et al.*, Voltage dips in distribution systems: load effects, measurements and theory. 7th International Conference on Electricity Distribution, Session 2, Paper 47, 2003.

[9] Cano J. M. *et al.*, Technical and economical assessment of the effect of voltage sags on adjustable speed drives. International Conference on Renewable Energy and Power Quality, 2004.

[10] Carbone R., Analyzing voltage background distortion effects on PWM adjustable-speed drives. *IEEE Transactions on Power Electronics*, vol. 19, no. 3, pp. 765–774, 2004.

[11] Carpinelli G. *et al.*, Single phase voltage source converters: analytical modelling for harmonic analysis in continuous and discontinuous current conditions. *International Journal of Power and Energy Systems*, vol. 23, no. 1, pp. 37–48, 2003.

[12] Chen S., Lipo T. A., Circulating type motor bearing current in inverter drives. *IEEE Industry Application Magazine*, January/February, pp. 32–38, 1998.

[13] Chindris M., Stefanescu S., A new zero-current-switching series resonant inverter topology. 6th European Conference on Power Electronics and Applications, pp. 2.796–2.800, 1995.

[14] Chindris M., Stefanescu S., Switched-mode power supplies using a new series resonant inverter topology. Mediterranean Electrotechnical Conference, pp. 1191–1194, 1996.

[15] Chindris M., Sudria A., *Harmonic pollution of electrical industrial networks*, Mediamira, Cluj Napoca, 1999.

[16] Chindris M. *et al.*, Active filters for improved power quality and voltage regulation at the customer bus. International Conference on Electromechanical Systems, I, pp. 265–269, 1999.

[17] Chindris M. *et al.*, *Mitigation of harmonic pollution in electrical industrial networks*, Mediamira, Cluj Napoca, 2003.

[18] Collombet C. *et al.*, *Harmonic disturbances in networks, and their treatment*, Cahier Technique Merlin Gerin No. 152, 1999.

[19] Delaballe J., *EMC: electromagnetic compatibility*, Cahier Technique Merlin Gerin No. 149, 2001.

[20] Delarue P. *et al.*, Reduction of perturbations induced by three-phase rectifiers in industrial power systems. Power Quality Applications, pp. 164–170, 1991.

[21] Desmet J. *et al.*, Influencing parameters on overvoltages at the terminals of inverter supplied induction motors. www.esat.kuleuven.be/electa/pub_1195.pdf.

[22] Desmet J. *et al.*, A practical approach to the influence of long lead cables on inverter supplied induction motors. www.esat.kuleuven.be/electa/pub_478.pdf

[23] Djokic S. Z. *et al.*, Sensitivity of AC adjustable speed drives to voltage sags and short interruptions. *IEEE Transactions on Power Delivery*, vol. 20, no. 1, pp. 494–505, 2005.

[24] Dugan R. C. *et al.*, *Electrical Power Systems Quality*, McGraw-Hill, New York, 1996.

[25] Duran-Gomez J. L. *et al.*, An approach to achieve ride-through of an adjustable-speed drive with flyback converter modules powered by super capacitors. *IEEE Transactions on Industry Applications*, vol. 38, no. 2, pp. 514–522, 2002.

[26] Dwyer R. *et al.*, C filters for wide-bandwidth harmonic attenuation with low losses. IEEE/PES Winter Meeting, 2000.

[27] Ferracci P., *Power quality*, Cahier Technique Merlin Gerin No. 199, 1993.

[28] Fiorina J. N., *Harmonics upstream of rectifiers in UPC*, Cahier Technique Merlin Gerin No. 160, 1993.

[29] Fiorina J. N., *Inverters and harmonics (case studies of non-linear loads)*, Cahier Technique Merlin Gerin No. 159, 2001.

[30] Golovanov C. (coordinator), Chindris M. *et al.*, *Modern measurements in power systems*, Editura Tehnica, Bucharest, 2001.

[31] Grundfos Motor Book, *8. Frequency Converter Operation*, 2004.

[32] Hanna R. A., Harmonics and technical barriers in adjustable speed drives. *IEEE Transactions on Industry Applications*, vol. 25, no. 5, pp. 894–900, 1989.
[33] Hanzelka Z, Bien A., *Harmonics. Interharmonics*, Power Quality Application Guide 3.1.1, Copper Development Association, 2004.
[34] Hanzelka Z., *Switch mode power supply*, Academia Gorniczo-Hutnicza & Leonardo Power Quality Initiative (www.lpqi.org), 2004.
[35] Heydt G. T., *Electric Power Quality*, Stars in a Circle Publications, 1991.
[36] IEC 60146-1-2, *Semiconductor convertors – General requirements and line commutated convertors – Part 1-2: Application guide*, Third Edition, 1991.
[37] IEC 61000-3-2, *Electromagnetic Compatibility: Limits for harmonic current emissions (equipment input current ≤ 16 A per phase).*
[38] IEC 61000-6-1, *Electromagnetic Compatibility: Generic standards – Immunity for residential, commercial and light-industrial environments.*
[39] IEC 61000-6-2, *Electromagnetic Compatibility: Generic standards – Immunity for industrial environments.*
[40] IEEE 519, *IEEE recommended practices and requirements for harmonic control in electrical power systems*, 1992.
[41] IEEE Interharmonic Task Force, *Interharmonics in power systems*, 1997.
[42] Krug H. P. *et al.*, Neutral-point clamped three-level general purpose inverter – features, benefits and applications. Academia Gorniczo-Hutnicza & Leonardo Power Quality Initiative (www.lpqi.org),
[43] Lee K. *et al.*, Closed-form analysis of adjustable-speed drive performance under input-voltage unbalance and sag conditions. *IEEE Transactions on Industry Applications*, vol. 42, no. 3, pp. 733–741, 2006.
[44] Lee K. *et al.*, Impact of inductor placement on the performance of adjustable-speed drives under input voltage unbalance and sag conditions. *IEEE Transactions on Industry Applications*, vol. 42, no. 5, pp. 1230–1240, 2006.
[45] Lee K. *et al.*, Source current harmonic analysis of adjustable speed drives under input voltage unbalance and sag conditions. *IEEE Transactions on Power Delivery*, vol. 21, no. 2, pp. 567–576, 2006.
[46] Leonardo Power Quality Initiative, http://www.lpqi.com, 2006.
[47] Luszcz J., Iwan K., Conducted EMI propagation in inverter-fed AC motor. *Electrical Power Quality and Utilisation Magazine*, vol. II, no. 1, pp. 47–51, 2006.
[48] Magueed F. A. *et al.*, Transient performance of voltage source converter under unbalanced voltage dips. 35th Annual IEEE Power Electronics Specialists Conference, 2004, pp. 1163–1168.
[49] Mohan N. T. *et al.*, *Power Electronics: Converters, Applications and Design*, John Wiley & Sons, Ltd, Chichester, 1989.
[50] Ollila J., Dimensioning a low cost LCL-filter for the PWM-rectifier, Power Quality Bremen, pp. 179–188, 1995.
[51] Paice D. A., *Power Electronic Converter Harmonics. Multipulse Methods for Clean Power*, IEEE Press, Piscataway, NJ, 1996.
[52] Peeran S. M., Cascadden C., Application, design, and specification of harmonic filters for variable frequency drives. *IEEE Transactions on Industry Applications*, vol. 31, no. 4, pp. 841–847, 1995.
[53] Radomski G., Analysis of Vienna rectifier. *Electrical Power Quality and Utilisation Journal*, vol. XI, no. 1, pp. 49–56, 2005.
[54] Rashid M. H. (ed.), *Power Electronics Handbook*, Academic Press, New York, 2001.
[55] Redl R., Electromagnetic environmental impact of power electronics equipment. *Proceedings of the IEEE*, vol. 89, no. 6, pp. 926–938, 2001.
[56] Sankaran C., *Power Quality*, CRC Press, Boca Raton, FL, 2002.

[57] Stefanescu S. *et al.*, Analysis and comparison between different methods of current reference generation for active filters control. International Conference on Renewable Energy and Power Quality, 2004, paper 211.

[58] Sudria A. *et al.*, Grid voltage sags effects on frequency converter drives and controlled rectifier drives. *Compatibility in Power Electronics*, June, pp. 39–45, 2005.

[59] Vermaelen C. *et al.*, An overview of techniques to reduce common mode EMI in speed drive systems. PVIM, 2003, paper pq_9_3.

[60] Vlatcovic V., Input filter design for power factor correction circuits. *IEEE Transactions on Power Electronics*, vol. 11, no. 1, pp. 199–205, 1996.

[61] Watanabe E. *et al.*, New inverter drive technology. *Yaskawa Technical Review*, vol. 65, no. 2, pp. 104–111, 2001.

[62] Yamanaka K. *et al.*, A novel neutral point potential stabilization technique using information of output voltage vector and currents. *IEEE Transactions on Industry Applications*, vol. 38, no. 6, pp. 1572–1579, 2002.

[63] Yaskawa, *Application Note G7.01*, 2005.

[64] Yaskawa, *Application Note G7.02*, 2005.

[65] Yaskawa, http://www.yaskawa.com, 2006.

[66] Zargari N. R. *et al.*, Input filter design for PWM current-source rectifiers. *IEEE Transactions on Industry Applications*, vol. 30, no. 6, pp. 1573–1578, 1994.

15

Compensation of Reactive Power

Stefan Fassbinder and Alan Ascolari

Capacitor banks, used for the compensation of fundamental reactive power, are essential for the economic operation of systems that include resistive–inductive loads. In fact resistive–inductive loads have been omnipresent since the beginning of electric power engineering. However, since substantial non-linear loads are now becoming similarly omnipresent, there are two new risks emerging around and inside the capacitor banks:

- current overload of capacitors;
- parallel resonance of capacitances with inductances in their (electrical) proximity.

Compensation capacitors continue to be indispensable, and it is technically fairly easy to design or upgrade them to cope with the new challenges, although it almost doubles the price. This chapter identifies the optimal approach to selection of new plant and the upgrade of existing capacitor banks in order to prevent problems caused by harmonics.

It is important first of all to understand the complementary behavior of L and C elements in order to understand the business of compensation. Compensation capacitors should always be detuned in order to avoid resonance with harmonics and overload by high-frequency current. Variable compensation units should be designed for rapid switching using semiconductor switches and intelligent control algorithms.

Very little additional effort and costs are incurred in mitigating dominant harmonics together with the compensation of fundamental reactive power, since compensation is done regardless and most compensators today already utilize detuning reactors. But in most cases, tuning the resonance frequencies of such systems to any possible harmonic frequencies in the power system is deliberately avoided. Sometimes the resonance frequencies are tuned

Handbook of Power Quality Edited by Angelo Baggini

close to the harmonics, which helps to some extent. But why not have it tuned exactly to a harmonic frequency? The risk of overloading the compensator with harmonics is not as high as generally assumed. A certain degree of reserve, of course, has to be installed. This is not a problem since this brings about a better cleaning effect and better energy efficiency at very little extra cost. Active conditioners are much more expensive and often either shoot beyond the target or bring about a minor mitigation effect despite better appliance performance. This is because, due to cost structure, they tend to be installed centrally.

It makes no sense that utilities charge for fundamental reactive power while not doing so for harmonics (wattless current). This is especially true as the latter bring with them many more disadvantages than the former. It is foreseeable that utilities will start to charge for harmonic dissipation as well as fundamental reactive power.

There is no case where the installation of filtering equipment, except filters installed with or even inside a load, can be used as a convincing argument for downsizing neutrals or any other conductors. Keeping the system's impedance low is even more important with filtering installed than where it is not, otherwise the filter's effects may be adverse!

15.1 BASICS

15.1.1 Characteristics of Inductances and Capacitances

Electrically, inductance is analogous to the inertia of mass in a mechanical system. A reactor, i.e. a component with a defined and intentional inductance value, represents the electrical equivalent of a flywheel which would have a defined inertia. Of course, anything that has a mass also has inertia, in the same way as any piece of conductor has a parasitic inductance.

Both the inductance L and the capacitance C represent reactive components with a reactance and a reactive power intake/output, whereas capacitive reactive power input is equivalent to inductive reactive power output and vice versa.[1] The reactances are calculated as follows:

$$X_L = 2\pi f L \quad \text{and} \quad X_C = \frac{1}{2\pi f C} \tag{15.1}$$

So the inductive reactance X_L is proportional to frequency while the capacitive reactance X_C is inversely proportional to frequency f. For any parallel combination of L and C there will be a frequency f_0, at which the reactances are equal – this is the resonant frequency. This frequency at which the LC combination oscillates is calculated as

$$f_0 = \frac{1}{2\pi\sqrt{LC}} \tag{15.2}$$

To be precise, it is any change of current that is lagging or leading in relation to the corresponding voltage change, e.g. the zero crossing. It originates from the energy being stored in the capacitance and from the special characteristics of the waveform.

[1] Reactive power has, in effect, no clearly definable direction of flow.

Electrical capacitance corresponds to the resilience (elasticity) of a mechanical component. A capacitor can be produced with a defined capacitance, corresponding to a spring in a mechanical system, but, just as any material is resilient (elastic) to some extent, so there is a certain amount of parasitic capacitance between any two pieces of conducting material.

The question is whether these parasitic reactances are large enough to play a role in practical engineering. At high voltages or high frequencies they often are, but this is not normally the case at low voltage levels and mains frequency.

The energy content in each of the two energy stores is given by

$$W_{spring} = \frac{D}{2} \cdot s^2 \qquad W_{mass} = \frac{m}{2} \cdot v^2 \tag{15.3}$$

where D is the elasticity constant (elongation per force, Hooke's law); s is elongation (instantaneous distance from point of relaxed state); m is the mass; and v is the speed of movement of this mass.

Here, s and v could be, and should be, written as functions of time $s(t)$ and $v(t)$, for that is what they are – periodically changing with time.

Now, combining the two, the inertial mass and the resilient spring, provides a system with two energy stores. Energy that is released from one of the components may flow right into the other one: the mass may be accelerated, with the force doing so coming from the relaxing spring. At the zero crossing of the force, where the spring reaches its relaxed state, the speed reaches its maximum, and the mass of course will not stop immediately but, while slowing down, rather deform the spring again, from the extended to a compressed state or vice versa. Nearly the same thing is valid for the energy in a capacitance C and in an inductance L. The tension of the extended/compressed spring correlates with the positive/negative voltage in the capacitor, and the speed of the mass is the current, also swapping polarity at regular intervals. All polarity swaps occur alternately and at constant intervals: once the voltage, once the current, one every quarter period, or every 90° we can also say because all changes of the two dimensions, tension and velocity in the mechanical model, voltage and current in the electrical model, follow a sine function. On account of the 90° phase shift it can also be said that one of the dimensions follows a cosine function, and since, assuming linear and loss-free components, at any point of time within the oscillation

$$\sin^2(\omega t) + \cos^2(\omega t) = 1 \tag{15.4}$$

then the internal energy

$$W = \frac{C}{2} \cdot u^2(t) + \frac{L}{2} \cdot i^2(t) = \text{constant} \tag{15.5}$$

at any point in time. With real components losses occur and the phase displacement of current against voltage in an inductive/capacitive component becomes a little bit less than ±90°, but if operated within the specified range, losses are low, and the influence of non-linearity in reactor core materials is largely negligible for technical purposes if the reactor is properly designed.

15.1.2 Reactive Power

In resistive loads, the instantaneous values of voltage and current are proportional to each other (Figure 15.1); in reactive components (Figure 15.3) and combinations of these (Figure 15.2) they are not. In the latter case, if one of the dimensions has a sinusoidal wave shape, so does the other but with a phase shift between the two; hence, during two sections of every a.c. period they have the same sign but during the other two sections their signs are different. During these periods of opposite voltage and current polarities their product, the power, is negative, so a power consumer in fact temporarily turns into a power supply.

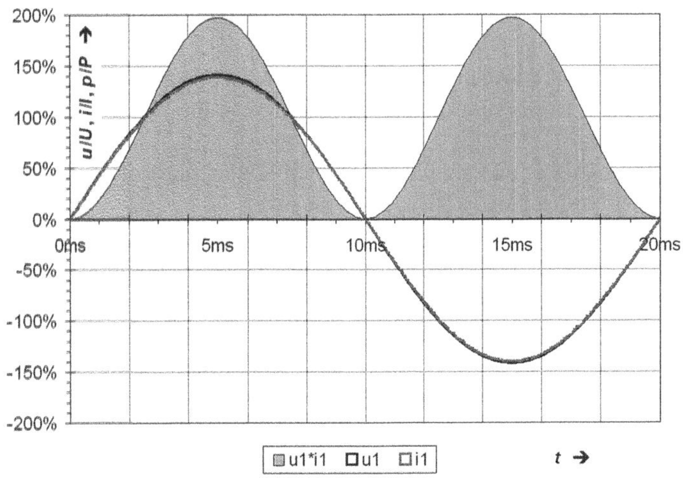

Figure 15.1 Ohmic load

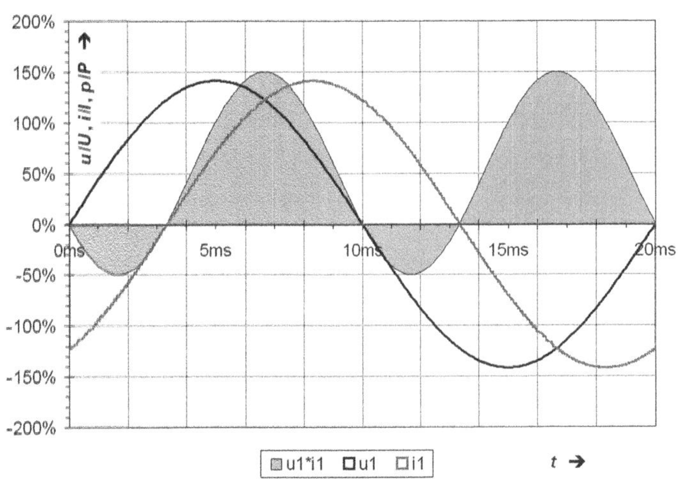

Figure 15.2 Ohmic–inductive load

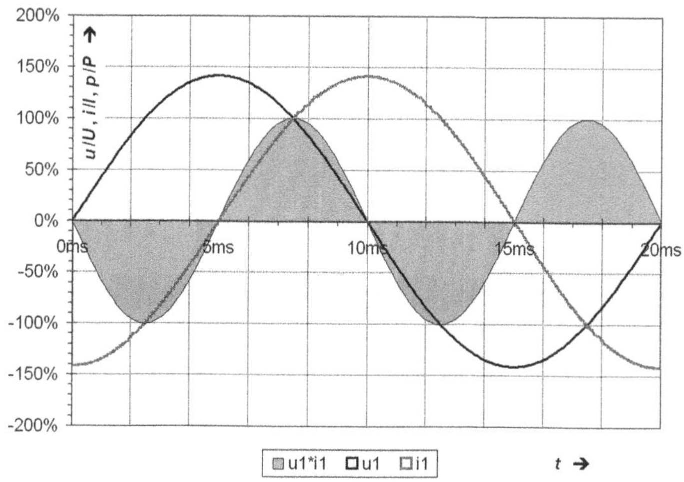

Figure 15.3 Inductive load

The electric energy absorbed a quarter period before was not consumed, e.g. converted into another form of energy (such as heat), but stored, and is now recovered and fed back into the network. The real active energy transferred during each full period equals the integral of power; that is, the area below the voltage times current curve, with the parts below the abscissa to be subtracted from those above. So fundamental reactive power is an oscillation of energy.

So far, the definition of reactive power as it relates to sinusoidal voltages and reactive loads is still relatively simple. However, reactive power is also present in phase-angle-controlled resistive loads. In a German electrical engineering journal, an author once claimed that such a load, e.g. an incandescent lamp with a dimmer, does not cause fundamental reactive power, since there are no periods of time within a full wave where voltage and current have opposite polarities. He provoked a flood of disagreement among readers, pointing out that in the Fourier analysis of such phase-angle-controlled current the fundamental wave did have a lagging phase shift against the voltage, so it was evident that there was fundamental reactive power. Both viewpoints sound logical but which one is correct?

Figure 15.4 provides an explanation. Looked at from the simple point of view of the load (top row in Figure 15.4), there is no reactive power – current is in phase with voltage (despite the distorted wave shape) and the displacement power factor is unity. But all loads exist in a common system, and should be examined from the system's perspective, shown in the bottom row of Figure 15.4. Now the voltage waveform is again sinusoidal and the displacement power factor is now 0.8 lagging (see the W, VA and var measurements).

15.1.3 Wattless current

Where reactive power occurs in a distribution system (usually inductive reactive power), part of the energy in the line is in effect not transferred from source to load. Rather it

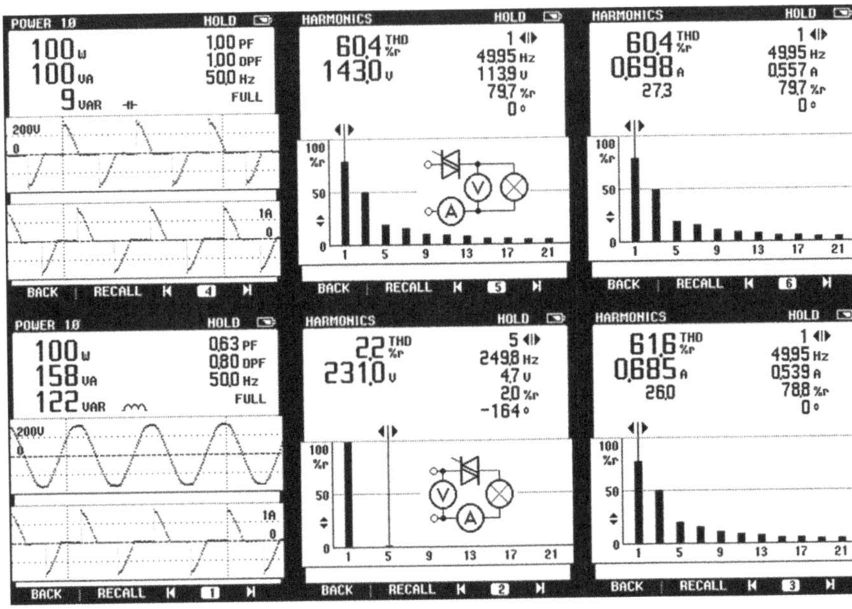

Figure 15.4 A phase-angle-controlled ohmic load can cause fundamental reactive power: from the utility viewpoint the additional impact on the mains is indeed there, while energy oscillation, which some experts see as a prerequisite for the existence of reactive power, does not take place

oscillates from a capacitance to a reactance and back again at a frequency of 100 Hz. For certain intervals of time, voltage and current have opposite polarities (Figure 15.5). Looking at the harmonics, the picture appears very similar. In Figure 15.6 the power of the third current harmonic has been plotted in isolation. The power transferred is the product of the

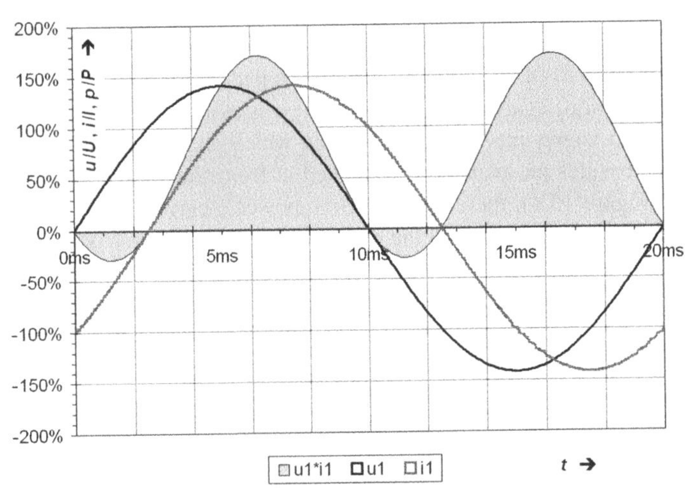

Figure 15.5 Fundamental reactive power

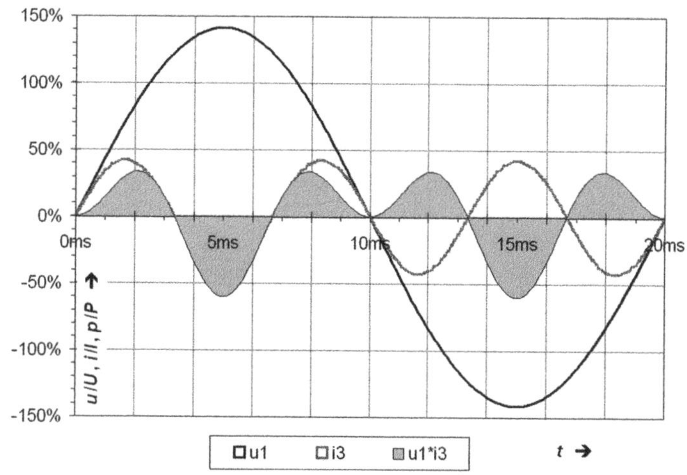

Figure 15.6 Wattless power

third current harmonic times the voltage present in the line, assuming the line voltage is still a pure sine wave. It can be shown that the areas above and below the abscissa cancel, meaning that on average no energy is transmitted. The third-harmonic current is therefore absolutely wattless.

But since harmonics cause additional losses, there must be some active power associated with them. This apparent contradiction originates from the incorrect assumption that the supply voltage was free of any harmonics. This is impossible, since the moment there is any 150 Hz current flowing, it will cause some – active and probably also reactive – 150 Hz voltage drop. This means that as soon as there is any additional frequency contained within the current, there will also be a certain amount of the same frequency in the voltage. Only when both voltage and current of the same frequency are present can active power occur at this frequency. It should be clear by this point that this will always be the case to some extent.

15.1.4 Reactive Power Compensator

In a normal network there are many simultaneously active loads. Many will be resistive while others have a capacitive component, where the current curve hurries a little ahead of the voltage curve (leading), and others have an inductive component where the current lags behind the applied voltage. In most networks the resistive–inductive loads prevail so that the overall current will have a resistive–inductive nature (Figure 15.5). This incessant but undesired oscillation of energy means an additional flow of currents in cables and transformers that adds to their loading, causes additional resistive losses and uses a potentially large part of their capacity. Therefore the basic reasons for compensating are to avoid:

- the undesired demand on transmission capacity;
- the energy losses caused by such;
- the additional voltage drops that the additional currents cause in the distribution system.

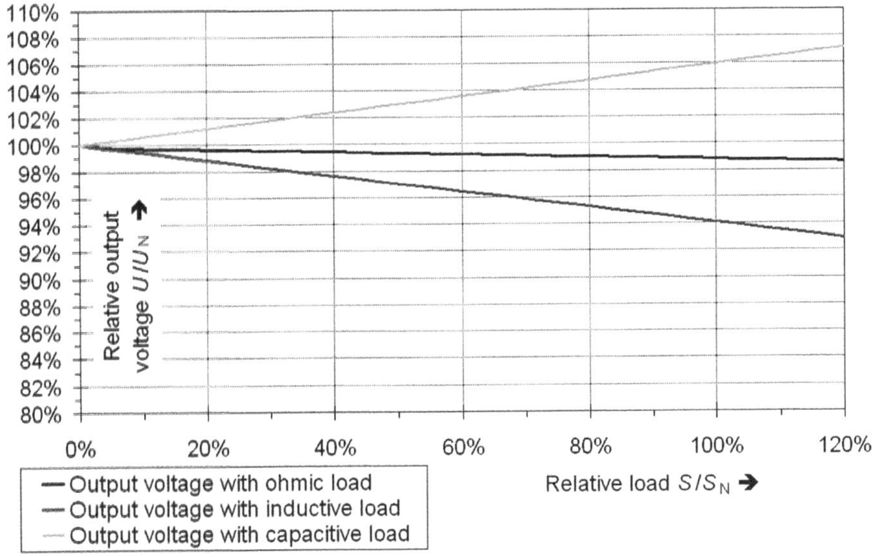

Figure 15.7 Voltage drop in a transformer (here 630 kVA according to HD 428 list C) is minor under resistive load, major with inductive load and even negative with capacitive load

These extra voltage drops in the system are important; a reactive current flowing in a resistance causes a real power loss. Even where the impedance is largely reactive, rapid changes of the reactive current may cause a flicker problem. A good example of this is a construction crane connected to a relatively small distribution transformer when a new home is erected in a residential area. The cranes are usually driven by contactor-controlled three-phase induction motors which are quite frequently switched from stop to start, from slow to fast and from downwards to upwards. The start-up currents of these motors are very high, several times the rated current, but these start-up currents have a very high inductive component, the power factor being around $\cos\varphi \approx 0.3$, or even smaller with bigger machines. The voltage drop in the transformer is also largely inductive, so this voltage drop has more or less the same phase angle as the start-up current of the motor and adds very much more to the flicker than the same current drawn by a resistive load (Figure 15.7). However, this also means that this flicker can be easily mitigated by adding a capacitor to compensate the inductive component of the motor's start-up current.

15.2 POWER FACTOR CORRECTION

15.2.1 Control and Regulation of Reactive Power

For these reasons, it is normally desirable to compensate reactive power. This is quite easy to achieve by adding an appropriate capacitive load parallel with the resistive–inductive loads so that the inductive component is offset. So while the capacitive element is feeding its stored energy back into the mains, the inductive component is drawing it, and vice versa, because the leading and the lagging currents flow in opposite directions at any point in time. In this way, the overall current is reduced by adding a load. This is called parallel compensation.

To do this properly requires knowledge of how much inductive load there is in the installation, otherwise overcompensation may occur. In that case, the installation would become a resistive–capacitive load which in extreme cases could be worse than having no compensation at all. If the load – more precisely its inductive component – varies, then a variable compensator is required. Normally this is achieved by grouping the capacitors and switching them on and off group-wise via relays. This of course causes current peaks with the consequent wear of the contacts, risk of contact welding and induced voltages in paralleled data lines. Care must be taken in timing the switch-on; when voltage is switched on to a fully discharged capacitor at the instance of line voltage peak, the inrush current peak is equal to that of a short circuit. Even worse, switching on a short time after switch-off, the capacitor may be nearly fully charged with the inverse polarity, causing an inrush current peak nearly twice as high as the plant's short-circuit current peak! If there are many switched mode power supply (SMPS) loads being operated on the same system, then a charged compensation capacitor, reconnected to the supply, may feed directly into a large number of discharged smoothing capacitors, more or less directly from capacitance to capacitance with hardly any impedance in between. The resulting current peak is extremely short but extremely high, much higher than in a short circuit! There are frequent reports about the failure of devices, especially the contacts of the relays controlling the capacitor groups, on account of short interruptions in the grid which are carried out automatically by for example auto-reclosers to extinguish a light arc on a high- or medium-voltage overhead line. It is often suggested that this doubling of peak value cannot occur with capacitors that are equipped with discharge resistors in accordance with IEC 831. However, the standard requires that the voltage decays to less than 75 V after 3 min, so they have little effect during a short interruption of a few tens of milliseconds up to a few seconds.

If, at the instant of reconnecting the capacitor to the line voltage, the residual capacitor voltage happens to equal the supply voltage, no current peak occurs. At least this is true if the compensator is viewed as a pure capacitance and the incoming voltage as an ideal voltage source, i.e. with zero source impedance. But if the self-inductance of the system is taken into account, certain resonances between that and the capacitance may occur. Assume the following case: the residual voltage of the capacitor is half the peak value and equal to the instantaneous line voltage, which would be the case 30° after the last voltage zero crossing, i.e.

$$u_C = \hat{u} \cdot \sin(30°) = 400\,\text{V}\frac{\sqrt{2}}{2} = 283\,\text{V} \tag{15.6}$$

At this point in time the current in the capacitor would be expected to be

$$i_C = -\frac{\hat{\imath}}{2} \tag{15.7}$$

but it is not because the capacitor has been disconnected from the supply up to this point in time. At the instant of connection, neglecting the system's inductance, the current would rise up to this value immediately, and nothing would happen that would not have happened anyway in the steady state. But a real system is not free of inductance, so the current will assume this value only hesitantly at first, then speed up and – again due to the inductance, its *inertia* – shoot beyond the target nearly up to double the expected value. Then it will come down again, and so on, and thus perform a short period of oscillation that may be attenuated down to zero well within the first mains cycle after connection. The frequency of such oscillation may be rather high, since the mains inductance is low, and may cause interference to equipment in the installation. Only if the instantaneous line and residual capacitor voltages are both at their positive or negative peaks, at which point in time the instantaneous current would be zero anyway, will the resistive–inductive current start without oscillation.

More precisely, there are two conditions to be fulfilled. Firstly, the sum of voltages across the capacitance and its serial reactance (be it parasitic or intentional detuning) must be equal to the line voltage. Secondly, the supposed instantaneous current, assuming connection had already taken place long before, has to equal the actual current, which of course is zero until the instant of switching. This second condition is fulfilled only at line voltage peak, which therefore has to equal the capacitor voltage. To achieve this, the capacitor is pre-charged from a supplementary power source. This practice has a secondary minor advantage in that it makes sure that there is always the maximum possible amount of energy stored in the capacitor while not in use, so that at the instant of turn-on it may help to mitigate some fast voltage dip and subsequent flicker which otherwise might occur.

Relays, however, are too slow and do not operate precisely enough for targeted switching at a certain point of the wave. When relays are used, measures have to be taken to attenuate the inrush current peak, such as inrush-limiting resistors or detuning reactors. The latter are frequently used anyway for other reasons (see Section 15.3 below), and are sometimes required by utilities. Although this series reactor replaces the inrush current peak at switch-on with a voltage peak (surge) at switch-off, it is still the lesser evil, since the reactive power rating of the reactor is just a fraction of the capacitor rating and so the energy available is less.

Electronic switches, such as thyristors, can be easily controlled to achieve accurate point-on-wave switching. It is also possible to control switching so as to mitigate a fast flicker caused by a large unstable inductive load, such as the crane motor mentioned previously, an arc furnace or a spot welder.

An alternative option frequently applied in some parts of Europe is FC/TCR compensation, the paralleling of a fixed capacitor with a thyristor-controlled reactor.

15.2.2 Centrally and Dispersed Compensation

The reason why commercial electricity users normally compensate is because some utilities charge for reactive power – not such a high charge as active power but still a significant charge – so that they are compensated for useless use of the distribution system. In some countries, the practice of charging for reactive power is reducing and power factor compensation is becoming less common. Electricity users see this as an advantage, but in fact it places an increased load on a system that is often working quite close to maximum.

The traditional approach is to place one large static compensator at the point of common coupling, the utility entry, and correct the power factor there to the level required to avoid charges, usually $\cos\varphi = 0.90$ or $\cos\varphi = 0.95$. The alternative approach is to disperse compensation near to resistive–inductive loads and, in the extreme, to an individual appliance that draws reactive current.

Centralized compensation is often believed to be cheaper because the central unit is less costly to purchase than spreading the same reactive power rating across the plant in dispersed small units. The installed compensation capacity can also be lower because it can be assumed that not every reactive current consumer will be simultaneously active. However, it must be remembered that reactive currents cause real losses within the installation – the voltage drop in a resistive element, such as a cable, is in phase with the current, so the product, the power loss, is always positive. Central compensation does nothing to reduce these losses, it merely reduces the power factor charge levied by the utility. On the other hand, when compensation is dispersed the total cost of the individual units will be greater than the cost of a single large centralized unit and the total installed compensation capacity will usually be greater – every device has compensation, whether in use or not. Losses are reduced because reactive current flows only between compensation and appliance, rather than back to a centralized compensator at the point of common coupling.

Apart from efficiency, there are technical arguments for and against centralized compensation. For example, if the aggregate load on a transformer is capacitive, the output voltage rises above nominal. This effect is sometimes used to offset the voltage drop in a heavily loaded transformer. The load is simply overcompensated so that the overall load appears capacitive to the transformer, so reducing the inductive voltage drop in the transformer [5]. In cases where a frequently switched heavy load causes a flicker problem, this may be a more sturdy and reliable solution than electronic flicker compensators and may also be considerably more cost efficient in cases where a degree of compensation would be needed in any case.

However, in general terms, the overvoltage of a transformer under capacitive load is a risk that should be avoided or must be adequately dealt with by, for example, using a slightly higher than nominal voltage rating ($\approx 6\,\%$). Sometimes it is necessary or desirable to apply compensation at the MV level and it is attractive to connect LV capacitors via an MV/LV transformer rather than paying the higher price for MV capacitors. In such a case, the transformer load is capacitive and the output voltage higher than expected. This can be dealt with by proper selection of components with adequate voltage rating or by selecting the transformer ratio, by the use of taps, to normalize it. The latter is preferable since it avoids running the transformer in an overexcited state with consequently higher losses. The idea may turn out to be a false economy, because although the installation cost is reduced, the running costs are increased. Reactive current in the installation is transformed twice – from the installation LV to the system MV and from the system MV to the capacitor LV – with two load losses to be paid for by the customer.

The other disadvantages of reactive power, transmission capacity demand and voltage drop, also occur inside the plant on any line and in any transformer between the inductive load and the compensator. It is better to spend 100 % of the budget on 100 % of the use than 75 % of the budget on only 50 % of the use.

In a decentralization scheme, each and every – even small – resistive–inductive load may be compensated by integrating a capacitor into it. This has been done quite successfully for instance in luminaires with one or two fluorescent lamps and magnetic ballasts. In Germany and Switzerland this is frequently implemented as serial compensation, where one out of every two lamp-and-ballast circuits is left uncompensated and the other one is (over)compensated by means of a series capacitor, dimensioned in such a way as to draw precisely the same amplitude of current as the uncompensated branch, but with the inverse phase angle.

Decentralization, however, has its limits in situations where an asynchronous induction machine is individually locally compensated. If the capacitor is located before the motor switch, then it may easily remain connected when the motor is off, leaving the system overcompensated. If the capacitor is located after the motor switch, so it is disconnected from the motor, then there is a risk of self-excitation in the machine as it decelerates. Voltage is generated although the device has been isolated from the supply, even overvoltage in case the capacitance is wrongly dimensioned.

It should be evident at this point that reactive power is not always undesired. Rather, the proper amount of capacitive reactive power needs to be generated to offset the inductive reactive power, and, in cases where ohmic–capacitive loads prevail, vice versa. Capacitive reactive power is also quite advantageous and loss reducing, e.g. for exciting asynchronous generators such as wind turbines and cogeneration plants if these are connected directly to the system without an inverter. It even becomes an absolute necessity in situations where such generators are supposed to feed an island network, otherwise there is no excitation, no voltage and no supply even while the machine is running.

15.2.3 Detuning

Detuning refers to the practice of connecting each compensation capacitor in series with a reactor. One reason for detuning, the attenuation of inrush currents, has already been mentioned. However, the basic reason why detuning is recommended by all compensators suppliers and most utilities – and why many consumers have already adopted it – is the problem of voltage disturbances on the network. Modern electronic loads draw harmonic currents, cause harmonic voltage disturbance and impose high-frequency noise on the network. As the reactance of a capacitor is inversely proportional to the frequency, these high frequencies can cause the current rating of the capacitor to be exceeded. This is prevented by the presence of a detuning inductor. The reactive power rating of the detuning reactor is usually 5, 7 or 11 % of the reactive power of the compensation capacitor. This percentage is also called the detuning factor.

When talking about ratings, there is scope for substantial confusion as to whether the reactive power indicated on the rating plate of a compensator refers to the rated mains voltage or to the rated capacitor voltage (which is higher), and whether or not the detuning factor has been taken into account. In fact, the stated reactive power should always refer to the combined unit – compensator plus detuning reactor – at the supply voltage and fundamental frequency.

As the reactance of a reactor rises proportionally to the frequency while that of the capacitor drops, an 11 % detuning factor at 50 Hz already becomes ≈ 100 % at 150 Hz,[2] meaning that the inductive and capacitive reactances are equal (in resonance with each other) and cancel out. This provides an option to design detuning factors in such a way as to suck out a particular harmonic from the network, while performing the basic compensation function as well. This is described in more detail in Section 15.3. Generally, however, in order to prevent capacitor (and reactor) overload, it is preferable to avoid detuning factors that place the resonant frequency on one of the predominant harmonic frequencies. Rather, the detuning factor is chosen so that the capacitor/inductor combination becomes inductive for frequencies just below the lowest occurring harmonic and above. This avoids resonances (Figure 15.8, Figure 15.9) that otherwise may occur between the capacitor and other elements of the system, especially the stray inductance of the nearest transformer (Figure 15.8), excited by one or another harmonic. In the figures the amplification factors are plotted versus the frequency. The amplification factor here is to be understood as the ratio of the system's behavior as it is, compared to the same system's behavior in absence of the compensator.

But this is not the only reason for detuning. Nowadays capacitors may also easily be overloaded by the higher frequencies omnipresent in the networks, higher than most common harmonics. Even small high-frequency voltages superimposed on the supply voltage – so small that they are not visible in the voltage recordings of a high-class network analyzer (Figure 15.10) – can drive high currents through the capacitors. On the left of this figure is an 11 W fluorescent lamp operated with a magnetic ballast but without compensation. However, the very high amount of reactive power requires compensation by capacitors. On the right of the figure, the lamp system current (the serial connection of lamp and ballast,

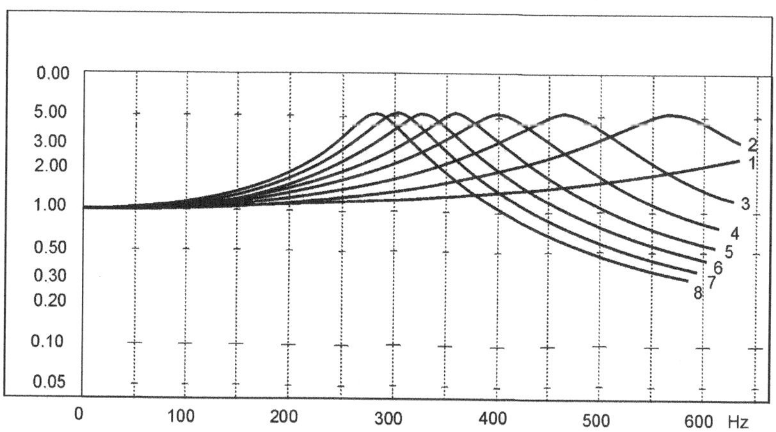

Figure 15.8 Resonance curves of different compensators ranging from 50 kvar (curve 1) to 400 kvar (curve 8), operated on a 1250 kVA transformer (Frako)

[2] X_L at 50 Hz = 11 %, so X_L at 150 Hz = 33 % (w.r.t. X_C at 50 Hz). X_C at 150 Hz = 33 %. Both are equal in magnitude, hence a 'detuning' factor of 100 %.

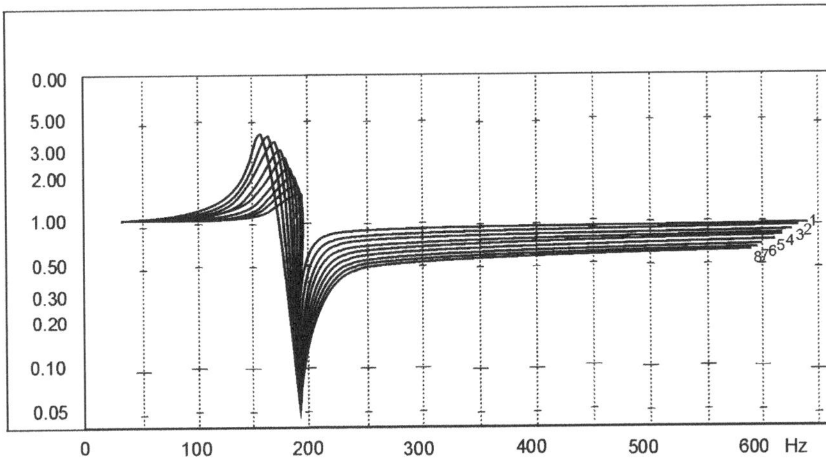

Figure 15.9 Resonance curves of different detuned compensators ranging from 50 kvar (curve 1) to 400 kvar (curve 8), operated on a 1250 kVA transformer (Frako)

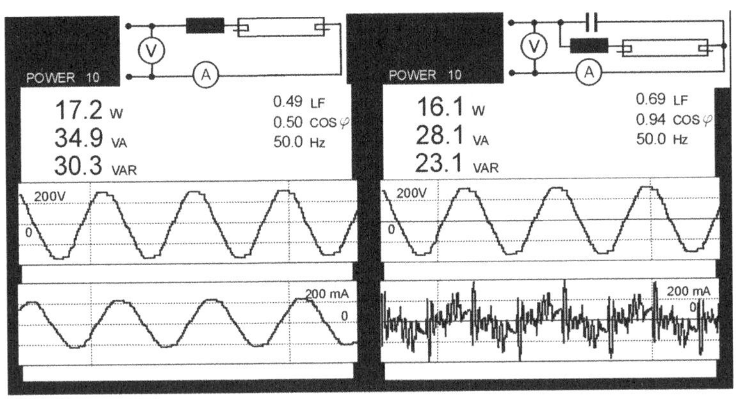

Figure 15.10 Current of fluorescent lamp (11 W) without (left) and with (right) parallel compensation

paralleled with the appropriate capacitor) is a bizarre zigzag rather than an approximate sine wave. This additional mixture of higher-frequency currents must be flowing through the capacitor, since nothing else has changed in the wiring. The measurements confirm this. As the current is almost sinusoidal on the left, the difference between the power factor, also called load factor LF, and $\cos\varphi$ (also called the displacement factor) is small, while on the right it is significant. The reason is that the power factor is the ratio of real (50 Hz) power to apparent power, including fundamental reactive power, harmonic power and noise power, while the good old $\cos\varphi$ only includes the fundamental reactive power caused by a phase shift between voltage and fundamental current. The capacitor, targeting only the latter (left), is nevertheless heavily affected by the former (right) if not detuned. This is the second reason for the widespread detuning practice of today and reveals how important it may be for the life of a capacitor designed for 50 Hz. The experiment can be repeated with similar results in nearly all modern networks. Simply connecting a capacitor to the line voltage and recording the current will give similar readings everywhere. It may be illustrative to make the capacitor current flow through an appropriately dimensioned loudspeaker. The noise is truly awful, but it changes back again to a calm and quiet 50 Hz hum as soon as the capacitor is detuned with a reactor.

The present example also makes the above-mentioned serial compensation practice for fluorescent lamps appear quite advantageous, as it represents a compensating capacitance with a detuning factor of 50 %, and this even with a reactor that is already there and need not be added.

15.3 PASSIVE FILTERS

Fundamental reactive power is always an onerous oscillation of energy. When considering harmonic currents it is not so clear that they can be addressed as a second type of reactive power. Harmonic currents may originate from systems in which there is no energy and where the sign of the (composite) current matches that of the voltage throughout the cycle (e.g. a phase angle controller for an incandescent lamp). The term **wattless current** is sometimes applied to the harmonic current when there are no substantial voltage harmonics of the same order to multiply them with – the product of current and voltage for an individual order being zero. However, harmonic currents have a lot in common with reactive currents:

- They both are undesirable in that they require part of the capacity of generators, cables and transformers, while contributing nothing to the generation and transport of electric energy.
- They both cause additional losses, because the voltage drop is phase related to the current so the product is real and non-zero.
- Harmonics originate mostly from the power-consuming load and flow back to the energy source, against the normal energy flow (Figure 15.11). (An exception is a renewable

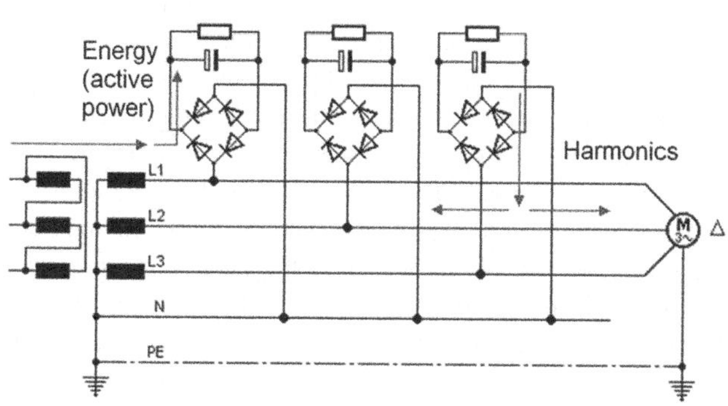

Figure 15.11 Harmonics originate from the load and flow upstream from there

energy source connected to the grid by a power electronics converter, where the harmonics flow from the source.) Fundamental reactive power does not have a defined direction – intake of inductive reactive power is synonymous with output of capacitive reactive power and vice versa.

It should therefore be possible to combat both reactive power and harmonics by similar means. This is indeed the case, as becomes evident considering the following points:

- Voltage wave shapes other than sine, when applied to reactive components, do not result in a current of similar wave shape. Rectangles become triangles, straight lines get bent, and slopes become straight. The reverse is also true because of the above-mentioned proportionality.
- The resistance of an ohmic element, sine or not, a.c. or d.c., is in principle constant if the skin effect is ignored. With inductive components, however, reactance rises proportionally to frequency. In capacitive components, reactance drops proportionally to rising frequency. This has consequences for their behavior under the impact of non-sinusoidal voltage and current waveforms, which both, as mentioned above, deviate from each other. These waveforms can be represented as an infinite multitude of different frequencies (so-called Fourier analysis). This behavior may incur certain risks such as capacitor overload, but can also be taken advantage of by the use of passive filters.

15.3.1 Dedicated Filtering Circuits for Individual Frequencies

A given L and a given C have equal absolute reactances at a well-defined frequency, the so-called resonance frequency (see Section 15.1.2).

Moreover, one of the components has a phase shift of $90°$ and the other of $-90°$, referring to the currents if both components are connected in parallel and to the voltage drops across each of them if they are connected in series. For harmonic filtering, the series LC connection (acceptor circuit) is normally used, while the parallel (rejection circuit) is applicable in only a few special cases. This book considers only the series connection. The two voltage drops (i.e. those across the inductance and capacitance) have a mutual phase shift of $180°$, i.e. inverse polarity. Even without resorting to complex geometry at this point, it is obvious that the L and C reactances in a serial LC filter subtract from each other rather than adding up, or in other words they do add up but have opposite signs, which is the same. At resonance frequency, where the magnitudes of their reactances are equal, they subtract to zero. Therefore an acceptor circuit is practically a short circuit at this particular frequency. Only the resistance, mostly of the reactor winding, remains to be considered, but which is very low compared to the reactance. The magnetizing and eddy current losses in the steel and the dielectric and ohmic losses in the capacitor are normally so low that they do not need to be taken into account as far as the behavior of the filter is concerned. However, all of these losses result in the generation of heat and are an important design consideration; they add up and form the reason for overheating and subsequent failure in overload conditions. Losses also influence the filtering quality; that is to say, the sharpness of separating the wanted from the undesired frequency is a lot better when losses are low.

In order to assess the quality a quality factor is defined, incurring the quotient of reactance by resistance. Note that if one wanted to reduce a certain voltage harmonic by for example more than 50 % with an acceptor circuit, this circuit must have a lower impedance than the network's short-circuit impedance at this specific frequency, since the harmonic current originates from the load and splits between the supplying source and the acceptor filter according to Kirchhoff's laws. Moreover, some producers of reactive power compensators and passive filters point out that in extreme cases the money saved through compensating reactive power is lost instead to active losses in the compensator if customers insist on the cheapest model. After all, the rate for reactive power is normally not as high as for active power. So if a decision is made to have a combined compensator and active filter installed, as hereby suggested, a duplicate profit is achieved by selecting a high-quality model.

Tuned to, for example, 150 Hz (the prevailing current harmonic in European office and commercial environments) whatever currents of that frequency are circulating will use this next-to-zero impedance path and disappear. At the fundamental 50 Hz, however, the C reactance is three times as high and the L reactance is only one-third, so the capacitive reactance is nine times that of the reactor.

In the considered case of a 150 Hz acceptor circuit, the 50 Hz voltage drop across L is 1/8 of the applied voltage. The voltage drop across C is 9/8. These subtract to the full voltage across an acceptor circuit. It is recommended by most utilities, and is indeed compulsory with many of them, to detune reactive power compensating capacitors. Detuning means connecting them in series with a reactor, which makes the capacitor look like an inductive element for harmonic frequencies, while for the fundamental supply frequency it still resembles a capacitor. In the above example only 12.5 % (1/8) of the capacitance is lost at 50 Hz through detuning the reactor. The purpose of detuning is to displace the resonance frequency to a range where no exciting oscillations are expected. This avoids resonance with other inductive components in the network, especially with stray inductance from transformers. That is where the term detuning comes from. Without detuning, resonance will lead to excessive harmonic currents and eventually to excessive voltage drops in the proximity of the transformers affected, since these form an acceptor circuit together with the compensator. The voltage drop across each component may considerably exceed the rated supply voltage or the transformer's no-load voltage. Detuned compensators, however, appear as inductive components to higher frequencies. The excessive voltages remain inside the compensator cubicle, say across the capacitors designed for these voltage values, but at its outside terminals no resonance can appear.

15.3.2 Central and Dispersed Solution

A further question accompanying the selection of the right model is the Y or Δ connection. Compensators usually come in Δ connection. For a passive filter this design provides only part of the effect, since the most essential harmonics in office environments flow between phase and neutral. There may also be some intermediate solutions with the capacitors connected in Δ but designing the detuning reactors as three-phase neutral reactors. Suppliers should be able to advise which design is best for the system.

Under no circumstances can the presence of any filtering equipment be used as a pretext to return to old TN-C wiring practice and downsized neutral conductor cross-sections.

It remains to be considered that filters will first of all provoke an increase of TRMS (True Root Mean Square) current between the source of harmonics and the filter. Any additional current here provokes additional active and reactive voltage drops, additional energy losses and in TN-C systems an increase of operating currents flowing along extraneous conductive parts. Only the sum of the load and filter currents is lower than the load current alone without a filter. The load current alone will be larger in the presence of a filter than without any filter operating in the proximity. Therefore the more dispersed the solution is, the more costly it is, but also the more efficient.

This goes for any kind of filter. Active harmonic conditioners suffer from a heavy cost disadvantage compared to passive filters. The smaller the unit, the more cost intensive it is. The price ratio shifts more or less from 1/2 for large filters to 1/3 for small units. Apart from that, their overwhelming effectiveness is sometimes doubted [1] Active filters make sense if integrated into UPS (Uninterruptible Power Supplies) used to bridge outages and also dips and sags. For this, the passive filter cannot replace the active one. But for a rough cleaning of waveforms, which is often sufficient, the passive filter is a much more cost effective, reliable and, if properly designed, lower-loss device. It offers the opportunity for dispersed installation. This avoids the unwelcome surprises that can occur through central filtering when harmonic currents – and the harmonic voltage drops they cause – increase, due to the installation of a large central unit.

But when decentralizing, care has to be taken to do so thoroughly and to connect detuned compensators (resp. passive filters) electrically remote from each other so that they can no longer be seen as strictly parallel. Even filters with equal ratings are never absolutely identical on account of tolerances and potentially different operating temperatures. So out of two filters with a rated resonance frequency of 250 Hz each, one may in fact resonate at 248 Hz and the other at 252 Hz. At 250 Hz the former appears capacitive and the latter inductive, and together they form an approximate or even perfect rejection circuit, which has the opposite of the desired effect. Moreover, a 250 Hz current will circulate between the two and may overload both of them simultaneously. Or, if one of the filters happens to hit 250 Hz precisely and the other one resonates at, say, 254 Hz, then the larger share of the 250 Hz pollution will use the former and may overload it, while the latter one is idling. Unfortunately this effect will be the more extreme, the higher is the mentioned quality factor. After all, a higher quality factor of an acceptor/rejection circuit means nothing more than a steeper decline/incline of impedance while approximating the resonance frequency. Therefore some impedance is needed between two units in order not to parallel them directly.

15.3.3 L/C Ratio

Yet, setting the resonance frequencies while detuning the compensators to one of the harmonics occurring in the network is usually avoided because compensators may easily be overloaded. The rating of the reactors is normally given as a percentage of the rated reactive power of the capacitors at 50 Hz. For example, a 5 % detuning rate means that 1/20 of the voltage drops across L and 21/20 drop across C, subtracting to 100 % overall. At 20 times the frequency, say 1000 Hz, the ratio would be reversed, so the resonance frequency where X_L and X_C are equal lies in the middle between these two frequencies, to be precise at $50\,\text{Hz} \cdot \sqrt{20} = 224\,\text{Hz}$.

Another common value, 7 %, yields a resonance frequency of 189 Hz, thus avoiding an approximate short circuit for any harmonic. In fact, such short circuits short out not only a harmonic within the premises, but also the equivalent one coming over from the neighboring premises. Therefore, to operate such a filter in an installation close to another one without it, it is necessary to oversize it. Apart from that, oversizing will not only avoid unforeseen overload but also improve the filter quality – that is, separate the desired from the undesired frequencies more sharply. On top of that it will reduce energy losses.[3]

For each frequency there is an infinite number of *LC* pairs with the same resonance frequency. The ratio of the capacitive reactive power by the harmonics' cleaning capability can be varied, but once the selection is made it is fixed forever after. Adding to this is the fact that fundamental capacitive reactive power cannot be selected to be zero. There will always be some. This is the disadvantage of passive filters. One hundred and fifty hertz and 250 Hz model filters, for example, drew 50 Hz currents of 100 mA, resp 37 mA. This is quite low compared to the measured harmonic currents, due to the fact that these filters were designed with much *L* and little *C*. A solution may be to arrange the filters in smaller groups and switched individually. This is done with controlled compensators anyway and when there is less need for fundamental compensation, there will also be less urgency to filter harmonics. Consideration should also be given whether to shut off the filters for the higher harmonics when less compensation is needed, as suggested in Figure 15.12. Though this is not a

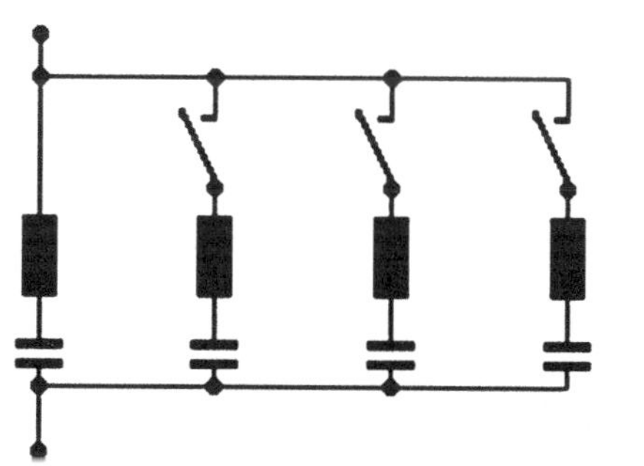

Figure 15.12 Combination of acceptor circuit filters

[3] Active harmonic conditioners are normally also operated as mains parallel (as shunts). Yet, the situation is slightly different. These electronic devices meter the current in a network, subtract the desired ideal sine wave from it and feed a negative image of this difference into the network, as far as their rating allows. If metering is done on one specific load, no more and – within the constraint of rating – no less than the pollution of this load is compensated for so that the power quality in this network remains unaffected, at least by this specific load. If this load is inactive then the active filter will also be inactive, although it could be used in the meantime to provide a better power quality. The passive filter, however, is in a way always active in that it is always on the alert, waiting for 'its' harmonic to show up. Thanks to electronic monitoring and control, overload of active electronic filters cannot occur. What will happen instead is an imperfect, partial cleansing effect when pollution exceeds the conditioner's ratings. Passive acceptor circuits tuned to resonance frequencies of, for example 150 Hz (11 % of detuning reactance) or 250 Hz (4 % of detuning reactance), accept any amplitude of the third and the fifth current harmonic, as the name suggests, well into the overload range. This depends on the amount of harmonics found in the mains and does not depend on the impact of one specific load. For this reason they should be generously dimensioned. This is not a cost issue when compared to active conditioners.

perfect solution, it is a very cost-effective one. The passive filter we are talking about is nothing more than a modified design or adequate selection of a compensator that is needed anyway. When applying this method, however, it must be ensured that the cut-out is done top down (from right to left in Figure 15.12.). Otherwise, one or the other of the higher-frequency acceptor circuits may resonate with an inductive network element at one of the lower harmonics.

15.3.4 Sound Frequency Signals

Individual utilities use different sound frequencies to control street lighting, night storage heating, and other systems for controlling the load in their system (demand side management). Care must be taken not to short out these signals and make them ineffective. The closer the signal frequency comes to the resonance point of an acceptor circuit, the lower the impedance of that circuit is at that particular frequency. It is better to use parallel *LC* filters, the so-called rejection circuits, in series with the serial filter arrangement (as shown in Figure 15.13 with a utility that uses 183.3 Hz signals, 13/3 of the mains frequency). Rejection circuits have the inverse characteristics compared to acceptor circuits. The equation for the resonance frequency is the same, so when the same components are used, the resonance points are equal. But note that the meaning of resonance is the opposite: impedance assumes a maximum in a rejection circuit, theoretically reaching infinity if no losses were incurred.

For case studies see web address

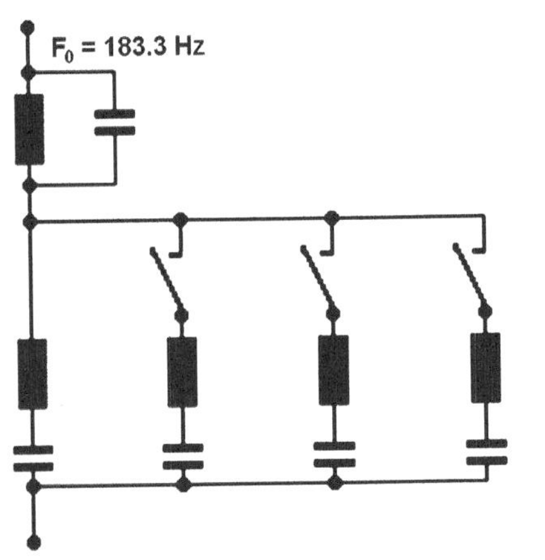

Figure 15.13 Combination of acceptor circuit filters with rejection circuit against losing sound frequency signals

BIBLIOGRAPHY

[1] Clewing M., Statische USV im Leistungsbereich unter 6 kVA, etz. *Elektrotechnishe Zeitschrift*, vol. 3–4, p. 26, 2000.
[2] Fender M., Vergleichende Untersuchungen der Netzrückwirkungen von Umrichtern mit Zwischenkreis bei Beachtung realer industrieller Anschluss-Strukturen, PhD Thesis, Wiesbaden, 1997.
[3] IEC 60076-1, *Power transformers. Part 1 General*, April 2000.
[4] IEEE 519, *Recommended Practices and Requirements for Harmonic Control in Electrical Power Systems*, 1992.
[5] Just W., Hofmann W., *Blindleistungs-Kompensation in der Betriebspraxis*, Fourth Edition, VDE Verlag, Offenbach, 2003.

16

Distributed Generation and Power Quality

Vu Van Thong and Johan Driesen

The electric power system is traditionally designed and operated to transport large amounts of power top down from the generation units, through the transmission and distribution networks, and finally to the electric energy consumers. Distribution systems are traditionally *passive* and designed to operate with only such unidirectional energy flow, in contrast to the transmission system, which is designed for two-way power flow. In planning and operating, it is assumed that electric power always flows from the secondary winding of the transformers in the substations to the end of feeders. With the introduction of distributed generation [1],[16], the distribution system becomes an *active* system with both energy generation and consumption at the formerly exclusive load nodes. Bidirectional power flows should now be incorporated in the hierarchical network design and its operating criteria [22].

Distributed generation (DG), or embedded generation, is a concept of installing and operating small electricity generators connected directly to the distribution network or at the customer side, with a rated power typically less than 10 MW [1],[3],[16],[22]. When also considering storage systems and even controllable loads, the name distributed energy resources (DER) is often used.

The different DER technologies can be divided through the primary resource used:

- **Fossil-fuel-based DG**: diesel or gas engines, gas turbines, Stirling engines, fuel cells – these technologies are often used in combined heat and power (CHP) systems.
- **Renewable-based DG**: wind turbines, photovoltaic panels, biomass (biofuels or biogas for engines and turbines), small hydro turbines, geothermal energy, tidal and wave generators.
- **Storage systems**: batteries, flywheels, supercapacitors, superconducting coils, reversible fuel cells, etc.

Handbook of Power Quality Edited by Angelo Baggini

These technologies are entering an era of rapid expansion and commercialization. The idea behind DG is to provide electricity to a customer at a reduced cost and more efficiently with lower overall losses than in the traditional utility central generating plant involving transmission and distribution losses. Additionally, local resources including renewable energies can be addressed.

DG may have a significant impact on the power flow, voltage profile, voltage stability, power system protection and power quality experienced by customers and electricity suppliers. This impact may manifest itself either positively or negatively, depending on the distribution system, distributed generator and load characteristics [3],[21],[24],[25].

Power quality has become a real problem over the last decade due to the ever-increasing use of power electronics and sensitive load equipment. The addition of DG can have a significant effect as well and surely increases the complexity of this problem [12]. The power injection from DG can raise the steady-state voltage level [13],[27]. Inverter-connected generators may inject non-sinusoidal currents into the network and cause additional harmonic distortion. The fluctuation of power injection of some DG technologies such as wind turbines and photovoltaics or heat-demand-driven CHP may cause voltage fluctuations [6],[27]. However, the connection of DG may improve the reliability of the distribution system [2]. Optimal placement of DG units may reduce the power losses and can improve the voltage profile of the system [10],[21].

In this chapter, the local impact of DG on the distribution system operation and control is addressed. Firstly, general points of DG and its possible impacts are discussed. Then, case studies with simulation results illustrate the subject. Modeling of the distribution network, distributed generators and loads is performed for both steady-state and dynamic situations, with only synchronous and induction machines used in DG technologies taken into consideration in the dynamic studies. The DG units are assumed to connect to an existing Belgian distribution network. Different DG technologies and load characteristics are studied.

16.1 DISTRIBUTION NETWORK MODELING

16.1.1 Steady-State Modeling

16.1.1.1 Transformer and Distribution Line Modeling

The distribution lines and transformers are represented by a general series model containing resistances and reactances [2],[7]. The shunt capacitances of the distribution lines are neglected as the feeders are short and distribution voltage level is used. The magnetizing current of transformers is small compared to normal load current and its magnetizing shunt impedance is large, hence negligible.

16.1.1.2 Static Load Modeling

Loads are modeled statically as active and reactive power consumers connected to the distribution network at any instant in time as functions of voltage magnitude and frequency [9],[14]. In the case of a small penetration of DG connected to the grid, the frequency of the power system is assumed to remain constant.

Loads are divided into industrial, commercial and residential types. Each load is made up of a number of subloads categorized by their voltage dependency. Accordingly, the load is divided into constant impedance, constant current and constant power loads. The total load at a node i is

$$S_i = \sum_{k=1}^{n} (P_{ik} + jQ_{ik}) \tag{16.1}$$

Each subload P_{ik} and Q_{ik} of n subloads at node i is described as a function of the node voltage U_i:

$$P_{ik} = P_{iko}\left(\frac{U_i}{U_{io}}\right)^{\alpha} \tag{16.2}$$

$$Q_{ik} = Q_{iko}\left(\frac{U_i}{U_{io}}\right)^{\beta} \tag{16.3}$$

where P_{iko}, Q_{iko} are power values at rated voltage U_{io}; α, β are constant values depending on the type of subload: $\alpha = \beta = 0$ represents a constant power load; $\alpha = \beta = 1$ represents a constant current load; and $\alpha = \beta = 2$ represents an impedance load.

16.1.2 Dynamic Modeling

16.1.2.1 Synchronous Generator

Modeling of synchronous generators is done according to Park's theory [8],[14]. The rotor is represented by the excitation winding in the direct axis and no windings along the quadrature axis. The generator is controlled through an automatic voltage regulator (AVR) to control the reactive power with limited excitation current and a governor to control the active power. The AVR is very important in islanding mode and when a very large number of DG units are connected to the grid. In that case DG has potential to play a larger role in contributing to ancillary services.

16.1.2.2 Induction Generator

The widely used technology for rotating DG is the induction generator, due to its construction simplicity and low costs. An induction generator in principle is an induction motor with torque applied to the shaft with some modifications to optimize its performance as a generator. It consists of a wound stator and generally a squirrel-cage rotor [12]. The full Park's model of an induction generator is used in which the dynamics of the rotor fluxes are calculated [8]. The rotor transient contributes to the short-circuit power during voltage dips and swells during disturbance.

16.2 POWER QUALITY AND DG

16.2.1 Voltage Rise

The active power produced by DG units increases the steady-state voltage, and the reactive power produced or consumed by DG can further increase or reduce the voltage rise depending on the type of DG technology. The value of voltage rise ΔU at the DG connection point of a radial feeder is approximately [23]

$$\frac{\Delta U}{U} \cong \frac{(P_{dg} - P_{Lj})R_{ij} + (Q_{dg} + Q_{Lj})X_{ij}}{U^2} \tag{16.4}$$

where P_{dg}, Q_{dg} are the real and reactive power of the generator; P_{Lj}, Q_{Lj} are the real and reactive power of the load; U is the line voltage at the DG connection point; and $Z_{ij} = R_{ij} + jX_{ij}$ is the impedance between the main substation and the DG connection point.

The synchronous generator can produce or absorb reactive power, but the induction generator only consumes reactive power. DG with inverter interface can change the reactive power output within a limited range. These outcomes, in combination with the system's R/X ratio or the distribution network characteristics and the load profiles, determine whether the voltage level at the connection point increases with a higher power production by DG or not.

16.2.2 Voltage Dips

A voltage dip may happen when opening a branch, starting up a motor, switching on a generator of a DG system away from the synchronous speed [5], energizing a transformer or when a short circuit occurs. Voltage dips are not desirable and may lead to malfunctioning of customer's equipment, especially with sensitive loads such as computers, process controllers and adjustable-speed drives. DG units may also cause a voltage dip in the system that may be experienced by other customers, such as a CHP unit that is often being switched on and off for economic reasons [21], or an induction-generator-based DG unit starting up, consuming reactive power for magnetization [12].

However, DG may also be a victim of voltage dips. The voltage dip may lead to a triggering of the islanding protection of DG, resulting in a cascade of undervoltage tripping of DG units when this propagates from the transmission system to the low-voltage grid. Tripping implies that the energy flow from the unit to the grid is interrupted immediately. The power input on the chemical side or prime mover of the unit may take some time to stop. The result is an energy surplus resulting in overvoltages or overspeed in the case of rotational machines. Obviously, safety measures are needed [5].

16.2.3 Voltage Fluctuations and Flicker

Rapid voltage variations cause a light to flicker, which affects the human eye. Conventional voltage flicker has traditionally been a concern when connecting large fluctuating loads such as arc furnaces. However, it is of considerable significance for some DG technologies as well.

The variation of the wind speed and the tower shadow effect of wind turbines may produce power variations. Moving clouds may cause the power output of photovoltaic systems to fluctuate. The operational modes of CHP are mostly based on the rather uncontrollable customer-driven heat demand, resulting in switching on or off of a CHP unit. These power variations cause voltage fluctuations in the power system, which in turn may cause voltage flicker. Connection and disconnection of induction generators also causes flicker emissions due to the magnetizing current.

On the one hand DG units may cause flicker problems, but on the other hand the connections of rotating DG units may raise short-circuit currents, resulting in an additionally mitigating flicker ability of the power system.

To evaluate the flicker impact caused by power fluctuations of a load or a DG unit, the flicker emission P_{lt} can be measured or calculated as in

$$P_{lt} = c_c \frac{S_{dg}}{S_{pcc}} \qquad (16.5)$$

where c_c is the capability of equipment to produce flicker; S_{dg} is the unit rated power; and S_{pcc} is the short-circuit power at the point of common coupling.

Many wind turbine manufacturers can provide information on the flicker coefficient, which is measured only once and then normalized, so that it becomes applicable in general [13].

The IEC 61400-21 standard [15], the standard applying power quality requirements for grid-connected wind turbines, recommends limiting the flicker emission from a single wind turbine to $P_{lt} = 0.25$. It also recommends limiting the flicker due to the total amount of wind turbines in a medium-voltage network to $P_{lt} = 0.5$ at any node.

16.2.4 Harmonics

DG may introduce harmonics and influence the harmonic performance of distribution systems. Like other end users, DG units have to limit harmonic current injections below the maximum allowable level. Standards such as IEEE 519-1992 impose limits for different order harmonics and the total harmonic distortion (THD) (Table 16.1). The limits for system voltage distortion are 5 % for THD and 3 % for any individual harmonics.

Table 16.1 Harmonic current injection limits of IEEE 519-1992

< 11th	4.0 %
< 11th to < 17th	2.0 %
< 17th to < 23rd	1.5 %
< 23rd to < 35th	0.6 %
< 35th or higher	0.3 %
Total harmonic distortion	5.0 %

DG units using power electronics converters may inject harmonic currents in the network. The type and severity depend on power electronics inverter technology and its interconnection configuration. Rotating generators can also be sources of harmonics, depending on the winding design, non-linear magnetizing (saturation), grounding, etc. [3]. They also alter the harmonic impedance of the network. In case triple harmonic currents are present, they add up in the neutral conductor of generators, when the neutral (star) point is directly grounded.

Another problem that might arise with a large penetration of DG is the occurrence of resonances of distribution system capacitance and reactance of DG units. The capacitors used to improve the power factor of induction generators and the capacitance of cables may be involved in series or parallel resonances with generator and transformer coils, which might cause amplification of harmonic distortions [5].

In general, harmonics injected by DG tend to become less severe due to the development of inverter technologies and mitigation methods. Most new inverters are capable of generating very near to a sine wave due to the high switching frequency [26].

16.2.5 Unbalance

The connection of single-phase DG units, such as small photovoltaic systems, to the low-voltage grid may generate an unbalance. The voltage unbalance might cause overheating of induction motors and induction generators, as induction machines have low negative-sequence impedances. Synchronous machines and electronic inverters are sensitive to voltage unbalance as well.

16.2.6 Direct Current

D.C. injection is an issue because of the economics of magnetic component design. The increased d.c. voltage has the potential to increase saturation of magnetic components, such as the cores of distribution transformers. This saturation, in turn, causes increased power system distortion [20]. There is a concern that inverters without a transformer interface may inject sufficient direct current into distribution circuits to cause the distribution transformer core to saturate. Following IEEE 1547, direct currents injected by DG must be smaller than 0.5 % of its rated current at the connection point [20]. As an example, in the Belgian technical guidelines, direct currents must be smaller than 1 % of the rated current.

For a case study see web address

BIBLIOGRAPHY

[1] Ackermann T., Andersson G., Soder, L., Distributed generation: a definition. *Electric Power System Research*, vol. 57, pp. 195–204, 2000.

[2] Arrillaga J., Watson N. R., *Computer modelling of electrical power systems*, John Wiley & Sons, Ltd, Chichester, 2001.

[3] Barker P. P., De Mello R. W., Determining the impact of distributed generation on power systems: Part 1 – Radial distribution systems. PES Summer Meeting, IEEE, Vol. 3, pp. 1645–1656, 2000.

[4] Belmans R., Michiels W., Vandenput A., Geysen W., Transient torques in wind turbine driven induction generators. Symposium on Electric Power Systems in Fast Developing Countries, Saudi Arabia, pp. 563–567, 1987.

[5] Bollen M. H. J. and Hager M., Power quality: integrations between distributed energy resources, the grid, and other customers. *Electrical Power Quality and Utilization Magazine*, vol. 1, no. 1, pp. 51–61, 2005.

[6] Brown R. E., *Electric power distribution reliability*, Marcel Dekker, New York, 2002.

[7] Chen T. H., Chen M. S., Inoue T., Kotas P., Chebli E. A., Three-phase cogenerator and transformer models for distribution system analysis. *IEEE Transactions on Power Delivery*, vol. 6, no. 4, pp. 1671–1681, 1991.

[8] *Eurostag, package document, part I*, Tractebel & Electricité de France, 2002.

[9] Grigsby L. L., *The electric power engineering handbook*, CRC Press & IEEE, Boca Raton, FL, 2001.

[10] Haesen E., Espinoza M., Pluymers B., Goethals I., Vu Van T., Driesen J., Belmans R., De Moor B., Optimal placement and sizing of distributed generator units using genetic optimization algorithms. *Electrical Power Quality and Utilisation Journal*, vol. xi, no. 1 pp. 97–104, 2005.

[11] IEEE P1547, *Standard for Interconnecting Distributed Resources with Electric Power Systems*, 2003.

[12] Jenkins N., Allan R., Crossley P., Kirschen D., Strbac G., *Embedded Generation*, IEE, London, 2000.

[13] Jensen K. K., Guidelines on grid connection of wind turbines. CIRED International Conference on Electric Power Distribution Engineering, France, June 1999.

[14] Kundur P., *Power system stability and control*; McGraw-Hill, New York, 1994.

[15] *Measurement and assessment of power quality characteristics of grid connected wind turbines*, IEC Standard, CIE/IEC 61400-21, 2001.

[16] Pepermans G., Driesen J., Haeseldonckx D., Belmans R., D'Haeseleer W., Distributed generation: definition, benefits and issues. *Energy Policy*, vol. 33, no. 6, pp. 787–798, 2005.

[17] Ranjan R. and Venkatech B., Voltage stability analysis of radial distribution networks. *Electric Power Components and Systems*, vol. 31, no. 5, pp. 501–511, 2003.

[18] Scott N. C., Atkinson D. J., Morrell J. E., Use of load control to regulate voltage on distribution networks with embedded generation. *IEEE Transactions on Power Systems*, vol. 17, no, 2, pp. 510–515, 2002.

[19] *Technical requirements for connection of dispersed generating systems operating in parallel on the distribution network*, Document C10/11 of the FPE/BFE, Belgium, 7 May 2002.

[20] The NRECA Guide to IEEE 1547, *Application guide for distributed generation interconnection*, 2006.

[21] Vu Van T., Impact of distributed generation on power system operation and control. PhD Thesis, Katholieke Universiteit Leuven, 2006.

[22] Vu Van T., Belmans R., Distributed generation overview: current status and challenges. *International Review of Electrical Engineering (IREE)*, vol. 1, no. 1, pp. 178–189, 2006.

[23] Vu Van T., Driesen J., Belmans R., Impacts of embedded generation on voltage stability of distribution system. Universities Power Engineering Conference, Thessaloniki, Greece, September 2003.

[24] Vu Van T., Driesen J., Belmans R., Interconnection of distributed generators and their influences on power system. *International Energy Journal*, vol. 6, no. 1, part 3, pp. 127–140, 2005.

[25] Vu Van T., Driesen J., Belmans R., Power quality and voltage stability of distribution system with distributed energy resources. *International Journal of Distributed Energy Resources*, vol. 1, no. 3, pp. 227–240, 2005.

[26] Woyte A., Design issues of photovoltaic systems and their grid integration. PhD Thesis, Katholieke Universiteit Leuven, 2003.

[27] Woyte A., Vu Van T., Belmans R., Nijs J., Voltage fluctuations on distribution level introduced by photovoltaic systems. *IEEE Transactions on Energy Conversion*, vol. 21, no. 1, pp. 202–209, 2006.

17

Electricity Market

Pieter Vermeyen and Johan Driesen

With energy markets being liberalized, companies and private consumers are free to choose an electricity retailer. Competition arises between different power retailers, which are forced to offer electric energy and good services at very keen prices. In order to create this situation, traditional power companies have been unbundled in independent entities and new players are allowed to enter the market. This has rendered the electricity market a complex affair. In this chapter the different market players are discussed briefly, an overview of contract types between retailers and consumers is given, load management in the electricity market is presented, and the evolution of power quality in the electricity market is proposed.

17.1 MARKET PLAYERS

In the past, most European countries had an electricity system controlled or dominated by a single company, *de facto* resulting in national monopolies. As these are assumed to be inefficient markets, they have to disappear, according to European legislation, and new companies are to be allowed to enter the electricity market for the activities of electricity production, supply and retail.

The different players or parties in the electricity sector are the following:

- Electricity producers.
- Grid operators: transmission system operators and distribution system operators.
- Retailers or suppliers which sell and supply electricity to consumers.
- Regulators: supervisors of the electricity market.

Handbook of Power Quality Edited by Angelo Baggini

17.1.1 Electricity Producers

Electricity producers own production capacity (e.g. wind farms, fossil-fuel-fired power plants, nuclear power plants, photovoltaic arrays, etc.), which they use to transform primary energy sources such as gas, coal, oil, biomass, uranium, wind, sunlight and hydro power into electric energy. This energy is sold to electricity retailers, which in turn sell it to electricity consumers, e.g. factories, hospitals, railways, supermarkets and residential consumers. Electric energy can be consumed in the same country it is generated in, but this is not a necessity. International electricity trade is an important activity, resulting in significant cross-border power flows.

17.1.2 Grid Operators

A grid operator manages a certain part of the electricity grid and is responsible for its operation, maintenance and development. Since there can only be one electricity grid in an area, grid operators have a natural monopoly. Grid operators should be independent of producers and retailers of electricity. Two kinds of grid operators exist: transmission system operators and distribution system operators.

17.1.2.1 Transmission System Operators

A transmission system operator (TSO) is in control of a section of the transmission system. Transmission systems are grids at high voltage, in general having a meshed structure, and used for transporting power over long distances, from large power plants to large industrial consumers and distribution grids. National transmission systems in Europe are interconnected.

The main tasks of a transmission system operator are:

- Maintenance and development of the grid and managing connections with other grids and large industrial consumers.
- Providing access to the transmission grid to producers and large consumers.
- Providing services related to transmission.
- Watching over the functioning of the grid.
- Managing the balance of production and consumption of electricity.
- Taking the initiative to improve the functioning of the electricity market.

17.1.2.2 Distribution System Operators

A distribution system operator (DSO) is in charge of the operation, maintenance and development of distribution grids. This part of the grid, traditionally in a radial topology, connects the end consumers of electricity at lower voltages. At the request of electricity retailers, they transport electricity to the consumers.

The main tasks of a distribution system operator are:

- Construction and operation of the distribution grid.
- Connecting new consumers to the grid, updating existing connections, installing electricity meters.

- Solving problems and malfunctions.
- Managing technical data of the grid and the consumers.

17.1.3 Electricity Retailers

Suppliers or retailers buy electric energy from producers and sell it to consumers. For their transactions they use the transmission grid and the distribution grids. They pay a certain tariff to the grid operators for the use of the grid. Depending on the characteristics of their customers (e.g. industrial or residential consumers), retailers offer a variety of contracts for energy supply, often including additional services. This is discussed in Section 17.2, concerning contract types.

17.1.4 Regulators

Regulators are independent bodies, appointed by governments to watch over the electricity market and to ensure that all activities occur in a fair and efficient way. They are also responsible for issuing supply permits and approving new production installations. Depending on national policy, regulators can also be in charge of support systems, such as stimulating the use of renewable energy and rational use of energy.[1]

Typical tasks of a regulator are:

- Verifying the independence of transmission and distribution system operators (TSOs and DSOs).
- Verifying that TSOs and DSOs perform their tasks in a proper way.
- Watching over compliance with regulations and permits by all grid users.
- Managing permit applications for the construction of new production and transmission infrastructure.
- Approving electricity prices and tariffs related to the use of infrastructure.
- Advising the government on electricity-related matters.
- Appointing grid operators (TSOs and DSOs).
- Acting as a mediator in disputes between grid users or market players.
- Verifying compliance with regulations and policy concerning renewable energy, generation of combined heat and power (CHP), rational use of energy and power supply in case of emergencies.

17.2 CONTRACT TYPES

In the restructured electricity market, retailers and producers have to be competitive in order to attract business. Because the relation between retailers and consumers is most significant to consumers in economic terms, this relationship is the subject of this section. Different possible contract types defining this relationship are discussed here.

[1] Rational use of energy is discussed in Chapter 19, including the impact of rational use of energy on power quality.

In order to attract customers, electricity retailers offer a variety of contracts, designed for different categories of consumers. Contracts vary from basic electricity delivery to a portfolio of energy and services. A contract can imply a fixed electricity tariff for a long period of time, offering the consumer price security and requiring little effort and interaction with the consumer. This type of contract is typical for residential consumers. On the other hand, large industrial consumers are interested in flexible, negotiated contracts, with prices varying in quasi-real time, requiring the consumer to follow up the price and consumption. This allows consumers to minimize their energy costs.

17.2.1 Residential Customers

The basic contract for residential consumers consists of supply of electricity at a certain tariff. The retailer can offer multiple tariffs, depending on the time of consumption, such as day and night tariff or weekend tariff. Consumers which wish to use renewable energy can opt for green electricity at a higher rate.

In order to differentiate from other retailers, certain services are offered to residential consumers. Certain retailers have a website with a section for customers, in which they can consult their account information. Here digital tools can be provided, offering personalized advice and tips for reducing energy consumption and indicating potential savings. Retailers also provide information concerning existing subsidies and tax benefits for measures for reducing energy consumption. Another service offered by certain retailers is a quick repair service in case of a malfunction in the consumer's electrical installation. Depending on the contract, the consumer can be entitled to, for example, two repairs per year free of charge.

In certain countries, electricity can be bought using prepaid cards. In this way, electricity is paid for before it is consumed. This system avoids the problem of consumers not paying their electricity bill because the retailer receives payments in advance. Because the administration for this system is minimal and there is no need for the electricity meter to be read, costs are reduced. There is also no need for disconnecting and reconnecting consumers. It places consumers in full control of their energy consumption and budget.

17.2.2 Small Businesses

For small businesses, e.g. shops, electricity retailers offer personalized contracts, taking into account the nature of the business and its consumption profile. The manager of the business can choose to buy electricity from renewable energy, at a higher rate.[2] A fixed electricity price can be offered, resulting in predictable energy costs over a certain period of time. Other service elements can be a quick repair service and maintenance of certain electrical equipment, e.g. climate control.

Another possible service is the energy audit. Energy measurements are conducted in the building and equipment and thermal insulation are inspected. Afterwards a report is

[2] A business manager may choose to pay more for electricity from renewable sources in order to help achieve a cleaner environment, or to make the business appear more environmentally-friendly, in order to attract more customers.

delivered to the business manager, with recommendations for energy savings, a description of relevant subsidies and a calculation of expected savings. Additional benefits can be added, such as a discount on the audit cost if the customer carries out the recommended improvements.

17.2.3 Small and Medium-Sized Companies

The larger a company's energy consumption, the more interesting it is for a retailer to attract or keep it as a customer. Therefore small and medium-sized companies are offered contracts tailored to their needs. Next to competitive electricity rates and the possibility to use renewable energy, extensive services are offered. The energy audit of a company's infrastructure is the most important element. It is also possible to have the retailer conduct continuous monitoring of a company's energy consumption, combined with regular contacts with energy experts. These experts can also provide advice concerning upgrades in electrical installations when the need for power increases. Other possible advice concerns lighting of buildings and workshops and heating, ventilation and air-conditioning (climate control). The possibility exists to offer a company access to its consumption and billing history on the retailer's website.

A loss of power has serious economic and technical consequences for most companies. In collaboration with insurance companies, power retailers can offer blackout insurance, covering damage and costs due to blackouts. Instead of compensating financial losses due to loss of power supply, the power retailer can assist in the design of uninterruptible power supplies (UPS), to prevent loss of supply in the first place.

High reliability of supply is one aspect of power quality. Other aspects can be taken care of by the supply company as well. Because good power quality is important to a company, offering power quality audits and carrying out projects for improving power quality are interesting business opportunities.

17.2.4 Large Companies

Large industrial companies are offered contracts tailored to their needs. For industrial sites it can be advantageous to have on-site generation capacity, which can be controlled according to the company's need for electricity. This is particularly interesting if certain production processes require heat. In this case, units for CHP production can be used, resulting in a higher overall energy efficiency. For the construction and operation of such units, collaboration with the supplying company can be considered. Next to flexible contracts and competitive electricity rates, extensive services are offered.

Power retailers can offer large companies energy audits and power quality audits of the entire company or concentrated on certain areas or sensitive processes. Supply companies can offer training courses for employees concerning rational use of energy, energy savings and power quality. In this way, a company's employees can be actively involved in projects concerning energy savings and power quality.

Excelis is a power quality service for companies, provided by the French power company Electricité de France (EDF) [5]. It consists of identification of the origin of disturbances, estimation of the financial damage due to the disturbances and the proposal of possible solutions. The following is an example of the implementation of this service. It was hired by the pharmaceutical company Pierre Fabre to secure power supply for a production plant situated in Pau (France), in which 135 persons were employed at that time. An investigation was necessary in order to prevent costly interruptions of critical production processes due to very short voltage interruptions. A new configuration for the plant's distribution network was proposed. In the new design critical production lines are connected to a high-quality grid, supplied by converters. The payback time of the investment was 18 months. The high-quality grid does not cover the entire plant. This would have cost five times more than the design that was implemented.

17.3 LOAD MANAGEMENT IN THE ELECTRICITY MARKET

In the electricity market, it is beneficial to all market players, including the consumers and even the environment, if the utility or retailers can exert a certain influence on the consumption behavior of consumers. This influence consists of incentives to consumers in order to decrease or defer energy consumption at certain times or under certain circumstances. One element is peak load management. Its goal is to reduce peak consumption and to level out daily and weekly consumption profiles. Managing or controlling loads can also be required during certain emergencies (load shedding). In the next subsections, the reasons for load management are explained, as well as possible techniques, consisting of tariff systems, contractual obligations, forced interruptions and intelligent power systems [6].

17.3.1 Reason for Load Management

At times when electricity consumption is at a peak, the marginal costs of electricity production are very high. In power generation, the least expensive power plants are always used first; they are used to supply the base load. Because consumption is not constant, standby units are necessary. Standby capacity can consist of gas turbines or diesel engines, which are inactive until additional power is required (off-line reserve), or it can be spinning reserve. This is stand-by capacity consisting of generators operating at partial capacity. When more power is required, the output of these generators is increased.

Generation units for off-line and spinning reserve are generally more expensive because they are rarely operated at optimal capacity, resulting in low energy efficiency. This also leads to higher emission of greenhouse gases and polluting substances. Because of the high cost of supplying additional power during peak load, consumers pay a high price. If peaks can be limited by reducing consumption or by spreading out consumption in time, costs and prices can be decreased. Depending on the characteristics of the consumers, different techniques can be used to try to influence their consumption profile. One possibility is to focus efforts on large consumers by imposing very high prices during consumption peaks. Because available power becomes scarce during peaks, electricity is already expensive at these moments.

When energy consumption does not change much with the price, demand is said to be inelastic.[3] As a consequence, prices are volatile: they increase strongly when demand increases or supply decreases. Demand elasticity increases when the number of responsive loads grows. These loads consume less when the price is high. With more responsive loads, price spikes are reduced. One way for making loads more responsive is to offer financial benefits to consumers which adapt their consumption to the price.

For certain loads, the time of consumption is not important. With certain heating applications, thermal energy is stored and can be used some time after the heat was produced. For these applications, electricity consumption can be deferred. Examples of systems that can be used are air-conditioners and water heaters. The use of certain industrial loads can be shifted to less convenient moments if the customer receives financial compensation.

Certain critical situations and emergencies are other cases in which load management is necessary. This occurs when the capacity of the grid is at its limits, when an amount of production or transport capacity is suddenly lost or when consumption suddenly surpasses production. Sudden loss of transport or production capacity can occur due to faults in the electricity grid or incidents in power plants. In order to keep the remaining components of the power system balanced, a quick decrease in consumption is required, i.e. load shedding. Because consumption has to be reduced immediately, certain loads and sections of the grid are disconnected, resulting in blackouts.

17.3.2 Multiple Electricity Tariff

The simplest billing system for residential consumers consists of a single tariff. With this system, the consumer always pays the same for the energy that is consumed, regardless of the time of day. Because the cost of generating electricity is not constant, the real cost and underlying technical aspects are not reflected in the price.

A better system consists of a double or triple tariff. In the case of a double tariff, the electricity price is lower during the night. For this system the meter is equipped with two energy counters. Switching from one counter to the other is done by means of a built-in timer or by means of communication with the utility. This can be achieved by means of low-frequency signals transmitted along the electricity grid (ripple control) [9] or by means of radio signals. Certain heavy loads, such as an electric water boiler and electric heating, can be automatically activated when the night tariff becomes effective. A triple tariff is another possibility. In this case two day tariffs are used. The highest tariff is activated during peak loading of the power system.

17.3.3 Real-Time Pricing

In order for a system of load response to work, some requirements have to be fulfilled. First of all, the real-time price must be communicated to the consumer or the load. A means of communication is needed, e.g. the Internet or radio waves. When the consumer or the load

[3] Inelastic demand or demand with low elasticity.

(equipped with a suitable control system) receives new information, it is decided whether or not to decrease consumption. In order to do this, there has to be a certain potential for the reduction of consumption. If the load is part of an important production process or a vital activity, response to a price signal is impossible. As a third requirement, the response of the loads must be measured. Without this, there is no incentive to respond. Measuring is also needed to create an accurate electricity bill (e.g. metering at 15 min intervals).

A special kind of contract is the buy-back contract. With such a contract the energy price is fixed. However, if the real-time market price is higher than the fixed price in the contract, the consumer can sell energy back to the retailer at the market price. In this way, the consumer receives the difference between the fixed price and the market price.

17.3.4 Contracts for Reducing Consumption

When the power system is operating at full capacity, there are opportunities for consumers which are willing to reduce consumption or accept not to have full access to electric power for 100 % of the time, in exchange for financial benefits. In this way power becomes available for other consumers which are willing to pay a certain amount extra in critical situations.

Consumers can agree to reduce consumption within a specific amount of time in case of reliability or capacity problems on the supply side. It is even possible to organize a market for reduction capacity. Consumers can also agree to shut down completely certain areas of their equipment. It is also possible for large, industrial consumers to make an agreement with the grid operator or the retailer, handing over control of a certain production process to the grid operator or the retailer, so it can be switched off when necessary. In this way reliability of supply can be traded.

An example of this system is the contract offered by the Otter Tail Power Company, which is operational in three states of the USA: North Dakota, South Dakota and Minnesota [8]. A consumer can choose to include energy control in the electricity contract. With this contract, the consumer pays about half the electricity price as with a regular contract. In exchange, the power company can switch off a number of the consumer's appliances. This is done by means of radio signals. The company installs the necessary control equipment. At the start of the contract the consumer decides which appliances will be controlled by the power company. The consumer's cost consists of an energy rate ($/kWh) and a charge for the power ($/kW) the consumer wants to dispose of during periods of energy control. E-mails are sent to notify consumers of the start or the end of a control period.

17.3.5 Intelligent Energy Meter

In the near future residential electrical installations will be equipped with intelligent energy meters. Intelligent meters will perform many functions. Next to a metering unit, it will have an information unit, consisting of a small computer, an information display and a communication link (Internet, power line communication or radio communication) to the DSO and/or the electricity retailer [1].

The energy meter measures consumption and monitors power quality. Readings are sent to the utilities. From the utility, service messages are sent, such as commands for disconnection or reconnection. Problems such as loss of measurements or unauthorized interference with the system are reported directly to the utility. The information system will offer the consumer a variable real-time electricity price. Using the interface, the user has access to real-time measurements, price information and historical data. There will also be the possibility for the information system to control the operation of large loads, such as air-conditioners, boilers, heaters, pumps, dishwashers, tumble driers and washing machines. These could be controlled directly by the utility or according to the preferences of the user, who can program the disconnection of certain loads when the price exceeds a certain value, or manually shut down certain devices.

If consumers receive transparent, real-time information concerning the energy price, they are stimulated to adapt consumption by shifting and reducing the use of certain devices. Price signals can motivate users to shift peak consumption by means of manual or automatic load controls. Promotion of voluntary energy savings in a liberalized electricity market is likely to increase in the future due to the need to conserve energy.

17.3.6 Other Techniques

Other techniques can be used. Consumers can simply be asked to consume less, e.g. by means of the media, as is often done for conservation of water during dry periods in summer. Another method is a contract between the utility and a consumer, giving the utility the ability to control the consumer's load. An example of this is a thermostat controlled by the retailer.

A simple system is *Cool Keeper*, from Rocky Mountain Power (USA). This is a voluntary system aimed at reducing consumption by air-conditioning [2]. A control device containing a radio receiver is connected to air-conditioning units. During peak demand, the power company broadcasts a signal, instructing the control devices to switch off the air-conditioning. Afterwards units are switched on again by another signal. *Cool Keeper* turns off air-conditioning units for a few minutes per half hour in the afternoon on summer weekdays. Participation in this system and installation of the control device are free of charge. Participating residential consumers receive a $20 discount on their electricity bill annually. For small businesses the discount is $40.

An extreme measure in case of a shortage in production capacity is the use of rolling blackouts. This consists of deliberately interrupting supply to a certain area for a limited amount of time, after which supply is restored and the blackout is shifted to another area. In this way the discomfort is divided among all consumers.

17.4 POWER QUALITY IN THE ELECTRICITY MARKET

In order for an electricity grid to operate in a satisfying way, a high power quality is required. Because power quality has its price, a trade-off between quality and cost has to be made. Electricity consumers request a certain level of power quality, adapted to their activity, at an attractive price. The clean electromagnetic spectrum can be considered a public asset that needs to be controlled and protected. It is an important resource and has to be used in a sustainable way. The question is: who should be responsible and accountable for improving poor power quality or maintaining existing quality?

In order to achieve durable improvements in power quality, the efforts of electricity retailers, DSOs, manufacturers of equipment and consumers have to be coordinated. Because improvements are very expensive, standards or operational rules are required to define each party's responsibilities. The specific implementation of mutual responsibilities between interacting parties can be established in contracts between the parties. Another possibility is to set up a market for power quality. In this case, permits for the emission of disturbances are traded, and private investments with the aim of reducing disturbance emissions become profitable. Power quality contracts and a possible outline of a future power quality market are discussed in the next subsections.

17.4.1 Power Quality Contracts

The restructuring of the electricity sector has replaced the monopolist utility with several independent companies. This evolution raises the following question: who takes care of power quality? In a competitive market, offering power quality services (e.g. evaluation and improvement of power quality at a customer's site) can be an interesting business opportunity. In this subsection, the important elements of power quality contracts are outlined as discussed in [7].

In the new structure of the electricity industry, contracts between the different participants are made, in which the expected system performance is described. The content of contracts concerning power quality depends on the parties involved and the characteristics of the system. The following issues are possible subjects of a power quality contract:

- List of power quality concerns to be evaluated.
- List of performance indices to be used.
- Expected level of power quality performance.
- Penalty for not complying with the expected level of performance.
- Reward for performing better than the expected level.
- Methods to verify performance (measurement, calculation).
- Responsibilities for each party in achieving the desired performance.
- Responsibilities of the parties for resolving problems.

The content of a contract between parties in the electricity market depends on the category to which the parties belong and their relationship. Here the most important elements of contracts are discussed for the following relationships, which are also shown in Figure 17.1:

- TSO and DSO, or TSO and large industrial consumers.
- DSO and consumers.
- Power retailer and consumers.
- DSO and small independent power producers (IPPs).

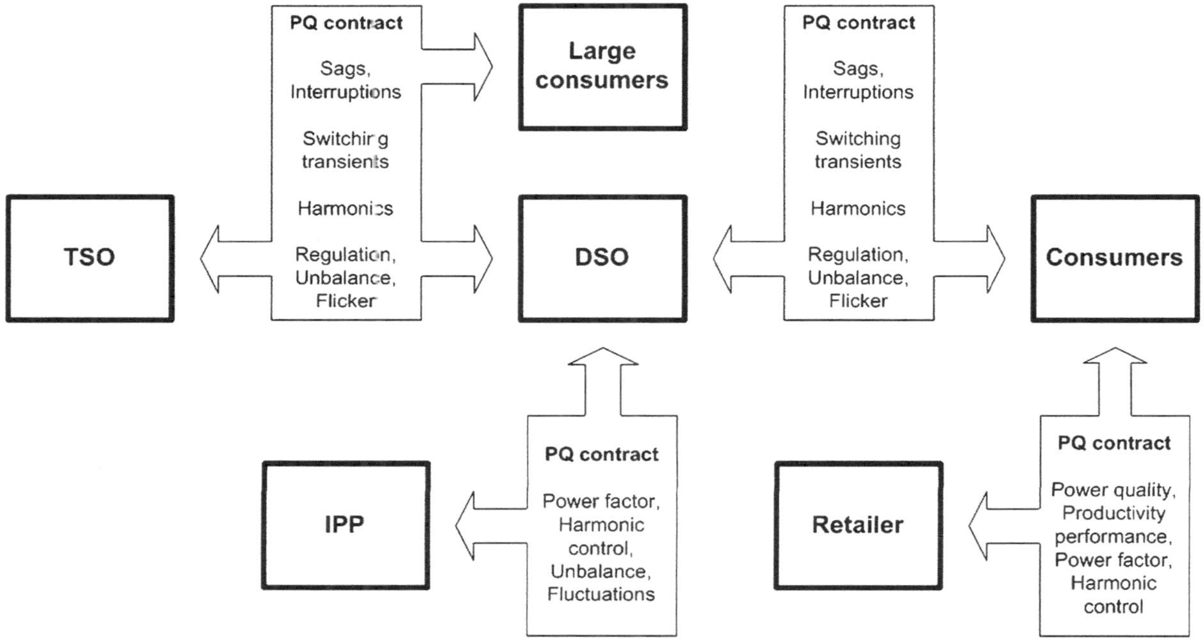

Figure 17.1 Power quality contracts in the restructured electricity market [7] (Reproduced from 1998 Proceedings of the 8th International Conference on Harmonics and Quality of Power, IEEE)

In Figure 17.1, the structure of the relationships between the TSO, the DSO, large consumers, (small) consumers and IPPs reflects the hierarchical structure of the power system. The relationship between the retailer and the consumer, however, is purely contractual; it cannot be linked to the structure of the power system.

17.4.1.1 Contracts Between the TSO and DSOs or Large Consumers

The rated voltage supplied by the transmission system has to be described, as well as the responsibility of voltage regulation.[4] The supplying transmission system is responsible for the overall flicker levels in the voltage. Flicker levels can be reduced by making short-circuit powers at interconnection points sufficiently high.[5] The DSO or the industrial consumer is responsible for limiting current fluctuations. These fluctuations can be caused by fluctuations of the load or by fluctuations in the energy produced in the distribution system (i.e. distributed generation). Stipulations concerning flicker are particularly important where industrial consumers with large arc furnaces are concerned. Concerning harmonics, in a similar way, the TSO is responsible for limiting the harmonic content of the voltage and the DSO has to limit harmonic current.

Switching capacitor banks in the transmission systems cause transient voltages. These voltages can be amplified by the impedances of the grid, which can result in malfunctions of distribution grids and sensitive loads. Transients can be limited by closing switches at the zero crossing of the voltage or by means of resistors. Limits for transients have to be defined at the point of connection between the transmission system and the distribution system, i.e. the point of common coupling (PCC), in the substation.

Expected performance concerning voltage sags and interruptions at the PCC has to be defined. Both parties have a certain responsibility concerning voltage sags because they can be caused by faults in both the transmission system and the distribution system. A penalty system can be used, according to which a fine has to be paid to the other party if voltage sags or interruptions occur more frequently than specified in the contract.

17.4.1.2 Contracts Between the DSO and Consumers

The characteristics of the voltage have to be defined. The consumer has to comply with requirements concerning fluctuating loads, unbalanced loads and the starting of motors. Concerning harmonics, the rule mentioned previously applies here as well: the DSO has to supply the voltage with limited distortion, while the consumer has to be mindful of the use of non-linear loads, in order to limit current harmonics.

The switching of capacitor banks for power factor correction is also important in the interaction between the distribution system and the consumers. The DSO has to limit transients arising from capacitors within the distribution grid. The consumer has to take care that transients that do occur are not amplified by capacitors within the consumer's own local grid. The switching of capacitors in the consumers' own grid may cause transients as well.

[4] Voltage regulation is discussed in Chapter 3.

[5] High short-circuit power corresponds to low grid impedance and, consequently, limited voltage fluctuations when current fluctuations occur.

The contract should also contain requirements and responsibilities concerning suppression of transients caused by lightning strokes.

The expected occurrence of voltage sags and interruptions should also be mentioned in the contract. As an optional service to the consumer, performance can be improved by means of modifications to the distribution system. The following case [3] is an example of the application of such a contract.

In January 2000 the utility company Detroit Edison (Michigan, USA) had a standing service contract with three automotive companies, namely Chrysler Corporation, Ford Motor Company and General Motors Corporation. This contract implied that Detroit Edison was the manufacturers' sole power supplier for a period of 10 years. The additional services included in the contract were assistance in improving energy efficiency and financial compensation in case the number of voltage interruptions and voltage sags exceeded performance targets. In order to monitor the power system's performance regarding interruptions and sags, Detroit Edison installed power quality measuring units in 56 locations on the manufacturers' premises. The targets for interruptions and sags were based on measurement data previously recorded on site.

Regarding voltage interruptions, the annual target (i.e. the number of allowed interruptions per year) and the corresponding compensation payment depend on the manufacturing process affected by the interruption. An example is shown in Table 17.1. If the assembly/stamping plants, for example, experience five interruptions during a year, the target is exceeded by two. If the compensation per interruption is \$50 000, the manufacturer receives \$100 000 from Detroit Edison. The targets for interruptions were decreased by 5 % each year (rounded to the nearest integer), in order to improve the quality of the service.

In order to reduce the number of interruptions, the utility company carried out additional maintenance on certain lines, replaced certain pieces of equipment and took measures to protect equipment against harmful animals. In 15 of the 56 locations, a second power supply was created, to serve as backup. As a result, the three manufacturers experienced fewer interruptions. Compared to 1995, the number of interruptions was 4 % lower in 1997 and 54 % lower in 1999. The corresponding payments were 40 % lower in 1997 and 78 % lower in 1999, compared to 1995.

System performance related to voltage sags was determined by means of calculated sag scores. Sag scores depend on the nature of voltage sags. The scores are summed for a

Table 17.1 Interruption targets for Ford Motor Company. The targets are decreased by 5 % each year, starting from their 1995 base value [3] (Reproduced from 2000 Proceedings of the 9th International Conference on Harmonics and Quality of Power, IEEE)

Group	Number of locations	Interruptions per calendar year: target for 1995	Payment per interruption (\$)
Assembly/stamping	6	3	50 000 to 88 000
Power train/glass	4	2	159 000 to 326 000
Components	9	9	16 000 to 152 000
Individual	1	1	30 000
Other	8	3	2000 to 11 000
Affiliates	1	1	85 000

certain group of locations. If the sum exceeds a preset value (the sag-score target) Detroit Edison makes a compensation payment. In 1999 the payment due to voltage sags was 9 % of the payment due to interruptions. Based on the sag-score data from 1995 to 1999, new, more stringent targets were defined for the year 2000, with target values between 50 and 95 % of the previous targets.

17.4.1.3 Contracts Between Power Retailers and Consumers

The main subject of the contract between retailers and consumers is the sale of electric energy. As was discussed before, retailers offer additional services. Advice about and improvement of power quality can be part of a retailer's portfolio of services.

The retailer's power-quality-related services are focused on the consumer's local grid. Therefore the required measures depend on the nature of the consumer's equipment. These requirements may be formulated in terms of the performance of the equipment, rather than voltages and currents. A possible service is taking responsibility for the interface with the distribution system, guaranteeing a certain level of power quality and reliability of the consumer's power supply. This may require the installation of a UPS system. In this way the consumer becomes almost entirely immune for poor power quality in the distribution grid. Payment for this service could be a share from the consumer's financial savings from improved productivity. The retailer can also offer services aimed at reducing emission, i.e. reducing the consumer's negative impact on the power quality in the distribution system. In general, DSOs impose limitations on the injection of harmonic current and poor power factor.

17.4.1.4 Contracts Between the DSO and Small IPPs

A recent evolution in power systems is the appearance of small IPPs, which operate generators connected to the distribution system. This is called distributed generation. As for consumers, mutual requirements concerning power quality are discussed in the contract. Important aspects, specific for distributed generation,[6] are the effects of power fluctuations due to the start-up of generators and variations in wind and sunlight, in case renewable energy sources are used.

17.4.2 Power Quality Market

A possible future development is the creation of a market for power quality. This is discussed in [4]. If a level of acceptable disturbance is defined or determined for a distribution grid, all users of this grid can be allotted a permit for emitting a certain amount of disturbance. This means that all users of the power grid have the right to cause a certain amount of power quality distortion. Users causing little disturbance would not make full use of their permits. A market could then be created, consisting of a trading system for emission permits. In this system users with high emission levels have to buy emission permits from other users or reduce their own emissions. In case users fail to comply, some kind of sanction

[6] Power quality problems caused by distributed energy sources are discussed in Chapter 16.

has to be applied, e.g. a fine. In this way investments in mitigating measures by grid users can be profitable because of the permit sale. The sum of the permits must be such that the worst-case allowed emission is not harmful. A control body or inspection body has to check whether emission levels remain below the limits of the corresponding permits. A threshold for a minimum, unavoidable emission has to be included. Otherwise, passive, linear customers would be penalized for emissions from their side, indirectly provoked due to the distorted voltage on the PCC. This system could be similar to the trade in permits for greenhouse gas emissions or the trade in green certificates, associated with electricity generation from renewable energy.

In contrast to permits for greenhouse gas emissions or green certificates, power quality permits should be limited in time and space. Their period of validity has to be short enough, in order to keep up with the evolution of technology, which has consequences for the desired level of power quality.

The area or region, for which a permit is valid, is determined by the electricity grid. There is no point in trading permits between customers that are connected to branches of the grid that are electrically independent or interconnected by means of stretches of high-voltage line. An emission can cause a power quality problem for customers connected to the same radial network, but not beyond this network. The electrical distance between customers determines the propagation of disturbances. Because the value of a permit should reflect the cost of the negative consequences of the distortion, the value decreases when the origin of a distortion is situated further away. In this way, it becomes less interesting to sell the permit.

Different distortion phenomena have different consequences, e.g. different time scales, different physical quantities. Therefore it would be difficult to have one permit system covering all types of power quality distortion. It is virtually impossible to compare a voltage sag of, for instance, a certain duration and depth to a current with a certain harmonic spectrum. A permit system should be divided into categories, according to different power quality problems, e.g. harmonic emission, unbalanced loads, surge currents, voltage sags and flicker.

In a permit system, customers would be stimulated to take mitigating measures. The investments are paid for with the savings resulting from reduced permit requirements or the sale of surplus permits. If some grid users make improvements, they can sell additional permits to other grid users. In order to keep track of every user's emissions and mitigating efforts, a permit system would require power quality metering, registration and communication. This can only be achieved by means of advanced IT and communication devices. Permits might be managed by an external party or offered as a service by the power retailer. A permit system for power quality would not be very suitable for purely residential areas, because all residential buildings emit more or less similar distortion and the financial benefits would be small compared to the efforts required from private persons. In areas with commercial and industrial activity, a permit system would be more interesting.

BIBLIOGRAPHY

[1] Chen Y., Hwang J. K., A reliable energy information system for promoting voluntary energy conservation benefits. *IEEE Transactions on Power Delivery*, vol. 21, no. 1, pp. 102–107, 2006.

[2] Cool Keeper, http://www.coolkeeper.net, 2007.

[3] Dettloff A., Sabin D., Power quality performance component of the special manufacturing contracts between power provider and customer. Proceedings of the 9th International Conference on Harmonics and Quality of Power, IEEE, 1–4 October 2000, 2, pp. 416–424.
[4] Driesen J., Green T., Van Craenenbroeck T., Belmans R., The development of power quality markets. IEEE Power Engineering Society Winter Meeting, New York, 27–31 January 2002.
[5] Excelis – Electricité de France, http://entreprises.edf.fr/102166i/GrandesEntreprises/temoignages/interviewFabre.html, 2007.
[6] Kirby B. J., Staunton R. H., Technical potential for peak load management programs in New Jersey. Oak Ridge National Laboratory, 2002.
[7] McGranaghan M., Kennedy B. W., Samotyj M., Power quality contracts in a competitive electric utility industry. Proceedings of the 8th International Conference on Harmonics and Quality of Power, IEEE, 14–16 October 1998, pp. 170–175.
[8] Otter Tail Power Company, http://www.otpco.com, 2007.
[9] Paessler E. R., *Rundsteuertechnik*, Publicis MCD Verlag, Erlangen, 1994.

18

Cost of Poor Power Quality

Roman Targosz and Jonathan Manson

It is increasingly being accepted that the impacts of poor power quality on organizations' operational efficiency are significant and disruptive – it is also being evermore clearly understood that financial impacts, where relevant, are far greater than had previously been recognized. Although PQ definitions commonly exclude continuity of supply phenomena, these impacts are included in this chapter as detailed in 18.1–18.3.

Whilst the consequences are not limited to electrical systems in non-residential environments and can harm domestic systems and appliances as well as mainly introducing electrical danger into the home, the residential sector is not addressed here.

It is only by knowing what the cost consequences are that effective plans can be drawn up to reduce or eliminate these unnecessary productivity and efficiency losses and to design in solutions that make economic as well as operational sense.

It is well known that although there are several methods to do it, quantifying the losses created by poor power quality costs is a complicated business and one that has to be carried out specifically for the site in question – survey results can give end users a good idea of the likely impacts of poor power quality on their activities, but in the end they should carry out their own investigations to ensure that both the diagnostics and any solutions that result from them are specifically tailored to their particular systems, environment and electrical supply conditions.

The economic impacts of power quality are usually divided into three broad categories:

- *Direct economics impacts:*
  - loss of production;
  - unrecoverable downtime and resources (e.g. raw material, labor, capital);
  - process restart costs;

Handbook of Power Quality Edited by Angelo Baggini

- spoilage of (semi-)finished production;
- equipment damage;
- direct costs associated with human health and safety;
- financial penalties incurred through non-fulfillment of contract;
- environmental financial penalties;
- utility costs associated with the interruption.

• *Indirect economic impacts:*

  - the costs to an organization of revenue/income being postponed;
  - the financial cost of loss of market share;
  - the cost of restoring brand equity.

• *Social economic impacts:*

  - uncomfortable building temperatures as related to reduction in efficient working/health and safety;
  - personal injury or fear, also as related to reduction in efficiency and health and safety;
  - evacuating neighboring residential buildings as an indirect social impact in the event of failure of industrial safety, as it relates to the additional costs incurred by an organization that has to carry out these measures.

In the field of power quality, which is still relatively underregulated and one in which lines of responsibility are often blurred, capital investment decisions become the more difficult to secure, the *softer* the problem areas become. This is especially so in cases where that capital investment decision is not seen as a priority by the final decision maker.

The consequence for this area of interest, the economic impact of poor power quality (PQ), is that the *indirect* and *social* impacts often fail to be recognized and go unaddressed.

It is a given that knowledge about PQ cost is essential for correctly planning activities to mitigate all these costs and so reduce or eliminate the avoidable wastage generated by them.

It is an important part of the end users' armory to arrive at the correct balance between what the organization's poor PQ costs are, the investment required for any PQ solution, as well as to assess whether unmitigated poor PQ can be tolerated.

The power sector and equipment manufacturers also need to assess how great an effort to place behind helping end users to reduce the impacts of poor PQ.

Currently the role of solving this issue lies in the hands of regulators who are responsible for optimizing the equation and balance between unmitigated poor PQ costs and the solutions associated with them on different mitigation levels:

• power generation (mainly reliability);
• power transmission and distribution (reliability and voltage quality);
• end users' electrical system level (reliability, voltage quality, PQ distortion emissions);
• Electric energy using equipment level (both immunity and emissions).

This issue is addressed on a purely societal level.

18.1 EXPLORING PQ COST

PQ costs are usually reported in the following categories:

- Voltage dips and swells
- Short interruptions
- Long interruptions
- Harmonics
- Surges and transients
- Flicker, unbalance, earthing and electromagnetic compatibility (EMC) problems

The harmonics category may be further subdivided to focus on particular impacts that were treated as separate cases, defined as follows:

- Overstressing insulation, effects on electrical equipment including transformers, capacitors, motors; the consequences of not measuring TRMS (True Root Mean Square); additional losses.
- Overheating of the neutral conductor (e.g. burn-off and subsequent disruption or damage to electrical equipment).
- Nuisance tripping of protective devices.
- Malfunction of equipment control systems due to additional zero crossing.

To calculate the total cost of each PQ disturbance the following PQ cost categories are to be considered for each disturbance category.

18.1.1 Staff Cost

Personnel rendered unproductive through disrupted workflow/process. This cost can be calculated either by: multiplying the total number of person-hours of staff who are unable to work and average person-hour rate of staff who are unable to work; or by estimating the percentage of plant activity (in terms of added value) which was stopped and multiplying by the idle time of such stoppage. This value is then compared to the total production time of the plant.

18.1.2 Work in Progress

This category includes:

- Costs of raw materials involved in the production of services which was inevitably lost.
- Labor involved in the production of services which was inevitably lost.
- Labor needed to make up for lost production, sales or services (such as overtime pay, extra shifts, etc.).

18.1.3 Equipment Malfunctioning

If the equipment is affected, the consequence can be the slowing down of the company's activity or part of the production running out of specification. In this case the percentage of such slowdown is calculated taking into account additional idle time, the value of products running out of specification and/or the value of insufficient quality of products.

18.1.4 Equipment Damage

If the operating equipment is affected, the consequence can be damage to it, the shortening of its lifetime, components wearing out prematurely and the need for additional maintenance or repair. The cost components include:

- Cost of equipment being damaged (completely and scrapped) or the cost of its repair. This category typically includes transformers, capacitors, motors, cables, contactors, relays, protective equipment, electronic equipment and lighting bulbs.
- Cost to run and/or rent backup equipment if necessary.
- Additional maintenance costs because of excessive equipment components wearing out. Usually this includes bearings and, if a machine is unbalanced due to distorted power, insulating, disconnecting any protective/signaling components and resets or reinstallation.

18.1.5 Other Costs

This category usually includes:

- Penalties due to contract non-delivery or late delivery.
- Environmental fines/penalties.
- Cost of evacuation of personnel and equipment (this can also include ensuring the safety of external communities).
- Costs of personnel injury (including the on-costs incurred through inability to work).
- Increased insurance rates (equipment, personnel, corporate liability).
- Compensation paid out.

18.1.6 Specific Costs

For some disturbances, some specific cost categories can be distinguished:

- Harmonics – electricity bill. By operating electrical equipment in a non-linear environment, additional eddy currents, heat dissipation and consequent energy losses may be experienced. It is, however, not easy to measure them as a full harmonic spectrum is needed. As an indication, a typical large-office environment generates about 50 % extra losses in transformers. Other possible problems which may arise from harmonic pollution refer to the correct measurement of electric energy consumption and problems related to the utility imposing penalties for harmonic pollution of the surrounding distribution network.

- Flicker can cause migraine and can be responsible for so-called sick building syndrome, which reduces personnel productivity. This can be defined for example as a comparison of staff error rates between flicker-free and flicker environments.

18.1.7 Savings

To make the total calculation fair and complete, savings also resulting from PQ disturbances are calculated. These usually include:

- Savings from unused materials or inventory.
- Savings from wages that were not paid.
- Savings on the energy bill.

Savings are deducted from the gross PQ cost to obtain the total cost of PQ for the plant in question.

18.2 STUDIES ON COST OF POOR PQ

At the beginning of the 1990s the costs of poor PQ to industry were estimated globally to be in the region of a few tens of billions of dollars. After conducting more comprehensive surveys the global cost of poor PQ was then estimated at a few hundreds of billions of dollars (ECI estimates 2003).

The soundest studies on the topic include:

- CIGRE – Methods to consider customer interruption costs in power system analysis. Interruption costs international surveys report/Task Force 38.06.01, Official Report from CIGRE by R. Billinton *et al.* from 2001 [2].
- The cost of power disturbances to industrial & digital economy companies prepared by Primen to EPRI's (CEIDS), by D. Lineweber and S. McNulty in 2001 [9].
- A European PQ Survey conducted by the Leonardo Power Quality Initiative team in 2005 to 2006. The study comprised 62 face-to-face interviews carried out in eight European countries. This allowed for an extrapolation of the overall wastage caused by poor PQ to the EU's 25 countries (EU-25) and for the analysis of many associated issues such as user perception of PQ problems, the causes of PQ problems, the equipment mainly affected and any solutions that were available and adopted [11].

Additional national studies have provided more precise findings. For example, in France EdF investigated the cost of interruptions to economically justify investment into improving network reliability. Also in the USA subsequent studies and summaries concluded by addressing the topic of the economic impact of PQ.

A summary of findings on the economic impacts by PQ cost category and by industrial sector are reviewed in the following and summarized in Table 18.1.

Table 18.1 Summary of PQ costs

Disturbance	Occurrence	Cost	Mitigation	Critical areas – sectors
Voltage dips	Joint responsibility of supplier, user and neighborhood Highly unpredictable although some occurrence indicators exist Dependency on location (neighborhood) Usually frequency from a few to a few tens a year Increasing trend	Usually between €2000 and €20 000 per event Average ~ €2000 and 25 % of 1 hour interruption. Cost highly dependent on dip depth and length. In the case of continuous manufacturing dips are usually responsible for more than 50 % of PQ cost.	Impossible to eliminate but mitigation techniques exist to reduce vulnerability level Depending on local conditions and type of equipment, different mitigation techniques and levels (equipment, user installation, network) are optimum	Continuous manufacturing, Electronics, control systems/equipment, automation and protection systems
Interruptions	Limited predictability, higher for long interruptions Average a few yearly, local network operator should provide information on SAIFI and MAIFI Trend stable for the amount of shorter-duration interruptions	From a dozen or so to tens of thousands of euros per event. Some are a few million euros Average for short interruption ~ €10 000 per event, 0.01 $/kWh, 10 $/kW Long interruption usually more costly, depending greatly on the end-user type Dependency (sustained) on notification and season	99.999 99 % reliability systems do exist but costs can be deemed excessive Emphasis on supplier–user relations, which may help to minimize consequences and selection of optimum PQ mitigation	Industry in general, especially high-energy-intensive users Cost varies greatly and highly dependent on duration Transaction sector (credit cards, brokers, banks), food and kindred products, farming, even if the interruption is relatively short

Harmonics	Impacts from both harmonic currents and voltages Directly related to harmonic currents, generation sites (power electronics and small SMPS, CFLs) Most often focus on 5th, 7th and 11th	Energy losses, up to 50 % above pure sine wave losses Lifetime shortening by a few to a dozen or so years Reliability decrease Costs expressed as mitigation cost at the level of €20–30bn in countries like France or Germany; Hydro-Quebec Canada source $\sim$ C$650m per year	Possibility to mitigate consequences locally to acceptable level exist (i.e. THD $< 5\,\%$) Costs will vary Due to changes in power system, active, more costly mitigation may be easier but will be more expensive The critical issue is the correct addressing of the current distortion origin	Neighborhood of harmonic polluting loads Losses, lifetime and reliability reduction in transformers, motors and cables Consequences for ICT equipment
Other	Consequences, especially in monetary value, not well covered in the technical literature. However, some observations exist that short and very short disturbances like transients and surges may play higher role in PQ system immunity Flicker, although some observations suggest that it affects equipment Flicker impact is still mainly considered as restricted to causing vision problems only. However, actual cost consequences can be up to 10 % percent of an organization's employment cost			

18.3 LONG INTERRUPTIONS

The first point to make is that long interruptions by all major definitions are treated completely separately; that is, outside the context of PQ as defined here. This is because long interruptions are currently referred to as continuity of supply and not as poor PQ. The main reason for this is that the causes of lack of continuity of power supply mainly originate from the energy sector, i.e. power generation, transmission and distribution. What often happens is that more mitigation of continuity of supply very often results in an increase in voltage dips, which, as will be seen later, generate some of the most troublesome poor PQ cost consequences. The reason for this is simple: to minimize the quantity and duration of these interruptions, utilities apply auto-reclosing techniques, which in turn result in increased number of voltage dips. The same thing happens when electric energy end users increase their power supply redundancy levels by connecting up to additional power lines. If no additional design measures are installed when doing this, the effect is usually an increase in the cost of voltage dips.

The costs of interruptions can be stated as the cost per event of a specific duration (most often 1 second, 1 minute, 1 hour, 4 hours, etc.). For comparisons and estimates these unit costs are usually specified by the following indices:

- The cost per peak power (based on maximum power demand).
- The cost per interrupted power (at the moment of the event).
- The cost per energy not supplied (level referred to the moment when an outage started or cost per peak energy not supplied when the cost is divided by the energy consumption at yearly peak demand level).
- The cost per energy consumption annually.
- The combination of both in one formula, i.e. the relative power cost plus relative energy cost.

Some examples are:

- The IEEE 493 [7] standard gives averages for assessing the cost of a single interruption of 1 h duration:
  - Industrial plants: 6.43 \$/kW + 9.11 \$/kWh.
  - Commercial facilities: 21.77 \$/kWh not delivered.
  - Office buildings containing computers: \$25.07 per peak kWh not delivered.
- The CEIDS/EPRI study found that the average cost of a 1 h outage to industrial and digital economy companies was \$7795. A 3 min outage cost \$2107. A 1 s outage cost \$1477. For continuous process manufacturing, the outage costs were \$12 654, \$18 476, \$14 746 for a 1 s, 3 min and 1 h outage respectively. For fabrication and essential services, the outage costs were below the continuous process manufacturing costs. Total average annual cost in the USA for zero-voltage events was \$23 318.
- The breakdown of outage cost components for three of these groups of companies is presented in Table 18.2.

Table 18.2 Outage cost components [9] (Reproduced from Report of Primen for Electric Power Research Institute, Consortium for Electric Infrastructure to Support a Digital Society)

	Average	Digital economy	Continuous process manufacturing	Finished product and ess. services
Costs ($)	8080	5061	14 837	13 463
Net Lost Production ($)	2922	3344	4 491	1 765
Labor ($)	1763	1046	2 919	3 114
Materials ($)	231	7	3 069	253
Equip. damage ($)	576	196	1 121	1 305
Backup ($)	449	119	886	1 085
Overhead ($)	364	74	691	933
Restart ($)	1457	59	1 332	4 475
Misc. ($)	318	216	329	534
Savings ($)	285	13	91	901
Materials ($)	263	0	36	863
Energy ($)	16	10	32	24
Labor ($)	6	2	23	13
Net cost ($)	*7795*	*5048*	*14 746*	*12 562*

The difference in costs between outages of different duration are presented in Table 18.3.

Clearly there are differences in the cost components for different sectors and duration of events. Production losses dominate in the digital economy sector and continuous manufacturing. The same applies but at lower costs in the case of equipment damage. Restart costs are very significant in the case of continuous manufacturing and are the dominant category for finished product and essential services. Very high material cost is very specific

Table 18.3 Duration characteristics of outage cost components [9] (Reproduced from Report of Primen for Electric Power Research Institute, Consortium for Electric Infrastructure to Support a Digital Society)

	1 second	Recloser	3 minutes	1 hour
Costs ($)	1489	2848	2124	8080
Net Lost Production ($)	284	274	466	2922
Labor ($)	433	263	590	1763
Materials loss or spoilage ($)	99	68	248	231
Equipment damage ($)	554	2011	500	576
Backup generation ($)	21	64	89	449
Overhead ($)	33	49	101	364
Other restart costs ($)	60	46	121	1457
Misc. ($)	5	73	9	318
Savings ($)	12	0	17	285
Unused materials ($)	10	0	14	263
Savings on energy bill ($)	2	0	2	16
Unpaid labor ($)	0	0	1	6

for continuous manufacturing. Labor cost is more significant for finished products and services.

As far as duration of events is concerned, recloser events result in doubling the cost of the preceding 1 s event only because of an increase in equipment damage cost. In the case of longer events the cost increases but much below the time linearity model and is higher mainly due to increased production and material cost.

Some other observations from the CEIDS/EPRI study are as follows:

- 74 % of companies can ride through a 1 s outage with no costs, whereas fewer (65 %) experience no costs from recloser events. Half of all users would not suffer any measurable costs from a 3 min outage and 26 % would not experience significant costs from a 1 h outage.
- The typical cost of a 1 h outage to most (56 %) companies is less than $500, whilst the remaining 44 % experience much higher than average costs.
- In 4 % of companies, costs generated by poor PQ are higher than 10 % of their turnover; in another 9 % these costs range between 1 and 9.99 % of turnover, in 27 % between 0.1 and 1 %, whilst 60 % of companies represent a cost of 0.1 % of turnover or below.
- Recovery periods of returning to normal production activity are summarized by this study as:

  - 30 min (56.8 % of establishments) for 1 s interruption;
  - 30 min (82 %) for 3 min event;
  - 2 h (82.1 %) for 1 h event.

- A general observation can be made, as may be imagined: organizations suffer less from fewer, longer outages than from more, shorter ones. However, the caveat is that there are many, great differences between cases.

When summarizing all existing studies the conclusions are as follows. The cost to the end user of a 1 h interruption per event can be estimated within an order of magnitude by commercial, digital economy, continuous manufacturing, fabrication and essential services and industrial sectors as being:

- commercial sector (basically different public and non-public services): €1000
- digital economy sector (e.g. firms that rely heavily on data storage and retrieval, data processing, or research and development operations; specific industries include telecommunications, data storage and retrieval services including collocation facilities or Internet hotels, biotechnology, electronics manufacturing, and the financial industry): €5000
- continuous manufacturing sector (specific industries include paper, chemicals, petroleum, rubber and plastic, stone, clay and glass, and primary metals): €13 000
- fabrication and essential services sector (e.g. all other manufacturing industries, plus utilities and transportation facilities such as railroads and mass transit, water and wastewater

treatment, and gas utilities and pipelines): slightly below (~ 10 %) continuous manufacturing, i.e. ±€11 500
- industrial sector in general: €3500

Today, information on reliability indices exists and should be made publicly available by network operators. This information includes three main indices (described in Chapter 2):

- SAIDI
- SAIFI
- MAIFI

Based on this information the estimated interruption cost index to a company can be calculated. As very rough estimate, SAIDI is usually between 1 and 2, whilst SAIFI is between 1.2 and 1.5, and these result in an average yearly cost factor that is 1.2 to 2 times higher than the cost per single long interruption.

Long interruptions have been the subject of many international studies worldwide and perhaps one generic finding is that the relation between the cost of a long interruption and its duration is not proportional and can be expressed as a logarithmic curve.

The industrial sector and type of activity can make a great difference and it may well be that some sectors hardly experience any cost until, that is, the power loss exceeds a critical time limit. Alternatively, other industries may experience almost the same damage from an interruption of any duration. This is, for example, typical in the case of most continuous manufacturing industries for which the difference between interruptions of 1 second, 1 minute and 1 hour is negligible.

There is a strong relation between the cost of an interruption and when it occurs. Such parameters as *weekday*, *working day* and *afternoon*, as one may imagine, all increase the cost estimate for industrial and commercial customers. Other studies have concluded that season is, not unsurprisingly, a key parameter for agriculture.

What is surprising is that whether end users are notified or not differs from study to study in terms of the effects that interruptions have on their activity. According to a CIGRE report, prior (usually two days, but much shorter advance period scenarios have also been investigated) notification about interruption could almost halve any financial loss, especially in the case of industry but less so for the commercial sectors. The CEIDS/EPRI study concludes that such notification does not necessarily create a significant difference or reduction in losses due to this PQ issue.

It is seen that the cost of interruptions has been thoroughly investigated. However, these costs refer to zero-voltage disturbances and it is unlikely that other disturbances would result in generating similar consequences.

So in making any PQ cost estimation the limitations of the continuity of supply referred to at the beginning of this section should be borne in mind. Although as stated at the beginning of this section it is not advisable to make PQ cost estimations for serious purposes like PQ cost mitigation investment analysis, in the specific case of long interruptions such estimates can be valuable and balanced, provided reliability indices like SAIDI and SAIFI are known and the business activity in question is not completely atypical.

18.4 SHORT INTERRUPTIONS

The question arises whether to treat short interruptions separately or together with voltage dips.

Some suggest that short interruptions lasting less than 1 min are specific types of voltage dips. Others make a clear distinction between a residual voltage presence in the case of dips and a zero-voltage condition where interruptions are concerned. In some US studies, especially that by CEIDS/EPRI, short interruptions were investigated and grouped together with long interruptions as zero-voltage events. Furthermore, although CEIDS/EPRI refers distinctly to zero-voltage conditions, the typical dip behavior of momentary interruption followed by recloser operation has been investigated. In fact, short interruptions, lasting less than 60 s could also be seen as a 100 % voltage dip, which would result in inevitable confusion when making such a classification. EN 50160 has adopted a conventional (at the time of writing) threshold of 1 % of the declared voltage (depth of 99 %). If the supply voltage value drops below 1 % of U_c the event is considered a short interruption, otherwise it is classified as a voltage dip.

From a cost point of view, short interruptions have already been included, at least in part, under long interruptions. However, some important comments need to be added.

In every instance of a short interruption, any kind of indirect cost estimates should be confirmed by an internal, organization-specific PQ cost survey. The reasons for this are as follows:

- Customers that do not have precise PQ monitoring may find it difficult if not impossible to differentiate between a short interruption and a voltage dip.
- The most disruptive and cost dynamic phase of a short interruption is the first few seconds which bias the analysis toward MAIFI-based estimates. Some commentators would have it that MAIDI (Momentary Average Interruptions Duration Index) might not be useful. This is not necessarily the case, but it probably would not be feasible to require or request this information from the network operator.

As MAIFI measurements are usually higher than SAIFI ones by a factor of 5 to 20 and as the cost of a single-event short interruption is between 20 and 90 % (70 % as a very rough average) of a long interruption, the total cost of short interruptions is at least similar to that of long interruptions.

On a somewhat speculative note, others maintain that short interruptions are at least three times higher than long interruptions. Both MAIFI and SAIFI in the CEIDS/EPRI survey were by default 1, whilst MAIFI is several times higher than SAIFI. This takes into account the fact that not all short events result in outages.

18.5 VOLTAGE DIPS

In the case of voltage dips, surveys are not useful for transferring their data to particular sites, for the following reasons:

- The data on the occurrence of voltage dips is not fully reliable as indices are not freely or easily available and network or user system reconfigurations may change it very significantly.
- A proportion of voltage dips are unlikely to be designated interruptions and may well be generated locally by user heavy loads, especially during start-up. This further complicates information relating to the occurrence of voltage dip and depth.
- Device/system immunity plays a much bigger role than in the case of interruptions; voltage dip immunization is more case specific and more sophisticated than the equivalent for interruptions.

When analyzing PQ costs there is always the question of whether this analysis is deterministic or stochastic – put simply, the balance between certainty, risk and uncertainty.

With interruptions, especially long interruptions, the level of certainty is reasonably high, whereas with voltage dips and even more so with harmonics and other voltage distortions/variations, the term probability is highly relevant. Consequently, to the equation for cost per event times number of events per time unit (usually one year), one should add the additional factor of probability as this answers the question of whether the event under consideration will result in cost and wastage.

The first thing to be determined is, as in the case of interruptions, the frequency of occurrence. Network operators may have some data, e.g. that derived from continuous measurement of voltage (at each phase), at different points and different voltage levels of the network.

To gain a better understanding of voltage dips that occur in the European MV supply networks, the former UNIPEDE group of experts DISDIP [12] carried out a coordinated measurement campaign over a period of three years in nine countries with different climatic conditions and network configurations.

The survey was carried out at 126 sites with standardized measurement and evaluation criteria, with the maximum duration of a dip set at 60 s in order to include less frequently occurring longer-duration dips. The measurements were taken at the LV busbars of distribution transformers and at various locations with the aim of ensuring that the results could be seen as representative of public LV networks in general. The results are presented in Table 18.4.

Further, for each site a so-called vulnerability study was performed. Devices like switching apparatus – namely, relays or contactors, electronic controls – have different

Table 18.4 Frequency of occurrence of dips per annum with a 95 % confidence interval [12] (Reproduced from Voltage Dips and Short Interruptions in Electricity Supply Systems. UNIPEDE Report 91)

Depth (% of rated voltage)		Duration (*d*)					
		(ms)	(ms)	(s)	(s)	(s)	(s)
From	To	10 < 100	100 < 500	0.5 < 1.0	1 < 3.0	3 < 20	20 < 60
10	30	111	68	12	6	1	0
30	60	13	38	5	1	0	0
60	99	12	20	4	2	1	0
99	100	1	12	16	3	3	4

Table 18.5 Cost of voltage dips [10]

Industry	Duration	Cost/sag
UK steelworks	30 % for 3.5 cycles	£250 000
US glass plant	Less than 1 s	$200 000
US computer center	2 s	$600 000
US car plant	Annual exposure	$10m
South Africa (country)	Annual exposure	$3bn

Table 18.6 Cost of voltage dips [3] (Reproduced from Leonardo Power Quality Application Guide – Part 2.1, Copper Development Association)

Industry	Typical financial loss per event (€000)
Semiconductor production	3800
Financial trading	6000 per hour
Computer center	750
Telecommunications	30 per minute
Steelworks	350
Glass manufacture	250

immunities described by immunity curves. This study was trying to determine at which dip duration and depth the process will stop or will be influenced. Determining the level of vulnerability of each single component to influence the tripping of the entire process would accomplish the task. Methodologies exist on how to carry out this part of the analysis, but it should be emphasized that such work requires knowledge, skill and experience.

Finally the potential consequences in terms of financial loss have to be determined. This is simply an answer to the question of what will happen if, for a precisely described scenario, a voltage dip occurs.

There have been numerous studies on voltage dips and their comparisons to interruptions: Table 18.5 presents the cost of voltage dips per event based on [10]; and Table 18.6 presents the financial losses caused by voltage sags [3].

In summary, a single voltage dip that is deep and long enough (usually referred to as longer than a few seconds and deeper than 60 %) may cost almost the same as short or long interruptions. Depending on the source, the overall effects of dips are between 50 % to more than 100 % of those attributed to interruptions. Individual companies and even industrial sectors may certainly experience a cost impact proportion of dips to interruptions that is much higher than 1. For example, a commonly acknowledged extreme case is the semiconductor sector, whilst for some other continuous manufacturing processes voltage dips are usually considered as the most financially damaging.

18.6 HARMONICS

The costs of harmonics are probably even more difficult to identify than any other PQ costs already considered. The effects of voltage and current distortions on any equipment or component fall into four categories:

- Additional energy losses compared to pure sine wave conditions.
- Premature equipment ageing.
- Incorrectly operating equipment performance compared to nominal conditions.
- Loss of reliability because of harmonics (this category is sometimes included under premature ageing or in the incorrectly operating category).

Methodologies and tools exist to quantify additional losses in cables, transformers, motors and capacitors. There are some on-line and off-line tools available to calculate these losses when the harmonics and load profiles are known. Customers usually complain that the load and especially the harmonics' profile are changing and encounter problems quantifying these losses.

Also, information from equipment manufacturers exists relating to the degree to which so-called hot spots and insulation working temperature (resulting from additional losses) negatively influence the equipment's effective lifetime. Again, lack of the necessary measurements is the main problem in applying this information.

Incorrectly operating equipment performance is an even more complex subject because of the absence of absolute references of the cause/effect link between harmonics and loss of performance of that equipment as the main issue that can identify harmonics as the only cause of the disturbance.

It makes sense only to consider aspects of reliability when sound referential information exists. Having said that, it must be admitted that there are few studies that satisfactorily describe how to deal with the reliability aspects of harmonics. Again the problem comes back to the question of whether and to what degree harmonics are responsible for an increased failure rate.

The following is a summary of a Eurelectric study and estimations of harmonics-related costs. Eurelectric in 2002 [6] reported that:

- Hydro Quebec assessed the cost of harmonics for a local community of 7.5 million people representing a GDP of C$150bn and energy use of 120 TWh at C$650m.
- EdF, instead of assessing harmonics-related PQ cost, calculated the cost of harmonics mitigation in system HV/MV substations by application of active filters and additional upsizing of neutral conductors in LV networks. These costs were at the level of €25bn.
- Similarly in Germany the cost of replacement of critical network components to fit the desired harmonic levels would cost approximately the same as in France.
- In Spain different costs to mitigate harmonics were estimated at the level of €6–7bn.

18.7 OTHER DISTURBANCES

There have been no other major surveys or studies on costs of PQ phenomena apart from the Leonardo PQ survey already mentioned above.

Individual case studies, however, do exist, which cover such phenomena as:

- Interharmonics
- Unbalance
- Overvoltages

- Transients
- Flicker
- Problems with earthing and high-frequency phenomena

These are responsible for substantial PQ costs within particular sites and regions. The approach here, however, is limited rather to analyzing historical events, recording the consequences mainly in terms of damaged equipment and associated costs.

The often quoted CEIDS/EPRI study grouped all PQ phenomena so far mentioned under one category, including voltage dips, though EPRI did not specify dips separately when enquiring about outages, clearly referring to total loss of electricity (zero voltage) and other PQ phenomena. However, when respondents were asked about 1 second interruptions and recloser effects, some deep dips were included by default. The summarized PQ cost in question amounted to €6.7bn; this accounted for approximately 10 % of the total surveyed PQ cost (dips plus interruptions) and for approximately 20 % of the total surveyed PQ cost attributed to the fabrication and essential services category.

Figure 18.1 and Figure 18.2 are two examples of non-interruption PQ cost findings from the EPRI/CEIDS survey. These examples include voltage dip impacts. In our opinion

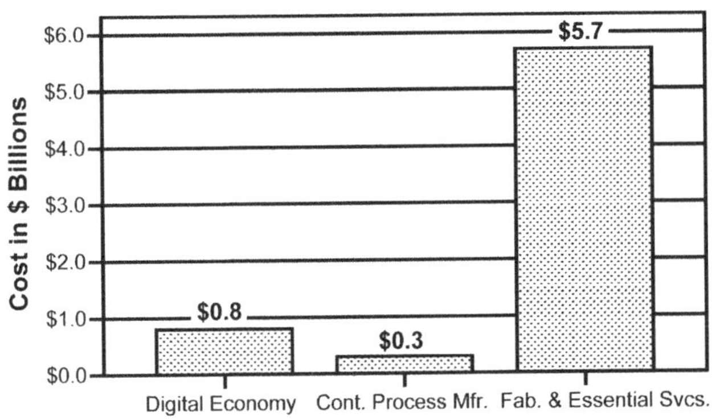

Figure 18.1 The division of non-interruption PQ cost per surveyed sector

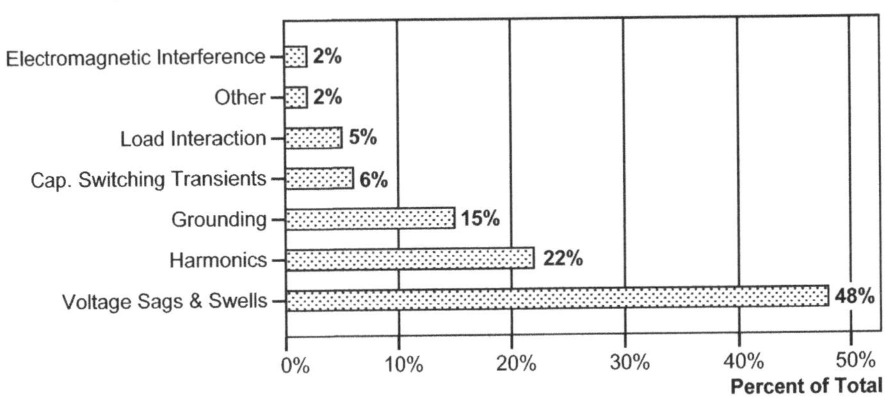

Figure 18.2 Root causes of non-interruption PQ cost

voltage dips, even if representing almost 50 % of root causes, are underestimated in the EPRI/CEIDS survey, especially for the continuous manufacturing sector.

18.8 PROFILES BY SECTOR

Apart from specifying the real PQ cost of wastage, respondents also defined those hypothetical costs that would be potential losses and risk avoided by power systems that had been immunized against the PQ disturbances under review.

The LPQI survey [11] is based on 11 individual cases per complete interview. The subsequent regression analysis was performed to estimate PQ cost across those sectors that offered a convergence of four specific indices. These indices initially were: employment, energy consumption, contractual power, annual turnover.

After refining them, the study concluded that annual turnover is a key indicator for a regression model (see Table 18.7 and Table 18.8 below).

To arrive at a statistically significant and acceptable model the survey sample was divided into two subsamples: *industry* and *services*. The banking sector was excluded because of its anomalous size and structure.

In the industry model, analysis shows that for the industry sector the estimation of how much wastage is caused by poor PQ is 4 % of annual turnover.

In the services model, the estimation of wastage caused by poor PQ is 0.1419 % of annual turnover.

Statistical bias is a real danger in research like this, especially in terms of how representative the study is of the target universe. This was resolved once the random and statistically based samples were checked. The regression analysis in the LPQI survey project proved that the samples and models were large and good enough to conclude that the variation explained by the model was not due to chance and that the relationship between the model and the dependent variable, which is annual PQ cost, was very strong.

The charts in Figure 18.3 present the cost extrapolations of wastage caused by the range of PQ phenomena throughout the sectors investigated in EU-25: PQ cost is characterized by disturbance type (absolute value in €bn and % value of total cost) and cost components.

The cost of wastage caused by poor PQ for EU-25 according to this analysis exceeds €150bn. Industry accounts for over 90 % of this wastage. That the proportion of these total PQ costs/losses accounted for relatively by services is possibly explained by cost underestimations by service sector organizations that often experience PQ problems in, say, an office environment, where distinguishing between the cause of a given PQ issue and other root causes may be difficult.

Furthermore, some service sectors like data centers, which probably experience high PQ costs, are not represented in the survey. Hospitals fit well into the services model and demonstrate slightly higher PQ costs than other service sectors.

Dips and short interruptions account for almost 60 % of the overall cost to industry and 57 % for the total sample.

Table 18.7 Statistics: frequencies; industry [11] (Reproduced from the 2007 Leonardo Power Quality Initiative Survey, R. Targosz)

	Total cost per kWh	Total cost per kW peak	Total cost per turnover	Total cost per electricity bill	Total cost per employment	Cost per kWh dip and short interruption	Cost per kWh long interruption	Cost per peak kW dip or short interruption	Cost per peak kW long interruption
	€/kWh	€/kW	Ratio	Ratio	€/employee	€/kWh	€/kWh	€/kW	€/kW
N Valid	41	41	41	41	41	41	41	41	41
Missing	0	0	0	0	0	0	0	0	0
Mean	0.625	730.37	0.0374	7.530	8 158	0.1352	0.0802	147	323
Std error of mean	0.459	298.24	0.0089	5.330	2 995	0.0922	0.0469	57	260
Median	0.031	95.57	0.0109	0.413	1 218	0.0098	0.0012	28	4.18
Std. deviation	2.939	1 909.72	0.0574	34 163	19 181	0.5903	0.3008	371	1 670
Variance	8	3 647 060	0.0030	1 167 000	367 922 790	0.3490	0.0910	137 717	2 789 088
Minimum	0	0	0	0	0	0	0	0	0
Maximum	18.861	10 681	0.2375	218 552	113 638	3.7720	1.7090	1 771	10 681

Table 18.8 Statistics: frequencies; services [11] (Reproduced from the 2007 Leonardo Power Quality Initiative Survey, R. Targosz)

	Total cost per kWh	Total cost per kW peak	Total cost per turnover	Total cost per electricity bill	Total cost per employment	Cost per kWh dip and short interruption	Cost per kWh long interruption	Cost per peak kW dip or short interruption	Cost per peak kW long interruption
	€/kWh	€/kW	Ratio	Ratio	€/employee	€/kWh	€/kWh	€/kW	€/kW
N Valid	21	21	21	21	21	21	21	21	21
Missing	0	0	0	0	0	0	0	0	0
Mean	9.330	746.23	0.01210	135.76	2 975.51	3.2080	3.8802	125.96	363.85
Std. Error of Mean	6.040	362.03	0.00370	89.26	813.53	2.3070	3.5747	63.09	276.99
Median	0.056	297.40	0.00470	0.98	1 510.70	0.0032	0.0046	7.72	10.89
Std. Deviation	27.700	1 659.07	0.01702	409.05	3 728.09	10.5700	16.3810	289.14	1 269.35
Variance	767	2 752 514	0	167 328	13 898 708	111	268	83 603	1 611 253
Minimum	0	0	0	0	0	0	0	0	0
Maximum	92.700	7 209.21	0.06980	1 603.80	11 853.18	46.3500	75.2850	133 692	5 872.29

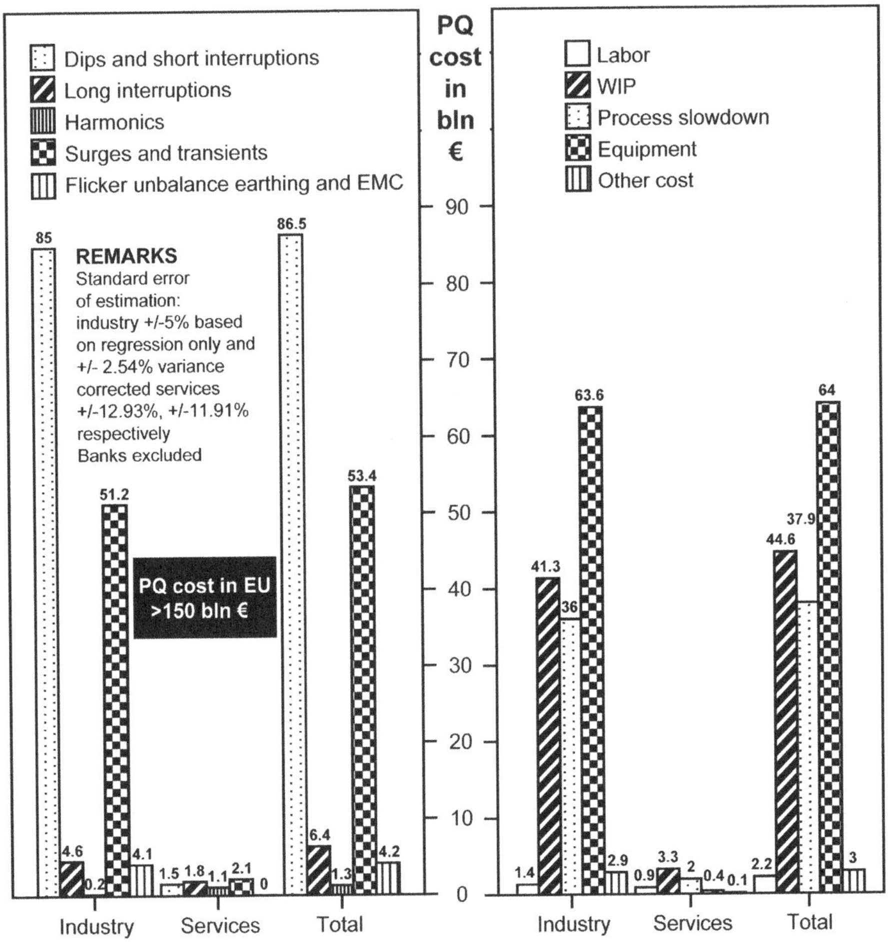

Figure 18.3 Extrapolation of PQ cost to EU economy in LPQI surveyed sectors [11] (Reproduced from the 2007 Leonardo Power Quality Initiative Survey, R. Targosz)

This extrapolation corresponds well with those levels indicated by the CEIDS survey [9] in 2000 which reports between \$119 and 188bn as the cost of poor PQ-generated wastage in the USA with 4 % of companies reporting annual costs of 10 % or more of annual revenue and 9 % reporting costs of between 1 and 9.99 %.

The LPQI survey shows that the economic impact of inadequate PQ costs industry some 4 % of turnover and services some 0.15 %. Whilst these values are extrapolations based on the sample interviewed (42 industrial companies and 21 service companies), it can be said with confidence that significant differences exist between the two.

Among industrial companies the highest values occur in typical continuous manufacturing industries and lower values in the metallurgical, food and beverages, and general manufacturing sectors.

Within the service sector hospitals are significantly higher than other groups.

Table 18.7 and Table 18.8 present the statistics relating to different PQ cost frequencies, grouped by industry and services.

The three industry histograms presented in Figure 18.4 show the distribution of frequencies for different types of PQ cost indices. They show that in the case of PQ cost per turnover, the frequency is closer to a normal distribution curve and the ratio of mean to standard deviation is far lower than in the case of other indices

Also, when comparing other statistics, especially variance, it is clear that the most accurate model would be based on PQ cost per company turnover.

When analyzing other indices, it is also apparent that the more appropriate representation is values that are closer to the median rather than the mean.

The general structure of PQ cost for each sector is presented in Figure 18.5.

It is noticeable that in typical continuous manufacturing sectors the losses incurred by lost work-in-progress (WIP) is quite significant and responsible for about one-third of the PQ costs recorded.

The slowing down of processes, which sometimes integrates WIP, and labor cost where these are not visible as independent components, is understandably even more significant.

In other sectors the situation is less clear with either labor cost or equipment-related costs being the most important source of economic losses.

Finally, in relation to public services like hotels and the retail sector, PQ impact is measured as slowing down their business activities, in terms of revenues that are irrevocably lost.

Figure 18.6, Figure 18.7, Figure 18.8, Figure 18.9 and Figure 18.10 respectively present PQ cost structures for five major groups of PQ disturbances – dips, short and long interruptions, harmonics, and surges and transients.

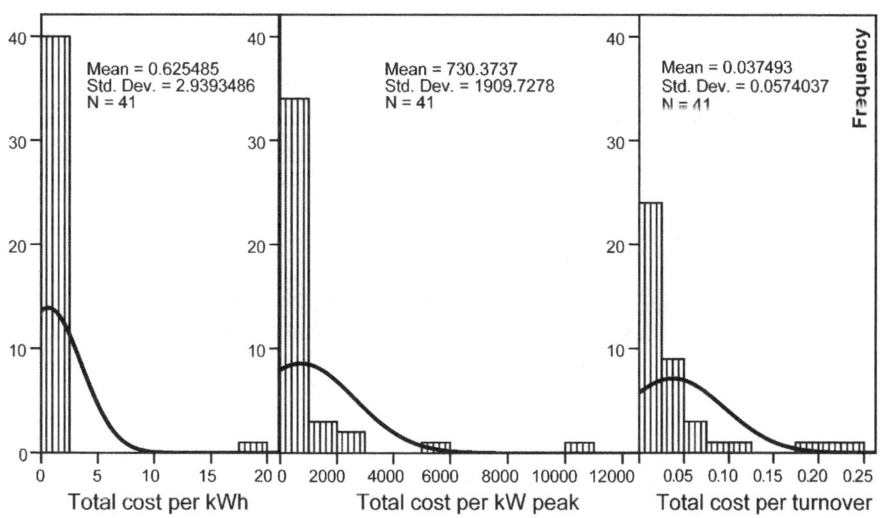

Figure 18.4 Frequency distribution of different PQ cost indices [11] (Reproduced from the 2007 Leonardo Power Quality Initiative Survey, R. Targosz)

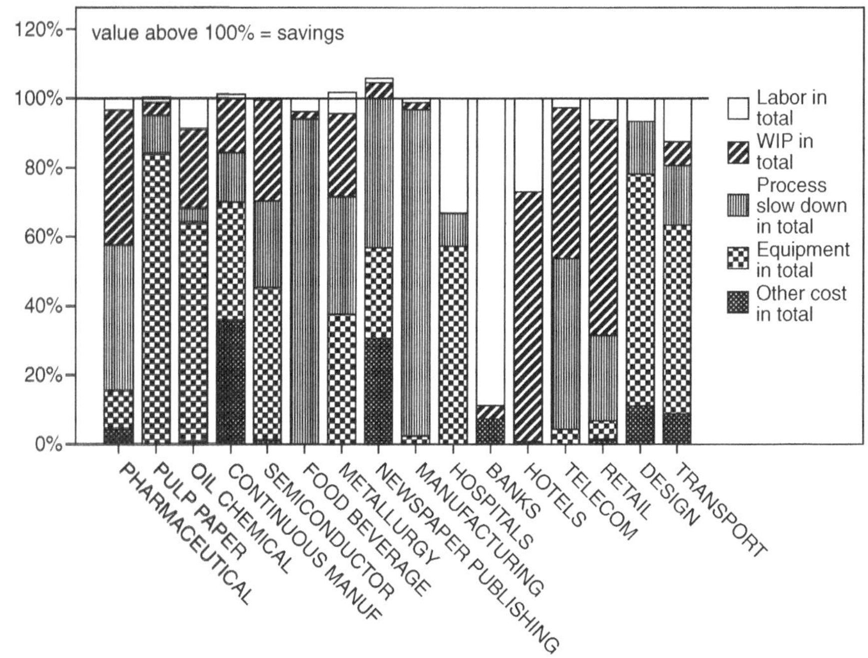

Figure 18.5 PQ cost components [11] (Reproduced from the 2007 Leonardo Power Quality Initiative Survey, R. Targosz)

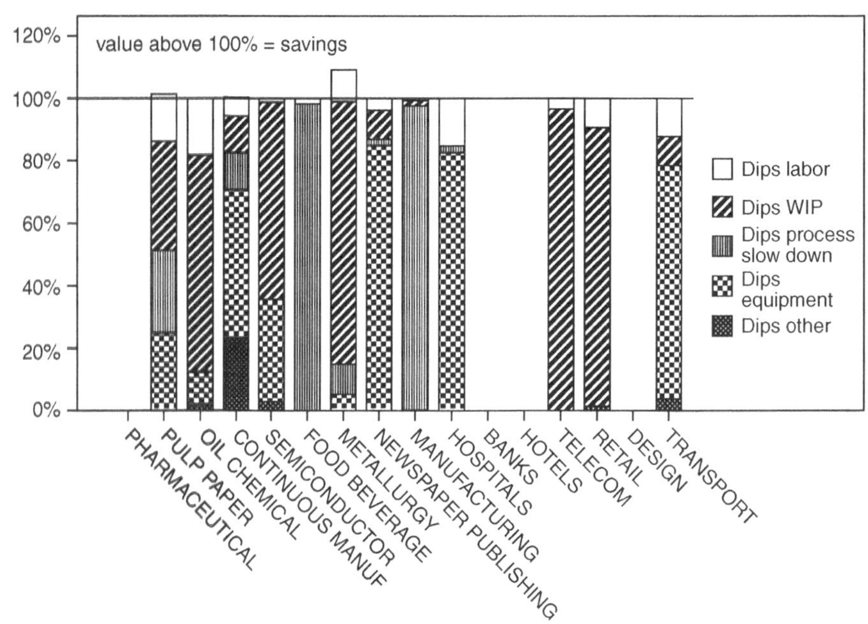

Figure 18.6 PQ cost components per disturbance type [11] (Reproduced from the 2007 Leonardo Power Quality Initiative Survey, R. Targosz)

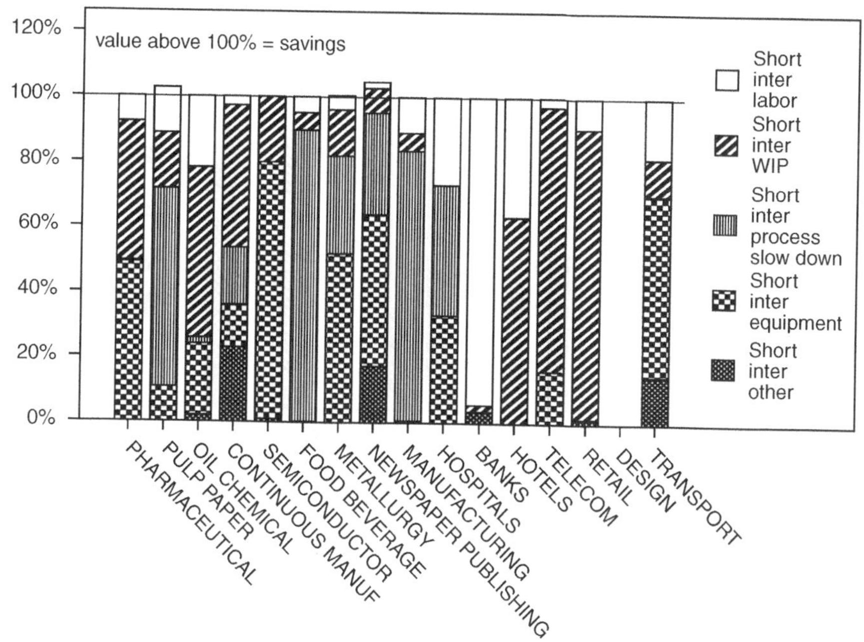

Figure 18.7 PQ cost components per disturbance type [11] (Reproduced from the 2007 Leonardo Power Quality Initiative Survey, R. Targosz)

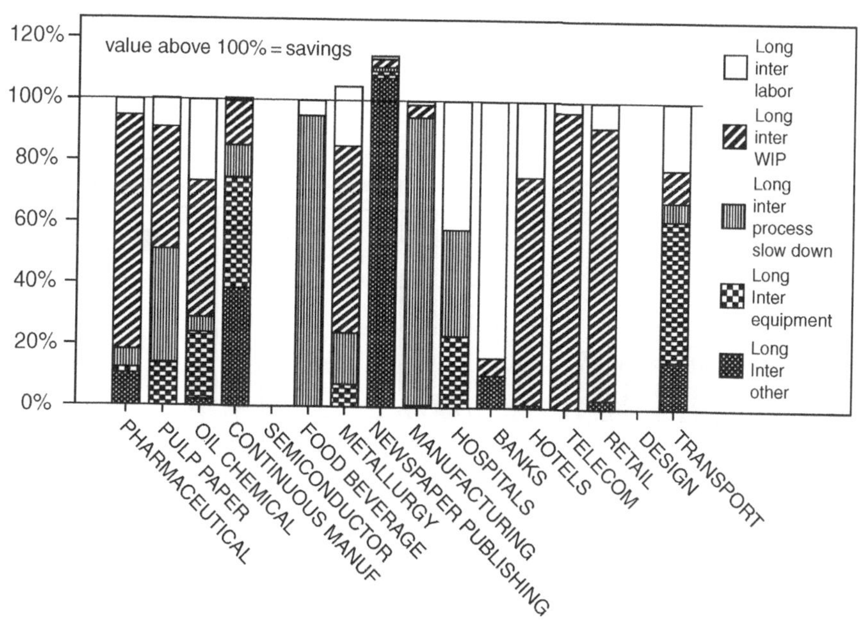

Figure 18.8 PQ cost components per disturbance type [11] (Reproduced from the 2007 Leonardo Power Quality Initiative Survey, R. Targosz)

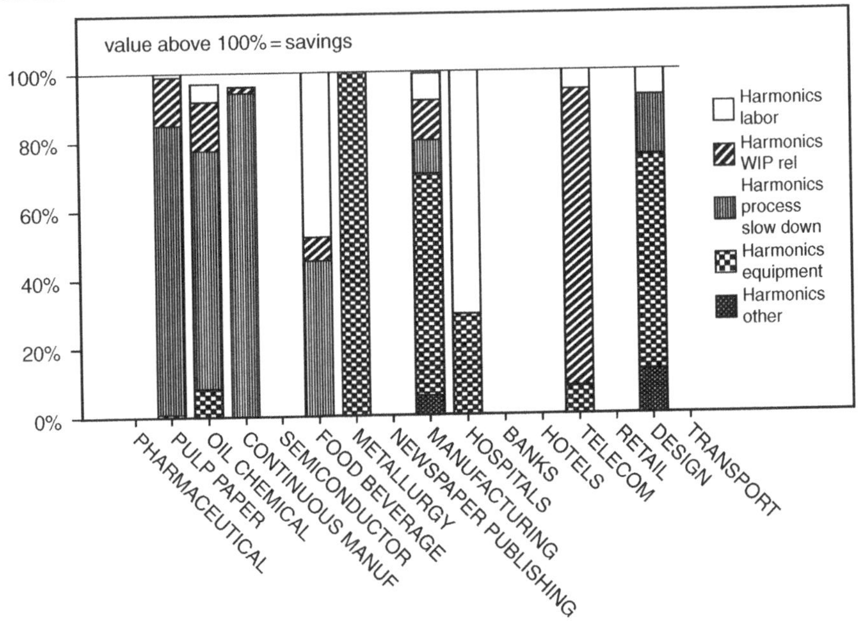

Figure 18.9 PQ cost components per disturbance type [11] (Reproduced from the 2007 Leonardo Power Quality Initiative Survey, R. Targosz)

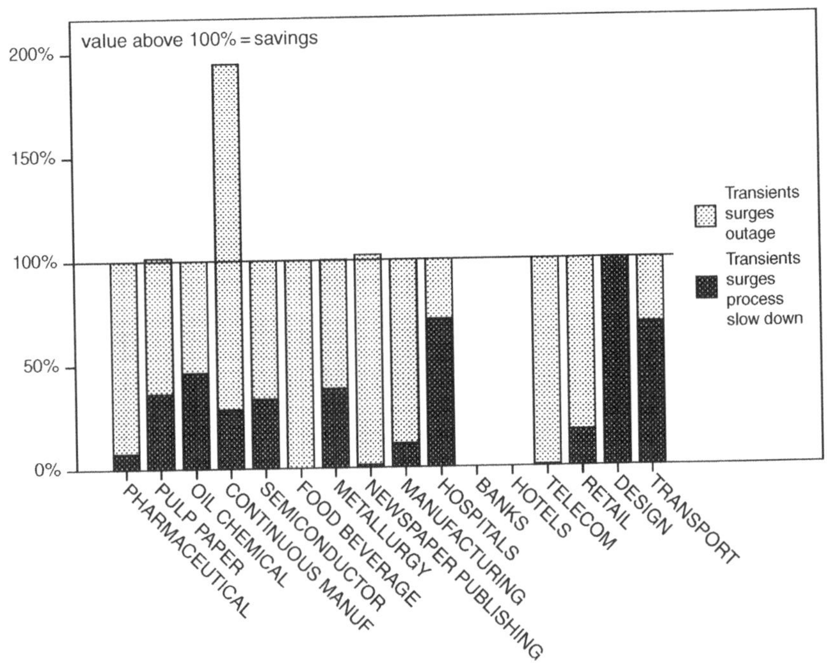

Figure 18.10 PQ cost components per disturbance type [11] (Reproduced from the 2007 Leonardo Power Quality Initiative Survey, R. Targosz)

The specific observations are:

- *Voltage dips* – WIP accounts for almost 50 % of PQ cost and the largest single source of PQ-caused economic losses, with process slowdown accounting for a further 30 %.
- *Short interruptions* – the cost structure is similar but costs relating to equipment failure/premature ageing are more significant. This is in part explained by the influence of the semiconductor sector. This sector claims to be fully immunized against dips; however, evidence suggests a lack of immunity to short interruptions, in which the category of equipment-related cost dominates.
- *Long interruptions* – the importance of labor costs increases, reaching some 20 % on average. This is twice as great as the equivalent figure for voltage dips and some 40 % higher than with short interruptions. In addition there are instances where some of the other cost categories take on greater importance and these tend to be related to the long-term economic consequences created by penalties (commercial or statutory), loss of brand equity or the need for unanticipated business investment to regain lost sales/market share.
- *Harmonics* – process slowdown generates almost two-thirds of all harmonics-related costs. Equipment-related costs represent about 25 % of these harmonics costs. However, only 8 out of 62 companies specified PQ cost of harmonics that related to additional energy losses. This cost in total is €186 000 and represents about 1 % of the total cost of harmonics in the sample.
- *Surges and transients* – production outage is again a major source of economic losses and is responsible for two-thirds of total PQ cost and consequential losses. An interesting observation from one of the continuous manufacturing companies (see Figure 18.10 – the bottom chart) was that 90 % of transient/surge cost was claimed successfully (presumably to the electricity supplier).

The breakdown of costs by different PQ disturbances is presented in Figure 18.11.

On average the absolute share of impacts (before sector grouping and extrapolation) of the six categories of disturbances taken from the total survey sample is as follows:

- Voltage dips 23.6 %
- Short interruptions 18.8 %
- Long interruptions 12.5 %
- Harmonics 5.4 %
- Surges and transients 29 %
- Other 10.7 %

These shares can be further summarized as:

- Voltage dips are the most important source of impacts in the continuous manufacturing sector.
- Short interruptions are most significant for food, metallurgy and newspaper publishing.
- Long interruptions are most costly for hotels and other public service sectors.
- Surges and transients are most destructive for the telecommunications sector and, perhaps surprisingly, for the pharmaceutical sector.

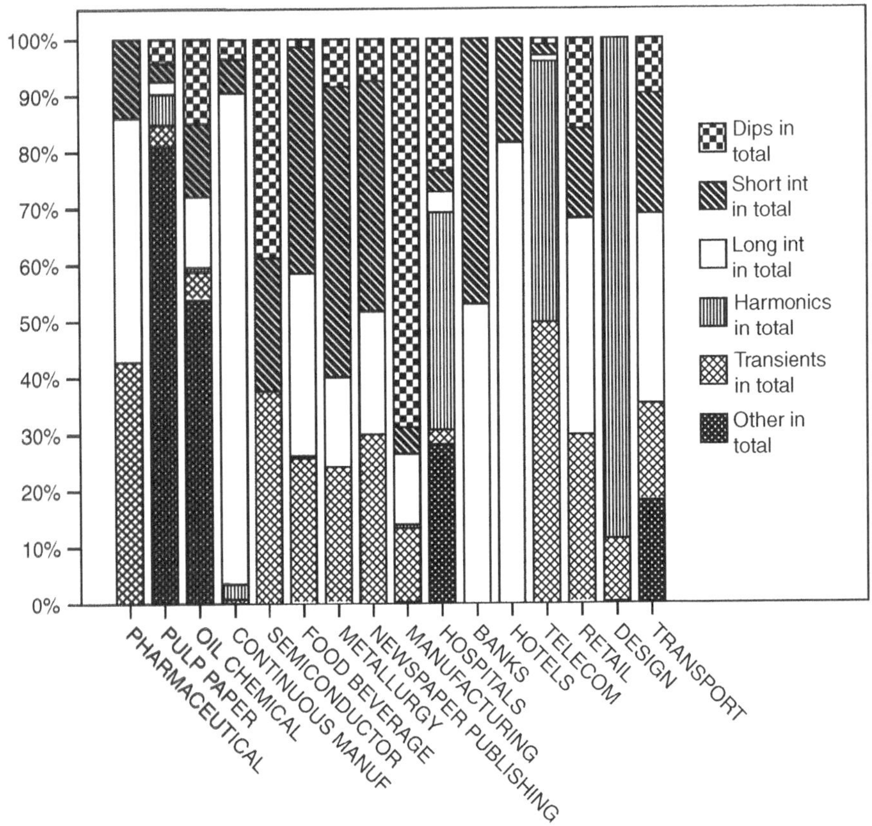

Figure 18.11 Cost structure of PQ disturbances in the survey sample [11] (Reproduced from the 2007 Leonardo Power Quality Initiative Survey, R. Targosz)

18.9 COST PER EVENT OF PQ DISTURBANCES

This section presents cost per event of PQ disturbances related to the frequency of different PQ disturbances by individual industrial sector as illustrated by the bar chart in Figure 18.12.

These are annualized data giving the frequency of disturbances per sector. In this figure one metallurgical company claiming short interruptions every day has been filtered to avoid distorting the overall picture.

Frequency of harmonics and flicker expressed in time percentage per year is illustrated by the chart in Figure 18.13.

The analysis of the average (yearly) values of disturbance frequencies broken down by the two categories of industry and services are shown in Table 18.9.

On average the MAIFI that has been measured by survey [11] is approximately 6; this is three times bigger than SAIFI. The number of recorded voltage dips identified by the respondents is approximately twice the number of short interruptions. This ratio is somewhat less for the service sectors.

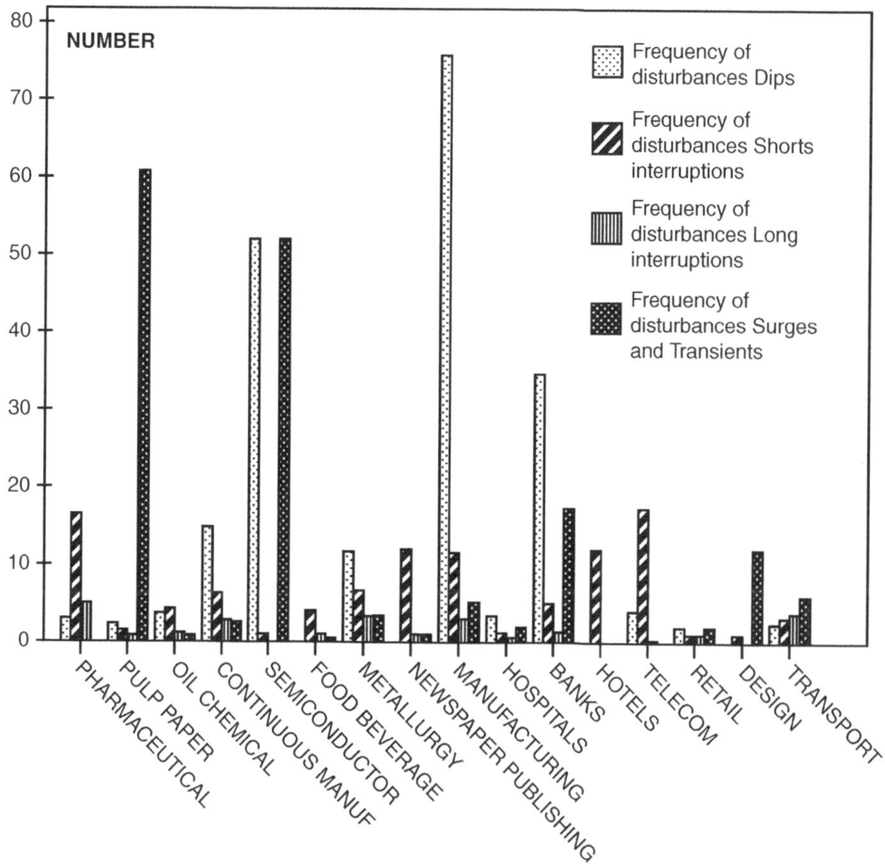

Figure 18.12 PQ disturbance frequency, dips, interruptions, surges and transients [11] (Reproduced from the 2007 Leonardo Power Quality Initiative Survey, R. Targosz)

The cost per event identified by the survey is shown in the chart in Figure 18.14.

To avoid potential distortion, 2 out of the 62 companies surveyed (semiconductors and retail) have been filtered out.

The average values of cost per event, i.e. the real losses incurred by the respondents as a result of PQ disturbances, are presented in Table 18.10.

These results can be summarized as follows:

- The cost per voltage dip event is between €2000 and €4000.
- Single short interruptions on average are 3.5 times more costly for industry and 9 times more costly for services.
- The average cost of long interruptions is €90 000 and is more homogeneous across the whole survey sample.

All these results are quite comparable to different studies described at the beginning of this chapter.

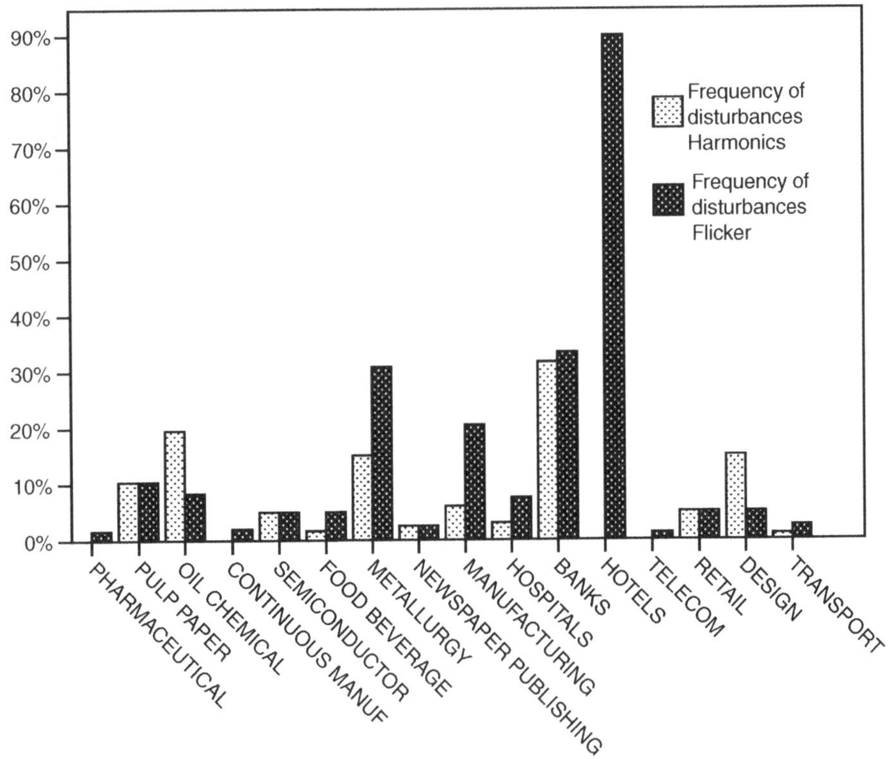

Figure 18.13 Frequency – annual time occurrence in %, harmonics, flicker [11] (Reproduced from the 2007 Leonardo Power Quality Initiative Survey, R. Targosz)

Table 18.9 Frequency of PQ disturbances [11] (Reproduced from the 2007 Leonardo Power Quality Initiative Survey, R. Targosz)

Frequency of PQ disturbances	Number of events				Annual (%)
	Voltage dips	Short interruptions	Long interruptions	Surges and transients	Harmonics
Industry	15.7	6.9	2.2	13.0	9.0
Services	7.7	5.4	2.1	6.7	7.5
Average	13.2	6.4	2.2	11.3	8.5

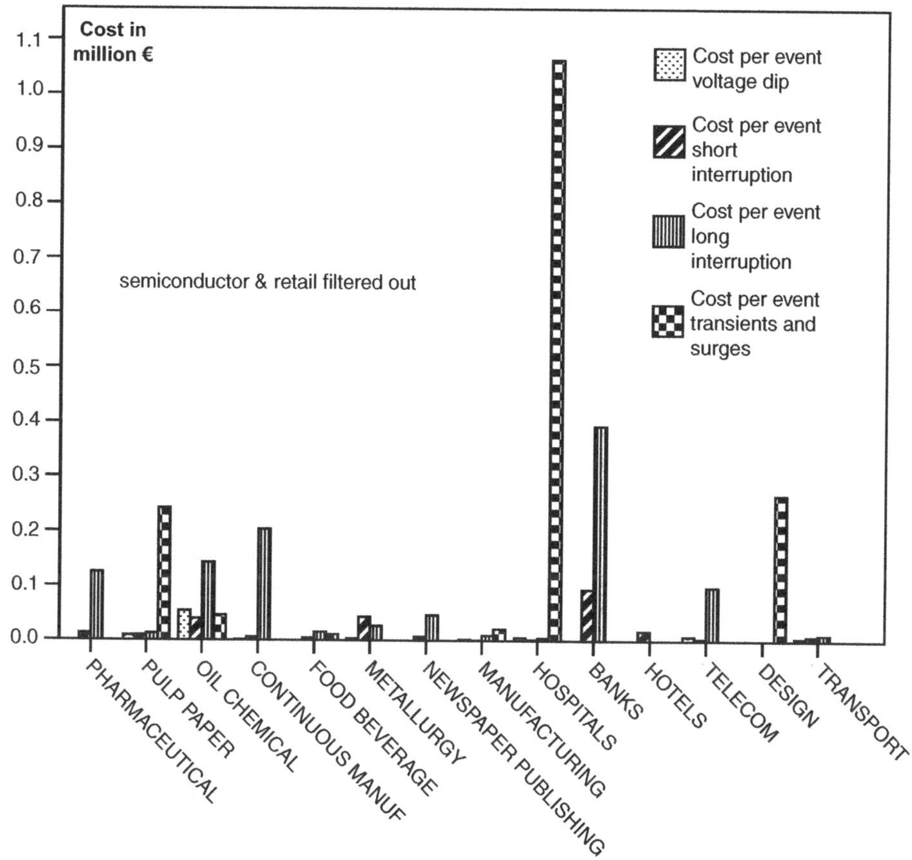

Figure 18.14 PQ cost per event for different disturbances [11] (Reproduced from the 2007 Leonardo Power Quality Initiative Survey, R. Targosz)

Table 18.10 PQ cost per event [11] (Reproduced from the 2007 Leonardo Power Quality Initiative Survey, R. Targosz)

Cost per event(€)	Voltage dips	Short interruptions	Long interruptions	Surges and transients
Industry	141 635	205 300	95 478	186 260
Services	22 064	47 762	272 916	122 602
Average	119 357	163 153	148 709	175 871
Industry[a]	4 682	15 484	95 478	186 260
Services[a]	2 120	19 447	80 326	122 602
Average[a]	4 177	16 539	91 021	175 871

[a] 2 out of the 62 companies surveyed (semiconductors and retail) filtered.

Assessing the generic cost per event for surges and transients is more problematic because of the lack of other research into these PQ phenomena. For survey [11] it is between €120 000 and €180 000.

18.10 PQ SOLUTIONS

PQ solution costs are to be evaluated case by case, but it is possible to provide some useful statistical data based on surveys carried out in a large number of cases.

Figure 18.15 charts the proportion of load per sector covered by different types of redundant or mitigating solutions.

The analysis of these solutions produced some interesting conclusions. Many of the correlations between solutions (both investment and load coverage) and PQ cost, frequency of events or sensitivity to PQ problems, which were thought to have been significant, have not been proven. Table 18.11 presents all significant relations, where 0.05 is used as reference threshold between PQ consequences and PQ solutions as confirmed by surveys.

Figure 18.16 shows a certain relation (although not proven by the linear regression model; R^2 linear $= 0.036$) between PQ investment and experienced PQ cost. Basically

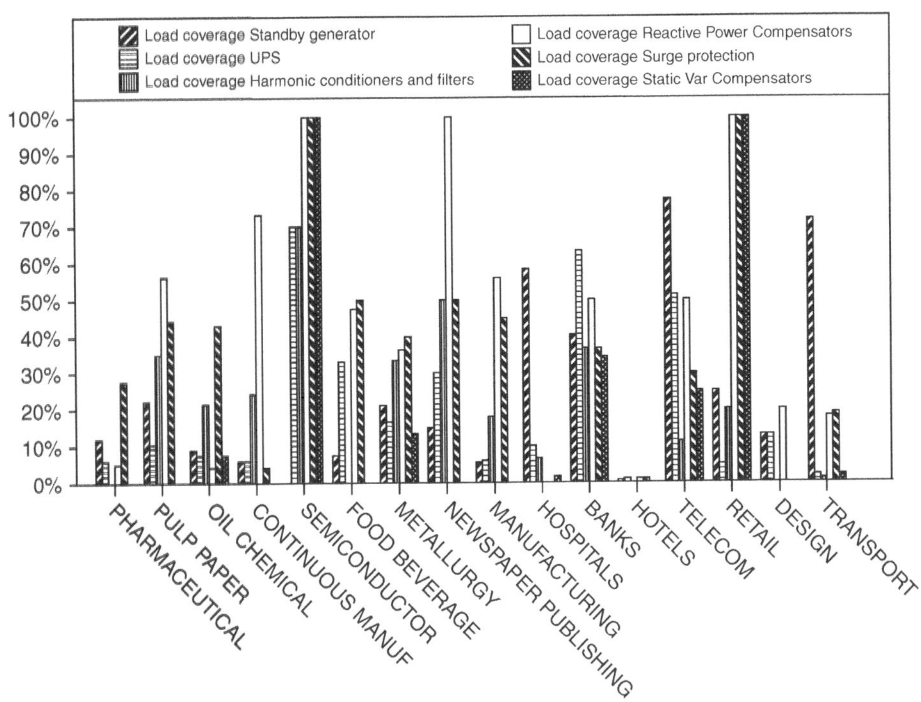

Figure 18.15 PQ solutions – installation coverage in % [11] (Reproduced from the 2007 Leonardo Power Quality Initiative Survey, R. Targosz)

Table 18.11 PQ consequence/PQ solution correlations [11]

PQ consequence	PQ solution	Sig. value[1]
Relay and contactor nuisance tripping	Harmonic filter (passive)	0.040
Noise interference in telecommunication lines	Multiple independent feeder	0.035
Motors or process equipment damaged	Shielding and grounding	0.046
Motor or other process equipment malfunctions	Backup generator	0.009
Loss of synchronization of processing equipment	Surge protection on key pieces of equipment	0.023
Loss of synchronization of processing equipment	Harmonic filter (passive)	0.010
Circuit-breakers or RCD nuisance tripping	Line conditioner or active filter	0.037
Circuit-breakers or RCD nuisance tripping	Backup generator	0.048
Capacitor bank failure	Shielding and grounding	0.007

[1] Relationship between PQ consequence and solution tested by chi-square test (one-sided): the lower the significance value [Sig.], the less likely it is that the two variables are independent (unrelated). Usually 0.05 is used as a reference threshold.

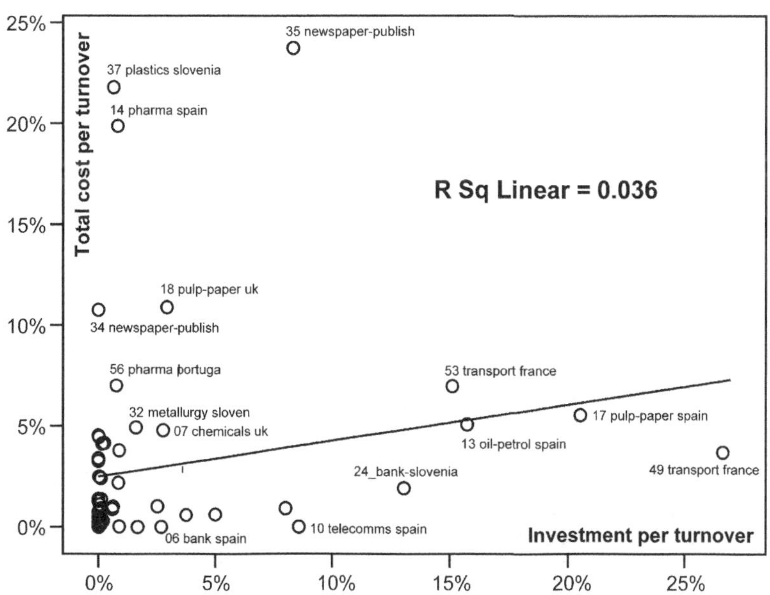

Figure 18.16 PQ cost/PQ solution investment relation [11] (Reproduced from the 2007 Leonardo Power Quality Initiative Survey, R. Targosz)

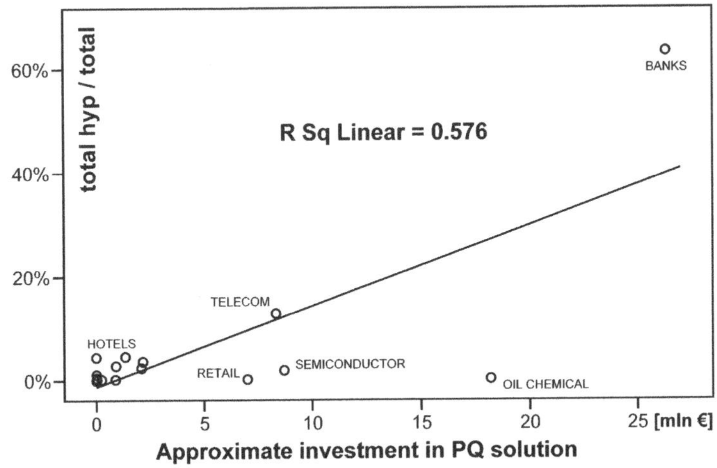

Figure 18.17 Mitigated and unmitigated PQ cost per unmitigated (real) PQ cost ratio as a function of PQ solution investment [11] (Reproduced from the 2007 Leonardo Power Quality Initiative Survey, R. Targosz)

the slope angle suggests a positive relation between PQ investment and PQ unmitigated cost.

Although there is no significant correlation between solutions and real cost, a strong correlation exists between investment in PQ solutions and the hypothetical to real cost ratio.

This results in an indirect but clear link between solutions and (real) consequences. See Figure 18.17.

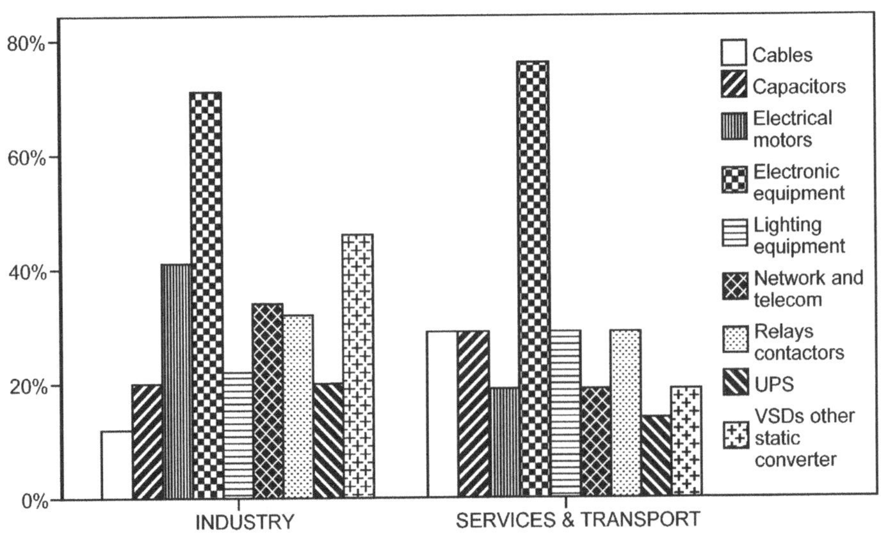

Figure 18.18 Occurrence of equipment affected by PQ in annual % [11] (Reproduced from the 2007 Leonardo Power Quality Initiative Survey, R. Targosz)

The following broad conclusions can be drawn:

- The increase in the ratio between hypothetical and real (mitigated/unmitigated) is very visible in the case of UPS.
- One side effect of UPS use is the increased cost of harmonics. This can be explained by suboptimal use of UPS systems that are based on diffused small units without active power wave modulation, which in turn generates significant input current distortion.
- There is a small but significant (positive) correlation between number of power lines and costs of short interruptions, whilst such a correlation is insignificant in so far as dips are concerned.

The study provided a number of additional conclusions regarding the occurrence of PQ problems, their sources and the equipment affected by them.

The occurrence of different equipment being affected by PQ is presented in Figure 18.18:

- Electronic equipment is most affected in the industry and service categories.
- Static converters and electric motors are the next most affected.
- All other equipment types are more evenly affected in the services category.

Below are some additional findings from the survey.

- The perceived level of presence of different PQ disturbances for all sectors is presented in Figure 18.19 and varies quite noticeably.
- The semiconductor respondents did not specify experiencing long interruptions, though they did record very intensive occurrence of voltage dips and short interruptions.
- For all sectors, on average, the presence of short interruptions is perceived as being the most intensive and disruptive.
- The same differences in perception between the industry and services categories also apply to the consequences of poor PQ (see Figure 18.20) and amount to:

  - Loss of synchronization of processing equipment, which is very common for continuous manufacturing and caused industry considerable problems for its activity.
  - Lock-ups of computers and switching equipment tripping were the second most problematic.
  - As far as services were concerned, circuit-breakers tripping and data loss cause the greatest problems.
  - According to survey [11], respondents affirmed that electric shocks are not relevant to the PQ issues investigated.

- The main sources or causes of PQ problems, see Figure 18.21, are defined as follows:

  - Motor-driven systems and, in general, static converters are the main sources of PQ problems for industry.
  - Electronic equipment and components are the equivalent main source for the services category.

- Regarding PQ solutions, Figure 18.22 presents the preferences of the two categories, industry and services.

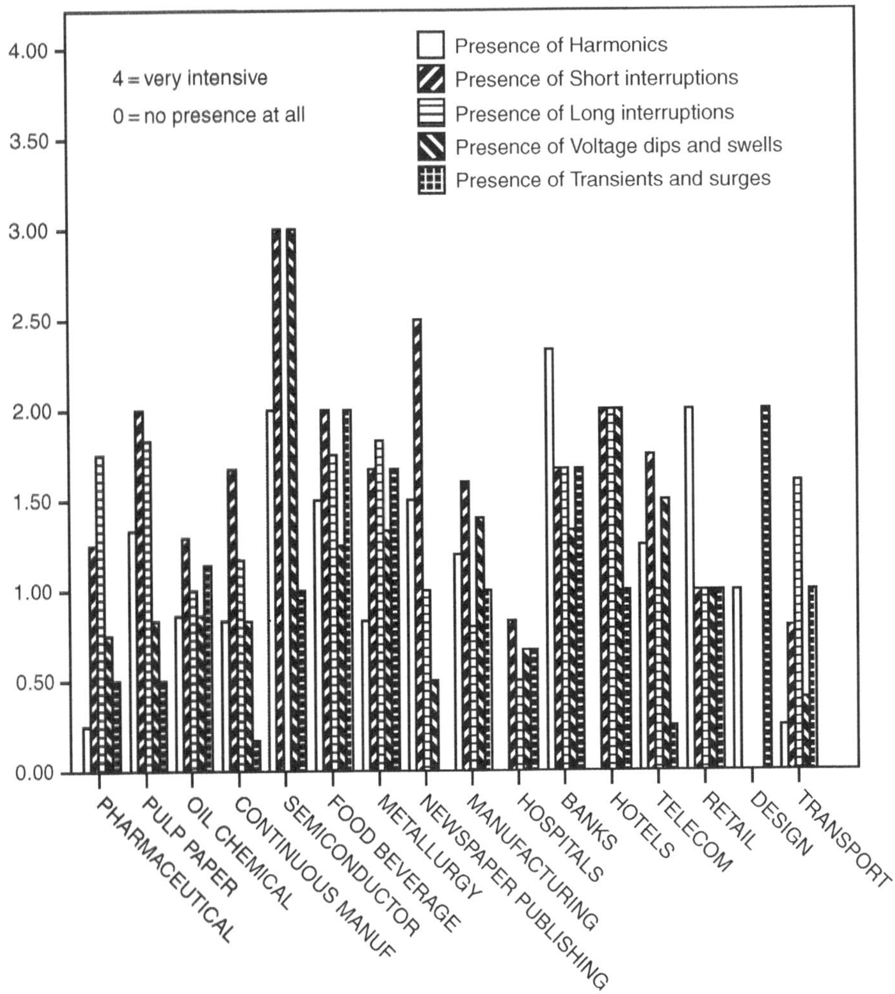

Figure 18.19 Presence (perceived) of PQ disturbances [11] (Reproduced from the 2007 Leonardo Power Quality Initiative Survey, R. Targosz)

- Both specify UPS most frequently.
- Backup generators, which prove to be most effective in the case of long interruptions, are dominant in services.
- Harmonic mitigation through harmonic filters is reported at 45 % to 65 % of the frequency.
- For industry, passive filters are almost three times more popular than active filters.
- For services, active filters are more popular but the difference is small.
- In general, services apply a higher frequency of different PQ solutions than found in industry.
- Industry tends to favor less costly, less universal solutions whenever possible.

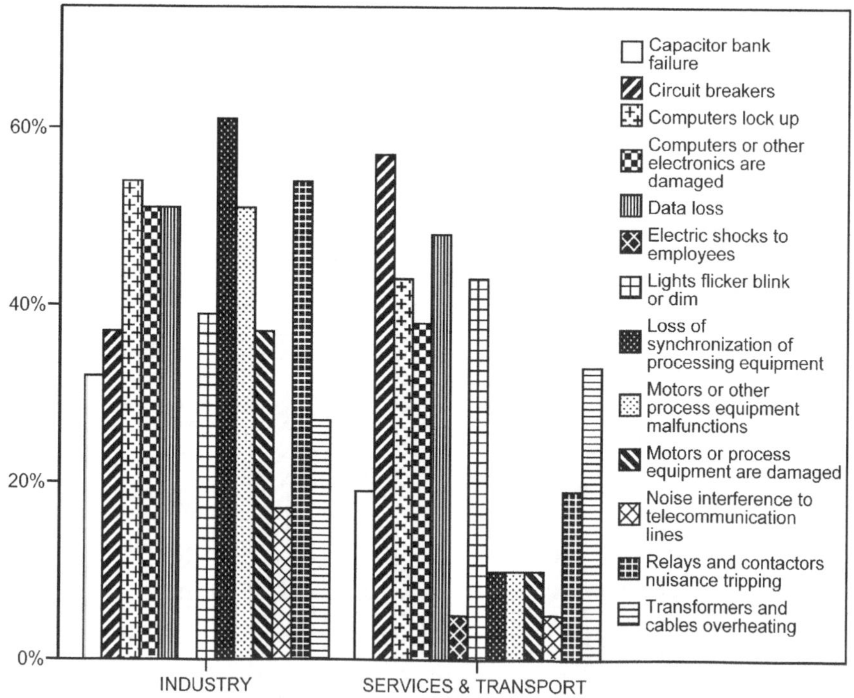

Figure 18.20 Frequency of PQ consequences as % of cases [11] (Reproduced from the 2007 Leonardo Power Quality Initiative Survey, R. Targosz)

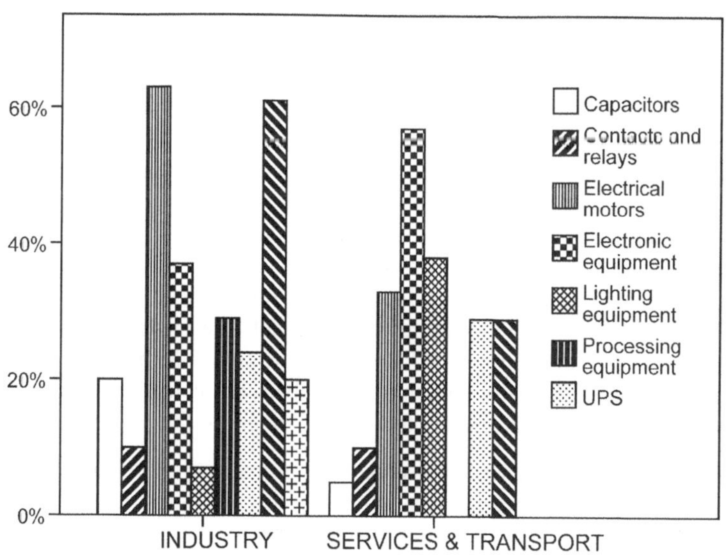

Figure 18.21 PQ problem source as % of cases [11] (Reproduced from the 2007 Leonardo Power Quality Initiative Survey, R. Targosz)

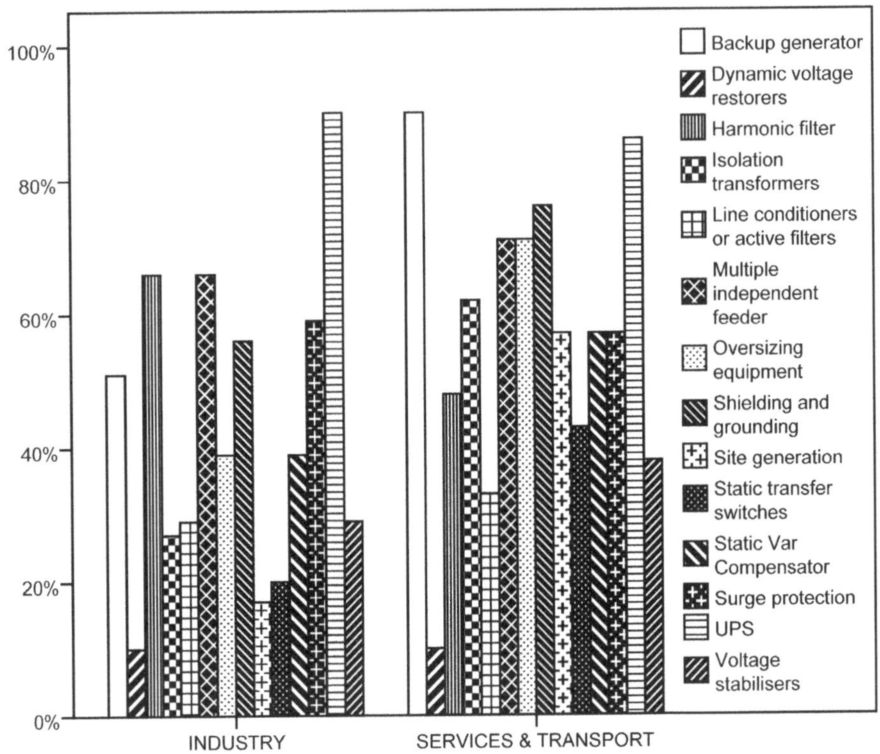

Figure 18.22 PQ solutions applied as % of cases [11] (Reproduced from the 2007 Leonardo Power Quality initiative Survey, R. Targosz)

- Looking at where the fault for poor PQ resides, in general the blame is usually placed at the foot of external causes. See Figure 18.23.
- Within that general statement, services more frequently admit that their installation could be the source.
- PQ measurement was of great concern because the survey [11] identified a different level of measurement of PQ parameters. Consequently the implication is that there exists an unequal level of understanding of and acceptance for the need for power-critical users to ensure consistently good PQ. Figure 18.24 presents the feedback to two questions – the ability to identify the sources of PQ events (the first four bars per category of the chart) and their frequency (the remaining eight bars per category) and the continuous monitoring of the key PQ parameters that further diagnose these issues:
  - For the identification data set, the average response across all information sources was 50 % – a level which, to repeat, is significantly low for industrial sectors that depend on good PQ. Within that, rather surprisingly the services' direct PQ measurement is much more frequent than that occurring in industry, where PQ data gathering is more reliant on the different PQ data acquisition components installed in its power systems.

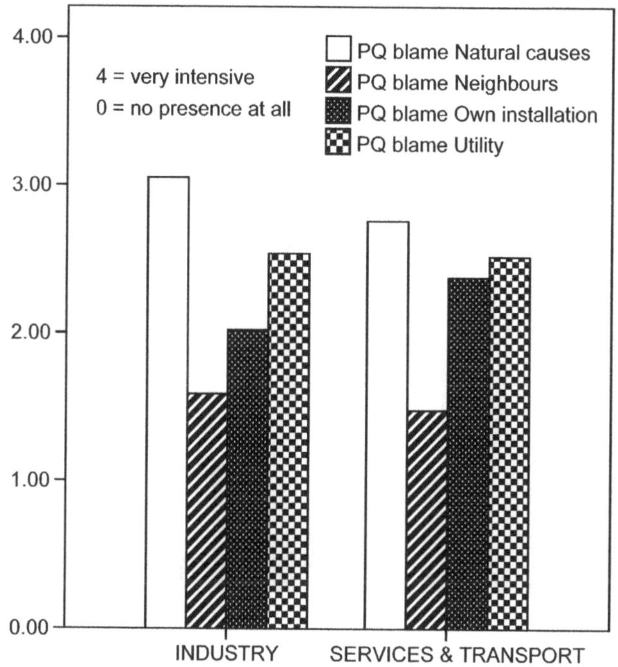

Figure 18.23 Poor PQ responsibility: 0, no; 4, high extreme [11] (Reproduced from the 2007 Leonardo Power Quality Initiative Survey, R. Targosz)

- Concerning continuous monitoring of key PQ parameters, this is more prevalent in industry than in services. Both reactive power and flicker are subject to continuous measurement several times more frequently by industry than by services. In 70 % of the industry cases, reactive power is subject to continuous measurement and this could be for financial reasons when reactive power is likely to be subject to separate accounting procedures.

- For the case of flicker, a high proportion of the companies interviewed in the survey (46 out of 62) agreed that flicker generates PQ costs in terms of losses generated in employee efficiency, which can amount to 10 % of annual employment cost. These costs are related to vision problems with symptoms like fatigue and increased error rate. These consequences relate to reduced productivity/inefficiency in work and in extreme cases to employee compensation. These costs amount to €167 m, which is equivalent to approximately 1.5 % of all hypothetical (mitigated) and real (unmitigated) costs. As this is an area of current and as yet inconclusive debate, and although respondents affirmed that their employees' efficiency was reduced by the levels of flicker experienced, the flicker cases at this stage have all been treated as hypothetical.

Finally, and in addition to the summary of these technical findings, as was stated earlier but merits repeating, it is astounding that industrial sectors, for which electric power is critical, are not fully aware of these issues.

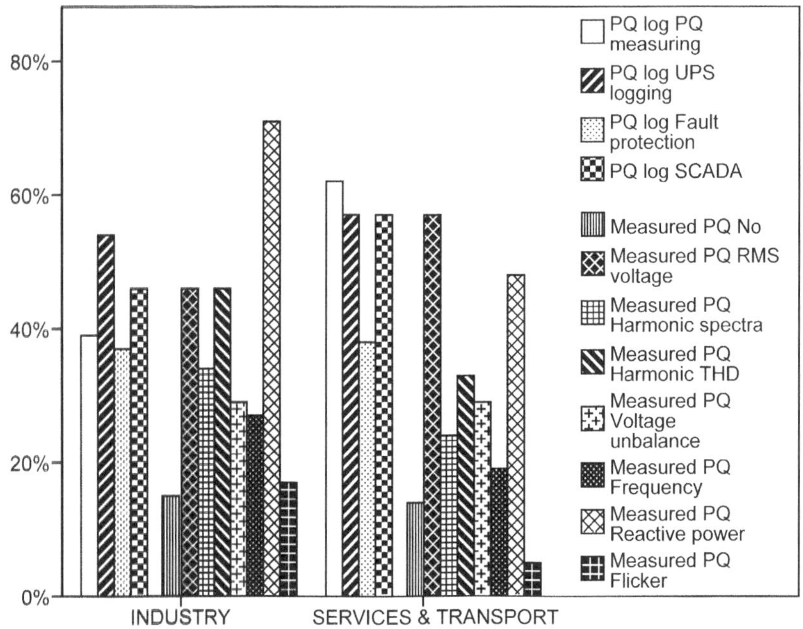

Figure 18.24 PQ monitoring: four left bars, source of PQ event information; remaining bars, measured PQ parameter [11] (Reproduced from the 2007 Leonardo Power Quality Initiative Survey, R. Targosz)

The main conclusion, however, remains that PQ costs in Europe are responsible for a serious reduction in industrial performance with an economic impact exceeding €150 bn.

18.11 INVESTMENT ANALYSIS TO MITIGATE COSTS OF PQ

18.11.1 Investment Analysis

Companies have several choices about how to spend capital in order to produce a return on their investment ranging from two possible extremes, on the one hand, of investing in a project, or, on the other, just depositing the money into an investment account.

Whatever the option, including PQ investments, it must compete for scarce capital with other investment opportunities. Therefore, the economic analysis of PQ investment should be conducted in the same way as the analysis of other capital investments to ensure that all options are compared on an equal basis. This decision process is called capital budgeting.

A particular problem arises for PQ investment, which is typical of any investment proposed in a cost reduction environment. In the capital budgeting process, some investments are earmarked as 'strategic', i.e. they are needed for the survival and growth of the firm, and hence receive priority. Another group of investments is required by law. They have little or no return on investment, but regulation requires them and the firm would never do them under purely economic criteria. Some typical examples of this are those investments made to reduce the environmental impact of a given operation.

Once the strategic investments and investments required by law are fulfilled, usually little capital budget remains for investment in cost reduction measures, such as PQ investments. These rely on specific business units, using operating not capital income, to see the light of day. Such investment planning is usually made within the confines of very short acceptable payback or time perspectives, which for PQ investments can be expected to be anything up to 1–2 years. Put another way, this represents the equivalent of a 50–100 % return rate, which is much higher than the average return on assets. Therefore, the scarcity of capital for PQ investments, and the requirement to finance from operating income, suboptimizes the firm's performance, and opens up opportunities for third-party financing.

A brief definition of capital budgeting principles and a summary of some useful definitions follow.

18.11.2 Capital Budgeting

The decision on whether to accept a project depends on the analysis of the cash flows resulting from the project. A capital budgeting decision rule should satisfy the following criteria:

- It must consider all of the project's cash flows (including working capital).
- It must consider the time value of money.
- It must always lead to the correct decision when choosing among mutually exclusive projects over different investment horizons.

The entire capital budgeting process relies on precise cash flow estimates. In general it is very important for the decision makers to obtain the most accurate forecasts possible. In order to do so they must do basically two things:

- identify all the variables that affect cash flows and determine which of those variables are critical to the success of the project;
- define the degree of forecasting accuracy required.

In the following subsections the most relevant capital budgeting decision rules will be presented. A distinction will be made between deterministic and stochastic methods. An evaluation method is considered to be deterministic if each cash flow can be precisely estimated, and it will be defined as stochastic when cash flows vary over a range and thus introduce a degree of uncertainty.

The focus will be on deterministic methods.

18.11.3 Project Classifications

When dealing with capital budgeting, projects can be classified as either *independent* or *mutually exclusive*.

An independent project is a project whose cash flows are not affected by the accept–reject decision of other projects. Thus, all independent projects which meet the company's capital budgeting criterion should be accepted.

Mutually exclusive projects are a set of projects among which only one will be accepted, e.g. a set of projects which accomplish the same task. Thus, when choosing among mutually exclusive projects, more than one project may satisfy the company's capital budgeting criterion, whilst only one, i.e. the best project, can be accepted.

18.11.4 Cost of Capital

As described below, discounted cash flow methods measure cash flows in terms of a required rate of return (hurdle rate) to determine their acceptability. This hurdle rate can be referred to as the firm's cost of capital.

The company's cost of capital is the discount rate which should be used in capital budgeting. The weighted average cost of capital (WACC) reflects the company's cost of obtaining capital to invest in long-term assets. Thus it reflects a weighted average of the company's cost of debt (long-term and short-term) and cost of equity (preferred stock, common stock).

Another way to define it is that the cost of capital represents the cost of funds used to acquire the total assets of the firm. Generally it refers to the rates of return expected by those parties contributing to the financial structure – preferred and common shareholders as well as creditors. Thus, it is generally calculated as a weighted average of the cost associated with each type of liability included in the financial structure of the enterprise.

With reference to capital budgeting, the concept underlying the definition of the cost of capital is that a firm must manage its assets and select capital projects with the goal of obtaining a yield at least sufficient to cover its cost of capital. Financial management separates the investment decision from the financing decision. A firm's financial structure is considered as fixed, and yields a WACC figure. Sometimes, the required rate of return for investment opportunities can be risk adjusted, i.e. low-risk projects have a lower hurdle rate, whereas high-risk projects must produce a return well above the WACC.

Another consideration is the debt-to-equity ratio. Firms may not wish to carry too much debt compared to equity, as this increases risk exposure of the firm. So projects may not be pursued, even if they provide an attractive return, because the firm wants to limit or reduce debt. Again, such a situation presents an opportunity for third-party financing.

18.11.5 The Time Value of Money

A given amount of money on hand today is worth more than the same amount to be received in the future because money available today can be invested to earn interest to yield more than the same amount in the future. The time value of money mathematics quantifies the value of a given amount of money through time. This, of course, depends upon the rate of return or interest rate which can be earned on the investment.

The time value of money concepts can be divided into two categories:

- Future value
- Present value

Future value describes the process of finding to which extent an investment today will grow to in the future. **Present Value** describes the process of determining what a given amount of money to be received in the future is worth in today's money.

18.11.6 Future Value of a Single Cash Flow

The future value of a single cash flow represents the amount, at some time in the future, that an investment made today will grow to if it is invested to earn a specific interest rate. For example, if you were to deposit €100 today in a bank account to earn an interest rate of 10 % compounded annually, this investment will grow to €110 in one year.

This can be shown as follows:

$$100(1+0.10) = €110.$$

The interest rate in the example is 10 % compounded annually. This implies that interest is paid annually. Thus the balance in the account was €110 at the end of the first year. Thus, in the second year the account pays 10 % on the initial principal of €100 and the €10 of interest earned in the first year. Thus, the €121 balance in the account after two years can be computed as follows:

$$110(1+0.10) = €121$$

or

$$100(1+0.10)(1+0.10) = €121 \text{ or } 100(1+0.10)^2 = €121$$

At the end of two years, the initial investment will have grown up to €121. Notice that the investment earned €11 in interest during the second year, whereas it only earned €10 in interest during the first year. Thus, in the second year, interest was earned not only on the initial investment of €100 but also on the €10 in interest that was paid at the end of the first year. This occurs because the interest rate in the example is a compound interest rate.

If the money were left in the account for one more year, interest would be earned on €121 and the balance in the account at the end of year 3 would be €133.10. This can be computed as follows:

$$121(1+0.10) = €133.10$$

or

$$100(1+0.10)(1+0.10)(1+0.10) = €133.10 \text{ or } 100(1+0.10)^3 = €133.10$$

A pattern should be becoming apparent. The future value of an initial investment at a given interest rate compounded annually at any point in the future can be found using the following equation:

$$FV_t = CF_0\,(1+r)^t \tag{18.1}$$

where FV_t is the future value at the end of year t; CF_0 is the initial investment; r is the annually compounded interest rate; and t is the number of years.

18.11.7 Present Value of a Single Cash Flow and of a Cash Flow Stream

Present value describes the process of determining what a cash flow to be received in the future is worth in today's money. Therefore, the present value of a future cash flow represents the amount of money today which, if invested at a particular interest rate, will grow to the amount of the future cash flow at that time in the future. The process of finding present values is called discounting and the interest rate used to calculate present values is called the discount rate. For example, the present value of €100 to be received one year from now is €90.91 if the discount rate is 10 % compounded annually.

This can be demonstrated as follows:

$$90.91(1+0.10) = €100 \text{ or } €90.91 = 100/(1+0.10)$$

Notice that the future value equation was used to describe the relationship between the present value and the future value. Thus, the present value of €100 to be received in two years can be shown to be €82.64 if the discount rate is 10 %.

This can be demonstrated as follows:

$$82.64(1+0.10)^2 = €100 \text{ or } €82.64 = 100/(1+0.10)^2$$

A pattern should be becoming apparent. The following equation can be used to calculate the present value of a future cash flow given the discount rate and number of years in the future that the cash flow occurs:[2]

$$PV = \frac{CF_t}{(1+r)^t} \tag{18.2}$$

where PV is the present value; CF_t is the future cash flow which occurs t years from now; r is the interest or discount rate; and t is the number of years.

The present value of a cash flow stream is equal to the sum of the present values of the individual cash flows:

$$PV = \sum_{t=0}^{T} \frac{CF_t}{(1+r)^t} \tag{18.3}$$

where PV is the present value of the cash flow stream; CF_t is the cash flow which occurs at the end of year t; r is the discount rate; t is the year, which ranges from 0 to T; and T is the last year in which a cash flow occurs.

18.11.8 Deterministic Approach to PQ Investment Analysis

The economic analysis of investments is one of the fundamental steps in any decision process because cost reduction is the main target for PQ investments.

[2] Note that this equation can be obtained algebraically from the future value equation.

The main elements of an investment to be investigated are:

- the capital cost or initial investment;
- the cost of capital;
- cost reduction;
- operating and maintenance expenses for the investment;
- the economic life of the investment.

Several evaluation methods can be used for investment, depending on the company's internal evaluation criteria. More or less sophisticated methods can be used as appropriate for the importance of the investment.

A distinction can be made between evaluation methods that use life cycle costing and those that do not. Evaluation methods that use life cycle costing are based on the conversion of investment and annual cash flows at various times to their equivalent present values. In other words, the *whole-life span* of the investment is taken into consideration. Typical examples of life cycle costing methods are: net present value (NPV) and internal rate of return (IIR).

Evaluation methods that do not use life cycle costing are for instance payback time (PBT) and break-even analysis. They do not take into consideration the life of the investment; they only define how long it will take to earn back the money spent on the project.

18.11.9 Discounted Cash Flow Methods

18.11.9.1 Net Present Value

The NPV of a project indicates the expected impact of the project on the value of the company.

Projects with a positive NPV are expected to increase the value of the company. Thus, the NPV decision rule specifies that all independent projects with a positive NPV should be accepted. If the NPV is greater than 0, the project is valid since the revenues are enough to pay the interest and to recover the initial capital cost before the end of the life of investment. When the NPV equals 0, the investment balances out at the end of the period and is consequently an unattractive proposition.

When choosing among mutually exclusive projects, the project with the largest (positive) NPV should be selected.

The NPV is calculated as the present value of the project's cash inflows minus the present value of the project's cash outflows. This relationship is expressed by the following formula:

$$NPV = \sum_{t=0}^{T} \frac{CF_t}{(1+r)^t} = CF_0 + \frac{CF_1}{(1+r)^1} + \frac{CF_2}{(1+r)^2} + \ldots + \frac{CF_T}{(1+r)^T} \tag{18.4}$$

where CF_t is the net cash flow at time t; r is the cost of capital; and T is life of the project.

The example in Table 18.12 illustrates the calculation of the NPV. Consider projects A and B which yield the following cash flows over their five-year lives. The cost of capital for the project is 10 %.

Table 18.12 Example that illustrates the calculation of NPV

Year	Project A cash flow (€)	Project B cash flow (€)
0	−1000	−1000
1	500	100
2	400	200
3	200	200
4	200	400
5	100	700
NPV	121.89	103.92

Thus, if projects A and B are independent projects then both projects should be accepted. On the other hand, if they are mutually exclusive projects then project A should be chosen since it has the larger NPV.

The NPV method takes all of the project's cash flows and the time value of money into consideration.

Projects can also be compared on the basis of the ratio between the present worth of the project and the related investment (NPV/I) being taken as a comparative parameter.

18.11.9.2 Internal Rate of Return

The IRR of a project is the discount rate at which the NPV of a project equals zero.

The IRR decision rule specifies that all independent projects with an IRR greater than the cost of capital should be accepted. When choosing among mutually exclusive projects, the project with the highest IRR should be selected as long as the IRR is greater than the cost of capital:

$$NPV = 0 = \sum_{t=0}^{T} \frac{CF_t}{(1+IRR)^t} = CF_0 + \frac{CF_1}{(1+IRR)^1} + \frac{CF_2}{(1+IRR)^2} + \ldots + \frac{CF_T}{(1+IIR)^T} \quad (18.5)$$

The example in Table 18.13 illustrates the determination of IRR. Consider projects A and B which yield the following cash flows over their five-year lives. The cost of capital for both projects is 10 %.

Thus, if projects A and B are independent projects then both projects should be accepted since their IRRs are greater than the cost of capital. On the other hand, if they are mutually exclusive projects then project A should be chosen since it has the higher IRR.

18.11.9.3 Annual Equivalent

If we assume the same cash flow every year, i.e. $CF_0 = CF_1 = \ldots = CF_T$, we can simplify Equation (18.3) to

$$PV = \frac{CF \cdot [(1+r)^T - 1]}{r(1+r)^T} \quad (18.6)$$

Table 18.13 Example that illustrates the calculation of IIR

Year	Project A cash flow (€)	Project B cash flow (€)
0	−1000	−1000
1	500	100
2	400	200
3	200	200
4	200	400
5	100	700
IRR (%)	17	13

This equation can be used to calculate annualized cash flows (ACFs) equivalent to an investment (I) made. For example, if an investment I is made in PQ mitigation, it is effective if the annual cost savings (ACS) are higher than this ACF + operating & maintenance expenses (OME):

$$ACF = \frac{I \cdot r \cdot (1+r)^T}{(1+r)^T - 1} \tag{18.7}$$

The annual cost of ownership (ACO) for this investment is ACS – OME – ACF. The investment decision should be positive when ACO is greater than 0.

A variant of this method is used in PQ Leonardo application note 5.5.1 [4] where the annual cost of poor power quality is added to annual investment (ACF) and operating and maintenance costs for various mitigation approaches. The minimum cost solution is proposed.

The ACO can be converted to total cost of ownership (TCO) through reuse of Equation (18.3):

$$TCO = \frac{ACO \cdot [(1+r)^T - 1]}{r \cdot (1+r)^T} \tag{18.8}$$

18.11.9.4 Comparison of These Methods

The decision process and the rules applied to both the NPV and IIR take all of the project's cash flows and the time value of money into consideration. The NPV and IRR differ with respect to their reinvestment rate assumptions.

The NPV decision rule implicitly assumes that the project's cash flows can be reinvested at the company's cost of capital, whereas the IRR decision rule implicitly assumes that the cash flows can be reinvested at the project's IRR. Since each project is likely to have a different IRR, the assumption underlying the NPV decision rule is more reasonable.

In general, engineering economic analysis presents the NPV as the most appropriate method on which to base investment decisions. The IRR has particular problems – for example, Equation (18.5) above does not always give a unique solution for the IRR. In addition, for a project where the IRR is high, e.g. 40 %, the assumption that the firm will be able to earn a 40 % return on the proceeds from the project is flawed. In today's information age, with the amount of desktop computing power available, there is no reason not to use the NPV systematically for making investment decisions.

18.11.10 Non Discounted Cash Flow Methods

18.11.10.1 Payback Time

The payback time (PBT) represents the amount of time that it takes for a project to recover its initial cost.

The use of the PBT as a capital budgeting decision rule specifies that all independent projects with a PBT less than a specified number of years should be accepted. When choosing among mutually exclusive projects, the project with the shortest payback is to be preferred.

The calculation of the PBT is best illustrated with an example (Table 18.14). Consider project A which yields the cash flows over its five-year life.

To begin the calculation of the PBT for project A an additional column to the above table is added (Table 18.15) and this represents the net cash flow (NCF) for the project in each year.

Notice that after two years the NCF is negative ($-1000+500+400=-100$) while after three years the NCF is positive ($-1000+500+400+200=100$).

Table 18.14 PBT project A

Year	Cash flow (€)
0	−1000
1	500
2	400
3	200
4	200
5	100

Table 18.15 PBT project A (NCF)

Year	Cash flow (€)	NCF (€)
0	−1000	−1000
1	500	−500
2	400	−100
3	200	100
4	200	300
5	100	400

Thus the PBT, or break-even point, occurs sometime during the third year. If we assume that the cash flows occur regularly over the course of the year, the PBT can be computed using the following equation:

$$PBT = Y_{LN} - \left(\frac{|NCF\left(Y_{LN}\right)|}{CF\left(Y_{LN+1}\right)} \right) \tag{18.9}$$

where Y_{LN} is the last year with a negative NCF; $NCF(Y_{LN})$ is the NCF in that year; and $CF(Y_{LN+1})$ is the total cash flow in the following year.

Thus in the example above, the last year with a negative NCF is year 2; the absolute value of the NCF in that year is equal to €100; the total cash flow in the following year (year 3) is equal to €200; therefore the project will recoup its initial investment in $2-(100/200)=2.5$ years.

Although widely used, the PBT suffers from several drawbacks. Firstly, it assumes that €200 received one year from today is equivalent to €200 received five years from today; in other words, it does not consider the time value of money. This issue can be resolved by calculating the discounted payback (DPBT), where cash flows are discounted to their present value based on the discount rate, making the DPBT consistent with life cycle costing methods such as the NPV and IIR.

Secondly, the PBT does not consider the effects of different lives of alternatives, penalizing projects that have long potential life. For example, if alternative investments A and B each cost €1000 and save €200 per year, then both would have a PBT of five years, making them seem equally acceptable. However, if investment A has an estimated useful life of five years and investment B has an estimated useful life of ten years, investment B would obviously be a better choice.

The third drawback is that the accept–reject criterion is often arbitrarily short. For example, many organizations require a one- to three-year payback period to consider a cost-saving project and place a higher priority on shorter payback projects. Therefore, the PBT will reject many interesting investment opportunities, though it may even accept projects that reduce a company's value. While it was used in the 1960s and 1970s before the computer era, today it should be avoided as much as possible. A fairly recent survey [3] shows that the NPV is by far the preferred tool among Fortune 1000 companies, with 85 % of respondents using it always or often.

18.11.11 Break-even Analysis

Break-even analysis can be used for projects where there is a gradual build-up of costs and benefits over time. For example, a production plant will need several years of investment in facilities, labor, training or services. After a certain time, the output of the plant will start to rise gradually as experience grows and output/sales develop. The point where accumulated costs equal accumulated benefits is called the break-even point. It typically applies for complex projects, and rarely applies for PQ investments.[3]

[3] More resources on this topic are to be found at http://www.lpqi.org and http://www.leonardo-energy.org/drupal.

BIBLIOGRAPHY

[1] Baggini A., Bua F., Investment analysis for PQ solutions. *Leonardo Power Quality Application Guide – Part 2.5*, 2004.
[2] Billinton R. *et al.*, 2001, Methods to consider customer interruption costs in power system analysis. Task Force 38.06.01, Official report.
[3] Chapman D., The cost of poor power quality. *Leonardo Power Quality Application Guide – Part 2.1*, Copper Development Association, November 2001.
[4] Didden M., Voltage sags in continuous processes – case study. *Leonardo Power Quality Application Guide – Part 5.5.1*, 2003.
[5] EN 50160, *Voltage characteristics of electricity supplied by public distribution systems*, November 1994 (revised version 1999).
[6] Gutierrez Iglesias J. *et al.*, Power quality in European electricity supply networks – NE Standardisation Specialist Group, N-E EMC & Harmonics, 2002.
[7] IEEE 493, *IEEE Recommended Practice for the Design of Reliable Industrial and Commercial Power Systems*, 1997.
[8] IEEE 1159, *Monitoring Electric Power Quality*, 1995.
[9] Lineweber D., McNulty S. R., The cost of power disturbances to industrial & digital economy companies. *Report of Primen for EPRI's (CEIDS)*, 2000.
[10] Milanovic J. V. and Jenkins N., Power quality cost for manufacturing industries. Paper presented at EdF Workshop, Paris, 2003.
[11] Targosz R., LPQI survey, Session 2, Block 4, Vienna, 2007.
[12] UNIPEDE, Voltage dips and short interruptions in electricity supply systems. Report 91, No. 50.02, 1991.

19

Power Quality and Rational Use of Energy

Pieter Vermeyen and Johan Driesen

Energy is too valuable to waste. Every unnecessary use should be avoided. This can be done in a simple way by being mindful of the use of equipment and infrastructure, such as switching off lights in offices after work or by not turning on the heating in spaces with open windows or open doors to the outside. Savings can also be accomplished by means of technology, such as adequate thermal insulation in buildings or efficient electrical drives. *Rational use of energy* is the concept of reducing energy consumption without sacrificing comfort.

Because of its environmental and economic importance, rational use of energy is of great interest to governments. Subsidies and tax benefits are granted for investments in certain energy-efficient technologies. Because certain groups of efficient devices are based on power electronics, the growing use of efficient devices can have a negative effect on power quality. The connection of power electronics systems to the distribution grid often leads to the injection of harmonic currents, although advanced power electronics can generate nearly perfect sinusoidal currents. On the other hand, power electronics devices can be more sensitive to poor power quality than classic lighting systems and drives, for example. This illustrates how the use of more power electronics within the scope of rational use of energy can be problematic from two points of view: emission of power quality problems and vulnerability (lack of immunity) to these problems.

The possible reasons for rational use of energy are discussed in the first section of this chapter. Next, possible techniques to implement rational use of energy are discussed. Afterwards the effects on power quality are discussed for the two most important domains: lighting with fluorescent lamps (including CFLs or energy-saving lamps) and variable speed drives (VSDs; or adjustable-speed drives, ASDs).

Handbook of Power Quality Edited by Angelo Baggini

19.1 REASONS FOR RATIONAL USE OF ENERGY

Organizing the use of electric energy in such a way that overall consumption becomes more efficient, i.e. using less energy while maintaining the same levels of comfort and technical performance, is beneficial for several reasons. It is a way of conserving exhaustible energy sources and dealing with energy shortages. It also reduces the impact of human activity on the environment. Because less energy has to be paid for, there is also an economic benefit, provided that the payback time of the investment cost is short. Regarding the electricity industry, higher energy efficiency of the consumers reduces the strain on the power system, which is often operated at its full capacity. These different aspects are briefly discussed in the next subsections.

19.1.1 Sustainable Development

Consuming less electricity contributes to the conservation of primary energy sources. Fossil fuels (coal, oil, gas) and nuclear fuel (i.e. fissionable material; uranium) will become depleted some day in the future. Decreasing electricity consumption postpones depletion of these sources and gives scientists more time to improve and develop existing and promising alternatives, such as renewable energy and nuclear fusion. Conserving fossil fuels is advisable for other reasons as well. Natural gas and oil are essential raw materials for the chemical industry. If possible, it is better to conserve them for this purpose rather than simply burning them for energy.

The use of fossil energy sources generates greenhouse gas emissions. Certain greenhouse gases escape during the handling and treatment of fossil fuels, but the main part of emissions consists of CO_2, which is the main residual product of the combustion of fossil fuels. Because there is considerable agreement among scientists that these emissions enhance the greenhouse effect and thereby affect the global climate, international climate agreements were made (e.g. the Kyoto Protocol) to reduce greenhouse gas emissions. Because fossil fuels represent an important part of the energy mix for electricity generation, reducing energy consumption is a possible way for contributing to emission reduction.

19.1.2 Economic Considerations

Rational use of electricity results in lower energy costs. However, in order to achieve energy savings, a considerable investment is often required. This can be the cost of rebuilding existing installations or replacing classic technology by modern technology. A financial analysis is required to determine whether the payback time of investments to improve efficiency is reasonable or not. Depending on the application, it can be important to replace a less efficient device by a more efficient device only at the end of the first device's lifetime, in order to have a profitable investment. Government support for efficient technology can be a decisive element.

Next to lower energy costs, increased energy efficiency is also beneficial for a company's electrical infrastructure. The ratings of power lines, transformers, capacitors, circuit-breakers and other electrical equipment will be sufficient for a longer period of time. In this way, major investments and downtime of equipment can be deferred or avoided.

19.1.3 Electricity Industry

The costs of production and trade of electricity fluctuate. Costs depend on factors such as fuel prices, technical restrictions and economic elements. Electricity retailers, who sell energy to consumers, try to minimize costs, as does every company. Influencing consumption patterns is a very useful tool to meet this purpose. This is the field of demand-side management. If consumers have a predictable and sufficiently level consumption profile, without high peaks, producers, traders and retailers are more in control of their budget and expenses.

For producers of electricity, which operate power plants, constant production profiles allow them to optimize fuel efficiency and minimize costs. If the demand profile contains high peaks, a large amount of standby production units are required, used only for a small fraction of the time. A reduction of standby capacity lowers the fixed and variable costs. If the demand profile is leveled, the contribution of base production units can be increased. These production units are not very flexible, but they are relatively cheap. Because they constantly operate about a certain operating point, they can be optimized for minimal fuel consumption and minimal emissions.

Likewise for the operators of the transmission and distribution grids, level consumption profiles result in more efficient operation and planning of the grid. Lower consumption peaks also result in lower currents and therefore lower losses in power lines. Influencing consumption profiles is done by means of variable electricity tariffs. A high tariff is used during consumption peaks and a low tariff in off-peak periods. In certain countries utilities control large loads on the consumers' premises, such as boilers and electric heating. This is done by means of control signals, broadcast by means of radio waves or ripple control (i.e. low-frequency power line communication). In the future fast communication and intelligent loads will be used to make power systems more flexible, in order to optimize load profiles.

19.2 TECHNIQUES FOR RATIONAL USE OF ENERGY

Because of the reasons mentioned, it is interesting for the different parties or players in the electricity market, including energy consumers, to reduce energy consumption or shift consumption in time. This can be achieved by adapting consumption habits and practices, by using energy-efficient technology and by means of variable electricity tariffs. These techniques are discussed in this section, including the available technology.

19.2.1 Adapted Use of Electrical Systems

Adapting the behavior of users of electrical systems is a simple, inexpensive, yet effective form of rational use of energy. Examples of adapted behavior are switching off devices that are not being used (e.g. lighting, computers, heating), unplugging switched-off devices to reduce standby losses, closing doors and windows of heated or cooled spaces, reducing climate control in buildings and instead adapting attire to the resulting temperature. People who do not wish to adapt their way of using energy, who do not care, or who are simply not aware of the need for it, should be informed or educated. This can be a part of the education of children and youths, training of employees or information campaigns by governments, for example.

19.2.2 Efficient Loads

In industry, a basic technique for keeping energy consumption under control or reducing consumption consists of good design and maintenance of the infrastructure. An important example is compressed air, which is often used in production processes to drive tools. By means of a compressor, air is pressurized and stored in a tank. Via a network of pipes the air is distributed over the factory. In such a system, air leaks are a major problem. Because energy is required to create pressure, leaks are equivalent to energy losses. Therefore a continuous leak-management program is required to save energy. This consists of constantly looking for leaks and repairing them. The system's air pressure should be limited, in order to reduce losses through leaks. Compressors should be accurately controlled, in order not to have excess air, which is vented into the atmosphere. Using compressors at partial capacity results in higher losses. This can be avoided by using multiple smaller compressors, one serving the base load, and the others switched on when additional capacity is required.

Conventional industrial equipment is often obsolete when it comes to efficiency. This equipment was installed because energy was not expensive and because it was simple to operate and maintain, or because there was no better alternative at the time. The typical example is flow control by means of a valve or speed control by means of a mechanical brake, while the pump or motor is working at full capacity. In this way control is achieved by converting excess mechanical energy into residual heat. Controlling flow or speed by applying speed control to the electric motor driving the system results in significant energy savings. In this way only the energy that is needed is converted from electrical into mechanical form. In general, switching to modern equipment can drastically decrease energy consumption.

Energy efficiency can also be increased by switching to an entirely different technology. The most important example for this is lighting. In classic incandescent lamps light is produced by sending a current through a filament to make it glow. These lamps have a low efficiency and a short lifetime. In compact fluorescent lamps (CFLs) light is generated by sending a current through ionized gas. CFLs are more efficient and last longer. Replacing incandescent lamps with CFLs reduces consumption and costs.

The principal technique for using energy in a more rational way is the (gradual) replacement of classical technology by technology that is more efficient. Important examples of possible improvements in electrical systems are:

- CFLs instead of incandescent lamps;
- motors with higher efficiency;
- VSDs instead of classical speed control;
- more efficient domestic appliances, e.g. refrigerators, washing machines;
- heating by means of microwaves and infrared radiation in industry.

In the next subsections, the following technologies are described: CFLs, high-efficiency motors and VSDs.

19.2.2.1 CFLs

Classical incandescent lamps have a very low efficiency. About 5 % of the consumed energy is emitted in the form of visible light. The remaining 95 % is turned into heat. They also have a short lifetime: 1000 hours on average. The advantages of incandescent lamps are the low price and the high quality of the light, whose spectrum resembles that of natural sunlight. Halogen lamps are incandescent lamps that are about 5 to 10 % more efficient than classic incandescent lamps. Halogen lamps with a heat-reflecting infrared coating (IRC) are 1.5 to 3 times more efficient than classic lamps. The lifetime of these lamps is 2000 hours or longer.

CFLs (also known as energy-saving lamps) are energy-efficient alternatives to incandescent lamps. They have similar shapes and can be placed in the same fittings. CFLs are compact versions of the fluorescent tube, with a built-in electronic ballast. Because of the different mechanism of light production (electrical discharge) the power required for equal light production is four to five times lower than that of incandescent lamps. The lifetime of a CFL is eight to twelve times longer than that of an incandescent lamp, i.e. 8000 to 12 000 hours. CFLs are more expensive, but due to the lower consumption and fewer replacements, the overall cost is lower. Topalis *et al.* [11] compare the costs of using incandescent lamps and using CFLs. The cost of using a 23 W CFL is at least 2.5 times lower than that of an equivalent (i.e. 100 W) incandescent lamp. Comparison of a 20 W CFL with a 75 W incandescent lamp shows the cost of the CFL is at least 2.2 times lower.

19.2.2.2 High-Efficiency Motors

Electric motors represent about 40 % of overall electricity consumption. Therefore improvements in motor efficiency have a significant effect on power consumption. A standard motor with a rated power below 10 kW has an efficiency of about 80 %. Efficiency increases with the power rating. Motors of over 75 kW have an efficiency of about 95 %. By means of better design and the use of new materials (e.g. better permanent magnets), high-efficiency motors are produced. Their efficiency is about 1 to 6 % higher than the standard equivalent. The improvement is larger for motors with lower power rating [4].

19.2.2.3 VSDs

In traditional motor-driven systems, control of speed or flow rate is often implemented mechanically: by means of a brake on the axis or a valve in a tube. In this way, the excess energy is converted into heat. With a VSD, the desired speed is imposed on the motor by means of the frequency of the supply voltage. Exactly the right amount of energy is delivered to the load, resulting in a more efficient system. Replacing a classic drive in a pump, fan or blower by a VSD typically results in energy savings of about 15 to 40 % [4]. Compressor drives have a smaller potential for energy savings of about 5 %.

19.2.3 Variable Tariffs

By shifting electricity consumption in time, a highly variable load profile can be made more level. As explained earlier, this is profitable for the consumer as well as the electricity retailer, the network operators and the electricity producers. Influencing consumers' load profile is attempted by means of variable electricity tariffs, e.g. day and night tariff for residential consumers. Cost-conscious consumers try to shift the use of certain appliances and devices to the off-peak hours as much as possible. Time-dependent tariffs can be realized by means of energy meters with a built-in timer or by means of signals from or communication with the utility.

19.3 IMPACT ON POWER QUALITY

Depending on the methods or technologies, rational use of energy can have a positive or a negative impact on power quality. If consumption is reduced by means of a more efficient use of electrical loads and by means of improved classic technology, power quality is improved. Due to the resulting lower currents in distribution systems, voltage drops are reduced. Load variations, including transient changes, become smaller, resulting in smaller variations of the supply voltage at the points of common coupling (PCCs). Because currents are lower, harmonic currents are lower as well. This results in lower harmonic content of the supply voltage. The impact of loads that emit distortion will be lower because currents are lower or because they are operated less frequently.

If efficient loads are connected to the grid by means of a power electronics interface, power quality may deteriorate. On the one hand, certain devices are a source of power quality problems because of the highly distorted current they draw; this is the problem of emission. On the other hand, there are devices that are sensitive to poor power quality. These devices will malfunction when power quality becomes too poor. This is the problem of (lack of) immunity. These problems are discussed for fluorescent lamps and variable drives. These technologies are key elements of rational use of energy, because lighting and drives represent the main part of electricity consumption.

19.3.1 Emission of Harmonic Distortion by Fluorescent Lamps

A fluorescent lamp is a gas-filled glass tube, through which a discharge current is sent to produce light. The impedance of the lamp (i.e. the impedance of the electric arc) is non-linear. If a sinusoidal voltage is applied directly to the lamp, a non-sinusoidal current flows. In order to limit and control this current, fluorescent lamps are equipped with a ballast. Two kinds of ballasts are used [12]: electromagnetic ballast and electronic ballast. Three types of fluorescent lighting systems exist: the fluorescent lamp with electromagnetic ballast, the fluorescent lamp with electronic ballast and the CFL.

19.3.1.1 Electromagnetic Ballast

Fluorescent lamps with electromagnetic ballast are inexpensive and have a simple design. The electromagnetic ballast is a series inductor, as shown in Figure 19.1. The combination of

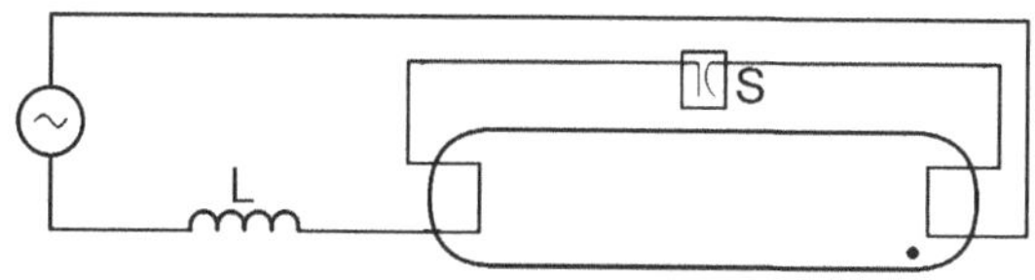

Figure 19.1 Fluorescent lamp with electromagnetic ballast (L) and starter (S)

the series inductor and the lamp results in an impedance that is more linear than the lamp's impedance. As a result, the current is more sinusoidal, with a small amount of harmonic distortion. The current lags the supply voltage. An example of the frequency spectrum of the current of a fluorescent lighting unit with electromagnetic ballast is shown in Figure 19.2 (third set of bars). The total harmonic distortion (THD) corresponding to this spectrum is 19.4 %. This particular lighting unit consists of two parallel fluorescent lamps, each with a ballast. The current waveform of a fluorescent lamp can contain more distortion if saturation and hysteresis occur in the ballast's iron core [1].

The disadvantages of the electromagnetic ballast are its weight, its large size and the occurrence of the stroboscopic effect. The frequency of the current is 50 Hz. As a result, the light output pulsates at a frequency of 100 Hz. If a single lamp is used the stroboscopic effect occurs. Because of the pulsating light output, rotating objects appear to be moving slower than they actually are. The stroboscopic effect is avoided by using two fluorescent lamps in parallel whose currents are not in phase. This is obtained by adding a series capacitor to one of the lamps. An alternative way is to feed the lamps from different phases.

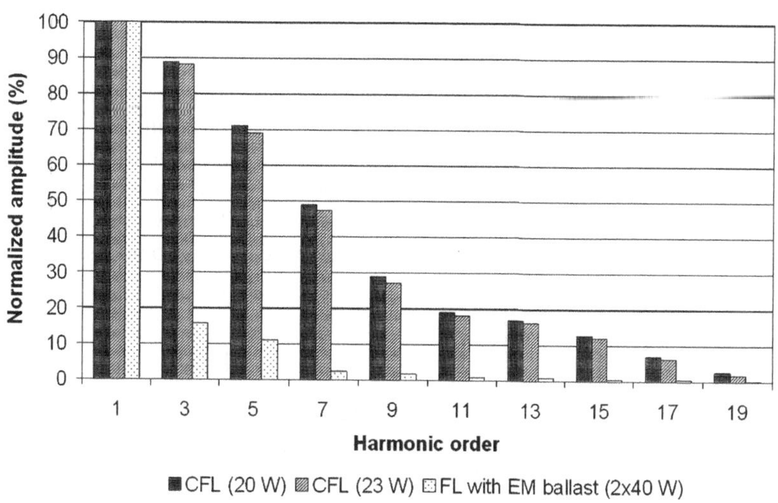

Figure 19.2 Harmonic spectrum of the current of CFLs (power: 20 W and 23 W) and a fluorescent lamp with electromagnetic ballast (power: 2×40 W) up to the 19th harmonic (in % of the fundamental component) [8],[10]

19.3.1.2 Electronic Ballast

The circuit of a fluorescent lamp with electronic ballast is shown in Figure 19.3. The principal elements of an electronic ballast are a rectifier, a d.c. filter, an oscillator (or high-frequency inverter), a capacitor and an inductor. The mains voltage is rectified and the d.c. voltage is converted to an alternating voltage with a high frequency, e.g. 30 kHz. This voltage is applied to the series circuit of the inductor and the lamp.

For large lamps, used in commercial and professional environments, separate electronic ballasts are used. Here it is possible to supply high-frequency voltage to multiple lamps. These external ballasts are equipped with a filter or wave-shaping controls to keep the THD minimal. In domestic lighting applications, however, considerations of cost, aesthetics and available space are more important than in professional environments. Therefore CFLs, with integrated ballast, are used in residential environments.

The main disadvantage of the electronic ballast is the distorted current drawn by the rectifier. The extent of this distortion depends on whether the rectifier is a standard diode-bridge rectifier [3], or a rectifier in combination with a passive filter or active control [12]. A passive filter can be added to the a.c. side to filter out harmonic current. In case the rectifier is equipped with active control, a nearly sinusoidal current is drawn from the grid, resulting in a high power factor. This type of ballast is more expensive than ballasts with a simple rectifier. Two systems for current control are possible. If the line voltage is used as the reference for the current waveform, the ballast and the lamp behave as a linear impedance. As a consequence distorted voltage results in distorted current. If a perfect sine wave is used as a reference, the current is sinusoidal under every circumstance.

In order to limit the size and the cost of CFLs, the integrated electronic ballast contains a standard rectifier, resulting in current spikes on the a.c. side, which correspond to a high level of harmonic currents. The fundamental current leads the voltage by a small angle. Examples of the frequency spectrum of this current are shown in Figure 19.2, for a CFL of 20 W and one of 23 W. The corresponding values of the THD are 130.5 % and 127.7 %, respectively [8].

Because of the high frequency of the current, the lamp's efficiency is higher than with an electromagnetic ballast. Additionally the inductor can be much smaller than the equivalent electromagnetic ballast, resulting in a lighter and more compact system. Other advantages, compared to the electromagnetic ballast, are the absence of the stroboscopic effect, high power factor, the possibility of dimming and quick start-up. A lamp with an electromagnetic ballast can also have a high power factor if it is equipped with a compensation capacitor. Dimming a lighting unit reduces its efficiency.

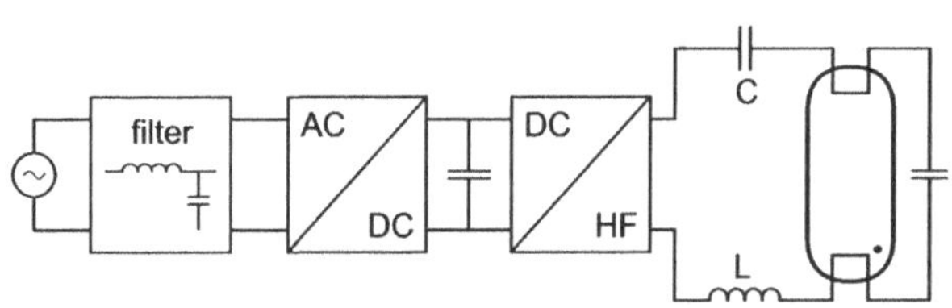

Figure 19.3 Fluorescent lamp with electronic ballast, consisting of an a.c. filter, a rectifier, a d.c. capacitor, a high-frequency oscillator, an inductor and a capacitor

19.3.1.3 Effect of CFLs on the Local Network

Due to the impedance of the network, harmonic currents lead to harmonic components in the voltage. In this way harmonic distortion is spread over the local network. If the number of CFLs in a grid section is small, the influence on the voltage waveform is acceptable. If large numbers of CFLs are used, harmonic distortion will become unacceptable. Radakovic *et al.* [9] analyze the effect of CFLs in a hotel in order to determine the allowable number of CFLs, taking into account the relevant standards. They conclude that the total active power of CFLs should not exceed 10 % of the rated power of the supply transformer. This number decreases if other non-linear loads are installed in the building. If the value of 10 % is exceeded, the voltage distortion in the hotel's network exceeds the limits stated in IEC/TR3 61000-3-6.

When several lamps are operated in parallel, the THD of this lighting system is lower than that of a single lamp. The harmonic currents from the CFLs are partially cancelled by each other. Cancellation occurs due to dispersion in the phase angles of the harmonic currents. This is called the diversity effect. Dispersion is caused by variations in the parameters of the network and the loads.

Another effect is the attenuation effect. Voltage harmonics, caused by harmonic currents, are a distortion of the supply voltage. The distortion results in a reduction of the voltage wave when the current spike occurs. Because of the voltage decrease, the current spikes are lower than when a sinusoidal voltage is applied. This results in an attenuation of the current harmonics [7].

19.3.2 Emission of Harmonic Distortion by VSDs

In the same way as fluorescent lamps with electronic ballasts, VSDs can be a source of harmonic distortion. VSDs with a diode-bridge rectifier generate harmonic currents. The outline of a VSD connected to a three-phase voltage source (i.e. the supply network) is shown in Figure 19.4. A VSD consists of a rectifier, a d.c. filter (e.g. a capacitor) and an inverter. With a three-phase rectifier, the current waveform consists of two peaks per half period, as is illustrated in Figure 19.5. This results in cancellation of the third harmonic if the supply voltage is balanced [3]. If the supply voltage is unbalanced, the two current peaks have a different magnitude, resulting in a large third-harmonic component in the current [13].

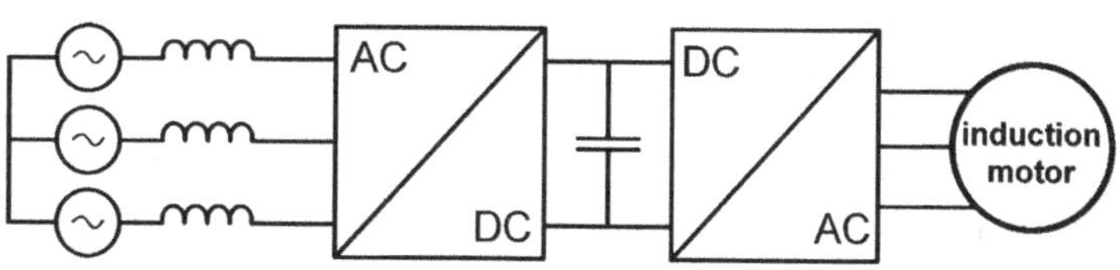

Figure 19.4 Schematic representation of a VSD. The VSD consists of a rectifier, a d.c. filter and an inverter. The frequency of the inverter's output voltage can be adjusted. This voltage is supplied to an induction motor. The rectifier is connected to the three-phase grid by means of inductors

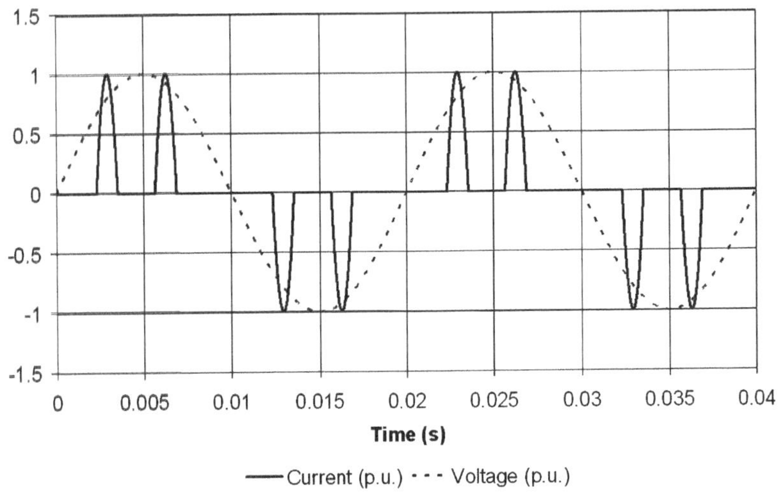

Figure 19.5 Input current of one phase of a three-phase rectifier

Both the rectifier and the inverter of a VSD contribute to the emitted harmonic and interharmonic currents [5],[13]. The order of the harmonic currents originating from the rectifier is calculated with Equation (19.1) below. The frequency of the interharmonic currents is calculated with Equation (19.2) [5]. Symbol p is the number of pulses per period (e.g. 6), k and n are integer numbers, f_N is the mains frequency and f_M is the fundamental frequency of the voltage supplied to the motor:

$$h = k \times p \pm 1 \tag{19.1}$$

$$f_{hi} = h \times f_N \pm n \times p \times f_M \tag{19.2}$$

Three main inverter types are used in VSDs: the inverter with pulse width modulation (PWM), the voltage source inverter (VSI) and the current source inverter (CSI) [13]. In a PWM inverter and a VSI the d.c. link usually consists of a large shunt capacitor, as in Figure 19.4. In the CSI the d.c. link is a series inductor. The a.c. input current of the rectifier supplying a PWM inverter or a VSI charges the capacitor. When the capacitor is fully charged, the input current is zero. This results in an alternating current consisting of a sequence of positive and negative spikes. This corresponds to strong harmonic distortion. In the case of a CSI, the direct current is always flowing. As a result, the alternating current resembles a square wave, which contains less harmonic currents. The attenuation effect mentioned in the discussion of fluorescent lamps occurs also with VSDs [13].

Harmonic distortion originating from VSDs may cause several problems, such as:

- Torque pulsations, reduced efficiency and overheating of a.c. motors connected directly to the mains voltage.
- Increased energy losses in transformers.
- Malfunction of protective relays and circuit-breakers.

In order to control the harmonics, recommended practices and requirements are specified in standards such as IEEE 519-1992 or IEC 61000-3-x. To check whether or not the installation of a VSD may result in problems concerning harmonics, certain elements have to be taken into account, such as the presence of capacitor banks without tuning reactors and the short-circuit capability of the supplying grid. A harmonic analysis may be required to determine the interaction of a VSD and the grid. If it is found that the harmonic distortion due to the installation of VSDs is increased beyond acceptable limits, one of the following measures can be taken [6]:

- Replacing the six-pulse rectifier by a twelve-pulse rectifier with a 30° phase shift.
- Replacing the three-phase bridge rectifier by a PWM-controlled inverter bridge, which generates a nearly sinusoidal current.
- Installing passive filters tuned to the most important harmonics (5th, 7th and 11th).
- Installing active filters with the ability to compensate current harmonics.
- Lowering the impedance of the main distribution transformer.

19.3.3 Immunity of Fluorescent Lamps

Because the use of CFLs is stimulated by utilities and governments, it is important that they have sufficient immunity to poor power quality originating elsewhere in the grid. Concerning lighting, an important type of distortion is voltage flicker. This is a low-frequency variation of the voltage. Flicker results in a visible variation of the light output of lamps.

Chang and Wu [2] compare the performance of incandescent lamps, fluorescent lamps with electromagnetic ballast and fluorescent lamps with electronic ballast (including CFLs). It is found that CFLs are less sensitive to voltage flicker than incandescent lamps and fluorescent lamps with electromagnetic ballast. The variation in light intensity of CFLs is lowest. This is due to the high frequency of the current flowing through the lamp. Different types of incandescent lamps show identical performance when flicker occurs, whereas a dispersion of performance is found for different types of fluorescent lamps and CFLs.

19.3.4 Immunity of VSDs

A voltage dip is a power quality phenomenon consisting of a voltage reduction by 10 to 99 % for a short period of time, typically less than 1 s. Voltage dips can occur in one, two or all three phases of the network. Dips are mainly caused by temporary, high currents, caused by the starting of large loads or by temporary short circuits.

VSDs are sensitive to voltage dips. This can be explained by analyzing the energy input and output of the d.c. bus during a voltage dip. A voltage dip causes the capacitor of the d.c. bus not to be loaded to its rated value. During normal operation of a VSD the inverter driving the motor is fed by the rectifier and the capacitor of the d.c. bus. When a dip occurs, the rectifier does not supply power because the amplitude of the a.c. voltage has dropped below the voltage of the d.c. bus. Because now only the energy stored in the capacitor

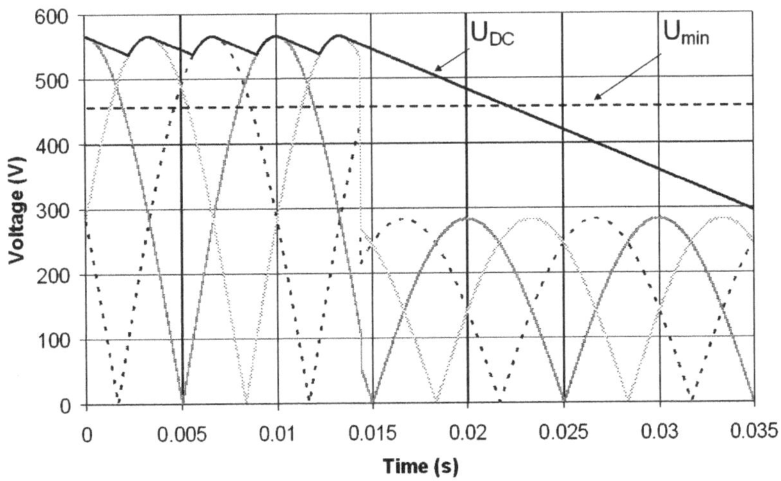

Figure 19.6 D.C. bus voltage of a VSD during a three-phase voltage dip. At about 0.022 s, U_{DC} becomes lower than $U_{\min}$

is available to the inverter, the d.c. voltage decreases as the capacitor is discharged. This is illustrated in Figure 19.6 for a three-phase voltage dip. If the d.c. voltage drops below the decreased a.c. voltage, supply through the rectifier is restored and a lower, steady d.c. voltage is established.

To prevent speed fluctuations and to protect the power electronics of the VSD against the high inrush currents that flow when the supply voltage is restored, most VSDs are equipped with undervoltage protection. If during the dip the d.c. voltage becomes lower than $U_{\min}$, the setting of the undervoltage protection, the VSD is switched off. Typical values for the protection device are 70 to 85 % of the rated d.c. voltage.

During an unbalanced dip (two-phase or single-phase dip) energy is still supplied by the unaffected phases. During a two-phase dip, the unbalance in the voltages causes the rectifier to operate in single-phase mode [6]. Whether or not the d.c. bus voltage will reach the undervoltage protection level $U_{\min}$ and consequently trip the drive depends on the load conditions and size of the capacitor of the d.c. bus. This is illustrated in Figure 19.7 for a two-phase dip. In this example, the capacitor is discharged sufficiently slowly by the inverter; the voltage remains higher than $U_{\min}$. This indicates that the probability that a VSD will not trip is higher when an asymmetrical dip occurs.

In most processes, the energy efficiency gain outweighs the costs of voltage dips since the tripped motor can be restarted easily and without any cost. In processes where synchronism of a large amount of motors has to be taken into account (e.g. extrusion processes), the failure of a motor can cause considerable losses. In these cases, dips can be countered by installing a backup d.c. power source connected to the d.c. bus, e.g. a boost converter. If the d.c. bus is not accessible, a.c. support systems can be installed, such as a dynamic voltage restorer, a dynamic sag corrector or a flywheel system.

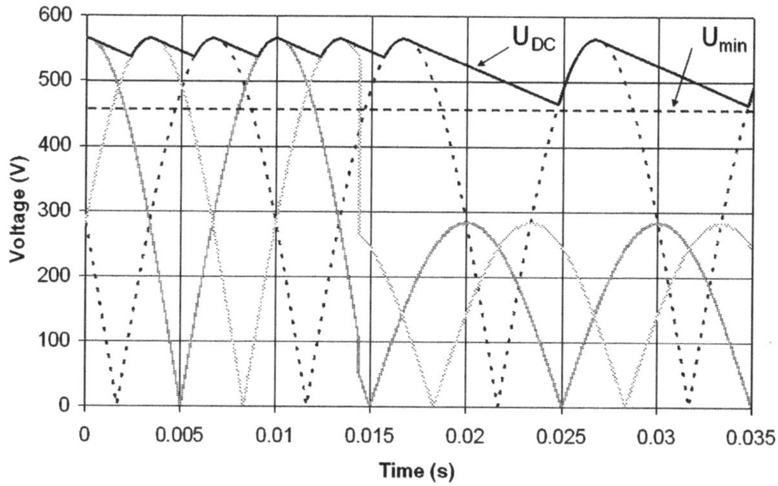

Figure 19.7 D.C. bus voltage of a VSD during a two-phase voltage dip. In this example, U_{DC} remains sufficiently high

BIBLIOGRAPHY

[1] Chang G. W., Characterizing harmonic currents generated by fluorescent lamps in harmonic domain. *IEEE Transactions on Power Delivery*, vol. 18, no. 4, pp. 1583–1585, 2003.

[2] Chang W. N., Wu C. J., The influence of voltage flicker on residential lamps. International Conference on Power Electronics and Drive Systems, 1997, pp. 392–396.

[3] Chen K., The impact of energy efficient equipment on system power quality. IEEE Industry Applications Conference, 2000, pp. 3240–3245.

[4] De Almeida A. T., Fonseca P., Characterisation of the electricity use in European Union and the savings potential in 2010. *Energy Efficiency Improvements in Electric Motors and Drives*, Springer, Berlin, 1997, pp. 19–36.

[5] De Capua C., Landi C., Quality assessment of electrical drives with strongly deformed supply waveform. *Measurement*, vol. 30, pp. 269–278, 2001.

[6] Didden M., Driesen J., Belmans R., Possible problems and solutions when introducing variable speed drives. EEMODS Conference, Treviso, Italy, September 2002.

[7] El-Saadany E. F., Salama M. M. A., Reduction of the net harmonic current produced by single-phase non-linear loads due to attenuation and diversity effects. *Electrical Power & Energy Systems*, vol. 20, no. 4, pp. 259–268, 1998.

[8] Korovesis P. N., Vokas G. A., Gonos I. F., Topalis F. V., Influence of large-scale installation of energy saving lamps on the line voltage distortion of a weak network supplied by photovoltaic station. *IEEE Transactions on Power Delivery*, vol. 19, no. 4, pp. 1787–1793, 2004.

[9] Radakovic Z., Topalis F. V., Kostic M., The voltage distortion in low-voltage networks caused by compact fluorescent lamps with electronic gear. *Electric Power Systems Research*, vol. 73, pp. 129–136, 2005.

[10] Teixeira M. D., de Oliveira J. C., Medeiros C. A. G., de Oliveira J. C., Teixeira G. S., A power quality comparative analysis related to electronic and electromagnetic fluorescent ballast operation. IEEE 10th International Conference on Harmonics and Quality of Power, 2, 2002, pp. 424–429.

[11] Topalis F. V., Kostic M. B., Radakovic Z. R., Advantages and disadvantages of the use of compact fluorescent lamps with electronic control gear. *Lighting Research and Technology*, vol. 34, no. 4, pp. 279–288, 2002.
[12] Wakabayashi F. T., Canesin C. A., Novel high-power-factor isolated electronic ballast for multiple tubular fluorescent lamps. IEEE Industry Applications Conference, 36th IAS Annual Meeting, 2001, pp. 230–237.
[13] Xu W., Dommel H. W., Hughes M. B., Chang G. W. K., Tan L., Modelling of adjustable speed drives for power system harmonic analysis. *IEEE Transactions on Power Delivery*, vol. 14, no. 2, pp. 959–601, 1999.

20

Perceived Power Quality

Maurizio Caciotta

The problems discussed in this chapter relate directly to the economic, sociological and psychological aspects of, and also the scientific–technical elements connected with, electric power production, distribution and fruition.

For this reason, only the international rules connected with the technical, economic and environmental fields will be considered, if possible, working in the industrial world. These are the set of ISO 9000 rules for the quality that will influence the ISO 14000 rules for the environment (not discussed here).

20.1 CUSTOMER DEFINITION

The ISO 9000 standard defines, in point 3.3.5, the customer as an 'organization or person that receives a product', and it is further defined in point 3.4.2 as a 'result of a process' that is defined in point 3.4.1 as a 'set of interrelated or interacting activities which transforms inputs into outputs'.

In other words, the customer is an organization or a person that receives the result of a complex of connected or interacting activities that transform entrance elements into exit ones.

In the note 1 to point 3.4.2 of ISO 9000 four product categories are defined:

- Services
- Software
- Hardware
- Direct product materials

Handbook of Power Quality Edited by Angelo Baggini

Let us consider the product, as in the above, of electric power. This is classified as a service, and note 2 to point 3.4.2 defines it as the result of an activity carried out necessarily to the interface between the customer and the supplier, defined in point 3.3.6 as an 'organization or person that supplies a product' which is generally intangible.

The customer receives the electric power, in a place called the delivery point, as an intangible product supplied by an organization or person, which transforms entrance elements into exit ones.

According to current technical knowledge, the intangibility of electric power is not borne out in practice, as the customer utilizes it directly on the interface, connecting with the network. Applying a load that interferes with its characteristics limits the distribution power.

There is a problem not only for the parameters, i.e. for supplying parameter values such as voltage, active or reactive power, components of a inverse sequence etc., but also in linear terms, for the presence of harmonics and interharmonics.

To limit the supplied power available at the customer interface points, these non-linear effects are diffused among all the distribution networks.

20.2 CUSTOMER REQUIREMENTS

The same rule as indicated in point 3.1.1 of the ISO 9000 standard defines quality as the 'degree to which a set of inherent characteristics fulfils requirements'. The characteristics are defined in point 3.5.1 as a 'distinguishing feature' while requirements are defined in point 3.1.2 as a 'need or expectation that is stated, generally implied or obligatory'. Generally implied means custom or common practice for the organization, its customers and other interested parties. To conclude, the ISO 9000 standard defines quality, when applied to electric power, as 'The Electric Power quality is all its inner and distinct elements that satisfy the clear or coercive expectations of the customers and the interested parties.'

Apart from the coercive rules concerning a service and that consider all the expectations of the interested parties, in the case of electric power, which is so important for the development of industrial companies, there is no sense in dealing with electric power quality without defining the clear and/or coercive expectations of the customers.

There have been numerous attempts to sound out a perception of customer quality, but this necessity arises from the academic world, which uses a correct and formal language, but not one scarcely diffused among the workers of small and medium-sized firms. This language and its technical nature achieves good results in all those structures that possess an energy manager, but it has practically no effect in other situations.

20.3 ANALYSIS PROCESS OF THE CUSTOMER WITH RESPECT TO THE REQUIREMENTS CONCERNING THE PRODUCT

All people with individual social and economic characteristics are customers, with regard to electric power, and they have very different expectations regarding this, so their evaluations of its levels of quality differ considerably.

To deal with such a complex reality, mathematical models have to be employed that consider homogeneous categories. For the economic and social importance of the problem the chosen category is the marketing one.

In the process of defining the customer, the economic dimensions of the organization become a fundamental parameter. The implications are of a social type either for the number of workers or for the products brought to market to satisfy needs.

Large firms are usually able economically and culturally to engage the services of experts – energy managers – who are able to enter into a dialogue at a technical level with productive organizations and/or energy suppliers to obtain the best service. This brings economic advantages to both and very substantial ones at that.

Small and medium-sized enterprises (SMEs), sometimes being the major part of the productive power (e.g. in Italy), are customers working in a variety of marketing categories and have, as far as energy and particularly electric power are concerned, a common psychological position.

In recent times, when many situations have arisen of almost absolute control by more or less important organizations that supply electric energy, dialogue with small customers has been impossible.

Usually the technical function of these organizations is a flattering one, but not for communication and transparency, because these organizations have rarely considered the problem of sedimentation and diffusion of an energy culture, particularly with regards to electric energy.

SMEs know little or nothing about electric energy, even if they do need it for their activities, except for the cost of the primary sources on which they depend for information from the media, on the ways to transform, carry or use it.

When an SME has something to say on the electric energy supply, it is usually along the lines of the popular saying 'It Rains Thieving Government!'

20.4 MULTIPLICITY OF GOODS: ACTIVE CATEGORIES IN THE TERRITORY OF ROME

The situation described above produces a sense of inability, and, moreover, an indifference to the working problems concerning production and transformation. Organizations have to satisfy their customers' primary needs, but for a long time there has been no information and no culture on electric energy.

In 2003, Roma Tre University and the Italian Federation for the Rational use of Electric Energy (FIRE) carried out research on a definition of electric energy quality as perceived by SMEs.

From the outset, the working problem was the way to carry out the research. There are two problem types:

- How to construct a dialogue with such different marketing categories?
- In which language?

The adopted methodology was to use a marketing filter (MF).

Electrical installations of SMEs are usually managed by professional designers or directly by installers (I/Pes). So their category can be one that is interfaced in order to obtain homogeneous information on the perceived quality of electric energy (PQEE).

A communication language with I/Pes was developed in two phases. To take the heuristic process further, for example, an electrician was consulted about who had more profound knowledge of technical matters, and who was capable of establishing a dialogue with academics but also of drawing up a list of the queries and needs common to all electricians which could then be compared to the technical regulations in force.

As a result, a questionnaire (Q) was prepared to obtain information on PQEE (Figure 20.1). The Q was given to a small number of I/Pes who were farther removed from the academic world. As there were a reduced number of the Q returned, the replies were examined along statistical lines according to Student's test and the Q planned again.

The characteristics of such an individualized Q were:

- A limited number of fields – only six.
- A limited number of questions – five at the most.
- Headings not corresponding to the rules.
- Answers in fields which do not allow determination of the best estimators from a statistical point of view, including the mean and standard deviation.
- Answers developed on substantial damage.

The technical standards issues considered in the Q were:

(a) Breaks and/or micro-breaks and lack of one or several phases
(b) Change of voltage
(c) Electrical disturbances
(d) Presence of harmonics
(e) Transients
(f) Scattered currents

First of all it is necessary to observe that the structure of the Q has been developed on negative aspects, as I/Pes give information on damage and have a very negative impression

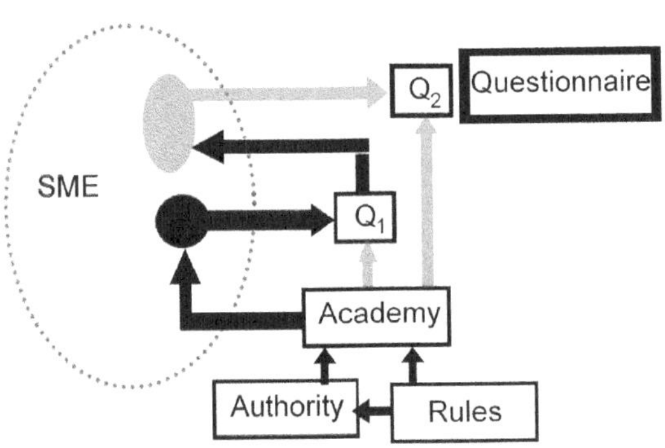

Figure 20.1 Flowchart of methodology

of electric energy quality, but it can and must be judged according to the fact that their electrical apparatus has a medium life as observed by the I/Pes in their work; they treat it like an inner value to be extracted during the values analysis (point (e) of the Q).

The quality categories indicated were only apparently similar to those of the EN 50160 standard (point (c) of the Q).

For example, the first category, 'Breaks and/or micro-breaks and lack of one or several phases', was developed over five questions (point (b) of the Q) on damage to machinery: namely, tools, electrical apparatus, electrical contact breaks, recharging of pad batteries, unsuccessful effects of system power factor correction.

Voltage changes were defined relating to incandescent lamps and to their replacement after faults in residual current devices, to transformers and motors, the latter without the intervention of protection, and to false alarms of anti-theft devices.

Electrical disturbances were defined by the damage noticed on the tuners of TV sets and by the troubles in data communication of electrical apparatus, which were explored through two questions only.

Curiously the electrical systems of cash registers, more precisely the higher input to ratings, were assigned to the category of harmonic presence. Also in this category there were only two proposed questions, which can scarcely define the world of perception.

Motors overheating was connected with the transients category, and also the damage to the conditioning apparatus, which was explored through three questions.

The scattered currents category covers corrosion effects. It was used substantially to amass and analyze data mostly in the macro marketing categories of the business, industry and agriculture.

Operationally, the Q asked for information on the presence of voltages in ground leakages and their number. Five questions were posed for this problem:

- Never
- 1–5
- 5–10
- 10–20
- More

They were easy to indicate, but there are no simple problems in the statistical treatment, as the moments and the central moment estimators are not excellent, going beyond the classic scientific treatment of probability distributions.

Working with a large schematization, evidence exists for the Never and More categories when separated from the other ones, while the intermediate categories have superimposed limits.

For a case study see web address

BIBLIOGRAPHY

[1] Caciotta M., Leccese F., Trifiro A., From Power Quality to Perceived Power Quality, National Congress.

[2] EN ISO 9000, *Quality management systems – Fundamentals and vocabulary*, 2005.

Index

Printed and bound in the UK by
CPI Antony Rowe, Eastbourne

Printed and bound by CPI Group (UK) Ltd, Croydon, CR0 4YY

16/09/2026

14963389-0002